Karl Menger

Ergebnisse eines Mathematischen Kolloquiums

Herausgegeben von/Edited by
E. Dierker, K. Sigmund

Mit Beiträgen von/With contributions by
J. W. Dawson jr., R. Engelking, W. Hildenbrand

Geleitwort von/Foreword by
G. Debreu

Nachwort von/Afterword by
F. Alt

Springer-Verlag Wien GmbH

Univ.-Prof. Dr. Egbert Dierker
Institut für Wirtschaftswissenschaften, Universität Wien
Hohenstaufengasse 9, A-1010 Wien

Univ.-Prof. Dr. Karl Sigmund
Institut für Mathematik, Universität Wien
Strudlhofgasse 4, A-1090 Wien

Gedruckt mit Unterstützung des Fonds zur
Förderung der wissenschaftlichen Forschung

Ursprünglich erschienen bei Springer-Verlag/Wien 1998
Softcover reprint of the hardcover 1st edition 1998

Satz: Vogel Medien GmbH, A-2102 Bisamberg

Gedruckt auf säurefreiem, chlorfrei gebleichtem Papier – TCF
SPIN:10637930

ISBN 978-3-7091-7330-5 ISBN 978-3-7091-6470-9 (eBook)
DOI 10.1007/978-3-7091-6470-9

Preface

The *Ergebnisse eines Mathematischen Kolloquiums* were published by Karl Menger (with the collaboration of Kurt Gödel, Georg Nöbeling, Abraham Wald and Franz Alt) in eight issues during the years 1929 to 1937. They constitute a major source text for the history of scientific ideas in the 'Thirties, containing path-breaking papers by Menger, Gödel, Tarski, Wald, Wiener, John von Neumann and many others. Due to political circumstances, the distribution of the Ergebnisse was very limited, and only few complete sets have survived to this day.

This volume contains the full text. We have reprinted the German original, adding English commentaries written by outstanding experts. John Dawson jr. comments the contributions on mathematical logic, Ryszard Engelking the works on topology and Werner Hildenbrand the papers on mathematical economics. Furthermore this volume contains a biography of Karl Menger, a survey written by Menger on the major topological and geometrical ideas of the Kolloquium, and an introduction by Nobel Prize winner Gerard Debreu. We have also listed those papers of the Ergebnisse which have been translated into English, and added short biographical notes on Menger and his collaborators. Prof Franz Alt, one of the co-editors of the Ergebnisse, was kind enough to provide us with an Afterword, and Mrs. Louis Golland with a photograph of Menger.

We wish to thank Doz. Friedrich Stadler for suggesting to reprint the Ergebnisse, Professors L. Schmetterer and E. Streissler for helpful advice, Dr. E. Köhler from the Vienna Circle and Professor R. Leonard for providing us with important informations. We also thank Dr. M. Stöltzner for his translation of Menger's survey article and the Fonds zur Förderung der wissenschaftlichen Forschung for support. We are greatly indebted to Professors Bert Schweizer and Abe Sklar, who were students, then collaborators of Menger and kindly volunteered to read the biographical introduction to this volume. Very special thanks, finally, go to Professor Franz Alt, one of the co-editors of the *Ergebnisse,* who kindly provided us with helpful suggestions and clarifications.

E. Dierker
K. Sigmund

Table of Contents

Foreword: Economics in a mathematics colloquium

Gérard Debreu

In stark contrast to the short eight years of its existence, the colloquium that met in Vienna from 1928 to 1936 had a long lasting influence on economic theory. At first, the reaction of the economics profession to the new ideas presented in Karl Menger's seminar was slow, and fifteen years elapsed before the strength of their impact was felt. Many explanations can be found for the delay that took place from Karl Schlesinger's paper given orally in 1934, to the Linear Programming Conference held in Chicago in 1949, a prelude to the *Activity Analysis of Production and Allocation* Monograph edited by Tjalling C. Koopmans, and published in 1951. Only a few of those explanations can be listed.

The world had already taken the path to a global conflict, which could be foreseen, with almost blinding clarity, in the abrupt end of the colloquium in 1936. In that world, the Proceedings was not common reading among economists, and it was easily available neither from specialized libraries, nor from colleagues. The title did not promise any economic contents. The language alone was a powerful deterrent. The place from which it came seemed distant. The main participants were, or were to become, mathematicians, and statisticians of the most brilliant magnitudes, but their names were little known to economists. A complete understanding of their mathematics required a high level of preparation. Their mode of reasoning was unforgivingly rigorous, at a time when logical rigor was not believed to be a condition for theorizing about economics. Partial as this list will remain, it must include as an especially forbidding obstacle to the rapid and full recognition of their contributions, the novelty of the questions they asked, and began to answer. Yet the ideas that Karl Schlesinger, Abraham Wald and John von Neumann published in volumes 6, 7, and 8 of the Proceedings in 1935, 1936, 1937, deserved immediate notice.

The starting signal for the development that would bring a profound transformation to mathematical economics was given by Schlesinger. In the first paper on economics of the Colloquium, presented on March 19, 1934 (and published in 1935), he raised a question aimed at the center of Walrasian theory. In his *Eléments d'Economie Politique Pure* (1874–77), Léon Walras did not inquire whether the equations with which he described the general equilibrium of an economy actually had a solution, let alone a solution in which all the variables had the proper signs. The mathematics available at that time did not permit a satisfactory answer, and he had to content himself with the giant step he had taken. But 60 years went by without any alteration of the argument for the existence of a General Equilibrium that the master of the School of Lausanne proposed, and that his disciples repeated for six decades. The reason given for existence was simply a count of equations and unknowns. Showing that their numbers were equal established no more, however, than the plausibility of a positive answer to Schlesinger's question. And yet, long before his query was formulated, one of the main mathematical tools with which it was eventually settled was provided in 1910–11 by L. E. J. Brouwer (*Mathematische Annalen*), in the form of his powerful fixed point theorem.

In his paper (on March 19, 1934), Schlesinger suggested a modification of Léon Walras' (and Gustav Cassel's) equations that soon turned out to be essential. Walras (and Cassel) assumed that the list of free commodities (i.e., for which demand is smaller than supply) was given a priori. Schlesinger wrote that this condition was inadmissible, that the list of free commodities should be determined by the system of equations itself, each one of the free commodities having a zero price.

Immediately after that correct formulation was given, Wald proved the existence of a GE (also on March 19, 1934), and, at the next session of the Colloquium, on November 6, 1934, established existence under much weaker conditions. His two papers, providing the first proofs of existence for a GE marked an important moment in the history of mathematical economics.

With a fifteen years lag, an echo was finally heard in Chicago in 1949, and from that time on, events began to unfold more rapidly. First, however, one must go back to Menger's Colloquium where still another major contribution was published in 1937 by von Neumann as the last article in the last volume of the Proceedings. His paper was presented orally first in English at the Mathematics Colloquium of Princeton University in the winter of 1932, next printed in German in the Ergebnisse, finally that version was translated into English and appeared in the 1945 *Review of Eco-*

nomic Studies as "A Model of General Equilibrium". Von Neumann's contribution reached many goals; one of the most important, in his own appraisal, was his "lemma" (in the English translation), or "Satz" (in the German original). That "lemma" concerns a generalization of a Brouwer Fixed Point to what is now known as a Kakutani Fixed Point. John von Neumann's "lemma", and Shizuo Kakutani's theorem can almost immediately be derived from each other. That von Neumann thought his "lemma" to be important can be read in his article's German title "Über ein ökonomisches Gleichungssystem und eine Verallgemeinerung des Brouwerschen Fixpunktsatzes". His generalization turned out to be essential for the existence proofs presented in December 1952 (and published in *Econometrica* in 1954) by Kenneth Arrow and Gérard Debreu, and by Lionel McKenzie. Yet his "lemma" was unnecessarily powerful for his purpose, in spite of John von Neumann's assertion that "the mathematical proof is possible only by means of a generalization of Brouwer's Fix-Point Theorem". Elementary proofs of von Neumann's central result that do not call on fixed point theorems were given later by several authors, in particular by David Gale (in "Linear Inequalities and Related Systems", edited in 1956 by H.W. Kuhn and A. W. Tucker). Thus the main mathematical tool for the proof of existence of a GE owes it origin to an accident.

Kakutani then stated von Neumann's "lemma" in a form that made it easier to use, and proved it in a simpler manner in 1941 (*Duke Mathematical Journal*). Nine years later, in a single page note, one of the shortest influential papers in science, John Nash defined, and proved the existence of, a "Nash equilibrium" for finite N-person games using Kakutani's generalization (*Proceedings of the National Academy of Sciences of the USA*, 1950). Nash's article was short, but it was also fundamental, as the Nobel committee in economics recognized in 1994. By itself it could destroy the myth that forceful economic ideas can be fully expressed only with great prolixity.

Nash's introduction of Kakutani's theorem in economics in 1950 played a major role in the solutions of the existence problem proposed in 1954 by Arrow and Debreu, and by McKenzie.

In the meantime, Morton Slater presented, in Chicago, a little known Cowles Commission Discussion Paper (Math 403, November 7, 1950) "Lagrange Multipliers Revisited". There he applied the von Neumann-Kakutani result to his study of a class of constrained maximum problems that had been considered by Kuhn and Tucker. In a different context from Nash's article with which it was simultaneous, Slater presented another introduction of a wide set of fixed point theorems to the group of scientists that was then doing research on mathematical economics, the theory of games, and mathematical programming.

The stage was now set for the proofs of existence of a GE published in 1954. Arrow and Debreu, and McKenzie, using as their major mathematical tool the fixed point theorem of Kakutani, weakened the conditions given by Abraham Wald nineteen years earlier. The uncompromising precision of existence proofs forced on the Walrasian model an extensive reformulation that provided a framework for the research that occurred in several directions in GE theory. Knowing that the central concept of Walrasian theory existed under mild assumptions validated that research.

Menger's *Ergebnisse* – a biographical introduction

Karl Sigmund

The Vienna Circle (the *Wiener Kreis*) was known in the Vienna of the 'twenties as the Schlick Circle, and was but one of many intellectual circles. As Karl Menger wrote in his reminiscences[1], there existed, among others, a large number of social-democratic discussion groups, for example, and psycho-analytic circles of various orientations, and groups debating educational reforms. There were discussion groups on constitutional law or economics, on art history, on the philosophy of law, on phenomenology, Kant, etc.

During the academic year 1928/29, several students at the *Seminar für Mathematik* (the forerunner of the institute of mathematics) of the University of Vienna asked their professor, Karl Menger, to direct a mathematical *Kolloquium*. Menger organised it in the form of a Circle[2] meeting regularly on Tuesdays during the academic semester. It had a flexible agenda including lectures by members or by invited guests, discussions of unsolved problems and reports on recent publications. From autumn 1929 onward, Menger kept a record of these meetings. These notes were published, first by Teubner Verlag and later by Deuticke, under the title *Ergebnisse eines mathematischen Kolloquiums*. Altogether eight issues appeared between November 1931 and November 1936. They are reprinted in the present volume. Together with the commentaries written by John Dawson Jr., Ryszard Engelking and Werner Hildenbrand, they offer a revealing insight into an intellectually fascinating period. Kurt Gödel was one of the main contributors and co-editors of the *Ergebnisse*, which also contain important papers by Alfred Tarski and other logicians. For topology (frequently called *mengentheoretische Geometrie* at the time), the *Ergebnisse* are a major source: in addition to Menger and his outstanding students, distinguished visitors like Borsuk, Čech and Whyburn lectured at the Colloquium. Other important visitors were John von Neumann, Norbert Wiener and Marston

Morse. Last but not least, the mathematical Colloquium played a key role in the emergence of mathematical economics, in particular through the work of Abraham Wald. Fittingly enough, the last paper published in the *Ergebnisse* is a break-through article by John von Neumann on equilibrium theory.

During this period – the closing years of the *gay apocalypse* – the cultural life in Vienna reached an astonishing intensity in almost every domain. The mathematicians did not lag behind. Indeed, many members of the Schlick Circle were mathematicians[3]: Hans Hahn, who is seen by many as the true originator of the Circle[4], his sister Olga, his students Kurt Gödel and Karl Menger, as well as Kurt Reidemeister, Gustav Bergmann and Theodor Radakovic. In addition, most of the philosophers of the Circle – in particular Carnap and Waismann – were well versed in mathematics and keenly interested in foundational questions. When Tarski, for instance, visited Vienna in 1929, he held three lectures in the Colloquium, of which two were on logic. Menger made sure that these were also attended by the philosophers of the Vienna Circle[5]. And when Karl Popper – who was not a member of the Vienna Circle, but strongly influenced by it – lectured at the Colloquium in 1935 on the concept of random sequences, his ideas led Abraham Wald to his important mathematical investigations on the concept of randomness[6].

This introduction attempts to briefly sketch the intellectual background of the Colloquium, based on the biography of its founder and promotor Karl Menger. This is followed by the commentaries of Dawson, Engelking and Hildenbrand on the main themes of the Colloquium: logic, topology, and mathematical economics. We also include the translation from Italian into English of a survey written by Menger himself, in 1936, on the main contributions of the Colloquium to geometry and topology.

A short biography of Karl Menger

Karl Menger was born in Vienna on January 13, 1902. His father was the famous economist Carl Menger, one of the founders of the Austrian School of Economics, whose path-breaking *Grundsätze der Volkswirtschaftslehre*, published in 1871, introduced marginalism at about the same time as Jevons and Walras[7]. Carl Menger was a professor at the University of Vienna from 1873 on and became the tutor of crown-prince Rudolf (of Mayerling fame). His brother Anton was a world-renowned professor of law[8], and his brother Max was, for more than thirty years, a liberal deputy in the *Reichsrat*. Karl's mother Hermione Andermann was a noted novelist. Any child born to such a family must have felt the pressure to succeed, and this pressure did not relax at school (the *Döblinger Gymnasium*): two of Menger's school mates, namely Wolfgang Pauli and Richard Kuhn, were destined for Nobel Prizes.

In the autumn of 1920, Karl Menger inscribed at the University of Vienna. He started in physics, but switched to mathematics after a few months. In March 1921, he attended the first seminar meeting of the newly appointed professor Hans Hahn. Hahn lectured on the concept of a curve, which no one had been able to define in a satisfactory way. Within a few days, Menger came up with a solution to this problem, which some well-known mathematicians had declared to be unsolvable[9].

Only weeks later, disaster struck. Menger was diagnosed with tuberculosis, an illness which, in the impoverished capital of an amputated state, was killing by the thousands. Menger was sent to a sanatorium in Aflenz (Styria). Before leaving Vienna, in feverish haste, he wrote down his preliminary results on curves, and deposited them in a sealed envelope at the Academy of Science. During his time in Aflenz, which lasted more than one year, both his eighty-year old father and his sixty-year old mother died. But when Karl Menger returned to the university, completely recovered, he had developed a full-fledged theory of curves which almost immediately earned him his doctorate with Hahn[10].

Soon afterwards, Menger, supported with a Rockefeller scholarship, went to Amsterdam, where he studied with L.E.J.Brouwer, one of the foremost topologists of the time, who quickly recognised his gifts and offered him a job as his assistant[11]. But their relationship soon became tense, in part because Menger disagreed with Brouwer's anti-French views and his legalistic mind. Their differences became increasingly acrimonious over a priority dispute. The young Russian mathematician Lew Urysohn had developed essentially the same dimension theory as Menger before perishing in a drowning accident. Brouwer edited the posthumous publication of Urysohn, stressing their link with a little-known note written by himself in 1913 which contained already the basic idea. Menger, who had known neither of Brouwer's nor of Urysohn's work, felt that his contribution was mis-represented by Brouwer[12].

In the meantime, the associate professor of geometry at the University of Vienna, Kurt Reidemeister, had left to take a chair in Königsberg. Hahn offered the job to Menger, who eagerly accepted. He had earned his habilitation in Vienna in the autumn of 1926 and was appointed *Extraordina-*

rius für Geometrie in February 1927. He immediately plunged into an ambitious program of lectures on all aspects of geometry, and published in quick succession two path-breaking books summarizing his work on curve theory and on dimension theory.

Like his predecessor Reidemeister, Menger was invited by Hahn to become a member of the Vienna Circle. Menger, who had an extensive knowledge of philosophy, and a first hand knowledge of Hahn's logistic and Brouwer's intuitionistic approach to the foundational problems of mathematics, became a valuable member of the group, but he did not sympathize with the more political aspects of the Circle s activities. This was the time when he decided to launch the Mathematical Colloquium and began to distance himself from the Circle. Another student of Hahn, the young Kurt Gödel, moved in unison.

At this time, some extraordinarily talented young topologists worked in Vienna, in particular Witold Hurewicz and Leopold Vietoris. Menger attracted a remarkable following of gifted young mathematicians to the Colloquium: Hans Hornich, Georg Nöbeling, Franz Alt, Olga Taussky etc. Hornich later turned to complex analysis and potential theory. He eventually became a professor at the Technical University of Vienna[13]. Franz Alt, who wrote his thesis under Menger's supervision, published many of his papers in the *Ergebnisse*. He became co-editor of the last volume. In 1938, he was forced to emigrate to the US. As a member of the US Army Ballistic Research Laboratory, and later the National Bureau of Standards, he did pioneering work on computers[14]. He was also a founding member of the Association for Computing Machinery and served as its President from 1950 to 1952. Georg Nöbeling, who kept working on geometry and topology, returned to Nazi Germany in 1935, to Menger's great dismay, and took up a chair in Frankfurt. Olga Taussky spent a year in Göttingen, where she took part in the publication of Hilbert's Collected Works. She later emigrated to England, where she married the mathematician John Todd. She went on to a most distinguished mathematical career in the United States[15].

But Menger's most promising student at the time was the Romanian Abraham Wald[16]. Quite in contrast to Menger, Wald was a latecome. Born (like Menger) in 1902, it was only in 1930 that he started serious mathematical work. He began by immediately solving a problem proposed to him by Menger (an axiomatization of the notion of 'betweenness' in metric spaces) and kept asking for more. In 1931 he obtained his doctorate, although he had only taken three courses so far. He became one of the major figures in the Colloquium, and one of the first co-editors of the *Ergebnisse*. Wald's main interest was at first geometry, but since there was no opening at the university, Menger arranged for Wald a job as private tutor to the banker Karl Schlesinger, and later obtained work for him at Oskar Morgenstern's *Institut für Konjunkturforschung*. This was a decisive turn in Wald's life. He became increasingly involved in mathematical economics, but also kept working with Menger on geometrical topics.

Menger interpreted geometry in its very widest sense, including 'set theoretic geometry', i.e. today's general topology. In addition, he wrote four long papers entitled 'Untersuchungen über allgemeine Metrik'[17] which laid the groundwork for much further research in the Colloquium and were the beginnings of distance geometry[18]. Among his results was the discovery of a 'universal curve' in three-space, which has the property that *every curve* can be topologically imbedded in it. Today, this curve is also known as 'Menger's sponge' in fractal theory[19]. For his lectures on euclidean geometry, Menger devised a completely new approach. His 'algebra of geometry' became another leading theme of the Colloquium, and soon played an important role in John von Neumann's mathematical foundation of quantum mechanics[20]. Through this work, Menger became one of the first to define and investigate lattice structures. He presented his work on curve theory and on dimension theory in two remarkable books.[21]

Moreover, already before and during his stay in Amsterdam, Menger had developed an intense interest in foundational problems. One of his first publications on this topic was a 'dictionary' translating Brouwer's somewhat obscure definitions into well-known set-theoretic concepts[22]. Menger was on excellent terms with Kurt Gödel (who was four years younger). When Gödel first mentioned his epoch-making discovery during a meeting in Königsberg, in September 1930, Menger was on a lecture tour in the United States. Nöbeling wrote him about it, and Menger, who (like John von Neumann) quickly grasped the significance of Gödel's incompleteness theorem, immediately started to lecture on it[23]. He usually couched the theorem in number theoretic terms: there exist propositions about natural numbers which can be neither proved by induction nor disproved, and which can (if they are valid) only be derived within a larger system, i. e., by means of techniques from analytic number theory.

Menger's family background made it almost inevitable that the social and economic sciences would also attract his attention. He started writing on these topics at the age of twenty-one, but published only much later, and sporadically[24]. Whereas his father (quite in contrast to his uncle Anton) had hardly ever felt any inclination to put his ideas into mathematical terms (thus achieving less impact than Léon Walras[25]), Menger kept calling attention to open mathematical problems concerning marginalism and economic equilibria. This became one of the major topics in the Colloquium. Furthermore, in 1934, Menger wrote a short, highly innovative book on formal problems in ethics[26].

At about this time, the Vienna Circle began to disintegrate[27]. Carnap had accepted a chair in Prague. Hahn died quite unexpectedly at the age of fifty-five. Neurath could not return to Austria for political reasons: the clerical-fascist regime which, after the civil war in February 1934, had outlawed the Social Democrats, would have arrested him as a leftist agitator. Authorities frowned upon the Vienna Circle as a citadel of free thinking. In 1936, Moritz Schlick was shot to death on the stairs of the University by a deranged student of philosophy and mathematics named Nelböck. Newspapers condemned the murder, but pontified on the demoralising effects of free-thinkers upon religiously educated people, and claimed that the proportion of Jewish members at the Institute of Philosophy was unduly high[28].

In 1935, Karl Menger had married Hilda Axamit, a student of actuarial mathematics. In the following year a son was born – Karl junior. The young father was deeply worried by the increasingly distressing political and academic situation in Austria. Hahn's vacant chair, for which Menger was the obvious candidate, was abolished. His hopes to succeed the mathematician Wilhelm Wirtinger as full professor were disappointed[29]. Menger's achievements were obviously much better appreciated abroad. In 1936, he was elected vice-president of the International Mathematical Congress in Oslo – a remarkable honour for someone who was barely thirty-four[30]. Soon after he first mentioned his plans to leave Vienna, he was offered a professorship in the USA, at the University of Notre Dame in Indiana.

When Hitler's troops invaded Austria, Menger's friend Schlesinger committed suicide. Menger managed to obtain, within a few weeks, an American immigration visa for Abraham Wald. Wald succeeded in an incredibly short time to attain a prominent position among American mathematical economists and statisticians[31]. However, his plans for continuing the joint work with Menger on a new differential geometry had to be postponed.

As soon as Menger arrived at Notre Dame, he resumed his journal, this time under the title *Reports of a Mathematical Colloquium, Second Series*.

Eight issues appeared between 1938 and 1946, but their impact did not compare with that of the *Ergebnisse*. When after the war Menger accepted a position at the newly founded Illinois Institute of Technology, he discontinued the series.

It seems that after Menger's emigration, his career lost its momentum. Within the American mathematical community, he certainly held a respected position, but he did not experience the tremendous success of fellow-emigrants John von Neumann, or Stanislas Ulam, or Abraham Wald, his former student, who dominated mathematical statistics within a few years. During the war, Menger published relatively little. This was not because his work was classified like that on computers or the bomb, but his enormous administrative and teaching load (he was engaged in the mathematical training of Navy cadets) made research almost impossible.

Menger's subsequent scientific work found less acclaim than his earlier work on curves and dimension; more precisely, its immediate impact was smaller. Nevertheless, his productivity remained impressive. He wrote on hyperbolic geometry and introduced a statistical notion of distance which he planned to analyse jointly with Wald[32]. But Wald died in 1950 during a plane crash, and Menger, thereafter, rarely returned to geometric problems. He devoted part of his work to an 'algebra of analysis', and in particular to the investigation of an algebra of functions which, today, is occasionally called a Menger algebra. His widespread teaching activity induced him to critically discuss the usual approaches towards teaching calculus, and to devise some more transparent notations. But in spite of his convincing and occasionally very witty arguments[33], Menger was unable to vanquish the inertia of tradition, and what he termed the '*x*-itis' of calculus curricula continues to this day to mar the classroom experiences of countless students. The textbook on calculus published by Menger in 1955 soon vanished from the market, in part because it was poorly produced and hardly advertised[34].

Menger's family increased by three more children in the US. It is unlikely that he would have seriously considered returning to the Vienna of after-war years[35]. But the University of Vienna certainly never did try to lure him back, and Menger was deeply hurt by this lack of interest. When he had accepted his position in Notre Dame, he had not given up his post in Vienna, but had only asked for an extended leave (the usual procedure in such a situation). Soon after the annexation of Austria, in 1938, Menger sent a telegram to Vienna resigning from his *Extraordinariat für Geometrie*, obviously in order to anticipate his dismissal by the Nazi authorities[36].

After the war, the dean wrote to the Austrian ministry that Menger had resigned his position of his own will, and hence that there was no cause, 'in the strict sense', for redressing a wrong.

In his later years, Karl Menger's enthusiasm for travel remained unbroken (in 1951–2, for instance, he spent a sabbatical at the Sorbonne in Paris); but only in 1963 did he return to Austria – as a guest of the *Institut für Höhere Studien* which had just been founded by Oskar Morgenstern. In 1971, the Austrian Academy of Science elected him as a corresponding member, and in 1975 he received – as his father had done many years ago – the Distinguished Cross for Science and Art. In 1979 Menger published a very readable volume of his selected papers[37], which displayed his two major qualities: the diversity of his interests and the clarity of his arguments. During his last years, Menger worked on a book about the Vienna of his memories. When he died in 1985, he left a manuscript which was incomplete, but fairly well advanced. It was published in 1994 under the title *Reminiscences of the Vienna Circle and the Mathematical Colloquium*[38], and offers an insider's picture of the two Circles which had played such a decisive role in his life.

Geometrical results

In 1936 Karl Menger described the work done in his Colloquium for the *Annali della Scuola Normali Superiore di Pisa.* This article had the title *Sull'indirizzo di idee e sulle tendenze prinzipali del colloquio matematico di Vienna*[39], and its English translation is contained in this volume. (Menger was fairly fluent in Italian; moreover, he was helped by the young Italian mathematician Ludovico Geymonat, who at that time spent a year in Vienna[40]). In the introduction, Menger mentions three main topics for the Colloquium: geometry, logic, and the mathematical investigation of social and economic problems. For the remainder of his article, Menger concentrates on geometry, where the bulk of the achievements were due to him.

Menger begins by stressing the remarkable progress due to the use of coordinate systems in geometry: by representing points in the plane by pairs of numbers, and points in space by triples, geometrical results could be verified computationally. This revealed hitherto undreamt of realms of geometrical problems; it became possible to introduce higher-dimensional spaces, to classify surfaces through equations, to derive theorems on volumes and curvature by means of calculus, etc. But this development placed such an emphasis upon coordinate systems, that geometry threatened to degenerate into an arithmetics of numerical tuples. To counter this, members of the Colloquium aimed to emphasise a *coordinate-free* geometry characterised by a more direct approach and more natural problems. Furthermore, analytic methods had a tendency to restrict the diversity of objects studied: in contrast, Menger and his co-workers were intent on developing a *set-theoretic* (rather than *algebraic* or *analytic*) geometry whose objects were not necessarily defined by equations.

The first great field of applications was, of course, dimension theory. At first glance this seems to offer no problems: a set is said to be n-dimensional if its points are described by n coordinates. Accordingly, the continuous image of an interval (traditionally defined as a *curve*) ought to be one-dimensional. But in 1890 the Italian mathematician Giuseppe Peano, and shortly thereafter the German David Hilbert, had produced examples of such 'curves' which contained all the points of a square, and hence were obviously two-dimensional[41]. This discovery played an essential role in the developement of topology.

One of the main theorems in this field was due to Hahn, who later became Menger's teacher, and to the Polish mathematician Stefan Mazurkiewicz. In 1913, they succeeded independently in characterising the continuous images of intervals as those sets which are compact, connected, and locally connected. These so-called 'Peano continua' showed strikingly how misleading reliance on intuition could be, and therefore became important object-lessons for philosophers, and in particular for the members of the Vienna Circle. (Hahn ended a lecture for a general audience with the words: 'Intuition is not, as Kant thought, a synthetic *a priori* means of knowledge, but merely force of habit rooted in psychological inertia.'[42])

We have seen how Menger had reacted to Hahn's problem of finding a workable definition for the concept of curve. His approach was based on a surprisingly simple definition of dimension, which had already been conjectured by Poincaré and explicitly proposed (albeit in an imperfect version) in 1913 by Brouwer. At that time, it did not find its way into the consciousness of mathematicians, and neither Hahn nor Menger had been aware of it[43]. The basic idea was to define dimension inductively upward, i.e. stepwise from zero- to one-, then to two-dimensional spaces etc. This approach was so simple that it still worked in spaces which were more general than euclidean space – for instance, in those metric spaces which the French mathematician Maurice Fréchet had introduced at about the time when Menger was born.

If M is a subset of a metric space, and p a point of M, then M is said to be *zero-dimensional* at p if there exist arbitrarily small open sets containing p whose boundaries contain no point of M. M is n-dimensional at p if M has no lower dimension at p, and if there exist arbitrarily small open sets whose boundaries intersect M in a set having dimension smaller than n at each point.

For manifolds, i.e., those sets which can be described by coordinates, this definition coincides with the usual one, but it is vastly more general. In his Italian paper, Menger quotes a few of the main theorems: If, for instance, a closed set has dimension n (i.e. contains no point with a larger dimension, and at least one point with dimension n), then the subset of all points where M has dimension precisely n is itself n-dimensional at every point. Every n-dimensional set can be written as the union of $n + 1$ zero-dimensional sets, but not as a union of fewer such sets (incidentally this theorem is not valid for Brouwer's definition of dimension). Furthermore, every n-dimensional subset M can be mapped bicontinuously onto a suitable subset N of a $(2n + 1)$-dimensional euclidean space. In euclidean space, every closed n-dimensional set can be deformed into an n-complex by a continuous mapping arbitrarily close to the identity (for $n = 2$, for instance, into a finite union of triangles).

The second geometric theme stressed by Menger in his survey concerns convexity. In euclidean space, a set is said to be convex if, with every two points p and q, it contains the entire *segment* joining p and q. A segment is easy to define in a linear space. In general, however, one must define a segment by using only the notion of distance. A subset M of a metric space is said to be convex if, with every two points p and q, it contains a point r lying in-between (i.e. such that the distance between p and q is the sum of the distances between p and r and between r and q). A closed subset of euclidean space is convex in the usual sense if and only if it is convex in the sense of the definition just given. If M is convex in this sense, and furthermore complete, then every two points p and q can be connected by a set which can be mapped in a distance-preserving way onto an interval. Such a set is called a *segment*. This notion allows one to transfer many classical problems into the setting of metric spaces, to develop a theory of geodesics (i.e. shortest paths) etc. Finding the corresponding axioms was Abraham Wald's first mathematical achievement. In his reminiscences, Menger mentions that this notion also attracted the interest of several philosophers[44]. Indeed, it is obviously related to the corresponding notion of *betweenness* in the axioms of euclidean geometry which are due to Pasch and Hilbert[45], but there are some interesting differences. In particular, it can happen that r lies between p and q, and q between r and s, but r does not lie between p and s. Menger illustrates this by means of the American railway net, using as distance the shortest time of travel. New York and Chicago are joined by two lines of equal length, one through Cleveland, the other through Pittsburgh. It follows that both Cleveland and Pittsburgh lie between New York and Chicago, but Youngstown, which lies on the shortest stretch between Pittsburgh and Cleveland, is not on the way between New York and Chicago.

Another paper on the notion of *metric convexity* is due to Gödel[46]. He associated to every triple (p, r, q) in a metric space a point (x, y, z) in the positive octant of euclidean three-space, such that x is the distance between p and r, y that between r and q, and z that between q and p. The triangle inequality implies $(x + y - z)(x - y + z)(-x + y + z) \geq 0$ and in addition $x + y = z$ whenever r lies between p and q. Wald's axioms for betweenness can therefore be translated into properties of subsets of euclidean three-space. Menger compared this translation with the famous *Gödelization* which translates logical into number-theoretical propositions[47].

The third geometrical theme mentioned by Menger, after those on dimension and convexity, dealt with congruence. Under what conditions, for instance, does there exist a distance-preserving map from a given metric space onto euclidean n-space? Menger's elegant solution is based on an intrepid generalisation. He first introduces *complex metric spaces*. The 'distance' between the two points p and q is a complex number which is required to be equal to the 'distance' between q and p (symmetry), but the triangle inequality is no longer required, and neither must p and q coincide if their 'distance' is zero. One assumes instead that the 'distance' is zero whenever $p = q$, and that if $p \neq q$, then there exists a point r whose 'distance' to p is different from its 'distance' to q. The most important examples are the space K^n of complex n-tuples, using as 'distance' between $(x_1, \ldots, x_n)$ and $(y_1, \ldots, y_n)$ the sum of the squares $(x_1 - y_1)^2 + \ldots + (x_n - y_n)^2$, and those subsets of K^n for which the first k coordinates are real and the others purely imaginary. Menger then derives an algebraic relation between the 'distances', which allows one to decide whether $n + 3$ given points of an arbitrary complex metric space can be imbedded in K^n by means of a congruence; and this yields the desired characterisation of the usual euclidean space. This example shows that in geometry, just as in number theory, the shortest path between two theorems may be via complex numbers.

For a partisan of coordinate-free geometry, it is of course eminently gratifying to see that euclidean space, i.e., the space underlying the usual analytic geometry, can be characterised by mere metric relations. The greatest challenge, however, is posed by differential geometry, which was born as an application of calculus and seems, at first, inextricably linked with coordinate systems. The challenge was taken up by Menger and Wald: they succeeded in deriving, without any use of calculus, a wealth of theorems containing propositions of classical differential geometry as special cases.

The most fundamental concept in this context is arguably the curvature of a surface. To four given points p_1 to p_4 of an arbitrary metric space, one can associate a curvature k if there exists a 'sphere' of curvature k containing four points $q_1, \ldots, q_4$ which are congruent to $p_1, \ldots, p_4$. (Here a 'sphere'of curvature $k > 0$ is an ordinary euclidean sphere of radius $1/\sqrt{k}$ with the usual great-circle distance, a 'sphere' of curvature *0* is a euclidean plane, and a 'sphere' of curvature $k < 0$ is a hyperbolic plane of curvature k.) Such a 'sphere' may not always exist (though Gödel, as Menger later wrote, 'found it not below his dignity' to prove that every non-planar quadruple in euclidean three-space is congruent to a quadruple on a sphere). When such a curvature exists, it need not be unique, but it is now possible to define the curvature at a given point p of a compact, metric space as the unique accumulation point, when such exist, of the curvatures of 'spheres' corresponding to points $p_1, \ldots, p_4$ arbitrarily close to p. If the metric space happens to be defined by a three times differentiable function in euclidean space, this reduces to the classical definition of Gaussian curvature. Amazingly, a converse is valid: if there exists a finite curvature at every point of a compact convex space, then this space is congruent to a Gaussian surface[48]. With this result, which is due to Wald, Menger concludes his Italian paper on the geometric investigations of the Colloquium.

It must be stressed that this manifesto of coordinate-free geometry certainly does not cover all the geometrical results published by Menger and his colleagues in the *Ergebnisse*. For instance, it does not even mention numerous papers on curves contained in this journal. Some of these contributions gained relevance in fields far removed from the original problems. A case in point is Menger's n-arc theorem, which became one of the fundamental theorems of graph theory[49]. A graph consists of finitely many vertices and connecting arcs. If A and B are two disjoint subsets of such a graph, each consisting of n vertices, then there either exist n disjoint paths, each connecting a point in A to a point in B, or else there exists a set of less than n vertices which separates A and B. Menger later described how he discovered this result, which is known today as Menger's theorem and is considered as *the* fundamental result on connectedness of graphs[50]. Menger also recalled that the Hungarian mathematician Denes König, who at that time was writing an encyclopedic work on graph theory, did not believe this theorem and told Menger, on taking leave from him in the evening, that he would not go to bed before he found a counterexample. Next morning he greeted him with the words: 'A sleepless night!'

Menger also became one of the founders of lattice theory, through his work on the algebra of geometric incidence relations. This was again a question of some philosophical importance, for it shed a new light on Euclid's famously obscure definition: 'A point is that which has no part'.

It is also worth to mention a short note by Gödel on projective mappings: every one-to-one transformation of the real projective plane onto itself which carries straight lines into straight lines is continuous and therefore, according to a fundamental theorem of projective geometry, a collineation[51].

Logical results

The first person to work on mathematical logic in Vienna seems to have been Olga Hahn, Hahn's younger sister, who had lost her eyesight at the age of twenty-two. Already before the first World War, she wrote several important papers in this field (in part jointly with her husband Otto Neurath)[52]. Hans Hahn himself never published on mathematical logic, but he was fascinated by it from an early age, mostly through the writings of Bertrand Russell. He frequently gave lectures, both to general and to academic audiences, on logical and set-theoretical questions. Popper conjectured that Hahn's seminar on the *Principia Mathematica* had a decisive impact on the young Kurt Gödel[53]. It was also Hahn who advised Menger to spend his postdoctoral years with Brouwer. As one of the leading figures in topology, Brouwer was obviously a very attractive choice for young dimension theorists like Menger or Hurewicz. But in the twenties, Brouwer's main interest had shifted to the foundations of mathematics. He had developed his own approach to the field, an *intuitionism* which contrasted sharply with Russell's *logicism* and Hilbert's *formalism*. It subsumed the early qualms of French mathematicians towards Cantor's set theory and Zermelo's axiom of choice, and was based on a fundamental distrust of operations involving infinity[54]. Brouwer defended the view that one was not allowed to use the principle of the excluded middle, the *tertium non datur*, for infinite sets. If one has shown, for instance, that the statement 'No three-digit numbers have property C' is false, then one can conclude that there exists at least one three-digit number with property C. But according to Brouwer, the negation of the statement 'No natural numbers have property C' does not imply that there exists a natural number with property C: this can only be proved by actually displaying such a number, or at least a method for constructing it. Intuitionists (a minority among mathematicians, but including such luminaries as Weyl) accept only a restricted part of classical analysis.

Inevitably, a philosophically inclined mathematician like Menger had to feel intensively challenged by Brouwer's position. He soon adopted a very skeptical stance towards it. This may have been caused in part by the aforementioned personal disappointement with Brouwer, but mostly, no doubt, by Menger's antipathy to vague or obscure claims. Nevertheless, after his return from Amsterdam, he seconded the invitation to Brouwer by Hahn and the physicist Ehrenhaft, which led, in March 1928, to two guest lectures at the University of Vienna. Wittgenstein listened to one of them[55]. It caused him to return to philosophy after several years of abstinence. The then completely unknown student Kurt Gödel also attended the lectures[56].

Accordingly, quite a few of the contributions to mathematical logic in the mathematical Colloquium dealt with intuitionism. The general trend consisted in *relativising* it. Menger's very first lecture on the subject was characteristic: it decribed an 'intutionistic-formalistic dictionary' which translated the rather awkward notions of Brouwer's set theory into familiar concepts of classical set theory, as found in Hausdorff's well-known treatise, among others. Brouwer's 'sets' corresponded to the analytic sets (an intensively studied notion at that time), his 'species' became subsets of analytic sets, his 'finitary sets' translated into compact sets and the particularly involved notion of 'individualised sets' turned out to be the counterpart of Borel sets[57].

Another basic idea in Menger's contributions to the f foundational debate later received great acclaim (in particular through Carnap's writings), as the 'logical tolerance principl[e]'58. According to this view, there does not exist jus*t o*ne logic, o*r o*ne notion of constructibility; there exist many of them. In the same vei there exists not just one intuitionistic viewpoint, but a whole range of them: after all, the creeds of Poincaré, Weyl, or Brouwer were far from congruent. Accordingly, a recourse to intuition (or the *Ur-Intuition*) was not well-defined: different mathematicians understood it differently.

Gödel elaborated this point in several contributions, with great success. Four years younger than Menger, he had also started out as a student of Hahn. In his Ph.D. thesis, which appeared in 1930 in the *Monatshefte*, he had solved a problem of Hilbert and proved the completeness of first order logic: every valid statement could be derived from the axioms[59]. However, this calculus concerns only a restricted range of formulas, which includes, roughly speaking, statements of the form '*f(x)* holds for all *x*', but not '*f(x)* holds for all *f*'. This Completeness Theorem was a great and widely noticed achievement, but in the following year, Gödel capped it with a historical breakthrough[60]. He proved that every mathematical theory encompassing the natural numbers was necessarily incomplete: there exists no finite system of axioms from which all valid statements can be derived. This was a heavy blow to Hilbert's program for the foundation of mathematics. Indeed, Hilbert had hoped that a consistency proof would be the core of his metamathematics, whereas Gödel showed that a system containing a proof of its own consistency is necessarily inconsistent!

It seems that Gödel first reported his incompleteness theorem in August 1930, in the 'Café Reichsrat', when he met with some members of the Vienna Circle to plan a joint trip to Königsberg[61]. In Königsberg, during the concluding discussion of a symposium on the foundations of mathematics, Gödel mentioned his theorem in public for the first time[62]. In November 1930, he finished writing it down in his famous *Habilitationsschrift*, which Hahn quickly published in the *Monatshefte*[63]. The first time that Gödel actually lectured on the Incompleteness Theorem was in February 1931. This was the lecture officially required for the habilitation (a lengthy academic procedure which essentially decided whether the candidate was deemed sufficiently qualified for a professorship).

In this lecture, which took place as part of the Colloquium, Gödel presented a simpler approach to the incompleteness results than in his written thesis for the habilitation[64]. Instead of Russell's theory of types, he used as basis Peano's axioms for the natural numbers. This became the canonical approach. Whereas some mathematicians had originally misunderstood Gödel's result as a statement concerning the axioms of *Principia Mathematica*, it now became clear that the Incompleteness Theorem holds for every system of axioms which includes natural numbers.

Gödels short report on his habilitation lecture was not his first contribution to the *Ergebnisse*. Before this, he had lectured on the decision problem[65]. Church and Turing had not yet clarified what actually characterises efficient algorithms, so that undecidability results in today's sense were not available. (In this context it may be mentioned that Gödel accepted Church's thesis only when it became clear that it was equivalent to characterisation via the Turing machine, the abstract forerunner of all modern computers[66]). This paper on the decision problem contains a gap which Gödel corrected one year later in the *Monatshefte*[67] (cf. W. Goldfarb's comment in volume I of Gödel's *Collected Works*[68]).

In 'A property of the realisations of the propositional calculus'[69], Gödel gave an answer to a question which Menger had asked him in a letter: even an uncountable set of statements representing a realisation of the axioms of the propositional calculus can be divided into two disjoint classes – the 'true' and the 'false' statements. In his next publication[70], Gödel deals with a question by Hahn: can every statement about independence be proved in the propositional calculus by means of a finite model? Gödel gives a counterexample. Obviously both Hahn and Menger were able to stimulate their young colleague even after his habilitation. Gödel's biographer Hao Wang conjectures that Gödel's readiness to publish his results was greatly enhanced by the opportunity to see them appear very quickly in print, an opportunity provided both through Hahn's *Monatshefte* and Menger's *Ergebnisse*[71]. In Gödel's American years this changed dramatically.

In his next two publications, Gödel deals with intuitionism and obtains several remarkable results. Since intuitionists do not accept many classical proofs, the theorems of intuitionistic number theory obviously constitute a proper subset of the theorems of classical number theory. But Gödel showed that a simple translation (having nothing to do with Menger's 'dictionary') transforms every classical theorem into an intuitionistic counterpart: according to this mapping, classical number theory is a subsystem of intuitionistic number theory[72]. In particular, the theorems of intuitionist mathematics are in no way more certain, or more consistent, than ordinary theorems. (This result had been partly anticipated in a paper written in 1925 by Kolmogoroff, which received little attention at the time and which Gödel had not known[73].) In his second note, Gödel provides an interpretation of intuitionistic propositional calculus by means of the supplementary concept '*p* is provable'[74]. (In this note Gödel refers to a related interpretation due to Kolmogoroff[75]).

The last paper by Gödel in the *Ergebnisse* (and the last that Gödel wrote in German) is concerned with the length of proofs[76]. Gödel claims (without proof) that the passage from an n-th order logic to an $(n + 1)$-th order logic not only allows one to prove statements that were formerly unprovable, it also allows one to shorten infinitely many of the proofs to a remarkable degree. In complexity theory (a branch of theoretical computer science) theorems of this type were later referred to as 'speed-up theorems'[77]. In this context it is interesting to remark that Gödel added, in proof, that his concept of 'computable' is independent of the underlying logic, quite in contrast to concepts like 'provable', 'definable' etc. In a paper submitted some ten years later[78], Gödel stressed that his notion coincides with Turing's notion of computability, which has come to play such an eminent role in computer science.

The Polish logician Tarski, who visited Vienna on several occasions, delivered some lectures on mathematical logic in the Colloquium[79]. In particular, he reported on the logical notation due to Lukasiewicz, which represented binary operations such as 'p implies q', not by '$p \to q$' as is usual, but by '*Cpq*', thereby placing the relevant symbol not *between*, but *in front of* the variables p and q. One of the advantages of this notation is that it enables one to write more complex formulas, e.g., $((p \to q) \to (q \to r)) \to (p \to r)$, without parenthesis (in this case, as *CCCpqCqrCpr*). In the *Tractatus*, Wittgenstein had stressed the 'apparently unimportant fact' that logical relations needed parenthesis, which was obviously falsified by this notation[80]. Menger did not fail to draw the attention of his Wittgenstein-addicted colleagues in the Vienna Circle, in particular Schlick and Waismann, to this error of their idol[81]. More importantly, the notation devised by Lukasiewicz (and independently by Post) yielded a simple computational criterion characterising all strings of symbols corresponding to meaningful logical propositions[82]. This provided a transparent connection between the structure and the meaningfulness of an expression. Menger was to return to this paper on several occasions[83].

During a later visit (in 1935), Tarski showed that under certain conditions, there exists a unique complete extension of a given contradiction-free set of propositions, and that most of the usual systems of propositional calculus satisfy these conditions[84].

Wald, Probability, Economics

The Vienna Circle was interested in the foundations, not only of set theory, number theory and analysis, but also of probability theory. Hilbert, too, on many occasions voiced his desire to see probability theory provided with a solid basis[85]. How can one characterise a 'random sequence' of zeros and ones (or of 'heads' and 'tails'), for instance? The first promising attempt at a solution was due to Richard von Mises[86]. (His brother, the world-renowned economist Ludwig von Mises, headed a group of economists who pursued the tradition of Carl Menger's marginalism[87]). Richard von Mises, who had studied at the Technical University in Vienna, where he made friends with both Hans Hahn and Robert Musil, became, as professor of mechanics in Berlin, one of the major applied mathematicians of his time. His textbook, 'Theory of Flight', influenced generations of aerodynamicists[88]. On top of this, Richard von Mises was an eminent philosopher (during the years of his exile in Turkey, he wrote a textbook on positivism[89]). From 1919 onward, in several papers and books he attempted to define the notion of a random sequence by means of 'collectives'[90]. A random sequence of zeros and ones must of course be 'disorderly': on the other hand, the relative frequency of ones must converge to 1/2; furthermore, all subsequences of blocks of digits obtained by selection rules – the so-called collectives – should also have frequencies converging to the proper limits, for otherwise, it would be possible to devise a winning strategy for the game 'heads or tails'. Soon, however, logical inconsistencies were discovered in the original definition of von Mises. This led to heated debates, among others between the philosophers Hans Reichenbach and Karl Popper[91]. When Popper elaborated upon the point of view which he had presented in his *Logik der Forschung* (which was to become *The Logic of Scientific Discovery*), he developed an improved version of von Mises' notion of a random sequence, and lectured on it in 1935 in the Colloquium[92]. As Menger later wrote in an obituary of Abraham Wald (which Popper quoted in his autobiography[93]), this lecture led to one of Wald's most important achievements, which he soon presented in the *Ergebnisse* under the title 'The consistency of the notion of collective in the theory of probability'[94]. It did not solve Popper's problem[95], but yielded a pathbreaking contribution to the foundations of probability. At this point, Kolmogoroff must again be mentioned: after having decisively shaped the development of probability theory in 1933 via his measure-theoretic approach, he returned in the early 'sixties to von Mises' and Wald's concept of 'collective' and gave the impetus for further important developments[96]. Wald himself found no more time, after his emigration, to deal with foundational questions; but the notion of a random sequence of observations played a major role in his later statistical investigations.

Two extremely important papers on mathematical economics which appeared in the *Ergebnisse* were also due to Wald. We have seen how his acquaintance with the banker Karl Schlesinger had drawn his attention to this subject. He started out with the so-called imputation problem, which played a central role in the theories of Carl Menger and Leon Walras: how do the prices of the means of production determine the prices of the products? The first to lecture in the Colloquium on the fundamental equations of Walras and Cassels was Schlesinger[97]. He stressed the fact that all the means of production which occured in the equations had to be 'scarce' in the sense that they were entirely used up in the production, i.e. that there remained no surplus. However, whether this holds or not is itself dependent on the production process. But any commodity which is in surplus has price zero. Similarly, if a commodity can only be sold for a price which is lower than its production costs, then it will not be produced at all. This yields a system of equations and inequalities instead of Cassel's system of equations. (The Danish economist Zeuthen discovered the same fact independently of Schlesinger.)

Taking Schlesinger's new system as his point of departure, Wald proved the existence of a unique positive solution, i.e. of a stable equilibrium[98]. This was a giant step forward: thus far, economists had been content with pointing out that the system of Walras and Cassels had the same number of equations as of unknowns, apparently under the impression that this implied the existence and uniqueness of a feasible solution. Someone like Menger could never accept such sloppy reasoning. He immediately understood the impact of Wald's paper and lectured on it on several occasions[99].

John von Neumann visited Vienna several times in the 'thirties (although he did not yet know Oskar Morgenstern[100]). He was greatly interested in Wald's results and, on Menger's urging, he published his own analysis of a model for an expanding economy in the *Ergebnisse*[101]. He had understood the role of inequalities in models for production at an even earlier date, during a seminar organised by Leo Szilard. In his dynamical economic model, he described a closed production loop: the supply is the output of the preceding period, and the demand is the input of the following period. John von Neumann proved the existence of an equilibrium solution by a generalisation of Brouwer's fixed point theorem. He underscored the similarities with his own minimax theorem, which was to become the basis of game theory. His Colloquium paper, which already foreshadows the essential ideas of George Dantzig's and Leonid Kantorovich's linear programming and of Koopman's activity analysis, is a precursor of the Leontieff model.

This paper by John von Neumann, a milestone in the history of mathematical economics, was to become the last article in the *Ergeb-*

nisse[102]. Wald, who (as a co-editor) allowed John von Neumann's paper to appear without any reference to his own contribution, had finished a further article on mathematical economics which he planned to publish in the following, ninth, volume of the series. In it he proved the existence of an equilibrium in a pure exchange economy, most probably also using Brouwer's fixed point theorem. We shall never know for sure: after the *Anschluss*, the forced annexation of Austria by Hitler's Germany in 1938, it became impossible to continue the *Ergebnisse*, and the manuscript disappeared. 'Wald's lost paper' became a legend among mathematical economists.

The book by von Neumann and Morgenstern on 'Game Theory and Economic Behaviour' appeared a few years later[103]. It stimulated some further important investigations by Wald[104]. But he was never to return to his equilibrium theorem for a pure exchange economy.

Menger himself participated only indirectly in this tremendous development, which dramatically increased the degree of mathematisation in the economic sciences and greatly affected the work of younger mathematicians, eventually leading to the 'general equilibrium thoery' for which Arrow and Debreu received their Nobel prizes[105]. But Menger published, in the *Ergebnisse* and elsewhere, highly original contributions to the social and economic sciences. Like John von Neumann, in these fields he tried to develop new mathematical techniques which were not simply transferred from the analytical techniques motivated by theoretical physics, but which were hand-tailored, so to speak, for the new problems. The other two main topics of the Colloquium – topology and logic – had already been championed in Vienna by Hans Hahn. The economic component, however, is entirely due to Menger's influence (notwithstanding the old friendship between Hahn and Schumpeter). Menger must, incidentally, have been struck by the fact that in this third field, as in the other two, Brouwer's influence was again inescapable; Brouwer's fixed point theorem turned out to be an essential tool for handling economic equilibria.

At the age of twenty, Menger had written an essay 'On the role of uncertainty in economics'. Ten years later, after being encouraged by Oskar Morgenstern, he published it in the *Zeitschrift für Nationalökonomie*. He also returned to this topic in the Colloquium[106]. The point of departure was the *St. Petersburg paradox*, which had provided probabilists with food for thought for more than two hundred years. It is a very simple game. Suppose that a casino offers you the following bet: throw a coin repeatedly, until the first time 'head' comes up. If this happens on the n-th throw, you will receive 2^n dollars. Of course you have to pay something for being admitted to the game. How much should you be prepared to pay? The first answer which comes to mind is: anything that is less than the expected value of your gain. But this value is infinitely large. Indeed, the probability that 'head' comes up for the first time at the n-th throw is 2^{-n}, and this yields a payoff of 2^n. Hence you should be prepared to pay any sum to be admitted to the game. In fact, you would of course not be prepared to do so, and neither would any reasonable person. This is the paradox. Bernoulli proposed an ingenious solution, but Menger recognised it as deficient. Whereas Bernoulli's approach had consisted in solving a probabilistic problem by means of the notion of subjective values, Menger tried to introduce subjective probabilities into the theory of values[107]. Morgenstern later reported that Menger's idea inspired von Neumann in his axiomatic treatment of utility[108].

In another paper, Menger dealt with the law of diminishing returns, which generally was considered to be the most secure part of mathematical economics[109]. Once again, all so-called proofs of this result failed to satisfy Menger's rigourous standards: painstakingly, he listed their gaps. Schumpeter wrote that with this paper, Menger was 'reading us (economists) the logician's riot act'[110].

Menger also undertook a particularly original attempt to apply exact logical thinking to ethical statements and value judgements. In a very moving part of his reminiscences, he describes how difficult it was for him to concentrate on pure mathematics during the politically uncertain situation in the 'thirties[111]. In this distressful time, he tried to sketch the principles of a formal ethics which should relate to traditional ethics just as mathematical logic does to traditional logic. The members of the Vienna Circle would of course never accept ethical statements as *knowledge*, since they could be neither justified nor falsified empirically. But Menger proposed to analyse statements about groups of adherents of ethical norms and about their preferences that certain norms were allowed, forbidden, required, unenforceable, or free. Among others, Menger defined 'well-rounded preferences' (for those who only wanted that *one* of these five possibilities held for some of n given norms), and he proved a related combinatorial theorem. It seems that Menger's approach to studying the compatibility of individual preference relations in groups had a strong influence on Morgenstern and, through him, on von Neumann and the developement of cooperative game theory[112].

Menger even wrote a little booklet on the logic of ethics, entitled 'Morality, Decision and Social Organization', and he lectured about it in the Colloquium[113]. In the politically supercharged atmosphere of the time, the University of Vienna was frequently closed, often for months at a time, but the Schlick Circle kept meeting, and so did the Colloquium. Menger, who as professor had a key to the building housing the Mathematics Institute, had to let the members in. A strange mood must have reigned in the room where the members of the Colloquium met: a small group lost in a huge, empty building, devotedly discussing economic equilibria, logical consistency and formal ethics while outside the police battled with illegal Nazis.

Menger assuredly would never have lectured on his personal ethical principles – this was something about which one could only remain silent, Wittgensteinian or not – but his passionate need for mathematical, irrefutable clarity broke through, more touchingly than in any confession, in his attempt to counter the disasters of a smouldering civil war by a calculus of theoretical ethics. Right after Menger's lecture, Wald proved that there exist exactly $2 \times 3^n + 2^{n-1}(n-2) - (n+1)^2 - 3\binom{n}{2}$ well-rounded preferences[114].

References

As an excellent biographical introduction, we recommend:
Seymour Kass (1996), Karl Menger, *Notices of the AMS*, **43**, 558–561.
In the references, we use the following abbreviations:
REM: K. Menger, *Reminiscences of the Vienna Circle and the Mathematical Colloquium*, Vienna Circle Collection vol. **20**, Kluwer, Dordrecht (1994).
SEL: K. Menger, *Selected Papers in Logic and Foundations, Didactics, Economics*, Vienna Circle Collection vol. **13**, Kluwer, Dordrecht (1979).
ERG: *Ergebnisse eines mathematischen Kolloquiums*.
MON: *Monatshefte für Mathematik und Physik.*
STAD: F. Stadler, *Studien zum Wiener Kreis*, Suhrkamp, Frankfurt a.M. (1997).

1. *REM* p.16–7.
2. Introduction to *ERG*, **1** (1931) p. 1.
3. STAD p. 627ff and F. Stadler, Wiener Kreis – Versuch einer Typologie, in *Jour Fixe der Vernunft*, (Ed. P. Kruntorad) Hölder-Pichler-Tempski, Wien (1991) p. 23–41.
4. K. Menger: Hans Hahn, *ERG* **6** (1934), p. 40–44 ; Ph. Frank: Hans Hahn, *Erkenntnis* **4** (1934), 315–6; K. Sigmund: A philosopher's mathematician – Hans Hahn and the Vienna Circle, *Mathematical Intelligencer*, **17** (4), 16–29 (1995); K. Popper, Hans Hahn – Reminiscences of a grateful student, in: *Hans Hahn – Collected Works* (Eds. L. Schmetterer und K. Sigmund), Vol 1, Springer, Vienna (1995).
5. *REM*, p. 147 ff.
6. K. Popper, *Unended Quest: an Intellectual Autobiography*, updated version (1992), Routledge, London, p.100–102.
7. J. Schumpeter, *History of Economic Analysis*, Oxford University Press (1954) p. 825 ff.

8. W.M. Johnston, *The Austrian Mind*, Univ. of California Press (1972), Chapters 4 and 5.
9. *REM* p. 38 ff; K. Menger, My memories of L.E.J. Brouwer, in *SEL* Ch. 21.
10. K. Menger, Über die Dimensionalität von Punktmengen Teil I, *MON* **33** (1923) 148–160; Teil II, *MON* **34** (1924), 137–161.
11. cf. 9.
12. L.E.J. Brouwer, *Proc. Acad. Amsterdam* **31** p.957. K. Menger, Antwort auf eine Note von Brouwer, *MON* **37** (1930) p. 175–182; Eine Zuschrift von Karl Menger an Hans Hahn betreffend Mengers Antwort auf eine Note von Brouwer in Band 37 der Monatshefte, *MON* **40** (1932), 233.
13. Hlawka, E., Nachruf auf E. Hornich, *MON* **89**, (1980), 1–8.
14. M. Pinl and A. Dick, Kollegen in einer dunklen Zeit, *Jahresberichte der DMV* **75** (1974), p. 183, and pages 850–859 of *ACM*, **30** (1987). See also H. Goldstine, *The computer from Pascal to von Neumann*, Princeton UP (1972), p. 115–6.
15. E. Luchins and M.A. McLoughlin, In memoriam: Olga Taussky–Todd, *Notices of the AMS* **43** (1996), 838–847.
16. Menger K., The formative years of Abraham Wald and his work in Geometry, *Annals of Math. Statistics* **23** (1952), 14–20.
17: K. Menger, Untersuchungen über allgemeine Metrik I-IV, *Mathematische Annalen* **100**, 75–163.
18 L.M. Blumenthal *Theory and Applications of Distance Geometry*, Clarendon Press, Oxford.
19.H. Sagan (1994) *Space-filling Curves*, Springer, New York.
20. J. von Neumann, Continuous geometry, *Proc. Nat. Acad. Sci.* **22** (1936); J.v. Neumann und G.D. Birkhoff, The logic of quantum mechanics, *Annals of Mathematics* **37** (1936), 823–43. Menger, K., Bemerkungen zu Grundlagenfragen IV: Axiomatik der endlichen Mengen und der elementargeometrischen Verknüpfungsbeziehungen, *Jahresbericht der Deutschen Mathematiker-Vereinigung* **37** (1929) p. 309–25. See also *REM* p. 220–224, and G. Grätzer, *General Lattice Theory*, Birkhäuser (1978).
21. K. Menger, *Dimensionstheorie*, Leipzig-Berlin (1928) and K. Menger, *Kurventheorie* (in collaboration with G. Nöbeling), Leipzig (1932), reprinted New York (1967).
22. Menger, K., Bemerkungen zu Grundlagenfragen I: Über Verzweigungsmengen, *Jahresbericht der Deutschen Mathematiker-Vereinigung* **37** (1929) 213–226.
23. *REM* p. 200–233.
24. K. Menger, Bernoullische Wertlehre und Petersburger Paradox, *ERG* **6** (1934), 26–27. For an expanded version see Ch. 22 in *SEL*.
25. cf. 7.
26. K. Menger, *Moral, Wille und Weltgestaltung*, Wien (1934).
27. See e.g. STAD, R. Haller, *Neopositivismus: eine historische Einführung in die Philosophie des Wiener Kreise*, Wissenschaftliche Buchgesellschaft Darmstadt (1993) and M. Geier, *Der Wiener Kreis*, rororo Monographie **508**, (1993).
28. See STAD.
29. E. Hlawka, Karl Menger, *Almanach der Österreichischen Akademie der Wissenschaften* **136** (1986) 342–61.
30. R. Einhorn, *Vertreter der Mathematik und Geometrie an den Wiener Hochschulen 1900–40*, Dissertation TU Wien (1985).
31. cf. the volume dedicated to A. Wald of *Annals of Math. Statistics* **23** (1952).
32. see B. Schweizer and A. Sklar, *Probabilistic metric spaces*, North-Holland, New York (1983).
33. K. Menger, Why Johnny hates Math, *The Mathematics Teacher* **49** (1956), 578–584. See also the Gulliver articles reprinted in Chapters 24–6 of SEL.

34. K. Menger, *Calculus. A Modern Approach*, Boston (1956).
35. cf.: In zwei Wiener Kreisen zugleich: Interview mit Karl Menger, *Die Presse*, 18.8. 1978.
36. Personalakte Menger, Archiv der Philosophischen Fakultät der Universität Wien.
37. *SEL.*
38. *REM.*
39. K. Menger, Sull'indirizzo di Idee e sulle Tendenze Principali del Colloquio Matematico di Vienna, *Annali di Pisa* **4** (1935) 1–13.
40. L. Geymonat, Persönliche Erinnerungen an den Wiener Kreis, in *Jour Fixe der Vernunft*, (Hrsg. P. Kruntorad) Hölder-Pichler-Tempski, Wien (1991), p. 42–48.
41. See 19.
42. H. Hahn, Die Krise der Anschauung (1934), in *Krise und Neuaufbau in den exakten Naturwissenschaften*, Fünf Wiener Vorträge, F. Deuticke, Leipzig und Wien (1933). A shortened translation appeared in the *Scientific American* (1954) (4), p. 84–91.
43. L.E.J. Brouwer, Über den natürlichen Dimensionsbegriff, *Crelles Journal für reine und angewandte Mathematik* **142** (1913), 146–52. For a very readable survey of dimension theory see H.C. Reichel, Zur Entwicklung des Dimensionsbegriffes, *Math. Semesterberichte* **27** (1980), 16–54.
44. *REM* p. 115
45. M. Pasch, *Vorlesungen über neuere Geometrie*, Teubner, Leipzig (1882); D. Hilbert, *Grundlagen der Geometrie*, Teubner, Leipzig (1899).
46. K. Gödel, Über die metrische Einbettbarkeit der Quadrupel des R^3 in Kugelflächen, *ERG* **4**, 16–7.
47. *REM* p. 208.
48. A. Wald, Begründung einer koordinatenlosen Geometrie der Flächen, *ERG* **7**, 24–45.
49. K. Menger, Zur allgemeinen Kurventheorie, *Fundamenta mathematica* **10** (1927), 96–115; K. Menger, Über plättbare Dreiergraphen und Potenzen nichtplättbarer Graphen, *ERG* **2**, 30–31.
50. K. Menger, On the origin of the *n*-arc theorem, *Journal of Graph Theory* **5** (1981), 341–50.
51. K. Gödel, Bemerkung über projektive Abbildungen, *ERG* **5** (1933) 1.
52. O. Hahn, Über die Koeffizienten einer logischen Gleichung und ihre Beziehungen zur Lehre von den Schlüssen, *Archiv für sys. Phil.* **15** (1909); O. Hahn und O. Neurath, Zum Dualismus der Logik, *Archiv für sys. Phil.* **15** (1909)
53. See K. Popper, Hans Hahn – Reminiscences of a grateful student, in: *Hans Hahn – Collected Works* (Eds. L. Schmetterer und K. Sigmund), Vol **1**, Springer Vienna (1995).
54. L.E.J. Brouwer, *Collected works* (ed. A. Heyting), Vol. **1** (1975).
55. L.E.J. Brouwer, Mathematik, Wissenschaft und Sprache, *MON* **36**, 153–64.
56. Hao Wang, *Reflections on Kurt Gödel* MIT Press, Cambridge, Mass (1987) p.49, and J.W. Dawson jr. *Logical Dilemmas: the life and work of Kurt Gödel*, Peters, Mass. (1997).
57. K. Menger, Ein intuitionistisch-formalistisches Wörterbuch, *ERG* **1** (1931) 30.
58. K. Menger, Logical tolerance in the Vienna Circle, *SEL*, p. 11 ff.
59. K. Gödel, Die Vollständigkeit der Axiome des logischen Funktionenkalküls, *MON* **37** (1930), 349–60.
60. K. Gödel, Über formal unentscheidbare Sätze der *Principia Mathematica* und verwandter Systeme I, *MON* **38**, 173–198.
61. E. Köhler, Gödel und der Wiener Kreis, in *Jour Fixe der Vernunft*, (Hrsg. P. Kruntorad), Hölder-Pichler-Tempski, Wien (1991) p. 127–158.
62. Diskussion zur Grundlegung der Mathematik, in *Erkenntnis* **2** *(1931), 147–51.*
63. K. Gödel, Über formal unentscheidbare Sätze der Principia Mathematica und verwandter Systeme I, *MON* **38** (1931), 173–98.

64. K. Gödel, Über Vollständigkeit und Widerspruchsfreiheit, *ERG* **3** (1932), 12–3.
65. K. Gödel, Ein Spezialfall des Entscheidungsproblems der theoretischen Logik, *ERG* **2** (1931) 27–8.
66. cf. p.18 of S. Feferman, Gödels Life and Work, in *Kurt Gödel – Collected Works* (ed. Feferman et al.) Oxford UP (1986), 1–36.
67. K. Gödel, Zum Entscheidungsproblem des logischen Funktionenkalküls, *MON* **40** (1933), 433–43.
68. W. Goldfarb, Introductory note, in *Kurt Gödel – Collected Works* Vol. I (ed. S. Feferman et al) Oxford UP (1986), 226–230.
69. K. Gödel, Eine Eigenschaft der Realisierungen des Aussagenkalküls, *ERG* **3** (1932) 20–21.
70. K. Gödel, Über Unabhängigkeitsbeweise im Aussagenkalkül, *ERG* **4** (1933) 9–10.
71. cf. Hao Wang, *Reflections on Kurt Gödel* MIT Press, Cambridge, Mass (1987), p.89.
72. K. Gödel, Zur intuitionistischen Arithmetik und Zahlentheorie, *ERG* **4** (1933) 34.
73. A.N. Kolmogoroff, *Matematicheskii Sbornik* **36** (1925), 646–67. Engl. Translation, On the Principle of the Excluded Middle, in: J. van Heijenoort (ed.) *From Frege to Gödel: a source book in mathematical logic, 1879–1931,* Harvard UP, Mass. (1967).
74. K. Gödel, Eine Interpretation des intuitionistischen Aussagenkalküls, *ERG* **4** (1933), 39–40.
75. A.N. Kolmogoroff, Zur Deutung der intuitionistischen Logik, *Mathematische Zeitschrift* **35**, 58–65.
76. K. Gödel, Über die Länge von Beweisen, *ERG* **7** (1936), 23–4.
77. R. Statman, Bounds for proof-search and speed-up in the predicate calculus, *Annals of mathematical logic* **15** (1978), 225–287.
78. K. Gödel, Remarks before the Princeton bicentennial conference on problems in mathematics, in M. Davis (ed.), *The Undecidable: basic papers on undecidable propositions, unsolvable problems and computable functions*, Hewlett NY, Raven Press (1965), p. 84–8.
79. viz. the 10th, 11th and 12th Kolloquium (*ERG* **2** (1930)) as well as the 91th und 96th Kolloquium (*ERG* **7** (1935)).
80. L. Wittgenstein, *Tractatus logico-philosophicus* (1921) Proposition 5.461.
81. *REM* p. 150–1.
82. K. Menger, Eine elementare Bemerkung über die Struktur logischer Formel, *ERG* **3** (1930). 22–3.
83. *SEL* 61–67.
84. A. Tarski, Über die Erweiterungen der unvollständigen Systeme des Aussagenkalküls, *ERG* **7** (1935) 51–57.
85. D. Hilbert, Mathematische Probleme, *Nachrichten von der Königlichen Gesellschaft der Wissenschaften zu Göttingen* (1900) 253–297.
86. F. Stadler, Richard von Mises – Wissenschaft im Exil. In Richard von Mises, *Kleines Lehrbuch des Positivismus* (F. Stadler, Hrsg.), suhrkamp taschenbuch wissenschaft, Frankfurt (1990), p. 7–48.
87. M. von Mises: *Ludwig von Mises. Der Mensch und sein Werk*, München, Philosophia (1981).
88. R. von Mises, *Fluglehre*, first edition Berlin (1918). Engl. title *Theory of Flight*, first edition New York (1945).
89. R. von Mises, *Kleines Lehrbuch des Positivismus*, first edition den Haag (1939). Cf. 86.
90. R. von Mises, Fundamentalsätze der Wahrscheinlichkeitsrechnung, *Mathematische Zeitschrift* **4** (1919) 1–97; *Wahrscheinlichkeit, Statistik und Wahrheit*, Vienna (1928); *Lectures on Probability and Statistics*, edited and supplemented by Hilde Geiringer, New York (1963).

91. K. Popper, 'Induktionslogik' und 'Hypothesenwahrscheinlichkeit', *Erkenntnis* **5** (1935), 170–2; H. Reichenbach, *Erkenntnis* **5** (1935), 367–84.
92. 84. Kolloquium, s. *ERG* **7** p.12.
93. cf. 6.
94. A. Wald, Die Widerspruchsfreiheit des Kollektivbegriffes der Wahrscheinlichkeitsrechnung, *ERG* **8** (1937) 38–72.
95. cf.6, p. 102.
96. A.N. Kolmogoroff, On tables of random numbers, *Sankhya* Ser A **25** *(1963), 369–76; C.P. Schnorr, Zufälligkeit und Wahrscheinlichkeit,* Springer Lecture Notes in Mathematics **218** (1971); G. Chaitin, *Information, Randomness and Incompleteness*, World Scientific, Singapore (1990).
97. K. Schlesinger, Über die Produktionsgleichungen der ökonomischen Wertlehre, *ERG* **6** (1934) p.10. see also F. Zeuthen, *Zeitschrift für Nationalökonomie* **4**, 1–24, (1932–3).
98. A. Wald, Über die eindeutige positive Lösbarkeit der neuen Produktionsgleichungen, *ERG* **6** *(1934) 12–18,* and A. Wald, Über die Produktionsgleichungen der ökonomischen Wertelehre (II), *ERG* **7**, 1–6 (1935); see also A. Wald, Über einige Gleichungssysteme der mathematischen Ökonomie, *Zeitschrift für Nationalökonomie* **7**, 637–70 (1936).
99. K. Menger, Österreichischer Marginalismus und mathematische Ökonomie, *Zeitschrift für Nationalökonomie* **32** (1972) 19–28.; expanded version in: Austrian Marginalism and Mathematical Economics, in *Carl Menger and the Austrian School of Economics* (ed. J.R. Hicks and W. Weber) Oxford UP (1973) p.38–60.
100. M. Shubik, Biographical article on Oskar Morgenstern, in *The new Palgrave: a Dictionary on Economics* (ed. Eatwell et al), Macmillan, London (1987) Volume 4, p.556.
101. J. von Neumann: Über ein ökonomisches Gleichgewichtssystem und eine Verallgemeinerung des Brouwer'schen Fixpunktsatzes, *ERG* **8** (1937) 73–83. This article is based on a lecture given by John von Neumann in 1932 at the departement of economics in Princeton. In 1945 this article appeared in translation in the *Review of Economic Studies*, with a commentary by Champernowne.
102. cf. J. Schumpeter, *History of Economic Analysis*, Oxford University Press (1954), p.968; see also the volume edited by M. Dore, P. Chakravarty and R. Goodwin 1989 with Oxford University Press *John von Neumann and Modern Economics*, and in particular the articles by K.J. Arrow, *Von Neumann and the Existence Theorem for General Equilibrium* (p. 15–28); P.A. Samuelson, *A Revisionist View of von Neumann's Growth Model* (p. 100–124); and L. F. Punzo, *Von Neumann and Karl Menger's Mathematical Colloquium*, p. 29–68.
103. J. von Neumann und O. Morgenstern, *Theory of Games and Economic Behaviour*, 2nd ed., Princeton UP (1947).
104. A. Wald, A review of J. von Neumann and O. Morgenstern, *Theory of Games and Economic Behaviour*, in: *Rev. Econ. Stat.* **29** (1947) p. 47–51; A. Wald, *Statistical Decision Functions*, John Wiley and Sons (1950).
105. cf. the article on 'General equilibrium' in *The new Palgrave: a Dictionary on Economics* (ed. Eatwell et al), Macmillan, London (1987), Vol. 2, p.498–511.
106. K. Menger, Bernoullische Wertelehre und Petersburger Spiel, *ERG* **6** (1934) p.26–7; K. Menger, Das Unsicherheitsmoment in der Wertelehre, *Zeitschrift für Nationalökonomie* **5** (1934) 459–485; Bemerkungen dazu, *Zeitschrift für Nationalökonomie* **6** (1935) p. 283–5.
107. see Chapter 22 in *SEL*.
108. H.W. Kuhn and A.W. Tucker, John von Neumann's work on the theory of games and mathematical economics, *Bull. Amer. Math. Soc.* **64** (1958) p. 100–122.

109. K. Menger, Bemerkungen zu den Ertragsgesetzen, *Zeitschrift für Nationalökonomie* **7** (1936) 25–26; Weitere Bemerkungen zu den Ertragsgesetzen, *ibid* p. 388–397.
110. J. Schumpeter *History of Economic Analysis*, Oxford University Press (1954) p. 587.
111. *REM*, p.181 ff.
112. See R.J. Leonard, From Parlor Games to Social Science, *J. of Economic Literature* **23**, 730–761.
113. cf. 26 as well as K. Menger, Ein Satz über endliche Mengen mit Anwendungen auf die formale Ethik, *ERG* **6** (1934), 23–6.
114. A. Wald, Über die Anzahl der abgerundeten Mengen, *ERG* **6** (1934) 26.

Logical contributions to the Menger colloquium

John W. Dawson jr.

Apart from articles on set-theoretical topics, the eight volumes of *Ergebnisse eines mathematischen Kolloquiums* contain reports of sixteen contributions to the field of mathematical logic. Three other presentations on logical topics, entitled "Über einige fundamentale Begriffe der Metamathematik", "Über die Vollständigkeit der Axiome des logischen Funktionenkalküls", and "Bemerkungen von Freges und Russells Definition der Zahl", are noted as having been delivered to the colloquium as well, during its eleventh, fifteenth and ninetieth sessions (20 February and 14 May, 1930, and 5 June 1935, respectively), but their contents were not published in the colloquium proceedings. In the first two cases, the speakers involved (Alfred Tarski and Kurt Gödel) presumably preferred that the results they discussed on those occasions appear in a journal having wider circulation[1]. Whether the third contribution (by Friedrich Waismann) ever appeared elsewhere is unknown to this commentator. The question is worth exploring, especially in view of the tantalizing comment (volume 7, page 15 of these *Ergebnisse*) that the following discussion included remarks by Gödel, Menger and Tarski.

In fact, those two presentations were significant events, not only for the colloquium, but for the developing careers of Tarski and Gödel. In his memoir of the colloquium (*Menger 1994*, pp. 146–157) Menger recalls that "Tarski eagerly accepted" his invitation to come from Warsaw to deliver a series of lectures, and that Tarski's "visit . . . was one of the first

1 Tarski's talk, a reprise of a lecture he had given two years earlier to the Warsaw Section of the Polish Mathematical Society (see the Bibliographical Note to the English translation, p. 30 in *Tarski 1983*), was first published in *Comptes Rendus des Séances de la Société et des Lettres de Varsovie* (*Tarski 1930a*). An expanded version (*Tarski 1930b*) appeared not long after in *Monatshefte für Mathematik und Physik*, in the same volume as Gödel's contribution (*Gödel 1930*).

important steps out of isolation for Polish logic and philosophy." Afterwards Tarski embraced Menger and "said he would never forget" the opportunity that Menger had provided him. As for Gödel, his presentation to the colloquium was the first occasion on which he lectured on the results of his dissertation: the completeness theorem for the first-order functional calculus[2]. (He had obtained his doctorate from the University of Vienna just three months before.)

Of the sixteen published contributions to the colloquium–half of them by Gödel–nine concern propositional logics (classical, intuitionistic, and modal), three are devoted to predicate logic of first or higher order, and three to formal number theory. The remaining article, and the only one published in the first volume of the proceedings, is a very brief and rather trivial note by Menger himself, in which four of the terms employed by Brouwer in his intuitionistic reconstruction of mathematics are (ostensibly) translated into their classical mathematical equivalents.

Presumably, Menger's intention was to make Brouwer's work comprehensible to a wider audience. The impression actually conveyed, however, is that Brouwer was an obscurantist who eschewed standard terminology; and though from a classical standpoint Menger's translations are literally correct, they are nonetheless highly misleading, since they take no account of the constructive context underlying Brouwer's intuitionistic program[3].

The first substantial discussion of logical issues in the colloquium was the aforementioned talk by Tarski, the second of the three that he gave there. In it Tarski introduced a number of basic metamathematical notions, including those of logical equivalence, consistency, completeness, independence, and finite axiomatizability, all of which he defined in terms of just two primitive concepts: that of *meaningful sentence* and that of the set *Cn(X)* of *consequences* of a set *X* of sentences under certain (unspecified) rules of inference. The primitives themselves were characterized by two groups of (informal) axioms, the first applying to "all deductive disciplines", including non-classical logics, the second to logics that include

[2] The completeness theorem is by now so well known, and the circumstances surrounding its discovery so thoroughly studied (see in particular the commentary by Dreben and van Heijenoort, pp. 44–59 in *Gödel 1986*), that I shall not discuss it further herein, except to note that it remains undetermined whether Gödel presented his compactness theorem in his lecture to the colloquium. It was not included in the dissertation itself, but did appear in its published revision (*Gödel 1930*).

[3] Menger's cavalier treatment of Brouwer and his philosophy was perhaps influenced by the well-known dispute between the two over priority for the creation of dimension theory.

symbols for negation and implication and "presuppose the [classical] sentential calculus". The five axioms of the first group stipulated that the set S of meaningful sentences be countable, that it be generated as the set of all consequences of a single sentence, and that it be closed under the consequence relation applied to any of its subsets; in addition, it was postulated that $Cn(Cn(X)) = Cn(X)$ for all $X \subseteq S$ and that $Cn(X)$ should coincide with the union of $Cn(Y)$ over all finite subsets Y of X.

Among the results Tarski announced as derivable from those five axioms were that $Cn(X) \subseteq Cn(Y)$ whenever $X \subseteq Y$ and that for every X there is a smallest $Y \supseteq X$ such that $Y = Cn(Y)$. Sets of sentences X for which $X = Cn(X)$ (theories, in modern terminology) Tarski called *systems*, and in terms of them he stated Lindenbaum's theorem that every consistent set X of sentences can be enlarged to a consistent, complete system[4]–another result provable from the first group of axioms alone. Then, having defined a set X of sentences to be *independent* if $Y \subseteq X$ and $Cn(Y) = Cn(X)$ entailed $Y = X$, he called an independent set of sentences X a *basis* for a set of sentences Z if $Cn(X) = Cn(Z)$, and noted that the first group of axioms imply that every finite set X of sentences has a subset that is a basis for it, and that the full set of axioms ensure that every (countable) set of sentences possesses a basis.

The choice of topics for the last two of Tarski's lectures was made by Menger, who had been impressed by the results of the Warsaw school of logicians and wanted to familiarize members of the Vienna Circle with them. To his great displeasure, however, the only members of the Circle who attended the session on 20 February were Hans Hahn and Rudolf Carnap. Before Tarski's final lecture Menger therefore contacted every Circle member by telephone and pointedly urged each one to come (as, indeed, they all did)[5].

As abstracted in the colloquium proceedings, Tarski's third lecture surveyed various formulations of classical two- and many-valued propositional calculi. Both syntactic and semantic methods for obtaining closed logical systems were mentioned, along with some recent results of the Warsaw school–most notably, Wajberg's theorem that every axiomatization of classical two-valued propositional logic must involve at least three distinct statement letters, and Łukasiewicz's discovery that all proposi-

[4] According to *Vaught 1986*, p. 873, Lindenbaum's theorem first appeared in print in the version of Tarski's talk published in *Monatshefte* (*Tarski 1930b*).

[5] *Menger 1994*, pp. 147–148.

tional statements can be written unambiguously without the use of parentheses.

Today, of course, "Polish notation" is widely known and used. At the time, however, it seemed a radical idea. Indeed, as Menger well knew, Wittgenstein had claimed in section 5.461 of his *Tractatus*–a work closely studied and highly esteemed by many in the Circle–that "the seemingly insignificant fact that . . . [the symbols] $\vee$ and $\supset$ require parentheses" was "highly significant" and showed that those signs were not truly primitive. So if Menger expected Tarski to mention Łukasiewicz's notation in his talk, he may have had an ulterior motive in inviting the Circle members to attend.

In any case, the lecture stimulated Menger himself to formulate a simple arithmetical criterion for determining whether an arbitrary sequence of symbols in Łukasiewicz's notation is a well-formed formula. The result was published in volume three of the colloquium proceedings, among the *Gesammelte Mitteilungen* (undated contributions that presumably were not presented at any of the colloquium sessions).

Both Gödel and Carnap had long conversations with Tarski during his sojourn in Vienna, and all three benefitted from the exchanges: Tarski was "greatly impressed" by Gödel's dissertation results and invited Carnap to lecture in Warsaw[6]; the methods and results of Tarski and the other logicians of the Polish school significantly influenced Carnap's work on the syntax of language, then in its early stages; and some of the ideas that Tarski broached in his survey lecture reappeared shortly thereafter in three contributions that Gödel made to the colloquium.

Specifically, Tarski had mentioned two results of Łukasiewicz in addition to his parenthesis-free system of notation: his axiomatization of the propositional calculus in terms of three independent axioms, and his use of many-valued truth tables in establishing formal independence results. Gödel adopted the former in the paper he delivered to the colloquium on 10 June 1931, entitled "Eine Eigenschaft der Realisierungen des Aussagenkalküls". Assuming nothing about the structure of well-formed formulas or their cardinality, but only that they satisfied Łukasiewicz's axioms, he there showed how, given an arbitrary realization of the propositional calculus, its formulas could be partitioned into two disjoint classes that behaved like the familiar classes of true and false propositions. Six months later, at the session held 2 December 1931, he presented an example illus-

6 *Menger 1994*, p. 156.

trating that independence proofs of propositional statements may require the use of infinitely many truth values; and in the *Gesammelte Mitteilungen* for 1931/32 (volume 4, p. 40) he demonstrated that no realization that employed only finitely many truth values could satisfy those and only those formulas provable in Heyting's intuitionistic propositional calculus.

On the preceding page of those *Gesammelte Mitteilungen* Gödel also formulated a version of the modal propositional logic now known as *S*4 (introduced by Clarence I. Lewis). By interpreting the modal operator not as meaning "necessity" (as Lewis had), but as expressing the informal notion "provability by any correct means", he then showed how Heyting's intuitionistic propositional calculus could be embedded within its classical counterpart.

Gödel's provability interpretation became the starting point for what is now called provability logic, a subject studied intensively by logicians in the 1970s and '80s; and, as A. S. Troelstra notes in his incisive commentary on Gödel's paper (pp. 296–299 in *Gödel 1986*), the embedding was also "instrumental in the development of . . . [the] semantics for intuitionistic logic." In particular, "once the semantics for modal [quantificational] logic . . . had been formulated, Gödel's interpretation . . . showed how one could obtain a semantics" for intuitionistic predicate logic as well.

Lewis's modal logic, under its standard interpretation, also figured in the colloquium session held 18 February 1932. The speaker then was the American visitor William Tuthill Parry, who outlined a decision procedure for the formulas of Lewis's calculus that are of first degree (those built up from variables, connectives and the necessity operator, in which no occurrence of the necessity operator lies within the scope of another). Parry had previously addressed the colloquium on 13 November 1931, at which time he proposed a system of axioms for what he called "analytic implication". The system was an example of what is now called relevance logic, a propositional logic in which a necessary condition for the provability of $A \to B$ is that every propositional variable that occurs in B also occurs in A.

One other paper concerning a decision procedure appeared in the *Gesammelte Mitteilungen* for volume 2. In it Gödel described a procedure for testing the validity of certain closed prenex formulas of "the restricted functional calculus"–in modern terminology, first-order quantification theory without equality. The formulas in question were those whose quantifier prefixes were of the form $\exists \ldots \exists\forall\forall\exists \ldots \exists$ (a sequence beginning and ending with nonzero numbers of existential quantifers, separated by exactly two universal quantifiers). The theorem, an extension of an earlier result of Ackermann (*Ackermann 1928*), was one of a number of positive solutions for restricted cases of the *Entscheidungsproblem* that were obtained in the years preceding the development of recursion theory. Following the demonstration by Church and Turing of the unsolvability of the general decision problem, further work on the solvability of special prefix classes continued. By 1962, a complete determination of which classes are or are not decidable had been achieved, for formulas not containing the equality symbol; but the corresponding determination for formulas *with* equality was not finally settled until 1983[7].

Rounding out the logical contributions to volumes 1–4 of the *Ergebnisse* are two more papers by Gödel. Both concern results in formal number theory. In the earlier of the two, delivered 22 January 1931 at the twenty-fourth meeting of the colloquium, Gödel announced certain extensions of the results that he had presented in his incompleteness paper (*Gödel 1931*), which was just then in press[8]. He had referred obliquely to the first incompleteness theorem in a discussion at Königsberg on 7 September 1930, a transcript of which (including a Postscript by him in which both theorems were described in somewhat more detail) was later published in the journal *Erkenntnis*[9]; but that too, had likely not yet appeared at the time of his address to the colloquium. If so, then that talk constituted Gödel's first forthright public presentation of the incompleteness results.

No details of the proofs are given in the published abstract, so it is unclear to what extent he may have discussed them. But the abstract does go beyond *Gödel 1931* in two respects: First, the results are stated in the context of Peano arithmetic, rather than the simple theory of types; second, Gödel notes that his methods can be extended to systems of higher type or order. In particular, he points out that in type-free systems such as set theory it is "axioms of cardinality ([those] . . . postulating the existence of sets of ever higher cardinality) that take the place of type extensions"–a remark foreshadowing his later advocacy of axioms of infinity as a promising focus for set-theoretical research.

Gödel's other number-theoretic contribution to the colloquium was his celebrated "negative translation" of classical first-order arithmetic into

[7] For an overview and further references, see the commentary by Warren Goldfarb in *Gödel 1986*, pp. 226–231. Detailed considerations may be found in *Ackermann 1954*, *Dreben and Goldfarb 1979*, *Goldfarb 1984*, and *Lewis 1979*.

[8] See the correction to the Chronology in *Gödel 1986*, noted on p. 518 of *Gödel 1994*.

[9] *Hahn et alii 1931*. See also the commentary accompanying the English translations in *Dawson 1984* and in *Gödel 1986*, pp. 196–199.

Heyting's intuitionistic system–a translation also discovered, independently and at nearly the same time, by Paul Bernays and Gerhard Gentzen. Gödel's own interpretation of his result was that it showed that intuitionistic arithmetic "is only apparently narrower" than classical arithmetic, so that intuitionism can be said "to introduce genuine restrictions only for analysis and set theory" (and even there, not because of the intuitionists' rejection of the Law of Excluded Middle, but because of their refusal to countenance impredicative definitions). Of course, the translation also provided a consistency proof for classical arithmetic relative to the intuitionistic version. Beyond that, it has proved quite useful in proof-theoretical investigations. And for Gödel himself, the presentation had an important effect on his subsequent career; for Menger had invited Oswald Veblen to attend the colloquium that day, and Gödel's talk made such an impression on the visitor that Veblen invited Gödel to come to the Institute for Advanced Study during its first year of operation (1933–34).

After volume 4, there is a hiatus: No further presentations on logic appear until volume 7. Gödel's absence in America, followed by his lengthy incapacitation due to depression, was no doubt one contributing factor, as was, perhaps, Hans Hahn's sudden death in the summer of 1934[10]. Of course, political tensions in Austria springing from the growing Nazi interference in Austrian affairs made concentration on any sort of intellectual work difficult at best. Still, the colloquium continued to meet, and Gödel did continue to make minor contributions–on geometry, however, rather than logic.

Logic reappeared as a topic of discussion in a rather insubstantial presentation by the Japanese visitor Tomoharu Hirano, who spoke to the colloquium on 30 November 1934. Hirano showed that if the symbols of the classical predicate calculus are reinterpreted so that the meanings of the universal and existential quantifiers are reversed and the connectives $\vee$ and $A \rightarrow B$ are taken to mean "and" and "A implies B is false", respectively, then the statements formally derivable within the system will be contradictions rather than tautologies.

Of much greater interest was the lecture that Tarski delivered on 12 June 1935, in which he reported on his joint work with Adolf Lindenbaum concerning limitations on the expressive power of deductive theories. Tar-

[10] Although Hahn himself proved no results in logic, in his later years he became quite interested in the subject and encouraged its study (as noted by Menger in his obituary of Hahn, volume 6, pp.43–44 of these *Ergebnisse*). Hahn was Gödel's dissertation advisor and a member of the committee that approved Gödel's *Habilitation*.

ski considered logics extending that of *Principia Mathematica* and containing a predicate symbol *R* representing the intuitive notion "R is a permutation of the set of individuals" constituting the universe of discourse. In terms of that formal predicate he stated a series of theorems establishing the invariance under such permutations of most logically definable notions. For example, the only classes definable by purely logical means are the empty set and the universal class; the only logically definable binary relations between individuals are the universal and empty relations, equality, and inequality; and any two classes of individuals of equal cardinality whose complements are also of equal cardinality are logically indistinguishable. Tarski went on to consider theories containing non-logical axioms (in particular, Euclidean geometry) and obtained a number of results in a similar vein (e.g., that no absolute unit of length is definable therein). Finally, he considered connections between the metamathematical notions "categorical set of axioms" and "non-ramifiable set of axioms"–notions that he defined syntactically but which may be expressed model-theoretically as "any two models of the axioms are isomorphic" and "given any well-formed formula $\sigma(x_1, \ldots, x_n)$ and any model of the axioms, either σ or its negation holds for all *n*-tuples of elements in the domain of the model". The principal conclusion was that the former notion implies the latter.

At the session immediately following Tarski's talk, held 19 June 1935[11], Gödel announced some results concerning the lengths of formal proofs in number theory, measured not in terms of the number of symbols involved, but the number of lines (separate inferences). The theorems in question, for which he supplied no proofs[12], are extensions of the incompleteness theorems and are regarded today as the first examples of so-called "speed-up theorems". For as Gödel pointed out, whereas the incompleteness theorems establish the existence of number-theoretic propositions undecidable within logic of a given order, his results on lengths of proofs show that the derivations of some propositions that are already provable can be "shortened extraordinarily" by passing to logics of higher order.

The eight volumes of the colloquium proceedings exhibit a certain symmetry. For example, the last, like the first, contains no dated entries.

[11] Erroneously printed as 1934.

[12] Only recently, in work by Samuel Buss, have proofs of some of Gödel's claims appeared in print. See *Buss 1994*.

Nor does it contain any articles on logic. Rather, the last such article comprised the subject matter of the last dated colloquium session (number ninety-six, held 3 July 1935). The speaker was again Alfred Tarski, who returned once more to the subject of propositional logic.

In that final talk Tarski proved a series of characterization theorems for various formulations of the classical two-valued propositional calculus. In particular, if the only connective symbol is that for implication, he showed that the corresponding calculus is the only consistent, complete system including the sentences $p \supset (q \supset p)$, $p \supset ((p \supset q) \supset q)$, and $(q \supset r) \supset ((p \supset q) \supset (p \supset r))$. If the symbol for negation is allowed as well, then the same conclusion holds with respect to the four sentences $p \supset \neg\neg\, p$, $q \supset (p \supset q)$, $\neg\, p \supset (p \supset q)$, and $p \supset (\neg\, q \supset (\neg\, (p \supset q)))$. He pointed out that the many-valued calculi of Łukasiewicz and the intuitionistic calculus of Heyting are examples of systems to which those theorems apply.

During its eight-year existence the Menger colloquium was one of the major loci for research in logic. Together with the Polish school in Warsaw, it produced some of the most outstanding work of the interwar period. The demise of both groups was part of the bitter legacy of Nazism.

References

Ackermann W (1928) Über die Erfüllbarkeit gewisser Zählausdrücke. *Mathematische Annalen* 100: 638–649

Ackerman W (1954) *Solvable Cases of the Decision Problem.* North-Holland Publishing Co, Amsterdam

Buss S R (1994) On Gödel's theorems on length of proofs, I: number of lines and speedup for arithmetics. *Journal of Symbolic Logic* 59: 737–756

Dawson, J W Jr (1984) Discussion on the foundation of mathematics. *History and Philosophy of Logic* 5: 111–129

Dreben B, Goldfarb WD (1979) *The Decision Problem: Solvable Classes of Quantificational Formulas.* Addison-Wesley, Reading, Mass.

Gödel K (1930) Die Vollständigkeit der Axiome des logischen Funktionenkalküls. *Monatshefte für Mathematik und Physik* 37: 349–360

Gödel K (1930) Über formal unentscheidbare Sätze der Principia Mathematica und verwandter Systeme, I. *Monatshefte für Mathematik und Physik* 38: 173–198

Gödel K (1986) *Collected Works.* Vol. I: Publications, 1929–1936. Ed. Solomon Feferman et alii. Oxford University Press, New York Oxford

Gödel K (1994) *Collected Works.* Vol. III: Unpublished Essays and Lectures. Ed. Solomon Feferman et alii. Oxford University Press, New York Oxford

Goldfarb WD (1984) The unsolvability of the Gödel class with identity. *Journal of Symbolic Logic* 49: 1237–1252

Hahn H, Carnap R, Gödel K, Heyting A, Reidemeister K, Scholz A, von Neumann J (1931) Diskussion zur Grundlegung der Mathematik. *Erkenntnis* 2: 135–151

Lewis HR (1979) *Unsolvable Classes of Quantificational Formulas*. Addison-Wesley, Reading, Mass.

Menger K (1994) *Reminiscences of the Vienna Circle and the Mathematical Colloquium*. Ed. Brian McGuinness, Louise Golland and Abe Sklar. Kluwer Academic Publishers, Dordrecht Boston London

Tarski A (1930a) Über einige fundamentale Begriffe der Metamathematik. *Sprawozdania z posiedzen Towarzystwa Naukowego Warszawskiego, wydzial III* 23: 22–29

Tarski A (1939b) Fundamentale Begriffe der Methodologie der deductiven Wissenschaften. I. *Monatshefte für Mathematik und Physik* 37: 361–404

Tarski A (1983) *Logic, Semantics, Metamathematics. Papers from 1923 to 1938*, 2nd ed. Translated by J. H. Woodger. Ed. John Corcoran. Hackett Publishing Company, Indianapolis

Vaught RL (1986) Alfred Tarski's work in model theory. *Journal of Symbolic Logic* 51(4): 869–882

Topology in the *Ergebnisse*

Ryszard Engelking

The first volume of the *Ergebnisse* opens with three lectures on dimension theory, two of them by Karl Menger. Clearly Menger wanted his Colloquium to continue the study of the notion of dimension. This notion has been introduced independently by P. S. Urysohn in 1922 and by Menger in 1923 and they both built the framework of the dimension theory for compact metric spaces. Since Urysohn's death in 1924, Menger was considered the foremost expert in the theory, and it was he who published the first monograph on the subject [M 1928].

Correspondingly, we shall start our presentation of the topological contents of the *Ergebnisse* with results involving dimensional properties of spaces.

I. Dimension

In his first lecture, Menger recalls the embedding theorem: every n-dimensional separable metric space is homeomorphic to a subspace of Euclidean $(2n+1)$-space R^{2n+1}, a theorem he formulated, and proved for $n = 1$ in 1926, and which was proved in its full strength only in 1931, by several topologists, G. Nöbeling among them (announcement in the *Ergebnisse,* volume 2, p. 29–30). He also puts forward the important question, whether the number $2n+1$ is an exact estimation, i. e., whether for every n there exists an n-dimensional space that is not embeddable in R^{2n}. The positive answer to this question was to be given in May 1933 by A. Flores (62. Kolloquium, 1932/33; see also 71. Kolloquium, 1933/34). Unfortunately, Flores' elegant note is probably the only publication from the *Ergebnisse* that is remembered and cited by contemporary topologists.

The second lecture, by H. Hornich, is devoted to Menger's five axioms for dimension: he proves that they are completely independent. In the third lecture Menger talks on some of his new dimension theoretical results and states a couple of problems. Two subjects there considered: weak n-dimensionality and the rational dimension, as well as the five axioms, were later studied by the participants of the Colloquium. We shall now briefly describe the results they obtained.

We start with Menger's axioms for dimension. They are formulated in the lecture by Hornich, and the question is if the only function f that satisfies them in the realm of all subspaces of Euclidean n-space is the Menger-Urysohn dimension dim. Menger proved that it is so for $n = 1$ and for $n = 2$. For higher n the problem is still open; it is even not known, if for n larger than 3 the dimension dim has the compactification property (axiom 3). Let us add that the dimension dim satisfies the five axioms in the realm of all subspaces of Euclidean spaces, but there are also other functions (cohomological dimensions) that satisfy them (see [E 1995]). By analysing his proof for the case of $n = 2$, Menger formulated (4. Kolloquium 1929/30) two hypothetical theorems, that would imply that the five axioms characterize dimension in the realm of all subspaces of R^n. It was shown, however, by Nöbeling (23. Kolloquium, 1930/31) that the second of these hypothetical theorems does not hold (see also 43. Kolloquium, 1931/32, for a positive result). Nöbeling must have been thinking on the subject further, because he claimed later (38. Kolloquium, 1931/32) that for any function f that satisfies the Menger axioms in the realm of all subspaces of Euclidean spaces we have $f(X) \leq \dim X$ for every separable metric space X; he promised a proof in *Monatshefte* but did not publish it. The result is valid nevertheless, and has been published later by Y. Hayashi [H 1990], who was not aware of Nöbeling's announcement. It is an easy consequence of the following lemma formulated by Hayashi: *Every G_δ-set in R^{2n+1} that contains all points with at least n+1 rational coordinates contains topologically all separable metric spaces of dimension $\leq n$.* The lemma might have been known to Nöbling. This is a special case of Menger's second hypothetical theorem, and it can be obtained by methods that Nöbeling was often using (such a proof is outlined in [E 1995], hint to Problem 1.11.F(c)).

The weakly n-dimensional spaces (i. e. separable metric spaces of dimension n that are n-dimensional only at the points of an $(n-1)$-dimensional subset) intrigued Menger, but they did not attract particular attention by his students, probably because the two interesting and hard problems that Menger recalled in his introductory talk were solved by S. Mazurkiewicz in 1928 and 1929. He returned, though, twice to the topic (5. Kolloquium, 1929/30, and 58. Kolloquium, 1932/33), to discuss the dimensional properties of Cartesian products of such spaces. His results were strengthened by B. Tomaszewski who proved in [T 1979] that if X is a weakly n-dimensional space and Y is a weakly m-dimensional space, then $\dim(X\times Y) \leq n+m-1 = \dim X+\dim Y-1$ (so that the Cartesian product of n weakly 1-dimensional spaces is at most $(n-1)$-dimensional).

The definition of the rational dimension is analogous to that of the Menger-Urysohn inductive dimension: (a) a space has rational dimension zero if and only if it is countable, and (b) a space has rational dimension $\leq n$, if every point of that space has arbitrarily small neighbourhoods with boundaries of rational dimension $\leq n-1$. Talking on his new dimension theoretical results, Menger discussed decomposition theorems for the rational dimension. Nöbeling studied the notion further: answering a question of Menger, he found a universal space (43. Kolloquium, 1932/32), and then characterized compact metric spaces with rational dimension $\leq n$ in terms of mappings to polyhedra (55. Kolloquium, 1932/33). These and other related results were published in [N 1934]. The announcement in 43. Kolloquium, and the paper in *Mathematische Annalen,* were forgotten. In [I 1988] S. D. Iliadis defined once more a universal space for all separable metric spaces with rational dimension ≤ 1 (it is true though, that it has some additional properties). Another construction of a universal space for this class of spaces was given by J. C. Mayer and E. D. Tymchatyn in [MT 1990].

We should also mention that Nöbeling was the first to study connections between the notion of dimension and that of Hausdorff measure (*Ergebnisse,* volume 3, p. 24–25), fully clarified by E. Marczewski (E. Szpilrajn) in [S 1937].

Nöbeling was undoubtedly the strongest topologist working at Menger's Colloquium. Unfortunately, some of his lesser results, published only in the *Ergebnisse,* went to oblivion, and have been rediscovered by others.

Thus the geometric measure introduced by Menger (6. Kolloquium, 1929/30) was rediscovered by K. Borsuk in [1983], and Nöbeling's result that every k-dimensional compact subset of R^n has a positive k-dimensional measure (22. Kolloquium, 1930/31), was reproved by K. Borsuk, S. Nowak and S. Spież in [BNS 1984]. Similarly, the nice result on projections of a compact n-dimensional subset of R^k (4. Kolloquium, 1930/31) was reproved by S. Mardešić in [Ma 1973].

One of Nöbeling's announcements is quite puzzling. He claims (26. Kolloquium, 1930/31) that if a normal family of sets (*Normalbereich*) is closed with respect to compactifications, then the derived family has the same property. This result was announced also in [N 1931a] with the promise of a forthcoming publication. It seems however that no proof has ever been published. It would be worthwhile to decide if the result is true.

In one case we know that a result he announced (*Ergebnisse,* volume 6, p. 46–47) and published [*Monatshefte* 42 (1935), p. 117–152] was false: he claimed having established that every separable n-dimensional manifold has a combinatorial triangulation and that the *Hauptvermutung* holds for such manifolds. As we know now, this is not true (see [Ra 1996] for further information).

II. Curves

This was another subject that Menger proposed to his Colloquium to work on. The need for a precise definition of a curve was one of the stimuli that gave rise to dimension theory and this theory, once developed, engendered curve theory. Menger was at the time working on his monograph on curves [M 1932], defined as one-dimensional continua. He had plenty of research problems to give to his students. Some of the solutions were included into the monograph. We shall list the most important curve theoretical results obtained at the Colloquium and published, or announced, in the *Ergebnisse.* They concern mainly curves that are regular (every point has arbitrarily small neighbourhoods with finite boundaries) or rational (every point has arbitrarily small neighbourhoods with countable boundaries).

The first volume of the *Ergebnisse* (p. 13–14) contains a summary of H. Reschovsky's results on rational curves published later in [R 1930]. She studied Menger's notion of gender (*Geschlecht*) of a point in a rational curve and obtained some results on the unions of rational curves as well as on the structure of such curves (as observed by R. G. Putnam, 73. Kolloquium, 1933/34, one of her proofs contains a gap).

In the same volume (p. 14–15), Nöbeling presented his results on representing regular curves as images of the circle under continous mappings with "small fibres". The results were published in [N 1933]. In the next issue he proved that there is no universal regular curve, i. e., that no regular curve contains topologically all possible regular curves (2. Kolloquium, 1929/30). From one of Reschovsky's results it follows that for rational curves the situation is similar.

Nöbeling also proved (20. Kolloquium, 1929/30) that in the class of locally connected curves, the dendrites are characterized by having the fixed point property (for each continuous mapping $f\colon X \to X$ there exists an

$x \in X$ such that $f(x) = x$). This result is often attributed in the literature to W. Scherrer or to W. L. Ayres, although they considered only homeomorphims $f: X \to X$ (a note in [A 1930] says that Borsuk extended Ayres' results to continuous mappings, see [B 1932]).

It is not hard to show that every regular curve is hereditarily locally connected (i. e., all its subcontinua are locally connected). G. T. Whyburn, one of the Colloquium's guests, proved as complementary result that every hereditarily locally connected continuum is a rational curve (the three classes being distinct). At first he had only a proof for continua in the plane (*Ergebnisse,* volume 2, p. 31–32), later he obtained the result in full generality (see [W 1931]).

III. Continua

Two results obtained at the Colloquium are among the fundamental facts of the theory of continua.

Whyburn described (8. Kolloquium, 1929/30) an example of a continuum in the plane that separates the plane hereditarily, i. e., has the property that any of its non-trivial subcontinua separates the plane. This is one of the basic examples in continua theory; its full description and an analysis of its properties can be found in [W 1930].

The most important result on continua obtained at the Colloquium is with no doubt Nöbeling's generalization of Menger's n-pod theorem (*Ergebnisse,* volume 2, p. 29, published as [N 1931]). It says that if a point x in a locally connected continuum X has order at least n (i. e., x has arbitrarily small neighbourhoods with boundaries of cardinality at least n) then X contains an n-pod with vertex x, the latter being the union of n arcs with only the point x in common. In the original Menger formulation (see [M 1927]) X was assumed to be regular curve and the order of x equal to n. In [N 1931] the notion of the frame (*Gerüst*) of a locally connected continuum is also studied, and it is proved that there always exists a frame which is the union of countably many arcs. The notion was first introduced by Nöbeling for regular curves (3. Kolloquium, 1929/30). In [M 1981] Menger gave an interesting account of the history of the n-pod theorem and of its relation to the n-arc theorem.

The *Ergebnisse* contains also many results on continua, curves and dimension which, although of lesser importance, are quite interesting and might serve as good working problems for students. They are due mainly to Nöbeling, Menger, Whyburn and Ayres. On the other hand, some results, connected with notions that proved nonviable, have lost their interest.

The *Ergebnisse* also contains quite interesting results that do not fit into the categories discussed above. Let us finish with commenting on three of them.

Following an idea of Menger's, A. Wald considered (23. Kolloquium, 1930/31) a notion of topological space defined without specifying the points of the space, but rather its "pieces" (some decreasing sequences of such pieces permitting eventually to obtain the points and the usual topological structure). For such objects, he found a set of axioms that characterize those corresponding to second countable regular topological spaces. The idea of "topology without points" reappears from time to time in various settings (see [M 1940], cf. [J 1983]).

Borsuk (49. Kolloquium, 1931/32) studied the notion of border-points and kernel-points of a compact metric space, the definition being based on the properties of continuous mappings of that space into Euclidean spheres. Another interesting notion studied by Borsuk (64. Kolloquium, 1932/33) is that of "spheroidal space". In [B 1936] Borsuk returned to this concept, and approached it in a different way.

The *Ergebnisse* closes with J. v. Neumann's article on equations of mathematical economy (volume 8, p. 73–83). In the article a topological theorem on intersections of sets with convex sections, related to the Brouwer fixed point theorem, is established and applied. The idea proved fruitful and was further developed (see [F 1966]).

Menger's Viennese Colloquium was in its time among the most active world research centers in topology. Its interests covered almost all contemporarily studied aspects of the topology of Euclidean spaces, the only notable exception being the theory of Borel and analytic sets. The methods applied in the Colloquium work were purely set theoretic. Combinatorial and algebraic methods appeared only accidentally in talks of some guests, such as K. Borsuk, A. Flores and E. Čech. Menger and his collaborators had close relations with other topology centers in the world. Many eminent topologists of the time visited the Colloquium and gave talks there. The most important results in topology obtained elsewhere were presented at the Colloquium by its participants. Menger reported what he had learned in his travels, and letters from his correspondents were read and discussed. Several topologists stayed longer in Vienna provided with various scholarships, among them: W. L. Ayres (Philadelphia), G. T. Whyburn (Baltimore), K. Zarankiewicz (Warsaw), K. Borsuk (Warsaw) and A. Flores (Madrid). In connection with the Colloquium, four Viennese PhD degrees in topology were conferred, namely to H. Hornich, H. Reschovsky, G. Nöbeling and K. Vanek.

Karl Menger was an inspiring leader. His influence worked by the notions he introduced and suggested to study, rather than by involving students in his own research. He had an inclination to bring forth new concepts, most often pertaining to concrete and classical mathematical objects. In this line, his notion of dimension was undoubtedly the greatest success.

The academic year 1935/36 was the last of the Colloquium. It had to suspend its work because of the political situation. Its participants were dispersed, tradition broken off, even the memory of the Colloquium and the *Ergebnisse* faded. Let Menger talk himelf ([M 1952]):

"Meanwhile the political situation in Austria deteriorated from month to month. The *Ergebnisse* was criticized (with specific reference to Wald) for its large number of Jewish contributions just when I felt that we ought to honor that journal by making Wald co-editor. Issue 7 was edited by Gödel, Wald and myself. But Issue 8 containing Wald's paper on collectives was destined to be the last of the series. Hahn was dead. Schlick had been assassinated. Viennese culture resembled a bed of delicate flowers to which its owner refused soil and ligth while a fiendish neighbour was waiting for a chance to ruin the entire garden. I left the country. A year later Hitler marched into Vienna."

I wish to express my thanks to Professors Z. Marciniak, R. Pol and H. Toruńczyk for their helpful comments and conversations.

References

[A 1930] Ayres, W. L., *Some generalizations of the Scherrer fixed-point theorem,* Fund. Math. 16 (1930), 332–336.

[B 1932] Borsuk, K., *Einige Sätze über stetige Streckenbilder,* Fund. Math. 18 (1932), 198–213.

[B 1936] Borsuk, K., *Über sphäroidale und H-sphäroidale Räume,* Mat. Sbornik 1 (43) (1936), 643–660.

[B 1983] Borsuk, K., *An alternative concept of the n-dimensional measure,* Ann. Polon. Math. 42 (1983), 17–24.

[BNS 1984] Borsuk, K., Nowak, S., Spież, S., *Remarks on the n-dimensional geometric measure of compacta,* Fund. Math. 121 (1984), 59–71.

[E 1995] Engelking, R., *Theory of dimensions, finite and infinite,* Heldermann Verlag, Lemgo 1995.

[F 1966] Fan, Ky, *Applications of a theorem concerning sets with convex sections,* Math. Ann. 163 (1966), 189–203.

[H 1990] Hayashi, Y., *An axiomatic characterization of the dimension of subsets of Euclidean spaces,* Topology and its Applications 37 (1990), 83–92.

[I 1988] Iliadis, S. D., *Rational spaces and the property of universality,* Fund. Math. 131 (1988), 167–184.

[J 1983] Johnstone, P. T., *The point of pointless topology,* Bull. Amer. Math. Soc. 8 (1983), 41–53.

[Ma 1973] Mardešić, Š., *Compact subsets of R^n and dimension of their projections,* Proc. Amer. Math. Soc. 41 (1973), 631–633.

[MT 1990] Mayer, J. C., Tymchatyn, E. D., *Universal rational spaces,* Diss. Math. 293 (1990).

[M 1927] Menger, K., *Zur allgemeinen Kurventheorie,* Fund. Math. 10 (1927), 96–115.

[M 1928] Menger, K., *Dimensionstheorie,* Teubner, Leipzig, 1928.

[M 1932] Menger, K., *Kurventheorie* (herausgegeben unter Mitarbeit von Georg Nöbeling), Teubner, Leipzig 1932.

[M 1940] Menger, K., *Topology without points,* Rice Inst. Pamphlet 27 (1940), 80–107.

[M 1952] Menger, K., *The formative years of Abraham Wald and his work in geometry,* Ann. Math. Statistics 23 (1952), 14–20.

[M 1981] Menger, K., *On the origin of the n-arc theorem,* Journ. Graph Theory 5 (1981), 341–350.

[N 1931] Nöbeling, G., *Eine Verschärfung des n-Beinsatzes,* Fund. Math. 18 (1931), 23–38.

[N 1931a] Nöbeling, G., *Die neuesten Ergebnisse der Dimensionstheorie,* Jahresbericht d. Deutschen Mathem. Vereinigung. 41 (1931), 1–17.

[N 1933] Nöbeling, G., *Reguläre Kurven als Bilder der Kreislinie,* Fund. Math. 20 (1933), 30–46.

[N 1934] Nöbeling, G., *Über die rationale Dimension,* Math. Ann. 109 (1934), 353–375.

[Ra 1996] Ranicki, A. A. (editor), *The Hauptvermutung book,* Kluwer, Dordrecht 1996.

[R 1930] Reschovsky, H., *Über rationale Kurven,* Fund. Math. 15 (1930), 18–37.

[S 1937] Szpilrajn, E., *La dimension et la mesure,* Fund. Math. 28 (1937), 81–89.

[T 1979] Tomaszewski, B., *On weakly n-dimensional spaces,* Fund. Math. 103 (1979), 1–8.

[W 1930] Whyburn, G. T., *A continuum every subcontinuum of which separates the plane,* Amer. Journ. of Math. 52 (1930), 319–330.

[W 1931] Whyburn, G. T., *Concerning hereditarily locally connected continua,* Amer. Journ. of Math. 53 (1931), 374–384.

An exposition of Wald's existence proof

Werner Hildenbrand

In the 74. Kolloquium of 19.3.1934 and in the 80. Kolloquium of 6.11.1934, both with the same title *"Über die Produktionsgleichung der ökonomischen Wertlehre"* Wald analysed the solvability of the following system of equations:

$$
(W) \qquad
\begin{array}{lll}
\text{(a)} & p = f(x), & x, p \in \mathbb{P}^n \\
\text{(b)} & r = Ax + u & \\
 & & q, u, \in \mathbb{R}^m_+ \\
\text{(c)} & p^T = q^T A & \\
\text{(d)} & q_i \cdot u_i = 0, & i = 1, \ldots, m
\end{array}
$$

where
$\mathbb{P}^n := \{x \in \mathbb{R}^n \mid x_j > 0, j = 1, \ldots, n\}$, f denotes a function from $\mathbb{P}^n$ into $\mathbb{P}^n$, $r \in \mathbb{P}^m$, $A = (a_{ij})$ is a matrix with n columns and m rows, m and n are arbitrary positive integers.

Abraham Wald has formulated conditions on the function f and the matrix A which insure the existence of a solution (x^*, p^*, q^*, u^*) of the system (W) with x^* and p^* in $\mathbb{P}^n$ and q^* and u^* in $\mathbb{R}^m_+$.

Before I explain Wald's existence proof[1] in Section 2, I shall describe in Section 1 the economic background which led Karl Schlesinger and Abraham Wald to the system of equations (W). Then, in Section 3, I shall discuss the impact of Wald's outstanding contribution on further developments in mathematical economics.

[1] I would like to thank Kurt Hildenbrand and Reinhard John for helpful discussions.

1 Economic background

In his *Eléments d'Economie Politique Pure* Léon Walras (1874) has formulated for the first time a closed model of a *private ownership economy with production*[2]:

- there are *m factors of production* $i = 1, \ldots, m$, the price of factor i is denoted by q_i.
- there are *n consumption goods* $j = 1, \ldots, n$, the price of consumption good j is denoted by p_j.
- the production of one unit of consumption good j requires a_{ij} units of factor i.
  Thus, to produce the vector $x \in \mathbb{R}^m_+$ of consumption goods one needs the vector $y = Ax \in \mathbb{R}^m_+$ of factors of production.
  The matrix A of production coefficients $a_{ij} \geq 0$ is exogenously given.
- There is a set K of *consumers* who own the factors of production. The factor endowment of consumer k is denoted by $r^k \in \mathbb{R}^m_+$; $r = \sum_{k \in K} r^k$ denotes the total availability of factors of production.
  The income of household k is $w^k = \sum_{i=1}^m q \cdot r^k$. Every household spends his income w^k on consumption for goods and (possibly) factors of production (e.g. land, labor).
  The consumption behavior of a consumer k is described by a *utility-function* $u^k(x, y)$ with $x \in \mathbb{R}^n_+$, $y \in \mathbb{R}^m_+$ (actually Walras assumes a specific form of a strictly concave function, but this does not matter here). Given (p, q), the prices of consumption goods and factors of production, the *demand function* $f^k(p, q)$ and $g^k(p, q)$ for consumption goods and for factors of production, respectively, is defined as a solution of the maximization problem

$$\text{(P)} \qquad \max \; u^k(x, y) \quad x \in \mathbb{R}^n_+, \; y \in \mathbb{R}^m_+, \; p \cdot x + q \cdot y = q \cdot r^k$$

  Note that the demand functions f^k and g^k are, by definition, homogenous in (p, q).

Finally, *total consumers' demand* is defined by

$F(p, q) = \sum_{k \in K} f^k(p, q)$ for consumption goods
$G(p, q) = \sum_{k \in K} g^k(p, q)$ for factors of production.

[2] I shall refer to Leçon 20 of the definitive edition (1926).

An *equilibrium* of the economy is defined by (x, y, p, q) which satisfies

$x = F(p, q)$	the supply x of consumption goods by the production sector equals total consumers' demand $F(p, q)$ for consumption goods
$y = r - G(p, q)$	the demand y for factors of production by the production sector equals total consumers' net supply $r - G(p, q)$ of factors of production
$y = Ax$	technological feasibility
$p_j = q \cdot a_j$	the price of consumption good j equals to cost of producing one unit (zero profit condition)

At the time of Walras' writing, he could not specify conditions on F, G and A which insure the existence of a positive solution of this non-linear system of equations. He could not do better, than observing that there are the same number of unknowns as equations.

Walras presentation of his general equilibrium theory was only accessible at the time to a very small number of economic theorists who have not been afraid by mathematical symbols. One among those was Gustav Cassel[3] who popularized the Walrasian theory by presenting a "simplified" version (1899) and (1918).

First, Cassel assumes that consumer demand is restricted to consumption goods, that is to say, $G(p, q) \equiv 0$.

Second, Cassel freed equilibrium analysis from utility theory; he based his presentation directly on demand functions for consumption goods.

Third, Cassel assumed that total consumers demand $F(p, q)$ does not depend on factor prices q.

These assumptions led Cassel to the system

$$\begin{aligned} x &= F(p), \quad x, p \in \mathbb{P}^n \\ r &= Ax \\ p^T &= q^T A, \quad q \in \mathbb{P}^m \end{aligned}$$

Cassel's simplification of the Walras model was heavily criticized, in particular, his third assumption which is, indeed, a severe restriction. It

3 "Von den Verfassern, die so gewissermassen als meine Vorgänger zu betrachten sind, sei hier nur Walras erwähnt. Dass dieser sich durch einen überaus schwerfälligen, mathematischen Apparat einen größeren Leserkreis selbst verschlossen hat, ist sehr zu beklagen." Cassel (1899), p. 396.

requires *either* that income of all consumers is exogenously given *or* that factor endowments r^k, $k \in K$ are colinear.

In the first case one leaves the Walrasian model of a private ownership economy where income is defined endogenously by $q \cdot r^k$.

In the second case, i.e., collinear factor endowments, many interesting situations are excluded. Yet this case is compatible with the Walrasian model since total demand then is a function in p and total income $q \cdot r$. Since $F(p, q)$ is homogeneous in (p, q) one can add the restriction $q \cdot r =$ constant, and thus obtains that total demand depends only on p.

In the late 20', and early 30' several authors realized that Cassel's system – in particular the equation $Ax = r$ – might have no positive solution. Frederic Zeuthen (1928) and later, yet independently, Karl Schlesinger (1935) proposed the following modification of Cassel's system:

(a') $x = F(p), \quad x, p \in I\!P^n$

(b) $Ax + u = r, \quad q, p \in \mathbb{R}^m_+$

(c) $p^T = q^T A$

(d) $u_i \cdot q_i = 0, \quad i = 1, \ldots, n$

It was this system, I assume, which was shown by the economist Karl Schlesinger to the young mathematician Abraham Wald (who gave private lessons in mathematics to Schlesinger). Wald offered a solution, yet a solution to a slightly different problem. The first equation was inverted to $p = f(x)$, which seems perhaps harmless to a mathematician but is a serious restriction from an economic point of view. Who made the inversion, Schlesinger or Wald? Who had a motive for doing so? I can hardly believe that Schlesinger posted the problem in this form to Wald since Schlesinger must have been aware of the problem, and furthermore, he had no need to invert the demand function. Wald, however, had a clear motive to write $p = f(x)$ instead of $x = F(p)$ as becomes obvious if one studies Wald's existence proof!

2 An exposition of Wald's existence proof

To prove the solvability of the system (W) Wald makes the following *Assumptions*

(1) $r \in I\!P^m$

(2) $a_{ij} \geq 0$ for all elements of the matrix A

(3) every column a_j of A has a positive element

(4) the function $f: I\!P^n \to I\!P^n$ is continuous

(5) *Boundary behavior*: if the sequence $(x^\nu)_{\nu=1,\ldots}$ with $x^\nu \in I\!P^n$ converges to $x \in I\!R^n_+$, then $f_j(x^\nu)$ tends to infinity if $x_j = 0$ and remains bounded if $x_j > 0$.

(6) *Wald's Axiom*: for every x and x' in $I\!P^n$ with $x \neq x'$

$$f(x) \cdot (x' - x) \leq 0 \text{ implies } f(x') \cdot (x' - x) < 0$$

Assumptions **(1)** to **(3)** need no comment. Assumption **(4)** seems harmless from a mathematical point of view, yet, as I explained in section 1, it is very restrictive from an economic point of view. The assumption requires implicitly that total consumers demand does not depend on factor prices. The inverse demand function f should exist and be defined on $I\!P^n$, that is to say, for every commodity vector $x \in I\!P^n$ there is a price vector at which x is demanded; a very restrictive assumption. Assumption **(5)** expresses a strong version of desirability for the consumption goods.

The most criticized assumption is **(6)**, what I called Wald's Axiom. The critique (or even the rejection) of **(6)** in the economic literature is mainly based on the argument that Wald's Axiom does not follow deductively from *individual rational behavior* (in form of utility maximization). Thus, the critique is not based on a lack of empirical content of Wald's Axiom.

Wald's Theorem: If assumption **(1)** *to* **(6)** *are satisfied, then there exists a solution* (x^*, p^*, q^*, u^*) *of the system (W) with* x^* *and* p^* *in* $I\!P^n$, q^* *and* u^* *in* $\mathbb{R}^m_+$. *Furthermore, the variables* x^*, p^* *and* u^* *are uniquely determined.*

The *idea* of Wald's proof is *very simple*, even though his presentation is not an easy reading[4]. I shall present Wald's proof in order to reveal its originality, simplicity and beauty and to make explicit the rôle that is played by assumption **(5)**, and in particular, by the much discussed assumption **(6)**. I have the impression that the rôle of assumption **(6)** is not well understood; there are some incorrect statements and even conflicting opinions on this point in the economic literature.

Proof:

Wald proves his theorem by induction on n, the dimension of the vector x.

For $n = 1$ the system (W) obviously has a solution. Indeed, define $x_1^* := \max\{\xi \mid \xi \cdot a_1 \leq r\} > 0$. Then $p_1^* := f_1(x_1^*) > 0$, $u^* := r - x_1^* a_1 \geq 0$ and $q^* \in \{q \in \mathbb{R}^m_+ \mid q_i = 0 \text{ if } u^i > 0 \text{ and } p_1^* = q \cdot a_1\}$, which is a non-empty convex set.

[4] Dorfman-Samuelson-Solow (1958), p.368: "Wald's proof . . . is extremely intricate and opaque", Arrow and Hahn (1971), p.10: "Wald's paper were of forbidding mathematical depth, not only in the case of sophisticated tools, but also in the complexity of the argument".

Assume now that every system W($n-1$) satisfying assumption **(1)** to **(6)** has a solution. Consider the system of equations

(1') $p_1 = f_1(x_1, \ldots, x_n), \ldots, p_{n-1} = f_{n-1}(x_1, \ldots, x_n)$

(2) $x_1 a_1 + \ldots + x_{n-1} a_{n-1} + u = r - x_n a_n$

(3') $p_1 = q \cdot a_1, \ldots, p_{n-1} = q \cdot a_{n-1}$

(4) $u_i \cdot q_i = 0, \quad i = 1, \ldots, m$

The only difference between this system, which I shall denote by W($n-1|x_n$), and the system W(n) – which one has to solve – is that one does not require

$$p_n = f_n(x_1, \ldots, x_n) \text{ and } p_n = q \cdot a_n.$$

If we choose $x_n > 0$ such that $r(x_n) := r - x_n a_n \in \mathbb{P}^n$, then W($n-1|x_n$) has a solution by the induction hypothesis. This is true for all x_n in the open interval $(0, \bar{x}_n)$ with $\bar{x}_n = \sup\{\xi | r - \xi a_n \in \mathbb{P}^m\}$.

For every solution of W($n-1|x_n$), i.e., $x_1^*(x_n), \ldots, x_{n-1}^*(x_n), p_1^*(x_n), \ldots, p_{n-1}^*(x_n)$ and $q_1^*(x_n), \ldots, q_m^*(x_n)$ we consider

$$q^*(x_n) \cdot a_n - f_n(x_1^*(x_n), \ldots, x_{n-1}^*(x_n), x_n) =: z^*(x_n).$$

If, by luck, $z^*(x_n) = 0$, then we have a solution of W(n) by setting $x_n^* = x_n$ and $p_n^* = q^*(x_n) \cdot a_n$.

In general, for every $x_n \in (0, \bar{x}_n)$ we define the set

$$\varphi(x_n) := \{ z^*(x_n) \mid \text{for every solution of W}(n-1|x_n) \},$$

which is non-empty by the induction hypothesis.

To complete the proof one has to show that there exists $x_n^* \in (0, \bar{x}_n)$ such that $0 \in \varphi(x_n^*)$.

Wald's argument on pages 15 and 16 of Kolloquium 74 provides a proof of the following (today well-known)

Lemma 1: "Intermediate Value Theorem for a correspondence"
Let ψ be a convex-valued and bounded correspondence from a closed interval I into $\mathbb{R}$ *that has a closed graph, i.e., the set*

$$\{(x, z) \in I \times \mathbb{R} | z \in \psi(x)\} \text{ is closed.}$$

If there are $\underline{x}$ and $\bar{x}$ in I such that $\psi(\underline{x})$ contains a non-positive number and $\psi(\bar{x})$ contains a non-negative number, then there is $x^* \in I$ with $0 \in \psi(x^*)$.

Clearly, if ψ is a function then Lemma 1 is the usual intermediate value theorem for a continuous function.

In order to apply the intermediate value theorem to the correspondence φ of $(0, \bar{x}_n)$ into $\mathbb{R}$ one has to show that

(α) φ is *bounded and has a closed graph* on every closed subinterval of $(0, \bar{x}_n)$

and

(β) $\varphi(x_n)$ is *convex* for $x_n \in (0,\bar{x}_n)$.

Property (α) follows from Hilfssatz 2 of Kolloquium 80. The essential steps of the proof are given in Hilfssatz 2 and 3 of Kolloquium 74.

Wald considers a sequence $(W^\nu)_{\nu=1,\ldots}$ of systems of equations:

(a) $\quad p = f^\nu(x), \quad x, p \in \mathbb{P}^n$

(b) $\quad r^\nu = A^\nu x + u \qquad\qquad (\nu = 1, \ldots)$

(c) $\quad p^T = q^T A^\nu, \quad q, u \in \mathbb{R}^m_+$

(d) $\quad q_i u_i = 0, \quad i = 1, \ldots, m.$

Lemma 2: Let the sequences (A^ν), (r^ν) *and* (f^ν) *converge to A, r and f, respectively. Assume that every system* (W^ν) *and* (W) *satisfies assumptions* **(1)** *to* **(5)**. *Furthermore, the boundary condition holds uniformly: if* (x^ν) *in* $\mathbb{P}^n$ *converges to* $x \in \mathbb{R}^n_+$, *then* $f^\nu_j(x^\nu)$ *tends to infinity if* $x_j = 0$ *and remains bounded if* $x_j > 0$.

If $(x^\nu, p^\nu, q^\nu, u^\nu)$ *is a solution of* (W^ν), $\nu = 1, \ldots$, *then the sequence of solutions is bounded and every limit point* (x, p, q, u) *is a solution of* (W).

Lemma 2 clearly implies property (α). Indeed, we have to show: if $x^\nu_n \to x_n \in (0, \bar{x}_n)$, $z^\nu \in \varphi(x^\nu_n)$, $\nu = 1, \ldots$; then the sequence (z^ν) is bounded and every limit point of (z^ν) belongs to $\varphi(x_n)$. This follows from Lemma 2 by setting $A^\nu = (a_1, \ldots, a_{n-1})$, $r^\nu = r - x^\nu_n a_n$, $f^\nu(x_1, \ldots, x_{n-1}) = f(x_1, \ldots, x_{n-1}, x^\nu_n)$.

We remark that Lemma 2 implies that it suffices to prove Wald's Theorem if one strengthens assumptions **(2)** and **(3)** to: all $a_{ij} > 0$. This we shall do in the remaining part of the proof.

Proof of Lemma 2:

Since A^ν, x^ν and u^ν are non-negative we have

$$0 \le u^\nu \le r^\nu \text{ and } 0 \le A^\nu x^\nu \le r^\nu.$$

Hence (u^ν) is bounded and assumptions **(2)** and **(3)** on A^ν and A imply that the sequence (x^ν) is bounded.

Assume now that (q^ν) is unbounded. Without loss of generality we can assume that (x^ν) converges to $x \in \mathbb{R}^n_+$ and

(1) $\quad q^\nu_1 \to \infty.$

We shall now show that for every $j = 1, \ldots, n$,

(2) $\quad x^\nu_j \cdot a^\nu_{1j} \to 0.$

This clearly holds if the sequence $a^\nu_{1j} \to 0$.

Otherwise we obtain by (1) that

$$p^\nu_j = q^\nu_1 \cdot a^\nu_{1j} + \ldots \to \infty.$$

Since $p_j^\nu = f_j^\nu(x_1^\nu, \ldots, x_j^\nu, \ldots)$, the uniform boundary condition implies that $(x_j^\nu) \to 0$, which implies (2).

Hence, $x_1^\nu a_{11}^\nu + \ldots + x_n^\nu a_{1n}^\nu < r_1$ for ν sufficiently large. But then equation (d) implies $q_1^\nu = 0$; a contradiction to (1).

Thus, the sequence (q^ν), and hence (p^ν), are bounded.

For every limit point (x, p, q, u) of a sequence of solutions of W^ν it follows by the uniform boundary condition that $x \in \mathbb{P}^n$. Thus, by continuity of f the limit point (x, p, q, u) is a solution of W.

Q.E.D.

To prove property (β), the convexity of $\varphi(x_n)$, Wald uses assumption **(6)**. This assumption is used nowhere else in the proof!

Obviously, by definition of the set $\varphi(x_n)$, one obtains that $\varphi(x_n)$ is convex if the solutions of $W(n-1|x_n)$ are unique in x^*. To obtain this, Wald proves (Hilfssatz 1, Kolloquium 80) the following general result.

Lemma 3: If there are solutions of the system (W) satisfying assumption **(6)**, *then the solutions are unique in x^*, p^* and u^*.*

Proof of Lemma 3:

Wald first observed that for two solutions (x, p, q, u) and (x', p', q', u') of the system (b), (c) and (d) it follows that

$$p \cdot (x' - x) \le 0 \le p' \cdot (x' - x) .$$

Indeed, (b) implies $A(x' - x) + (u' - u) = 0$. Then one obtains from (c) that $p \cdot (x' - x) + q \cdot (u' - u) = 0$ and $p' \cdot (x' - x) + q' \cdot (u' - u) = 0$. But (d) implies $q \cdot (u' - u) \ge 0$ and $q' \cdot (u' - u) \le 0$, which proves the claimed inequalities.

Consequently, if $p = f(x)$ and $p' = f(x')$, which holds for any solution of (W), we obtain

$$f(x) \cdot (x' - x) \le 0 \le f(x') \cdot (x' - x) .$$

Now assumption **(6)** comes into play; it implies $x = x'$, hence $u = u'$, and $p = p'$.

Q.E.D.

Now we are ready to apply the intermediate value theorem to the correspondence φ.

If $x_n \in (0, \bar{x}_n)$ is sufficiently small, then $\varphi(x_n)$ contains a negative number. Indeed,

$$z^*(x_n) = q^*(x_n) \cdot a_n - f_n(x_1^*(x_n), \ldots, x_{n-1}^*(x_n), x_n)$$

becomes negative since $f_n(x_1^*(x_n), \ldots, x_n)$ becomes large (assumption **(5)**) and $q^*(x_n) \cdot a_n$ remains bounded (Lemma 2). Finally, if x_n is sufficiently near to $\bar{x}_n$, then $\varphi(x_n)$ contains a positive number. Indeed, there is at least one i such that

$$r(x_n)_i = r_i - x_n a_{in}$$

becomes arbitrarily small (recall we assume that all $a_{ij} > 0$). But this implies that $x_1^*(x_n), \ldots, x_{n-1}^*(x_n)$ become arbitrarily small. Hence by assumption **(5)**, $p_1^*(x_n), \ldots, p_{n-1}^*(x_n)$ become large and hence also $q^*(x_n)$. But then $q^*(x_n) \cdot a_n$ becomes large and $f_n(x_1^*(x_n), \ldots, x_n)$ remains bounded (assumption **(5)**).

Q.E.D.

Wald's contribution in retrospect

The economic profession is in agreement that Wald's existence proof is an outstanding contribution to mathematical economics. After all, this was the first proof of the logical coherence of Walrasian equilibrium theory; an open problem since 1874.

However, due to the restrictiveness of Wald's assumptions, his existence proof is mainly a mathematical achievement. Put in short, and somewhat exaggerated, Wald solved, *from an economic point of view*, the wrong system of equations. The existence proofs by Arrow and Debreu (1954) and McKenzie (1954) made Wald's theorem obsolete, since they did not need restrictive assumptions on the demand side of the economy. This achievement was possible by using Kakutani's fixed point theorem. Wald's intermediate value theorem for a correspondence (Lemma 1, above) is, of course, a one dimensional version of Kakutani's fixed point theorem.

Also Wald's method of proof, i.e., induction on the number of consumption goods, which – as it seems to me – requires the inverse demand function, has not been pursued in mathematical economics. What remains of interest up to our days is Wald's assumption **(6)** even though this assumption is not needed any more for proving existence. I would like to conclude with a remark on Wald's Axiom.

First, Wald's assumption **(6)** and the strengthening **(6')** reappears in the *mathematical* literature (in connection with the non-linear complementarity problem) under the names "strict pseudomonotonicity" and "strict monotonicity", respectively, e.g. Ortega and Rheinboldt (1970) and Harker and Pang (1990).

Second, Wald's assumption **(6)** reappears in the *economic* literature under the name "Weak Axiom of Revealed Preference" in Samuelson (1938).

Wald considers the inverse $p = f(x)$ of a market demand function $x = F(p)$ (with a fixed income distribution). In terms of the market demand function F, Wald's assumption **(6)** is:

$$\text{for any } p \text{ and } p' \text{ with } F(p) \neq F(p'),$$
$$p \cdot (F(p') - F(p)) \leq 0 \text{ implies } p' \cdot (F(p') - F(p)) < 0.$$

Wald also considers assumption **(6)** for individual demand functions $g(p)$ (with fixed income) and denotes in this case his assumption by **(6_w)**; w stands for "Wirtschafter" (economic agent). Wald emphasizes the difference between **(6)** and **(6_w)** on page 4, bottom, of Kolloquium 80.

If one considers now an individual demand function $g(p, y)$ in *prices* and *income*, which is *homogenous* in p and y (i.e., $g(\lambda p, \lambda y) = g(p, y)$, $\lambda > 0$) and satisfies the *budget identity* $p \cdot g(p, y) \equiv y$, (demand function of this type underly Walrasian analysis) then *Wald's assumption* **(6_w)** for $g(p, \bar{y})$ with some fixed income $\bar{y}$ *is equivalent to*: for every (p, y) and (p', y') with $g(p, y) \neq g(p', y')$,

$$p \cdot g(p', y') \leq y \text{ implies } p' \cdot g(p, y) > y',$$

which is Samuelson's Weak Axiom of Revealed Preferences.

This formal equivalence, however, does not imply that Wald anticipated the *theory* of revealed preferences, which definitely has been initiated by Samuelson.

It is very astonishing that Wald did not realize that a demand function $g(p, y)$ derived from utility maximization satisfies always his assumption **(6_w)** (see p. 5 of Kolloquium 80). Wald interpreted assumption **(6_w)** as a mild restriction on neoclassical individual demand functions.

Samuelson, however, formulated his Axiom as a postulate of *consistent* (rational) *individual behavior*, which is implied by utility maximization.

Although in neo-classical economics Wald's assumption **(6_w)** is considered as an acceptable assumption for every economic agent it does not follow that Wald's assumption **(6)** is harmless. Wald is very explicit on this point. But then (p. 5 of Kolloquium 80) Wald makes a puzzling remark: "Aber wenn **(6_w)** für jeden W gilt und **(6)** nicht gilt, so müssen, wie in einer folgenden Mitteilung ausgeführt werden soll, besondere Beziehungen zwischen den Nachfragen der einzelnen Wirtschafter W bestehen, welche statistisch unwahrscheinlich sind, so daß mit einer statistischen Wahrscheinlichkeit aus der Annahme, daß **(6_w)** für alle W gilt, die Gültigkeit von **(6)** folgt." The announcement of such a result is repeated in Wald (1936), but it was never actually published. Perhaps Wald had a much better solution to the aggregation problem of deriving **(6)** from **(6_w)** than the one which I offered in my book *Market demand* (1994).

References

Arrow K, Debreu G (1954) Existence of an equilibrium for a competitive economy. *Econometrica 22*: 265–290

Arrow K, Hahn FH (1971) *General competitive analysis.* Holden Day, San Francisco

Cassel G (1899) Grundriss einer elementaren Preislehre. *Zeitschrift f. die gesamte Staatswissenschaft 55*: 395–458

Cassel G (1918) *Theoretische Sozialökonomie.* Winter, Leipzig

Dorfmann R, Samuelson PA, Solow RM (1958) *Linear programming and economic analysis.* McGraw-Hill, New York

Harker PT, Pang J-S (1990) Finite-dimensional variational inequality and nonlinear complementary problems: a survey of theory, algorithms and applications. *Mathematical Programming 48*: 161–220

Hildenbrand W (1954) *Market Demand: Theory and Empirical Evidence.* Princeton University Press, Princeton

McKenzie LW (1954) On equilibrium in Graham s model of the world trade and other competitive systems. *Econometrica 22*: 147–161

Ortega JM, Rheinboldt WC (1970) *Iterative solution of nonlinear equations in several variables.* Academic Press, New York

Samuelson PA (1938) A note on the pure theory of consumer's behaviour. *Economica 5*: 61–71

Schlesinger K (1935) Uber die Produktionsgleichungen der ökonomischen Wertlehre. In: Menger K (Ed) *Ergebnisse eines mathematischen Kolloquiums, 74. Kolloquium 1934,* Vol 6. Deuticke, Leipzig Wien, pp 1–6

Wald A (1936) Über die Produktionsgleichungen der ökonomischen Wertlehre. In: Menger K (Ed) *Ergebnisse eines mathematischen Kolloquiums, 80. Kolloquium 1934,* Vol 7. Deuticke, Leipzig Wien, pp 1–6

Walras L (1874) *Eléments d'économie politique pure.* L. Corbaz, Lausanne

Zeuthen F (1928) *Den okonomiske fordeling.* Nyt Nordisk Forlag, Kopenhagen

Zeuthen F (1932–33) Das Prinzip der Knappheit, technische Kombination und ökonomische Qualität. *Zeitschrift für Nationalökonomie 4*: 1–24

On the direction of ideas and the principal tendencies of the Vienna Mathematical Colloquium

by Karl Menger (Vienna)*

1 Coordinate-free geometry and the geometry of sets

This article is not so much intended as a recapitulation of all the results achieved by the Vienna Mathematical Colloquium[1] during its six years of existence—such a recapitulation would in fact require a much larger space—but intends instead to expound the direction of the ideas that were pursued in this colloquium. Moreover, we limit ourselves to shedding light on the tendencies of the colloquium in the direction of *geometry*, although it certainly did not deal exclusively with geometrical studies. And in truth, apart from the fact that the geometric researches cultivated in this colloquium often required investigations into other fields of mathematics, other studies, above all in logic[2], but also regarding new applications of the exact sciences to problems of a sociological character, were carried out.[3]

---

* *I thank Dr. L. GEYMONAT for having helped me in writing the present article in Italian.*

[1] See: Ergebnisse eines mathematischen Kolloquiums. Unter Mitwirkung von K. GÖDEL und G. NÖBELING herausgegeben von K. MENGER, volumes 1–6 (Deuticke, Vienna). In the sequel we cite them as "Ergebnisse".

[2] Apart from a short lecture of K.GÖDEL (Ergebnisse 3, p.12) on his fundamental results concerning the completeness and the consistency of axiomatic systems [that is: in every theory that contains the arithmetic of PEANO one finds propositions that cannot be proved by means of this theory itself, and in particular it is impossible to prove the consistency of such a theory without stepping outside of the axiomatic system on which it is based], there are investigations of classical mathematical logic, of many-valued logic (GÖDEL, TARSKI, PARRY) and of intuitionistic logic, among them a proof of the surprising fact that the so-called intuitionistic arithmetic is only apparently more restrictive than classical arithmetic, whereas in an appropriate interpretation it contains the latter (GÖDEL, Ergebnisse 4, p.34).

[3] For instance on the existence and uniqueness of the equations of production in mathematical economics (See SCHLESINGER-WALD, Ergebnisse 6, p.10).

And the rapprochement of geometry with logic in not merely accidental because in studying geometrical problems one of the most characteristic tendencies was precisely the transition from a treatment exclusively based on calculation (rechnende Geometrie) to a logico-deductive one (schließende Geometrie). But in order to clarify this idea we have to step back a bit.

One of the greatest advances geometry has made since the first days of its existence was the discovery of analytic methods that took place at the beginning of the modern epoch: the representation of the points of the plane by coordinate pairs and of the points of space by coordinate triples, the definition of linear manifolds by linear equations, of conics by quadratic equations, etc. and the resulting possibility of applying the whole of algebra to geometry.

With this discovery not only a new treatment, by means of calculations, of the objects already known to geometry became possible, but also a considerable extension of the domain of this science.

In fact, there began a systematic study of geometrical objects defined by equations of degree higher than two, whereas before the discovery of analytic geometry only few such objects were known, and these few were studied with completely non-systematic methods (essentially only as a means for the solution of the three classical problems of geometry). With the discovery of analytic methods, however, one obtained a systematic principle for the classification of these various objects by the degrees of the equations defining them. Against the tendency, which prevailed during the first period of analytic geometry, to restrict geometrical research to objects definable by algebraic relations, one afterwards proceeded rather quickly to the study of objects defined by means of transcendental functions. The discovery of the notion of coordinates revealed a new field of geometrical entities, whose properties, mutual relations, transformations, etc. could be treated with the methods of analytic geometry. In particular, all those theorems which the earlier axiomatic geometry had proven only by limiting itself to the simple geometrical objects that constituted its domain were proven by mere calculation. Hence, it seemed that synthetic geometry could at best barely keep pace with analytic geometry. Moreover, the local properties of objects defined by equations, their curvature properties, their contact relations, etc. also turned out to be susceptible to a purely analytic treatment by means of infinitesimal calculus, and this so-called differential geometry even became one of the strongest incentives for the foundation of differential and integral calculus. In this way, the discovery of analytic methods—through the application of algebra and calculus to geometrical objects represented by coordinates—provided geometry for centuries with a series of problems which are still not completely exhausted today.

In the wake of this enormous progress, the habituation to purely analytical methods, which lasted several centuries, had—it is true—its shortcomings. First of all, the concept of coordinates, certainly one of the most fruitful tools introduced into geometry, but only a *tool* for treating geometrical problems, was given such importance that many researchers came to see in geometry just an arithmetic of n-tuples of numbers and a theory of equations among these n-tuples. In contrast, one of the most characteristic tendencies of the Vienna Mathematical Colloquium is precisely to develop a geometry *without the help of coordinates*.—Secondly, a great number of geometers, because of their natural disposition to analytic methods, restricted geometry to the study of only those objects which can be defined by means of equations between the coordinates of their points. Certainly, the realm of the equations considered was vastly extended in the course of time. As already mentioned, only objects which were definable by *algebraic* equations were originally recognized as geometrical entities; later on, those definable by means of *analytic* functions and even those definable by means of *continuous* relations were added to the list; but even after these extensions an essential limitation of the domain of geometrical research still prevailed, a limitation that was necessitated solely by the restriction to the given methods, i.e. the analytic methods. Accordingly, another basic tendency of the Vienna Colloquium was the desire to consider all subsets of Euclidean spaces and of still more general spaces, not only sets given by equations involving the coordinates of their points. In other words, we are studying *"set-theoretical geometry"* (mengentheoretische Geometrie).[4]

2 Dimension theory

A first example illustrating the two basic tendencies mentioned above—amd one that is better suited for this purpose than other general considerations—is the concept of dimension. In analytic geometry the plane is said to have "two dimensions" because its points can be represented by two coordinates; our usual space is said to have "three dimensions" because its

[4] See MENGER: *Neuere Methoden und Probleme der Geometrie*, Verhandl. d. Intern. Math. Kongr., Zürich 1932, vol. I, p. 310.

points can be represented by three coordinates; circles have "one dimension" because the coordinates of their points can be expressed as continuous functions of a single parameter; spherical surfaces are of "two dimensions" because the coordinates of their points can be expressed as continuous functions of two parameters, etc. In the last analysis, in analytic geometry, the number of coordinates necessary for the continuous representation of an object therefore constitutes the criterion for its dimension. But this conception runs into some fundamental difficulties: If one denotes all objects for which the coordinates of their points are functions of a single parameter as one-dimensional, then, as G. PEANO showed, quadratic surfaces and cubic bodies belong to the class of one-dimensional figures. Therefore, for continuous transformations one sometimes loses what one commonly understands as the dimension of a figure. One has to restrict oneself to transformations which are *topological*, i.e. one-to-one and bicontinuous; and in this way one arrives at the following concept of an n-dimensional manifold: an object, each of whose points has a neighborhood which can be put into a one-to-one and bicontinuous correspondence with an n-dimensional ball. But this concept is extremely restricted, or rather there exist a large number of spatial figures to which, according to this definition, no dimension at all can be attributed.

Today it is well-known how dimension theory[5] resolves this problem. It states that a non-empty subset M_0 of Euclidean space has *dimension zero*, if every point of M_0 is contained in arbitrarily small open sets whose frontiers do not have any point in common with M_0. Of course, here, as in the theory of functions, a subset U of an Euclidean space is said to be *open* if every point of U is the center of a sphere of that Euclidean space which is entirely contained in U; and by the *frontier* of U one understands the set of all accumulation points of U which do not lie in U. The set M_1 is said to be *one-dimensional*, if it is not zero-dimensional and if each point of M_1 is contained in arbitrarily small open sets whose frontiers intersect M_1 in a set of points that is either empty or of dimension zero. And in general, a set M_n is said to be *n-dimensional* if it does not have a dimension lower than n, and if every point of M_n is contained in arbitrarily small open sets whose frontiers intersect M_n in sets of dimension lower than n. In this way one can

[5] See MENGER: *Dimensionstheorie*, Leipzig 1928, where one finds the complete references up to 1928, including in particular the works of URYSOHN. See also NÖBELING: *Die neuesten Ergebnisse der Dimensionstheorie*, Jahresbericht d. deutsch. Math. Ver., 41, 1931, p. 1.

attribute to each subset of an Euclidean space, no matter how complicated it may be, a finite number as its dimension, and one can prove that each k-dimensional manifold, in the sense explained above, has precisely dimension k.

The importance of this definition lies in the fact that it allows one to develop an extensive theory describing the essential properties and relations of geometric figures. To list only a few examples: to every set M one can associate a dimension to each point P such that the set M is said to be k-dimensional at P if k is the minimal number that enjoys the property that P is contained in arbitrarily small open sets whose frontiers intersect M in sets of dimension lower than k. The set M has dimension n if, and only if, it has dimension $\leq n$ at every point and there exists at least one point at which it has exactly dimension n. The set of all those points of an n-dimensional set M, at which M has exactly dimension n, is either n-dimensional or $(n-1)$-dimensional. If M is a set of dimension n which is closed (i.e. contains all its accumulation points), then the set of all points at which M has dimension n is itself a set of dimension n at each point. Another theorem states that every set of dimension n—be it closed or not—is the sum of $n+1$, but not fewer than $n+1$, sets of dimension zero. Moreover, it is possible to prove that every closed n-dimensional set can be transformed into a n-dimensional complex (i.e., a set that is the sum of a finite number of simplices among which at least one is of dimension n) by a continuous deformation that moves the single points by as little as one wishes; and thus every closed n-dimensional set can, so to speak, be approximated by n-dimensional complexes. Of particular interest are *curves*, i.e. one-dimensional continua, for which one obtains a special and extensive theory.[6]

It is now clear: Dimension theory and the theory of curves belong to coordinate-free geometry and also to set-theoretical geometry because they treat point sets of Euclidean space without referring to a representation of these points by means of coordinates, and deal with entirely arbitrary sets, including sets not given by equations for the coordinates of their points.

This being said, the impression could remain that when considering subsets of Euclidean space one makes implicit use of their representation by means of coordinates. But this is absolutely not the case. And in fact, parallel to the treatment of arbitrary subsets of Euclidean space, one even obtains a generalization of the space concept itself. Just as the Riemannian

[6] See MENGER: *Kurventheorie*, Teubner, Leipzig, 1932.

generalization of the space concept is (according to an observation of WEYL) adequate for the treatment of the figures studied in differential geometry (because the RIEMANNian space is as general as the spatial objects studied in differential geometry), so to the geometry of general sets in Euclidean space there corresponds the even broader generalization of the space concept due to FRÉCHET which is completely detached from the representation of points by means of coordinates. Hence set-theoretical geometry, in particular dimension theory and the theory of curves, can be handled appropriately by means of arbitrary subsets of such generalized spaces.

FRÉCHET calls a set of arbitrary elements a *metric space*, if to every pair of elements p, q there corresponds a real number $pq = qp$ that is $= 0$ if $p = q$ and > 0 if $p \neq q$, and in addition, for three elements p, q, r the triangle inequality $pq + qr \geq pr$ is always satisfied. Using a very suggestive expression, the elements of such sets are called *points*, and the number pq is said to be the *distance* of the point p to the point q. In no way, however, can one attribute n-tuples of coordinates to these elements, as in the special case of Euclidean spaces of dimension n.

It is clear that spheres can be defined in a metric space. A sphere with center p and radius ϱ is the set of all points of the space that have distance $< \varrho$ from p. Therefore, in a metric space, open sets and their frontiers can be defined just as easily as in Euclidean space, and the previous definition of dimension can accordingly be applied to arbitrary subsets of metric spaces: the subset M of a metric space is called n-dimensional if n is the lowest number such that every point of M is contained in arbitrarily small open sets whose frontiers intersect M in a set of dimension lower than n; as initial point of this recursive definition we can choose the empty set if we view it as a set of dimension -1.

All the above-mentioned theorems for n-dimensional subsets of Euclidean spaces still hold for subsets of arbitrary metric spaces, provided that these metric space are separable, i.e., that they possess some countable and dense set. Beyond that, one also obtains the following fundamental result: every n-dimensional subset of a separable metric space can be transformed into a subset of an Euclidean space of *dimension* $(2n + 1)$ by means of a topological (i.e. one-to-one and bicontinuous) mapping.[7] Hence, we see that from the viewpoint of topological representations separable metric spaces of finite dimension are no more general than subsets of Euclidean

[7] For the literature see NÖBELING, loc.cit., footnote 5.

spaces, but identical with them. Finally, as an aside we ne that the term $2n + 1$ turns out to be irreducible, insofar as any n-dimensional set can be represented topologically as a $(2n + 1)$-dimensional subset of Euclidean space, but for each n there are n-dimensional sets that do not permit topological representations on $2n$-dimensional subsets of Euclidean space.[8]

3 Convexity

Perfecting dimension theory and curve theory was by no means the only problem that the geometers of the Vienna Mathematical Colloquium posed for themselves. One of their most important aims is the systematic treatment of those properties and relations of subsets of metric spaces that are studied in classical geometry by means of definite equations for subsets of Euclidean spaces. One of the first chapters of this *"metric geometry"* was the development of a general theory of convexity.[9] We say that a point q of a metric space lies *between* the two points p and r, if $p \neq q \neq r$ and $pq + qr = pr$. A metric space is then called *convex* if for each pair of points p and r there exists at least one point q that lies between p and r. In this theory we restrict our considerations a priori to *closed* metric spaces, i.e. those metric spaces in which every CAUCHY sequence possesses a limit point; a sequence of points $p_1, p_2, \ldots, p_n, \ldots$ is called CAUCHY sequence, if for every ε there exists an N such that for $n > N$ and $m > N$ the distance $p_n p_m$ is $< \varepsilon$. One can prove[10] that a closed convex space must, with any pair of points p and q, contain a *segment* connecting them, i.e. a subset that can be represented, one-to-one and without changing the distance between the different points, by a line segment (in fact a segment of length pq). If, in addition, the space is *externally convex*, i.e. if for every pair of points p and r there exists at least one point s such that r lies between p and s, then any two points of the space are contained within at least one straight line, i.e. a set that can be mapped onto a straight line by means of a transformation which is one-to-one and distance-preserving.

The relations between these investigations and the axiomatics of elementary geometry, and in particular the axioms of order, are evident. Beyond that, the above-mentioned concept of "lying between" leads to the definition of many structural properties of metric spaces, the development

[8] For a short proof see FLORES, Ergebnisse 5, p.17 and vol.6, p.4.

[9] See MENGER, Mathem. Annalen, 100, p.76, and for an abstract Ergebnisse 1, p.20.

[10] For a simple new proof of this theorem, see ARONSZAJN, Ergebnisse vol. 6, p.45.

of a theory of geodesic lines in metric spaces, etc. It is worth noting that these metric properties are not preserved by topological transformations; for example, it is clear that a closed convex space can be represented topologically by a non-convex space. Hence some concepts of metric geometry are linked to profound and often not yet resolved problems of topology. For example, what properties must a compact metric space (i.e. a metric space in which every infinite set has an accumulation point) possess in order to be topologically representable by a convex space?

Pathwise connectedness is certainly necessary. Is this also sufficient? If not, what further conditions have to be added? This constitutes a question that is not yet resolved and seems extraordinarily difficult.

4 The spaces of analytic geometry within the metric spaces

One of the most important metric examples is R_n, the n-dimensional Euclidean space in which every point P is given by n real numbers $x_1, x_2, \ldots, x_n$ and the distance pq between the point p and the point q with coordinates $y_1, y_2, \ldots, y_n$ is defined by the formula $\sqrt{(x_1 - y_1)^2 + (x_2 - y_2)^2 + \ldots + (x_n - y_n)^2}$. What properties must a metric space have in order to be congruent with R_n, or with a subset of R_n? Here we call two metric spaces *congruent* if they can be mapped onto each other by a distance-preserving one-to-one transformation. After this question had been answered[11], the Vienna Mathematical Colloquium was immediately able to attack and resolve a new, much more general problem that will now be briefly expounded.[12]

First of all, if all the spaces studied in analytic geometry are to fall under the concept of space, then this latter must be generalized even beyond the concept of a metric space. In a metric space only non-negative real numbers are admitted as distances, while complex and negative distances enter in the spaces of complex analytic geometry. Furthermore, in a metric space any two non-coincident points have distance $\neq 0$, whereas in complex analytic geometry one considers isotropic figures consisting of distinct points having pairwise distance zero. A generalization of the concept of a metric space that encompasses the spaces considered in analytic geometry is obtained by means of the following definition[13]: a set R of arbitrary elements (cal-

[11] See MENGER, Math. Annalen 100, p.113 and, for an abstract, Ergebnisse 1, p.20.

[12] See WALD, Ergebnisse 5, pp.32-42.

[13] See MENGER, Ergebnisse 2, p.34; WALD-FLEXER, Ergebnisse 5, p.32 and, for an even more farreaching generalization, TAUSSKY, Ergebnisse 6, p. 20.

led points) forms a *complex metric space*[14] if to each pair of points p and q there corresponds a complex number $pq = qp$ that is $= 0$ if $p = q$, but if $p \neq q$ there exists at least one point r of R for which $pr \neq qr$. This concept encompasses, for example, n-dimensional complex space, each of whose points is given by n complex numbers $x_1, x_2, \ldots, x_n$ and where the distance of the point p with the coordinates $x_1, x_2, \ldots, x_n$ from the point q with the coordinates $y_1, y_2, \ldots, y_n$ is defined by $(x_1 - y_1)^2 + (x_2 - y_2)^2 + \ldots + (x_n - y_n)^2$. Among the most important subsets of this K_n, which are complex metric spaces on their own (and precisely with real, but not necessarily positive distances), we mention the sets $R_{n,s}$ consisting of those points of K_n among whose n coordinates the first $\frac{n+s}{2}$ are real and the last $\frac{n-s}{2}$ purely imaginary. For $s = n$ we obtain $R_{n,n}$ which is closely related to the n-dimensional Euclidean space R_n, insofar as the distance between two arbitrary points in this $R_{n,n}$ is equal to the square of the distance of the points of R_n with the same coordinates. The space $R_{n,s}$ is even a *real metric space*, i.e. a complex metric space for which the distances between any two points are real.

The fundamental metric property of the spaces K_n and $R_{n,s}$, which is directly related to homogeneity and the existence of groups of congruences, is the following: If R is a complex metric spaces, such that for any $n+3$ of its points $p_1, p_2, \ldots, p_{n+3}$ there exist $n+3$ points $p'_1, p'_2, \ldots, p'_{n+3}$ (not necessarily pairwise distinct) of K_n, resp. $R_{n,s}$, with the property that $p_i p_j = p'_i p'_j$ $(i, j = 1, 2, \ldots n+3)$, then there exists a subset R' of K_n, resp. $R_{n,s}$, to which R is *applicable* (in the sense that to each point p in R one can associate a point p' in R', such that $pq = p'q'$ holds for the corresponding pairs of points p, q and p', q'. (In order that R and R' be congruent, it is necessary and sufficient that R is applicable to R' or R' is applicable to R by means of a one-to-one transformation).

The following definition serves to characterize K_n and $R_{n,s}$: Let $p_1, p_2, \ldots, p_n$ be arbitrary points of a complex metric space, and let $\Delta(p_1, p_2, \ldots, p_n)$ denote the determinant

$$\begin{vmatrix} 0 & 1 & 1 & 1 & \cdots & 1 \\ 1 & 0 & p_1p_2 & p_1p_3 & \cdots & p_1p_k \\ 1 & p_1p_2 & 0 & p_2p_3 & \cdots & p_2p_k \\ \vdots & \vdots & \vdots & \cdots & \ddots & \vdots \\ 1 & p_1p_k & p_2p_k & p_3p_k & \cdots & 0 \end{vmatrix}$$

[14] This definition and those of a real metric space and a positive metric space given on the following pages should not be confused with the concept of a metric space in the sense of FRÉCHET (See p.2).

Obviously, one has

$$\Delta(p_1) = \begin{vmatrix} 0 & 1 \\ 1 & 0 \end{vmatrix}, \quad \Delta(p_1, p_2) = \begin{vmatrix} 0 & 1 & 1 \\ 1 & 0 & p_1p_2 \\ 1 & p_1p_2 & 0 \end{vmatrix} = 2p_1p_2,$$

etc. To ensure that a real metric space R is applicable, in the sense defined above, to a subset of K_n it is necessary and sufficient that for any $n + 3$ points $p_1, p_2, \ldots p_{n+3}$ of R one has $\Delta\,(p_1, p_2, p_{n+3}) = 0$ and for any $n + 2$ points $p_1, p_2, \ldots p_{n+2}$ of R one has $\Delta\,(p_1, p_2, \ldots p_{n+2}) = 0$. Every metric space containing no more than $n + 1$ points is applicable to a subset of K_n.

For a real metric space R to be applicable to a subset of $R_{n,s}$ it is necessary and sufficient that for any $n + 3$ and $n + 2$ points of R the relations $\Delta = 0$ are valid and moreover that any $n + 1$ points of R are applicable to points of $R_{n,s}$. If for $n + 1$ points $p_1, p_2, \ldots p_{n+1}$ of a real metric space one has $\Delta\,(p_1, p_2, \ldots, p_{n+1}) \neq 0$, then the $n + 1$ points can be renumbered as $p_{i_1}, p_{i_2}, \ldots, p_{i_{n+1}}$ in such a way that within the sequence

$$\Delta(p_{i_1}), \quad \Delta(p_{i_1}, p_{i_2}), \ldots, \quad \Delta(p_{i_1}, p_{i_2}, \ldots, p_{i_{n+1}}) \qquad (*)$$

two consecutive numbers cannot both be zero. For $n + 1$ points to be applicable to points of $R_{n,s}$ it is necessary and sufficient that in the sequence (*), possibly after excluding zero terms, $\frac{n+s}{2}$ changes of sign occur. Thus it becomes clear that if $n + 1$ points with $\Delta \neq 0$ are applicable to points of $R_{n,s}$, they are not applicable to points of a $R_{n,t}$ $(t \neq s)$.

Hence, in order that $n + 1$ points $p_1, p_2, \ldots, p_{n+1}$ with $\Delta(p_1, p_2, \ldots, p_{n+1}) \neq 0$ are applicable to points of $R_{n,n}$ (in which case they are congruent with $n + 1$ points of $R_{n,n}$ because any two distinct points of $R_{n,n}$ have non-zero distance) it is necessary and sufficient that there are n changes of signs in (*). But since $\Delta(p_{i_1}) = -1$, this is possible only if

$$\operatorname{sgn}\Delta(p_{i_1}, p_{i_2}, \ldots, p_{i_k}) = (-1)^k \quad \text{for } k = 1, 2, \ldots, n + 1.$$

Using this it is also possible to obtain a characterization of R_n, the n-dimensional Euclidean space[15], where the distance between any two points is the square root of the points of $R_{n,n}$ with the same coordinates. In this case, for k given points $p_1, p_2, \ldots, p_k$, instead of Δ, we introduce the determinant

[15] loc.cit. footnote 11.

$$D(p_1, p_2, \ldots p_k) = \begin{vmatrix} 0 & 1 & 1 & 1 & \ldots & 1 \\ 1 & 0 & (p_1p_2)^2 & (p_1p_3)^2 & \ldots & (p_1p_k)^2 \\ 1 & (p_1p_2)^2 & 0 & (p_2p_3)^2 & \ldots & (p_2p_k)^2 \\ \vdots & \vdots & \vdots & \ldots & \ddots & \vdots \\ 1 & (p_1p_k)^2 & (p_2p_k)^2 & (p_3p_k)^2 & \ldots & 0 \end{vmatrix}$$

and observe that we are restricting ourselves to real metric spaces with non-negative distances, or as we say *positive metric spaces*. For a positive metric space R to be congruent to a subset of R_n it is necessary and sufficient that for any $n+2$ points $p_1, p_2, \ldots, p_{n+2}$ of R, one has $D(p_1, p_2, \ldots, p_{n+2}) = 0$ and that for each k $(2 \leq k \leq n+1)$ and for any k points $p_1, p_2, \ldots, p_k$,

$$\operatorname{sgn} D(p_1, p_2, \ldots, p_k) = (-1)^k \text{ or } 0.$$

If R consists of exactly $n+3$ points $p_1, p_2, \ldots, p_{n+3}$, then one has to add the further condition $D(p_1, p_2, \ldots, p_{n+3}) = 0$. If R contains more than $n+3$ points, then R is already congruent to a subset of R_n whenever any $n+2$ points of R are congruent to some $n+2$ points of R_n (an assertion that would be false for K_n or $R_{n,s}$, $0 \neq s \neq n$). But for each n there exist positive metric spaces containing exactly $n+3$ points which are not congruent to $n+3$ points of R_n, although any $n+2$ of their points are congruent to some $n+2$ points of R_n. We call spaces of this type pseudo-euclidean $n+3$-tuples. For $n = 1$, one obtains pseudo-euclidean quadruples which can be characterized in the following way: For 4 points p_1, p_2, p_3, p_4 to be congruent to 4 points on a straight line it is necessary and sufficient that the following relations among the six distances hold:

$$\begin{aligned} p_1p_2 &= p_3p_4 > 0, \\ p_1p_3 &= p_2p_4 > 0, \\ p_1p_4 &= p_2p_3 = p_1p_2 + p_1p_3. \end{aligned} \tag{2}$$

For $n = 2$ one obtains pseudo-euclidean quintuples whose theory is connected with the theory of isogonal transformations of the plane; finally if $n > 2$, then the theory of $(n+3)$-tuples is related to the theory of some Cremona transformations of R_n.[16]

It follows easily from the theory of convexity that for a positive metric space R to be congruent with R_n itself, it is necessary and sufficient that R is closed, internally and externally convex, and moreover satisfies conditions ensuring its congruence with a subset of R_n.

[16] See MENGER, Mathem. Annalen, 100, pp.120-128.

5 Coordinate-free differential geometry

As shown above, the spaces of analytic geometry can be included in the general space concept of coordinate-free metric geometry. On the other hand, it seemed that one particularly important domain of geometrical research was indissolubly linked to the coordinate concept and to the numerical calculations of analytic geometry, namely differential geometry. Until now the unique way of treating the local metric properties of spatial figures seemed to be infinitesimal calculus, which in the course of centuries had developed a complete symbolism for this purpose. This symbolism and its rich methods required the unavoidable hypothesis that the points and figures studied were given by coordinates. It is therefore noteworthy that coordinate-free metric geometry also succeeded in treating local metric properties of figures whose points are not given by coordinates and which are not given by equations. When applied to the special case of Euclidean spaces, some of these theorems contain propositions of classical differential geometry, now proven without differential calculus, i.e., solely through metric geometry.[17]

Here I want to exhibit only one particularly important example of this coordinate-free differential geometry: the theory of curvature of surfaces recently developed in the Vienna Mathematical Colloquium.[18] Let 4 points p_1, p_2, p_3, p_4 of a metric space, together with their distances p_ip_j $(i, j = 1, 2, 3, 4)$, be given; we define the *radius of curvature* of the points p_1, p_2, p_3, p_4 as a real number ϱ having the property that the surface of a sphere of radius ϱ contains 4 points p'_1, p'_2, p'_3, p'_4 which are congruent with the 4 given points. As distance between two points on a spherical surface we take the length of the shortest of the two arcs of the great circle through these points. By $P(p_1, p_2, p_3, p_4)$ we denote the set of all radii of curvature of p_1, p_2, p_3, p_4. It turns out that this set may be empty, which means that 4 given points cannot be congruent to 4 points of any spherical surface. On the other hand, it may happen that the set $P(p_1, p_2, p_3, p_4)$ contains more than one number, and hence that the 4 given points are congruent to 4 points on surfaces of different radii.

Now let R be a convex compact metric space and p a given point of R; we define a p-sequence as a sequence of quadruples of points $\{p_n^1, p_n^2, p_n^3, p_n^4\}$ $(n = 1, 2, 3, \ldots)$, such that, first of all, these quadruples converge to the

[17] See MENGER: Zur Theorie der Bogenkrümmung, Mathem. Annalen, 103, p.480; Jahresber. d. deutsch. Math. Ver., 50, pp. 211-213, 218.

[18] See WALD, Ergebnisse 6, p.29.

point p, i.e. that for the distances $p_i^n p$, $\lim_{n\to\infty} p_i^n p = 0$ $(i = 1, 2, 3, 4)$ holds, and secondly that they satisfy the following conditions: for each n let d_n be the largest of the six distances $p_i^n p_j^n$ $(i, j = 1, 2, 3, 4)$ and consider the sequence of sextuples of numbers

$$\frac{p_1^n p_2^n}{d_n}, \quad \frac{p_1^n p_3^n}{d_n}, \quad \frac{p_1^n p_4^n}{d_n}, \quad \frac{p_2^n p_3^n}{d_n}, \quad \frac{p_2^n p_4^n}{d_n}, \quad \frac{p_3^n p_4^n}{d_n},$$

and if this sequence of sextuples possesses the six numbers

$$\pi_{12}, \ \pi_{13}, \ \pi_{14}, \ \pi_{23}, \ \pi_{24}, \ \pi_{34},$$

as accumulation points, then we require that each of these six numbers π_{ij} is larger than zero and that there do not exist four collinear points q_1, q_2, q_3, q_4 for which

$$q_1q_2 = \pi_{12}, \quad q_1q_3 = \pi_{13}, \quad q_1q_4 = \pi_{14}, \quad q_2q_3 = \pi_{23}, \quad q_2q_4 = \pi_{24}, \quad q_3q_4 = \pi_{34}.$$

Using the concept of p-sequence, the surface curvature is defined as follows: the real number κ is called the *surface curvature of the space R at the point p* if, for any p-sequence $\{p_1^n, p_2^n, p_3^n, p_4^n\}$, there exists an index N with the property that for each $n > N$ one can find a number $\bar{\varrho}_n$ in the set $P(p_1^n, p_2^n, p_3^n, p_4^n)$ such that $\lim_{n\to\infty} \frac{1}{\bar{\varrho}_n^2} = \kappa$. Although it may happen, as already mentioned, that for each n the set $P(p_1^n, p_2^n, p_3^n, p_4^n)$ of the radii of curvature contains different numbers, the surface curvature as defined turns out to be *unique*. In other words, if κ is the surface curvature of R at p, then one can prove the following: whatever numbers $\bar{\rho}_n$ one choses from the set $P(p_1^n, p_2^n, p_3^n, p_4^n)$ for each p-sequence, the relation $\lim_{n\to\infty} \frac{1}{\bar{\rho}_n^2} = \kappa$ is always valid. Furthermore, one can also prove: if, in particular, the space R is a GAUSSian surface within an Euclidean space, given by functions with third derivatives, then at each point p there exists a surface curvature of R, and this coincides precisely with the GAUSSian curvature of the surface at the point p.

Hence, in this way the concept of a surface curvature without coordinates was successfully introduced. But the most noteworthy and surprising result of this theory is a sort of converse to the previous theorem. This is the following[19]: if R is a convex and compact metric space that admits a

[19] See WALD, Ergebnisse volume 7, that will appear in autumn 1935. This article will also contain a simplification of the definition of curvature stated above.

surface curvature (in the sense just made precise) at each point, then R is congruent to a two-dimensional Riemannian space, i.e. a GAUSSian surface. Observe that apart from compactness, convexity, and the existence of a surface curvature one does not make any assumptions about the metric space: but one can prove, based on these three conditions, that it is actually a GAUSSian surface. One need not even assume that the space is two-dimensional. Incidentally, the two-dimensionality is a rather immediate consequence of the existence of a finite surface curvature. The main difficulty in the proof of the theorem resides in showing that every point of a compact convex space with finite surface curvature is connected with each other point of a sufficiently small neighborhood by a unique segment (that two points of a compact convex space are connected by at least one segment is a result of the theory of convexity). This uniqueness of segments between neighboring points of the space reflects the fundamental fact that each point of a GAUSSian surface is joined to each point of a sufficiently small neighborhood by one and only one geodesic. The final difficulty in the proof then consists in introducing a system of polar coordinates on the segments originating from a point [segments which correspond to the geodesics lines of the GAUSSian surface], and in checking that the classical formulas of the differential geometry of surfaces apply to it.

It is evident that the methods outlined above allow one to deduce new rigorous proofs of the most important theorems of the classical differential geometry of surfaces under extraordinarily general hypotheses, e.g., to obtain a new development of the theory of surfaces of constant curvature, etc. In the opinion of the geometers of the Vienna Mathematical Colloquium, these are the first steps towards a new foundation of the whole of differential geometry; accordingly, this part, which seemed to be most remote from the methods of coordinate-free geometry, offers—after the fundamental results reached by these methods in other branches of geometry—to become one of the most important domains of application.

Translation: Michael Stöltzner

From: *Annali della R. Scuola Normale Superiore di Pisa*, Scienze Fisiche a Matematiche, Serie II, Volume IV (1935), pp. 327-339. Original title: *Sull'indirizzo di idee e sulle tendenze principali del colloquio matematico di Vienna*

A table of contents of the Ergebnisse

1928/29

1929/30

Gesammelte Mitteilungen des Jahres 1929/30

1930/31

Gesammelte Mitteilungen des Jahres 1931/32

1932/33

Gesammelte Mitteilungen des Jahres 1932/33

1933/34

1934/35

Gesammelte Mitteilungen des Jahres 1934/35

1935/36

Gesammelte Mitteilungen der Jahre 1928/29.

Jene Notizen dieses Heftes, die durch einen Stern vor dem Titel gekennzeichnet sind, stellen bearbeitete Abdrücke von Notizen dar, die im Anzeiger der Wiener Akademie der Wissenschaften (Mathem.-naturwiss. Klasse) Bd. 65 (1928) und 66 (1929), erschienen sind.

*Ein Theorem der Topologie. Von Karl Menger.

Zwei Grundfragen der Topologie lauten:

I. Welche Räume besitzen eine endliche Dimension?

II. Welche topologischen Räume sind mit endlich vielen Koordinaten beschreibbar, d. h. homöomorph mit einer Teilmenge eines Zahlenraumes?

Als höchstens n-dimensional bezeichnet man[1]) einen Raum, wenn zu jedem seiner Punkte beliebig kleine Umgebungen mit höchstens $(n-1)$-dimensionalen Begrenzungen existieren, und man nennt endlichdimensional einen Raum, der für irgendeine natürliche Zahl n höchstens n-dimensional ist. Dabei ist unter „Raum" eine abstrakte Menge zu verstehen, die gewissen inneren Bedingungen (z. B. den Hausdorffschen Umgebungsaxiomen) genügt. Nicht im mindesten wird also durch die angegebene Definition für die endlichdimensionalen Räume irgendeine Beschreibbarkeit mit Koordinaten postuliert. Es gilt nun aber der fundamentale Satz, daß jeder separable n-dimensionale Raum homöomorph ist mit einer Teilmenge des $(2n+1)$-dimensionalen Zahlenraumes, also mit $2n+1$ Koordinaten beschrieben werden kann. Aus diesem Theorem folgt insbesondere als Antwort auf die beiden eingangs formulierten Fragen: Dieselben Räume sind es, die eine endliche Dimension besitzen und die mit endlich vielen Koordinaten beschreibbar sind. Die für die endlichdimensionalen Räume nicht postulierte Beschreibbarkeit mit endlich vielen Koordinaten kommt ihnen von selbst zu. Denn die Zahlenräume sind dem Theorem zufolge von einer so überraschenden Allgemeinheit, daß sie zu jedem endlichdimensionalen separabeln Raum eine homöomorphe Teilmenge enthalten. Natürlich ist nicht jeder n-dimensionale Raum mit n Koordinaten beschreibbar. Es lassen sich für jedes n ohne weiteres n-dimensionale Räume angeben,

---

[1]) Vgl. mein Buch Dimensionstheorie, bei Teubner 1928, und die Literaturangaben daselbst. Dieses Buch wird in allem folgenden kurz als „Dimensionstheorie" zitiert.

die zu ihrer Beschreibung mindestens $n+2$ Koordinaten erfordern. Jedenfalls kommt man dem Theorem zufolge zur Beschreibung der n-dimensionalen Räume mit $2n+1$ Koordinaten aus. Die Frage, ob diese Schranke für $n > 1$ verringert werden kann, ist noch offen. [Für $n \leqq 1$ kann sie natürlich nicht verringert werden, denn es gibt ja sogar Kurven, die Summe von endlich vielen Strecken sind, welche nicht mit Teilkurven der Ebene homöomorph sind. Hingegen ist noch nicht entschieden, ob es Flächen gibt, die Summe von endlich vielen Dreiecken sind, welche nicht homöomorph mit Teilmengen des vierdimensionalen Zahlenraumes sind.]

***Über die vollständige Unabhängigkeit der Mengerschen Dimensionsaxiome.** Von Hans Hornich.

Mengers Postulate an die Dimension unter den Funktionen f der Teilmengen des R_n lauten[1]:

1. *Ist M' Teilmenge von M, so gilt $f(M') \leqq f(M)$.*

2. *Ist $\{A_k\}$ eine Folge von abzählbar vielen abgeschlossenen Mengen, so ist $f(\sum_k A_k) \leqq Max.\ f(A_k)$.*

3. *Sind M und M' homöomorphe Mengen, so ist $f(M) = f(M')$.*

4. *Jede Menge M ist homöomorph mit einer Teilmenge M' einer kompakten Menge A, so daß $f(M) = f(A)$ ist.*

5. *Für die k-dimensionalen Intervalle I_k ist $f(I_k) = k$.*

Unter den Funktionen der Teilmengen des R_1 und des R_2 sind diese Eigenschaften für die Dimension charakteristisch[1]). Es beruht dies auf der Tatsache, daß unter allen Systemen $\mathfrak{S}$ von Teilmengen dieser Räume nur die Systeme aller höchstens k-dimensionalen Mengen die vier folgenden Eigenschaften besitzen:

A. Ist M eine Menge des Systems $\mathfrak{S}$ und $M' \subset M$, so ist M' eine Menge von $\mathfrak{S}$.

B. Ist $\{M_k\}$ $(k = 1, 2, \ldots$ ad inf.) eine Folge von abgeschlossenen Mengen von $\mathfrak{S}$, so ist $\sum_{k=1}^{\infty} M_k$ eine Menge von $\mathfrak{S}$.

C. Ist M eine Menge von $\mathfrak{S}$ und M' mit M homöomorph, so ist M' eine Menge von $\mathfrak{S}$.

D. Zu jeder Menge M von $\mathfrak{S}$ existiert eine kompakte Menge von $\mathfrak{S}$, welche eine mit M homöomorphe Menge als Teilmenge enthält.

Menger hat gezeigt[2]), daß sowohl unter den 5 Forderungen an die Dimension als auch unter den 4 Forderungen an Mengensysteme jede von den übrigen unabhängig ist. In Verallgemeinerung dieses gewöhnlichen Unabhängigkeitsbegriffes nennt E. H. Moore ein System von Eigenschaften vollständig unabhängig[3]) (com-

---

[1]) Monatshefte f. Math. u. Phys., 36, S. 193.

[2]) A. a. O. S. 202.

[3]) New Haven Mathematical Colloquium, Yale University Press (1910), S. 82.

1*

pletely independent), wenn es für jede Gruppe von Eigenschaften aus diesem System ein Ding gibt, welches alle Eigenschaften dieser Gruppe besitzt und alle anderen Eigenschaften des Systems nicht besitzt. Es soll nun gezeigt werden, daß **Mengers Postulate in diesem Sinn vollständig unabhängig sind.** Die den folgenden Beispielen vorangestellten ungestrichenen Buchstaben deuten das Vorhandensein, die gestrichenen das Fehlen der entsprechenden Eigenschaften im betreffenden Beispiel an.

$\bar{A}\ B\ C\ D$. Das System der Borelschen Mengen des R_n[1]).

$A\ \bar{B}\ C\ D$. Das System der endlichen Mengen des R_n.

$A\ B\ \bar{C}\ D$. Das System aller Mengen von erster Kategorie im R_n.

$A\ B\ C\ \bar{D}$. Das System der abzählbaren Mengen des R_n.

$\bar{A}\ \bar{B}\ C\ D$. Das System der G_δ-Mengen des R_n.

$\bar{A}\ B\ \bar{C}\ D$. Das System der unendlichen (d. h. unendlichviele Punkte enthaltenden) Teilmengen eines Bogens B im R_n.

$\bar{A}\ B\ C\ \bar{D}$. Das System der abzählbaren insichdichten Mengen des R_n.

$A\ \bar{B}\ \bar{C}\ D$. Das System der beschränkten Mengen des R_n.

$A\ \bar{B}\ C\ \bar{D}$. Das System der abzählbaren Mengen des R_n, die keinen ihrer Häufungspunkte enthalten.

$A\ B\ \bar{C}\ \bar{D}$. Das System der Mengen von rationalen Punkten des R_n.

$\bar{A}\ \bar{B}\ \bar{C}\ D$. Das System der beschränkten insichdichten Mengen des R_n.

$\bar{A}\ \bar{B}\ C\ \bar{D}$. Das System der Mengen des R_n, welche topologische Bilder des offenen Intervalls $(0,1)$ sind.

$\bar{A}\ B\ \bar{C}\ \bar{D}$. Das System der mehrpunktigen Mengen von rationalen Punkten des R_n.

$A\ \bar{B}\ \bar{C}\ \bar{D}$. Das System der abzählbaren beschränkten Mengen des R_n.

$\bar{A}\ \bar{B}\ \bar{C}\ \bar{D}$. Das System der abzählbaren insichdichten beschränkten Mengen des R_n.

In derselben Symbolik beweisen wir die vollständige Unabhängigkeit der Postulate 1—5.

$\bar{1}\ 2\ 3\ 4\ 5$. Für die Borelschen Mengen des R_n sei f gleich der Dimension, für die nichtborelschen gleich n.

$1\ \bar{2}\ 3\ 4\ 5$. Im R_1 sei f für die leere Menge gleich —1, für die endlichen Mengen 0, für die unendlichen Mengen ohne Teilintervalle $\frac{1}{2}$ und für die ein Teilintervall enthaltenden Mengen 1.

[1]) Hier und im folgenden bedeutet n irgend eine ganze Zahl $\geqq 1$.

$1\ 2\ \bar{3}\ 4\ 5$. Für die Mengen von erster Kategorie im R_n sei f gleich der Dimension, für alle anderen gleich n.

$1\ 2\ 3\ \bar{4}\ 5$. Im R_n sei f für die leere Menge -1, für die abzählbaren Mengen 0, für die irrational $(k-1)$-dimensionalen und die rational k-dimensionalen Mengen $k\,(0 < k \leqq n)$.

$\bar{1}\ \bar{2}\ 3\ 4\ 5$. Für die kompakten Mengen des R_n sei f gleich der Dimension, für alle anderen gleich n.

$\bar{1}\ 2\ \bar{3}\ 4\ 5$. Sei D ein im R_n konstruiertes Cantorsches Diskontinuum. Dann sei f für D gleich -1, für alle anderen Mengen gleich der Dimension.

$\bar{1}\ 2\ 3\ \bar{4}\ 5$. Für die unabzählbaren, total imperfekten Teilmengen des R_1 sei f gleich 2, für alle anderen gleich der Dimension.

$1\ \bar{2}\ \bar{3}\ 4\ 5$. Sei P die Menge der rationalen Punkte des R_n. Für eine Menge $M \supset P$ sei f der größere der Werte 1 und der Dimension von M, für die anderen Mengen gleich der Dimension.

$1\ \bar{2}\ 3\ \bar{4}\ 5$. Im R_1 sei f für die leere Menge -1, für alle endlichen Mengen 0, für die abzählbar unendlichen $\frac{1}{2}$ und für die unabzählbaren Mengen 1.

$1\ 2\ \bar{3}\ \bar{4}\ 5$. Sei D eine insichdichte unabzählbare, total imperfekte Menge des R_n. Für jede Menge $M \subset D$ sei f gleich -1, für alle anderen Mengen gleich der Dimension.

$\bar{1}\ \bar{2}\ \bar{3}\ 4\ 5$. Für die k-dimensionalen Intervalle des R_n sei f gleich k, für alle anderen Mengen 0.

$\bar{1}\ \bar{2}\ 3\ \bar{4}\ 5$. Für die kompakten Mengen des R_n sei f gleich der Dimension, für alle anderen Mengen gleich der um 1 vermehrten Dimension.

$\bar{1}\ 2\ \bar{3}\ \bar{4}\ 5$. Für den gesamten R_n sei f gleich $n-1$, für alle echten Teile gleich der Dimension.

$1\ \bar{2}\ \bar{3}\ \bar{4}\ 5$. Für beschränkte Mengen des R_n sei f gleich der Dimension, für unbeschränkte gleich $n+1$.

$\bar{1}\ \bar{2}\ \bar{3}\ \bar{4}\ 5$. Für die k-dimensionalen Intervalle des R_n sei f gleich k, für alle kompakten Mengen, die nicht Intervalle sind, gleich $\frac{1}{2}$ und für alle anderen Mengen gleich 0.

In allen diesen Beispielen erfüllt f das Postulat 5. Setzt man in jedem statt f die Funktion $f+1$ ein, so entstehen neue Funktionen, die sämtlich das Postulat 5 nicht erfüllen. Zusammen ergeben dann diese Beispiele den Nachweis der vollständigen Unabhängigkeit.

*Dimensionstheoretische Notizen. Von Karl Menger.

I. Ein Satz zur Ergänzung der Dimensionsdefinition.

Die mit Umgebungsbegrenzungen operierende Einführung des Dimensionsbegriffes, — derzufolge ein Raum höchstens n-dimensional heißt, wenn jeder Punkt in beliebig kleinen Umgebungen mit höchstens $(n-1)$-dimensionalen Begrenzungen enthalten ist, und derzufolge eine Teilmenge M eines Raumes höchstens k-dimensional heißt, wenn jeder Punkt von M in beliebig kleinen Umgebungen enthalten ist, deren Begrenzungen mit M höchstens $(k-1)$-dimensionale Durchschnitte haben, — legt folgende Frage nahe: Ist jeder Punkt einer höchstens k-dimensionalen Teilmenge M eines höchstens n-dimensionalen Raumes in beliebig kleinen Umgebungen enthalten, deren Begrenzungen zugleich höchstens $(n-1)$-dimensional sind und mit M höchstens $(k-1)$-dimensionale Durchschnitte haben? Die Antwort ist positiv und es gilt allgemeiner der folgende (übrigens noch in mehrfacher Hinsicht verallgemeinerbare) Satz: *Sind in einem höchstens n-dimensionalen kompakten Raum A eine höchstens $(n-1)$-dimensionale abgeschlossene Menge A_1, eine höchstens $(n-2)$-dimensionale abgeschlossene Teilmenge A_2 von A_1, ... usw. ... endlich eine höchstens $(n-k)$-dimensionale abgeschlossene Teilmenge A_k von A_{k-1} gegeben, so existieren zu jedem Punkt von A_k beliebig kleine Umgebungen, deren Begrenzungen höchstens $(n-1)$-dimensional sind, mit A_1 höchstens $(n-2)$-dimensionale Durchschnitte, mit A_2 höchstens $(n-3)$-dimensionale Durchschnitte, ... usw. ... endlich mit A_k höchstens $(n-k-1)$-dimensionale Durchschnitte haben.* Der Beweis dieses Satzes kann ganz analog dem Beweis des Summensatzes (Monatshefte f. Math. u. Phys., *34*, S. 147) mit Hilfe der Methode der Modifikation der Umgebungen in der Nähe ihrer Begrenzungen geführt werden. Der erwähnte Satz kann zum Beweise des allgemeinen Zerlegungstheorems herangezogen werden.

II. Über schwachdimensionale Mengen.

Ein endlichdimensionaler Raum, dessen Dimension größer ist als die Dimension der Menge aller Punkte höchster Dimension des Raumes, heißt *schwachdimensional.* Auf Grund des zweiten Fundamentaltheorems über dimensionelle Raumstruktur (Dimensionstheorie, S. 135) kann die Dimension eines endlichdimensionalen Raumes die Dimension der Menge aller Punkte höchster Dimension höchstens um 1 übersteigen. Ein n-dimensionaler Raum heißt demgemäß *schwach n-dimensional,* wenn die Menge aller Punkte, in denen der Raum n-dimensional ist, die Dimension $n-1$ besitzt. Ein Raum, welcher bloß in den Punkten einer höchstens $(n-1)$-dimensionalen Menge n-dimensional ist, heiße *höchstens schwach n-dimensional.* So heißt also ein Raum, welcher schwach n-dimensional oder höchstens $(n-1)$-dimensional ist. Sierpiński hat (Fund. Math., *2*, S. 81) einen eindimensionalen Raum konstruiert, welcher bloß in abzählbar vielen Punkten eindimensional ist. Derartige Räume mögen kurz als äußerst schwach eindimensional bezeichnet werden. Die Frage, ob für $n > 1$ schwach n-dimensionale Mengen existieren, ist noch ungelöst[1]). Es gilt nun der Satz: *Wenn die n-te Potenz eines äußerst schwach eindimensionalen Raumes n-dimensional ist, so ist sie schwach n-dimensional*[2]). Offen bleibt die Frage, ob die Summe von endlich oder abzählbar vielen schwach n-dimensionalen Mengen, von denen jede in der Summe abgeschlossen ist, schwach n-dimensional ist[3]).

III. Über n-dimensionale Gewächse.

Unter den Kurven (den eindimensionalen Kontinua) spielen eine besondere Rolle die Baumkurven (vgl. „Grundzüge einer Theorie der Kurven“, Mathem. Annalen, *95*, insbes. S. 302, und *96*, S. 572). Dieselben können definiert werden als jene kompakten zusammenhängenden Räume, in denen je zwei Punkte durch einen Punkt trennbar sind. Es sei hier nun darauf hingewiesen, daß dieser Begriff der Baumkurve Verallgemeinerungen für beliebige Dimensionen gestattet. Betrachten wir zunächst den Fall $n=2$. Wir können als zweidimensionales Gewächs im weiteren Sinn jedes zweidimensionale Kontinuum bezeichnen, in dem je zwei Punkte entweder durch eine Baumkurve oder durch einen Punkt trennbar sind. Derartige Gewächse in weiterem Sinn enthalten keine geschlossenen Teilflächen (keine Kugelflächen, Torusflächen usw.), können dagegen z. B. von der Art einer Zylinderfläche sein. Wollen wir derartige Gebilde ausschließen, so können wir als zweidimensionale Gewächse die zweidimensionalen Kontinua, in denen je zwei Punkte und je zwei fremde Teilbäume durch einen Punkt oder durch einen Teilbaum trennbar sind, bezeichnen. Dabei heißen zwei fremde abgeschlossene Teilmengen A und B eines Raumes R durch eine (eo ipso abgeschlossene) Menge C getrennt, wenn zwei offene Mengen U und V existieren, so daß $A \subset U, B \subset V, \overline{U} + \overline{V} = R, \overline{U} \cdot \overline{V} \subset C$ gilt. Allgemein können wir, indem wir als nulldimensionales Gewächs jede aus einem einzigen Punkt bestehende Menge bezeichnen, definieren: Ein n-dimensionales Kontinuum heißt Gewächs (beziehungsweise Gewächs in weiterem Sinn), wenn je zwei fremde höchstens $(n-1)$-dimensionale Teilgewächse (beziehungsweise je zwei Punkte) durch ein höchstens $(n-1)$-dimensionales Teilgewächs trennbar sind. Die Verallgemeinerungen des Begriffes Baumkurve, von denen die beiden angeführten Definitionen Beispiele darstellen, sind für verschiedene topologische Fragen von Interesse:

a) für das Studium der Verteilung der mehrdimensionalen Verzweigungspunkte (derartige Sätze über das Verhalten im kleinen gelten für n-dimensionale Gewächse im kleinen);

---

[1]) Diese Frage wurde inzwischen von Mazurkiewicz (Fund. Math., *13*, 1929, S. 210) positiv entschieden.

[2]) Vgl. Dimensionstheorie, S. 148.

[3]) Diese Frage wurde von Mazurkiewicz (Fund. Math., *12*, S. 111) negativ entschieden.

b) für das Studium stetiger Abbildungen;
c) für Sätze über die Existenz von Fixpunkten bei topologischen Selbstabbildungen der Räume;
d) für die topologische Charakterisierung des n-dimensionalen Simplexes.

IV. Ein Zerlegungssatz für rational dimensionale Mengen.

Ein endlichdimensionaler Raum heißt *rationaldimensional*, wenn seine Dimension durch Tilgung einer abzählbaren Teilmenge verringert werden kann. Da die Dimension eines endlichdimensionalen Raumes durch Tilgung einer nulldimensionalen Menge höchstens um 1 vermindert werden kann, so heißt dieser Definition zufolge rational n-dimensional ein n-dimensionaler Raum, welcher Summe einer $(n-1)$-dimensionalen und einer abzählbaren Menge ist. Ein Raum, welcher entweder rational n-dimensional oder höchstens $(n-1)$-dimensional ist, heiße *höchstens rational n-dimensional.* Gleichbedeutend ist (Dimensionstheorie S. 122) folgende rekursive Definition: Ein Raum heißt höchstens rational n-dimensional, wenn jeder Punkt in beliebig kleinen Umgebungen mit höchstens rational $(n-1)$-dimensionalen Begrenzungen enthalten ist. Rational nulldimensional sind die abzählbaren Mengen und nur diese. Ein Raum, der nicht höchstens rational n-dimensional ist, heißt *mindestens irrational n-dimensional.*

Der allgemeine dimensionstheoretische Zerlegungssatz (Dimensionstheorie, S. 156) besagt, wenn man ihn speziell für Teilmengen kompakter Räume und in metrischer Form ausspricht: Ist M eine n-dimensionale Menge, so ist M für jedes $\varepsilon > 0$ Summe von endlich vielen in M abgeschlossenen Mengen $< \varepsilon$, die zu je k höchstens $(n-k+1)$-dimensionale Durchschnitte haben $(k=2, 3, \ldots n+2)$. Mit (S) wollen wir folgende Umkehrung dieser Behauptung bezeichnen: Ist M n-dimensional, so existiert ein $\eta > 0$, so daß bei jeder Zerlegung von M in endlich viele in M abgeschlossene Mengen $< \eta$ k-Tupel mit mindestens $(n-k+1)$-dimensionalem Durchschnitt existieren. (Bewiesen ist Satz (S) bloß für den Fall, daß M kompakt ist. Im allgemeinen ist bloß bewiesen, daß eine Menge M höchstens n-dimensional ist, falls M für ein k $(2 \leqq k \leqq n+1)$ und für jedes Überdeckungssystem von M Summe ist von endlich vielen in M abgeschlossenen Mengen, von denen jede in einer Menge des Überdeckungssystems enthalten ist und die zu je k einen höchstens $(n-k+1)$-dimensionalen Durchschnitt haben.) Für die rational n-dimensionalen Mengen gilt nun entsprechend:

Ist M eine rational n-dimensionale Menge, so ist M für jedes $\varepsilon > 0$ *Summe von endlich vielen in M abgeschlossenen Mengen* $< \varepsilon$, *die zu je k höchstens rational* $(n-k+1)$*-dimensionale Durchschnitte haben* $(k=2, 3, \ldots n+2)$. *Umgekehrt folgt aus Satz (S): Ist M irrational n-dimensional, so existiert ein* $\eta > 0$, *so daß bei jeder Zerlegung von M in endlich viele in M abgeschlossene Mengen* $< \eta$ *k-Tupel mit mindestens irrational* $(n-k+1)$*-dimensionalem Durchschnitt auftreten.*

Speziell besagen die letzten Teile dieser Behauptungen: Eine rational n-dimensionale Menge kann in endlich viele beliebig kleine relativ abgeschlossene Mengen zerlegt werden, die zu je $n+1$ abzählbare Durchschnitte haben, während bei jeder Zerlegung einer irrational n-dimensionalen Menge in endlich viele hinreichend kleine relativ abgeschlossene Mengen $(n+1)$-Tupel mit unabzählbarem Durchschnitt auftreten.

Zum Beweise der ersten Hälfte der allgemeinen Aussage kann jede der beiden zum Beweise des allgemeinen Zerlegungssatzes verwendbaren Methoden (Dimensionstheorie, S. 161 u. 170) herangezogen werden. Nach der zweiten Methode (mit der a. a. O. ein viel weitergehender Hilfssatz bewiesen wird) verläuft der Beweis des rationalen und des allgemeinen Zerlegungssatzes unter der Annahme der Gültigkeit des letzteren für $(n-1)$-dimensionale Mengen so: Es sei eine [rational] n-dimensionale Menge M und ein $\varepsilon > 0$ gegeben. Wir überdecken M (Dimensionstheorie, S. 158) mit endlich vielen offenen Mengen $U_1, U_2, \ldots U_m$, die $< \varepsilon$ sind und deren abgeschlossene Hüllen zu je $n+2$ leere Durchschnitte haben. Sodann spalten wir M in eine $(n-1)$-dimensionale Menge M' und eine [abzählbare] nulldimensionale Menge M'', welche keine nichtleere in M offene Teilmenge enthält. Hierauf zerlegen wir nach dem allgemeinen Zerlegungssatz (für $n-1$) die Menge M' in endlich viele in M' abgeschlossene Mengen $A'_1, A'_2, \ldots A'_m$, so daß $A'_i \subset U_i (i=1, 2, \ldots m)$ gilt, daß also insbesondere jedes $A'_i < \varepsilon$ ist, und daß der Durchschnitt von je k Mengen A'_i höchstens $(n-k)$-dimensional ist $(k=2, 3, \ldots n+1)$. Setzen wir $A_i = \bar{A}'_i \cdot M$, so sind die Mengen A_i abgeschlossen in M und es ist $M = \sum_{i=1}^{m} A_i$, weil M'' keine nichtleere in M offene Teilmenge enthält. Wir behaupten, daß der Durchschnitt von je k Mengen A_i höchstens [rational] $(n-k+1)$-dimensional ist. Es sei Π_k der Durchschnitt von k Mengen A_i, wobei erstens $k < n+2$ sei. Es gilt $\Pi_k = \Pi'_k + N$, wo Π'_k den Durchschnitt der k entsprechenden Mengen A'_i bezeichnet und N eine Teilmenge von M', also [abzählbar] nulldimensional, ist. Da die Menge Π'_k höchstens $(n-k)$-dimensional ist, ist Π_k höchstens [rational] $(n-k+1)$-dimensional. Handelt es sich zweitens um den Durchschnitt von $n+2$ Mengen A_i, so bedenken wir, daß wegen $A'_i \subset U_i$ jede Menge A_i in der entsprechenden $\bar{U}_i$ enthalten ist und daß je $n+2$ $\bar{U}_i$ einen leeren Durchschnitt haben. Damit ist der Beweis beendet[1]).

Zum Beweise der Umkehrung seien n und k zwei gegebene natürliche Zahlen, für die $2 \leq k \leq n+1$ gilt, und M eine Menge, welche für jedes $\varepsilon > 0$ Summe ist von endlich vielen in M abgeschlossenen Mengen $< \varepsilon$, die zu je k einen höchstens $(n-k+1)$-

[1]) Pontrjagin (Ann. of Math. *32*, S. 761) bezeichnet denselben als „sehr kompliziert“.

dimensionalen Durchschnitt haben. Wir haben zu zeigen, daß M höchstens rational n-dimensional ist. Wir stellen für jedes natürliche m die Menge M als Summe von endlich vielen in M abgeschlossenen Mengen $A_1^m, A_2^m, \ldots A_{r_m}^m$ dar, die $< \frac{1}{m}$ sind und zu je k einen höchstens $(n-k+1)$-dimensionalen Durchschnitt haben. Wir nennen D^m die Menge aller Punkte, die in mindestens k von den r_m Mengen A_i^m $(i=1, 2, \ldots r_m)$ enthalten sind. Als Summe von endlich vielen relativ abgeschlossenen höchstens rational $(n-k+1)$-dimensionalen Mengen ist D^m höchstens rational $(n-k+1)$-dimensional. Dasselbe gilt daher für die Menge $D = \sum_{m=1}^{\infty} D^m$. Setzen wir nun $D' = M - D$, so ist die Menge D' für jedes natürliche m Summe von endlich vielen in D' abgeschlossenen Mengen $< \frac{1}{m}$, die zu je k leere Durchschnitte haben. (Durch die Formel $D' = \sum_{i=1}^{r_m} D' . A_i^m$ wird eine solche Zerlegung geliefert.) Also ist D', falls Satz (S) gilt, höchstens $(k-2)$-dimensional. Die Menge M ist demnach Summe der höchstens $(k-2)$-dimensionalen Menge D' und der höchstens rational $(n-k+1)$-dimensionalen Menge D, also höchstens rational n-dimensional, wie behauptet.

Es sei noch erwähnt, daß die eben zum Beweis der zweiten Hälfte verwendete Methode unter Voraussetzung der Gültigkeit von Satz (S) auch eine neue Zurückführung der Umkehrung des allgemeinen Zerlegungssatzes auf die Umkehrung des letzten Teiles ergibt. Ferner bemerken wir, daß man in derselben Weise wie das obige Theorem für beliebige Normalbereiche Zerlegungseigenschaften der Mengen der iteriert übergeordneten Normalbereiche beweisen könnte.

V. Über die nirgendsdichten Teilmengen des R_n.

Eine der Fragen, welche durch die mit Umgebungsbegrenzungen operierende Einführung des Dimensionsbegriffes nahegelegt werden (vgl. S. 6 dieses Heftes), bezieht sich auf die Möglichkeit der Beschränkung auf Umgebungsbegrenzungen besonderer Art. Beispielsweise ist (Dimensionstheorie, S. 260), wenn M eine Teilmenge des R_n ist, deren Komplement zum R_n im R_n dicht ist, jeder Punkt des Raumes in beliebig kleinen n-dimensionalen Minkowskischen Strahlumgebungen (also in n-dimensionalen topologischen Kugeln) enthalten, deren Begrenzungen mit M höchstens $(n-1)$-dimensionale Durchschnitte haben. Für die nirgendsdichten Teilmengen des R_n beweisen wir nun folgenden schärferen Satz:

Ist M eine nirgendsdichte Teilmenge des R_n, so ist jeder Punkt p von M in beliebig kleinen n-dimensionalen Kugeln mit dem Mittelpunkt p enthalten, in deren Begrenzungen die Durchschnitte mit M nirgendsdicht sind.

Es genügt offenbar, die Behauptung unter der Voraussetzung, daß M abgeschlossen ist, zu beweisen. Wir können ihr dann die Form geben: ist M eine Teilmenge des R_n und p ein Punkt von M, so daß die Begrenzung von jeder hinreichend kleinen n-dimensionalen Kugel um den Punkt p (etwa von jeder, deren Radius $\leqq \rho$ ist) mit M einen in der Begrenzung nicht nirgendsdichten Durchschnitt, also eine $(n-1)$-dimensionale sphärische Kreisfläche gemein hat, — dann enthält M eine nichtleere offene Teilmenge des R_n. Wir wollen eine n-dimensionale Kugel um den betrachteten Punkt p, deren Begrenzung mit M eine $(n-1)$-dimensionale sphärische Kreisfläche gemein hat, welche mindestens den k-ten Teil der Begrenzungssphäre ausmacht, kurz als k-Kugel bezeichnen. Wir nennen nun S_k die Menge aller reellen Zahlen $\leqq \rho$, welche Radius einer k-Kugel sind. Da jede Kugel mit einem Radius $\leqq \rho$ mit M eine sphärische Kreisfläche gemein hat, so ist jede reelle Zahl, die > 0 und $\leqq \rho$ ist, für irgendein k (und daher für alle größeren k) in S_k enthalten, d. h. es ist $(0, \rho] = \sum_{k=1}^{\infty} S_k$. Da $(0, \rho]$ von zweiter Kategorie ist, ist also für ein gewisses k die Menge S_k nicht nirgendsdicht in $(0, \rho]$, enthält demnach, da offenbar jede Menge S_k abgeschlossen ist, ein Teilintervall $[r', r'']$ von $(0, \rho]$. Es hat also die Begrenzung von jeder Kugel um p, die dem Ring zwischen den Radien r' und r'' angehört, mit M eine sphärische Kreisfläche gemein, welche mindestens den k-ten Teil ihrer Fläche ausmacht. Dann muß aber offenbar ein Sektor des genannten Ringes Teil von M sein. Also enthält M einen offenen Teil.

Statt n-dimensionaler Kugeln hätten in derselben Weise auch andere einparametrige Umgebungsscharen betrachtet werden können. Z. B. lautet unsere Behauptung für $n = 2$: Jeder Punkt einer nirgendsdichten Teilmenge M der Ebene (z. B. einer ebenen Kurve) ist in beliebig kleinen Kreisen mit dem betrachteten Punkt als Zentrum enthalten, deren Begrenzungen mit M diskontinuierliche Durchschnitte haben. Ebenso läßt sich auch zeigen, daß jede ein Flächenstück überdeckende einparametrige Schar von Bogen oder geschlossenen Kurven ein Element enthält, welches mit M einen diskontinuierlichen Durchschnitt hat. (Hingegen ist ein Punkt p einer Kurve K des R_3 nicht notwendig in einer Kugel mit p als Mittelpunkt enthalten, deren Begrenzung mit K einen diskontinuierlichen Durchschnitt hat.)

VI. Über die Summe regulärer Kurven.

Eine nichtleere, nicht nulldimensionale Menge M heißt eindimensional, beziehungsweise rational eindimensional, beziehungsweise regulär eindimensional, falls jeder Punkt von M in beliebig kleinen Umgebungen enthalten ist, deren Begrenzungen mit M höchstens nulldimensionale, beziehungsweise abzählbare, beziehungsweise endliche Durchschnitte haben. Die Summe von abzählbar vielen eindimensionalen, beziehungsweise rational eindimensionalen F_σ eines separabeln Raumes ist eindimensional, beziehungsweise rational eindimensional. Hingegen ist die Summe von endlich vielen, ja von zwei regulär eindimensionalen abgeschlossenen Mengen eines kompakten Raumes nicht notwendig regulär eindimensional. (Z. B. kann die von mir Mathem. Annalen, *95*, S. 301, angegebene nicht reguläre Kurve als Summe zweier regulärer Kurven dargestellt werden.) Es gilt jedoch folgender

Satz 1. *In einem kompakten Raum ist die Summe von endlich vielen abgeschlossenen regulär eindimensionalen Mengen, die paarweise diskontinuierliche Durchschnitte haben, regulär eindimensional.*

Andernfalls enthielte die Summe ein Kontinuum, bestehend aus irregulären Punkten. Da sie aber in jedem Punkte, der bloß einer der abgeschlossenen regulär eindimensionalen Summanden angehört, offenbar regulär eindimensional ist, widerspricht dies der Voraussetzung, daß keine zwei Summanden ein Kontinuum gemein haben.

Wir wollen eine Menge M *beständig regulär eindimensional* nennen, *wenn die Summe von M mit jeder regulär eindimensionalen abgeschlossenen Menge regulär eindimensional ist.*

Satz 2. *Damit die abgeschlossene Menge M eines separablen Raumes beständig regulär eindimensional sei, ist hinreichend, daß zu jeder Teilmenge A von M, welche Differenz einer abgeschlossenen und einer endlichen Menge ist, und zu jeder Umgebung U von A eine Umgebung U' von A existiert, die $\subset U$ ist und deren Begrenzung mit M einen endlichen Durchschnitt hat.*

Diese Bedingung verschärft die für jede regulär eindimensionale Menge M giltige Tatsache, daß zu jeder abgeschlossenen Teilmenge A von M und zu jeder Umgebung U von A eine Teilumgebung U' von A existiert, deren Begrenzung mit M einen endlichen Durchschnitt hat.

Zum Beweise von Satz 2 seien in einem separabeln Raum irgend eine abgeschlossene regulär eindimensionale Menge A_1 und eine abgeschlossene Menge A_2, welche die in Satz 2 genannte Bedingung erfüllt, gegeben. Es sei p ein Punkt von $A_1 + A_2$, $Z(p)$ eine Umgebung von p. Um nachzuweisen, daß $A_1 + A_2$ regulär eindimensional ist, haben wir zu zeigen, daß eine Umgebung $U(p) \subset Z(p)$ existiert, deren Begrenzung B mit $A_1 + A_2$ einen endlichen Durchschnitt hat. Es sei V eine Umgebung von p, die $\Subset Z(p)$ ist und deren Begrenzung mit A_1 einen endlichen Durchschnitt hat. Wir bilden eine Umgebung $V^* \subset Z(p)$ von $B—A_1 . B$, welche zu A_1 fremd ist und deren Begrenzung mit A_1 höchstens solche Punkte gemein hat, die in $B . A_1$ liegen. Die Menge $B . A_2—B . A_1 . A_2$ ist eine Teilmenge von A_2, welche Differenz einer abgeschlossenen und einer endlichen Menge ist, V^* ist eine Umgebung dieser Menge, also existiert nach der Voraussetzung über A_2 eine Umgebung $V' \subset V^*$ von $B . A_2—B . A_1 A_2$, deren Begrenzung mit A_2 einen endlichen Durchschnitt hat. Die Menge $U(p) = V + V'$ ist dann offenbar eine Umgebung von p, die $\subset Z(p)$ ist und deren Begrenzung mit $A_1 + A_2$

einen endlichen Durchschnitt besitzt. Offenbar ließe sich die Bedingung von Satz 2 erheblich abschwächen.

Es sei noch erwähnt, daß mit Hilfe der Methode der Modifikation der Umgebungen in der Nähe ihrer Begrenzungen folgende allgemeinere Sätze bewiesen werden können: Es sei in einem kompakten Raum ein System $\mathfrak{S}$ von abgeschlossenen Mengen gegeben, welches neben jeder Menge M auch alle abgeschlossenen Teilmengen von M und neben je zwei Mengen M und N auch $M+N$ enthält. Wir bezeichnen mit $\mathfrak{S}'$ das *(„übergeordnete")* System aller abgeschlossenen Mengen A mit folgender Eigenschaft: Jeder Punkt von A ist in beliebig kleinen Umgebungen enthalten, deren Begrenzungen mit A Durchschnitte haben, die dem System $\mathfrak{S}$ angehören. *Dann gehört die Summe von je endlich vielen Mengen von $\mathfrak{S}'$, die paarweise diskontinuierliche Durchschnitte haben, dem System $\mathfrak{S}'$ an. Damit für die Menge A aus $\mathfrak{S}'$ die Summe von A und jeder Menge aus $\mathfrak{S}'$ dem System $\mathfrak{S}'$ angehört, ist hinreichend, daß A folgende Eigenschaft besitzt: Ist D irgend eine Teilmenge von A, welche Differenz einer abgeschlossenen Menge und einer Menge aus $\mathfrak{S}$ ist, und U irgend eine Umgebung von D, so existiert eine Umgebung V von D, die $\subset U$ ist und deren Begrenzung mit A einen dem System $\mathfrak{S}$ angehörigen Durchschnitt besitzt.*

***Über rationale Kurven.** Von Helene Reschovsky.

Ein Kontinuum K heißt nach Menger[1]) eine *rationale Kurve*, wenn jeder Punkt von K in beliebig kleinen Umgebungen mit abzählbaren Begrenzungen enthalten ist. Ein rationaler Kurvenpunkt p besitzt[2]) ein *Geschlecht* α, wo α eine Ordinalzahl der ersten oder zweiten Zahlenklasse ist, wenn α die kleinste Zahl der Eigenschaft ist, daß p in beliebig kleinen Umgebungen enthalten ist, deren Begrenzungen leere α-te Ableitungen haben. Wenn α eine isolierte Ordinalzahl ist, so heißt p vom *Typus* (α, n), wenn zu p beliebig kleine Umgebungen existieren, so daß die $(\alpha-1)$-ten Ableitungen ihrer Begrenzungen n Punkte enthalten und n die kleinste natürliche Zahl dieser Eigenschaft ist. Es gilt der Satz: *Zu jeder rationalen Kurve K gibt es eine Zahl α der ersten oder zweiten Zahlenklasse derart, daß das Geschlecht eines jeden Kurvenpunktes $\leqq \alpha$ ist.* Die kleinste Ordinalzahl, die nicht übertroffen wird von dem Geschlecht irgend eines Punktes, kann daher als *Geschlecht der Kurve* bezeichnet werden.

Über die Summe rationaler Kurven gelten folgende Sätze: *Wenn die Summe endlich vieler rationaler Kurven ein Kontinuum ist, so ist das Geschlecht der Summe höchstens so groß wie die Summe der Geschlechter der einzelnen Kurven.* Existiert zu zwei rationalen Kurven mit dem Geschlecht α, respektive β $(\beta \leqq \alpha)$, eine isolierte Zahl $\gamma \leqq \beta$ und eine endliche Zahl n derart, daß der

---

[1]) Menger, Grundzüge einer Theorie der Kurven. Math. Ann., *95*, S. 277; vgl. auch Urysohn, Mémoire sur les multiplicités Cantoriennes II. Verhandl. Ak. Amst., XIII, Nr. 4.

[2]) Menger, Zur allgemeinen Kurventheorie. Fund. Math., *10*, S. 111.

Durchschnitt der Kurven in jedem seiner Punkte von einem Typus $\leqq (\gamma, n)$ ist, so ist ihre Summe von einem Typus $\leqq (\alpha + \gamma, n)$.

Ist eine rationale Kurve in einem Punkt p von einem Geschlecht $\geqq \alpha$, *so ist die Menge aller Kurvenpunkte von einem Geschlecht* $> \beta$, *für* $\beta < \alpha$, *in p von einem Geschlecht* $\geqq -\beta + \alpha$.

In einer nichtregulären isolierttypigen rationalen Kurve, d. h. in einer Kurve, die in jedem ihrer Punkte höchstens vom Typus (α, n) ist, wobei α eine isolierte Ordinalzahl und n endlich ist, *liegen die Punkte von einem Typus* $\leqq \left(\alpha, \frac{n}{2}\right)$ *dicht*[1]. Daraus ergibt sich, da der topologische Kreis die einzige reguläre Kurve homogener Ordnung ist: *Unter den isolierttypigen Kurven ist der topologische Kreis die einzige homogentypige Kurve.*

In einem Teilkontinuum einer rationalen Kurve K, in dessen sämtlichen Punkten K höchstens vom Typus (α, n) ist, wobei α eine isolierte Ordinalzahl und n endlich ist, liegen die Punkte von einem Typus $\leqq \left(\alpha, \frac{n}{2} + 1\right)$ dicht. *In einer rationalen Kurve ist die Menge aller Punkte vom Typus* (α, n) *für eine isolierte Zahl* α *und eine endliche Zahl* $n > 2$ *diskontinuierlich. Sind in einer isolierttypigen Kurve alle Punkte von einem Typus* $\leqq (\alpha, n)$, *so ist die Menge aller Punkte von einem Typus* $> \left(\alpha, \frac{n}{2} + 1\right)$ *nulldimensional.*

*** Über die stetige Abbildung des Kreises auf reguläre Kurven.**

Von Georg Nöbeling.

Satz: *Es gibt eine eindeutige und stetige Abbildung des Kreises auf eine beliebige reguläre Kurve, die jedem Kurvenpunkt von endlicher Ordnung n höchstens n, jedem Kurvenpunkt von wachsender Ordnung höchstens abzählbar viele Urbildpunkte zuordnet.*

Beweisskizze: Wir konstruieren für jedes natürliche i eine aus endlich vielen Bogen bestehende Baumkurve B_i, so daß B_i in B_{i+1} liegt und die gegebene Kurve die abgeschlossene Hülle der Summe $B_1 + B_2 + \ldots$ ist. Bedeutet $n_i = n_i(p)$ die Ordnung des Punktes $p \subset B_i$ bezüglich B_i, so bezeichnen wir mit α_i eine eindeutige und stetige Abbildung des Kreises auf den Baum B_i, die allen Punkten $p \subset B_i$, für die $n_i = n_{i+1} = \ldots$ gilt, genau n_i Urbildpunkte zuordnet, während die Urbildmenge jedes der höchstens abzählbar vielen übrigen Punkte von B_i aus $n_i - 1$ einzelnen Punkten und einem Kreisbogen von einer Länge $< \frac{1}{i}$ besteht. Ausgehend von einer beliebigen Abbildung α_1 konstruieren wir durch Induktion eine Folge

---

[1]) Dieser und die folgenden Sätze verallgemeinern Ergebnisse über reguläre Kurven von Urysohn, Mémoire sur les multiplicités Cantoriennes II. Verhandl. Ak. Amst., XIII, S. 105. Whyburn, On regular points of continua and regular curves of at most order n. Bulletin of the Am. Math. Soc., *35,* S. 218. Künneth, Ein Theorem der Kurventheorie. Monatsh. f. Math. u. Physik, *36,* S. 149.

von Abbildungen $\alpha_1, \alpha_2, \ldots$ eines festen Kreises auf die Kurve mit der Eigenschaft, daß die n_i, beziehungsweise $n_i - 1$ einzelnen Urbildpunkte, die einem beliebigen Punkte $p \subset B_i$ durch die Abbildung α_i zugeordnet werden, auch bei der Abbildung α_{i+1} in der Urbildmenge desselben Punktes p liegen. Jedem Kreispunkt, der bei fast allen Abbildungen α_i unserer Folge denselben Bildpunkt auf der Kurve hat, ordnen wir diesen Bildpunkt endgültig zu. Die so entstehende Abbildung einer Teilmenge des Kreises auf eine Teilmenge der Kurve läßt sich auf genau eine Weise zu einer eindeutigen und stetigen Abbildung des Kreises auf die Kurve vervollständigen, und diese Abbildung besitzt die im Satze behaupteten Eigenschaften.

Ist die Kurve keine Baumkurve, so kann man die Konstruktion dahin modifizieren, daß entweder mindestens ein Kurvenpunkt endlicher Ordnung n weniger als n oder mindestens ein Kurvenpunkt wachsender Ordnung nur endlich viele Urbilder bekommt. Dies ist charakteristisch für diejenigen regulären Kurven, die keine Baumkurven sind, denn bei einer beliebigen eindeutigen und stetigen Abbildung des Kreises auf eine reguläre Baumkurve hat jeder Punkt von endlicher Ordnung n mindestens n, jeder Punkt von wachsender Ordnung mindestens abzählbar unendlich viele Urbildpunkte.

* **Über Verallgemeinerungen der Jordanschen Kontinua.** Von William Leake Ayres.

Eine Punktmenge heißt ein Jordansches Kontinuum, wenn sie ein stetiges Bild eines abgeschlossenen Intervalles ist. Wir betrachten die Punktmengen, die stetige Bilder einer Halbgeraden sind, und diejenigen, welche stetige Bilder einer ganzen Geraden sind. Damit eine Punktmenge M stetiges Bild einer Halbgeraden sei, ist notwendig und hinreichend, daß M die Summe von abzählbar vielen Jordanschen Kontinuen ist, und je zwei verschiedene Punkte von M durch einen in M liegenden Bogen verbunden werden können. Damit eine Punktmenge M stetiges Bild einer Halbgeraden sei, ist ferner notwendig und hinreichend, daß $M = \sum_{n=1}^{\infty} A_n$, wo jede Menge A_n ein Jordansches Kontinuum und eine Teilmenge von A_{n+1} ist. Damit eine Punktmenge M ein stetiges Bild einer ganzen Geraden sei, ist notwendig und hinreichend, daß M ein stetiges Bild einer Halbgeraden ist.

* **Über geschlossene Kurven in Peanoschen Räumen.** Von William Leake Ayres.

Wir setzen eine endliche Teilmenge K eines Peanoschen Raumes M voraus. Ein Peanoscher Raum = Jordansches Kontinuum ist ein metrischer Raum, der stetiges Bild einer Strecke ist; also kompakt, zusammenhängend und zusammenhängend im kleinen. Ein wichtiges Problem der Theorie des Peanoschen Raumes ist, die Bedingungen ausfindig zu machen, unter denen die Menge K durch eine geschlossene Kurve oder einen Bogen von M verbunden werden kann.

In dieser Note geben wir eine allgemeine Lösung dieses Problems. Sie lautet, wenn die Menge K aus m Punkten besteht, folgendermaßen: *Wenn keine zwei Punkte von K durch eine Menge von $m-1$ Punkten von M zerspalten werden können, dann existiert eine geschlossene Kurve von M, die K enthält. Und wenn keine Menge von $m-2$ Punkten von M zwei Punkte von K zerspaltet, dann existiert ein Bogen, der K enthält.*

Spezielle Fälle dieses ersten Satzes haben G. T. Whyburn ($m=2$) (Bull. Amer. Math. Soc., 1927, S. 308 und 661) und W. L. Ayres ($m=3$) (Bull. Acad. Polonaise, 1928, S. 132) schon bewiesen. Für $m=2$ haben Kaluzsay, Mazurkiewicz, Moore, Tietze und Vietoris den zweiten Satz bewiesen. Die allgemeinen Beweise findet man durch Induktion und „chain-wise" Erreichbarkeit. Durch Beispiele sieht man, daß die Zahlen $(m-1)$ und $(m-2)$ nicht vermindert werden können. Aber in der Ebene gilt die folgende Verschärfung der vorangehenden Sätze: *Jede endliche Teilmenge eines Peanoschen Raums, der in der Ebene liegt und durch keine zwei Punkte zerspalten wird, kann durch eine geschlossene Kurve* (und daher durch einen Bogen) *verbunden werden.* Dieser Satz ist unrichtig, wenn der Peanosche Raum nicht in der Ebene liegt. Für unendliche Teilmengen gilt er auch nicht.

* **Über die Dichtigkeit der Zerschneidungspunkte und der Endpunkte.** Von William Leake Ayres.

Aus der Menger-Urysohnschen Definition der Endpunkte folgt, daß die Zerschneidungspunkte eines Peanoschen Raumes M dicht in M sind, wenn die Endpunkte dicht in M sind. Es gilt ferner: Damit die Menge Z aller Zerschneidungspunkte dicht in M sei, ist notwendig und hinreichend, daß die Endpunkte von M dicht in $M-Z$ sind. Damit die Endpunkte von M dicht in M seien, ist notwendig und hinreichend, daß die Zerschneidungspunkte von M von einer Ordnung ≥ 3 überall dicht sind. Wenn die Endpunkte überall dicht sind, dann ist die Menge aller Zerschneidungspunkte von M von der Ordnung 2 nulldimensional. Und umgekehrt, wenn die Menge aller Zerschneidungspunkte von der Ordnung 2 nulldimensional ist und die Zerschneidungspunkte überall dicht sind, dann sind die Endpunkte überall dicht.

* **Ein allgemeiner Trennungssatz.** Von Karl Menger.

Sind A und B irgend zwei Teilmengen eines metrischen Raumes und U und V zwei offene Mengen, so daß

$$(+) \qquad A-A\,.\,B \subset U \text{ und } B-B\,.\,A \subset V$$

gilt, dann existieren zwei offene Mengen U' und V' gemäß den Bedingungen:

$$1. \qquad A-A\,.\,\overline{B} \subset U' \subset U \text{ und } B-B\,.\,\overline{A} \subset V' \subset V$$

$$2. \qquad \overline{U'}\,.\,\overline{V'} \subset \overline{A}\,.\,\overline{B}.$$

Sind $\bar{A}$ und $\bar{B}$ fremd, so sind U' und V' Umgebungen von A und B von vorgeschriebener Kleinheit (d. h. $\subset U$ und $\subset V$), mit fremden abgeschlossenen Hüllen, und von solchen Umgebungen sagt man, daß sie die Mengen A und B trennen. Sind $\bar{A}$ und $\bar{B}$ nicht fremd, so sind U' und V' Umgebungen von vorgeschriebener Kleinheit von $A—A.\bar{B}$ und $B—B.\bar{A}$, deren abgeschlossene Hüllen gleichsam möglichst fremd sind, d. h. nur solche Punkte gemein haben, welche in $\bar{A}.\bar{B}$ liegen. Von diesen offenen Mengen U' und V' kann man also sagen, daß sie A und B möglichst trennen.

Zum Beweise bezeichnen wir für jeden nicht in $\bar{A}.\bar{B}$ liegenden Punkt p des Raumes mit $\tau(p;\ A, B)$ das Verhältnis des Abstandes zwischen p und A zum Abstand zwischen p und B. Dann ist $\tau(p;\ A, B)=0$, wenn p in $\bar{A}—\bar{A}.\bar{B}$ liegt, und $\tau(p;\ A, B)=\infty$, wenn p in $\bar{B}—\bar{B}.\bar{A}$ liegt. Für jeden nicht in $\bar{A}.\bar{B}$ liegenden Punkt p ist $\tau(p;\ A, B)$ eine (eventuell unendliche) reelle Zahl und es gilt offenbar

(++) $$\tau(p;\ A, B)=\frac{1}{\tau(p;\ B, A)}.$$

Wir bezeichnen nun mit U^* die Menge aller Punkte p des Raumes, für welche $\tau(p;\ A, B)<\frac{1}{3}$. Man bestätigt (auf Grund der Dreiecksungleichung) leicht, daß U^* eine offene Menge ist, deren abgeschlossene Hülle (außer etwa Punkten von $\bar{A}.\bar{B}$) gerade jene Punkte des Raumes enthält, für die $\tau(p;\ A, B)\leqq\frac{1}{3}$. Entsprechend sei V^* die Menge aller Punkte, für die $\tau(p;\ B, A)<\frac{1}{3}$. Dann gilt $\bar{A}—\bar{A}.\bar{B}\subset U^*$ und $\bar{B}—\bar{B}.\bar{A}\subset V^*$. Aus dieser Tatsache im Verein mit (+) folgt, wenn wir $U'=U.U^*$ und $V'=V.V^*$ setzen, die behauptete Bedingung 1. Für einen Punkt p von $\bar{U}'.\bar{V}'$ müßte sowohl $\tau(p;\ A, B)\leqq\frac{1}{3}$ als auch $\tau(p;\ B, A)\leqq\frac{1}{3}$ gelten, was wegen (++) für eine reelle Zahl unmöglich ist. Also muß p in $\bar{A}.\bar{B}$ liegen, womit auch die Bedingung 2 erwiesen ist.

Sind A und B beide in A + B abgeschlossen, so gilt $A—A.\bar{B}=$ $=A—A.B$ und $B—B.\bar{A}=B—B.A$. In diesem Falle läßt sich also die Bedingung 1 ersetzen durch

1'. $A—A.B\subset U'\subset U$ *und* $B—B.A\subset V'\subset V$.

* **Über zweigradigen Zusammenhang.** Von Hans Hornich.

Menger hat (Dimensionstheorie, S. 312) auf eine neue Verallgemeinerung des Zusammenhangsbegriffes, den n-gradigen Zusammenhang hingewiesen; es soll nun der Fall $n=2$ erörtert werden.

2

Ein System σ von drei abgeschlossenen zusammenhängenden Mengen α, β, γ eines topologischen Raumes, welche zu je zweit nichtleeren, zu dritt aber leeren Durchschnitt haben, heißt ein *Mengendreieck,* in Zeichen $\sigma = (\alpha, \beta, \gamma)$. Enthält eine Menge M die drei Mengen eines Mengendreieckes $\sigma = (\alpha, \beta, \gamma)$ als Teil, so heißt M *zwischen diesem Mengendreieck zusammenhängend,* wenn bei jeder Zerlegung von M in drei in M abgeschlossene Mengen $M = A + B + C$ derart, daß $\alpha \subset A$, $\beta \subset B$, $\gamma \subset C$ gilt, der Durchschnitt $A . B . C$ nichtleer ist.

Eine Menge M heißt *zweigradig zusammenhängend,* wenn sie erstens ein Mengendreieck enthält und zweitens zwischen jedem in M enthaltenen Mengendreieck zusammenhängend ist. Eine zweigradig zusammenhängende Menge ist im allgemeinen nicht zusammenhängend. Sind alle Komponenten einer Menge M zweigradig zusammenhängend, so ist auch M zweigradig zusammenhängend. Ist umgekehrt M zweigradig zusammenhängend und im kleinen zusammenhängend, so ist jede Komponente von M, welche ein Mengendreieck enthält, auch zweigradig zusammenhängend.

Vietoris (Proc. Acad. Amsterdam *29*) nennt ein Kontinuum K, für welches bei jeder Zerlegung von K in zwei Kontinua A und B der Durchschnitt $A . B$ ein Kontinuum ist, ein Kontinuum „*ohne Henkel*"; wir wollen dabei auch die Möglichkeit zulassen, daß $A . B$ aus einem einzigen Punkt besteht.

Ein zweigradig zusammenhängendes bogenverknüpftes[1]) *Kontinuum hat keine Henkel. Ein im kleinen zusammenhängendes Kontinuum K ohne Henkel ist, sofern K ein Mengendreieck enthält, zweigradig zusammenhängend.*

Für ein im kleinen zusammenhängendes Kontinuum K sind folgende zwei Eigenschaften identisch:

a) K ist henkellos;

b) wenn O ein Gebiet (d. h. eine offene zusammenhängende Menge) *von K ist, so daß K—O nichtleer ist, so zerschneidet jede Komponente H der Begrenzung B(O) von O, welche mit der Begrenzung eines Gebietes V von K—O nichtleeren Durchschnitt hat, die Menge K,* d. h. $K - H$ ist unzusammenhängend, und zwar so, daß O und V in verschiedenen Komponenten von $K - H$ liegen.

In beiden angeführten Sätzen kann überall mit leichter Modifikation die Voraussetzung der Kompaktheit durch die der Separabilität ersetzt werden.

Ist $\sigma = (\alpha, \beta, \gamma)$ ein Mengendreieck eines kompakten Raumes R, so ist die Eigenschaft einer abgeschlossenen Menge F, daß die Durchschnitte von F mit den Mengen α, β, γ ein Mengendreieck (αF, βF, γF) bilden, *induzibel.* Es gibt also nach dem Brouwerschen Reduktionssatz zu jedem Mengendreieck $\sigma = (\alpha, \beta, \gamma)$ in R eine abgeschlossene Menge F^* derart, daß $\sigma^* = (\alpha F^*, \beta F^*, \gamma F^*)$ ein Mengendreieck

[1]) Ein Kontinuum K heißt *bogenverknüpft* (*arcwise connected*), wenn je zwei Punkte von K in einem Teilbogen von K enthalten sind.

ist und F^* hinsichtlich dieser Eigenschaft *irreduzibel* ist; das Mengendreieck σ^* heiße ein zu σ gehöriges irreduzibles Mengendreieck. Die Menge F^* ist dabei zusammenhängend und es gibt keine Zerlegung von F^* in zwei abgeschlossene echte Teilmengen A und B, so daß $A . B$ einpunktig ist. Um von einem kompakten Raum R zu zeigen, daß er zwischen jedem in ihm enthaltenen Mengendreieck zusammenhängend ist, genügt es, dies für die irreduziblen Mengendreiecke zu zeigen.

In einem kompakten, zweigradig zusammenhängenden Raum R ist für jedes Mengendreieck $\sigma = (\alpha, \beta, \gamma)$ die Eigenschaft einer abgeschlossenen Menge, σ zu enthalten und zweigradig zusammenhängend zu sein, eine induzible. Es gibt also zu jedem Mengendreieck σ in R eine abgeschlossene, σ enthaltende, zweigradig zusammenhängende und hinsichtlich dieser Eigenschaften irreduzible Teilmenge K von R. Die Menge K ist dabei zusammenhängend und, falls das Mengendreieck ein irreduzibles ist, gibt es keine Zerlegung von K in zwei abgeschlossene echte Teilmengen mit einpunktigem Durchschnitt.

Ein Kontinuum K heiße *spaltbar,* wenn K Summe von zwei echten Teilkontinua ist, deren Durchschnitt aus einem Punkt besteht. Ist K ein im kleinen zusammenhängendes Kontinuum und zweigradig zusammenhängend, so nennen wir eine größte (d. h. in keiner andern enthaltene) nichtspaltbare, zusammenhängende Teilmenge von K, welche den Punkt p von K enthält, *eine zu p gehörige zweigradige Komponente von K.* Diese (durch den Punkt p nicht notwendig eindeutig bestimmte) zweigradige Komponente kann sich auf p selbst reduzieren; es gibt jedoch zweigradige Komponenten von K, die mehr als einen Punkt enthalten. Jede von ihnen ist selbst ein im kleinen zusammenhängendes Kontinuum und zweigradig zusammenhängend; außerdem ist sie zweistufig zusammenhängend und daher in jedem Punkt mindestens zweidimensional. Zwei zweigradige Komponenten von K haben höchstens einen Punkt gemein.

Der euklidische R_n und die Intervalle des R_n sind für $n \geq 2$ zweigradig zusammenhängend. Der R_1 und seine sämtlichen Teile sind nicht zweigradig zusammenhängend.

Der zweite Teil folgt daraus, daß der R_1 kein Mengendreieck enthalten kann. Der Beweis für die erste Tatsache führt diese auf den Satz zurück, daß bei jeder Zerlegung eines endlichen Intervalls des R_n in zwei zusammenhängende abgeschlossene Intervallsummen A und B der Durchschnitt $A . B$ zusammenhängend ist. Jede kompakte Teilmenge der Ebene E, welche die Ebene nicht zerschneidet, ist, sofern sie ein Mengendreieck enthält, zweigradig zusammenhängend. Denn jedes endliche Intervallsystem der Ebene, welches diese nicht zerschneidet, ist zweigradig zusammenhängend. — Weiter folgt (ohne Anwendung des Jordan'schen Kurvensatzes): Jedes beschränkte Mengendreieck der Ebene zerschneidet diese [1]).

---

[1]) Einen etwas allgemeineren Satz hat Janiszewski bewiesen (Prace matem.-fizyczne *26*, 1913): Sind A und B zwei Kontinua der Ebene und ist der Durchschnitt $A . B$ unzusammenhängend, so zerschneidet $A + B$ die Ebene.

2*

Für die beschränkten Gebiete der Ebene ergeben sich die zweigradig zusammenhängenden und die mit zusammenhängender Begrenzung als identisch.

*** Metrische Untersuchungen.** Von Karl Menger.

I. Theorie der Konvexität[1]).

Es liege ein metrischer Raum vor, d. h. eine Menge, in der je zwei Elementen (Punkten) p und q eine Abstandszahl $pq = qp > 0$ für $p \neq q$ und $= 0$ für $p = q$ zugeordnet ist, wobei für je drei Punkte p, q, r die Dreiecksungleichung $pq + qr \geqq pr$ erfüllt ist. Wir sagen, daß q zwischen p und r liegt, falls $p \neq q \neq r$ und $pq + qr = pr$ gilt. Es sind für diesen Zwischenbegriff die wichtigsten Relationen, denen der elementargeometrische Zwischenbegriff genügt, erfüllt (aber nicht immer alle: Es kann in einem metrischen Raum sein, daß sowohl p als auch r zwischen zwei Punkten s und t liegen und q zwischen p und r liegt, ohne zwischen s und t zu liegen). Wir nennen einen metrischen Raum konvex, falls er zu je zweien seiner Punkte einen Zwischenpunkt enthält. Man kann zeigen, daß ein vollständiger metrischer Raum (d. h. ein Raum, in dem jede Cauchysche Folge einen Limespunkt besitzt), wenn er konvex ist, zusammenhängend und im kleinen zusammenhängend ist und zu je zweien seiner Punkte p und q eine Strecke, d. h. einen diese beiden Punkte verbindenden („geodätischen“) Bogen enthält, welcher abstandstreues Bild einer Strecke ist, deren Länge gleich dem Abstand pq ist. Eine abgeschlossene Teilmenge eines euklidischen Raumes ist dann und nur dann konvex im klassischen Sinn (in dem Sinn nämlich, daß sie zu je zweien ihrer Punkte die durch sie bestimmte Strecke als Teil enthält), wenn sie konvex im Sinne unserer allgemeinen Definition ist.

Die Betrachtung von Zwischenpunkten und geodätischen Bögen ermöglicht für allgemeine metrische Räume einerseits die Definition von metrischen Singularitäten (Spitzen, Kanten usw.), anderseits den sukzessiven Übergang zu den linearen Räumen. Von besonderer Wichtigkeit auf dem Wege zu normaleren Räumen sind jene, in denen zu je zwei Punkten genau eine Strecke existiert, und jene, in welchen je zwei Punkte eindeutig einen gewissen saturierten, offenen, halboffenen oder abgeschlossenen geodätischen Bogen (eine „Gerade“) bestimmen, der durch je zwei seiner Punkte bestimmt ist. Die Räume der letzteren Art sind unter den konvexen vollständigen Räumen durch folgende Eigenschaft gekennzeichnet: Wenn von irgend vier Punkten zwei Tripel auf einer Strecke liegen, so liegen alle vier Punkte auf einer Strecke.

II. Die euklidische Metrik[2]).

Wie sind die euklidischen Räume und die Teilmengen euklidischer Räume unter den metrischen Räumen charak-

[1]) Vgl. Mathem. Annalen, *100*, S. 76.
[2]) Vgl. Mathem. Annalen, *100*, S. 113.

terisiert? M. a. W.: Welche metrischen Räume sind auf den R_n, beziehungsweise auf Teilmengen des R_n abstandstreu abbildbar? Die vollständige Lösung dieses Problems ist in den folgenden Theoremen enthalten, in denen die euklidischen Mengen sogar unter den halbmetrischen Räumen gekennzeichnet werden, d. h. unter den Mengen, in denen je zwei Punkten p und q eine Abstandszahl $pq=qp$ zugeordnet ist, die > 0 für $p \neq q$ und $= 0$ für $p=q$ ist (aber nicht notwendig der Dreiecksungleichung genügt).

I. Ein halbmetrischer Raum, von dem je $n+3$ Punkte auf $n+3$ Punkte des R_n abstandstreu abbildbar sind, ist abstandstreu auf eine Teilmenge des R_n abbildbar.

II. Ein mehr als $n+3$ Punkte enthaltender halbmetrischer Raum, von dem je $n+2$ Punkte auf $n+2$ Punkte des R_n abstandstreu abbildbar sind, ist auf eine Teilmenge des R_n abstandstreu abbildbar.

III. Für jede natürliche Zahl n existieren halbmetrische Räume, welche genau $n+3$ Punkte enthalten, von denen je $n+2$ auf $n+2$ Punkte des R_n abstandstreu abbildbar sind, ohne daß alle $n+3$ Punkte auf $n+3$ Punkte des R_n abstandstreu abbildbar sind. (Derartige $(n+3)$-Tupel von Punkten heißen pseudoeuklidisch.)

Sind $p_1, p_2, \ldots, p_k$ k Punkte eines halbmetrischen Raumes $(k>1)$ und bezeichnen wir mit r_{ij} den Abstand der Punkte p_i und p_j, so führen wir folgende Größe ein:

$$D(p_1, p_2, \ldots, p_k) = \begin{vmatrix} 0 & 1 \\ 1 & r_{ij}^2 \end{vmatrix}_{i,j=1,2,\ldots k} = \begin{vmatrix} 0 & 1 & 1 & 1 & . & . & 1 \\ 1 & 0 & r_{12}^2 & r_{13}^2 & . & . & r_{1k}^2 \\ 1 & r_{12}^2 & 0 & r_{23}^2 & . & . & r_{2k}^2 \\ 1 & r_{13}^2 & r_{23}^2 & 0 & . & . & r_{3k}^2 \\ . & . & . & . & . & . & . \\ . & . & . & . & . & . & . \\ 1 & r_{1k}^2 & r_{2k}^2 & r_{3k}^2 & . & . & 0 \end{vmatrix}$$

Es gilt dann:

IV. Ein aus $n+2$ Punkten $p_1, p_2, \ldots, p_{n+2}$ bestehender halbmetrischer Raum ist dann und nur dann abstandstreu in den R_n einbettbar $(n \geq 0)$, wenn

1. $D(p_1, p_2, \ldots, p_{n+2})=0$ gilt,

2. für jede Zahl k, für die $1<k<n+2$ ist, und für je k Punkte $p_{i_1}, p_{i_2}, \ldots, p_{i_k}$ des Raumes die Beziehung gilt:

$$\operatorname{sign} D(p_{i_1}, p_{i_2}, \ldots, p_{i_k}) = \operatorname{sign} (-1)^k \text{ oder } 0.$$

V. Ein Raum, der genau $n+3$ Punkte enthält, ist dann und nur dann in den R_n abstandstreu einbettbar, falls je $n+2$ seiner

Punkte abstandstreu in den R_n einbettbar sind und überdies (um pseudoeuklidische $(n+3)$-Tupel auszuschließen) die Beziehung gilt:

$$D(p_1, p_2, \ldots p_{n+3}) = 0.$$

VI. Damit ein halbmetrischer Raum R abstandstreu auf den R_n selbst abgebildet werden könne, ist notwendig und hinreichend,

1. daß R vollständig ist,

2. daß je $n+2$ Punkte von R in den R_n abstandstreu einbettbar sind (d. h. den Bedingungen von IV genügen),

3. daß $n+1$ Punkte $p_1, p_2, \ldots, p_{n+1}$ von R existieren, für die $D(p_1, p_2, \ldots, p_{n+1}) \neq 0$.

4. daß zu je zwei Punkten a und b von R drei Punkte c_1, c_2, c_3 existieren, so daß die Beziehungen gelten

$$ac_1 + c_1b = ab, \quad c_2a + ab = c_2b, \quad ab + bc_3 = ac_3.$$

Es sei noch erwähnt, daß sich in analoger Weise die konvexen Körper des R_n und der Halb-R_n kennzeichnen lassen.

III. Eine n-dimensionale Metrik[1]).

Der Begriff des metrischen Raumes ist eine Präzisierung der für die Theorie der reellen Funktionen und für die Topologie wichtigsten Eigenschaften der euklidischen Abstandsmetrik. Man kann nun zur Begründung einer Theorie der n-dimensionalen Metrik die wichtigsten Eigenschaften der für euklidische Räume definierten n-dimensionalen Volumsmetrik für beliebige Mengen postulieren unter Abstraktion von der Tatsache, daß die Punkte eines k-dimensionalen euklidischen Raumes k-Tupel reeller Zahlen sind und die Volumsfunktion in bestimmter Weise von den Koordinaten der Punkte abhängt. Es kommen insbesondere folgende allgemeine Eigenschaften der euklidischen Volumsfunktion in Betracht: Es ist je $n+1$ Punkten $p_1, p_2, \ldots, p_{n+1}$ eine reelle Zahl $p_1 p_2 \ldots p_{n+1}$ zugeordnet, welche als das n-dimensionale Volumen des durch die $n+1$ Punkte bestimmten Simplexes bezeichnet wird und folgenden Bedingungen genügt:

S_n (Symmetrieaxiom). Für je $n+1$ Punkte $p_1, p_2, \ldots, p_{n+1}$ gilt, wenn $i_1, i_2, \ldots, i_{n+1}$ irgendeine Permutation der Zahlen $1, 2, \ldots, n+1$ bezeichnet:

$$p_1 p_2 \ldots p_{n+1} = p_{i_1} p_{i_2} \ldots p_{i_{n+1}}.$$

O_n (Nullbedingung). Für je $n+1$ Punkte $p_1, p_2 \ldots, p_{n+1}$, welche in einer $(n-1)$-dimensionalen Ebene liegen, gilt $p_1 p_2 \ldots p_{n+1} = 0$.

P_n (Positivitätsbedingung). Für je $n+1$ Punkte, welche nicht in einer $(n-1)$-dimensionalen Ebene liegen, gilt $p_1 p_2 \ldots p_{n+1} > 0$.

[1]) Vgl. Mathem. Annalen, *100*, S. 142.

U_n (Simplexungleichung). Von den $n+2$ n-dimensionalen Simplexen, welche durch irgend $n+2$ Punkte gegeben sind, hat keines ein größeres Volumen als die Summe der $n+1$ übrigen, m. a. W.: für je $n+2$ Punkte $p_1, p_2, \ldots, p_{n+2}$ des Raumes gilt

$$p_1 p_2 \ldots p_{n+1} \leqq \sum_{i=1}^{n+1} p_1 \ldots p_{i-1} p_{i+1} \ldots p_{n+1} p_{n+2}.$$

Man kann nun in abstrakten Mengen eine n-dimensionale Metrik definieren als Zuordnung von reellen Zahlen zu den $(n+1)$-Tupeln von Elementen, so daß die Symmetriebedingung S_n und die Simplexungleichung U_n erfüllt sind. Die Schwierigkeit liegt darin, einen Ersatz für die Bedingungen O_n und P_n zu finden, da ja in abstrakten Mengen keine $(n-1)$-dimensionalen Ebenen definiert sind. Diese Schwierigkeiten lassen sich lösen, indem man zunächst postuliert:

O_n^s (Schwache Nullbedingung). Für je $n+1$ Punkte, unter denen identische Punkte vorkommen, gilt $p_1 p_2 \ldots p_{n+1} = 0$.

Sodann kann man, falls eine $(n-1)$-dimensionale Metrik definiert ist, einführen:

O_n^m (Mittlere Nullbedingung). Gilt für $n+1$ Punkte $p_1, p_2, \ldots, p_{n+1}$

$$p_1 p_2 \ldots p_n = \sum_{i=1}^{n} p_1 \ldots p_{i-1} p_{i+1} \ldots p_n p_{n+1},$$

oder, wie wir dies auch ausdrücken können, liegt p_{n+1} zwischen den Punkten $p_1, p_2, \ldots, p_n$, dann ist $p_1 p_2 \ldots p_{n+1} = 0$.

Für den euklidischen R_1 und den R_2 sind die Bedingungen O_1^m und O_2^m mit O_1 beziehungsweise O_2 identisch und daher zugleich hinreichende und notwendige Bedingungen für das Verschwinden der Volumsfunktion. In allgemeineren Räumen ist schon für $n=2$ die Annahme, daß O_2^m auch notwendig für das Verschwinden der Volumsfunktion ist, mit der Simplexungleichung U_2 unverträglich. Um eine auch notwendige Bedingung zu erhalten, muß man zu O_n^m die Forderungen hinzunehmen:

a) Bestimmen für $n+2$ Punkte $n+1$ von den $n+2$ aus ihnen bildbaren $(n+1)$-Tupeln das n-dimensionale Volumen Null, dann bestimmt auch das $(n+2)$-te $(n+1)$-Tupel das Volumen Null.

b) Bestimmt jedes der $(n+1)$-Tupel einer gegen ein gewisses $(n+1)$-Tupel konvergenten Folge das Volumen Null, dann auch das Grenz-$(n+1)$-Tupel.

IV. Konvexitätstheoretische Notizen.

1. Über konvexe Hüllen.

In der Konvexitätstheorie (Mathem. Annalen, *100*, S. 76) wurde bewiesen: Damit in einem vollständigen metrischen Raum der Durchschnitt von je zwei konvexen abgeschlossenen Teilmengen konvex sei, ist notwendig und hinreichend, daß der Raum ein *Streckenraum* sei, d. h. daß zu je zwei Punkten des Raumes höchstens eine in ihnen endende Strecke (d. h. mit einer Strecke kongruente Teilmenge des Raumes) existiere. Ist der betrachtete Raum konvex, so kann diese Bedingung auch dahin ausgesprochen werden, daß je zwei Punkte des Raumes eindeutig eine Strecke bestimmen. Diese Bedingung erwies sich zugleich als notwendig und hinreichend dafür, daß zu jeder Teilmenge M des Raumes eindeutig eine konvexe Hülle existiert, d. h. eine konvexe abgeschlossene Menge, die M als Teilmenge enthält, aber keine echte Teilmenge besitzt, die konvex und abgeschlossen ist und M als Teilmenge enthält.

Wir beweisen nun zunächst den Satz: Ist in einem kompakten metrischen Raum $\{A_k\}$ $(k=1, 2, \ldots \text{ ad inf.})$ eine monoton abnehmende Folge von konvexen abgeschlossenen Mengen, so ist die Menge $A = \prod_{k=1}^{\infty} A_k$ konvex. Um dies zu zeigen, ist nachzuweisen, daß die Menge A, die sicher abgeschlossen ist, zu je zweien ihrer Punkte einen Mittelpunkt enthält. Seien a' und a'' zwei vorgegebene Punkte von A. Wir bezeichnen mit M_k die Menge aller in A_k enthaltenen Mittelpunkte von a' und a''. Für jedes natürliche k ist die Menge M_k, wie man leicht zeigt, abgeschlossen und, da A_k als konvex vorausgesetzt ist, nicht leer. Die Mengen $\{M_k\}$ $(k=1, 2, \ldots \text{ ad inf.})$ bilden also eine monoton abnehmende Folge von nichtleeren abgeschlossenen Teilmengen eines kompakten Raumes. Nach dem Cantorschen Durchschnittssatz ist also die Menge $M = \prod_{k=1}^{\infty} M_k$ nichtleer, d. h. es existiert ein in A enthaltener Mittelpunkt von a' und a'', was zu beweisen war.

Aus dem bewiesenen Satz folgert man mit Hilfe des Brouwerschen Reduktionstheorems unmittelbar: zu jeder Teilmenge M eines konvexen kompakten Raumes existiert eine Menge, welche konvex und abgeschlossen ist und M als Teilmenge enthält und welche hinsichtlich dieser drei Eigenschaften irreduzibel ist, d. h. zu jeder Teilmenge eines konvexen kompakten Raumes existiert eine konvexe Hülle (für deren Eindeutigkeit, wie erwähnt, die Eindeutigkeit der Strecken zwischen den Punktepaaren notwendig und hinreichend ist).

2. Über Vollkonvexität.

Eine Teilmenge eines Raumes heißt konvex, wenn sie zu je zweien ihrer Punkte einen Zwischenpunkt enthält. Wir wollen eine Teilmenge eines Raumes vollkonvex nennen, wenn sie zu je zweien ihrer Punkte alle (im Raum vorhandenen) Zwischenpunkte enthält. Im Gegensatz zur Konvexität ist die Vollkonvexität eine Beziehung der Menge zum Raum, keine innere Eigenschaft der Menge. Offenbar gilt der Satz: Damit in einem konvexen vollständigen Raum jede konvexe Teilmenge vollkonvex sei, ist notwendig und hinreichend, daß der Raum ein Streckenraum sei. Der

Durchschnitt aller eine vorgelegte Menge M enthaltenden vollkonvexen Teilmengen des Raumes ist offenbar vollkonvex und irreduzibel hinsichtlich der Eigenschaften, vollkonvex zu sein und M zu enthalten. Wir wollen diese eindeutig bestimmte Menge als die vollkonvexe Hülle von M bezeichnen. In Streckenräumen und nur in diesen ist für jede Menge die eindeutig bestimmte konvexe Hülle zugleich vollkonvex. Es sei darauf hingewiesen, daß in beliebigen Räumen die Menge, welche aus einer Menge M entsteht, indem zu ihr alle Punkte hinzugefügt werden, die zwischen zwei Punkten von M liegen, nicht notwendig vollkonvex ist; denn ein Punkt, welcher zwischen zwei Punkten liegt, deren jeder Zwischenpunkt zweier Punkte von M ist, muß ja (vgl. S. 20 dieses Heftes) nicht zwischen den letzteren zwei Punkten von M liegen.

3. Über Verallgemeinerungen des Zwischenbegriffes.

Die Ergebnisse der beiden vorangehenden Notizen gestatten, den Zwischenbegriff in doppelter Weise zu verallgemeinern. Die Definition des Zwischenpunktes von zwei Punkten, welche sich auf die Dreiecksungleichung stützt, kann auf Grund konvexitäts-theoretischer Sätze folgendermaßen ausgesprochen werden: Der Punkt c eines konvexen vollständigen Raumes heißt Zwischenpunkt von a und b, wenn er auf einem a und b verbindenden geodätischen Bogen liegt. Nun ist jeder a und b verbindende geodätische Bogen eine konvexe Hülle der aus den Punkten a und b bestehenden Menge. Dies legt folgende Definition nahe: Der Punkt p eines vollständigen konvexen Raumes heißt Zwischenpunkt der Menge M, wenn er in einer konvexen Hülle von M enthalten ist. Besteht M aus zwei Punkten, so reduziert sich diese Definition auf die der Konvexitätstheorie zugrunde liegende. Und man kann ferner den Punkt p einen Zwischenpunkt von M im weiteren Sinn nennen, wenn p in der vollkonvexen Hülle von M liegt. Zufolge der Bemerkung am Ende der vorangehenden Notiz ist ein Zwischenpunkt im weiteren Sinn von M nicht notwendig Zwischenpunkt von M; in Streckenräumen ist dies ausnahmslos der Fall. Und allgemein kann man sagen: Die Punkte der Menge M liegen im weiteren Sinn zwischen den Punkten der Menge M', oder kurz: M liegt im weiteren Sinn zwischen M', wenn M Teilmenge der vollkonvexen Hülle von M' ist. Diese allgemeine Zwischenrelation ist transitiv: Wenn M im weiteren Sinn zwischen M' und M' im weiteren Sinn zwischen M'' liegt, so liegt M im weiteren Sinn zwischen M''. Speziell in Streckenräumen ist diese Tatsache eine äußerst weitgehende Verallgemeinerung des in Streckenräumen gültigen Satzes: Wenn die Punkte q und r zwischen den Punkten s und t liegen und p zwischen q und r liegt, so liegt p zwischen s und t; ein Satz, aus dem durch Identifikation von r und t unmittelbar die für die ganze Konvexitätstheorie fundamentale Beziehung folgt: Wenn b zwischen a und c liegt und c zwischen a und d liegt, so liegt b zwischen a und d und liegt c zwischen b und d.

4. Über den Begriff der Konkavität.

Während die Konvexitätsvorstellung geradezu zwangsläufig die anschauliche und fruchtbare Präzisierung, welche der Konvexitätstheorie zugrunde liegt, fordert, sind für die verschwommenere Konkavitätsvorstellung, wie im folgenden gezeigt werden soll, mehrere verschiedene Präzisierungen möglich: Man kann erstens als konkav jene Räume bezeichnen, in denen zu keinem Punktepaar ein Zwischenpunkt existiert, also Räume, die in einem extremen Gegensatz zu den konvexen stehen, in denen zu jedem Punktepaar ein Zwischenpunkt existiert. Man kann zweitens als konkav einen Raum bezeichnen, wenn zu je zweien seiner Punkte a und b zwei Punkte c und d existieren, so daß a zwischen b und c liegt und b zwischen a und d liegt. Bei dieser Ausdrucksweise sind unter den abgeschlossenen Teilmengen der euklidischen Räume die ebenen Teilräume dadurch gekennzeichnet, daß sie sowohl konvex als auch konkav sind. Man bezeichnet diese Räume vielleicht besser als konvex nach außen. Man kann endlich drittens eine Definition geben, welche nur für Teilmengen von Räumen formulierbar ist, nicht aber für Räume selbst, indem man eine Teilmenge M eines Raumes konkav nennt, wenn ihr Komplement konvex ist, oder was gleichbedeutend ist, wenn kein Punkt von M zwischen Punkten des Komplements von M liegt.

V. Die Metrik des Hilbertschen Raumes.

I. *Damit ein separabler halbmetrischer Raum R in den Hilbertschen Raum abstandstreu einbettbar sei, ist notwendig und hinreichend, daß für jede natürliche Zahl k und für je k Punkte $p_1, p_2, \ldots, p_k$ von R die Beziehung gilt:*

$$\operatorname{sign} D\,(p_1, p_2, \ldots, p_k) = \operatorname{sign}\,(-1)^k \text{ oder } 0,$$

wobei $D\,(p_1, p_2, \ldots, p_k)$ die oben S. 21 angegebene Bedeutung hat.

Während zur Charakterisierung der euklidischen Räume merkwürdigerweise keine Voraussetzungen über die Separabilität oder auch nur die Mächtigkeit von R erforderlich sind, ist die Voraussetzung der Separabilität von R im Theorem I unentbehrlich. Denn bezeichnet R eine Menge von hinreichend großer Mächtigkeit, in der je zwei verschiedene Punkte den Abstand 1 haben, so ist die Bedingung des Theorems I erfüllt, ohne daß R abstandstreu in den Hilbertschen Raum einbettbar wäre. Dieses Beispiel zeigt, daß die Methode zum Beweis der Charakterisierung euklidischer Räume zum Beweis von Theorem I nicht ohne weiteres verwendbar ist.

Der Beweis von Theorem I erfolgt so, daß man erstens die Gültigkeit des Theorems für den Fall zeigt, daß R eine abzählbare Punktefolge ist. Dieselbe kann, wenn die Bedingung des Theorems erfüllt ist, unter Benutzung der Theoreme über den euklidischen Raum sukzessive in den Hilbertschen Raum abstandstreu eingebettet werden. Man führt sodann, wenn R ein beliebiger separabler

Raum ist, für eine in R dichte Punktefolge die abstandstreue Einbettung in den Hilbertschen Raum durch und kann mit Rücksicht auf die erfüllten Stetigkeitsforderungen die Abbildung zu einer von ganz R auf eine Teilmenge des Hilbertschen Raumes ergänzen.

Analog wie die Charakterisierung der euklidischen Räume selbst ergibt sich aus dem Theorem I das Theorem

II. *Damit ein halbmetrischer Raum mit dem Hilbertschen Raum abstandsgleich sei, ist notwendig und hinreichend,*

1. *daß $\mathring{R}$ separabel und vollständig sei,*

2. *daß für jedes natürliche k und für je k Punkte $p_1, p_2, \ldots, p_k$ von R* sign $D(p_1, p_2, \ldots, p_k) = \text{sign}\,(-1)^k$ *oder* 0 *gilt,*

3. *daß für jedes natürliche k der Raum R k Punkte $p_1, p_2, \ldots, p_k$ enthält, für die $D(p_1, p_2, \ldots, p_k) \neq 0$ gilt,*

4. *daß R konvex und nach außen konvex sei.*

Eine metrische Charakterisierung des Simplexes. Von M. M. Biedermann (Amsterdam).

In einem metrischen Raume heiße der Punkt b ein *k-dimensionaler Zwischenpunkt* der Punkte $a_0, a_1, \ldots, a_k$, wenn ein Punkt c existiert, welcher $(k-1)$-dimensionaler Zwischenpunkt von $a_0, a_1, \ldots, a_{k-1}$ ist und so daß $cb + ba_k = ca_k$ gilt.

Als Ausgangspunkt dieser Rekursion diene die Festsetzung, daß b eindimensionaler Zwischenpunkt von a_0 und a_1 heißt, falls $a_0 b + b a_1 = a_0 a_1$ gilt. Zum Unterschied von der üblichen Definition des Zwischenpunktes (siehe oben S. 20) wird dabei nicht außerdem die Verschiedenheit des Punktes b von a_0 und a_1 gefordert. Überdies heiße nulldimensionaler Zwischenpunkt eines Punktes a_0 dieser Punkt a_0 selbst.

Dann gilt der Satz: Der metrische Raum T^n ist mit dem n-dimensionalen Simplex homöomorph, wenn er folgende Eigenschaften besitzt:

I. *Der Raum T^n ist kompakt.*

II. *Der Raum T^n ist konvex und enthält zu je zwei Punkten höchstens einen Mittelpunkt* (und folglich genau eine sie verbindende Teilstrecke).

III. *Der Raum enthält $n+1$ Punkte $a_0, a_1, \ldots, a_n$ derart, daß a_{k+1} kein k-dimensionaler Zwischenpunkt von $a_0, a_1, \ldots, a_k$ ist $(k=1, 2, \ldots, n-1)$ und jeder Punkt des Raumes n-dimensionaler Zwischenpunkt von $a_0, a_1, \ldots, a_n$ ist.*

IV. *Die Menge T^k aller k-dimensionalen Zwischenpunkte von $a_0, a_1, \ldots, a_k$ ist abgeschlossen und konvex $(k=1, 2, \ldots, n-1)$.*

V. *Sind e und e' zwei verschiedene Punkte von T^k, so gibt es keinen von a_{k+1} verschiedenen Punkt, welcher sowohl zwischen a_{k+1} und e als auch zwischen a_{k+1} und e' liegt.*

Da die Behauptung für $n=1$ trivial ist, beweisen wir sie unter der Annahme ihrer Gültigkeit für $n-1$. Ist dann ein Raum T^n mit den fünf erwähnten Eigenschaften gegeben, so erfüllt jede der Teilmengen T^k ($k=1, 2, \ldots, n-1$), als Raum betrachtet, dieselben Bedingungen. Es ist demnach T^{n-1} auf ein $(n-1)$-dimensionales Simplex E^{n-1} des R_n topologisch abbildbar. Wir bilden ferner den Punkt a_n auf irgend einen Punkt a'_n des R_n ab, der nicht in der E^{n-1} enthaltenden $(n-1)$-dimensionalen Ebene des R_n liegt. Ist dann a irgend ein Punkt von T^n, so enthält T^n auf Grund seiner fünf Eigenschaften genau eine Strecke zwischen a_n und einem Punkt b von T^{n-1}, welche a enthält. Ist dem Punkt b bei der Abbildung von T^{n-1} auf E^{n-1} der Punkt b' von E^{n-1} zugeordnet worden, so ordnen wir a jenen Punkt der a'_n mit b' verbindenden Strecke zu, welcher in bezug auf a'_n und b' dasselbe Teilverhältnis hat, wie a in bezug auf a_n und b. Die so definierte Abbildung von T^n auf das durch E^{n-1} und a'_n bestimmte n-dimensionale Simplex ist offenbar topologisch, womit die Behauptung bewiesen ist.

Zur Axiomatik der endlichen Mengen und der elementargeometrischen Verknüpfungsbeziehungen. Von Karl Menger.

Es sei ein System von Dingen gegeben, in welchem *(I. „Summen- und Durchschnittsaxiom")* je zwei Dingen A und B zwei Dinge $A+B$ und $A.B$ entsprechen, wobei Summen- und Durchschnittsbildung assoziativ, kommutativ, totallinear sind (d. h. $A+A=A=A.A$) und (eo ipso eindeutige) Einheitselemente besitzen ($A+V=A=A.T$). Es mögen *(II. „Absorptionsaxiom")* für je zwei Dinge A und B die Beziehungen $A+B=A$ und $A.B=B$ äquivalent sein (d. h. gleichzeitig gelten oder nicht gelten). Wenn sie gelten, nennen wir B ein Teilding von A und, falls überdies $A\neq B$ gilt, ein echtes Teilding von A. Es möge *(III. „Subtraktionsaxiom")*, wenn A Teilding von B ist, ein Ding A' existieren, so daß $A+A'=B$ und $A.A'=V$ gilt. Es sei *(IV. „Betragsaxiom")* jedem Ding A eine ganze Zahl, bezeichnet mit $|A|$ und genannt der Betrag von A, zugeordnet derart, daß 1.) $|V|=-1$, daß 2.) ein Ding, dessen sämtliche echten Teildinge Beträge $\leq n-1$ haben, einen Betrag $\leq n$ und 3.) ein Ding mit einem echten Teilding, dessen Betrag $\geq n-1$ ist, einen Betrag $\geq n$ hat und daß 4.) für je zwei Dinge A und B die „Betragsgleichung" $|A|+|B|=|A+B|+|A.B|$ gilt.

Diese Axiome sind erfüllt für das System aller linearen Teilräume eines projektiven Raumes, wenn wir unter dem Betrag eines Dinges seine Dimension verstehen. Verschärfen wir das Subtraktionsaxiom durch die Forderung der Eindeutigkeit von A', so ist das Dingsystem isomorph mit dem System aller Teilmengen einer Menge, deren Mächtigkeit dem Betrage des Dinges T gleich ist.

***Zur Axiomatik der Elementargeometrie.** Von Gustav Bergmann.

Ein System von Dingen heiße *Feld*, wenn es gegenüber zwei Operationen abgeschlossen ist, die je zwei Dingen A und B zwei Dinge $A+B$ und $A.B$ gemäß folgenden Bedingungen zuordnen:

I. Axiom der Verknüpfung.

1. $A+(B+C)=(A+B)+C$, $\quad A.(B.C)=(A.B).C$;
2. $A+B=B+A$, $\quad A.B=B.A$;
3. $A+A=A$ $\quad A.A=A$;

Es existieren zwei Elemente V und T, so daß für jedes A

4. $A+V=A$ $\quad A.T=A$.

II. Axiom der Komplementbildung.

Zu je zwei Elementen A, B gibt es zwei Elemente C_1 und C_2, so daß gilt:

1*a*. $A.B+C_1=A$, $\quad$ 1*b*. $(A+B).C_2=A$.

2*a*. $A.B.C_1=V$, $\quad$ 2*b*. $A+B+C_2=T$.

Aus diesen Axiomen folgt der in Mengers vorangehender Note postulierte Absorptionssatz: Jede der beiden Relationen

$$(T) \qquad A.B=B. \qquad A+B=A$$

ist eine Folge der andern. Bestehen für A und B die beiden Relationen (T), dann heiße B ein Teil von A, $B \subseteq A$. Für diese Teilrelation folgen alle Eigenschaften der elementargeometrischen Teilrelation (Transitivität; aus $A \subseteq C$, $B \subseteq C$ folgt $A+C \subseteq B$, usw.) aus dem Axiomensystem.

Schon im Schröderschen Klassenkalkül wurde die Frage nach dem Zusammenhang zwischen den Distributivgesetzen und der eindeutigen Umkehrbarkeit der beiden Grundoperationen aufgeworfen. Es gilt nun der Satz: Die Gültigkeit der beiden Distributivgesetze:

III*a*. $A+B.C=(A+B).(A+C)$. $\quad$ III*b*. $A.(B+C)=A.B+A.C$

ist äquivalent mit der eindeutigen Bestimmtheit von C_1 und C_2 in II.

Zunächst folgt jede der beiden Hälften von III aus I., II. und der anderen Hälfte; ebenso folgen aus I., II. mit Hilfe der Eigenschaften der Relation (T) die distributiven Ungleichungen:

$$A+B.C \subseteq (A+B).(A+C). \qquad A.(B+C) \supseteq A.B+A.C$$

Beim Beweis, daß III aus der eindeutigen Umkehrbarkeit folgt, handelt es sich vor allem darum, zu zeigen, daß aus

$$X \subseteq B+C,\ B.C=V \text{ die Beziehung } X=B.X+C.X$$

folgt. Daraus ergeben sich dann die Distributivgesetze mit Hilfe der distributiven Ungleichungen und der übrigen Eigenschaften der Teilrelation. Ebenso leicht gelingt der Beweis der anderen Hälfte.

Was Mengers Betrags- oder „Dimensionsaxiom" betrifft, so stellen wir nun die Frage, ob seine explizite Formulierung sich vermeiden läßt, indem man an seiner statt Eigenschaften der Verknüpfungsrelationen postuliert, die scharfe Bedingungen für die Dimensionierbarkeit abstrakter Felder darstellen. Es ergibt sich, daß eine solche interne Charakterisierung der dimensionierbaren Felder innerhalb der abstrakten Felder in der Tat möglich ist.

Ein Element $\neq V$, das außer V kein echtes Teilding enthält, nennen wir einen Punkt. Enthält jedes Element $\neq V$ mindestens einen Punkt als Teilding, dann kann man jedes Element $\neq V$ durch gewisse — im allgemeinen nicht kommutative — Ketten $(P_1, P_2, \ldots)$ in ihm enthaltener Punkte charakterisieren. Dementsprechend postulieren wir:

D 1. Jedes Element enthält einen Punkt und die Gliederzahl sämtlicher Ketten liegt unter einer endlichen Schranke.

D 1 ist eine scharfe Bedingung für die Einführbarkeit einer Funktion mit den Eigenschaften IV, 1, 2. Ferner postulieren wir:

D 2. Wenn A ein beliebiges Element ist und P ein Punkt, dann gibt es kein Element D, so daß gilt: $A \subset D \subset A + P$ (echtes Teilzeichen!).

D 2 ist eine scharfe Bedingung dafür, daß alle Ketten kommutativ und gleich lang sind. Nennt man die Punkte einer Kette linear unabhängig, so ist jetzt ein n-dimensionales Ding durch genau $n+1$ linear unabhängige Punkte eindeutig bestimmt.

D 3. Wenn $A+P=A+X$ ist, P ein Punkt, $A.P=A.X=V$, dann ist auch X ein Punkt.

D 1, 2, 3 zusammen sind äquivalent mit IV, 1, 2, 3.

Schließlich kann man durch Beispiele zeigen, daß jedes dieser drei sukzessiven Dimensionsaxiome eine wirkliche Verschärfung der vorausgehenden ist.

Für eine ausführliche Darstellung dieser Ergebnisse vgl. Monatshefte f. Math. u. Phys., *36*, S. 269.

Intuitionistisch-formalistisches Wörterbuch. Von Karl Menger.

Eine Übersetzung der (überflüssig verwickelten) Definitionen, welche Brouwer (Math. Annalen, *93*ff.) an die Spitze des sog. intuitionistischen Aufbaues der Mathematik stellt, in die allgemein übliche Ausdrucksweise ergibt folgende Beziehungen:

Was Brouwer bezeichnet als:	heißt in üblicher Ausdrucksweise:
Finite Menge	Kompakte Menge
Individualisierte Menge	Borelsche Menge
Menge	Analytische Menge
Spezies	Teilmenge einer analytischen Menge.

Von Knoten berandete singularitätenfreie Flächen.

F. Frankl bewies im Studienjahre 1928/29 in Wien, daß im R_3 jedes einfach geschlossene Polygon (auch wenn es beliebig verknotet ist) Rand einer singularitätenfreien (d. h. sich nicht selbst durchdringenden) orientierbaren Fläche ist (vgl. Frankl und Pontrjagin, Math. Ann., 102, S. 785). Insbesondere bildet also die Kleeblattschlinge den Rand einer zweiseitigen Fläche, die man z. B. folgendermaßen konstruieren kann: Man zeichne auf einem Blatt Papier drei Strecken mit gemeinsamem Endpunkt, die paarweise einen Winkel von 120° einschließen, und schneide sodann die sich aus drei Streifen zusammensetzende Figur aus, bestehend aus allen Punkten, welche von mindestens einer Strecke einen Abstand $\leqq \rho$ haben (für eine Zahl $\rho > 0$, welche im Verhältnis zur Länge der drei Strecken klein ist). Die Enden der drei Streifen sind Halbkreisflächen vom Durchmesser 2ρ. Diese Halbkreisflächen ersetze man durch gleichschenklige Dreiecke mit einer Basis der Länge 2ρ und einem gegenüberliegenden Winkel von 120°. Nun tordiere man jeden der drei Streifen (so, wie wenn man in bekannter Weise aus einem Streifen durch Torsion und darauffolgendes Aneinanderfügen der Enden ein Möbiussches Band herstellt), und zwar führe man die Torsionen der drei Streifen im selben Drehungssinne durch und füge dann die drei Enden der Streifen so zusammen, daß die drei Winkel von 120° sich zu einem vollen Winkel zusammenschließen. Die so entstehende Fläche ist zweiseitig und hat die Kleeblattschlinge zum Rande. — Nach Knaster kann eine geschlossene Fläche vom Geschlecht 2 durch eine Kleeblattschlinge in zwei (natürlich zweiseitige) Flächen zerlegt werden, welche beide die Kleeblattschlinge zum Rande haben. — Eine einseitige, von der Kleeblattschlinge berandete Fläche gab Nöbeling an: man tordiere einen rechteckigen Streifen dreimal im selben Sinne und füge die Enden zusammen. — Diese topologischen Tatsachen legen das Problem nahe, in Knoten oder wenigstens in gewisse Knoten sich nicht durchsetzende Minimalflächen einzuspannen. Dieselben wären mit einer Kreisscheibe nicht homöomorph.

Angelegenheiten des Kolloquiums und Nachrichten.

Wiener Dissertationen im Zusammenhange mit dem Kolloquium 1928/29:

H. Hornich, „Über einen zweigradigen Zusammenhang" (vgl. S. 17 dieses Heftes).

H. Reschovsky, „Über das Geschlecht von Kurven" (vgl. S. 13 dieses Heftes und Fund. Math., 15, S. 18).

Stipendien.

Dr. W. L. Ayres, Philadelphia, arbeitete mit einer International Research Fellowship 1928/29 im Kolloquium.

Druck von Gottlieb Gistel & Cie., Wien.

Bericht über das Kolloquium 1929/30.

1. Kolloquium (24. X. 1929).

Menger: *Eine dimensionstheoretische Bemerkung von Otto Schreier.* Es sei $S_{k,\,n-k}$ die Menge aller Punkte des R_n mit k rationalen und $n-k$ irrationalen Koordinaten. Dann ist für jedes natürliche n für jede der drei Mengen $S_{0,\,n}$, $S_{n-1,\,1}$, $S_{n,\,0}$ jeder Punkt in beliebig kleinen n-dimensionalen Würfeln enthalten, deren Begrenzungen zur betreffenden Menge fremd sind. Dagegen hat, wenn $n \geqq 3$ ist, für $1 \leqq k \leqq n-2$ die Menge $S_{k,\,n-k}$, obwohl sie nulldimensional ist, mit jeder stetigen Parameterfläche, d. h. mit jeder Hyperfläche des R_n, welche dadurch definiert ist, daß eine Koordinate eine stetige Funktion der $n-1$ übrigen Koordinaten ist, einen nichtleeren Durchschnitt. (Ausführung: Monatshefte f. Math. u. Phys., *37*, S. 7. Über die Frage nach der Möglichkeit, die mit Umgebungsbegrenzungen operierende Dimensionsdefinition auf Umgebungen bestimmter Art zu beschränken, siehe auch Heft *1* dieser Ergebnisse, S. 10.)

Nöbeling: Man sagt, ein Kontinuum K besitze eine Eigenschaft E erblich, wenn jedes echte oder unechte Teilkontinuum von K die Eigenschaft E hat. Nennt man ein Kontinuum zerlegbar, wenn es als Summe zweier echter Teilkontinua dargestellt werden kann, so ist z. B. der Bogen erblich zerlegbar.

Knaster hat gezeigt (Fund. Math., *3*, S. 247), daß es in der Ebene erblich unzerlegbare Kontinua gibt, Kontinua also, die keine zerlegbaren Teilkontinua, insbesondere keine Bögen enthalten. Da ein Teilkontinuum der Summe von abzählbarvielen Kontinua nicht mit allen Summanden einen diskontinuierlichen Durchschnitt haben kann, besitzt jedes Teilkontinuum der Summe von abzählbarvielen Knasterschen Kontinua seinerseits ein erblich unzerlegbares Teilkontinuum.

Wir wählen nun im Einheitsquadrat der x, y-Ebene ein Knastersches Kontinuum K, das die Punkte (0, 1) und (1, 0), aber nicht den Punkt (0, 0) enthält (übrigens besitzt das von Knaster konstruierte Kontinuum selbst diese Eigenschaften). Das Spiegelbild von K an der x-Achse heiße K', das Spiegelbild von $K+K'$ an der y-Achse heiße K''. Dann ist $K^* = K + K' + K''$ Summe von vier Knasterschen Kontinua und es gibt eine im Quadrat $-1 \leqq x, y \leqq 1$ liegende Umgebung von (0, 0), deren Begrenzung Teilmenge von K^* ist, nämlich die Menge aller mit (0, 0) durch einen zu K^* fremden Bogen verbindbaren Punkte der Ebene.

Hieraus folgert man mühelos, daß sich abzählbarviele Knastersche Kontinua K_1, K_2, . . . der Ebene so finden lassen, daß jeder Punkt der Ebene in beliebig kleinen Umgebungen liegt, deren Begrenzungen Teilmengen der Summe $S = K_1 + K_2 + \ldots$ sind. Das Komplement der F_σ-Menge S ist demnach eine nulldimensionale Menge, deren Komplement kein erblich zerlegbares Kontinuum enthält. Wir haben damit den Satz:

Es gibt in der Ebene eine nulldimensionale G_δ-Menge, die mit jedem erblich zerlegbaren Kontinuum, insbesondere mit jedem Bogen, einen nichtleeren Durchschnitt hat.

Wenn man im Einheitswürfel W des R_n ein Knastersches Kontinuum konstruieren kann, das die Punkte (0, . . . , 0) und (1, . . . , 1) in W trennt, so läßt sich unser Satz durch dieselben Überlegungen auch für den R_n beweisen.

Menger: Jeder n-stufig zusammenhängende kompakte Raum R bleibt auch nach Tilgung einer nicht-abgeschlossenen $(n-2)$-dimensionalen Menge M zusammenhängend. Zerfiele nämlich die Menge $R-M$, so müßte M nach einem bekannten Satz auch eine abgeschlossene Teilmenge enthalten, nach deren Tilgung R zerfiele. — Ein n-dimensionaler kompakter Raum R enthält auch nach Tilgung irgend einer höchstens $(n-2)$-dimensionalen Teilmenge M Kontinua. Tilgt man nämlich aus R eine (sicher existierende) höchstens $(n-2)$-dimensionale M enthaltende G_δ-Menge M', so ist $R-M'$ ein mindestens eindimensionales F_σ eines kompakten Raumes, also nicht diskontinuierlich.

2. Kolloquium (6. XI. 1929).

Nöbeling: Lemma. Zu jeder Zahlenfolge $\{a_k\}$ ($k=2, 3, \ldots$ ad inf.) existiert eine reguläre Kurve K, die in k paarweise fremde Teilmengen bloß durch Tilgung von mindestens a_k Punkten zerlegt werden kann. Hieraus folgt der Satz: Zu jeder regulären Kurve K existiert eine reguläre Kurve K', die mit keiner Teilmenge von K homöomorph ist. *Es gibt also keine reguläre Universalkurve.* (Vgl. Math. Ann. *104*, S. 81.)

Menger: Aus Satz I von Reschovsky (vgl. diese Ergebnisse Heft *1*, S. 13, und Fund. Math., *15*, S. 19) folgt unmittelbar, daß auch *keine rationale Universalkurve existiert.* Es drängt sich daher die Frage auf: Bleibt die Eigenschaft eines Mengensystems, eine kompakte Universalmenge zu besitzen bzw. nicht zu besitzen, beim Übergang zum übergeordneten Mengensystem[1]) erhalten, wenigstens unter gewissen Additivitätsvoraussetzungen über die Mengensysteme, z. B. für Normalbereiche? Für die Systeme der endlichen, der abzählbaren und der nulldimensionalen Mengen, von denen die beiden ersten keine kompakte Universalmenge besitzen und das dritte eine kompakte Universalmenge besitzt, trifft dies zu, denn von den drei übergeordneten Mengensystemen haben die beiden ersten, d. i. das System der reguläreindimensionalen und das System der rationaleindimen-

[1]) Vgl. Menger, Dimensionstheorie, bei Teubner 1928, S. 123. Im folgenden mit „Dimensionstheorie" zitiert.

sionalen Mengen keine kompakte Universalmenge, das dritte, d. i. das System der höchstens eindimensionalen Mengen, enthält eine Universalmenge, nämlich die Universalkurve[1]). In diesem Zusammenhang ergibt sich auch die Frage, ob für jedes natürliche $n \geqq 2$ reguläre Kurven existieren, die alle Kurven n^{ter} Ordnung topologisch enthalten (für $n=2$ ist der Kreis eine Universalkurve zweiter Ordnung).

Entsprechende Fragen erheben sich für die Kompaktifizierbarkeit. Überhaupt sollten in Bearbeitung des Programmes einer Überblickung und Aufzählung von Mengensystemen[2]) vor allem jene Eigenschaften von Mengensystemen, die beim Übergang zum übergeordneten Mengensystem erhalten bleiben, systematisch untersucht werden.

3. Kolloquium (20. XI. 1929).

Menger: *Eine schwach irrationale Kurve von Mazurkiewicz.* (Fund. Math., *15*, S. 222.)

Nöbeling: Roberts hat (Fund. Math., *14*, S. 327) gezeigt: Es existiert eine reguläre Kurve K mit folgender Eigenschaft: Zerlegt man K in eine Summe von endlichvielen Teilkontinua, deren Durchmesser kleiner als ein gewisses $\varepsilon > 0$ sind, so haben mindestens zwei dieser Summanden mehr als einen Punkt gemein. Aus dem im 2. Kolloquium erwähnten Lemma ergibt sich unmittelbar folgende Verschärfung dieser Aussage: *Zu jedem natürlichen k existiert eine reguläre Kurve K, so daß bei jeder Zerlegung von K in endlichviele abgeschlossene Teilmengen zwei von diesen Teilmengen mindestens k Punkte gemein haben.*

Nöbeling: In jeder regulären Kurve K existiert ein *Gerüst*, d. h. eine Summe G von abzählbarvielen Teilbogen, so daß zu jedem Punkte p von K, der die Ordnung n bzw. wachsende Ordnung besitzt, n bzw. abzählbarviele in p endende und sonst fremde Teilbogen der Menge $(p)+G$ existieren. Zum Beweise wählen wir eine in K dichte Punktefolge $\{p_k\}$ $(k=1, 2, \ldots$ ad inf.) und verbinden je zwei Punkte p_i und p_j durch abzählbarviele Teilbogen $\{B_{i,j}^k\}$ $(k=1, 2, \ldots$ ad inf.) von K derart, daß zu jedem in p_i und p_j endenden Teilbogen B von K und zu jeder Zahl $\varepsilon > 0$ ein Bogen $B_{i,j}^k$ in $U(B; \varepsilon)$ existiert. Die Summe G der abzählbarvielen Bogen $B_{i,j}^k$ $(i, j, k=1, 2, \ldots$ ad inf.) ist dann ein Gerüst von K.

4. Kolloquium (27. XI. 1929).

G. T. Whyburn: Damit die Kurve K beständig regulär sei, d. h. (vgl. diese Ergebnisse, Heft *1*, S. 12) damit die Summe $K+K'$ von K und einer beliebigen regulären Kurve K' regulär sei, ist notwendig und hinreichend, daß die abgeschlossene Hülle der Menge aller Verzweigungspunkte von K diskontinuierlich ist. (Ausführung: Monatshefte f. Math. u. Phys., *38*, S. 1.)

---

[1]) Dimensionstheorie, S. 314.

[2]) Vgl. Menger, C. R. du I. Congr. Math. Slav., S. 60.

1*

Beer: *Über Kompaktifizierbarkeit.* Das einem kompaktifizierbaren Mengensystem übergeordnete Mengensystem ist nicht notwendig kompaktifizierbar. (Vgl. die Fragestellung im 2. Kolloquium.) Beispielsweise ist das System aller aus höchstens vier Punkten bestehenden Mengen kompaktifizierbar (es enthält sogar bloß kompakte Mengen). Das übergeordnete Mengensystem, d. h. das System aller Mengen, die in jedem ihrer Punkte von höchstens vierter Ordnung sind, ist dagegen nicht kompaktifizierbar, wie folgendes Beispiel zeigt: Es sei in der Ebene M die Summe aller Geraden $x=k$ und $y=k$ ($k=\pm 1, \pm 2, \ldots$ ad inf.). Diese Menge vierter Ordnung ist nicht homöomorph mit einer Teilmenge einer kompakten Menge vierter Ordnung. Wenn man nämlich die Ebene stereographisch auf die Kugeloberfläche ohne den Nordpol n abbildet, so ist, wenn M' die Menge bezeichnet, in die M bei dieser Abbildung übergeht, $M'+(n)$ eine reguläre Kurve, die in n von wachsender Ordnung ist. Ist umgekehrt K eine reguläre kompakte Menge, die eine mit M homöomorphe Teilmenge M^* enthält, so besteht die Menge $\overline{M^*}-M^*$ aus genau einem Punkt n'. Denn enthielte sie noch einen zweiten Punkt n'', so enthielte M^* zwei gegen n' bzw. n'' konvergente Punktefolgen $\{n'_i\}$ bzw. $\{n''_i\}$ derart, daß n'_i und n''_i für jedes natürliche i durch einen Bogen $B_i \subset M^*$ verbunden werden könnten, so daß B_i und B_j für $i \neq j$ fremd sind, was der Regularität widerspricht. Dann ist aber $\overline{M^*}$ mit $M'+(n)$ homöomorph und enthält im Bildpunkt von n' einen Punkt von wachsender Ordnung. Da jede M topologisch enthaltende kompakte Menge mindestens einen Punkt wachsender Ordnung besitzt, ist zugleich bewiesen: *Das System aller Mengen, die in jedem Punkte eine endliche Ordnung besitzen, ist nicht kompaktifizierbar.*

Nöbeling wirft die Frage auf, ob es eine ganzzahlige Funktion $f(n)$ mit folgender Eigenschaft gibt: Ist K irgend eine reguläre Kurve von höchstens n^{ter} Ordnung, so gibt es ein K erzeugendes finites Umgebungssystem $\{U^k_m\}$ ($m=1, 2, \ldots m_k$, $k=1, 2, \ldots$ ad inf.) derart, daß für jeden Punkt p von K und für jedes natürliche k die Summe aller p enthaltenden Mengen U^k_m ($m=1, 2, \ldots m_k$) des k^{ten} Schrittes eine aus höchstens $f(n)$ Punkten bestehende Begrenzung besitzt, daß m. a. W. jeder Punkt p von K in bezug auf das System $\{U^k_m\}$ höchstens die Ordnung $f(n)$ besitzt.

Menger: *Zur Axiomatik der Dimension.* Die Eigenschaften I—V der Dimension (vgl. diese Ergebnisse, Heft *1*, S. 3) sind für dieselbe unter den Funktionen der Teilmengen des R_n charakteristisch, wenn die beiden folgenden Sätze gelten:

I^n_k. Zu jeder kompakten k-dimensionalen Teilmenge M des R_n existiert eine Folge $\{M_i\}$ ($i=1, 2, \ldots$ ad inf.) von mit M homöomorphen Mengen derart, daß $\mathit{dim}\,(R_n - \sum_{i=1}^{\infty} M_i) = n-k-1$.

II_k^n. Jedes G_δ des R_n, dessen Komplement höchstens $(n-k-1)$-dimensional ist, enthält topologisch alle k-dimensionalen Mengen des R_n. (Aus einem Briefwechsel mit Kuratowski, Monatshefte f. Math. u. Phys. *37*, S. 172.)

5. Kolloquium (8. I. 1930).

Hornich: *Über Zusammenhangseigenschaften.*

Es sei $\mathfrak{S}$ ein System von Teilmengen eines topologischen Raumes, welches *monoton* ist, d. h. neben jeder Menge alle ihre Teilmengen enthält. Eine Menge M des Raumes heißt *zusammenhängend bezüglich* $\mathfrak{S}$, wenn M durch Tilgung keiner zu $\mathfrak{S}$ gehörigen Menge in zwei getrennte (d. h. fremde, relativ abgeschlossene) nichtleere Mengen zerfällt. Zusammenhängend im üblichen Sinne sind die in bezug auf das aus der leeren Menge bestehende Mengensystem zusammenhängenden Mengen. Die Menge M heißt im kleinen zusammenhängend bezüglich $\mathfrak{S}$, wenn jeder Punkt von M in beliebig kleinen Umgebungen enthalten ist, die bezüglich $\mathfrak{S}$ zusammenhängend sind. Es gilt dann der Satz: *Ein bezüglich* $\mathfrak{S}$ *im kleinen zusammenhängendes Kontinuum ist bezüglich* $\mathfrak{S}$ *zusammenhängend.* Zerfällt nämlich das Kontinuum K durch Tilgung der Menge S von $\mathfrak{S}$, welche (vgl. S. 2 dieses Heftes) wegen der Monotonie von $\mathfrak{S}$ als abgeschlossen angenommen werden kann, in zwei nichtleere getrennte Summanden A und B, so ist die Menge S' aller nicht in A gelegenen Häufungspunkte von A eine Teilmenge von S, nach deren Tilgung K in zwei getrennte nichtleere Summanden A und B' zerfällt. Da K im üblichen Sinn zusammenhängend ist, so ist S' nichtleer. Es existiert ein Punkt p von S', der Häufungspunkt sowohl von A als auch von B' ist. Ist nun U eine beliebige Umgebung von p, so zerfällt U durch Tilgung der Menge $U.S'$ in die beiden getrennten nichtleeren Mengen $U.A$ und $U.B'$. Dann ist also K bezüglich $\mathfrak{S}$ nicht zusammenhängend im kleinen. Durch analoge Überlegungen zeigt man: Ist der Raum zusammenhängend und zusammenhängend im kleinen (im üblichen Sinn), so ist, damit er in bezug auf das monotone Mengensystem $\mathfrak{S}$ zusammenhängend im kleinen sei, notwendig und hinreichend, daß jedes Gebiet (d. h. jede zusammenhängende offene Teilmenge des Raumes) zusammenhängend in bezug auf $\mathfrak{S}$ sei.

Menger: *Bemerkungen über Produktsatz und dimensionelle Raumstruktur.*

Das Produkt
- eines höchstens n-dimensionalen und eines höchstens m-dimensionalen
- eines höchstens rational n-dimensionalen und eines höchstens rational m-dimensionalen
- eines höchstens schwach n-dimensionalen und eines höchstens schwach m-dimensionalen

Raumes ist
- höchstens $(m+n)$-dimensional
- höchstens rational $(m+n)$-dimensional
- höchstens schwach $(m+n)$-dimensional.

Die Menge aller Punkte eines schwach n-dimensionalen Raumes, in denen der Raum mindestens eindimensional ist, ist mindestens stark $(n-1)$-dimensional. (Vgl. Prace mat. fis., Warschau 1930.)

6. Kolloquium (15. I. 1930).

Menger: *Entwurf einer neuen Theorie des Maßes*[1]).

I. Axiomatik von Dimension und Maß.

Die Axiome für die Dimension (vgl. Heft *1* dieser Ergebnisse, S. 3) weisen weitgehende Analogien mit Lebesgues Axiomen für das Maß auf. Sowohl für die Dimension, wie für das Maß gilt das *Monotonie*-Axiom I, daß der Funktionswert eines Teiles niemals größer als der Funktionswert des Ganzen ist. Für beide Funktionen gilt ein *Additions*-Axiom II, welches sich sowohl für Maß als auch für Dimension in zwei Teile spalten läßt:

II*a*) *Für jede Mengenfolge* $\{M_k\}$ ($k=1, 2, \ldots$ *ad inf.*) *gilt*

$$f(\sum_{k=1}^{\infty} M_k) = \lim_{n=\infty} f(\sum_{k=1}^{n} M_k).$$

II*b*) *Für je zwei Mengen A und B gilt*

$$f(A+B)=\begin{cases} f(A)+f(B)-f(A\,.\,B) & \textit{für das Maß.} \\ Max.\,\big(f(A),\ f(B)\big) & \textit{für die Dimension.} \end{cases}$$

Beide Mengenfunktionen genügen ferner einem *Invarianz*-Axiom III: die Dimension ist invariant gegenüber topologischen, das Maß gegenüber kongruenten Transformationen. Beide Mengenfunktionen werden endlich durch die Festsetzung eines Funktionswertes für den k-dimensionalen Einheitswürfel *normiert*, indem für denselben die Dimension k und das Maß 1 in einem Axiom V festgesetzt wird.

Diesen Analogien stehen folgende Unterschiede gegenüber: Die Forderungen II*a*) und II*b*) werden von Lebesgue hinsichtlich des Maßes für *jede* Mengenfolge bzw. für *je* zwei Mengen, hinsichtlich der Dimension aber nur für Folgen von *abgeschlossenen* Mengen bzw. für zwei *abgeschlossene* Mengen gemacht; für Folgen bzw. Paare von Mengen, die nicht wenigstens *in der Summe abgeschlossen sind,* gelten II*a*) und II*b*) hinsichtlich der Dimension gar nicht. Für die Dimension wird ein Axiom IV postuliert, welches für jede Menge die Existenz einer sie topologisch enthaltenden kompakten Menge mit gleichem Funktionswert fordert, woraus insbesondere für jede Menge die Existenz einer sie als Teilmenge enthaltenden G_δ-Menge mit gleichem Funktionswert folgt (welch letztere Tatsache nebenbei nach einer Bemerkung von Kuratowski, Monatshefte, *37*, S. 169, als Ersatz für das Axiom IV nicht ausreichend wäre). Für das Maß existiert zwar zu jeder Menge eine maßgleiche Hülle, d. h. eine sie als Teilmenge enthaltende G_δ-Menge mit gleichem Funktionswert,

[1]) Bearbeitung einer im Wiener akad. Anz., *67* (16. I. 1930) erschienenen Notiz.

aber diese Tatsache ist eine Folge aus den übrigen Axiomen, während ein Analogon zum schärferen Axiom IV in den bisherigen Maßaxiomatiken nicht vorhanden war.

Während die Dimensionsaxiome im Systeme *aller* Teilmengen euklidischer Räume realisierbar sind (*jede* Menge eines euklidischen Raumes *„dimensionierbar"* ist), sind die bisherigen Maßaxiome nur im Systeme *gewisser* (der sogenannten *meßbaren*) Mengen realisierbar. Zur Lösung des wichtigen Problems *einer anschauungsgemäßen Messung aller Mengen euklidischer Räume* liegt es daher nahe, *die Forderungen an das Maß restlos an die Forderungen an die Dimension anzugleichen,* d. h. die Additivitätsforderungen IIa) und IIb) auch für das Maß auf Summen von *relativ abgeschlossenen* Mengen zu beschränken und statt dessen (weil durch die Axiome I, III, V im Verein mit diesem eingeschränkten Axiom II keine Mengenfunktion eindeutig festgelegt wird) eine dem Dimensionsaxiom IV entsprechende Forderung an das Maß zu stellen, welche das Maß beliebiger Mengen in irgend einer Weise auf das Maß kompakter Mengen zurückführt.

Die Beschränkung der Additivität des Maßes auf Summen relativ abgeschlossener Mengen liegt durchaus im Geiste der Dimensionstheorie; denn wenn bei Addition von zwei fremden nicht abgeschlossenen Mengen die *Dimension* der Summe die *Dimensionen* beider Summanden übertrifft, so kann nicht in natürlicher Weise gefordert werden, daß das *Maß* der Summe gleich der Summe der *Maße* der Summanden sein soll. Überhaupt zwingt die dimensionstheoretische Auffassungsweise dazu, die neuen Forderungen von vornherein so einzurichten, daß sie sich auf ein m-dimensionales Maß im n-dimensionalen euklidischen Raum beziehen, wie dies hinsichtlich der Lebesgueschen Forderungen bekanntlich von Carathéodory durchgeführt worden ist.

Indem ich mich hinsichtlich der neuen *impliziten* (axiomatischen) Einführung des Maßbegriffes hier auf die vorliegenden Andeutungen beschränke, versuche ich im folgenden vor allem eine den angegebenen Gesichtspunkten gemäße neue *explizite* Maßdefinition aufzustellen, welche sämtliche Mengen euklidischer Räume erfaßt, und entwickle zunächst die topologischen Grundlagen derselben für den Fall der Ebene.

II. Die topologischen Grundlagen der neuen Maßdefinition in der Ebene.

Von Alexandroff wurden für beliebigdimensionale euklidische Räume die beiden folgenden Sätze bewiesen:

(A_k). *Eine k-dimensionale kompakte Teilmenge eines euklidischen Raumes kann für jedes $\varepsilon > 0$ in einen k-dimensionalen Komplex ε-defomiert werden,* d. h. auf ihn stetig abgebildet werden derart, daß jeder Punkt von seinem Bildpunkt einen Abstand $< \varepsilon$ hat.

(B_k). *Eine k-dimensionale kompakte Teilmenge eines euklidischen Raumes kann nicht für jedes $\varepsilon > 0$ in einen weniger als k-dimensionalen Komplex ε-deformiert werden.*

Wir beweisen zunächst eine Verschärfung der Sätze A_0 und B_0, indem wir die Beschränkung auf kompakte Teilmengen euklidischer Räume fallen lassen und statt dessen beliebige (auch nicht abgeschlossene) Teilmengen beliebiger kompakter Räume in Betracht ziehen. Es gilt nämlich der Satz:

Damit die Teilmenge M eines kompakten Raumes nulldimensional sei, ist notwendig und hinreichend, daß M für jedes $\varepsilon > 0$ in eine endliche Menge ε-deformiert werden kann.

Notwendig: Man zerlege eine vorgelegte nulldimensionale Menge M bei gegebenem $\varepsilon > 0$ in endlich viele, paarweise fremde, relativ abgeschlossene nichtleere Teilmengen $M_1, M_2, \ldots M_n$, deren Durchmesser $< \varepsilon$ sind, wähle in jeder Menge M_i einen Punkt p_i und ordne jedem Punkt von M_i den Punkt p_i zu. Dann hat man eine ε-Deformation von M in die aus den Punkten $p_1, p_2, \ldots p_n$ bestehende endliche Menge. Hinreichend: Sei M eine mehr als nulldimensionale Menge eines kompakten Raumes. Es existiert (vgl. Dimensionstheorie, S. 207) ein (eventuell in M enthaltener) Punkt p des Raumes, so daß die Menge $M + p$ eine p enthaltende echte (d. h. nicht nur aus p bestehende) Quasikomponente besitzt. Es sei q ein von p verschiedener Punkt von M, welcher derselben Quasikomponente von $M + p$ angehört wie p, und es sei $\{p_k\}$ $(k = 1, 2, \ldots$ ad inf.) eine gegen p konvergente Folge von Punkten aus M. Bezeichnet 3ε den Abstand der Punkte p und q, so führt die Annahme der Existenz einer ε-Deformation von M in eine endliche Menge E auf einen Widerspruch. Denn bei einer ε-Deformation haben fast alle Punkte p_k einen Bildpunkt in E, der vom Bildpunkt des Punktes q (derselbe heiße e) verschieden ist. Betrachten wir nun die Zerlegung von E in die beiden fremden abgeschlossenen Mengen, von denen die eine aus dem Punkte e und die andere aus den übrigen Punkten von E besteht. Die Urbildmengen dieser beiden Mengen müssen fremd und abgeschlossen sein, die eine muß q enthalten, die andere fast alle Punkte p_k. Dies aber widerspricht der Annahme, daß p und q zu derselben Quasikomponente von $M + p$ gehören, d. h. bei jeder Zerlegung von $M + p$ in zwei relativ abgeschlossene fremde Mengen demselben Summanden angehören. Denn ein solcher Summand ist, da sein Komplement in $M + p$ abgeschlossen ist, in $M + p$ offen und enthält daher zugleich mit p fast alle Punkte p_k. Eine mehr als nulldimensionale Teilmenge eines kompakten Raumes kann also nicht für jedes $\varepsilon > 0$ in eine endliche Menge ε-deformiert werden, womit unser Satz bewiesen ist.

Jede eindimensionale beschränkte Teilmenge der Ebene (dieselbe mag abgeschlossen sein oder nicht) *ist für jedes $\varepsilon > 0$ in einen eindimensionalen Komplex ε-deformierbar.*

Sei M eine beschränkte eindimensionale Menge der Ebene. Wenn ihre abgeschlossene Hülle $\overline{M}$ eindimensional ist, folgt die Behauptung aus Satz A_1. Andernfalls wählen wir, was wegen der Eindimensionalität von M möglich ist, in der Menge $U = U\left(M; \frac{\varepsilon}{4}\right)$

eine dichte abzählbare Folge $\{p_k\}$ ($k=1, 2, \ldots$ ad inf.) von nicht zu M gehörigen Punkten. Wir wählen ferner eine Folge $\{\varepsilon_k\}$ von Zahlen >0, für die $\sum_{k=1}^{\infty} \varepsilon_k < \frac{\varepsilon}{8}$. Wir bilden nun um den Punkt p_1 einen Kreis K_1 von einem Radius $<\varepsilon_1$, der $\subset U$ ist, und bilden $U-p_1$ topologisch auf $U-K_1$ ab, derart, daß jeder Punkt eine Verrückung um höchstens $2\varepsilon_1$ erfährt. Wir bezeichnen den dem Punkt q von $U-p_1$ durch diese Deformation zugeordneten Punkt mit $\alpha_1(q)$. Angenommen, es sei bereits $U-\sum_{i=1}^{n-1} p_i$ topologisch abgebildet auf eine Menge $U-\sum_{i=1}^{n-1} K_i$, wo die K_i paarweise fremde abgeschlossene Kreise $\subset U$ sind, und zwar derart, daß jedem Punkt q von $U-\sum_{i=1}^{n-1} p_i$ ein Punkt $\alpha_{n-1}(q)$ in einem Abstand $<2\sum_{i=1}^{n-1} \varepsilon_i$ zugeordnet wird. Wir betrachten dann den Punkt $\alpha_{n-1}(p_n)$, welcher, da die Abbildung α_{n-1} topologisch ist, nicht in $\alpha_{n-1}(M)$ liegt, bilden um ihn einen abgeschlossenen Kreis K_n, dessen Radius $<\varepsilon_n$ und überdies so klein ist, daß K_n in U_n liegt und zu $\sum_{i=1}^{n-1} K_i$ fremd ist. Wir können nun

$$\alpha_{n-1}\left(U-\sum_{i=1}^{n} p_i\right) \text{ auf } U-\sum_{i=1}^{n} K_i$$

topologisch abbilden derart, daß jedem Punkt $\alpha_{n-1}(q)$ von $U-\sum_{i=1}^{n-1} K_i$ ein Bildpunkt $\alpha_n(q)$ in einem Abstand $<2\varepsilon_n$ zugeordnet wird, welcher also von q einen Abstand $<2\sum_{i=1}^{n} \varepsilon_i$ hat. Wir haben damit eine Folge topologischer, also stetiger Abbildungen $\{\alpha_n\}$ definiert, welche gleichmäßig konvergiert und daher eine stetige Grenzabbildung liefert, die $U-\sum_{i=1}^{\infty} p_i$ auf eine Menge $U-\sum_{i=1}^{\infty} K_i$ abbildet derart, daß jeder Punkt der Menge $U-\sum_{i=1}^{\infty} p_i$ von seinem Bildpunkt einen Abstand $<2\sum_{i=1}^{\infty} \varepsilon_i$ hat. Die Menge $U-\sum_{i=1}^{\infty} K_i$ ist, wie man leicht einsieht, nirgends dicht in der Ebene. Daher ist ihre abgeschlossene Hülle eindimensional. Wir haben also M $\frac{\varepsilon}{2}$-deformiert in eine Teilmenge der eindimensionalen kompakten Menge H und können diese nach Satz A_1 $\frac{\varepsilon}{2}$-deformieren in einen eindimensionalen Komplex, wodurch wir eine ε-Deformation von M in eine Teilmenge eines eindimensionalen Komplexes gewonnen haben.

Eine (auch nicht abgeschlossene) *zweidimensionale Menge der Ebene* enthält eine Quadratfläche und *kann* daher nach Satz B_2 *nicht für jedes* $\varepsilon > 0$ *in eine Teilmenge eines eindimensionalen Komplexes* ε*-deformiert werden.* Hingegen ist die abgeschlossene Hülle einer beschränkten zweidimensionalen Menge M der Ebene eine kompakte zweidimensionale Menge und kann daher nach Satz A_2 für jedes $\varepsilon > 0$ in einen zweidimensionalen Komplex ε-deformiert werden, so daß also *eine beschränkte zweidimensionale Menge M für jedes $\varepsilon > 0$ in eine Teilmenge eines zweidimensionalen Komplexes ε-deformiert werden kann.*

Zusammenfassend können wir also sagen: *In der Ebene kann jede beschränkte Menge für jedes $\varepsilon > 0$ in eine Teilmenge eines gleichdimensionalen Komplexes ε-deformiert werden, ist aber nicht für jedes $\varepsilon > 0$ in eine Teilmenge eines niedrigerdimensionalen Komplexes ε-deformierbar.*

III. Eine Definition des Maßes.

Wir nehmen den im vorigen Abschnitt für $n=2$ bewiesenen Satz an, daß jede Teilmenge eines n-dimensionalen euklidischen Raumes für jedes $\varepsilon > 0$ in eine Teilmenge eines gleichdimensionalen Komplexes ε-deformiert werden kann, aber nicht für jedes $\varepsilon > 0$ in eine Teilmenge eines niedrigerdimensionalen Komplexes ε-deformierbar ist. Ist dann eine k-dimensionale Menge M vorgelegt und eine Zahl $\varepsilon > 0$ gegeben, so kann M auf verschiedene Weisen in Teilmengen k-dimensionaler Komplexe ε-deformiert werden. Jeder dieser k-dimensionalen Komplexe hat einen k-dimensionalen Inhalt, der elementargeometrisch (d. h. auf Grund der Normierung des Maßes für k-dimensionale Simplexe durch Addition endlichvieler solcher Simplexe) berechnet werden kann. Wir bezeichnen mit $\mu_\varepsilon^k(M)$ die untere Schranke der Maße aller k-dimensionalen Komplexe, welche Teilmengen enthalten, in die M ε-deformiert werden kann, und nennen diese Zahl *das k-dimensionale ε-Maß von M.* Als k-dimensionales Maß von M bezeichnen wir die Zahl

$$\mu^k(M) = \lim_{\varepsilon=0} \mu_\varepsilon^k(M).$$

7. Kolloquium (18. I. 1930).

v. Neumann (Berlin): *Über Maßtheorie.*

Beer: Es sei im R_n eine kompakte Menge M eindeutig und stetig auf eine Menge M' abgebildet. Besitzt der Punkt $x_1, x_2, \ldots, x_n$ von M das Bild $x_1', x_2', \ldots, x_n'$ in M', so wollen wir ihm für die natürliche Zahl k den Punkt des R_{2n} mit den Koordinaten $x_1', x_2', \ldots, x_n'$, $\frac{x_1}{k}, \frac{x_2}{k}, \ldots, \frac{x_n}{k}$ zuordnen. Bezeichnet M_k die Menge aller den Punkten von M auf diese Art zugeordneten Punkte des R_{2n}, so ist die Abbildung von M auf M_k topologisch. Wenn k die natürlichen Zahlen durchläuft, so konvergieren die Abbildungen von M auf M_k gleichmäßig gegen die Abbildung von M auf M'.

Nöbeling: Die Bemerkung von Beer liefert für den Hilbertschen Raum H folgendes: Ist eine kompakte Menge $M \subset H$ eindeutig und stetig auf eine Menge $M' \subset H$ abgebildet, so existieren zu M homöomorphe Mengen $M_k \subset H$ derart, daß die topologischen Abbildungen von M auf M_k gleichmäßig gegen die Abbildung von M auf M' konvergieren. Diese Tatsache ermöglicht die Konstruktion einer *n-dimensionalen Universalmenge im Hilbertschen Raum.*

8. Kolloquium (20. I. 1930).

Beer: Bericht über die Theorie der Fréchetschen Dimensionstypen.

G. T. Whyburn zeigt, daß Kontinua existieren, deren sämtliche Teilkontinua die Ebene zerlegen, und daß jedes solche die Ebene erblich zerlegende Kontinuum in jedem seiner Teilkontinua topologisch enthalten ist, also einen Fréchetschen Dimensionstypus erblich besitzt, so wie die Strecke. Jedes solche Kontinuum enthält ferner ein die Ebene erblich zerlegendes Teilkontinuum, welches überdies die gemeinsame Begrenzung zweier Gebiete ist. Jedes Kontinuum K mit diesen beiden Eigenschaften kann oberhalb stetig zerlegt werden in ein System von abgeschlossenen zusammenhängenden Mengen (darunter abzählbarviele mehrpunktige Mengen), in bezug auf welche K ein topologischer Kreis ist. Es enthält überdies zwei zueinander nicht homöomorphe Teilkontinua. Es enthält kein unzerlegbares Kontinuum, aber auch keinen Bogen, so wie ein Knastersches Kontinuum J (C. R. I. Congr. d. pays Slaves, 1929, S. 290).

9. Kolloquium (5. II. 1930).

L. Whyburn: Ist in einer regulären Kurve K eine dichte Menge D gegeben, welche alle lokalen Zerschneidungspunkte von K, deren Ordnung > 2 ist, enthält, so hat jeder Punkt p von K in bezug auf D dieselbe Ordnung, wie in bezug auf K, d. h. wenn p ein Punkt n^{ter} Ordnung von K ist, so ist p in beliebig kleinen Umgebungen enthalten, deren Begrenzungen aus höchstens n der Menge D angehörigen Punkten bestehen. Da die Menge aller lokalen Zerschneidungspunkte, deren Ordnung > 2 ist, abzählbar ist (vgl. G. T. Whyburn, Monatshefte *36*, S. 305), so kann die Menge als abzählbare Menge gewählt werden.

Menger: *Eine neue Definition der Bogenlänge.* Die Länge von B ist die obere Schranke aller Zahlen $\lambda(E)$, wo E alle endlichen Teilmengen von B durchläuft und $\lambda(E)$ die Länge des kürzesten Polygones durch E bezeichnet (Ausführung: Math. Ann., *103*).

Beer: Es sei irgend eine bestimmte „ausgezeichnete" Folge $\{E_n\}$ ($n = 1, 2, \ldots$ ad inf.) von endlichen Teilmengen von B gegeben, d. h. eine Folge, für welche, wenn E_n in B δ_n-dicht ist, $\lim\limits_{n=\infty} \delta_n = 0$ gilt. Dann ist die Länge von B gleich $\lim\limits_{n=\infty} \lambda(E_n)$.

Menger: Die Länge des Bogens B erweist sich (vgl. Math. Ann., *103*) als die obere Schranke der Zahlen $\mu(E)$, wo E die endlichen Teilmengen von B durchläuft und $\mu(E)$ die Länge des kürzesten zusammenhängenden Graphen, dessen Scheitel in E liegen, bezeichnet. Ist E eine endliche Teilmenge des Raumes R, so bezeichnen wir mit $\nu_R(E)$ die untere Schranke der Längen aller zusammenhängenden Graphen, welche E enthalten (aber auch nicht in E gelegene Scheitel besitzen können). Solcher Graphen gibt es unendlichviele, während nur endlichviele Graphen existieren, deren Scheitel in E liegen. Es ist also $\nu_R(E) = \inf\limits_{E \subset E' \subset R} \mu(E')$. Es gilt stets $\nu_R(E) \leqq \mu(E)$, aber nicht immer das Gleichheitszeichen. Es fragt sich, ob die Länge $l(B)$ eines Bogens B die obere Schranke der Zahlen $\nu_R(E)$ für alle Teilmengen von B ist, ob also

$$(+) \qquad l(B) = \sup_{E \subset B} \inf_{E \subset E' \subset R} \mu(E')$$

gilt? Im Falle $R = B$ ist die Frage wegen $l(B) = \sup\limits_{E \subset B} \mu(E)$ offenbar zu bejahen. Wenn sie nicht für jeden metrischen Raum R zu bejahen sein sollte, so würde also die rechte Seite von (+) nicht nur von B, sondern auch von R abhängen.

Wir bezeichnen als *Botenproblem* (weil diese Frage in der Praxis von jedem Postboten, übrigens auch von vielen Reisenden zu lösen ist) die Aufgabe, für endlichviele Punkte, deren paarweise Abstände bekannt sind, den kürzesten die Punkte verbindenden Weg zu finden. Dieses Problem ist natürlich stets durch endlichviele Versuche lösbar. Regeln, welche die Anzahl der Versuche unter die Anzahl der Permutationen der gegebenen Punkte herunterdrücken würden, sind nicht bekannt. Die Regel, man solle vom Ausgangspunkt erst zum nächstgelegenen Punkt, dann zu dem diesem nächstgelegenen Punkt gehen usw., liefert im allgemeinen nicht den kürzesten Weg.

10. Kolloquium (19. II. 1930).

Tarski (Warschau): *Neue Resultate und unentschiedene Probleme der Kardinalzahlarithmetik.*

Die Hauptrichtungen, in denen sich die Kardinalzahlarithmetik gegenwärtig entwickelt, beziehen sich erstens auf eine möglichst weitgehende Begründung der Theorie ohne Verwendung des Auswahlaxiomes und zweitens auf die Feststellung der Unentbehrlichkeit des Auswahlaxiomes zum Beweise gewisser Sätze. Beispielsweise ist der Satz: „Aus $2\mathfrak{m} > \mathfrak{m} + \mathfrak{p}$ folgt $\mathfrak{m} > \mathfrak{p}$“ ohne das Auswahlaxiom beweisbar, dagegen der Satz: „Aus $2\mathfrak{m} < \mathfrak{m} + \mathfrak{p}$ folgt $\mathfrak{m} < \mathfrak{p}$“ mit dem Auswahlaxiom äquivalent. — Ungelöste Probleme der Kardinalzahltheorie beziehen sich vor allem auf die Frage nach der Cantorschen *Alefhypothese,* die sich auf zwei verschiedene Arten formulieren läßt:

H_1: Es gibt kein x, so daß $\mathfrak{m} < x < 2^{\mathfrak{m}}$.

H_2: $2^{\aleph_\alpha} = \aleph_{\alpha+1}$.

Die Hypothese H_1 ist äquivalent mit H_2 + Auswahlaxiom. Aus H_1 folgt das Auswahlaxiom. Tarski vermutet, daß das Auswahlaxiom schwächer ist als H_1. — Weitere Probleme betreffen die Existenz großer Kardinalzahlen, insbesondere der von Kuratowski *unerreichbar* genannten Zahlen. Bemerkenswert ist die logische Einfachheit und der prinzipielle Charakter der unentschiedenen Probleme der Mengenlehre im Gegensatz zu den bekannten unentschiedenen Problemen der klassischen Mathematik. Vermutlich werden sie erst auf Grund metamathematischer Untersuchungen eine definitive Entscheidung finden.

11. Kolloquium (20. II. 1930).

Tarski: *Über einige fundamentale Begriffe der Metamathematik.* (Vgl. Monatshefte f. Math. u. Phys., *37,* S. 361.)

12. Kolloquium (21. II. 1930).

Tarski: *Untersuchungen über den Aussagenkalkül.*

Zunächst lassen sich für den Aussagenkalkül die Begriffe der sinnvollen Aussage und der Folgerung präzisieren. Eine Aussagenmenge heiße ein *abgeschlossenes System,* wenn sie neben jeder in ihr enthaltenen Aussage auch alle Folgerungen derselben enthält. Zur Konstruktion solcher abgeschlossener Systeme stehen zwei Methoden zur Verfügung. Erstens kann man ein *Axiomensystem,* d. h. eine beliebige (meist endliche) Aussagenmenge X angeben und die Menge aller Folgerungen von X, d. i. das kleinste abgeschlossene System über X, bilden. Die zweite Methode besteht in der Konstruktion einer „Werttabelle", der sog. *logischen Matrix;* die Menge aller eine solche Matrix erfüllenden Aussagen ist unter einer gewissen Voraussetzung über die Matrix ein abgeschlossenes System. Beispielsweise können im gewöhnlichen Aussagenkalkül die Implikation und Negation durch folgende Matrizen definiert werden:

$\supset$	0	1
0	1	1
1	0	1

$\sim$	
0	1
1	0

Die erste Matrix bedeutet folgendes: Jeder Aussage sei der Funktionswert 1 oder 0 zugeordnet, je nachdem die Aussage wahr oder falsch ist. Der Funktionswert der Aussage „aus a folgt b" befindet sich in der quadratischen Matrix an der Schnittstelle der Zeile, vor welcher der Funktionswert der Aussage a steht, und der Spalte, oberhalb welcher der Funktionswert der Aussage b steht. — Die Definition von abgeschlossenen Systemen des Aussagenkalküls durch Matrizen ist, wie Lindenbaum gezeigt hat, äquivalent mit Definitionen durch Axiomensysteme. Ein (zum Unterschiede von den Systemen von Frege und Whitehead-Russell *unabhängiges* Axiomensystem von Łukasiewicz besteht aus den drei Sätzen (in der Symbolik der Principia):

$$p \supset q \,.\supset:\, q \supset r \,.\supset.\, p \supset r, \quad p \supset .\sim p \supset q, \quad \sim p \supset p \,.\supset p.$$

Wajsberg hat bewiesen, daß in jedem Axiomensystem des gewöhnlichen Aussagenkalküls mindestens drei verschiedene Aussagenvariablen vorkommen müssen. Tarski hat ein aus dem einzigen Axiom

$$p \supset .\, q \supset p :\supset ::: r \supset .\, s \supset t : \supset : r \supset s \,.\supset.\, r \supset t :.\supset :.\sim s \supset \sim r \,.\supset.\, r \supset s : \supset$$
$$\supset u :: \supset u :\cdot: \supset v ::: .\supset .\, w \supset v.$$

bestehendes Axiomensystem angegeben. Eine erhebliche Vereinfachung der Schreibweise, nämlich eine Vermeidung von Klammern und Punkten in den Formeln[1]), ergibt sich, wenn man mit Łukasiewicz Cpq statt $p \supset q$ und Np statt $\sim p$ schreibt. Z. B. lauten in dieser Schreibweise die drei oben angeführten Axiome

$$C\,Cpq\,C\,Cqr\,Cpr, \quad Cp\,C\,Npq, \quad C\,C\,N\,p\,p\,p.$$

Eine interessante Klasse abgeschlossener Systeme bilden die von Łukasiewicz stammenden sog. *mehrwertigen Systeme des Aussagenkalküls.* Im dreiwertigen Aussagenkalkül wird jeder Aussage einer der Werte 0, $\frac{1}{2}$, 1 zugeordnet (nicht bloß wie beim gewöhnlichen Aussagenkalkül einer der Werte 0, 1), Implikation und Negation durch die beiden Matrizen

$\supset$	0	$\frac{1}{2}$	1
0	1	1	1
$\frac{1}{2}$	$\frac{1}{2}$	1	1
1	0	$\frac{1}{2}$	1

$\sim$	
0	1
$\frac{1}{2}$	$\frac{1}{2}$
1	0

erklärt entsprechend der oben beschriebenen Matrizendefinition im gewöhnlichen (zweiwertigen) Aussagenkalkül. Die Sätze des dreiwertigen Aussagenkalküls bilden eine Teilklasse der Sätze des zweiwertigen Kalküls. Viele Sätze dieses letzteren, die von Intuitionisten angegriffen werden, z. B. die Reductio ad absurdum, fehlen im dreiwertigen Kalkül, weshalb Łukasiewicz vorschlägt, auch die zum dreiwertigen Kalkül gehörige Mathematik zu studieren. Entsprechend kann ein n-wertiger, ja ein ω-wertiger Aussagenkalkül definiert werden, welch letzterer vielleicht auf die Wahrscheinlichkeitsrechnung anwendbar ist.

[1]) Klammern in logischen Formeln wurden für unvermeidbar gehalten. Vgl. Wittgenstein, Tractatus logico-philosophicus, London 1922, S. 126, wo es heißt: „Bedeutungsvoll ist die scheinbar unwichtige Tatsache, daß die logischen Scheinbeziehungen, wie $\vee$ und $\supset$, der Klammern bedürfen — im Gegensatz zu den wirklichen Beziehungen."

13. Kolloquium (14. III. 1930).

Knaster (Warschau): *Über eindimensional unvergleichbare Kontinua.* Nennt man mit Menger zwei n-dimensionale Kontinua *k-dimensional unvergleichbar,* wenn keine k-dimensionale Teilmenge eines der beiden Kontinua mit einer Teilmenge des anderen homöomorph ist, so gibt es unter den gegenwärtig bekannten ebenen Gebilden vier paarweise eindimensional unvergleichbare Kontinua, und zwar:

1. Die Strecke.

2. Das von mir (Fund. Math., *3*, S. 275) konstruierte erblich unzerlegbare Kontinuum K_3.

3. Die von mir (C. R. Congr. Slav., Warschau 1929, S. 290) konstruierte Kurve J, die weder einen Bogen noch ein unzerlegbares Kontinuum enthält und die Ebene nicht zerschneidet.

4. Das die Ebene erblich zerschneidende Kontinuum M von G. T. Whyburn (Am. Journ. of Math., *52*, 1930, S. 319, vgl. auch 8. Kolloquium).

Da offenbar kein Teilkontinuum einer dieser Mengen mit einem Teilkontinuum einer der drei anderen homöomorph ist, so ist nur zu zeigen, daß auch keine diskontinuierliche eindimensionale Teilmenge einer dieser Mengen mit einer Teilmenge einer der drei anderen homöomorph ist, was sich aus folgenden Sätzen ergibt:

Jede eindimensionale Teilmenge

1. der Strecke enthält ein beiderseits stetiges Linienstück;

2. von K_3 enthält kein Linienstück;

3. von J enthält einseitig stetige Linienstücke und nur solche;

4. von M enthält beiderseits unstetige Linienstücke und nur solche.

(Dabei heißt Linienstück [nach Vietoris, Monatshefte f. Math. u. Phys., *31*, S. 179, vgl. Knaster, Fund. Math., *10*, S. 276] eine Menge, die irreduzibel ist in bezug auf die Eigenschaft, zwei feste Punkte p_0 und p_1 zu enthalten und zusammenhängend zu sein. Das Linienstück $L = \sum_{0 \leq x \leq 1} (p_x)$ heißt im Punkt p_x linksseitig stetig, wenn aus $x = \lim_{n=\infty} x_n$ und $x_n < x$ stets $p_x = \lim_{n=\infty} p_{x_n}$ folgt. Das beiderseits stetige Linienstück ist abgeschlossen und daher nach Lennes ein Bogen.)

Menger: Gibt es eine beliebige endliche oder gar unendliche Anzahl von paarweise eindimensional unvergleichbaren Kontinua?

Knaster weist auf das Problem hin, Eigenschaften einer Menge A, die als Durchschnitt $\prod_{k=1}^{\infty} A_k$ offener Mengen gegeben ist, aus den Eigenschaften und gegenseitigen Beziehungen der A_k herzuleiten. Z. B. ist (Knaster, Fund. Math., *3*, S. 248) die Unzerlegbarkeit von A eine Folge gewisser Oszillationen von A_k in A_{k-1}. Dies ist eines der wenigen Beispiele einer Lösung des Problems.

14. Kolloquium (7. V. 1930).

Menger: *Bemerkungen von Mazurkiewicz und Tarski über die Axiomatik des Dimensionsbegriffes.*

Läßt man in der Dimensionsaxiomatik in der Forderung, daß die Dimension einer Summe von abzählbarvielen (bzw. endlichvielen) abgeschlossenen Mengen gleich dem Maximum der Dimensionen der Summanden sein soll, die Beschränkung auf abgeschlossene Mengen sowie das Kompaktifizierungsaxiom weg, so erhält man (vgl. Kolloquium 6) ein dem Lebesgueschen System von Maßaxiomen völlig analoges System von Axiomen, welche wir als *falsche Dimensionsaxiome* bzw. (wenn die Summenforderung bloß für endlichviele Summanden postuliert wird) als *falsche restringierte Dimensionsaxiome* bezeichnen wollen. Die falschen Dimensionsaxiome sind im R_1 für das System aller Teilmengen realisierbar, wenn man der leeren Menge den Wert -1, den abzählbaren Mengen den Wert 0, den unabzählbaren Mengen den Wert 1 als Dimension zuordnet. Für den $R_n (n > 1)$ ist aber schon das falsche restringierte System von Dimensionsaxiomen im System aller Teilmengen unlösbar. Denn der R_2 ist Summe von drei nulldimensionalen Mengen, also von drei Mengen, die mit Teilmengen des R_1 homöomorph sind und daher einen Funktionswert $\leqq 1$ haben, während die Ebene den Funktionswert 2 haben soll. Auch das dem Maßproblem analoge Dimensionsproblem wird also erst durch Beschränkung auf relativ abgeschlossene Summanden allgemein lösbar.

Menger: *Über Mengensysteme.* Wenn es n bzw. $\aleph_0$ paarweise eindimensional unvergleichbare Kurven gibt (vgl. 13. Kolloquium), so gibt es einen kompakten eindimensionalen Raum, der mindestens $2^n + 1$ bzw. mindestens $2^{\aleph_0}$ „*ausgezeichnete*" Systeme von Teilmengen besitzt, d. h. monotone, F_σ-additive, topologische und kompaktifizierbare Mengensysteme (vgl. die Schlußweise Monatshefte f. Math. u. Phys., *36*, S. 207). Ein ausgezeichnetes Mengensystem $\mathfrak{S}$ eines Raumes heiße *vollkommen*, wenn jedes andere ausgezeichnete Mengensystem $\mathfrak{S}'$ des Raumes das System $\mathfrak{S}$ entweder als Teilsystem enthält oder in $\mathfrak{S}$ als Teilsystem enthalten ist. Man könnte nun erwarten, eine Dimensionsaxiomatik für beliebige kompakte eindimensionale Räume analog jener des R_1 auf Grund der Behauptung zu erhalten, daß jeder eindimensionale kompakte Raum genau drei vollkommene Mengensysteme enthält. Diese Behauptung ist aber unrichtig. Man bestätigt nämlich leicht den Satz: *Wenn ein eindimensionaler kompakter Raum genau vier ausgezeichnete Mengensysteme besitzt, so ist jedes von ihnen vollkommen.* Anderseits *existieren* eindimensionale kompakte Räume, die *genau vier* ausgezeichnete Systeme besitzen, wie folgendes Beispiel zeigt: Man addiere zum Knasterschen Kontinuum K_3 abzählbarviele Strecken, deren Summe erstens keinen topologischen Kreis enthält, zweitens in den Punkten einer abzählbaren in K_3 dichten Menge Verzweigungspunkte von paarweise verschiedener Ordnung besitzt und drittens, zu K_3 addiert, einen kompakten Raum R

ergibt. Beer beweist, daß in R jede kompakte Menge, die die Summe der abzählbarvielen Strecken topologisch enthält, ein Teilkontinuum von K_3 enthält. Demnach enthält R außer den drei Dimensionssystemen noch genau ein viertes ausgezeichnetes Mengensystem, nämlich das System aller Mengen, welche Summe einer Teilmenge von K_3 und einer höchstens nulldimensionalen Menge sind.

Nöbeling: Sind die Dimensionssysteme beliebiger kompakter Räume kompaktifizierbar?

Menger: Der Inhalt einer konvexen Fläche des R_3 scheint gleich der oberen Schranke der Flächeninhalte der Minimaltriangulierungen aller endlichen Teilmengen der Fläche zu sein.

15. Kolloquium (14. V. 1930).

Gödel: *Über die Vollständigkeit der Axiome des logischen Funktionenkalküls.* (Vgl. Monatshefte f. Math. u. Phys., *37,* S. 349.)

16. Kolloquium (28. V. 1930).

Wald: *Axiomatik des metrischen Zwischenbegriffes.*

Damit eine für die Punktetripel jedes metrischen Raumes definierte Dreipunkterelation $a\,b\,c$ mit der Mengerschen Zwischenrelation identisch sei, d. h. dann und nur dann gelte, wenn $ab + bc = ac$ und $a \neq b \neq c$ gilt, ist notwendig und hinreichend, daß die Relation folgenden Forderungen genügt:

1. Aus abc und $ab = a'b'$, $bc = b'c'$, $ca = c'a'$ folgt $a'b'c'$.
2. Aus abc folgt cba.
3. Aus abc folgt non acb.
4. Aus abc und acd folgt abd und bcd.
5. Für je zwei Punkte a, b des metrischen Raumes R ist die Menge $Z_R(a, b)$ bestehend aus a, b und allen Punkten x, für welche axb gilt, abgeschlossen in R.
6. Ist R ein aus zwei Punkten a, b bestehender metrischer Raum, so existiert ein metrischer Raum R', der ein mit a, b kongruentes Punktepaar a', b' und außerdem einen Punkt x' enthält, so daß $a'\,x'\,b'$ gilt.

Taussky: In einem beliebigen halbmetrischen Raum kann man ebenfalls festsetzen: abc soll für je drei Punkte gleichbedeutend sein mit $ab + bc = ac$, $a \neq b \neq c$. Diese Relation besitzt die Eigenschaften 1—6, bloß gilt statt 4 nur die schwächere Eigenschaft: Aus abc, acd und abd folgt bcd.

Wald: Diese sechs Eigenschaften sind für die Zwischenrelation in halbmetrischen Räumen nicht charakteristisch. Sie kommen z. B. auch folgender Relation zu: Es sei abc gleichbedeutend mit

$$ab\,.\,bc = ac,\ a \neq b \neq c,\ ab \neq 1,\ bc \neq 1,\ ca \neq 1.$$

Menger: Man könnte als *Dreieckskontraungleichung* die Beziehung $ab + bc \leq ac$ bezeichnen und einen halbmetrischen Raum

2

kontrametrisch nennen, wenn für je drei seiner Punkte a, b, c in einer Anordnung (in mehr als einer Anordnung ist dies für drei paarweise verschiedene Punkte unmöglich) die Dreieckskontraungleichung gilt.

Priebsch: *Ein mehrdimensionaler Zwischenbegriff.*

Wir sagen von den m Punkten $a_1, a_2, \ldots, a_m$ des R_k, sie liegen zwischen den n Punkten $b_1, b_2, \ldots, b_n$ des R_k $(m + n \leqq k + 2)$ und schreiben $a_1, a_2, \ldots, a_m ; b_1, b_2, \ldots, b_n$, wenn das durch die m ersten Punkte bestimmte Simplex mit dem durch die n zweiten Punkte bestimmten Simplex genau einen Punkt gemein hat, welcher für beide Simplexe innerer Punkt ist (dabei gilt der einzige Punkt eines nulldimensionalen Simplexes als innerer Punkt). Dieser höherdimensionale Zwischenbegriff besitzt folgende Eigenschaften:

I. Aus

$$a_1, a_2, \ldots, a_m ; b_1, b_2, \ldots, b_n \text{ folgt } b_1, b_2, \ldots, b_n; a_1, a_2, \ldots, a_m.$$

II. Ist $i_1, i_2, \ldots, i_m$ eine Permutation der Zahlen $1, 2, \ldots, m$ und $j_1, j_2, \ldots, j_n$ eine Permutation der Zahlen $1, 2, \ldots, n$, so folgt aus $a_1, a_2, \ldots, a_m; b_1, b_2, \ldots, b_n$, daß $a_{i_1}, a_{i_2}, \ldots, a_{i_m}, b_{j_1}, b_{j_2}, \ldots, b_{j_n}$ gilt.

III. Jede aus $a_1, a_2, \ldots, a_m; b_1, b_2, \ldots, b_n$ durch eine nicht auf I und II zurückführbare Permutation der $m + n$ Punkte hervorgehende Zwischenbeziehung ist falsch. (Insbesondere gibt es keine zwischen zwei echten Teilmengen der $m + n$ Punkte bestehende Zwischenrelation.)

IV. Je $n + 2$ Punkte des R_n zerfallen in zwei Punktgruppen, die zueinander in einer Zwischenrelation stehen.

Für $n = 2$ erhält man, wenn man zu I, II, III, IV folgende Sätze hinzunimmt

1. Aus $a, b, c; d$ und $a; b, d, e$ folgt $a; b, c, e$,

2. Aus $a, b; c, d$ und $a, b, d; e$ folgt $a, b; c, e$,

3. Aus $a, b; c, d$ und $a, b; d, e$ folgt $a, b, c; e$ oder $a, c; b, e$ oder $b, c; a, e$ oder $c; a, b, e$ oder $a, c; e$ oder $c; a, e$ oder $b, c; e$ oder $c; b, e$ oder $c; e$,

4. Aus $a, c; b, d$ und $a, d; b, e$ folgt $a, c; b, e$,

5. Aus $a; b, c, d$ und $a, b, d; e$ folgt $a, b; c, e$ oder $a; b, c, e$ oder $a; c, e$,

ein unabhängiges System von Forderungen, welches auch vollständig ist in dem Sinne, daß aus ihm jeder für die Ebene gültige Satz der obigen Art folgt. Für $n = 0$ ist der Zwischenbegriff mit dem Identitätsbegriff äquivalent. Aus den Huntington-Klineschen Axiomen, welche die lineare Anordnung durch Axiome charakterisieren, die analog zum Falle $n = 1$ der obigen Sätze sind, erhält man bekanntlich durch Hinzunahme von Separabilitäts- und Vollständigkeitsforderungen die Kennzeichnung der Geraden. *Problem:* Man ergänze die obigen Sätze derart, daß jedes ihnen genügende

System von Dingen auf die Ebene topologisch bzw. ähnlich (d. h. zwischen-treu) abgebildet werden kann, wobei als Umgebungen jene Mengen gelten, welche zwischen irgend drei Punkten a, b, c liegen, für welche keine lineare Zwischenrelation besteht.

17. Kolloquium (17. VI. 1930).

Grell (Jena): *Ein Beweis des Dedekindschen Fundamentalsatzes* von der eindeutigen Zerlegbarkeit der Ideale in Primideale in algebraischen Zahlkörpern, der sich aus den Arbeiten von E. Noether (Math. Ann., *96*), kombiniert mit Artins Untersuchungen zur Arithmetik hyperkomplexer Systeme (Abh. d. Hamb. math. Sem. *5*, § 2) ergibt, indem man in der letzteren Arbeit die Entwicklungen auf den kommutativen Fall spezialisiert.

18. Kolloquium (18. VI. 1930).

Grell: *Allgemeine Verzweigungstheorie kommutativer und nichtkommutativer hyperkomplexer Systeme.* Verzweigungstheorie maximaler Ordnungen in Dedekindschen hyperkomplexen Systemen nach E. Noether, Brandt, Grell und Verzweigungstheorie beliebiger Ordnungen endlicher algebraischer Zahlkörper von Grell.

19. Kolloquium (25. VI. 1930).

Grell: *Allgemeine Idealtheorie.*

20. Kolloquium (2. VII. 1930).

Nöbeling: *Eine Fixpunkteigenschaft der Baumkurven.*

Unter den im kleinen zusammenhängenden Kurven sind die Baumkurven dadurch gekennzeichnet, daß sie bei jeder stetigen Abbildung einen Fixpunkt besitzen. (Scherrer hatte, Math. Zeitschr., *24*, S. 125, die Existenz von Fixpunkten bei jeder *topologischen* Abbildung eines Baumes auf eine Teilmenge bewiesen. Diese Eigenschaft ist aber für Bäume nicht charakteristisch, wie man am Beispiel der Lemniskate erkennt.) Zum Beweise sei *erstens* ein Baum B auf $B' \subset B$ eindeutig und stetig abgebildet. Ist dann p ein Punkt von mindestens zweiter Ordnung, der kein Fixpunkt ist, so existiert, da p Zerlegungspunkt von K ist, ein Punkt q, der durch p vom Bildpunkt p' von p getrennt wird. Es sei A die Menge aller Punkte p, welche ihr Bild von diesem festen Punkt q trennen. Ist a irgend ein Punkt von A, so liegen auch alle hinreichend nahe bei a gelegenen Punkte des in a' und q endenden Teilbogens von B in A. Die Menge A ist also in jedem ihrer Punkte von mindestens zweiter Ordnung. Es sei A' eine Komponente von A. Die Menge $\bar{A}'$ ist ein Baum und besitzt daher einen von q verschiedenen Endpunkt f. Dieser Punkt f liegt nicht in A, denn sonst müßte er als Endpunkt eines ganz in A' gelegenen Bogens in A' liegen, was nicht möglich ist, da die Menge A' in jedem ihrer Punkte von mindestens zweiter Ordnung ist. Dieser Punkt f ist nun Fixpunkt. Andernfalls würde er, da er nicht in A liegt, seinen Bildpunkt f' von q nicht trennen.

2*

Dann würden auch alle hinreichend nahe bei f gelegenen Punkte ihre Bildpunkte von q nicht trennen. Dies aber ist unmöglich, da f Häufungspunkt von A ist. Der Punkt f ist also Fixpunkt.

Sei *zweitens* eine im kleinen zusammenhängende Kurve R gegeben, die einen topologischen Kreis K enthält. Durch eine Modifikation der Mengerschen Methode für die Einbettung von Kurven in den R_3 (vgl. Dimensionstheorie, S. 296) kann man die Kurve R derart auf eine Kurve R' des R_3 topologisch abbilden, daß erstens der Kreis K' bestehend aus den Punkten $x^2+y^2=1$, $z=0$ das Bild von K ist und daß zweitens die z-Achse Z zu R' fremd ist. Nun läßt sich die Menge R_3-Z und damit auch R' eindeutig und stetig so auf K' abbilden, daß jeder Punkt von K' Fixpunkt dieser Abbildung ist. Führt man nun noch eine kleine Drehung des Kreises K' aus, so entsteht eine fixpunktfreie Abbildung von R' auf K' und damit eine fixpunktfreie Abbildung von R auf K.

Nöbeling: *Eine reguläre Kurve, die jede Kurve beschränkter Ordnung topologisch enthält.*

Es existieren (vgl. die Vermutung von Kolloquium 4) für jedes natürliche $n>1$ zwei ganzzahlige Funktionen $f_n(k)$ und $g_n(k)$ der ganzen Zahlen k derart, daß für jede Kurve K von höchstens n^{ter} Ordnung eine erzeugende Doppelfolge $\{A_m^k\}$ $(m=1, 2, \ldots, m_k$, $k=1, 2, \ldots$ ad. inf.) von Stücken mit endlichen Begrenzungen existiert, in welcher für jedes k die Beziehung $m_k \leq f_n(k)$ gilt und höchstens $g_n(k)$ Punkte von K in mehr als einer der Mengen A_m^k $(m=1, 2, \ldots, m_k)$ enthalten sind. Für jedes natürliche s gibt es endlichviele Kurven höchstens n^{ter} Ordnung, so daß die Systeme $\{A_m^k\}$ $(m=1, 2, \ldots, m_k, k=1, 2, \ldots, s)$ für jede Kurve höchstens n^{ter} Ordnung mit dem entsprechenden System einer dieser endlichvielen Kurven in bezug auf Inzidenzrelationen äquivalent sind. Durch Konstruktion einer erzeugenden Doppelfolge, die für jedes s zu jedem dieser endlichvielen Systeme ein äquivalentes enthält, gelangt man (vgl. die Fragestellung von Kolloquium 2) zu einer regulären Kurve K_n, die zu jeder Kurve von höchstens n^{ter} Ordnung eine homöomorphe Teilmenge enthält. Durch Addition der Kurven K_n $(n=1, 2, \ldots$ ad inf.) konstruiert man nun leicht *eine reguläre Kurve K, welche für beliebiges n zu jeder Kurve von höchstens n^{ter} Ordnung eine homöomorphe Teilmenge enthält.*

Menger: *Probleme der allgemeinen metrischen Geometrie.*

Es ist so gut wie nichts über die algebraischen Eigenschaften der Determinanten der Form

$$\begin{vmatrix} 0 & 1 \\ 1 & r_{ij}^2 \end{vmatrix}_{i,j=1,2,\ldots,k}, \quad (r_{ij}=r_{ji},\ r_{ii}=0)$$

bekannt. Man durchblickt dieselben bloß im Falle $k=3$ auf Grund der Faktorenzerlegung der Determinante, die der Heronischen Formel für den Flächeninhalt des Dreieckes entspricht. Auf geometrischem

Wege sind aber mehrere Relationen zwischen Determinanten dieser Form und ihren Unterdeterminanten bewiesen. Z. B. gilt (vgl. Math. Ann., *100*, S. 116) der Satz: Ein mehr als 5 Punkte enthaltender Raum, von dem je vier Punkte mit vier Punkten der Ebene kongruent sind, ist mit einer Teilmenge der Ebene kongruent; für einen aus 5 Punkten bestehenden Raum ist dies nicht notwendig richtig. Diese Behauptung besagt für Determinanten der erwähnten Form folgendes: Die 36 Zahlen $r_{ij}^2 (i, j = 1, 2, \ldots, 6)$ mögen folgenden Bedingungen genügen:

1. Für jedes der $\binom{6}{3}$ Tripel k_1, k_2, k_3, die sich aus den Zahlen $1, 2, \ldots, 6$ bilden lassen, gilt $\begin{vmatrix} 0 & 1 \\ 1 & r_{ij}^2 \end{vmatrix}_{i, j = k_1, k_2, k_3} \leqq 0.$

2. Für jedes der $\binom{6}{4}$ Quadrupel k_1, k_2, k_3, k_4, die sich aus den Zahlen $1, 2, \ldots, 6$ bilden lassen, gilt $\begin{vmatrix} 0 & 1 \\ 1 & r_{ij}^2 \end{vmatrix}_{i, j = k_1, k_2, k_3, k_4} = 0.$

Dann gilt $\begin{vmatrix} 0 & 1 \\ 1 & r_{ij}^2 \end{vmatrix}_{i, j = 1, 2, \ldots, 6} = 0.$

Hingegen ist nicht notwendig $\begin{vmatrix} 0 & 1 \\ 1 & r_{ij}^2 \end{vmatrix}_{i, j = 1, 2, \ldots, 5} = 0,$

wenn für je $\binom{5}{3}$ Tripel und für je $\binom{5}{4}$ Quadrupel, die sich aus den Zahlen $1, 2, \ldots, 5$ bilden lassen, die Bedingungen 1. und 2. erfüllt sind.

Es wäre wünschenswert, diese und ähnliche (geometrisch bewiesene) algebraische Beziehungen (und damit zugleich ihre geometrischen Anwendungen) auch algebraisch beweisen zu können. Ein systematisches Studium dieser Determinanten würde zugleich Bedingungen dafür ergeben, daß $\binom{n+1}{k+1}$ vorgelegte reelle Zahlen als Volumina der k-dimensionalen Randsimplexe eines n-dimensionalen Simplexes auftreten können. Die in Frage kommenden Bedingungen wären Verallgemeinerungen der Dreiecksungleichung. Speziell ist zu vermuten, daß das Volumen eines n-dimensionalen Simplexes nicht nur durch die Längen seiner Kanten, d. h. durch die Volumina seiner $\binom{n+1}{2}$ eindimensionalen Randsimplexe, sondern auch durch die Volumina seiner $\binom{n+1}{n-1} = \binom{n+1}{2}$ Randsimplexe der Dimension $n-2$ eindeutig bestimmt ist, u. zw. durch eine Formel analog (oder in gewissem Sinne dual) zur bekannten Volumsformel, welche das Quadrat des Volumens des n-dimensionalen Simplexes mit den Kanten r_{ij} $(i, j = 1, 2, \ldots, n+1)$ der Determinante

$$\begin{vmatrix} 0 & 1 \\ 1 & r_{ij}^2 \end{vmatrix}_{i,j=1,2,\ldots,k}$$

proportional setzt. Auch erscheint eine Axiomatisierung der Eigenschaften dieser metrischen Volumsdeterminanten analog der bekannten Axiomatik der Koordinaten-Volumsdeterminanten nicht ausgeschlossen.

Weitere Probleme beziehen sich auf den *Aufbau einer Differentialgeometrie ohne Differentialrechnung*, d. h. einer koordinatenlosen Behandlung lokaler metrischer Eigenschaften von Raumgebilden. Ein erstes Ergebnis dieses Aufbaues ist der (in der vierten metrischen Untersuchung, Math. Ann., *103*) bewiesene Satz, daß im R_3 und in jedem Raum, von dem je vier Punkte mit vier Punkten des R_3 kongruent sind, die Bogen mit überall verschwindender Krümmung mit der Strecke kongruent sind. Weder in der Formulierung dieses Satzes, noch in der (auch für andere Probleme verwendbaren) Beweismethode werden irgendwelche Koordinatendarstellungen der Punkte des Bogens oder Funktionendarstellungen des Bogens oder gar Differenzierbarkeitsvoraussetzungen gemacht. Von weiteren Problemen wären mit diesen Methoden zunächst zu behandeln: Die Kennzeichnung der Kreise als ebene Kurven konstanter Krümmung, die Definition von Scheiteln, der allgemeine Beweis des Vierscheitelsatzes, das isoperimetrische Problem, ferner die von Mayrhofer aufgeworfene Frage, ob zwei ebene Kurven, die längentreu derart aufeinander abgebildet werden können, daß die Krümmung in entsprechenden Punkten denselben endlichen Wert hat, notwendig kongruent sind. Weiter wäre die Torsion rein metrisch zu definieren und zu zeigen, daß unter den Bogen des R_3 durch überall verschwindende Torsion die ebenen Kurven gekennzeichnet sind. Sodann wäre der allgemeine metrische Kern der Frenétschen Formeln zu untersuchen. Bezüglich der Kurven auf Flächen wäre zunächst festzustellen, ob auf der Kugelfläche durch überall verschwindende Krümmung die Großkreisbogen gekennzeichnet sind. Ferner wären Bogenscharen koordinatenlos zu behandeln.

Sobald die Räume konstanter Krümmung unter den metrischen Räumen gekennzeichnet sein werden, wie dies in der zweiten metrischen Untersuchung für den R_n geschehen ist[1]), wird die Möglichkeit einer metrischen Definition der Krümmung n-dimensionaler Gebilde und damit die Basis einer koordinatenlosen n-dimensionalen Differentialgeometrie gegeben sein.

[1]) Dabei ergibt sich (besonders im zweidimensionalen Fall) die Frage, ob nicht die Kennzeichnungen der Gebilde positiver und negativer Krümmung durch Einführung imaginärer Größen auf einander zurückführbar sind in der Art der Formel von Taurinus.

Gesammelte Mitteilungen des Jahres 1929/30.

Bericht über die mengentheoretischen Überdeckungssätze. Von Karl Menger.

Bezeichnet man als *Überdeckungssystem* einer Menge M ein System von offenen Mengen derart, daß jeder Punkt von M in mindestens einer Menge des Systems enthalten ist, so lauten die beiden ältesten (insbesondere auf Borel zurückgehenden) Überdeckungstheoreme folgendermaßen:

1. *Überdeckungssatz für kompakte Mengen.* Jedes Überdeckungssystem eines kompakten Raumes oder einer abgeschlossenen Teilmenge eines kompakten Raumes enthält ein endliches Überdeckungssystem des Raumes bzw. der Menge.

2. *Überdeckungssatz für separable Mengen.* Jedes Überdeckungssystem eines separablen Raumes oder einer beliebigen Teilmenge eines separablen Raumes enthält ein abzählbares Überdeckungssystem des Raumes bzw. der Menge.

Ferner ist bekannt folgender (von Vitali stammender)

3. *Maßtheoretischer Überdeckungssatz.* Ist M eine Teilmenge eines euklidischen Raumes, U eine Umgebung von M und $\mathfrak{M}$ ein System von abgeschlossenen Mengen, das zu jedem Punkt p von M eine Folge $\{M_k(p)\}$ ($k=1, 2, \ldots$ ad inf.) von (p nicht notwendig enthaltenden) Mengen enthält, derart, daß für jeden Punkt p von M die Menge $M_k(p)$ Teil der Kugel $U_k(p)$ mit dem Zentrum p und dem Radius $\frac{1}{k}$ ist und eine endliche Zahl $\varepsilon(p) > 0$ existiert, so daß für jedes k die Beziehung $\frac{\mu\left(U_k(p)\right)}{\mu\left(M_k(p)\right)} \leq \varepsilon(p)$ gilt, wo $\mu(A)$ das Maß der Menge A bezeichnet — dann enthält $\mathfrak{M}$ eine Folge von paarweise fremden Mengen $\subset U$, welche M bis auf eine Menge vom Maße 0 überdecken. (Ohne die Annahme der Existenz einer Zahl $\varepsilon(p) > 0$ für jeden Punkt p wäre die Behauptung im allgemeinen unrichtig.)

Weitere Fragestellungen hat die Dimensionstheorie mit sich gebracht, durch welche insbesondere der Begriff des *unbegrenzt feinen Überdeckungssystems* eingeführt wurde. Wir sagen, der Punkt p sei in beliebig kleinen Mengen des Mengensystems $\mathfrak{U}$ enthalten, wenn zu jeder Umgebung U von p eine p enthaltende Teilmenge von U

in $\mathfrak{U}$ existiert. Ein System $\mathfrak{U}$ von offenen Mengen heißt ein unbegrenzt feines Überdeckungssystem der Menge M, wenn jeder Punkt von M in beliebig kleinen Mengen von $\mathfrak{U}$ enthalten ist. Dann gilt (Dimensionstheorie, S. 64) folgender

4. *Verfeinerter Überdeckungssatz für separable Mengen.* Jedes unbegrenzt feine Überdeckungssystem eines separablen Raumes oder einer beliebigen Teilmenge eines separablen Raumes enthält ein abzählbares unbegrenzt feines Überdeckungssystem des Raumes bzw. der Menge.

Da ein Überdeckungssystem eines separablen Raumes nach Satz 2 eine überdeckende Mengenfolge enthält, so kann man ferner erwarten, daß ein unbegrenzt feines Überdeckungssystem überdeckende Mengenfolgen besonderer Art enthalte. Bezeichnet man als *halbkompakt* einen Raum, der Summe abzählbarvieler kompakter Mengen ist, so gilt (Dimensionstheorie, S. 46, und Wien. Ber., *133*, S. 423) in dieser Richtung auf Grund des Überdeckungssatzes für kompakte Mengen vor allem folgender

5. *Überdeckungssatz für halbkompakte Mengen.*

a) Metrische Form. Jedes unbegrenzt feine Überdeckungssystem eines halbkompakten Raumes oder einer F_σ-Menge eines halbkompakten Raumes enthält eine den Raum bzw. die Menge überdeckende Nullfolge von Mengen (d. h. endlichviele oder abzählbarviele Mengen, deren Durchmesser gegen Null konvergieren).

b) Topologische Form. Jedes unbegrenzt feine Überdeckungssystem einer F_σ-Menge M eines halbkompakten Raumes enthält eine M überdeckende Folge von Mengen, von denen für jede vorgelegte Umgebung U von $\overline{M}$ fast alle Mengen Teilmengen von U sind.

c) Abstrakte Form. Ist M ein halbkompakter Raum oder eine F_σ-Menge eines halbkompakten Raumes und ist eine Folge von M überdeckenden Folgen offener Mengen gegeben, d. h. ein System $\{U_i^k\}$ ($i=1, 2, \ldots$ evt. ad inf., $k=1, 2, \ldots$ ad inf.) derart, daß für jedes k die Beziehung $M \subset \sum_{i=1}^{\infty} U_i^k$ gilt — dann existiert eine Zahlenfolge $\{n_k\}$ ($k=1, 2, \ldots$ ad inf.), so daß $M \subset \sum_{k=1}^{\infty} \sum_{i=1}^{n_k} U_i^k$ gilt.

Hieran schließt eine Reihe negativer Resultate: Ein Beispiel des Referenten (Dimensionstheorie, S. 49, und Wien. Ber., *133*, S. 424) dafür, daß für nicht halbkompakte separable Räume die Aussage des Überdeckungssatzes 5 im allgemeinen unrichtig ist; daß nämlich für die G_δ-Menge aller dyadisch irrationalen Punkte der Strecke ein unbegrenzt feines Überdeckungssystem existiert, aus dem sich keine die Menge überdeckende Nullfolge herausgreifen läßt; ferner der von Hurewicz (Math. Zeitschr., *24*, S. 401) stammende Satz, daß zu jeder Borelschen Menge eines kompakten Raumes, die kein

F_σ ist, eine Folge von überdeckenden Mengenfolgen $\{U_i^k\}$ mit $M \subset \sum_{i=1}^{\infty} U_i^k$ ($k = 1, 2, \ldots$ ad inf.) existiert, für welche keine Zahlenfolge $\{n_k\}$ existiert, so daß $M \subset \sum_{k=1}^{\infty} \sum_{i=1}^{n_k} U_i^k$ gelten würde. Andrerseits hat Sierpiński (Fund. Math., *8*, S. 223) bemerkt, daß, falls die Kontinuumshypothese $2^{\aleph_0} = \aleph_1$ gilt, nicht-Borelsche Mengen existieren, für welche jedes unbegrenzt feine Überdeckungssystem eine überdeckende Nullfolge enthält.

So wie der Überdeckungssatz für kompakte Mengen mit Sätzen über die Wertebereiche stetiger Funktionen mit kompakten Definitionsbereichen zusammenhängt, so hängen, wie Hurewicz (Fund. Math., *9*, S. 193) gezeigt hat, die eben besprochenen Probleme mit Fragen über Wertebereiche von Folgen stetiger Funktionen zusammen.

Wichtig ist ferner die Frage nach Eigenschaften eines unbegrenzt feinen Überdeckungssystems einer beliebigen (eventuell nicht-halbkompakten) separablen Menge, welche gewährleisten, daß das System eine die Menge überdeckende Nullfolge enthält. Bezeichnet man als *inneren Durchmesser* $d(U, p)$ einer offenen Menge U in bezug auf den Punkt p von U den Abstand des Punktes p von der Begrenzung von U, so beweist man zunächst mühelos, daß ein Überdeckungssystem einer total beschränkten Menge M, welches für eine Zahl $\varepsilon > 0$ zu jedem Punkt p von M eine Umgebung enthält, die in bezug auf p einen inneren Durchmesser $> \varepsilon$ hat, ein endliches Überdeckungssystem von M enthält. Daraus folgt, daß, wenn $\mathfrak{U}$ ein Überdeckungssystem der total beschränkten Menge M ist, eine Punktfolge $\{p_n\}$ ($n = 1, 2, \ldots$ evt. ad inf.) und eine M überdeckende Mengenfolge $\{U_n\}$ aus $\mathfrak{U}$ existiert, so daß $\lim_{n=\infty} d(U_n, p_n) = 0$ gilt. Hieraus ergibt sich, wenn man als *Exzentrizität* der offenen Menge U in bezug auf ihren Punkt p den Quotienten aus dem Durchmesser von U gebrochen durch den inneren Durchmesser von U in bezug auf p bezeichnet, folgender

6. *Nullfolgenüberdeckungssatz für Teilmengen halbkompakter Räume.* Ist M eine beliebige Teilmenge eines halbkompakten Raumes und $\mathfrak{U}$ ein System von offenen Mengen, welches zu jedem Punkt p für eine endliche Zahl $\varepsilon(p)$ beliebig kleine Umgebungen enthält, die alle in bezug auf p eine Exzentrizität $\leqq \varepsilon(p)$ besitzen, dann enthält $\mathfrak{U}$ eine M überdeckende Nullfolge.

Trotz aller Verschiedenheiten liegt eine große Analogie dieses Satzes mit dem maßtheoretischen Überdeckungssatz 3 auf der Hand, die auf die Gültigkeit allgemeinerer diesbezüglicher Sätze deutet.

Die Überdeckungssätze für kompakte und separable Mengen sind in gewissem Sinne umkehrbar. Kompaktheit bzw. Separabilität eines Raumes ist nämlich nicht nur hinreichend, sondern auch notwendig, damit jedes Überdeckungssystem des Raumes ein endliches bzw. abzählbares Überdeckungssystem des Raumes enthält (vgl. W. Gross, Wien. Ber., *123*, S. 810). Die Gültigkeit dieser Überdeckungsaussagen ist also mit Kompaktheit bzw. Separabilität äquivalent. Allgemeiner haben Kuratowski und Sierpiński (Fund. Math., *2*, S. 172) folgendes bewiesen: Ist M eine Teilmenge einer Limesklasse, so möge ein System $\mathfrak{U}$ von Mengen der Limesklasse ein Überdeckungssystem von M heißen, wenn jeder Punkt von M in mindestens einer Menge von $\mathfrak{U}$ als innerer Punkt enthalten ist. Damit jedes Überdeckungssystem von M ein abzählbares bzw. endliches Überdeckungssystem von M enthalte, ist notwendig und hinreichend, daß jede unabzählbare Teilmenge A von M mindestens einen Kondensationspunkt von A enthält, d. h. einen Punkt, von dem jede Umgebung mit A einen unabzählbaren Durchschnitt hat bzw., daß M von jeder unendlichen Teilmenge A von M, die kompakt ist und eine kompakte Ableitung besitzt, mindestens einen Punkt enthält, von dem jede Umgebung mit A einen Durchschnitt hat, dessen Mächtigkeit gleich der von A ist.

Eine Analyse der Beweise von Kuratowski und Sierpiński zeigt nun aber, daß die bewiesenen Überdeckungssätze keinerlei geometrischen, sondern einen abstrakt mengentheoretischen oder relationstheoretischen Inhalt haben. Es seien zwei Mengen M und N gegeben und eine Relation R derart, daß, wenn m irgend ein Element von M und n irgend ein Element von N ist, die Beziehung $m R n$ entweder gilt oder nicht gilt. Ist $A \subset M$ und n ein Element von N, so bezeichnen wir mit $A(n)$ die Menge aller Elemente a von A, für die $a R n$ gilt. Entsprechend ist, wenn $B \subset N$ und m ein Element von M ist, die Menge $B(m)$ zu erklären. Wenn $A \subset M$ und $B \subset N$ gilt und zu jedem Element a von A mindestens ein Element b von B existiert, so daß $a R b$ gilt, so schreiben wir $A \underset{(R)}{\rightarrow} B$ und nennen B ein *R-Bild von A*. Die Beziehung $A \underset{(R)}{\rightarrow} B$ ist offenbar äquivalent mit der Beziehung $A \subset \sum_{b \, r \, B} A(b)$. Ist A eine Teilmenge von M, so heiße ein Element a von M mit der Eigenschaft, daß aus $a R n$ stets $|A(n)| > \aleph_\alpha$ bzw. $|A(n)| \geqq \aleph_\alpha$ folgt, wobei wir mit $|X|$ die Mächtigkeit der Menge X bezeichnen, ein *R-Verdichtungselement* einer Mächtigkeit $> \aleph_\alpha$ bzw. $\geqq \aleph_\alpha$ von A. Ist beispielsweise M eine Menge von Punkten eines Raumes, N eine Menge von offenen Mengen dieses Raumes und R die Relation des Enthaltenseins eines Punktes in einer Menge, so ist eine Teilmenge B von N dann und nur dann R-Bild einer Teilmenge A von M, wenn B ein Überdeckungssystem der Menge A ist. Die R-Verdichtungselemente einer Mächtigkeit $\geqq \aleph_\alpha$ sind identisch mit den Häufungspunkten von A. *Der abstrakte Kern der Überdeckungstheoreme für kompakte und separable Mengen* sind dann die beiden folgenden Sätze:

I. *Ist $M \underset{(R)}{\rightarrow} N$, so sind die beiden folgenden Aussagen gleichwertig:*

a) Aus $M \underset{(R)}{\rightarrow} N'$, *wo* $N' \subset N$ *gilt, folgt die Existenz einer endlichen Menge* N'', *die* $\subset N'$ *ist und so daß* $M \underset{(R)}{\rightarrow} N''$ *gilt.*

b) Zu jeder Menge $M' \subset M$ *mit der Mächtigkeit* $\aleph_\alpha$ *existiert in* M *ein R-Verdichtungselement der Mächtigkeit* $\aleph_\alpha$.

II. *Ist* $M \underset{(R)}{\rightarrow} N$ *und* $\aleph_\alpha$ *eine gegebene Kardinalzahl, so sind die beiden folgenden Aussagen gleichwertig:*

a) Aus $M' \subset M$, $M' \underset{(R)}{\rightarrow} N'$, $N' \subset N$ *folgt die Existenz einer Menge* N'' *mit den Eigenschaften* $N'' \subset N'$, $|N''| \leq \aleph_\alpha$, $M' \underset{(R)}{\rightarrow} N''$.

b) Ist $M' \subset M$ *und* $|M'| > \aleph_\alpha$, *so enthält die Menge* M' *ein R-Verdichtungselement einer Mächtigkeit* $> \aleph_\alpha$ *von* M'.

Daß die Sätze in dieser Form nicht mehr den Charakter punktmengentheoretischer Überdeckungssätze besitzen, geht daraus hervor, daß man in ihnen, statt M als Punktmenge und N als System offener Mengen zu deuten, umgekehrt M als System offener Mengen und N als Punktmenge interpretieren kann, eine Interpretation, bei der man (in Beantwortung einer von Hurewicz gestellten Frage) Bedingungen dafür erhält, daß eine Punktmenge, von der in jeder Menge eines Systems offener Mengen mindestens ein Punkt enthalten ist, eine endliche bzw. abzählbare Teilmenge von Punkten enthält, so daß in jeder offenen Menge des Systems bereits mindestens ein Punkt dieser Teilmenge entbalten ist.

Ein Spezialfall des Entscheidungsproblems der theoretischen Logik. Von Kurt Gödel.

Im Anschluß an die für den Beweis der Vollständigkeit des Funktionenkalküls verwendete Methode (Monatshefte f. Math. u. Phys., *37,* S. 349) kann man ein Verfahren entwickeln, welches für jede Formel mit einer Normalform der Gestalt:

$$(Ex_1 \ldots x_n)(y_1 y_2)(Ez_1 \ldots z_m) A(x_i y_i z_i)$$

zu entscheiden gestattet, ob sie erfüllbar ist. Man führt das Problem zunächst in bekannter Weise[1]) auf Formeln der Gestalt:

$$(x_1 x_2)(Ex_3 \ldots x_n) F(x_1 \ldots x_n) \qquad (1)$$

zurück. Wir setzen zunächst voraus, daß in F nur zweistellige Funktionsvariable, und zwar $\varphi_1, \varphi_2 \ldots \varphi_s$ vorkommen. Sei A die Menge der Formeln $\varphi_i(x_1 x_1)$ $i = 1, 2 \ldots s$; B die Menge der Formeln $\varphi_i(x_p x_q)$ $i = 1, 2 \ldots s$; $p = 1, 2$; $q = 1, 2$. Es gilt $A \subset B$. Eine eindeutige Zuordnung von Wahrheitswerten $+$, $-$ zu den Formeln aus A (bzw. B) bezeichnen wir allgemein mit $\mathfrak{A}_i$ ($i = 1, 2 \ldots$ usw.) bzw.

[1]) Vgl. M. Ackermann, Über die Erfüllbarkeit gewisser Zählausdrücke, § 4, Math. Ann., *100,* S. 647.

$\mathfrak{B}_i$ ($i = 1, 2 \ldots$ usw.) [1]) (es gibt nur endlichviele $\mathfrak{A}_i$ und $\mathfrak{B}_i$). Wenn eine Realisierung der φ_i in einem Individuenbereich $\mathfrak{J}$ gegeben ist, so liefert jedes Element $a \varepsilon \mathfrak{J}$ bzw. je zwei Elemente $a, b \varepsilon \mathfrak{J}$ ein bestimmtes $\mathfrak{A}_i$ bzw. $\mathfrak{B}_j$, wenn man a mit x_1 und b mit x_2 identifiziert. Dieses $\mathfrak{A}_i$, $\mathfrak{B}_j$ bezeichnen wir auch mit $\mathfrak{A}_a$, $\mathfrak{B}_{a,b}$. Falls ein $\mathfrak{A}_k$ mit einem $\mathfrak{B}_i$ für die Menge A übereinstimmt, schreiben wir $\mathfrak{A}_k \subset \mathfrak{B}_i$. Mit $\mathfrak{B}'_i$ bezeichnen wir dasjenige $\mathfrak{B}_j$, welches aus $\mathfrak{B}_i$ durch Vertauschung von x_1 mit x_2 hervorgeht.

Sei ein System von Funktionen $\psi_1 \ldots . \psi_s$ in einem Individuenbereich $\mathfrak{J}$ gegeben, welches die Formel (1) erfüllt. Wir bilden die Menge $\mathfrak{P}$ bzw. $\mathfrak{Q}$ derjenigen $\mathfrak{A}_i$ bzw. $\mathfrak{B}_j$, welche man erhält, wenn man in $\mathfrak{A}_a$ bzw. $\mathfrak{B}_{a,b}$ a, b alle Individuen aus $\mathfrak{J}$ durchlaufen läßt. Die (endlichen und nicht leeren) Mengen $\mathfrak{P}$, $\mathfrak{Q}$ genügen offenbar folgenden Bedingungen:

I. Wenn $\mathfrak{A}_i \varepsilon \mathfrak{P}$ und $\mathfrak{A}_k \varepsilon \mathfrak{P}$, so gibt es ein $\mathfrak{B}_j \varepsilon \mathfrak{Q}$, so daß $\mathfrak{A}_i \subset \mathfrak{B}_j$ und $\mathfrak{A}_k \subset \mathfrak{B}'_j$.

II. Wenn $\mathfrak{B}_j \varepsilon \mathfrak{Q}$, so gibt es ein für den Bereich von n Individuen $a_1, a_2 \ldots a_n$ definiertes System von Funktionen $\chi_1, \chi_2 \ldots \chi_s$, so daß:

1. $\mathfrak{B}_j = \mathfrak{B}_{a_1 a_2}$,
2. $F(x_1 \ldots x_n)$ den Wahrheitswert $+$ erhält, wenn man darin die φ_i durch die χ_i und die x_i durch die a_i ersetzt.
3. $\mathfrak{B}_{a_i a_k} \varepsilon \mathfrak{Q}$ und $\mathfrak{A}_{a_i} \varepsilon \mathfrak{P}$ für alle $i, k = 1, 2 \ldots n$.

Ob es für eine vorgegebene Formel der Gestalt (1) ein Paar $\mathfrak{P}$, $\mathfrak{Q}$ nicht leerer Mengen gibt, welches den Bedingungen I, II genügt, kann man offenbar immer in endlichvielen Schritten entscheiden (weil es nur endlichviele $\mathfrak{A}_i$, $\mathfrak{B}_k$ gibt). Existiert ein solches Paar n i c h t, so ist die vorgegebene Formel n i c h t erfüllbar. Existiert ein solches Paar, dann ist die Formel erfüllbar. Denn es gilt folgendes: Wenn $\mathfrak{P}$, $\mathfrak{Q}$ den Bedingungen I, II genügen und wenn für die in einem endlichen Bereich $\mathfrak{R}$ definierten Funktionen $\varphi_1 \ldots \varphi_s$ $\mathfrak{A}_x \varepsilon \mathfrak{P}$ und $\mathfrak{B}_{x\,y} \varepsilon \mathfrak{Q}$ für beliebige $x, y \varepsilon \mathfrak{R}$, dann kann man, wenn zwei beliebige Individuen a, b aus $\mathfrak{R}$ vorgegeben sind, die Funktionen φ_i auf einen durch eine Menge $\mathfrak{R}'$ von $n-2$ Individuen $(r'_1 \ldots r'_{n-2})$ vermehrten Bereich $\mathfrak{R} + \mathfrak{R}'$ so erweitern, daß $F(a, b, r'_1 \ldots r'_{n-2})$ richtig ist und daß auch für den erweiterten Bereich immer $\mathfrak{A}_x \varepsilon \mathfrak{P}$ und $\mathfrak{B}_{x\,y} \varepsilon \mathfrak{Q}$.

Denn man kann die φ_i zunächst wegen II für die Elementepaare der Menge $\{a, b\} + \mathfrak{R}'$ und dann wegen I für die Paare $(x\,y)$ mit $x \varepsilon \mathfrak{R} - \{a, b\}$, $y \varepsilon \mathfrak{R}'$ so definieren, daß die geforderten Bedingungen erfüllt sind.

Falls in F mehr als zweistellige Funktionen vorkommen, hat man unter B (bzw. A) die Menge der Formeln zu verstehen, welche man aus den φ_i erhält, indem man auf alle möglichen Arten die Variablen x_1, x_2 (bzw. x_1) an die Leerstellen einsetzt, und kann dann ebenso schließen.

---

[1]) D. h. wir numerieren die obigen Zuordnungen irgendwie und bezeichnen die i-te mit $\mathfrak{A}_i$ (bzw. $\mathfrak{B}_i$).

Eine Verschärfung des n-Beinsatzes. Von Georg Nöbeling[1]).

In Verallgemeinerung eines Begriffes von Menger (Fund. Math., *10,* S. 96) nennen wir einen Raum K zwischen zwei fremden abgeschlossenen Teilmengen P und Q *von mindestens n-ter Ordnung,* falls P und Q nicht durch eine (zu P und Q fremde) höchstens $(n-1)$-punktige Menge in K getrennt werden können. Dann gilt folgender

Satz: Ist ein im kleinen zusammenhängender kompakter Raum K zwischen zwei fremden abgeschlossenen Teilmengen P und Q von mindestens n-ter Ordnung, so enthält K mindestens n Bögen, die P und Q verbinden und zu je zwei höchstens Endpunkte gemein haben.

Zum Beweise überzeugt man sich zunächst leicht davon, daß es genügt, P und Q als endlich anzunehmen. Sodann wählt man zwei Folgen $U_1, U_2, \ldots; V_1, V_2, \ldots$ von Umgebungen von P und Q, die sich monoton fallend auf P und Q zusammenziehen. Nun spannt man zwischen U_1 und V_1 endlichviele Bögen $\subset K$, — ihre Summe heiße B_1 —, so daß $K_1 = U_1 + V_1 + B_1$ zwischen P und Q von mindestens n-ter Ordnung ist. Sodann spannt man zwischen U_2 und V_2 endlich viele Bögen $\subset K_1$, — ihre Summe heiße B_2 —, so daß $K_2 = U_2 + V_2 + B_2$ zwischen P und Q von mindestens n-ter Ordnung ist, usw. Die Menge $K' = P + Q = \sum_{i=1}^{\infty} B_i$ ist kompakt, regulär-eindimensional und zwischen P und Q von mindestens n-ter Ordnung. Sie enthält daher nach einem bekannten Satz von Menger (Fund. Math., *10*, S. 100) n Bögen, die P und Q verbinden und zu je zwei höchstens Endpunkte gemein haben, womit unser Satz bewiesen ist.

Ist P einpunktig, so ergibt er folgende

Verschärfung des n-Beinsatzes. Zu jedem Punkte p von mindestens n-ter Ordnung eines im kleinen zusammenhängenden Kontinuums C existieren mindestens n in p endende sonst fremde Teilbögen von C. (Ausführung: Fund. Math., *18,* S. 23.)

Der allgemeine Einbettungssatz[2]). Von Georg Nöbeling.

Menger hat (Dimensionstheorie S. 295, vgl. auch Heft *1* dieser Ergebnisse, S. 2) den folgenden Satz ausgesprochen:

Jeder n-dimensionale Raum R ist mit einer Teilmenge des $(2n+1)$*-dimensionalen Cartesischen Raumes* R_{2n+1} *homöomorph.*

Für $n=0$ folgt dies aus einem Satz von Sierpiński. Ein Beweis wurde von Menger für $n=1$ erbracht und für $n>1$ skizziert (Dimensionstheorie, S. 295).

Zum Beweise kann man nach Sätzen von Hurewicz und Urysohn den Raum R als n-dimensionale kompakte Teilmenge des Hilbertschen Raumes annehmen. Nach Alexandroff läßt sich die

[1]) Abdruck aus dem Wiener akad. Anz., *67* (20. II. 1930).

[2]) Abdruck aus dem Wiener akad. Anz., *67* (20. III. 1930).

Menge R für jedes natürliche i auf einen n-dimensionalen Komplex K_i eindeutig und stetig derart abbilden, daß jeder Punkt aus R von seinem Bildpunkt in K_i einen Abstand $< \frac{1}{i}$ hat. Diese n-dimensionalen Komplexe K_i lassen sich topologisch auf Teilkomplexe K'_i des R_{2n+1} abbilden. Die hiedurch entstehenden eindeutigen stetigen Abbildungen der Menge R auf Teilmengen K'_i $(i = 1, 2, \ldots$ ad inf.) des R_{2n+1} konvergieren bei Beachtung gewisser Vorsichtsmaßregeln gegen eine topologische Abbildung von R auf eine Teilmenge R^* des R_{2n+1}, womit der Beweis beendet ist. — Man kann es außerdem noch erreichen, daß die zu R homöomorphe Menge $R^* \subset R_{2n+1}$ in der Menge aller derjenigen Punkte des R_{2n+1} enthalten ist, die höchstens n rationale Koordinaten haben. Damit ist folgender von Menger vermutete Satz bewiesen:

Für jedes n gibt es im R_{2n+1} eine n-dimensionale Universalmenge.

(Ausführung: Math. Ann., *104*, S. 71.)

Über plättbare Dreiergraphen und Potenzen nichtplättbarer Graphen. Von Karl Menger.

Ausgehend von Überlegungen über kombinatorisch-topologische Probleme habe ich vor einigen Jahren die Frage untersucht, unter welchen Bedingungen ein Dreiergraph (d. h. eine Menge, die Summe ist von endlichvielen Bogen, die zu je zweien höchstens Endpunkte gemein haben, und die keinen Punkt enthält, in welchem mehr als drei bis auf den Punkt fremde Bogen zusammenstoßen) *plättbar,* d. h. mit einem ebenen Graphen homöomorph sei. Der mich leitende Gedanke war, *daß jeder nicht plättbare Graph einen hinsichtlich dieser Eigenschaft irreduziblen Graphen enthalten muß.* Es ist leicht, einen irreduzibel unplättbaren Dreiergraph anzugeben: Man nehme die sechs Kanten eines Tetraeders und verbinde zwei innere Punkte von gegenüberliegenden Tetraederkanten durch einen bis auf seine Endpunkte zu den Tetraederkanten fremden Bogen. Die Summe dieser sieben Bogen ist offenbar irreduzibel unplättbar, d. h. mit keiner ebenen Menge homöomorph, während jede echte Teilmenge mit einer ebenen Menge homöomorph ist. Diese „*Tetraederkonfiguration*" läßt sich auch folgendermaßen beschreiben: Man wähle zwei Punktetripel a, b, c und a', b', c' im R_3 und verbinde jeden Punkt des einen Tripels mit jedem Punkt des anderen, und zwar so, daß die neun verwendeten Bogen, abgesehen höchstens von ihren Endpunkten, paarweise fremd sind. Ich habe nun damals (1925) gefunden, *daß jeder nicht plättbare Dreiergraph meine Tetraederkonfiguration als Teilmenge enthält.* Der Beweis erfolgt so, daß man zu jedem Dreiergraphen ohne Tetraederkonfiguration durch vollständige Induktion nach der Anzahl der im Graphen enthaltenen einfachen geschlossenen Kurven einen homöomorphen ebenen Graphen konstruiert. Ist die Anzahl der im Graphen enthaltenen geschlossenen Kurven $= 0$, liegt also ein Baum vor, so ist die Existenz eines homöomorphen ebenen Graphen bekannt. Ist der Satz bewiesen für alle Dreiergraphen, die höchstens $m-1$ geschlossene Kurven enthalten, und liegt ein Dreiergraph G vor, der m geschlossene Kurven enthält, so tilge man eine von ihnen aus dem Graphen. Der Rest kann dadurch in mehrere Komponenten $G_1, G_2, \ldots G_n$ zerfallen, die aber alle nach der induktiven Annahme plättbar sind. Läßt man nun der getilgten geschlossenen Kurve eine Kreislinie K in der Ebene entsprechen, so kann man unter Verwendung des Jordanschen Kurvensatzes und der Sätze über zyklische Anordnung zeigen, daß, falls der gegebene Dreiergraph keine Tetraederkonfiguration enthält, ebene Bilder $G'_1, G'_2, \ldots G'_n$ von $G_1, G_2, \ldots G_n$ derart an die Kreislinie angefügt werden können, daß $K + G'_1 + + G_2' + \ldots + G'_n$ mit G homöomorph ist.

Analog zur Tetraederkonfiguration lassen sich offenbar für jedes n Komplexe der Dimension n angeben, welche nicht mit Teilmengen des R_{n+1} homöomorph sind. Von besonderer Wichtigkeit erscheint mir nun gerade im gegenwärtigen Moment folgende Frage: Ist der Produktraum von zwei Tetraederkonfigurationen, welcher sicher zweidimensional und daher mit einer Teilmenge des R_5 homöomorph ist, schon mit einer Teilmenge des R_4 homöomorph oder nicht?[1])

(Vgl. die Fragestellung in Heft *1* dieser Ergebnisse, S. 3.)

Ein Satz über erblich im kleinen zusammenhängende Kontinua der Ebene. Von G. T. Whyburn.

In der Ebene ist jedes erblich im kleinen zusammenhängende Kontinuum eine rationale Kurve.

Da ein im kleinen zusammenhängendes Kontinuum K genau dann rational[2]) ist, wenn jede maximalzyklische Teilkurve von K rational ist, können wir uns auf die Betrachtung eines zyklisch zusammenhängenden und erblich im kleinen zusammenhängenden Kontinuums M beschränken. Da zu jedem Punkte p eines erblich im kleinen zusammenhängenden Kontinuums K der Ebene E ein erblich im kleinen zusammenhängendes Kontinuum K' existiert, das $\supset K$ ist und so daß p Begrenzungspunkt keiner Komponente von $E-K'$ ist, genügt es zu zeigen, daß ein erblich im kleinen zusammenhängendes Kontinuum M in jedem Punkte p, der Begrenzungspunkt keiner Komponente von $E-M$ ist, rational ist. Bekanntlich ist[3]) ein solcher Punkt p in beliebig kleinen Umgebungen enthalten, deren Begrenzungen J_i topologische Kreise sind, so daß jeder Punkt von J_i Begrenzungspunkt mindestens einer Komponente von $E-M$ ist.

---

[1]) Die vorstehende Notiz ist ein Abdruck aus dem Wien. akad. Anz. *67* (20. III. 1930). Damals erschien auch ein hochinteressanter Artikel von C. Kuratowski (Fund. Math., *15*, S. 271), in welchem gezeigt wird, daß für die plättbaren Graphen überhaupt (nicht bloß für die plättbaren Dreiergraphen) kennzeichnend ist, daß sie weder die oben behandelte Tetraederkonfiguration noch folgende Kuratowskische Konfiguration enthalten: Von fünf Punkten wird jeder mit den vier anderen durch einen Bogen so verbunden, daß die zehn Bogen abgesehen von ihren Endpunkten paarweise fremd sind.

[2]) Menger, Math. Ann., *95*, S. 277.

[3]) Vgl. G. T. Whyburn, Ann. of Math., *29*, S. 409—411.

Weiter ist es bekannt[1]), daß die Begrenzung jeder Komponente von $E—M$ ein topologischer Kreis ist. Wir bezeichnen mit G_i die Summe aller Komponenten von $E—M$, deren Begrenzung einen Punkt von J_i enthält. Dann können wir für jedes J_i mit hinreichend kleinem Durchmesser einen topologischen Kreis C_i konstruieren, so daß die folgenden Beziehungen gelten:

a) $C_i \subset G_i + J_i$;

b) für jede Komponente D von G_i mit der Begrenzung E enthält $C_i . E$ höchstens zwei Punkte;

c) $\delta(C_i) < \delta(J_i) + \varepsilon/2$;

d) p ist im Innern von C_i enthalten.

Hieraus folgt, daß C_i beliebig klein gewählt werden kann und daß $C_i . M$ abzählbar ist.

Die Frage, ob unser Satz im $R_n (n > 2)$ gilt, bleibt hier offen.

Über möglicherweise reguläre Punktmengen. Von G. T. Whyburn.

Ein separabler metrischer Raum R heiße möglicherweise regulär, wenn zu jedem Punkt p von R eine monoton abnehmende Folge von Umgebungen $\{U_i\}$ von p mit aus endlich vielen Punkten bestehenden Begrenzungen existiert, so daß $p = \prod_1^\infty \overline{U_i}$ ist, wobei aber die Durchmesser der U_i nicht gegen Null zu konvergieren brauchen. Falls R zusammenhängend ist, so ist R dann und nur dann möglicherweise regulär, wenn je zwei Punkte von R durch eine aus endlich vielen Punkten bestehende Menge getrennt werden können. Offenbar ist jeder im Menger-Urysohnschen Sinne regulär eindimensionale Raum R auch möglicherweise regulär, aber nicht umgekehrt. Hingegen gilt der Satz. *Ist R ein separabler metrischer Raum, der zusammenhängend und möglicherweise regulär ist, dann gibt es eine eineindeutige und stetige Abbildung T von R auf einen separablen metrischen Raum R^*, der zusammenhängend und im Menger-Urysohnschen Sinne regulär-eindimensional ist. Die Eigenschaft einer endlichen Menge, zwei gegebene Punkte von R zu trennen, ist bei der Abbildung T invariant.* Aus dem zweiten Teil dieses Satzes folgt, daß die Zahl $\liminf_{i=\infty} |B(U_i)|$ eines Punktes P von R (wo $|B(U_i)|$ die Mächtigkeit von $B(U_i)$ ist) gleich der Menger-Urysohnschen Ordnung seines Bildpunktes P^* in R^* ist.

Vollständig zerschneidbare zusammenhängende Mengen. Von G. T. Whyburn.

Eine zusammenhängende Menge M heiße vollständig zerschneidbar, wenn je zwei Punkte von M durch einen Punkt von M getrennt werden können. Damit in einem separablen metrischen Raum

[1]) Vgl. G. T. Whyburn, Proc. Ntl. Acad. Sci. (Amerika), *13*, 31—38.

eine zusammenhängende Menge vollständig zerschneidbar sei, ist jede der beiden folgenden Bedingungen notwendig und hinreichend:

a) es gibt eine eineindeutige und stetige Abbildung von M auf eine Teilmenge einer regulären Baumkurve;

b) je zwei Punkte von M können durch eine einzige in M abgeschlossene irreduzibel zusammenhängende Menge verbunden werden und jede Komponente des Komplements einer in M abgeschlossenen zusammenhängenden Menge ist in M offen.

Eine solche Menge M ist möglicherweise regulär und für jeden Punkt p von M ist $\liminf_{i=\infty} |B(U_i)|$ gleich der Anzahl der Komponenten von $M-p$, falls diese Zahl endlich ist; immer ist $|B(U_i)|$ gleich der Anzahl der Komponenten von $M-U_i$. *Damit eine zusammenhängende separable Menge M in einer regulären Baumkurve topologisch enthalten sei, ist notwendig und hinreichend, daß (1) M vollständig zerschneidbar ist und (2) für jede kompakte Teilmenge K von M und jedes* $\varepsilon > 0$ *höchstens endlichviele Komponenten von M—K einen Durchmesser* $> \varepsilon$ *haben.*

Über Zerlegungseigenschaften im kleinen zusammenhängender Kurven. Von Karl Vanek.

Durch Kombination kurventheoretischer Sätze und Methoden von Menger[1]) und Urysohn[2]) ergibt sich folgender

Zerlegungssatz für im kleinen zusammenhängende Kurven. Eine im kleinen zusammenhängende Kurve K ist für jedes $\varepsilon > 0$ *Summe von endlichvielen im kleinen zusammenhängenden Kontinua* $< \varepsilon$, *die zu je zweien einen diskontinuierlichen und zu je dreien einen leeren Durchschnitt haben.* Ist in K überdies eine abgeschlossene diskontinuierliche Teilmenge D gegeben, so können die Summanden so gewählt werden, daß jeder Punkt von D in einem einzigen Summanden liegt. Hieraus folgt: *Zu jeder im kleinen zusammenhängenden Kurve existiert ein aus im kleinen zusammenhängenden Kontinua bestehendes erzeugendes System*[3]), *in dem je zwei Mengen gleichen Schrittes einen diskontinuierlichen und je drei Mengen gleichen Schrittes einen leeren Durchschnitt haben.* Ist K speziell eine im kleinen zusammenhängende rationale, beziehungsweise reguläre Kurve, so kann in diesem Theorem das Wort „diskontinuierlich" durch die Worte „abzählbar", beziehungsweise „endlich" ersetzt werden. (Ausführung: Monatshefte, *38*, S. 5.)

[1]) Vgl. K. Menger, Math. Annalen, *95*, S. 256 ff.

[2]) Vgl. P. Urysohn, Fund. Math., *8*, S. 301 ff.

[3]) Vgl. K. Menger, Dimensionstheorie, S. 287.

Einführung des Komplexen in die allgemeine Metrik und metrische Untersuchungen in abstrakten Gruppen[1]. Von Karl Menger.

I. Problemstellung.

Den bisherigen Untersuchungen über allgemeine Metrik lag stets die Annahme reeller Abstandszahlen zugrunde. Man untersuchte *halbmetrische Räume,* das sind Mengen, in denen je zwei Elementen p und q eine reelle Abstandszahl pq zugeordnet ist (*Realitäts*forderung), so daß für je 2 Elemente p und q die Beziehung $pq = qp$ gilt (*Symmetrie*forderung), daß ferner für je zwei verschiedene Elemente $pq \neq 0$ gilt (*Verschiedenheits*forderung), und zwar $pq > 0$ (*Positivitäts*forderung), während für jeden Punkt p die Beziehung gilt $pp = 0$ (*Identitäts*forderung). Ferner untersuchte man *metrische Räume,* das sind halbmetrische Räume, die der *Dreiecksungleichung* genügen, d. h. für je drei Elemente p, q, r der Beziehung genügen $pq + qr \geqq pr$. Zu den metrischen Räumen gehört insbesondere der R_n, der n-dimensionale euklidische Raum, d. h. die Menge aller geordneten n-Tupel reeller Zahlen mit pythagoreischer Abstandsfestsetzung.

Zahlreiche Tatsachen der elementaren analytischen Geometrie werden indes durch den Begriff des halbmetrischen Raumes nicht erfaßt: Im $\overrightarrow{R_1}$, dem orientierten eindimensionalen euklidischen Raum, d. h. in der Menge aller reellen Zahlen, in der je zwei Zahlen ihre Differenz als Abstand zugeordnet ist, treten negative Abstandszahlen auf; es ist also die Positivitätsforderung nicht erfüllt und auch nicht die Symmetrieforderung, es gilt vielmehr $pq = -qp$. — Im $\overrightarrow{K_1}$, im orientierten eindimensionalen euklidisch-komplexen Raum, d. h. in der Menge aller komplexen Zahlen, in der je zwei Zahlen ihre Differenz als Abstand zugeordnet ist, treten komplexe Abstände auf; es ist also nicht einmal die Realitätsforderung erfüllt. — Im K_n, im n-dimensionalen euklidisch-komplexen Raum, d. h. in der Menge aller geordneten n-Tupel komplexer Zahlen, in der je zwei Punkten $x_1, x_2, \ldots x_n$ und $y_1, y_2, \ldots y_n$ die komplexe Zahl $(x_1 - y_1)^2 + (x_2 - y_2)^2 + \ldots + (x_n - y_n)^2$ als Abstandsquadrat zugeordnet ist, ist zwar die Symmetrieforderung erfüllt, aber weder die Realitätsforderung noch die Verschiedenheitsforderung, denn es gibt auf isotropen Geraden Paare verschiedener Punkte mit dem Abstand 0. Dabei ist die Relation von Punktepaaren, das Abstandsquadrat 0 zu besitzen, nicht transitiv; im K_2 hat z. B. der Punkt 0, 0 von jedem der Punkte i, 1 und $-i$, 1 das Abstandsquadrat 0, diese beiden Punkte aber haben -4 als Abstandsquadrat.

II. Die Teilmengen orientierter Gruppen.

Es sei G eine Gruppe. Das den Elementen a und b zugeordnete Element bezeichnen wir mit $a+b$, das Einheitselement mit 0, das zu a inverse Element mit $-a$. Statt $a + (-b)$ schreiben wir

[1]) Bearbeitung von Notizen im Wien. akad. Anz., *67*.

kurz $a-b$. Eine Menge M heiße G-**orientiert**, wenn jedem geordneten Paar p, q von Elementen von M ein Element von G, welches wir mit $p\,q$ bezeichnen, zugeordnet ist gemäß der Bedingung $p\,q \neq 0$ für $p \neq q$ und $p\,q = 0$ für $p = q$. Zwei G-orientierte Mengen M und M' heißen *kongruent*, wenn sie derart eineindeutig aufeinander abgebildet werden können, daß je zwei Elementen p und q von M zwei Elemente p' und q' von M' entsprechen, für welche die Beziehungen gelten $p\,q = p'\,q'$ und $q\,p = q'\,p'$. Insbesondere wollen wir die Gruppe G selbst und ihre Teilmengen *orientiert* nennen, wenn jedem geordneten Paar a, b von Elementen von G das Element $b-a$, welches wir auch mit $a\,b$ bezeichnen werden, als „**Abstand**" zugeordnet ist. Offenbar genügt dieser Abstand für je zwei Elemente von G der Beziehung

$$a\,b = -\,b\,a \begin{cases} \neq 0, \text{ wenn } a \neq b \\ = 0, \text{ wenn } a = b \end{cases};$$

z. B. ist der $\overrightarrow{R_1}$ bzw. $\overrightarrow{K_1}$ die orientierte additive Gruppe der reellen bzw. komplexen Zahlen.

Dann gilt folgender auf den $\overrightarrow{R_1}$ und $\overrightarrow{K_1}$ anwendbarer Satz:

Damit die G-orientierte Menge M mit einer Teilmenge der orientierten Gruppe G kongruent sei, ist notwendig und hinreichend, daß für je drei Elemente p, q, r von M die Beziehung gilt $p\,q + r\,p + q\,r = 0$. (Vgl. Mathem. Zeitschr., *33*, S. 398).

III. Die Teilmengen metrischer Gruppen.

Mit $|G|$ bezeichnen wir die Menge aller Paare von zueinander inversen Elementen der Gruppe G.

Eine Menge M heiße *G-metrisch*, wenn je zwei Elementen p und q von M ein Element $p\,q$ von $|G|$ als Abstand zugeordnet ist gemäß der Bedingung

$$p\,q = q\,p \begin{cases} \neq 0, \text{ wenn } p \neq q \\ = 0, \text{ wenn } p = q \end{cases}.$$

G selbst und jede Teilmenge von G ist G-metrisch, wenn je zwei Elementen p und q von G das Elementepaar $(p-q,\ q-p)$ zugeordnet wird. In Verallgemeinerung der metrischen Kennzeichnung der Teilmengen des R_1 (vgl. Heft *1* dieser Ergebnisse, S. 20) gilt dann: Wenn je vier Elemente der G-metrischen Menge M mit vier Elementen der Abelschen Gruppe G kongruent sind, so ist M mit einer Teilmenge von G kongruent. Wenn die Abelsche Gruppe G keine Elemente der Ordnung 3 oder 2 enthält und je drei Elemente der mehr als vier Elemente enthaltenden G-metrischen Menge M mit drei Elementen von G kongruent sind, so ist M mit einer Teilmenge von G kongruent. Ist G eine Abelsche Gruppe ohne Elemente der Ordnung 2, so ist eine aus vier Elementen (p_1, p_2, p_3, p_4) beste-

3*

hende Menge, von der je drei Elemente mit drei Elementen von G kongruent sind, dann und nur dann ein *Pseudo-G-Quadrupel* (d. h. nicht mit vier Elementen von G kongruent), wenn

$$p_1 p_2 = p_3 p_4, \quad p_1 p_3 = p_2 p_4, \quad p_1 p_4 = p_2 p_3.$$

IV. Über pseudokomplexe Quadrupel und den K_2.

Ist G die additive Gruppe der komplexen Zahlen, so wollen wir jede G-metrische Menge *komplex-metrisch* nennen. Drei Punkte p_1, p_2, p_3 einer komplex-metrischen Menge sind offenbar dann und nur dann mit drei Punkten des K_1 kongruent, wenn einer der Abstände Summe der beiden anderen ist. Eine aus vier Punkten p_1, p_2, p_3, p_4 bestehende komplex-metrische Menge ist (dem in der vorigen Notiz erwähnten Satz zufolge) dann und nur dann ein pseudokomplexes Quadrupel (d. h. nicht mit einem Quadrupel des K_1 kongruent, während je drei Punkte mit einem Tripel des R_1 kongruent sind), wenn gegenüberliegende Punktepaare gleiche Abstände haben. Wir beweisen nun: *Ein pseudokomplexes Quadrupel ist nicht kongruent mit vier Punkten des K_2.* Wir nehmen an, p_1, p_2, p_3 seien kongruent mit drei Punkten $(x_1, 0)$, $(x_2, 0)$, $(x_3, 0)$ des K_2, wo x_1, x_2, x_3 paarweise verschiedene komplexe Zahlen sind und die zweite Koordinate der drei Punkte $=0$ angenommen werden konnte, weil p_1, p_2, p_3 mit drei Punkten des K_1 (insbesondere also auch des im K_2 enthaltenen durch die Gleichung $y=0$ gegebenen K_1) kongruent sind, und haben zu zeigen: Es gibt keinen Punkt (x_4, y_4) des K_2, dessen Abstände

$$\text{von } (x_1, 0), \text{ bzw. von } (x_2, 0), \text{ bzw. von } (x_3, 0)$$

gleich sind

$$p_4 p_1 = p_2 p_3, \text{ bzw. } p_4 p_2 = p_1 p_3, \text{ bzw. } p_4 p_3 = p_1 p_2.$$

D. h. es soll kein x_4, y_4 existieren, so daß die Gleichungen gelten

$$(x_4-x_1)^2+y_4^2=(x_3-x_2)^2, \quad (x_4-x_2)^2+y_4^2=(x_3-x_1)^2, \quad (x_4-x_3)^2+y_4^2=(x_2-x_1)^2.$$

Nun folgt aus der zweiten und dritten Gleichung $x_2^2-x_3^2= =(x_2-x_3)(x_1+x_4)$, also wegen $x_2 \neq x_3$

$$x_2+x_3=x_1+x_4. \tag{1}$$

Aus der ersten und dritten Gleichung folgt $x_1^2-x_3^2= (x_1-x_3)(x_2+x_4)$, also wegen $x_1 \neq x_3$

$$x_1+x_3=x_2+x_4. \tag{2}$$

Aus (1) und (2) aber folgt $x_1=x_2$ im Widerspruch zur Annahme. Die drei Gleichungen besitzen also keine Lösung x_4, y_4, womit die Behauptung bewiesen ist.

Wenn je fünf Punkte einer komplex-metrischen Menge mit fünf Punkten des K_2 kongruent sind, dann ist M mit einer Teilmenge von K_2 kongruent — oder M ist eine Menge, deren Mächtigkeit *größer als die des Kontinuums ist und in der je zwei Elemente den Abstand Null haben.*

Sei M erstens eine komplex-metrische Menge, von der je fünf Punkte mit fünf Punkten des K_2 kongruent sind und die zwei Punkte p_1 und p_2 enthält, deren Abstand $\neq 0$ ist. Wir ordnen ihnen zwei kongruente Punkte p'_1 und p'_2 des K_2 zu. Zu jedem Punkt p_3 von M existiert ein Punkt p'_3 des K_2, dessen Abstände von p'_1, beziehungsweise p'_2 gleich $p_3 p_1$, beziehungsweise $p_3 p_2$ sind. Ist für jeden Punkt p_3 von M das Tripel p'_1, p'_2, p'_3 auf einer Geraden des K_2 gelegen, dann sind je drei Punkte von M mit einer Teilmenge des K_1 kongruent, also ist M mit einer Teilmenge des K_1 und daher erst recht mit einer Teilmenge des K_2 kongruent (wenn M mehr als vier Punkte enthält, folgt dies daraus, daß der K_1 die Quasikongruenzordnung 3 besitzt; wenn M genau vier Punkte enthält, aus dem vorher bewiesenen Satz über pseudokomplexe Quadrupel). Enthält M einen Punkt p_3, so daß p'_1, p'_2, p'_3 nicht auf einer Geraden liegen, so stellt man ganz wie im Reellen eine eindeutige Abbildung von M auf eine Teilmenge des K_2 her, von der man auf Grund der vorausgesetzten Einbettbarkeit von je fünf Punkten von M in den K_2 einsieht, daß sie abstandstreu ist. Haben zweitens je zwei Punkte von M den Abstand Null, dann bilden wir M, wenn die Mächtigkeit von M nicht größer als die des Kontinuums ist, eineindeutig auf eine Teilmenge der isotropen Geraden $x = iy$ ab. Da je zwei Punkte derselben den Abstand Null haben, ist diese Abbildung abstandstreu. Hat dagegen M eine größere Mächtigkeit als das Kontinuum, so ist schon eine eineindeutige, geschweige denn eine kongruente Abbildung von M auf eine Teilmenge von K_2 unmöglich. Damit ist unsere Behauptung bewiesen.

V. Über Stufengruppen.

Die Kennzeichnung der Teilmengen einer Gruppe G unter den G-metrischen Mengen enthält die metrische Kennzeichnung der Teilmengen des R_1. Zur Verallgemeinerung der metrischen Kennzeichnung des R_n würde es sich um die Betrachtung von Potenzen Abelscher Gruppen handeln. Für dieselben läßt sich ein Abstand aber nur dann definieren, wenn in der Gruppe überdies jedem Element ein Quadrat oder irgend eine Potenz zugeordnet ist. Man gelangt auf diese Weise zu einer Begriffsbildung, welche sich zwischen die Begriffe der Abelschen Gruppe und des Ringes einschiebt. Eine Abelsche Gruppe G heiße eine *Stufengruppe*, wenn jedem Element g von G ein Element $g^s = \varepsilon(-g)^s$ zugeordnet ist, wo ε entweder $+1$ oder -1 für alle Elemente g von G ist und $g^s = 0$ dann und nur dann gilt, falls $g = 0$ ist. G möge *quadratisch gestuft* heißen, wenn für zwei Elemente g und h von G die Beziehung gilt

$$(+) \quad (g+h)^s + (g-h)^s = 2(g^s + h^s).$$

In diesem Falle schreiben wir g^2 statt g^s. Man beweist nämlich leicht, daß, wenn für eine natürliche Zahl n mit $n.g$ die Summe von n

Elementen g bezeichnet wird, stets die Beziehung gilt $(n.g)^2 = n^2.g^2$. In einer quadratisch gestuften Gruppe sind kommutative Doppelprodukte der Elemente definierbar durch die Festsetzung

$$2.g.h = 2.h.g = (g+h)^2 - (g^2+h^2).$$

Diese Doppelproduktbildung ist hinsichtlich der Addition distributiv, d. h. es gilt stets $2.f.(g+h) = 2.f.g + 2.f.h$, wenn für je drei Elemente f, g, h folgende Beziehung gilt, aus der sich auch (+) ergibt:

$$(f+g)^2 + (g+h)^2 + (h+f)^2 = (f+g+h)^2 + (f^2+g^2+h^2).$$

In der n-ten Potenz einer quadratisch gestuften Gruppe kann den Elementen $g_1, g_2, \ldots, g_n$ und $g'_1, g'_2, \ldots, g'_n$ als pythagoreischer Abstand das Element $(g_1 - g_1)^2 + (g_2 - g'_2)^2 + \ldots (g_n - g'_n)^2$ zugeordnet werden. (Es fragt sich übrigens auch, inwieweit die Artin-Schreier-sche Realitätsdefinition für beliebige quadratisch gestufte Gruppen zu einer Ordenbarkeit derselben führt.)

Zur Metrik der Gruppen. Von Olga Taussky.

Damit das G-metrische Quadrupel p_1, p_2, p_3, p_4 ein Pseudo-G-Quadrupel sei, ist, wenn $(p_{ij}, -p_{ij})$ das den Elementen p_i und p_j zugeordnete Element von $|G|$ bezeichnet, notwendig und hinreichend,

1) daß $2p_{ij} \neq 0 \ (i, j = 1, 2, 3, 4, i \neq j)$ gilt,

2) daß sechs Zahlen ε_{ij} existieren, die alle $= \pm 1$ sind und den Beziehungen genügen

$$\varepsilon_{12}p_{12} + \varepsilon_{23}p_{23} + \varepsilon_{13}p_{13} = 0,$$

$$\varepsilon_{12}p_{12} + \varepsilon_{24}p_{24} + \varepsilon_{14}p_{14} = 0,$$

$$\varepsilon_{13}p_{13} + \varepsilon_{14}p_{14} + \varepsilon_{34}p_{34} = 0,$$

$$\varepsilon_{23}p_{23} + \varepsilon_{24}p_{24} + \varepsilon_{34}p_{34} = 0.$$

(Ausführung: Mathem. Zeitschr., *33*, S. 412.)

Angelegenheiten des Kolloquiums und Nachrichten.

Wiener Dissertationen im Zusammenhange mit dem Kolloquium 1929/30.

J. Priebsch, Der n-dimensionale Zwischenbegriff.

A. Wald, Über das Hilbertsche Axiomensystem.

Stipendien.

Prof. G. T. Whyburn, Baltimore (Ml.), arbeitete mit einer Guggenheim Memorial Fellowship, G. Nöbeling mit einem Stipendium der österreichisch-deutschen Wissenschaftshilfe 1929/30 im Kolloquium.

Druck von Gottlieb Gistel & Cie., Wien.

Bericht über das Kolloquium 1930/31.

Während des Studienjahres 1930/31 wurde das Kolloquium von G. Nöbeling geleitet.

21. Kolloquium (5. XI. 1930).

Nöbeling: *Zum Mengerschen Maßbegriff.* Es sei M eine beschränkte k-dimensionale Menge des R_n. Wir nehmen an, daß sich M für jedes $\varepsilon > 0$ in einen k-dimensionalen Komplex ε-deformieren läßt (was z. B. der Fall ist, wenn M kompakt ist). Dann läßt sich das Mengersche Maß der Menge M (vgl. diese Ergebnisse Heft 2, S. 10) definieren als die kleinste Zahl m mit der Eigenschaft, daß M sich für jedes $\varepsilon > 0$ in einen k-dimensionalen Komplex K mit einem (elementargeometrischen) Inhalt $< m + \varepsilon$ durch ε-Deformation überführen läßt. Dabei ist es übrigens unerheblich, ob man K als Teilmenge des R_n annimmt oder eines höherdimensionalen euklidischen Raumes; denn wenn K in $R_{n'}$ mit $n' > n$ liegt und K' die Projektion von K auf den R_n ist, so ist die Menge M auch in K' durch ε-Deformation überführt und K' hat einen nicht größeren Inhalt als K; hieraus folgt, daß das Maß m der Menge M in jedem M enthaltenden euklidischen Raume dasselbe ist.

Nun sei $\mathfrak{S}$ irgendeine Unterteilung des Komplexes K in Simplexe $< \frac{\varepsilon}{2}$ und es seien $q_1, \ldots, q_s$ die bei $\mathfrak{S}$ auftretenden Simplexeckpunkte. Dann kann man K in s Teilmengen $K_i < \varepsilon$ zerlegen, von denen jede genau einen der s Punkte q_j, etwa K_i den Punkt q_i, enthält und sodaß je $k+2$ Mengen K_i einen leeren Durchschnitt haben. Beispielsweise kann man dies erreichen, indem man zu $\mathfrak{S}$ die baryzentrische Unterteilung $\mathfrak{S}'$ konstruiert und mit K_i den Stern von q_i bezüglich $\mathfrak{S}'$ bezeichnet. Die Urbildmengen M_i von K_i haben Durchmesser $< 3\varepsilon$ und zu je $k+2$ einen leeren Durchschnitt. Bezeichnet man mit p_i einen beliebigen Urbildpunkt von q_i, so muß p_i in M_i liegen. Macht man jetzt die Punkte $p_{i_0}, \ldots, p_{i_t}$ $(t \leqq k)$ genau dann zu Eckpunkten eines t-dimensionalen Simplexes, wenn die Mengen $M_{i_0}, \ldots, M_{i_t}$ einen nichtleeren Durchschnitt haben, so bildet die Summe $\overline{K}$ aller dieser Simplexe einen k-dimensionalen Komplex, in den M sich 3ε-deformieren läßt. Damit haben wir gefunden: Wenn die beschränkte k-dimensionale Menge M des R_n

1

sich für jedes $\varepsilon > 0$ in einen k-dimensionalen Komplex ε-deformieren läßt, so läßt sie sich insbesondere ε-deformieren in einen Komplex $\overline{K}$, der entsteht, wenn man M in endlich viele abgeschlossene Teilmengen $M_i < \varepsilon$ zerlegt, die zu je $k+2$ einen leeren Durchschnitt haben, in jedem M_i einen Punkt p_i wählt und die Punkte $p_{i_0}, \ldots, p_{i_t}$ genau dann zu den Eckpunkten eines Simplexes macht, wenn der Durchschnitt $M_{i_0} \ldots M_{i_t}$ nichtleer ist (vgl. hierzu Alexandroff, Ann. of Math., *30*, S. 101). Es erhebt sich nun die Frage, ob das Mengersche Maß m die kleinste Zahl ist mit der Eigenschaft, daß M sich für jedes $\varepsilon > 0$ in einen Komplex $\overline{K}$ mit einem Inhalt $< m + \varepsilon$ ε-deformieren läßt.

Falls diese Frage positiv zu beantworten ist, ist die folgende Definition eine naturgemäße Verallgemeinerung der Mengerschen Maßdefinition für einen beliebigen k-dimensionalen total-beschränkten metrischen Raum: Es sei Z_ε eine Zerlegung von M in endlichviele abgeschlossene Teilmengen M_i $(i = 1, 2, \ldots, s)$ mit Durchmessern $< \varepsilon$. Man wähle in jedem M_i einen Punkt p_i. Für jedes $(k+1)$-Tupel $i_0, i_1, \ldots, i_k$ von natürlichen Zahlen $\leqq s$ definieren wir eine Zahl $v_{i_0, i_1, \ldots, i_k}$ durch die Festsetzung:

$$v_{i_0, i_1, \ldots, i_k} = \begin{cases} 0, \text{ falls der Durchschnitt } M_{i_0} \,.\, M_{i_1} \ldots M_{i_k} \text{ leer ist} \\ \sqrt{\dfrac{(-1)^{k+1}}{2^k (k!)^2} \,.\, D(p_{i_0}, p_{i_1}, \ldots, p_{i_k})} \text{ andernfalls,} \end{cases}$$

wobei wir, wenn $(m\,,\,n)$ den Abstand der Punkte p_{i_m} und p_{i_n} bezeichnet, unter $D(p_{i_0}, p_{i_1}, \ldots, p_{i_k})$ die symmetrische Determinante

$$\begin{vmatrix} 0 & 1 & 1 & . & . & 1 \\ 1 & 0 & (0\,,1)^2 & . & . & (0\,,k)^2 \\ 1 & (1\,,0)^2 & 0 & . & . & (1\,,k)^2 \\ . & . & . & . & . & . \\ . & . & . & . & . & . \\ 1 & (k\,,0)^2 & (k\,,1)^2 & . & . & 0 \end{vmatrix}$$

verstehen, so daß, falls M eine Teilmenge eines euklidischen Raumes ist, $v_{i_0, i_1, \ldots, i_k}$ der elementargeometrische Inhalt des durch die Punkte $p_{i_0}, p_{i_1}, \ldots, p_{i_k}$ bestimmten Simplexes ist (vgl. Menger, Math. Ann. *100*, S. 119, und diese Ergebnisse Heft *1*, S. 20). Wir setzen

$$v(Z_\varepsilon;\, p_1, p_2, \ldots, p_s) = \Sigma v_{i_0, i_1, \ldots, i_k},$$

wobei über alle $(k+1)$-Tupel von Zahlen $\leqq s$ zu summieren ist, und nennen v_ε das Infimum aller Zahlen $v(Z_\varepsilon;\, p_1, p_2, \ldots, p_s)$ für alle Wahlen von Z_ε und der p_i. Dann ist $\lim\limits_{\varepsilon=0} v_\varepsilon$ das Mengersche Maß von M.

22. Kolloquium (19. XI. 1930).

Nöbeling: *Dimension und Mengersches Maß.*

Jede k-dimensionale kompakte Teilmenge M des R_n hat ein positives Mengersches Maß, m. a. W.: Eine k-dimensionale kompakte Menge M kann nicht für jedes $\varepsilon > o$ in einen k-dimensionalen Komplex K mit einem Inhalt $< \varepsilon$ ε-deformiert werden.

Zum Beweise sei M eine kompakte Menge des R_n, die für jedes $\varepsilon > o$ in einen k-dimensionalen Komplex mit einem Inhalt $< \varepsilon$ sich ε-deformieren läßt. Wir behaupten, daß M höchstens $(k-1)$-dimensional ist. Es sei W ein n-dimensionaler Würfel des R_n, der M enthält und so groß ist, daß M von $R_n - W$ einen Abstand $\geqq 2$ hat. Wir zerlegen W in endlichviele Teilwürfel $W_1, \ldots, W_t < \delta$, die zu je $k+1$ einen höchstens $(n-k)$-dimensionalen Durchschnitt haben, und bezeichnen die $(n-k)$-dimensionalen Hyperebenen des R_n, welche diese Durchschnitte enthalten, mit $E_1, \ldots, E_s$. Nun wählen wir $\varepsilon \leqq (\sqrt{n})^{-n} \cdot s^{-1}$ und $\leqq \delta$ und ε-deformieren M in einen k-dimensionalen Komplex K mit einem Inhalt $< \varepsilon$, der wegen $\varepsilon \leqq 1$ in W enthalten ist und von $R_n - W$ einen Abstand $\geqq 1$ hat. Wir nehmen die Existenz einer Parallelverschiebung des Würfels W um einen Betrag $\leqq 1$, die jede Ebene $E_\sigma (\sigma = 1, \ldots, s)$ in eine zu K fremde Ebene E'_σ überführt, als bereits bewiesen an. Sind W' und W'_τ $(\tau = 1, \ldots, t)$ die Würfel, in welche W und W_τ übergehen, so ist K in W' enthalten und je $k+1$ der Würfel W'_τ haben einen zu K fremden Durchschnitt. Mithin haben die Mengen $K \cdot W'_\tau$ $(\tau = 1, \ldots, t)$ zu je $k+1$ einen leeren Durchschnitt. Diese Zerlegung von K in die Mengen $K \cdot W'_\tau < \delta$ induziert eine Zerlegung von M in abgeschlossene Teilmengen (die Urbildmengen der $K \cdot W'_\tau$) mit Durchmessern $< 3\delta$, die zu je $k+1$ einen leeren Durchschnitt haben. Weil dies für jedes $\delta > o$ gilt, ist M höchstens $(k-1)$-dimensional, was wir behauptet hatten. Es bleibt noch die Existenz einer Parallelverschiebung $\leqq 1$ zu beweisen, welche jede Ebene E_σ in eine zu K fremde Ebene E'_σ überführt. Wir bezeichnen mit S eine beliebige Einheitskugel des R_n und bilden die Punkte von S und die Parallelverschiebungen $\pi \leqq 1$ eineindeutig aufeinander ab, indem wir jedem π denjenigen Punkt zuordnen, in welchen π das Zentrum von S überführt. Bezeichnen wir mit N_σ $(\sigma = 1, \ldots, s)$ die Menge aller Punkte von S, deren entsprechende Parallelverschiebungen die Ebene E_σ in eine zu K nicht fremde Ebene E'_σ überführt, so genügt es zu zeigen, daß die Menge $S - (N_1 + \ldots + N_s)$ nichtleer ist. Zunächst behaupten wir die Ungleichung

$$m(N_\sigma) < \frac{1}{s} J,$$ wo $m(N_\sigma)$ bzw. J das n-dimensionale Maß von N_σ bzw. S bedeutet. Wir legen im R_n ein Koordinatensystem so, daß E_σ die Gestalt $x_1 = \ldots = x_k = o$ hat. Mit K' bezeichnen wir die Projektion von K auf den Teilraum $x_{k+1} = \ldots = x_n = o$ und mit K'' die Menge aller Punkte der n-dimensionalen Einheitskugel S' um den

1*

Ursprung, deren Projektionen auf $x_{k+1} = \ldots = x_n = o$ in K' liegen. Nun hat K' einen k-dimensionalen Inhalt $\leqq \varepsilon$ und daher K'' einen n-dimensionalen Inhalt $\leqq 2^n \cdot \varepsilon$, weil S' im Würfel $|x_i| \leqq 1$ $(i = 1, \ldots, n)$ liegt, und da $\varepsilon \leqq (\sqrt{n})^{-n} \cdot s^{-1}$ ist, gilt $m(K'') \leqq \frac{1}{s} \cdot \left(\frac{2}{\sqrt{n}}\right)^n$. Nun ist $\left(\frac{2}{\sqrt{n}}\right)^n$ der Inhalt des in S' enthaltenen Würfels $|x_i| \leqq \frac{1}{\sqrt{n}}$ $(i = 1, \ldots, n)$ und daher $< J$. Also ist $m(K'') < \frac{1}{s} J$, und weil K'' zu N_σ kongruent ist, auch $m(N_\sigma) < \frac{1}{s} \cdot J$. Dies gilt für jedes $\sigma \leqq s$. Mithin gilt $m(N_1 + \ldots + N_s) < J$, und daher ist die Menge $S - (N_1 + \ldots + N_s)$ nicht leer, was zu zeigen war. Damit ist unser Satz vollständig bewiesen.

Beer: Der Satz gilt auch für die nichtkompakten Mengen der Ebene. Für $k = o$ und $k = 2$ ist dies trivial. Im Falle $k = 1$ verläuft der Beweis genau wie der Mengersche Beweis für die Tatsache, daß eine eindimensionale Menge der Ebene nicht für jedes $\varepsilon > o$ in einen nulldimensionalen Komplex ε-deformiert werden kann (vgl. diese Ergebnisse, Heft *2*, S. 8).

O. Haupt (Erlangen; nach brieflicher Mitteilung) und Alt: *Zum Krümmungsbegriff.*

Wenn ein ebener Bogen B in einem Punkte P eine (endliche) Mengersche Krümmung k besitzt, so ist er in P zweimal differenzierbar (und besitzt daher in P eine Krümmung im klassischen Sinne).

Man beweist zunächst das Lemma: In P existiert die Tangente t und benachbarte Sehnen konvergieren gegen t. Die Existenz von t folgt aus Mengers Überlegungen (Math. Ann. *103*, S. 482). Gäbe es nun Sehnen RT des Bogens, deren Winkel mit t eine fest positive Schranke nicht untenschreiten, so gilt das gleiche für den Winkel zwischen PT und RT, falls T hinreichend nahe bei P liegt (denn t ist Limes der Sehnen PT). Dann könnte aber die Mengersche Krümmung nicht endlich sein.

Wir legen nun ein Koordinatensystem so, daß P mit dem Nullpunkt und t mit der x-Achse zusammenfällt. Bedeuten $(x_i y_i)$ $(i = 1, 2, 3)$ Punkte von B, so kann man die Mengersche Krümmung k in der Form

$$k = \lim_{x_i y_i \to 0} 2 \frac{\dfrac{\dfrac{y_3 - y_2}{x_3 - x_2} - \dfrac{y_2 - y_1}{x_2 - x_1}}{x_3 - x_1}}{\sqrt{\left[1 + \left(\dfrac{y_2 - y_1}{x_2 - x_1}\right)^2\right]\left[1 + \left(\dfrac{y_3 - y_2}{x_3 - x_2}\right)^2\right]\left[1 + \left(\dfrac{y_1 - y_3}{x_1 - x_3}\right)^2\right]}}$$

schreiben. Da hierin der Nenner nach dem Lemma gegen 1 konvergiert, existiert also

$$(1)\qquad \lim_{x_i y_i \to 0} \frac{\dfrac{y_3-y_2}{x_3-x_2} - \dfrac{y_2-y_1}{x_2-x_1}}{x_3-x_1}.$$

Analog ergibt sich die Existenz von

$$(2)\qquad \lim_{x_i y_i \to 0} \frac{\dfrac{y_2-y_1}{x_2-x_1} - \dfrac{y_1-y_3}{x_1-x_3}}{x_2-x_3}.$$

Aus (1) folgt erstens, daß B in einer Umgebung des Ursprungs differenzierbar ist; denn wäre B in beliebig nahe beim Ursprung gelegenen Punkten $(x_2 y_2)$ nicht differenzierbar, so ließen sich $(x_1 y_1)$ und $(x_3 y_3)$ nahe bei $(x_2 y_2)$ so finden, daß der Ausdruck hinter dem Limeszeichen in (1) beliebig groß wird, was mit der Existenz eines endlichen Grenzwertes unvereinbar ist. Man setze jetzt $x_1 = y_1 = 0$. Wegen (2) existiert dann $\lim\limits_{x_i, y_i \to 0} \frac{1}{x_2 - x_3}\left(\frac{y_2}{x_2} - \frac{y_3}{x_3}\right)$, also wegen $\lim\limits_{x_3, y_3 \to 0} \frac{1}{x_2 - x_3}\left(\frac{y_2}{x_2} - \frac{y_3}{x_3}\right) = \frac{y_2}{x_2^2}$, auch $\lim\limits_{x_2, y_2 \to 0} \frac{y_2}{x_2^2}$.

Zufolge (1) existiert ferner $\lim\limits_{x_i, y_i \to 0} \frac{1}{x_3}\left(\frac{y_3 - y_2}{x_3 - x_2} - \frac{y_2}{x_2}\right)$.

Wählt man dabei x_2 so nahe bei x_3, daß $\frac{y_3 - y_2}{x_3 - x_2} = y'(x_3) + \varepsilon x_3$ und $\frac{x_3}{x_2} \to 1$, so ergibt sich die Existenz von $\lim\limits_{x_3, y_3 \to 0} \frac{y'(x_3)}{x_3} = y''(0)$.

Alt: *Zur Theorie der Krümmung.*

Ein Beispiel einer Kurve, bei der überall die klassische Krümmung existiert, die metrische dagegen nicht, ist $y = x^4 \sin \frac{1}{x}$. Abgesehen vom Nullpunkt hat diese Kurve überall eine metrische und daher auch eine klassische Krümmung; diese letztere ist aber im Ursprung unstetig und daher kann die metrische Krümmung im Ursprung nicht existieren. Denn wenn die metrische Krümmung in allen Punkten eines Bogens existiert, so ist sie stetig. Es sei nämlich p ein Punkt dieses Bogens mit der Krümmung k; dann haben alle in einer hinreichend kleinen Umgebung U von p liegenden Punktetripel eine sich von k beliebig wenig unterscheidende Krümmung; daher unterscheiden sich auch die Krümmungen der Punkte aus U von k um beliebig wenig.

Zwei ebene Kurven endlicher Länge, die längentreu derart aufeinander abgebildet werden können, daß in entsprechenden Punkten die (von Null verschiedene) metrische Krümmung denselben Wert hat, sind kongruent[1]). Dieser Satz folgt aus dem bekannten klassischen Satz, auf Grund der Tatsache, daß, wo die metrische Krümmung existiert, auch die klassische existiert und gleich der metrischen ist. Doch läßt sich auch ein rein metrischer Beweis erbringen, der hier kurz skizziert sei. Es seien p, q zwei Punkte des ersten Bogens und p', q' ihre Bildpunkte auf dem andern. Wir haben zu zeigen, daß die Abstände $\delta(p, q)$ und $\delta(p', q')$ gleich sind. Die Punkte p, q bzw. p', q' sind Endpunkte zweier Teilbogen B bzw. B' mit gleicher Länge l. Wir schreiben den beiden Bogen B und B' je ein n-Gitter[2]) ein, dessen Seitenlänge a_n bzw. a_n' sei. Da $n \,.\, a_n$ und $n \,.\, a_n'$ mit wachsemdem n gegen l konvergiert, ist der Quotient $\frac{a_n'}{a_n}$ für großes n sehr nahe bei 1. Es sei etwa $a_n' \geqq a_n$. Führen wir nun den Bogen B' durch eine Ähnlichkeitstransformation in einen Bogen B'' über derart, daß das B'' eingeschriebene n-Gitter die Seitenlänge a_n bekommt, so ändert sich dabei für hinreichend großes n der Abstand der beiden Punkte p' und q' beliebig wenig. Da die Krümmung stets von Null verschieden ist, haben die beiden n-Gitter in B und B'' die Eigenschaft, daß die Winkel $< \pi$ alle auf einer Seite liegen. Ändert man nun die Winkel des zweiten n-Gitters derart ab, daß sie den entsprechenden Winkeln im ersten n-Gitter gleich werden, so läßt sich zeigen, daß sich dabei der Abstand der beiden Endpunkte p'' und q'' von B'' abermals um beliebig wenig ändert. Da aber dann die beiden n-Gitter kongruent geworden sind, so haben wir also die beiden Punkte p', q' durch eine beliebig kleine Verrückung einen Abstand bekommen, welcher gleich $\delta(p, q)$ ist. Also muß $\delta(p, q) = \delta(p', q')$ sein, w. z. b. w.

23. Kolloquium (3. XII. 1930).

Wald: *Über den allgemeinen Raumbegriff.*

Zur Einführung des allgemeinen Raumbegriffes geht Menger (Dimensionstheorie, S. 15 ff.) von einer abzählbaren Menge $\mathfrak{A}$ von Dingen, bezeichnet mit $U, V \ldots$, aus, für welche eine transitive Relation, ausgedrückt durch das Symbol $U \subset V$, derart erklärt ist, daß zwei Dinge U und V von $\mathfrak{A}$ dann und nur dann identisch sind, falls jedes Ding von $\mathfrak{A}$, welches $\subset U$ bzw. $\subset V$ ist, auch $\subset V$ bzw. $\subset U$ ist. Zwei Dinge U und V von $\mathfrak{A}$ heißen fremd, wenn kein Ding W von $\mathfrak{A}$ existiert, welches $\subset U$ und $\subset V$ ist. Eine Folge $\{U_k\}$ ($k = 1, 2, \ldots$ ad inf.) heißt monoton abnehmend, wenn $U_{k+1} \subset U_k$ für jedes natürliche k gilt. Gewisse monoton abnehmende Folgen von Dingen aus $\mathfrak{A}$ werden Punkte genannt, wobei zwei Punkte $\{U_k\}$ und $\{V_k\}$ dann und nur dann identisch heißen, falls $U_n \subset V_m$ für jedes m und fast alle n sowie $V_m \subset U_n$ für jedes n und fast alle m gilt. Der Punkt $\{U_n\}$ heißt Punkt von V, wenn $U_n \subset V$ für fast

[1]) Vgl. diese Ergebnisse, Heft *2*, S. 22.
[2]) Zur Definition des n-Gitters siehe Menger, Math. Ann. *103*, S. 487.

alle n gilt. Es wird sodann gefordert: 1. Wenn $U \subset V$ nicht gilt, so existiert ein Punkt, der in U, aber nicht in V liegt. 2. Liegt der Punkt $\{U_n\}$ in V, so existiert ein V' derart, daß p in V' liegt, während jeder Punkt, der nicht in V liegt, in einem zu V' fremden Ding liegt. Die Menge aller Punkte ist dann ein regulärer topologischer Raum im Sinne von Hausdorff, der dem zweiten Abzählbarkeitsaxiome genügt. Die Menge aller Punkte, die in einem Dinge U von $\mathfrak{A}$ liegen, heiße die Menge U. Dann ist das System der Mengen U (für alle Dinge von $\mathfrak{A}$) ein den Raum definierendes Umgebungssystem. Die Relation $U \subset V$ gilt im mengentheoretischen Sinne dann und nur dann, wenn sie für die entsprechenden Dinge von $\mathfrak{A}$ im symbolischen Sinne gilt. Der Zweck dieser Art, den topologischen Raum einzuführen zum Unterschiede von Hausdorffs Einführung als Punktmenge, in welcher gewisse Teilmengen, die Umgebungen, gemäß gewissen Bedingungen definiert sind, besteht darin, daß bei der obigen Einführungsart die Analogie zur Einführung der reellen Zahlen gewahrt bleibt, welche ja auch als monoton abnehmende Folgen von Intervallen mit rationalen Endpunkten eingeführt werden, wobei aber zunächst diese Intervalle nicht als Punktmengen betrachtet werden, sondern als Paare rationaler Zahlen, für welche symbolisch eine Relation des Enthaltenseins definiert ist. Ein von Menger gesprächsweise betonter Mangel seiner speziellen Durchführung dieses Gedankens für beliebige Räume ist jedoch, daß bei Menger kein Gesetz angegeben wird, *welche* monoton abnehmenden Folgen von Dingen aus $\mathfrak{A}$ Punkte heißen, so daß außer der $\subset$-Relation in $\mathfrak{A}$ noch für jede monoton abnehmende Folge erklärt sein muß, ob sie einen Punkt darstellt oder nicht. Dieser Mangel soll im folgenden dadurch behoben werden, daß wir statt der $\subset$-Relation des Teilseins die $\Subset$-Relation des scharfen Teilseins axiomatisieren. Eine Menge U heißt nach Menger scharfe Teilmenge von V, in Zeichen $U \Subset V$, *wenn sie mit allen Häufungspunkten Teil von V ist,* d. h. wenn $\overline{U} \subset V$ gilt. Allerdings werden wir die $\Subset$-Relation nicht nur für je zwei Elemente einer abzählbaren Menge als erklärt annehmen, sondern für je zwei *Teilmengen* einer abzählbaren Menge. In dem zu definierenden Raume wird hinterher der abzählbaren Menge ein abzählbares den Raum definierendes Umgebungssystem entsprechen und jeder Teilmenge der abzählbaren Menge die Summe jener offenen Mengen, welche den Elementen der Teilmenge entsprechen. Dadurch wird also dem System aller Teilmengen der abzählbaren Menge das System aller offenen Mengen des Raumes entsprechen und der Umstand, daß wir für je zwei Teilmengen der abzählbaren Menge die $\Subset$-Relation als definiert annehmen, läuft also darauf hinaus, daß wir für je zwei offene Mengen U und V des Raumes wissen, ob $U \Subset V$ gilt oder nicht.

Es sei $\mathfrak{A}$ eine abzählbare Menge von Dingen, die mit $U, V, W, \ldots$ bezeichnet seien, $\mathfrak{B}$ die Menge aller Teilmengen von $\mathfrak{A}$, welche mit $X, Y, Z, \ldots$ bezeichnet seien. Die aus einem einzigen Element U von

$\mathfrak{A}$ bestehende Menge bezeichnen wir als Element von $\mathfrak{B}$ gleichfalls mit U, da Verwechslungen ungefährlich sind. Es sei in $\mathfrak{B}$ eine transitive Relation $\Subset$ definiert derart, daß zwei Dinge X und Y von $\mathfrak{B}$ dann und nur dann identisch sind, wenn jedes Ding U von $\mathfrak{A}$, welches $\Subset X$ bzw. $\Subset Y$ ist, auch $\Subset Y$ bzw. $\Subset X$ ist. *Fremd* heißen zwei Dinge X und Y von $\mathfrak{B}$, wenn kein Ding von $\mathfrak{A}$ existiert, welches $\Subset X$ und $\Subset Y$ ist. *Scharf abnehmend* heißt eine Folge $\{X_k\}$ ($k=1, 2, \ldots$ ad inf.) von Dingen von $\mathfrak{B}$, wenn $X_{k+1} \Subset X_k$ für jedes k gilt.

Punkt heißt eine scharf abnehmende Folge $\{M_k\}$ ($k=1, 2, \ldots$ ad inf.) von $\mathfrak{A}$, *wenn für je zwei Dinge X und Y von $\mathfrak{B}$, für welche $X \Subset Y$ gilt und für welche $U_k \Subset Y$ für kein k gilt, fast alle U_k zu X fremd sind.* Zwei Punkte $\{U_k\}$ und $\{V_k\}$ heißen *identisch* dann und nur dann, wenn $U_n \Subset V_m$ für jedes m und fast alle n sowie $V_m \Subset U_n$ für jedes n und fast alle m gilt. Wir sagen, der Punkt $\{U_k\}$ liegt im Dinge X von $\mathfrak{B}$ oder sei Punkt von X, wenn $U_k \Subset X$ für fast alle k gilt, und wir sagen, der Punkt $\{U_k\}$ sei Hüllenpunkt von X, wenn U_k und X für kein k fremd sind. Es gilt dann folgende *Bemerkung:* Gilt $U \Subset V$, so ist jeder Hüllenpunkt von U Punkt von V. Denn wenn $\{U_k\}$ Hüllenpunkt von U ist, so ist U zu keinem U_k fremd. Dann aber muß, da die Folge $\{U_k\}$ einen Punkt definiert und $U \Subset V$ gilt, $U_k \Subset V$ für fast alle k gelten und daher der Punkt $\{U_k\}$ in V liegen. Wir nehmen nun an:

I. *Jedes Ding von $\mathfrak{A}$ enthält mindestens einen Punkt.*

II. *Wenn $U \Subset V$ für zwei Dinge von $\mathfrak{A}$ nicht gilt, so existiert mindestens ein in V nicht enthaltener Hüllenpunkt von U.* Auf Grund der obigen Bemerkung folgt hieraus: *$U \Subset V$ gilt dann und nur dann, wenn jeder Hüllenpunkt von U Punkt von V ist.*

III. *Damit der Punkt $\{U_k\}$ im Dinge X von $\mathfrak{B}$ enthalten sei, ist notwendig und hinreichend, daß er in mindestens einem Dinge U von $\mathfrak{A}$ enthalten ist, welches Element der Teilmenge X von $\mathfrak{A}$ ist.*

Wir bezeichnen nun die Menge aller Punkte als *Raum* und nennen für jedes Ding X von $\mathfrak{B}$ die Menge aller Punkte von X eine *offene Menge,* und zwar die offene Menge X. Aus der Annahme I folgt, daß zwei offene Mengen X und Y dann und nur dann fremd sind, wenn die entsprechenden Dinge X und Y des Systems $\mathfrak{B}$ im symbolischen Sinne fremd sind. Wegen der Transitivität der $\Subset$-Relation folgt aus $X \Subset Y$ für zwei Dinge von $\mathfrak{B}$ zunächst, daß die entsprechende Menge X Teilmenge der Menge Y ist. Auf Grund der Definition der offenen Mengen sieht man ferner, daß jeder Hüllenpunkt eines Dinges X von $\mathfrak{B}$ Hüllenpunkt (d. h. Punkt oder Häufungspunkt) der entsprechenden Menge X im mengentheoretischen Sinne ist und umgekehrt; daher folgt aus der Konsequenz der Annahme II, daß die Relation $X \Subset Y$ für zwei Dinge von $\mathfrak{B}$ dann und nur dann besteht, wenn sie für die entsprechenden Mengen im mengentheoretischen Sinne besteht, d. h. wenn jeder Punkt und Häufungspunkt von X Punkt von Y ist. Man bestätigt nunmehr mühelos, daß für die abzählbar vielen offenen Mengen, welche den Dingen von $\mathfrak{A}$ entsprechen, die Hausdorffschen Umgebungsaxiome erfüllt sind und der Raum überdies regulär ist, d. h. zu jeder Umgebung U eines Punktes p eine Umgebung existiert, die $\Subset U$ ist. Auf Grund von Annahme III ist jede offene Menge Summe von offenen Mengen des Systems $\mathfrak{A}$ und umgekehrt und es ist das System aller offenen Mengen mit dem abzählbaren System der den Dingen von $\mathfrak{A}$ entsprechenden offenen Mengen im Hausdorffschen Sinne gleichwertig. Wir haben also bewiesen: *Die Menge aller Punkte ist, wenn wir die den Dingen von $\mathfrak{B}$ entsprechenden Mengen als Umgebungen bezeichnen, ein regulärer topologischer Raum, der dem zweiten Abzählbarkeitsaxiome genügt.* Umgekehrt beweisen wir nun: *Jeder reguläre topologische Raum, der dem zweiten Abzählbarkeitsaxiome genügt, kann auf die obige Art eingeführt werden.* Dies ist offenbar eine unmittelbare Konsequenz von folgendem mengentheoretischen

Satz: *Damit eine scharf abnehmende Folge $\{U_k\}$ von offenen Mengen eines regulären topologischen Raumes sich auf einen Punkt zusammenziehe, ist notwendig und hinreichend, daß zu jeder offenen Menge V, welche keine Menge U_k als Teil enthält, jede offene Menge W, die $\Subset V$ ist, zu fast allen Mengen U_k fremd ist.* Dabei sagen wir, die Folge $\{U_k\}$ ziehe sich auf den Punkt p zusammen, wenn $(p) = \prod_{k=1}^{\infty} U_k$ ist und jede Umgebung von p fast alle Mengen U_k als Teil enthält.

Notwendig: Es sei $\{U_k\}$ eine auf den Punkt p sich zusammenziehende Folge von offenen Mengen und V eine offene Menge, welche keine Menge U_k als Teil enthält. Wenn p_k irgend ein Punkt von U_k ist, so ist $p = \lim p_k$. Wäre nämlich p nicht Limes der Folge $\{p_k\}$, so existierte eine Umgebung W von p und eine Zahlenfolge $\{j_k\}$ ($k = 1, 2, \ldots$ ad inf.) derart, daß alle Punkte p_{j_k} im Komplement von W liegen. Dann würde also W nicht fast alle U_k enthalten im Widerspruch zur Voraussetzung, daß die Folge $\{U_k\}$ sich auf p zusammenzieht. Es sei nun W eine offene Menge $\Subset V$. Wir haben zu zeigen, daß W zu fast allen U_k fremd ist. Andernfalls existierte eine Zahlenfolge $\{i_k\}$ derart, daß W mit jeder der Mengen U_{i_k} einen Punkt p_{i_k} gemein hätte. Der Punkt p ist wegen $p = \lim p_{i_k}$ Punkt oder Häufungspunkt von W, also wegen $W \Subset V$ Punkt von V. Dann aber enthält, da die $\{U_k\}$ sich auf p zusammenziehen, V fast alle Mengen U_k im Widerspruch zur Voraussetzung.

Hinreichend: Wir setzen voraus, die Bedingung sei erfüllt, und zeigen

a) die Menge $\prod_{k=1}^{\infty} U_k$ enthält höchstens einen Punkt. Denn enthielte sie zwei verschiedene Punkte p und q, so existierte eine Umgebung U von p, welche q nicht enthält, und eine Umgebung U' von p, welche $\Subset U$ ist. Da q nicht in U liegt, enthält U keine der Mengen U_k. Dennoch ist U' zu keiner der Mengen U_k fremd, da p in U' liegt. Dies widerspricht der vorausgesetzten Bedingung.

b) Die Menge $\prod_{k=1}^{\infty} U_k$ enthält mindestens einen Punkt. Wir machen die Annahme, die Menge sei leer, und leiten einen Widerspruch her. Es sei p_k ein Punkt von U_k. Existierte ein Häufungspunkt p der Folge $\{p_k\}$, so wäre für jedes natürliche k, da U_k alle Punkte $p_k, p_{k+1} \ldots$ enthält, p Punkt oder Häufungspunkt von U_k, also für $k \geqq 2$ wegen $U_k \Subset U_{k-1}$ Punkt von U_{k-1} und daher Punkt von $\prod_{k=1}^{\infty} U_k$ entgegen der Annahme, diese Menge sei leer. Da die Folge $\{p_k\}$ keinen Häufungspunkt besitzt, ist die Menge aller in ihr enthaltenen Punkte unendlich. Sie enthält demnach eine Teilfolge $\{p_{i_j}\}$ $(j=1, 2, \ldots$ ad inf.) von paarweise verschiedenen Punkten derart, daß je zwei Punkte dieser Folge verschieden sind und daß unendlich viele Punkte der Folge $\{p_k\}$ in der Teilfolge nicht enthalten sind. Da insbesondere kein Punkt p_{i_j} Häufungspunkt der Folge $\{p_k\}$ ist, existiert eine Folge $\{V_{i_j}\}$ von offenen Mengen derart, daß V_{i_j} von der Folge $\{p_k\}$ bloß den Punkt p_{i_j} enthält und $\Subset U_{i_j}$ ist. Ferner bestimmen wir zu jedem Punkte p_{i_j} eine Umgebung W_{i_j}, die $\Subset V_{i_j}$ ist. Wir setzen nun $V = \sum_{j=1}^{\infty} V_{i_j}$, $W = \sum_{j=1}^{\infty} W_{i_j}$. Dann gilt:

1. V enthält keine der Mengen M_k $(k=1, 2, \ldots$ ad inf.). Denn V enthält von der Punktefolge $\{p_k\}$ bloß die Punkte $\{p_{i_j}\}$ $(j=1, 2, \ldots$ ad inf.); also enthält V die unendlich vielen Punkte der Folge $\{p_k\}$ nicht, welche in dieser Teilfolge nicht enthalten sind.

2. W ist zu keiner der Mengen U_k $(k=1, 2, \ldots$ ad inf.) fremd. Denn W enthält die Punkte p_{i_j} $(j=1, 2, \ldots$ ad inf.).

3. $W \Subset V$. Ist p ein von allen p_{i_j} verschiedener Punkt, so existiert, da die Folge $\{p_k\}$ keinen Häufungspunkt, insbesondere also nicht p als Häufungspunkt besitzt, eine Umgebung U von p, in der kein p_{i_j} liegt, die also keine Menge U_k als Teil enthält. Nach der vorausgesetzten Bedingung existiert daher eine Umgebung U' von p, die $\Subset U$ ist und zu fast allen U_{i_j} fremd ist. Wenn p ein von allen Punkten p_{i_j} verschiedener Punkt von $\overline{V}$ ist, so muß also p Häufungspunkt von einer der Mengen V_k sein. Erst recht muß ein von allen p_{ij} verschiedener Punkt von $\overline{W}$ Punkt einer der Mengen $\overline{W_k}$ sein, so daß $\overline{W} \subset \sum_{k=1}^{\infty} \overline{W_k} + \sum_{j=1}^{\infty} (p_{i_j})$ gilt. Wegen $W_k \Subset V_k$ und $(p_{i_j}) \subset V_{i_j}$ folgt daraus die Behauptung 3.

Das Zusammenbestehen von 1, 2, 3 ist der gesuchte Widerspruch gegen die vorausgesetzte Bedingung, womit *b)* bewiesen ist. Hieraus folgt wegen *a)*, daß $\prod_{k=1}^{\infty} U_k$ aus einem Punkte p besteht.

c) Die Folge $\{U_k\}$ zieht sich auf den Punkt p zusammen. Es sei nämlich U eine vorgelegte Umgebung von p. Wir haben zu zeigen: $U_k \subset U$ für fast alle k. Wir machen die Annahme, diese Behauptung sei falsch und leiten einen Widerspruch her. Aus der Annahme folgt, daß das Komplement U^* von $\overline{U}$ mit unendlich vielen und daher mit fast allen Mengen U_k nichtleere Durchschnitte hat. Wenden wir auf die abgeschlossenen Mengen (p) und $\overline{U^*}$ den Trennungssatz an, so ergibt sich die Existenz einer offenen Menge V^*, die $\supseteq U^*$ ist und p nicht enthält, also keine Menge U_k als Teil enthält. Dies widerspricht, da die Menge $U^* \subseteq V^*$ zu keiner Menge U_k fremd ist, der vorausgesetzten Bedingung.

Nöbeling: Die Widerlegung des Produktsatzes der Dimensionstheorie durch Pontrjagin.

Nöbeling: *Zur Axiomatik der Dimension.*

Menger hat (Monatshefte *37,* S. 172, vgl. diese Ergebnisse, Heft *2,* S. 4) gezeigt, daß die fünf Dimensionsaxiome (vgl. diese Ergebnisse, Heft *1,* S. 3) für die Dimension unter den Mengenfunktionen des R_n sicherlich dann kennzeichnend sind, wenn zwei Sätze gelten, von denen der zweite lautet: (II_k^n): Jede G_δ-Menge des R_n, deren Komplement höchstens $(n$-k-$1)$-dimensional ist, ist eine k-dimensionale Universalmenge des R_n, d. h. enthält zu jeder k-dimensionalen Teilmenge des R_n eine homöomorphe Menge als Teil. In dieser Allgemeinheit ist der Satz jedoch unrichtig, wie folgendes Beispiel gegen den Satz II_2^3 zeigt: Es sei A die Antoinesche nulldimensionale kompakte Menge des R_3, die mit jeder topologischen Kugelfläche, für welche sowohl im Innern als auch im Äußeren Punkte von A liegen, einen nichtleeren Durchschnitt hat. Es sei $\{p_i\}$ ($i = 1, 2, \ldots$ ad inf.) eine im R_3 dichte Punktefolge. Dann existieren offenbar abzählbar viele zu A homöomorphe Mengen $\{A_k\}$ ($k = 1, 2, \ldots$ ad inf.) derart, daß jedes Punktepaar p_i, p_j der Folge $\{p_i\}$ in mindestens einer Menge A_k enthalten ist. $R_3 - \sum_{k=1}^{\infty} A_k$ ist dann eine G_δ-Menge, deren Komplement Summe abzählbarvieler nulldimensionaler kompakter Mengen, also nulldimensional ist, die aber nicht jede Fläche des R_3 topologisch enthält; sie enthält z. B. keine topologische Kugel, da jede solche Fläche irgend zwei Punkte p_i und p_j trennt und daher mit der dieses Paar enthaltenden Menge A_k einen nichtleeren Durchschnitt hat, also nicht Teilmenge von $R_3 - \sum_{k=1}^{\infty} A_k$ ist. Es bleibt die Frage offen, ob nicht der Satz II_n^{2n+1} für jedes n gilt.

24. Kolloquium (22. I. 1931).

Gödel: *Über Vollständigkeit und Widerspruchsfreiheit.*

Sei Z das formale System, welches man erhält, indem man die Peanoschen Axiome durch das Schema der rekursiven Definition (nach einer Variablen) und die logischen Regeln des engeren Funktionenkalküls ergänzt. Z soll also keine anderen Variablen als solche für Individuen (d. h. natürliche Zahlen) enthalten und das Prinzip der vollständigen Induktion muß daher als Schlußregel formuliert werden. Dann gilt:

1. Jedes Z umfassende[1]) formale System S mit endlich vielen Axiomen und der Einsetzungs- und Implikationsregel als einzigen Schlußprinzipien ist unvollständig, d. h. es gibt darin Sätze (u. zw. Sätze aus Z), die aus den Axiomen von S unentscheidbar sind, vorausgesetzt, daß S ω-widerspruchsfrei ist. Dabei heiße ein System ω-widerspruchsfrei, wenn für keine Eigenschaft F natürlicher Zahlen zugleich $(Ex)\overline{F(x)}$ und sämtliche Formeln $F(i)$, $i=1, 2 \ldots$ usw. beweisbar sind.

2. In jedem solchen System S ist insbes. die Aussage, daß S widerspruchsfrei ist (genauer die mit ihr äquivalente arithmetische Aussage, welche man erhält, indem man die Formeln ein-eindeutig auf natürliche Zahlen abbildet), unbeweisbar.

Satz 1 und 2 gelten auch für Systeme mit unendlich vielen Axiomen und anderen Schlußprinzipien als den genannten, vorausgesetzt, daß bei einer Durchnumerierung der Formeln (nach wachsender Länge und bei gleicher Länge lexikographisch) die Klasse der den Axiomen zugeordneten Nummern, sowie die folgende Relation $R(x_1\, x_2 \ldots x_n)$ zwischen natürlichen Zahlen: „Die Formel mit der Nummer x_1 folgt aus den Formeln mit den Nummern $x_2 \ldots x_n$ durch einmalige Anwendung einer der Schlußregeln" im System Z enthalten und entscheidungsdefinit sind. Dabei heißt eine Relation (Klasse) $R(x_1 \ldots x_n)$ entscheidungsdefinit in Z, wenn für jedes n-tupel von nat. Zahlen $(k_1 \ldots k_n)$ entweder $R(k_1 \ldots k_n)$ oder $\overline{R}(k_1 \ldots k_n)$ in Z beweisbar ist (bisher ist keine entscheidungsdefinite Zahlrelation bekannt, die nicht schon in Z enthalten und entscheidungsdefinit wäre).

Denkt man sich das System Z sukzessive dadurch erweitert, daß man Variable für Klassen von Zahlen, Klassen von Klassen von Zahlen usw. sowie die zugehörigen Komprehensionsaxiome einführt, so erhält man eine (ins Transfinite fortsetzbare) Reihe von formalen Systemen, die den obigen Voraussetzungen genügen, und es zeigt sich, daß die Widerspruchsfreiheit (ω-Widerspruchsfreiheit) der vorhergehenden innerhalb der folgenden beweisbar ist. Auch die zum Beweise von Satz 1 konstruierten unentscheidbaren Sätze werden

[1]) Ein formales System S umfaßt ein anderes T, soll bedeuten: Jeder in T ausdrückbare (beweisbare) Satz ist auch in S ausdrückbar (beweisbar).

durch Hinzunahme höherer Typen und der zugehörigen Axiome entscheidbar; doch kann man in den höheren Systemen nach demselben Verfahren andere unentscheidbare Sätze konstruieren usf. Alle so konstruierten Sätze sind zwar in Z ausdrückbar (also zahlentheoretische Sätze), aber nicht in Z, sondern erst in höheren Systemen, z. B. dem der Analysis entscheidbar. Falls man die Mathematik typenlos aufbaut, wie es im Axiomensystem der Mengenlehre geschieht, so treten an Stelle der Typenerweiterungen Mächtigkeitsaxiome (d. h. Axiome, welche die Existenz von Mengen immer höherer Mächtigkeit fordern) und es ergibt sich daraus, daß gewisse in Z unentscheidbare arithmetische Sätze durch Mächtigkeitsaxiome entscheidbar werden, z. B. durch das Axiom, daß es Mengen gibt, deren Mächtigkeit größer ist als jedes α_n, wenn $\alpha_0 = \aleph_0$, $\alpha_{n+1} = 2^{\alpha_n}$.

25. Kolloquium (4. III. 1931).

Taussky: *Über ähnliche Abbildungen von Gruppen.*

In einer Gruppe G, deren Verknüpfungssymbol wir durch $+$ symbolisieren, bezeichnen wir (vgl. diese Ergebnisse *2*, S. 35, und Menger, Math. Zeitschr. *33*, S. 396) als rechtsseitigen Abstand der Elemente a und b das Elementepaar $(a-b,\ b-a)$ und als linksseitigen Abstand von a und b das Elementepaar $(-b+a,\ -a+b)$. Als rechtsseitige bzw. linksseitige Spiegelung von G am Elemente s bezeichnen wir die Transformation $x' = s - x$ bzw. $x' = -x + s$. Liegt s im Zentrum von G, so sind links- und rechtsseitige Spiegelung an s dieselbe Transformation, welche wir dann kurz die Spiegelung an s nennen. Insbesondere wollen wir die Spiegelung am Einheitselement 0 von G kurz als Spiegelung bezeichnen. Wir sagen, *G gestatte Spiegelung,* falls für je zwei Elemente der Abstand gleich dem ihrer (an 0) gespiegelten Elemente ist.

Daß G Spiegelung gestatte, ist offenbar gleichbedeutend damit, *daß für je zwei Elemente rechts- und linksseitiger Abstand identisch sind.* Insbesondere gestattet jede Abelsche Gruppe Spiegelung. Damit die Gruppe G Spiegelung gestatte, ist notwendig und hinreichend, *daß je zwei nicht vertauschbare Elemente a und b von G eine Quaternionengruppe erzeugen,* d. h. den Relationen genügen

$$2a = 2b,\quad 4a = 4b = 0,\quad -a = -b + a + b.$$

Eine eineindeutige Abbildung einer rechtsmetrischen bzw. linksmetrischen Gruppe G auf eine rechts- bzw. linksmetrische Gruppe G' möge nach einem Vorschlage von Menger *ähnlich* heißen, wenn für je vier Elemente a, b, c, d von G, falls a', b', c', d' ihre Bildelemente bezeichnen, aus $ab = cd$ die Beziehung $a'b' = c'd'$ folgt. Eine eineindeutige Abbildung einer rechts- bzw. linksmetrischen Gruppe G auf eine links- bzw. rechtsmetrische Gruppe G', bei welcher aus $ab = cd$ stets $a'b' = c'd'$ folgt, heiße *spiegelbildlich ähnlich.*

Jede isomorphe bzw. reziprok isomorphe Abbildung ist ähnlich bzw. spiegelbildlich ähnlich, wobei eine Abbildung von G auf G' isomorph bzw. reziprok isomorph heißt, wenn für je zwei Elemente a und b von G die Beziehung $(a+b)' = a' + b'$ bzw. $(a+b)' = b' + a'$ gilt. Dann gelten folgende Sätze:

Enthält die Gruppe G keine Quaternionengruppe als Untergruppe, so ist eine eineindeutige Abbildung von G auf G' dann und nur dann ähnlich bzw. spiegelbildlich ähnlich, wenn sie aus einer isomorphen bzw. reziprok isomorphen Abbildung von G auf G' und einer Translation von G' in sich zusammengesetzt ist. Enthält G eine Quaternionengruppe als Untergruppe, so ist eine eineindeutige Abbildung von G auf G' dann und nur dann ähnlich bzw. spiegelbildlich ähnlich, falls sie aus einer isomorphen bzw. reziprok isomorphen Abbildung von G auf G', einer Translation von G' in sich und eventuell der Spiegelung von G' (am Einheitselement von G') zusammengesetzt ist. Nennen wir zwei Gruppen ähnlich bzw. spiegelbildlich ähnlich, wenn eine ähnliche bzw. spiegelbildlich ähnliche Abbildung zwischen ihnen existiert, so ergibt sich aus den erwähnten Sätzen: *Zwei Gruppen sind dann und nur dann ähnlich, wenn sie isomorph sind, und dann und nur dann spiegelbildlich ähnlich, wenn sie reziprok isomorph sind.*

Nöbeling: *Zum Einbettungssatz für Kurven.*

In Verschärfung einer von Menger brieflich ausgesprochenen Vermutung läßt sich folgender Satz beweisen:

Ein eindimensionaler kompakter Raum R kann topologisch auf eine Teilmenge R' des dreidimensionalen Cartesischen xyz-Raumes R_3 derart abgebildet werden, *daß die senkrechte Parallelprojektion von R' auf eine Teilmenge K der xy-Ebene die folgenden Eigenschaften hat: K ist regulär-eindimensional, die Menge K_0 aller Punkte mit mehrpunktiger Urbildmenge in R' ist abzählbar; die Urbildmenge R'_0 von K_0 ist nulldimensional.* Ist R *regulär*-eindimensional, so hat jeder Punkt von K_0 eine *endliche* Urbildmenge in R'.

Der Beweis wird geführt, indem die Polyeder, mit deren Hilfe Menger[1]) den eindimensionalen Raum in den R_3 einbettet, speziell so gewählt werden, daß erstens der Durchschnitt zweier Polyeder des n-ten Schrittes entweder leer oder eine zur z-Achse parallele Strecke ist und so, daß zweitens gilt: ist S_n die Summe aller Polyeder des n-ten Schrittes und G eine beliebige Parallele zur z-Achse, so ist der Durchschnitt $G . S_n$ entweder zusammenhängend oder alle Komponenten dieses Durchschnittes sind Strecken, in welcher zwei Polyeder des n-ten Schrittes zusammenstoßen.

[1]) Dimensionstheorie, S. 296.

26. Kolloquium (18. II. 1931).

Nöbeling: *Zur Theorie der Mengenbereiche.*

Es sei $\mathfrak{B}$ ein Bereich separabler metrisierbarer Räume und $\mathfrak{T}$ ein Teilbereich von $\mathfrak{B}$. Der Bereich $\mathfrak{T}$ heiße *monoton in* $\mathfrak{B}$, wenn für jeden Raum R aus $\mathfrak{T}$ jeder zu $\mathfrak{B}$ gehörige Raum R', welcher $\subset R$ ist, zu $\mathfrak{T}$ gehört. $\mathfrak{T}$ heiße *additiv* bzw. *σ-additiv in* $\mathfrak{B}$, wenn jeder Raum R aus $\mathfrak{B}$, welcher Summe ist von endlich bzw. abzählbar vielen, in R abgeschlossenen und zu $\mathfrak{T}$ gehörigen Teilräumen, zu $\mathfrak{T}$ gehört. Ein in $\mathfrak{B}$ monotoner und σ-additiver Bereich heiße ein *Normalbereich in* $\mathfrak{B}$. Ist $\mathfrak{B}$ speziell der Bereich aller separablen metrisierbaren Räume, so lassen wir in allen diesen Definitionen den Zusatz „in $\mathfrak{B}$" weg. Einen Bereich nennen wir *abgeschlossen gegenüber Kompaktifizierung* oder kurz *abgeschlossen,* wenn er zu jedem seiner Räume R einen kompakten Raum mit einer zu R homöomorphen Teilmenge enthält[1]). Mit $\mathfrak{K}$ wollen wir den Bereich aller kompakten Räume bezeichnen. Dann gilt:

Satz 1. *Der einem in $\mathfrak{K}$ monotonen und in $\mathfrak{K}$ additiven Bereich übergeordnete Bereich ist abgeschlossen.*

Beispielsweise ist der Bereich aller endlichpunktigen Räume in $\mathfrak{K}$ monoton und additiv; also ist der übergeordnete Bereich, d. h. der Bereich aller höchstens regulär-eindimensionalen Räume abgeschlossen[2]).

Satz 2. *Ist $\mathfrak{N}$ ein abgeschlossener Normalbereich, so ist jeder $\mathfrak{N}$ übergeordnete Raum R enthalten in einem $\mathfrak{N}\,.\,\mathfrak{K}$ übergeordneten Raum R'.*

Da $\mathfrak{N}\,.\,\mathfrak{K}$ in $\mathfrak{K}$ monoton und additiv ist und da der $\mathfrak{N}\,.\,\mathfrak{K}$ übergeordnete Bereich Teil des $\mathfrak{N}$ übergeordneten Bereiches ist, folgt aus Satz 1 und 2 unmittelbar[3])

Satz 3. *Der einem abgeschlossenen Normalbereich übergeordnete Normalbereich ist abgeschlossen.*

Bedeutet $\mathfrak{S}_n$ ($n \geqq 0$) den Bereich aller höchstens n-dimensionalen (separablen und metrisierbaren) Räume, so ist der Bereich $\mathfrak{S}_{n+1}$ dem Bereich $\mathfrak{S}_n$ übergeordnet. Da $\mathfrak{S}_0$ ein abgeschlossener Normalbereich ist, gilt dasselbe nach Satz 3 für jeden Bereich $\mathfrak{S}_n$[4]).

---

[1]) Zur Terminologie vgl. Hurewicz, Math. Ann. *96*, S. 736, und Menger, Monatshefte f. Math. u. Phys. *36*, S. 197. Menger gebraucht dort statt „abgeschlossen gegenüber Kompaktifizierung" den Ausdruck „kompaktifizierbar".

[2]) Nöbeling, Math. Ann. *104*, S. 81.

[3]) Die Aufspaltung des Beweises von Satz 3 in die Sätze 1 und 2 erfolgte erst bei der Herausgabe dieses Heftes.

[4]) Hurewicz, l. c. [1]), und Monatshefte f. Math. u. Phys. *37*, S. 199.

27. Kolloquium (4. III. 1931).

Nöbeling: *Ein dimensionstheoretischer Isotopiesatz.*

Es seien R ein Raum und $f(R, 0)$ und $f(R, 1)$ zwei eindeutige stetige Bilder von R in einem Raume E. Wenn dann für jedes t mit $0 < t < 1$ ein topologisches Bild $f(R, t)$ von R im Raume E existiert derart, daß die Funktion $f(p, t)$ stetig in den Punkten p von R und dem Parameter t mit $0 \leqq t \leqq 1$ ist [also aus $p_i \to p$ und $t_i \to t$ folgt $f(p_i t_i) \to f(p, t)$], so heißen die Abbildungen $f(R, 0)$ und $f(R, 1)$ in E isotop[1]. Es gilt der

Satz: *Ist R ein kompakter n-dimensionaler Raum und E ein mindestens $(2n + 2)$-dimensionaler Euklidischer Raum* (oder der Fundamentalquader Q_ω des Hilbertschen Raumes), *so sind in E je zwei eindeutige stetige Bilder von R isotop.*

Für jedes p aus R ist die Menge $\{f(p, t)\}$ $(0 \leqq t \leqq 1)$ unseres Satzes ein stetiges Streckenbild. Es wäre denkbar, daß man sie sogar als Bogen wählen kann. In jedem höchstens $(2n+1)$-dimensionalen Euklidischen Raum sind natürlich je zwei Bilder $f(R, 0)$ und $f(R, 1)$ von R homotop[2], d. h. es läßt sich eine in $p \subset R$ und t $(0 \leqq t \leqq 1)$ stetige Funktion $f(p, t)$ finden, welche für $t = 0$ und $t = 1$ in die gegebenen Funktionen $f(p, 0)$ und $f(p, t)$ übergeht. Vermutlich kann man nun in Vervollständigung des obigen Satzes behaupten, daß man im R_{2n+2-k} die Funktion $f(p, t)$ so finden kann, daß für jedes t mit $0 < t < 1$ die Menge aller Punkte von $f(R, t)$ mit mehreren Urbildpunkten höchstens $(k—1)$-dimensional ist.

28. Kolloquium (11. III. 1931).

Nöbeling: *Eine Verschärfung des Mengerschen Einbettungssatzes.*

Nach einer brieflichen Mitteilung von Hurewicz gilt:

Jeder n-dimensionale Raum R läßt sich eindeutig und stetig auf eine n-dimensionale Teilmenge R' des R_{n+k} so abbilden, daß die Menge R'_0 aller Punkte von R' mit mehreren Urbildpunkten höchstens $(n—k)$-dimensional ist. Überdies kann man erreichen, daß auch die Urbildmenge von R'_0 höchstens $(n—k)$-dimensional ist[3].

Beweisskizze: Das System V aller eindeutigen stetigen Abbildungen von R auf Teilmengen des R_{n+k} wird durch Metrisation zu einem vollständigen metrischen Raum gemacht. Die Menge M aller Abbildungen, deren Existenz im Satze behauptet wird, läßt sich darstellen als Durchschnitt von in V offenen Mengen. Von diesen offenen Mengen wird gezeigt, daß sie in V dicht liegen. Also ist die Menge M nicht leer[4], d. h. es gibt eine Abbildung mit den verlangten Eigenschaften.

---

[1]) Vgl. z. B. Lefschetz, Topology (1930), S. 76.

[2]) Ausführung wird in den Monatsh. f. Math. u. Phys. erscheinen.

[3]) Vgl. die inzwischen erschienenen Noten von Hurewicz, Proc. Amsterdam *34*, Nr. 3, und Wiener akad. Anz. 1931, Nr. 12.

[4]) Hausdorff, Grundzüge der Mengenlehre, 1914, S. 327.

Kann man den Raum R topologisch in den R_{2n+1} so einbetten, daß sogar die Projektion von R auf eine Teilmenge R' des R_{n+k} die im Satze genannten Eigenschaften hat?[1])

29. Kolloquium (20. V. 1931).

Beer: Der Beweis von W. L. Ayres (Trans. Am. Math. Soc. *33*, S. 252) für den von G. T. Whyburn vermuteten Satz, daß in jedem Kontinuum K die Punkte n-ter Ordnung ($n > 2$) eine höchstens nulldimensionale Menge bilden und daß ferner die Menge der gewöhnlichen Punkte von K, d. h. der Punkte zweiter Ordnung von K, Summe einer höchstens nulldimensionalen Menge von lokalen Zerlegungspunkten von K und von höchstens abzählbarvielen, nach Tilgung ihrer Endpunkte in K offenen Bogen ist. In einer anschließenden Arbeit (Trans. Am. Math. Soc. *33*, S. 663) zeigt Menger, daß die Menge M aller gewöhnlichen Punkte von K Summe ist von der höchstens nulldimensionalen Menge aller Punkte, in denen M nulldimensional ist, von der höchstens abzählbaren Menge aller Endpunkte von M und von dem *gewöhnlichen Kern* von K, d. i. die Menge aller gewöhnlichen Punkte von M. Dabei ist der gewöhnliche Kern von K entweder mit R identisch (und zwar dann und nur dann, wenn K mit einem Kreis homöomorph ist) oder die Summe höchstens abzählbarvieler paarweise fremder Mengen, von denen jede mit einer Geraden homöomorph und in K offen und in Bezug auf diese beiden Eigenschaften saturiert ist.

30. Kolloquium (10. VI. 31).

Nöbeling: *Über die Kurven, deren Punkte die Ordnung n oder 2n—2 haben.*

Eine solche Kurve wurde von Urysohn (Verhand. Ak. Amsterd. *13*, 1927) als eindeutiges stetiges Bild einer nulldimensionalen Menge konstruiert. Diese Konstruktion läßt sich formal so abändern, daß die Kurve als Durchschnitt einer monoton fallenden Folge von Polygonflächen erscheint. Wir diskutieren nur den Fall, daß n ($\geqq 4$) ungerade ist. Wir bezeichnen in der Ebene als n-Stern eine Fläche S, deren Rand ein Polygon ist, welche Summe von $2n$ Strecken ist und folgende Eigenschaften hat: Erstens sind, wenn wir den Rand durchlaufen, die inneren Winkel an den Eckpunkten abwechselnd $\geqq \pi$ und $< \pi$, zweitens gibt es einen inneren Punkt z von S, so daß jeder Strahl von z mit dem Rand von S genau einen Punkt gemein hat. Wir bezeichnen diesen Punkt z als ein Zentrum von S und nennen diejenigen n Eckpunkte von S, für welche der innere Winkel $< \pi$ ist, Ecken von S, die n übrigen Eckpunkte, für welche der innere Winkel $\geqq \pi$ ist, Pseudoecken von S. Durchlaufen wir den Rand in positivem Sinne, beginnend bei einer Ecke e_1, so bezeichnen wir die Ecken und Pseudoecken, in derselben Reihen-

[1]) Vgl. das 25. Kolloquium (S. 13 dieses Heftes).

folge, wie wir sie passieren, mit $e_1, p_1, e_2, p_2, \ldots, e_n, p_n$; diese Numerierung wollen wir die normale Numerierung nennen.

Wir definieren nun induktiv ein System $K_{i_1 i_2 \ldots i_k}$ $(k=1, 2, \ldots)$ von n-Sternen. Mit K_1 bezeichnen wir ein regelmäßiges n-Eck von einem Umfange <1, welches dadurch zu einem n-Stern gemacht wird, daß wir die Mittelpunkte der Seiten zu Pseudoecken und den Mittelpunkt des n-Ecks zum Zentrum erklären. Dieser n-Stern besitzt die Eigenschaft, daß jedes Dreieck $(p_{i-1}\, e_i\, p_i)$ (normale Numerierung!) einen Umfang $<\frac{1}{2}$ hat. Wir machen nun die für $k=1$ soeben realisierte Induktionsvoraussetzung, es liege ein n-Stern $K_{i_1 i_2 \ldots i_k}$ mit einem Umfange $<\frac{1}{2^{k-1}}$ vor, derart, daß jedes Dreieck $(p_{i-1}\, e_i\, p_i)$ einen Umfang $<\frac{1}{2^k}$ besitzt, und konstruieren n Teil-n-Sterne $K_{i_1 i_2 \ldots i_k i}$ $(i=1, 2, \ldots, n)$ folgendermaßen: Wir wählen auf jeder der n Verbindungsstrecken eines Zentrums z von $K_{i_1 i_2 \ldots i_k}$ mit der Pseudoecke p_i $(i=1, 2, \ldots, n)$ $\frac{n-1}{2}$ voneinander und von z verschiedene Punkte $e_i^1 = p_i$, $e_i^2, \ldots, e_i^{\frac{n-1}{2}}$ so numeriert, wie wir sie beim Durchlaufen der Verbindungsstrecke von p_i nach z passieren. Da jeder von z ausgehende Strahl mit dem Rande von $K_{i_1 i_2 \ldots i_k}$ genau einen Punkt gemein hat, so ist die kleinste die n Punkte mit unterem Index $i-1$ oder i (einschließlich e_i) enthaltende konvexe Polygonfläche F_i ein Fünfeck. Alle Punkte e_i, e_{i-1}^j, e_i^j liegen auf dem Rande von F_i. Da wir vorausgesetzt haben, daß jedes Dreieck $(e_{i-1}^1\, e_i\, e_i^1)$ einen Umfang $<\frac{1}{2^k}$ hat, so können wir die Punkte e_i^j so nahe bei e_i^1 gewählt denken, daß auch die Fünfecke F_i einen Umfang $<\frac{1}{2^k}$ haben. Wir wollen nun in F_i einen n-Stern $K_{i_1 i_2 \ldots i_k i}$ konstruieren. Die (auf dem Rande von F_i liegenden) n Punkte e_i, e_{i-1}^j, e_i^j erklären wir zu seinen Eckpunkten. Durch Tilgung dieser n Punkte zerfällt der Rand von F_i in n Strecken. Den Mittelpunkt der Strecke $\left(e_{i-1}^{\frac{n-1}{2}}, e_i^{\frac{n-1}{2}}\right)$ erklären wir zur Pseudoecke. Ebenso erklären wir die Mittelpunkte der beiden Strecken (e_{i-1}^1, e_i) und (e_i, e_i^1) zu Pseudoecken. Die Mittelpunkte der übrigen n-3 Strecken $(e_{i-1}^{j-1}, e_{i-1}^j)$ und (e_i^{j-1}, e_i^j) verbinden wir mit einem inneren Punkt z_i von F_i und erklären nun auf jeder dieser Strecken einen von z_i verschiedenen Punkt zur Pseudoecke. Auf diese Weise entsteht ein n-Stern $K_{i_1 i_2 \ldots i_k i}$. Indem wir die zuletzt definierten n-3 Pseudoecken hinreichend nahe an den Streckenmittelpunkten wählen, können wir, da F_i einen Umfang $< \frac{1}{2^k}$ hat, erreichen, daß erstens auch der n-Stern $K_{i_1 i_2 \ldots i_{ki}}$ einen Umfang $< \frac{1}{2^k}$ hat und zweitens alle Dreiecke $(p_{i-1}\, e_i\, p_i)$ (normale Numerierung) in $K_{i_1 i_2 \ldots i_k i}$ einen Umfang $< \frac{1}{2^{k+1}}$ haben. Damit ist die induktive Konstruktion abgeschlossen.

Konstruieren wir auf diese Weise für jedes k ein System von n-Sternen $K_{i_1 i_2 \ldots i_k}$, so ist die Menge $C = \prod_k \sum K_{i_1 i_2 \ldots i_k}$ aller Punkte, die in unendlich vielen n-Sternen liegen, eine reguläre Kurve, deren Punkte (abgesehen von den Eckpunkten des Ausgangs-n-Ecks K_1) die Ordnung n oder $2n-2$ haben.

31. Kolloquium (24. VI. 31).

Beer: *Zwei Kurven, deren sämtliche Punkte die Ordnungen 3 und 6 bzw. 3 und 8 besitzen.*

Es sei Δ ein gleichseitiges Dreieck. Man verbinde je zwei Seitenmittelpunkte von Δ durch ein Kreisbogenzweieck, welches ganz innerhalb von Δ verläuft, zur Verbindungsstrecke seiner beiden Ecken symmetrisch ist und so, daß die drei Zweiecke bis auf die Ecken paarweise fremd sind. Nun tilge man aus Δ alle inneren Punkte der Zweiecke. Der Rest ist Summe von vier Dreiecken Δ_{i_1} ($i_1 = 1, 2, 3, 4$), deren Seiten Strecken oder Kreisbogen sind. Jedes dieser Dreiecke Δ_{i_1} zerlege man analog wie Δ durch Tilgung der inneren Punkte von drei schmalen, die Seitenmittelpunkte von Δ_{i_1} verbindenden Kreisbogenzweiecken in vier Dreiecke $\Delta_{i_1 i_2}$ ($i_2 = 1, 2, 3, 4$). Dieses Verfahren setze man in infinitum fort. Die Menge aller Punkte von Δ, die nicht getilgt werden, ist eine reguläre Kurve K, deren Punkte, mit Ausnahme der Eckpunkte e_1, e_2, e_3 von Δ, die Ordnung 3 oder 6 haben; die Punkte e_j sind von zweiter

Figur zu S. 20.

2*

Ordnung. Fügt man nun im R_3 drei zu K homöomorphe Kurven K^i $(i=1, 2, 3)$ so zusammen, daß für die den Eckpunkten e_j entsprechenden Punkte $\overset{i}{e_j}$ die Gleichungen $\overset{1}{e_j}=\overset{2}{e_j}=\overset{3}{e_j}$ $(j=1, 2, 3)$ gelten, daß aber im übrigen die Kurven K^i paarweise fremd sind, so ist $K^*=K^1+K^2+K^3$ eine Kurve, die in jedem ihrer Punkte die Ordnung 3 oder 6 besitzt.

Zur Konstruktion einer Kurve, deren sämtliche Punkte die Ordnungen 3 und 8 besitzen, tilgt man aus einem gleichseitigen Dreieck Δ die inneren Punkte der in der Figur schraffierten Kreisbogenzwei- bzw. -dreiecke. Der Rest ist Summe von 13 Dreiecken Δ_{i_1}, deren Seiten Strecken oder Kreisbogen sind. Jedes dieser Dreiecke zerlegt man analog wie Δ in 13 Dreiecke $\Delta_{i_1 i_2}$ usw. Die Menge aller nicht getilgten Punkte von Δ ist eine reguläre Kurve K, deren Punkte, mit Ausnahme der Eckpunkte von Δ, die Ordnung 3 oder 8 haben. Durch Zusammenfügung von 4 homöomorphen Kurven sowie bei der Konstruktion von K^* erhält man eine Kurve der gewünschten Art.

Beer: *Ein Problem über reguläre Kurven.*

Es sei K eine reguläre Kurve. Dann existiert bekanntlich in K ein erzeugendes Mengensystem[1]) $\{K_{i_1 \ldots i_n}\}$, das aus Kontinuen besteht und die Eigenschaft besitzt, daß je zwei Kontinua desselben Schrittes nur endlich viele Punkte gemein haben. Ist p ein beliebiger Punkt von K und $S_n(p)$ die Summe aller p enthaltenden Mengen $K_{i_1 \ldots i_n}$ des n-ten Schrittes, so ist die Mächtigkeit $s_n(p)$ der Menge $S_n(p) \cdot \{\overline{K-S_n(p)}\}$ endlich und die Ordnung $o(p)$ des Punktes p ist $\leq \liminf_{n=\infty} s_n(p)$. Welche speziellen Struktureigenschaften des erzeugenden Mengensystems sind nun dafür hinreichend, daß für jeden Punkt p von K in dieser Ungleichung das Gleichheitszeichen gilt? Ein Beispiel für ein erzeugendes Mengensystem, bei welchem das Gleichheitszeichen gilt, ist das erzeugende Dreiecksystem der Sierpińskischen Dreieckskurve. Die Kenntnis solcher Bedingungen würde bei Konstruktionen spezieller Kurventypen (z. B. der Urysohnschen Kurven mit den Ordnungen n und $2n-2$) den vielfach komplizierten Nachweis, daß die konstruierte Kurve die verlangten Ordnungen besitzt, überflüssig machen.

Gödel: *Eine Eigenschaft der Realisierungen des Aussagenkalküls.*

In Beantwortung einer mir von Menger brieflich gestellten Frage kann man folgenden Satz beweisen:

Es sei irgend eine Menge $\mathfrak{S}$ von Dingen $p, q, r, \ldots$ gegeben, in der eine einstellige Operation $\sim p$ und eine zweistellige Operation $p \supset q$ definiert sind, die nicht aus der Menge $\mathfrak{S}$ herausführen. Es sei ferner $\mathfrak{T}$ eine echte Teilmenge von $\mathfrak{S}$, die folgenden Bedingungen genügt:

[1]) Vgl. Dimensionstheorie, S. 287.

I. Falls p, q, r beliebige Dinge aus $\mathfrak{S}$ sind, gehören immer die folgenden drei Dinge zu $\mathfrak{T}$:

a) $(\sim p \supset p) \supset p$,

b) $p \supset (\sim p \supset q)$,

c) $(p \supset q) \supset [(q \supset r) \supset (p \supset r)]$.

II. Wenn sowohl p als $p \supset q$ zu $\mathfrak{T}$ gehören, dann gehört auch q zu $\mathfrak{T}$.

Unter diesen Voraussetzungen gibt es immer *eine Einteilung von* $\mathfrak{S}$ *in zwei elementefremde Klassen:* $\mathfrak{S} = \mathfrak{W} + \mathfrak{F}$, $\mathfrak{T} \subseteq \mathfrak{W}$, *so daß*

a) von den beiden Dingen p und $\sim p$ *immer genau eines zu* $\mathfrak{W}$ *und eines zu* $\mathfrak{F}$ *gehört,*

b) das Ding $p \supset q$ *dann und nur dann zu* $\mathfrak{F}$ *gehört, wenn p zu* $\mathfrak{W}$ *und q zu* $\mathfrak{F}$ *gehört.*

Das heißt: Wenn eine beliebige Realisierung der Axiome des Aussagenkalküls gegeben ist, so können die Elemente („Aussagen") in zwei fremde Klassen eingeteilt werden, welche sich ganz so wie die Klassen der wahren und der falschen Aussagen des gewöhnlichen Aussagenkalküls verhalten.

Beweisskizze: Man denke sich $\mathfrak{S}$ wohlgeordnet; dann ordne man jeder Ordinalzahl α eine Teilklasse $\mathfrak{T}_\alpha$ von $\mathfrak{S}$ zu durch die Festsetzungen:

1. $\mathfrak{T}_0 = \mathfrak{T}$.

2. Wenn α Limeszahl ist, $\mathfrak{T}_\alpha = \sum\limits_{\mu<\alpha} \mathfrak{T}_\mu$.

3. Wenn a das erste Element aus $\mathfrak{S}$ ist, für welches weder a noch $\sim a$ zu $\mathfrak{T}_\alpha$ gehört: $\mathfrak{T}_{\alpha+1}$ = Menge derjenigen $x \varepsilon \mathfrak{S}$, für welche $(a \supset) x \varepsilon \mathfrak{T}_\alpha$. Falls ein solches a nicht existiert, sei $\mathfrak{T}_{\alpha+1} = \mathfrak{T}_\alpha$.

Durch transfinite Induktion zeigt man:

A. Jedes $\mathfrak{T}_\alpha$ genügt den beiden Forderungen I und II, die oben für $\mathfrak{T}$ aufgestellt wurden.

B. $\mathfrak{T}_\alpha \subseteq \mathfrak{T}_\beta$ für $\alpha \leqq \beta$.

C. x und $\sim x$ gehören niemals beide zu $\mathfrak{T}_\alpha$.

Wenn ν die kleinste Ordinalzahl ist, für welche $\mathfrak{T}_\nu = \mathfrak{T}_{\nu+1}$, so hat $\mathfrak{W} = \mathfrak{T}_\nu$, $\mathfrak{F} = \mathfrak{S} - \mathfrak{T}_\nu$ die verlangten Eigenschaften.

B zeigt man z. B. folgendermaßen: Weil für $\mathfrak{T}_\alpha$ nach induktiver Annahme *A* erfüllt ist, gilt für beliebige $x \varepsilon \mathfrak{S}: x \supset . a \supset x \varepsilon \mathfrak{T}_\alpha$ und daher folgt aus $x \varepsilon \mathfrak{T}_\alpha$ $a \supset x \varepsilon \mathfrak{T}_\alpha$, d. h. $x \varepsilon \mathfrak{T}_{\alpha+1}$, also $\mathfrak{T}_\alpha \subseteq \mathfrak{T}_{\alpha+1}$. Daß für $\mathfrak{W} = \mathfrak{T}_\nu$ Bedingung (*b*) erfüllt ist, folgt daraus, daß wegen *A* gilt: $x \supset . y \supset x \varepsilon \mathfrak{T}_\nu$, $\sim y \supset . y \supset x \varepsilon \mathfrak{T}_\nu$, $\sim x \supset [y \supset . \sim (y \supset x)] \varepsilon \mathfrak{T}_\nu$ und zwar für beliebige $x, y \varepsilon \mathfrak{S}$.

Gesammelte Mitteilungen 1930/31.

Eine elementare Bemerkung über die Struktur logischer Formeln. Von Karl Menger.

Bedient man sich im Aussagenkalkül der klammerlosen Schreibweise von Łukasiewicz (vgl. diese Ergebnisse, Heft *2*, S. 14), indem man Aussagen mit kleinen lateinischen Buchstaben $p, q, \ldots$ bezeichnet, Cpq gleichbedeutend mit „p impliziert q" und Np gleichbedeutend mit „Non p" schreibt, so wird jede Aussage eine geordnete Menge von C, N und kleinen Buchstaben. Z. B. wird der Syllogismus durch $CCCpqCqrCpr$ symbolisiert. Es gilt dann der Satz:

Damit eine endliche geordnete Menge $\mathfrak{M}$ von C, N und kleinen Buchstaben eine Aussage darstelle, ist notwendig und hinreichend, daß folgende Bedingungen erfüllt sind:

1. *$\mathfrak{M}$ enthält genau um einen kleinen Buchstaben mehr als C.*

2. *Jeder echte* (d. h. nicht alle Elemente enthaltende) *Anfangsabschnitt der geordneten Menge $\mathfrak{M}$ enthält mindestens so viele C als kleine Buchstaben.*

3. *Das letzte Element von $\mathfrak{M}$ ist kein N.* (Aus 1. und 2. folgt überdies, daß das letzte Element von $\mathfrak{M}$ kein C ist, also muß es ein kleiner Buchstabe sein.)

Der Beweis der Notwendigkeit ergibt sich durch vollständige Induktion nach der Mächtigkeit von $\mathfrak{M}$. Enthält die Aussage $\mathfrak{M}$ nur ein Element, so ist dasselbe notwendig ein kleiner Buchstabe, so daß die Bedingungen 1., 2., 3. erfüllt sind. Ist die Notwendigkeit für alle Aussagen, die weniger als n Elemente enthalten, bewiesen und $\mathfrak{M}$ eine Aussage mit n Elementen, so besteht $\mathfrak{M}$ entweder aus $n-1$ Symbolen N, gefolgt von einem kleinen Buchstaben, in welchem Falle die Bedingungen 1., 2., 3. erfüllt sind, oder $\mathfrak{M}$ hat die Form CPQ, wo P und Q Aussagen mit weniger als n Elementen sind, die daher nach induktiver Annahme den Bedingungen genügen, woraus man mühelos folgert, daß auch $\mathfrak{M}$ den Bedingungen 1., 2., 3. genügt. Daß die Bedingungen hinreichend sind, ergibt sich daraus, daß man durch vollständige Induktion nach der Mächtigkeit von $\mathfrak{M}$ jede den drei Bedingungen genügende Menge $\mathfrak{M}$ in eine Aussage übersetzen kann. Der bewiesene Satz enthält, wie man unmittelbar bestätigt, das

Korollar. *Ist E irgend ein Element der eine Aussage darstellenden geordneten Menge $\mathfrak{M}$, so enthält $\mathfrak{M}$ genau einen mit dem*

Element E beginnenden Abschnitt, welcher eine Aussage darstellt. Wir bezeichnen diesen (nicht notwendig bis ans Ende von $\mathfrak{M}$ reichenden) Abschnitt mit $\mathfrak{M}_E$ und nennen ihn *die mit E beginnende, in* $\mathfrak{M}$ *eingeschachtelte Aussage.* Es sei z. B. $\mathfrak{M}$ der Syllogismus $CCCpqCqrCpr$; ist dann E das zweite C, so ist $\mathfrak{M}_E = CCpqCqr$; ist E das dritte C, so ist $\mathfrak{M}_E = Cpq$; ist E ein kleiner Buchstabe, etwa p, so ist die aus ihm bestehende Aussage die Aussage $\mathfrak{M}_E$. Es ist zu beachten, daß zu einem Element E von $\mathfrak{M}$ kein mit E endender Abschnitt von $\mathfrak{M}$, der eine Aussage darstellt, existieren muß (es existiert keiner, wenn $E = C$ oder N), und daß anderseits mehrere in E endende Abschnitte von $\mathfrak{M}$ existieren können (z. B. enden mit dem achten Element des Syllogismus die beiden Abschnitte $CCpqCq$r und Cqr).

Der Begriff der eingeschachtelten Aussage ermöglicht auch eine Präzisierung dessen, was im Funktionenkalkül als „*Wirkungsbereich*“ eines Quantifikators bezeichnet wird, und der Forderung, „daß die Wirkungsbereiche von Quantifikatoren, die zu gleichbezeichneten Variablen gehören, sich nicht überschneiden“ (Hilbert-Ackermann, Grundz. d. theor. Logik, S. 52).

Über das Hilbertsche Axiomensystem der Geometrie[1]. Von A. Wald.

Das Axiom I_5: „Irgend drei Punkte einer Ebene, die nicht auf ein und derselben Geraden liegen, bestimmen diese Ebene“ ist eine Folge der Verknüpfungs- und Anordnungsaxiome, falls postuliert wird: Eine Ebene ist durch die Menge aller in ihr liegenden Punkte bestimmt (d. h. zwei Ebenen, welche sämtliche Punkte gemein haben, sind identisch)[2].

Das Axiom I_7: „Wenn zwei Ebenen α, β einen Punkt A gemein haben, so haben sie wenigstens noch einen weiteren Punkt B gemein“ kann bei Benützung der Anordnungsaxiome durch folgende schwächere Fassung ersetzt werden: Es gibt mindestens einen Punkt 0 derart, daß je zwei Ebenen, die den Punkt 0 gemein haben, wenigstens noch einen weiteren Punkt gemein haben.

Das Axiom I_2: „Irgend zwei voneinander verschiedene Punkte einer Geraden bestimmen diese Gerade“ kann mit Hilfe des Parallelenaxioms ohne Benützung von Axiom I_5 bewiesen werden[3].

Sind A, B, C, D, E fünf paarweise verschiedene Punkte einer Geraden, über welche folgende Konventionen getroffen seien:

---

[1]) D. Hilbert, Grundlagen der Geometrie. 4. Auflage.

[2]) S. Weinlös (Fund. Math. *11*, S. 218) leitet diesen Satz unter Berufung auf A. Lindenbaum aus den übrigen Axiomen, aber mit Benützung des Parallelenaxioms her.

[3]) S. Weinlös beweist (a. a. O., S. 214) dasselbe mit Verwendung des Axioms I_5.

A	liege zwischen	B	und	C	und auch zwischen	D	und	E
B	„ „	A	„	D	„ „ „	C	„	E
C	„ „	A	„	E	„ „ „	B	„	D
D	„ „	A	„	C	„ „ „	B	„	E
E	„ „	A	„	B	„ „ „	C	„	D,

dann genügen diese fünf Punkte sämtlichen linearen Axiomen der Anordnung (II_1—II_3).

Das Axiom III_3 lautet: „Es seien AB und BC zwei Strecken ohne gemeinsame Punkte auf der Geraden a und ferner $A'B'$ und $B'C'$ zwei Strecken auf derselben oder auf einer anderen Geraden a' ebenfalls ohne gemeinsame Punkte: wenn dann $AB \equiv A'B'$ und $BC \equiv B'C'$ ist, so ist auch stets $AC \equiv A'C'$." Dieser Satz ist eine Folge der übrigen Kongruenzaxiome und der Axiomgruppen I und II.

Das Axiom III_4 lautet: „Es sei ein Winkel $\sphericalangle(hk)$ in einer Ebene α und eine Gerade a' in einer Ebene α', sowie eine bestimmte Seite von a' in α' gegeben. Es bedeute h' einen Halbstrahl der Geraden a', der vom Punkte $0'$ ausgeht: dann gibt es in der Ebene α' einen und nur einen Halbstrahl k', so daß der Winkel $\sphericalangle(hk)$ kongruent dem Winkel $\sphericalangle(h'k')$ ist, und zugleich alle inneren Punkte des Winkels $\sphericalangle(h'k')$ auf der gegebenen Seite von a' liegen, in Zeichen $\sphericalangle(hk) \equiv \sphericalangle(h'k')$." Statt dessen genügt es, für einen einzigen Winkel $\sphericalangle(hk)$ in α zu fordern, daß zu jedem Halbstrahl h' in α' mindestens ein Halbstrahl k' in α' existiert derart, daß $\sphericalangle(hk) \equiv \sphericalangle(h'k')$ ist.

Hausdorffsche und mengentheoretische Dimension. Von Georg Nöbeling.

Es sei M eine Teilmenge des Euklidischen R_n. Man überdecke M durch endlich- oder abzählbar unendlichviele Kugeln K_j mit Durchmessern $d_j < \rho$, bilde für irgendeine nichtnegative Zahl $p \leqq n$ die untere Grenze $L_p(\rho, M)$ aller Summen $\sum_j d_j^p$ für alle solche Kugelfolgen $\{K_j\}$ und setze $L_p(M) = \lim_{\rho \to 0} L_p(\rho, M)$. Es gibt offenbar genau eine Zahl $p = p(M)$ derart, daß für jedes $q > p$ die Gleichung $L_q(M) = 0$ und für jede Zahl $q < p$ die Gleichung $L_q(M) = \infty$ gilt. Diese eindeutig definierte Zahl p nennen wir die Hausdorffsche Dimension[1]) der Menge M und behaupten den

Satz. *Für jede Menge M ist die Hausdorffsche Dimension mindestens gleich der mengentheoretischen Dimension.*

Der Beweis geschieht durch Induktion nach der mengentheoretischen Dimension dim M. Für alle mindestens nulldimensionalen Mengen M ist auch $p(M) \geqq 0$. Sei für mindestens k-dimensionale Mengen auch $p \geqq k$ und M eine mindestens $(k+1)$-dimensionale

[1]) Hausdorff, Math. Ann. 79, S. 157. Übrigens nennt Hausdorff p nur dann die Dimension von M, wenn $0 < L_p(M) < \infty$ ist.

Menge. Wir behaupten, daß auch $p(M) \geqq k+1$ ist. Da M mindestens $(k+1)$-dimensional ist, gibt es einen Punkt P von M und eine Zahl r_0 derart, daß für jede $(n-1)$-dimensionale Sphäre S_r mit dem Radius $r \leqq r_0$ der Durchschnitt $M \cdot S_r$ mindestens k-dimensional ist, und daher nach Induktionsvoraussetzung $p(M \cdot S_r) \geqq k$ ist. Es sei nun $\{K_{ij}\}$ eine Doppelfolge von n-dimensionalen Kugeln mit der Eigenschaft, daß für jedes j die Menge M von den Kugeln $K_{i1}, K_{i2}, \ldots$ überdeckt wird und K_{ij} einen Durchmesser $d_{ij} < \frac{1}{i}$ hat. Für eine beliebige Zahl $q < k+1$ und $0 < x \leqq r_0$ setzen wir nun $f_{ij}(x) = d_{ij}^{q-1}$, falls $S_x \cdot K_{ij}$ nicht leer ist, andernfalls $f_{ij}(x) = 0$. Außerdem setzen wir $s_i(x) = \Sigma f_{ij}(x)$. Dann folgt aus der Induktionsvoraussetzung, daß für $i \to \infty$ die Zahlen $s_i(x)$ für jedes x gegen ∞ konvergieren. Also kann man für jedes C ein i_0 so finden, daß für $i > i_0$

$$\int_0^{r_0} s_i(x)\,dx = \sum_j \int_0^{r_0} f_{ij}(x)\,dx > C$$

ist. Nun ist $f_{ij}(x) = d_{ij}^{q-1}$ für die Zahlen x eines Intervalls, dessen Länge höchstens d_{ij} ist, und im übrigen ist $f_{ij}(x) = 0$. Also ist $\int_0^{r_0} f_{ij}(x)\,dx \leqq d_{ij}^q$ und daher $\sum_j d_{ij}^q > C$. Hieraus folgt $L_q(M) = \infty$. Dies gilt für jedes $q < k+1$. Also ist $p(M) \geqq k+1$, wie wir behauptet hatten. Damit ist die Induktion abgeschlossen und unser Satz bewiesen.

Über Ecken und Spitzen abgeschlossener Mengen der Ebene.

Von Georg Nöbeling.

Ist M eine abgeschlossene Menge der Ebene E, so nennen wir einen nicht isolierten Punkt p von M eine Spitze von M, wenn sich zu jedem $\varepsilon > 0$ ein zu M fremder offener Kreissektor mit dem Mittelpunkt p finden läßt, dessen Winkelöffnung ω der Bedingung $2\pi - \varepsilon < \omega < 2\pi$ genügt. Ist der nicht isolierte Punkt p von M keine Spitze von M, existiert aber ein zu M fremder offener Kreissektor mit dem Mittelpunkt p, dessen Winkelöffnung ω der Bedingung $\pi < \omega < 2\pi$ genügt, so heißt p eine Ecke von M[1]). Wir behaupten den

Satz. *Die Menge A aller Ecken und Spitzen einer abgeschlossenen ebenen Menge ist höchstens abzählbar.*

Beweis. Ist p ein Punkt von A, so existiert ein zu M fremder Kreissektor $K(p)$ mit dem Mittelpunkt p, dessen Winkelöffnung ω der Bedingung $\pi < \omega < 2\pi$ genügt. Ohne Beschränkung der Allgemeinheit kann man annehmen, daß ω und der Radius von $K(p)$ rational sind und die Winkelhalbierende von $K(p)$ mit einer festen Richtung der Ebene einen rationalen Winkel einschließt. Hieraus folgt aber, daß die Menge A in endlich- oder abzählbarviele Teilmengen

[1]) Rosenthal, Math. Ann. *73*, S. 480.

$A_i (i=1, 2, \ldots)$ zerfällt, so daß die Sektoren $K(p)$ und $K(p')$ zweier Punkte p und p' derselben Menge A_i durch Translation in einander übergehen. Es genügt, für beliebiges i zu zeigen, daß die Menge A_i höchstens abzählbar ist. Nun enthält die Menge A_i nur isolierte Punkte; denn wäre etwa p ein Häufungspunkt von A_i und p' ein hinreichend benachbarter Punkt, so müßte entweder p in $K(p')$ oder p' in $K(p)$ enthalten sein.

Angelegenheiten des Kolloquiums und Nachrichten.

Wiener Dissertationen im Zusammenhange mit dem Kolloquium 1930/31.

G. Nöbeling, „Über eine n-dimensionale Universalmenge im R_{2n+1}“. (Vgl. diese Ergebnisse *2*, S. 29, und Math. Ann. *104*, S. 71.)

K. Vanek, „Über Zerlegungseigenschaften im kleinen zusammenhängender Kurven“. (Vgl. diese Ergebnisse *2*, S. 33 und Monatshefte f. Math. u. Phys. *38*, S. 5.)

Druck von Gottlieb Gistel & Cie., Wien.

Bericht über das Kolloquium 1931/32.

32. Kolloquium (27. X. 1931).

Menger: *Über die lokale Dimension von Mengensummen.*

Es scheint bisher nichts Positives darüber bekannt zu sein, wie sich die Dimension der Summe zweier Mengen in einem Punkt zu den Dimensionen der Summanden in dem betreffenden Punkt verhält. Klar ist nur, daß die Summe zweier im Punkte p nulldimensionaler Mengen, selbst wenn dieselben kompakt sind, in p von beliebig hoher Dimension sein kann. Im folgenden wird bewiesen:

Sind M und N zwei in ihrer Summe abgeschlossene separable Mengen, so gilt:

$$Max\,(dim_p M, dim_p N) \leqq dim_p (M+N) \leqq Max\,(dim_p M, dim_p N, 1+dim\, M.N).$$

Zum Beweis legen wir $M+N$ als Raum zugrunde. Es sind dann M und N abgeschlossene Mengen, d. h. es gilt $M=\overline{M}$ und $N=\overline{N}$. Sind U und V irgend zwei den Punkt p enthaltende, in M bzw. in N offene Mengen, so lehren einfache Beispiele, daß $U+V$, selbst wenn M und N kompakt sind, nicht notwendig in $M+N$ offen ist; doch bestätigt man:

a) Der (im Raume $M+N$) offene Kern W von $U+V$ enthält jeden Punkt von $U.V$, insbesondere also den Punkt p.

Ferner gilt $(U-U.N)+(V-V.M)\subset W, \overline{W}\subset \overline{U}+\overline{V}$ und daher

$$\begin{aligned}\overline{W}-W &\subset \overline{U}+\overline{V}-(U-U.N)-(V-V.M)\\ &\subset (\overline{U}-U)+(\overline{V}-V)+U.N+V.M.\end{aligned}$$

Wegen $U\subset M$ und $V\subset N$ ergibt sich also für die Begrenzung der im Raum $M+N$ offenen Menge W

b) $B(W)=\overline{W}-W\subset(\overline{U}-U)+(\overline{V}-V)+M.N.$

Wegen der Abgeschlossenheit von M und N in $M+N$ folgen aus $U\subset M$ und $V\subset N$ die Beziehungen $\overline{U}\subset M$ und $\overline{V}\subset N$; da U in M offen ist, ist $\overline{U}-U$ in M, also im Raume $M+N$ abgeschlossen; ebenso ist $\overline{V}-V$ in $M+N$ abgeschlossen. Also gilt:

c) Die Durchschnitte von $B(W)$ mit den drei Summanden auf der rechten Seite von *b)* sind abgeschlossen.

Es sei nun $\dim_p M=m$ und $\dim_p N=n$. Offenbar gilt dann $\dim_p(M+N)\geqq Max\,(m,n)$. Um die zweite Hälfte der eingangs be-

1

haupteten Ungleichung zu beweisen, geben wir irgend eine Umgebung Z von p vor. Wegen $\dim_p M = m$ existiert eine p enthaltende, in M offene Menge $U \subset Z$, so daß $M.(\bar{U}-U) = \bar{U}-U$ höchstens $(m-1)$-dimensional ist. Entsprechend existiert eine p enthaltende, in N offene Menge $N \subset Z$, so daß $\dim(\bar{V}-V) \leqq n-1$. Der offene Kern von $U+V$ ist $\subset Z$, wegen *a)* eine Umgebung von p und besitzt wegen *b)* eine Begrenzung, welche Summe dreier Mengen mit Dimensionen $\leqq m-1$, $n-1$, $\dim M.N$ ist, wobei wegen *c)* jeder der drei Summanden in $B(W)$ abgeschlossen ist. Da zu jeder Umgebung Z von p eine derartige Umgebung $W \subset Z$ von p existiert, ist auch die zweite Hälfte der behaupteten Ungleichung bewiesen.

Zusätze. Durch vollständige Induktion läßt sich unsere Formel mühelos auf eine endliche Anzahl von Summanden verallgemeinern. Für abzählbar unendlichviele Summanden ist eine Abschätzung der lokalen Dimension der Summe durch die lokalen Dimensionen der Summanden und die Dimensionen der Durchschnitte unmöglich. Da im Beweis der Formel für zwei Summanden M und N die Dimension von $M.N$ nur in einer Umgebung von p eine Rolle spielt, können wir dagegen unsere Formel offenbar folgendermaßen verschärfen:

$\mathrm{Max}(\dim_p M, \dim_p N) \leqq \dim_p (M+N) \leqq \mathrm{Max}(\dim_p M, \dim_p N, 1+\mathrm{Min}\,[\dim_p Z.M.N])$, wobei das Minimum sich auf alle Umgebungen Z von p bezieht.

Ist p ein isolierter Punkt von $M.N$ und sind U und V hinreichend klein, so gilt $\bar{W}.M.N = p$; und wir können, da p nicht in $B(W)$ liegt, statt *b)* schreiben $B(W) \subset (\bar{U}-U)+(\bar{V}-V)$, woraus nach den oben verwendeten Schlüssen die Ungleichung $\dim_p(M+N) \leqq \mathrm{Max}(\dim_p M, \dim_p N)$ folgt. Ist p ein isolierter Punkt von $M.N$, so gilt also $\dim_p(M+N) = \mathrm{Max}(\dim_p M, \dim_p N)$.

Schließlich lassen sich Analoga unserer Formeln für rationaldimensionale Mengen beweisen, wovon hier nur erwähnt sei: *Die Summe zweier im Punkte p rational eindimensionaler kompakter Mengen mit höchstens abzählbarem Durchschnitt ist in p rational eindimensional.*

Taussky: *Zur Axiomatik der Gruppen.*

Ein gegenüber einer Kompositionsvorschrift abgeschlossenes System ist dann und nur dann eine Gruppe, wenn folgende vier Axiome erfüllt sind:

I. Die Kompositionsvorschrift ist assoziativ.

II. Es gibt ein Element u in dem System, so daß für jedes Element a mindestens ein linksseitig inverses Element a_l in bezug auf u existiert: $a_l a = u$.

III. Für dasselbe u existiert zu jedem a höchstens ein rechtsseitig inverses Element a_r in bezug auf u, so daß $a a_r = u$.

IV. Jede Lösung der Gleichung $eu = u$ ist idempotent ($e^2 = e$).

Um zu beweisen, daß das System eine Gruppe bildet, haben wir für jedes a nachzuweisen, daß 1. die Beziehung $ea = a$ gilt und daß 2. die Gleichung $xa = e$ eine Lösung besitzt.

1. folgt, indem man zu $ea = e^2 a$ das zufolge Axiom II existierende linksseitig inverse Element a' in bezug auf u bildet. Es ist also $a' e^2 a = a' ea = u$; daher gilt zufolge Axiom III die Behauptung: $ea = a$.

Um 2. zu beweisen, zeigen wir zuerst, daß die Beziehung $ue = u$ gilt. Bilden wir nämlich zum Element ae das linksseitig inverse Element a_l in bezug auf u, so gilt: $a_l ae = u = a_l a e^2 = ue$, woraus obendrein folgt, daß e das einzige idempotente Element ist. Ist dann $\bar{a}$ das linksseitig inverse Element in bezug auf u zu au, so gilt: $\bar{a}au = u$. Zufolge Axiom IV ist $\bar{a}a$ idempotent, also $\bar{a}a = e$.

Wir beweisen noch mit Hilfe der folgenden vier Gegenbeispiele, daß das vorliegende Axiomensystem unabhängig ist: I. Die Menge der positiven Zahlen ohne die Null, wobei als Komposition die Bildung des geometrischen Mittels genommen wird. II. Die Menge bestehend aus den positiven reellen Zahlen und der Null mit der Addition als Komposition. III. Die Felder[1]), wenn man sich auf eine der beiden Kompositionsvorschriften beschränkt. IV. Die Menge der eineindeutigen Abbildungen einer abzählbar unendlichen Menge auf echte unendliche Teilmengen, deren Komplement auch unendlich ist[2]).

Axiom IV kann nicht durch das schwächere Axiom: „es existiert ein idempotentes Element e“ ersetzt werden, wie das unter IV angeführte Gegenbeispiel zeigt, wenn man zu der betrachteten Menge von Abbildungen auch die identische Abbildung hinzunimmt. Daß das Element u in Axiom II mit demjenigen in Axiom III übereinstimmen muß, zeigt die Menge, bestehend aus den sämtlichen Elementen des nur die Null und Eins enthaltenden Körpers, mit der Multiplikation als Kompositionsvorschrift. Setzen wir nämlich $u = 0$ für Axiom II, $u' = 1$ für Axiom III, so wären alle vier Axiome erfüllt, ohne daß das System eine Gruppe bildete.

Wir zeigen schließlich, daß für Abelsche und endliche Gruppen Axiom IV aus den anderen drei Axiomen folgt. Ist das System nämlich kommutativ, so folgt aus $eu = u$ die Beziehung $ue = u$, was zufolge Axiom III besagt, daß e idempotent ist. — Für Systeme aus endlich vielen Elementen können wir schließen, daß unter den Potenzen von u zwei gleiche Elemente vorkommen müssen, etwa $u^m = u^n$, $n > m$. Es ist also $u^m = u^m u^{n-m}$ und daher $e = u^{n-m}$ Lösung der Gleichung $eu = u$ und zufolge Axiom III idempotent. Denn ist $\bar{u}$ das linksseitig inverse Element von u^m in bezug auf u, so gilt $\bar{u}\, u^m = \bar{u}\, u^m u^{n-m} = u^{n-m}\, u = u$. Dieses $e = u^{n-m}$ ist die einzige Lösung von $eu = u$.

---

[1]) Menger, Jahresber. d. D. M. V., *37*, S. 312, sowie diese Ergebnisse, *1*, S. 28 f.

[2]) Dieses Beispiel findet sich auch bei Baer und Levi, S. B. Heidelberg. Akad. Wiss., *18*, 1932.

1*

33. Kolloquium (7. XI. 1931).

Alt: *Zur Theorie der Krümmung.*

Ist P ein Häufungspunkt eines metrischen Raumes, so wollen wir die Zahl $\varkappa(P) = \lim\limits_{P', P'' \to P} \varkappa(PP'P'')$, falls sie existiert, als *Krümmung* des Raumes in P bezeichnen, wobei $\varkappa(PP'P'')$ der Mengersche Ausdruck für die Krümmung der drei Punkte P, P' und P'' ist[1]). In jedem Punkt P, in welchem die Mengersche Krümmung[1]) d. i. die Zahl $\lim\limits_{P', P'', P''' \to P} \varkappa(P'P''P''')$ existiert, existiert auch die Krümmung $\varkappa(P)$; jedoch braucht das umgekehrte nicht zu gelten. Für ebene Kurven existiert die Krümmung $\varkappa(P)$ in jedem Punkt, in welchem die klassische Krümmung existiert, und ist ihr gleich. Hingegen kann $\varkappa(P)$ existieren, ohne daß die klassische Krümmung existiert. Eine ebene Kurve hat in jedem Punkt P, in welchem sie eine Krümmung $\varkappa(P)$ besitzt, eine Tangente. Notwendig und hinreichend für die Existenz der Krümmung $\varkappa(P)$ in einem Punkte x_0 einer ebenen Kurve $y = y(x)$ ist die Existenz der Grenzwerte

$$\lim_{x \to x_0} \frac{\bar{y}(x) - y'(x_0)}{x - x_0} \quad \text{und} \quad \lim_{x \to x_0} \frac{\underline{y}(x) - y'(x_0)}{x - x_0},$$

wo $\bar{y}(x)$, $\underline{y}(x)$ die obere und untere Ableitungsgrenze und $y'(x)$ die Ableitung bedeuten. Diese beiden Grenzwerte sind, wenn sie existieren, gleich $\varkappa(P)$. Ein euklidischer Bogen, der in jedem Punkt die Krümmung $\varkappa = 0$ hat, ist eine Strecke.

Gödel hat vorgeschlagen, die in der obigen Definition der Krümmung $\varkappa(P)$ auftretenden Punkte P' und P'' der Einschränkung zu unterwerfen, daß sie auf verschiedenen Seiten von P liegen. Notwendig und hinreichend für die Existenz der Krümmung einer ebenen Kurve $y = y(x)$ in diesem Sinne ist, daß es (für ein geeignetes Koordinatensystem) zu jedem $\varepsilon > 0$ eine Umgebung des betrachteten Punktes gibt, in welcher $\left| y - \frac{1}{2} \varkappa x^2 \right| < \varepsilon$.

34. Kolloquium (13. XI. 1931).

Mimura (Tokio): *Über die Stetigkeit des Inhaltes konvexer geschlossener Flächen.*

Ist $\{F_n\}$ *eine gegen die geschlossene Fläche* F *konvergente Folge von konvexen Flächen, so gilt für die Inhalte die Gleichung* $\lim\limits_{n=\infty} m(F_n) = m(F)$. Der Beweis ergibt sich unmittelbar durch Kombination der folgenden beiden Tatsachen: 1) Ist p ein Punkt im Innern des von der konvexen geschlossenen Fläche F begrenzten Körpers und führt man F durch zwei Ähnlichkeitstransformationen mit dem Zentrum p in eine größere Fläche F' und in eine kleinere Fläche F'' über, so ist das Gebiet zwischen F' und F'' eine Um-

[1]) Menger, Math. Ann., *103*, S. 480.

gebung von F (was für nicht konvexe Flächen nicht notwendig der Fall ist) und enthält daher fast alle Flächen F_n; sind die Ähnlichkeitstransformationen hinreichend klein, so weichen die Inhalte von F' und F'' beliebig wenig vom Inhalt der Fläche F ab. 2) Der Inhalt einer konvexen geschlossenen Fläche, die im Gebiet zwischen zwei konvexen geschlossenen Flächen liegt, ist kleiner als der Inhalt der äußeren und größer als der Inhalt der inneren Fläche.

W. T. Parry (Cambridge, Mass.): *Ein Axiomensystem für eine neue Art von Implikation (analytische Implikation).*

Grundzeichen: $p.q$ (p und q), $\sim p$ (nicht p), $p \rightarrow q$ (p impliziert analytisch q). Aus . und $\sim$ werden $\vee$ (oder) und $\supset$ (impliziert), aus $\rightarrow$ wird $\leftrightarrow$ (analytisch äquivalent) in bekannter Weise definiert.

Axiome:

1. $p.q \rightarrow q.p$
2. $p \rightarrow p.p$
3. $p \rightarrow \sim\sim p$
4. $\sim\sim p \rightarrow p$
5. $p.(q \vee r) \rightarrow p.q \vee p.r$
6. $[p \vee (q.\sim q)] \rightarrow p$
7. $(p \rightarrow q).(q \rightarrow r) \rightarrow (p \rightarrow r)$
8. $(p \rightarrow q.r) \rightarrow (p \rightarrow q)$
9. $(p \rightarrow q)(r \rightarrow s) \rightarrow (pr \rightarrow qs)$
10. $(p \rightarrow q)(r \rightarrow s) \rightarrow (p \vee r \rightarrow q \vee s)$
11. $(p \rightarrow q) \rightarrow (p \supset q)$
12. $p \leftrightarrow q.f(p) : \rightarrow f(q)$
13. $f(p) \rightarrow (p \rightarrow p)$.

Schlußregeln: Einsetzungs- und Implikationsregel.

Das Charakteristische der analytischen Implikation ist, daß keine Formel $A \rightarrow B$ beweisbar ist, welche in B eine Aussagevariable enthält, die in A nicht vorkommt. Daher gilt z. B. $p \rightarrow p \vee q$ nicht, während $p.q \rightarrow p$ gilt (Aufhören des Dualismus). Auch „Transposition" $p \rightarrow q . \rightarrow . \sim q \rightarrow \sim p$ gilt nicht, wohl aber: $(p \rightarrow q . \sim q) \rightarrow \sim p$. Nach Ansicht des Vortragenden soll der Begriff des logischen Folgens die angegebene Eigenschaft haben, da die Konklusion keine anderen Begriffe enthalten kann als die Prämissen, während z. B. in der (sowohl für die Russellsche als für die Lewissche Implikation gültigen) Formel $p \supset p \vee q$ der Satz $p \vee q$ andere Begriffe enthalten kann als p. Der Beweis, daß keine derartige Implikation beweisbar ist, wird mit Hilfe der folgenden Matrix geführt:

p	0′	1′	0	1
$\sim p$	1′	0′	1	0

$p.q$	0′	1′	0	1
0′	0′	0′	0	0
1′	0′	1′	0	1
0	0	0	0	0
1	0	1	0	1

$p \rightarrow q$	0′	1′	0	1
0′	1′	1′	0	0
1′	0′	1′	0	0
0	1	1	1	1
1	0	1	0	1

ausgezeichnete Elemente: 1, 1′.

Sei $A \rightarrow B$ eine Formel, die in B eine in A nicht vorkommende Variable p enthält. Man setze 0 für p und setze 0′ für alle übrigen Variablen; dann wird A ein gestrichener, B ein ungestrichener Wert (weil jede der Operationen aus gestrichenen Werten gestrichene, dagegen aus einem ungestrichenen und einem beliebigen anderen einen ungestrichenen Wert erzeugt). Daher wird $A \rightarrow B = 0$, während jede beweisbare Formel 1 oder $1'$ ergibt.

Gödel: „p impliziert q analytisch", kann man vielleicht so interpretieren: „q ist aus p und den logischen Axiomen ableitbar und enthält keine anderen Begriffe als p" und es wäre, nachdem man diese Definition genauer präzisiert hat, ein Vollständigkeitsbeweis für die Parryschen Axiome zu erstreben, in dem Sinn, daß alle Sätze, welche für die obige Interpretation von $\rightarrow$ gelten, ableitbar sind.

In der Diskussion wird die Frage aufgeworfen, wieviele verschiedene Wahrheitswerte es im Heytingschen Aussagenkalkül gibt, d. h. wieviele nicht äquivalente Funktionen einer Variablen ($p \vee \neg p$ z. B. ist weder mit p noch mit $\neg p$ noch mit $\neg\neg p$ äquivalent). Bisher ist nicht einmal bekannt, ob es endlich oder unendlich viele sind.

Wald: *Zur Axiomatik der Verknüpfungsbeziehungen* (nach brieflicher Mitteilung).

In seiner den Klassenkalkül und die elementargeometrischen Verknüpfungsbeziehungen zusammenfassenden Axiomatik betrachtet Menger ein System von Dingen, für welche zwei assoziative, kommutative, totallineare Operationen der Addition und Multiplikation mit Einheitselementen (V für die Addition, T für die Multiplikation) definiert sind. Machen wir noch folgende weitere Annahme über eindeutige Umkehrbarkeit der Addition: *Zu je zwei Elementen A und B existiert genau ein Element C derart, daß $A.B + C = A$ und $A.B.C = V$ gilt*, so gilt der Satz: *Zu je zwei Elementen A und B existiert mindestens ein Element D, so daß $(A + B).D = A$ und $A + B + D = T$ gilt.* Einem Satz von Bergmann (diese Ergebnisse, *1*, S. 28 und Monatshefte f. Math. u. Phys., *36*, S. 269) zufolge existiert dann *nur ein* solches Element D. Auf Grund der Voraussetzungen gilt nämlich der Absorptionssatz und folgender Hilfssatz: Aus $(A + B).C = V$ folgt stets $(A + B).(A + C) = A$. Betrachtet man dann das zu den Elementen $A + B$ und T existierende Element C, für das $(A + B).T + C = T$ und $(A + B).T.C = V$ gilt, so genügt mit Rücksicht auf den Hilfssatz das Element $D = A + C$ der Behauptung.

Menger: Die im Hilfssatz von Wald aus der Eindeutigkeit von C gefolgerte Implikation $(A + B).C = V \rightarrow (A + B).(A + C) = A$ gilt offenbar in jedem System, in dem für je drei Elemente A, B, C die Beziehung $A + (A + B).C = (A + B).(A + C)$ besteht. Da hier die rechte Seite in B und C symmetrisch ist, die linke aber nicht, so liegt es nahe, Systeme zu betrachten, in denen für je drei Elemente A, B, C die Formel

$$A + (A + B).C = A + (A + C).B$$

gilt. Beispielsweise gilt sie für das System aller linearen Teilgebilde eines projektiven Raumes, dagegen nicht allgemein für die linearen Teilgebilde eines affinen Raumes. Z. B. ist sie für die euklidische Ebene falsch, wie man erkennt, wenn man für A und B zwei Punkte und für C eine zu ihrer Verbindungsgeraden parallele Gerade wählt.

Die erwähnte Formel und ihr duales Gegenstück ist eine vielleicht nicht unwichtige Abschwächung der Distributivgesetze.

35. Kolloquium (19. XI. 1931).

Menger: *Eine NZ-Kurve.* Teilt man ein Quadrat in 9 kongruente Teilquadrate und tilgt im oberen und unteren mittleren Neuntel die beiden offenen Dreiecke, welche oberhalb bzw. unterhalb der von links oben nach rechts unten laufenden Diagonale liegen, so entsteht ein *N*-förmiges Band, welches Summe von 6 Quadraten und einem Parallelogramm ist. Werden in jeder dieser 7 Flächen durch eine ähnliche Konstruktion abwechselnd *N*-förmige und *Z*-förmige Bänder eingezeichnet und iteriert man dieses Verfahren ad infinitum, so entsteht eine im kleinen zusammenhängende Kurve von einem bisher nicht untersuchten Typus. Ob alle Kurven, die man durch verwandte Konstruktionen erhalten kann, konvexifizierbar sind, ist fraglich.

36. Kolloquium (26. XI. 1931).

G. C. Evans (Houston, Texas): *Zur Dimensionsaxiomatik.* (Nach einer brieflichen Mitteilung vorgetragen von Menger).

Es existiert genau eine Funktion $D_1(M)$ und genau eine Funktion $D_2(M)$ der Teilmengen M des R_n mit folgenden Eigenschaften:

1. Die Funktionen sind monoton, d. h. aus $M' \subset M$ folgt $D_i(M') \leqq D_i(M)$, $i=1, 2$.

2. Die Funktionen sind invariant gegenüber Antoineschen Transformationen, d. h. wenn M und M' erweiterbar homöomorph sind, so ist $D_i(M) = D_i(M')$, wobei M und M' erweiterbar homöomorph heißen, falls eine topologische Abbildung zwischen ihnen existiert, die zu einer topologischen Abbildung zwischen zwei Umgebungen von M und M' erweitert werden kann.

3. Jede Menge M ist erweiterbar homöomorph mit einer Teilmenge eines Komplexes C, bzw. eines unendlichen Komplexes C', für welchen $D_1(M) = D_1(C)$, bzw. $D_2(M) = D_2(C')$ gilt.

4. Für Komplexe, bzw. unendliche Komplexe stimmt der Funktionswert von D_1, bzw. von D_2 mit der elementargeometrischen Dimension überein. Dabei heißt Komplex bzw. unendlicher Komplex eine Menge, die Summe von endlichvielen, bzw. von abzählbarvielen Simplexen ist. Daher gilt für jede Menge M die Beziehung $\dim M \leqq D_2(M) \leqq D_1(M)$.

Etwas verallgemeinert lassen sich die Begriffsbildungen, Sätze und Beweismethoden von Evans folgendermaßen formulieren: Es sei $\mathfrak{M}$ ein System von Teilmengen eines Raumes R und $\mathfrak{T}$ ein System von Paaren von Teilmengen von R. Wir betrachten eine Funktion $E(M)$ aller Teilmengen von R mit folgenden Eigenschaften:

1. Monotonie, d. h. aus $A' \subset A$ folgt $E(A') \leqq E(A)$.

2. $\mathfrak{T}$-Invarianz, d. h. für jedes Mengenpaar A, A' aus $\mathfrak{T}$ gilt $E(A) = E(A')$.

3. $\mathfrak{M}$—$\mathfrak{T}$-Reduzierbarkeit, d. h. zu jeder Menge A existiert eine Menge M_A aus $\mathfrak{M}$ und eine Menge B gemäß den Bedingungen:

A, B ist ein Paar aus $\mathfrak{T}$; $B \subset M_A$; $E(A) = E(M_A)$.

Bei Evans ist R der R_n, $\mathfrak{M}$ das System der Komplexe, bzw. unendlichen Komplexe und $\mathfrak{T}$ das System aller Paare von mit einander erweiterbar homöomorphen Mengen. Allgemein gilt folgender

Unizitätssatz. *Es gibt keine von $E(A)$ verschiedene Mengenfunktion, welche den Bedingungen 1) 2) 3) genügt und für alle Mengen von $\mathfrak{M}$ mit $E(A)$ identisch ist.*

Zum Beweise sei $E'(A)$ irgend eine den drei Bedingungen genügende Mengenfunktion, so daß $E'(M) = E(M)$ für jede Menge M von $\mathfrak{M}$ gilt. Sei dann A irgend eine Teilmenge des Raumes. Da die Funktion E' die Bedingung 3. erfüllt, existiert eine Menge M_A^* von $\mathfrak{M}$ und eine Menge B' derart, daß A, B' ein Paar von $\mathfrak{T}$ ist, $B' \subset M_A^*$ und $E'(A) = E'(M_A^*)$ gilt. Da die Funktion E die Bedingung 2. erfüllt und A, B' ein Paar von $\mathfrak{T}$ ist, gilt $E(A) = E(B')$. Da E monoton ist und $B' \subset M_A^*$ gilt, ist $E(B') \leqq E(M_A^*)$. Da M_A^* in $\mathfrak{M}$ liegt, gilt $E(M_A^*) = E'(M_A^*)$. Die vier Beziehungen zusammengenommen ergeben $E(A) \leqq E'(A)$ und umgekehrt, d. h. die Funktionen E und E' sind identisch, w. z. b. w.

Wir wollen die Menge M_A von $\mathfrak{M}$ zur Menge A $\mathfrak{T}$-zugehörig nennen, wenn eine Menge B existiert derart, daß A, B ein Paar von $\mathfrak{T}$ ist und $B \subset M_A$ gilt. Dann gilt folgender

Existenzsatz. *Es seien in einem Raume R ein Mengensystem $\mathfrak{M}$ und ein System $\mathfrak{T}$ von Mengenpaaren gegeben gemäß folgenden Bedingungen:*

$\mathfrak{M}$ *sei $\mathfrak{T}$-vollständig,* d. h. zu jeder Menge A des Raumes existiert eine zu A $\mathfrak{T}$-zugehörige Menge von $\mathfrak{M}$, *und $\mathfrak{T}$-abgeschlossen,* d. h. ist A eine Menge von $\mathfrak{M}$ und A, B ein Paar von $\mathfrak{T}$, so ist auch B eine Menge von $\mathfrak{M}$.

$\mathfrak{T}$ *sei reflexiv,* d. h. für jede Menge A sei das Paar A, A in $\mathfrak{T}$ enthalten; *symmetrisch,* d. h. wenn A, B in $\mathfrak{T}$ enthalten ist, so auch B, A; *transitiv,* d. h. wenn A, B und B, C in $\mathfrak{T}$ liegen so auch A, C; *erblich,* d. h. wenn A, B in $\mathfrak{T}$ liegt und $A' \subset A$ gilt, so existiert eine Menge B', derart, daß A', B' in $\mathfrak{T}$ liegt und $B' \subset B$ gilt.

Ist dann $F(M)$ irgend eine monotone, $\mathfrak{T}$-invariante Funktion der Mengen von $\mathfrak{M}$ derart, daß jede Menge von Funktionswerten eine kleinste Zahl enthält, dann existiert eine Funktion $E(A)$ aller Teilmengen des Raumes, welche monoton, $\mathfrak{T}$-invariant, $\mathfrak{M}$—$\mathfrak{T}$-reduzierbar ist und so daß für jede Menge M von $\mathfrak{M}$ die Beziehung $E(M) = F(M)$ gilt.

Ist A irgend eine Teilmenge des Raumes, so existieren wegen der $\mathfrak{T}$-Vollständigkeit von $\mathfrak{M}$ zu A $\mathfrak{T}$-zugehörige Mengen von $\mathfrak{M}$. Wegen der Voraussetzung über F existiert unter ihnen mindestens eine Menge M_A^* derart, daß $F(M_A^*) \leqq F(M_A)$ für alle zu A $\mathfrak{T}$-zugehörigen Mengen M_A von $\mathfrak{M}$ gilt. Wir setzen $E(A) = F(M_A^*)$ und behaupten, daß die so definierte Funktion aller Teilmengen des Raumes unseren Forderungen genügt.

Zum Beweise der Monotonie sei $A' \subset A$. Sind beide Mengen in $\mathfrak{M}$ enthalten, so folgt die Ungleichung $E(A') \leqq E(A)$ aus der Monotonie von F auf $\mathfrak{M}$. Ist A in $\mathfrak{M}$ enthalten, A' nicht, so folgt die Ungleichung daraus, daß wegen der Reflexivität von $\mathfrak{T}$ das Paar A', A' in $\mathfrak{T}$ liegt und daher A eine zu A' $\mathfrak{T}$-gehörige Menge ist. — Ist A nicht in $\mathfrak{M}$ enthalten, so folgt sie daraus, daß wegen der Erblichkeit von $\mathfrak{T}$ jede zu A $\mathfrak{T}$-zugehörige Menge erst recht zu A' $\mathfrak{T}$-zugehörig ist.

Zum Beweise der $\mathfrak{T}$-Invarianz von E sei A, A' ein Paar von $\mathfrak{T}$. Sind A und A' in $\mathfrak{M}$ enthalten, so folgt die Gleichheit $E(A) = E(A')$ aus der Invarianz von F. Daß genau eine der Mengen A, A' zu $\mathfrak{M}$ gehört, ist wegen der $\mathfrak{T}$-Abgeschlossenheit von $\mathfrak{M}$ und der Symmetrie von $\mathfrak{T}$ unmöglich. Ist weder A noch A' in $\mathfrak{M}$ enthalten, so existiert eine Menge M_A^* von $\mathfrak{M}$ und eine Menge B derart, daß A, B in $\mathfrak{T}$ liegt, $B \subset M_A^*$ und $E(A) = E(M_A^*)$ gilt. Wegen der Transitivität von $\mathfrak{T}$ ist auch A', B ein Paar von $\mathfrak{T}$, so daß M_A^* auch eine zu A' $\mathfrak{T}$-zugehörige Menge ist. Daher ist $E(A') \leqq E(M_A^*) = E(A)$. Analog zeigt man $E(A) \leqq E(A')$, womit die $\mathfrak{T}$-Invarianz bewiesen ist. — Die $\mathfrak{M}$-$\mathfrak{T}$-Reduzierbarkeit von E ist evident, womit der Existenzsatz bewiesen ist.

Das System $\mathfrak{M}$ aller kompakten Mengen des R_n und das System $\mathfrak{T}$ aller Paare von homöomorphen Mengen erfüllen die Voraussetzungen des Existenzsatzes, so daß derselbe zusammen mit dem Unizitätssatze ergibt: Zu jeder monotonen topologisch invarianten Funktion aller kompakten Mengen, für welche jede Menge von Funktionswerten eine kleinste Zahl enthält, existiert genau eine monotone topologisch invariante kompaktifizierbare Mengenfunktion, welche für alle kompakten Mengen mit der gegebenen Funktion übereinstimmt. Die Bedingung, daß jede Menge von Funktionswerten eine kleinste Zahl enthält, ist insbesondere für jede ganzzahlige nichtnegative Funktion erfüllt.

Menger: *Über eine Limesklasse.*

Nennt man in einem euklidischen Raume R eine Folge $\{A_k\}$ von kompakten Mengen gegen die kompakte Menge A konvergent, wenn A für jedes $\varepsilon > 0$ in fast alle A_k ε deformiert werden kann, so wird das System aller kompakten Mengen von R zu einer Limesklasse L. Dieselbe ist separabel, da in ihr die Menge aller Komplexe mit rationalen Eckpunkten dicht liegt. Die Dimension ist in L halbstetig: dim $A \leqq \underline{\lim}$ dim A_k.

37. Kolloquium (2. XII. 1931).

Gödel: *Über Unabhängigkeitsbeweise im Aussagenkalkül.*

Läßt sich jeder Unabhängigkeitsbeweis innerhalb des Aussagenkalküls mit Hilfe von endlichen Modellen [Matrizen[1])] führen? Diese von Hahn aufgeworfene Frage ist zu verneinen. Z. B. ist die Aussage $p \supset \sim\sim p$ von folgenden drei Axiomen:

[1]) Über den Begriff der logischen Matrix vgl. Lukasiewicz und Tarski, Untersuchungen über den Aussagenkalkül, Comptes Rendus de la Soc. d. Sciences et d. Lettres de Varsovie, *23*, 1930.

1) $p \supset p$ 2) $p \supset \sim\sim q . \supset . p \supset q$ 3) $\sim\sim p \supset \sim\sim q . \supset . p \supset q$

mit Einsetzungs- und Implikationsregel als Schlußschemata unabhängig.

Dies zeigt folgende unendliche Matrix: Elemente seien die ganzen Zahlen $\geqq 0$; ausgezeichnetes Element sei 0; $\sim p = p + 1$; $p \supset q = 0$ für $p \geqq q$; $p \supset q = 1$ für $p < q$.

Anderseits aber gilt: Für jede endliche normale Matrix, für welche 1) 2) 3) erfüllt sind, ist auch $p \supset \sim\sim p$ erfüllt. Beweis: Es sei $\mathfrak{A}$ die endliche Menge der Elemente, $\mathfrak{A}^* \subset \mathfrak{A}$ die der ausgezeichneten Elemente einer normalen Matrix, für welche 1) 2) 3) erfüllt sind, d. h. $\sim$ und $\supset$ sind als Operationen innerhalb $\mathfrak{A}$ gemäß folgenden Bedingungen definiert:

a) Die Axiome 1) 2) 3) ergeben für beliebige $p, q \varepsilon \mathfrak{A}$ immer Elemente aus $\mathfrak{A}^*$.

b) Aus $a \varepsilon \mathfrak{A}^*$ und $a \supset b \varepsilon \mathfrak{A}^*$ folgt immer $b \varepsilon \mathfrak{A}^*$.

Es ist zu zeigen, daß dann auch $p \supset \sim\sim p \varepsilon \mathfrak{A}^*$ für jedes $p \varepsilon \mathfrak{A}$. Da $\mathfrak{A}$ nur endlich viele Elemente enthält, muß es zwei Zahlen k, n geben, so daß $\sim^{2k} p = \sim^{2n} p$ und $k < n$. Wegen 1) gilt dann:

$$\sim^{2k} p \supset \sim^{2n} p \, \varepsilon \, \mathfrak{A}^*;$$

wegen 3) gilt:

$$(\sim^{2k} p \supset \sim^{2n} p . \supset . \sim^{2k-2} p \supset \sim^{2n-2} p) \, \varepsilon \, \mathfrak{A}^*$$

und daher: $\sim^{2k-2} p \supset \sim^{2n-2} p \, \varepsilon \, \mathfrak{A}^*$. Durch mehrmalige Anwendung dieses Schlusses folgt schließlich $p \supset \sim^{2(n-k)} p \, \varepsilon \, \mathfrak{A}^*$. Wendet man darauf $(n-k-1)$-mal 2) an, so folgt $p \supset \sim\sim p \, \varepsilon \, \mathfrak{A}^*$, w. z. b. w. In den obigen Axiomen könnte man statt $\sim\sim$ überall $\sim$ setzen. Das durch die Axiome 1) 2) 3) definierte logische System wäre aber dann kein Teilsystem der gewöhnlichen Logik (trotzdem wäre es widerspruchsfrei in dem Sinn, daß A und $\sim A$ niemals beide beweisbar sind).

Zu jeder gegenüber der Einsetzungsregel abgeschlossenen Menge M von Formeln des Aussagenkalküls gibt es eine kleinste Kardinalzahl m ($\leqq \aleph_0$), so daß Modelle mit m, aber nicht mit weniger Elementen existieren. Unter einem Modell wird dabei eine Matrix verstanden, für welche alle und nur die Formeln aus M erfüllt sind. Für die Menge der aus obigen Axiomen ableitbaren Formeln ist diese Zahl $\aleph_0$; ebenso z. B. für den Heytingschen Aussagenkalkül[1]). Die Formelmengen, für die $m = 2$ ist, sind isomorph mit solchen Teilsystemen des gewöhnlichen Aussagenkalküls, welche alle und nur die Tautologien enthalten, die sich unter alleiniger Verwendung gewisser Begriffe (z. B. mit $\supset$ allein) ausdrücken lassen.

Klanfer: Über d-zyklische (d. h. mit Punkten eines konvex metrisierten Kreises vom Umfang d kongruente) Quadrupel (vgl. Blumenthal, Amer. Journ. of Math., *54*, S. 387). —

Ist jedes Quadrupel des R_3 mit vier Punkten einer konvex metrisierten Kugelfläche von geeignetem Durchmesser kongruent? (Sicher gibt es Quadrupel auf konvex metrisierten Kugelflächen, die nicht mit Quadrupeln des R_3 kongruent sind.)

---

[1]) Vgl. Gödel, S. 40 dieses Heftes.

38. Kolloquium (16. XII. 1931).

Nöbeling: Genügt die für alle separablen Mengen M definierte Mengenfunktion $f(M)$ den fünf Mengerschen Dimensionsaxiomen (vgl. diese Ergebnisse *1*, S. 3), so gilt $f(M) \leqq \dim M$ für jedes M. (Erscheint in den Monatsheften f. Math. u. Phys.)

N. Wiener (Cambridge, Mass.) Bemerkungen über Kapazität.

39. Kolloquium (14. I. 1932).

Diskussion über die Verzweigungsordnung von Flächenpunkten.

Menger: Es wäre wichtig, für Flächenpunkte einen zur Ordnung von Kurvenpunkten analogen Begriff einzuführen. Sicher kann in Analogie zu den rationalen Kurvenpunkten ein Flächenpunkt als *rational* bezeichnet werden, wenn er in beliebig kleinen Umgebungen enthalten ist, deren Begrenzungen mit der Fläche rational eindimensionale Durchschnitte haben. Großen Schwierigkeiten begegnet aber die feinere zahlenmäßige Beschreibung der Flächenpunkte. Man hat hier die Möglichkeit, die *Komponentenanzahl* kleiner Umgebungsbegrenzungen oder deren kurventheoretische *Ordnung* in Betracht zu ziehen. Aber schon bei jenen Flächen, welche Summe von endlichvielen Dreiecken sind, führt keines dieser Prinzipien zu einer befriedigenden Klassifikation. Einer Beschreibung der Flächenpunkte durch eine ganze Zahl müßte ja eine Klassifikation der linearen Graphen vorangehen, deren Möglichkeit, selbst wenn man für eine gewisse Zahl $n > 1$ sämtliche Graphen durch n ganze Zahlen beschreiben wollte, zweifelhaft ist. Ein anderer naheliegender Weg zur Definition einer Ordnung von Flächenpunkten bestünde in der Betrachtung von *Zerlegungen der Fläche in drei Teilmengen.* Man könnte etwa versuchen, den Punkt p der Fläche F von höchstens n-ter Ordnung zu nennen, wenn für jedes $\varepsilon > 0$ eine Zerlegung von F in drei abgeschlossene Summanden existiert, von denen genau einer, dessen Durchmesser $< \varepsilon$ ist, p als inneren Punkt enthält und die höchstens n Punkte gemein haben. Insbesondere ergäben sich dann Punkte der Ordnung 0. Indes führt diese Definition, selbst wenn man voraussetzt, daß die drei Summanden Kontinua sind und Kontinua als Durchschnitte haben, nicht zum Ziel. Denn wenn man aus einer Quadratfläche eine gegen einen Randpunkt p des Quadrates konvergente Folge von paarweise fremden offenen Kreisflächen tilgt, so ist die verbleibende Fläche F sicher für jedes $\varepsilon > 0$ in drei Kontinua ohne gemeinsamen Punkt zerlegbar, die paarweise Kontinua als Durchschnitt haben und von denen genau eines einen Durchmesser $< \varepsilon$ besitzt und p als inneren Punkt enthält. Es müßte also F in p die Ordnung 0 haben, während F anderseits eine p am Rande enthaltende mit einer Quadratfläche homöomorphe Teilmenge enthält. Ferner sei F' Summe zweier Quadratflächen, die nur einen gemeinsamen Eckpunkt p' als Durchschnitt haben. F' ist topologisch auf eine Teilmenge einer Quadratfläche so abbildbar, daß p' einem Randpunkt entspricht. Wenn wir also an die Flächenordnung die naheliegenden Forderungen stellen, 1. daß homöomorphe Flächen in entsprechenden Punkten dieselbe Ordnung besitzen sollen, 2. daß die Ordnung einer Teilfläche die der Gesamtfläche in keinem Punkt übertreffen solle, 3. daß eine Quadratfläche in jedem Randpunkte die Ordnung 1 besitzen solle, so zeigen die erwähnten Flächen F und F', daß eine solche Flächenordnung nicht durch Zerlegungen in drei Summanden oder durch die Komponentenanzahl kleiner Umgebungsbegrenzungen erklärt werden kann, — es sei denn, daß einem Flächenpunkt als Ordnung ein System von ganzen Zahlen zugeordnet wird.

Zarankiewicz (Warschau): Für die Betrachtung von Zerlegungen in drei Summanden spräche, da die Ebene und eine Kugelfläche jedenfalls in jedem Punkt die Ordnung 2 haben müssen, immerhin folgender Satz über die Zerlegung der Ebene:

Ist die euklidische Ebene E Summe der drei abgeschlossenen echten Teilmengen A, B und C, so daß erstens A beschränkt und $B \neq C$ ist und zweitens keine der Mengen AB, AC und BC die Ebene zerschneidet, dann enthält der Durchschnitt ABC mindestens zwei Punkte.

Beweis: Zunächst behaupten wir, daß die Menge $A - A(B+C)$ nicht leer ist. Nehmen wir an, $A - A(B+C)$ sei leer. Dann ist $A \subset B+C$, also $E = B+C$. Da B und C echte Teilmengen von E sind, existieren zwei Punkte $p \subset B$ und $q \subset C$, so daß weder $p \subset C$, noch $q \subset B$ gilt. Die Begrenzung $F(B)$ von B zerschneidet die Ebene E zwischen p und q. Mithin zerschneidet auch die Menge $BC \supset F(B)$, da sie keinen der Punkte p und q enthält, die Ebene im Widerspruch zur Voraussetzung.

Die nichtleere Menge $A - A(B+C)$ ist offen. Sei a ein Punkt dieser Menge und A' die a enthaltende Komponente. Die Begrenzung $F(A')$ von A' ist wegen der Beschränktheit von A beschränkt und enthält daher einen irreduziblen Schnitt F der Ebene zwischen a und einem Punkt x aus $E - A$[1]). Aus $F(A') \subset B+C$ folgt $F \subset B+C$. Wäre nun $F \subset B$ oder $F \subset C$, so auch $F \subset AB$ bzw. $F \subset AC$; daher würde die Ebene durch AB bzw. AC zwischen a und x zerschnitten, im Widerspruch zur Voraussetzung. Es ist also $FB \neq 0$ und $FC \neq 0$ und diese Mengen sind echte Teilmengen von F, zerschneiden also E nicht; ihre Summe $FB + FC = F$ hingegen zerschneidet die Ebene E. Demnach ist der Durchschnitt $FB \,.\, FC$ nicht zusammenhängend[2]), enthält demnach mindestens zwei Punkte. Dieselben sind wegen $F \subset A$ auch im Durchschnitt ABC enthalten, womit der Satz bewiesen ist.

Ein analoger Satz gilt für jedes einfach zusammenhängende Gebiet E der Ebene, wobei nur statt der Beschränktheit von A die Beziehung $F(A) \subset E$ vorausgesetzt werden muß. Man erhält diesen Satz aus dem oben bewiesenen, indem man das Gebiet E topologisch auf die ganze Ebene abbildet. Für die Kugeloberfläche gilt unser Satz ebenfalls.

---

[1]) Mazurkiewicz, Fund. Math., *1*, S. 63.

[2]) Satz von Janiszewski, vgl. Straszewicz, Fund. Math., *7*, S. 159.

40. Kolloquium (21. I. 1932).

Menger: Existiert eine rational n-dimensionale Universalmenge? Offenbar könnte dieselbe nicht kompakt sein, da ja schon keine kompakte rational nulldimensionale (d. h. abzählbare) Menge existiert, welche alle rational nulldimensionalen Mengen topologisch enthält. Existiert eine rational eindimensionale Universalmenge, welche Summe von abzählbarvielen Bogen ist? Wie Knaster inzwischen bemerkt hat, ist die letztere Frage zu verneinen. Wenn eine Menge S Summe von abzählbarvielen Bogen B_i $(i=1, 2, \ldots)$ ist, so enthält sie keine mit der Sierpińskischen Dreieckskurve D homöomorphe Teilmenge D', da $D'.B_i$ für jedes i in D' nirgends dicht wäre, während D' kompakt, also von zweiter Kategorie ist.

Nöbeling: Zusammenhänge des kurventheoretischen n-Beinsatzes mit n-Bogensätzen (vgl. Menger, Kurventheorie, S. 213—221).

41. Kolloquium (4. II. 1932).

Über komplexmetrische Quadrupel.

Eine Menge, deren Elemente Punkte heißen mögen, wird *komplexmetrisch* genannt (vgl. diese Ergebnisse, *2*, S. 36), wenn je zwei Elementen p und q ein Paar von entgegengesetzten komplexen Zahlen (d. h. von komplexen Zahlen mit der Summe 0), genannt der Abstand pq, so zugeordnet ist, daß p und q das Paar 0,0 entspricht, falls p und q identisch sind. J. Groiß hat rechnerisch bewiesen, *daß jede aus vier Elementen bestehende komplexmetrische Menge mit vier Punkten des K_3 kongruent ist,* wobei K_n die Menge aller n-Tupel komplexer Zahlen ist, für welche den n-Tupeln

$x_1, x_2, \ldots, x_n$ und $y_1, y_2, \ldots, y_n$ das Zahlenpaar $\pm\sqrt{\sum_{i=1}^{n}(x_i-y_i)^2}$

als Abstand zugeordnet wird. Insbesondere ist jedes pseudokomplexe Quadrupel (d. h. jedes Quadrupel, welches nicht mit vier Punkten des K_1 kongruent ist, obwohl je drei seiner Punkte mit drei Punkten des K_1 kongruent sind) mit vier Punkten des K_3 kongruent; gelten für das Quadrupel p, q, r, s die für pseudokomplexe Quadrupel charakteristischen Relationen (vgl. diese Ergebnisse, *2*, S. 36)

$$(pq)^2=(rs)^2=a^2, \; (pr)^2=(qs)^2=b^2, \; (ps)^2=(qr)^2=(a+b)^2,$$

so sind, wenn für c eine beliebige komplexe Zahl $\neq 0$ gewählt wird,

$$p'=(0,0,0), \; q'=(a,0,0), \; r'=(b, ic, c),$$
$$s'=\left(a-b, \; -i\frac{b(b-a)}{c}, \; \frac{-b(b-a)}{c}\right)$$

vier kongruente Punkte des K_3. Es ist bemerkenswert, daß je drei von den Punkten p', q', r', s' mit drei Punkten des K_1 kongruent sind, ohne auf einer Geraden des K_3 zu liegen. Allgemein können von vier Punkten des K_3, wenn sie mit p, q, r, s kongruent sein sollen, keine drei auf einer mit dem K_2 kongruenten Ebene des K_3 liegen. Es gibt, wie besonders bemerkt sei, im K_3 Mengen, welche mit einer Geraden des K_3 kongruent sind, ohne Gerade zu sein (ein Vorkommnis, das im K_2 nicht auftreten kann).

Ein Quadrupel heißt pseudolinear, wenn es nicht mit vier Punkten des R_1 (der reellen Geraden) kongruent ist, obwohl je drei seiner Punkte mit drei Punkten des R_1 kongruent sind. Ein pseudolineares Quadrupel ist für kein natürliches n mit vier Punkten des R_n kongruent (vgl. Menger, Amer. Journ. of math., *53*, S. 744). Menger bemerkt, daß ein pseudolineares Quadrupel auch nicht mit vier Punkten des K_1 kongruent ist, da vier Punkte des K_1 mit paarweise reellen Abständen auf der reellen Teilgeraden des K_1 liegen müßten, was bei Punkten, die mit einem pseudolinearen Quadrupel kongruent sind, unmöglich ist. *Ein pseudolineares Quadrupel ist also pseudokomplex und daher auch nicht mit vier Punkten des K_2 kongruent, aber, wie jedes pseudokomplexe Quadrupel, mit vier Punkten des K_3.*

Menger: Wie ist der K_n unter den komplexmetrischen Räumen gekennzeichnet? —

Bemerkung über konvexe ebene Kurven.

Die ebene Menge P bestehe aus den n Punkten $p_1, p_2, \ldots, p_n$ die ein konvexes Polygon bilden mögen, in dem je zwei konsekutive Winkel eine Summe $> 180^0$ haben (also $n > 4$). Dann ist jede konvexe Kurve $\supset P$ kürzer als jenes Polygon, dessen Ecken die Schnittpunkte der Geraden durch p_i und p_{i+1} und durch p_{i+2} und p_{i+3} sind ($i = 1, 2, \ldots, n$, wobei $n + k = k$ zu setzen ist). Es existiert daher ein Supremum der Längen aller P enthaltenden konvexen Kurven, welches wegen der Kompaktheit der Menge aller P enthaltenden konvexen Kurven als Länge mindestens einer von ihnen auftreten muß, und zwar als Länge eines P enthaltenden *Polygones,* da zu jeder konvexen Kurve $\supset P$, die kein Polygon ist, ein längeres Polygon $\supset P$ sich angeben läßt.

Schreiber: *Zur Axiomatik der Verknüpfungsbeziehungen.*

In der Axiomatik der elementargeometrischen Verknüpfungsbeziehungen nimmt Bergmann (vgl. diese Ergebnisse, *1*, S. 28) zu den Mengerschen Voraussetzungen über Addition und Multiplikation (vgl. dieses Heft, S. 6) folgendes Umkehrungsaxiom: Zu je zwei Elementen A und B existiert ein Element C_1, so daß $A \,.\, B + C_1 = A$ und $A \,.\, B \,.\, C_1 = V$ gilt, und ein Element C_2, so daß $(A + B) \,.\, C_2 = A$ und $A + B + C_2 = T$ gilt. Statt dieser Voraussetzung über die Zuordnung von zwei Elementen zu je zwei Elementen gemäß vier Bedingungen genügt es, folgende Zuordnung eines Elementes zu je drei Elementen gemäß zwei zu einander dualen Bedingungen zu postulieren: *Zu je drei Elementen A, B, C existiert ein Element K, so daß $(A+B) \,.\, (A+C) \,.\, (C+B) + K = A + B$ und $(A \,.\, B + A \,.\, C + C \,.\, B) \,.\, K = A \,.\, B$ gilt.*

42. Kolloquium (18. II. 1932).

Parry: *Zum Lewisschen Aussagenkalkül.*

Grundzeichen sind: $p . q$ (p und q), $\sim p$ (nicht p), $\mathsf{U}(p)$ (p ist unmöglich). Daraus definiert man $\vee$, $\supset$, $\equiv$ (oder, impliziert, äquivalent) in bekannter Weise, ferner „Notwendigkeit": $Np \equiv \mathsf{U} \sim p$ und „strikte Implikation": $p < q \equiv \mathsf{U}(p \sim q)$, was mit $N(p \supset q)$ äquivalent ist.

Ein U-Zeichen in einer Formel F soll einem anderen übergeordnet heißen, wenn das zweite in der mit dem ersten beginnenden Teilformel von F enthalten ist. Eine nur aus Grundzeichen und Variablen aufgebaute Formel, die U-Zeichen enthält, in der aber kein U einem anderen übergeordnet ist, soll von erstem Grad heißen. Im folgenden wird ein Verfahren angegeben, um zu entscheiden[1]), *ob eine Formel ersten Grades aus den Lewisschen Axiomen ableitbar ist.* Jede solche Formel hat die Gestalt:

$$F\,\{\mathsf{U}(h_1),\ \mathsf{U}(h_2) \ldots \mathsf{U}(h_m),\ p_1, p_2 \ldots p_n\},$$

wobei die h_i kein U mehr enthalten, p_i die sämtlichen in der Formel vorkommenden Aussagevariablen sind und F eine nur mittels „und" und „nicht" aufgebaute Funktion mit $m+n$ Leerstellen ist. Die konjunktive Normalform von F (die auch im Lewisschen Kalkül mit F äquivalent ist, da alle Formeln des gewöhnlichen Aussagenkalküls auch dort gelten) ist ein Produkt von Faktoren der Gestalt:

$$\mathsf{U}(h_{i_1}) \vee \ldots \vee \mathsf{U}(h_{i_l}) \vee \sim \mathsf{U}(_{i_{l+1}}) \vee \ldots \vee \sim \mathsf{U}(h_{i_{l+k}})$$
$$\vee\, p_{j_1} \vee \ldots \vee p_{j_r} \vee \sim p_{j_{r+1}} \vee \ldots \vee \sim p_{j_{r+s}}$$

und es genügt, für jeden Faktor ein Entscheidungsverfahren anzugeben. Indem man statt U das Zeichen N verwendet und die von U freien Glieder unter einem Zeichen (f_{k+1}) zusammenfaßt, erhält man eine Formel der Gestalt:

$$N(f_1) \vee \ldots \vee N(f_k) \vee \sim N(g_1) \vee \ldots \vee \sim N(g_l) \vee f_{k+1}, \qquad (1)$$

wobei f_i, g_i Ausdrücke des gewöhnlichen Aussagenkalküls sind. Beachtet man $N(a) . N(b) \equiv N(a . b)$ und bezeichnet $g_1 . g_2 \ldots g_l$ mit g, so kann man (1) umformen in:

$$N(g) \supset . N(f_1) \vee N(f_2) \vee \ldots \vee N(f_k) \vee f_{k+1}, \qquad (2)$$

g sowie jedes f_i bringe man auf die „ausgezeichnete" disjunktive Normalform (vgl. Hilbert-Ackermann, Theoretische Logik, p. 16). Die Menge der dabei in g (bzw. in f_i) auftretenden Konstituenten bezeichnen wir mit G (bzw. F_i). Dann gilt: (2) ist dann und nur dann beweisbar, wenn G in mindestens einem F_i enthalten ist. Sei nämlich erstens $G \subseteq F_i$; dann gilt $g < f_i$ im Lewisschen Kalkül

[1]) Der unten stehende Beweis weicht in einigen Punkten von dem von Parry vorgetragenen ab.

(wegen $p < p \vee q$). Wegen $p < q . < . Np < Nq$ und $Np < p$ gilt weiter $N(g) \supset N(f_i)$ und $N(g) \supset f_i$ (für den Fall $i = k+1$) und daher auch (2). Wenn zweitens $G \subseteq F_i$ für kein i gilt, dann gibt es für jedes i eine Konstituente c_i, so daß $c_i \varepsilon G$, $c_i \sim \varepsilon F_i$. Um zu zeigen, daß in diesem Fall die Formel (2) im System of Strict Implication nicht ableitbar ist, betrachte man folgende Realisierung für das Lewissche System: Elemente seien sämtliche Teilmengen einer Menge M. Es sei ferner $p . q$ der Durchschnitt der Mengen p und q, $\sim p$ sei $M - p$,

$$\mathsf{U}\,(p) = M \text{ für } p = \Lambda \text{ und } \mathsf{U}\,(p) = \Lambda \text{ für } p \neq \Lambda,$$

wo Λ die leere Menge bezeichnet; daher gilt:

$$p \vee q = p + q,\ p \supset q = (M - p) + q,\ N(p) = M \text{ für } p = M,$$
$$N(p) = \Lambda \text{ für } p \neq M.$$

Sämtliche im Lewisschen System beweisbaren Formeln geben immer den Wert M (dies bleibt auch richtig, wenn man das Axiom von Becker $\sim \mathsf{U}(p) < \mathsf{U}\,\mathsf{U}(p)$ zu den Lewisschen hinzunimmt). Wenn unter den c_i genau r verschiedene vorkommen, nehme man für M eine Menge aus r Elementen $e_1, e_2 \ldots, e_r$ und bestimme die p_i als Teilmengen von M so, daß jedes c_i gleich wird einem $\{e_j\}$ und alle übrigen Konstituenten leer werden; (bekanntlich kann man zu jeder Aufspaltung einer Menge M in 2^n paarweise fremde Klassen Teilmengen $p_1, p_2 \ldots p_n$ von M so bestimmen, daß die 2^n Konstituenten der p_i gleich werden den 2^n gegebenen Klassen). Setzt man die angegebenen Werte für p_i in (2) ein, so erhält man $g = M$, $f_i \neq M\ (i = 1, 2 \ldots k+1)$; daher $N(f_i) = \Lambda\ (i = 1, 2 \ldots k)$. Die Formel (2) erhält also den Wert: $M \supset f_{k+1}$ und dies ist $= f_{k+1} \neq M$.

Der Fall, daß kein g_i (bzw. f_i) in der Formel (1) vorkommt, kann durch Hinzufügung eines Disjunktionsgliedes der Gestalt $\sim N(p_1 \vee \sim p_1)$ [bzw. $N(p_1 \sim p_1)$] auf den früheren zurückgeführt werden.

Vermöge des Beckerschen Zusatzaxioms wird jede Formel mit einer ersten Grades äquivalent[1]) und man hat daher für das erweiterte System ein für beliebige Formeln gültiges Entscheidungsverfahren.

Gödel: *Über die metrische Einbettbarkeit der Quadrupel des R_3 in Kugelflächen.*

In Beantwortung einer im 37. Kolloquium (dieses Heft, S. 10) aufgeworfenen Frage gilt: *Ein metrisches Quadrupel, das mit vier Punkten des R_3 kongruent ist, ist, wenn es nicht mit vier Punkten der Ebene kongruent ist, auch kongruent mit vier Punkten einer Kugelfläche, in welcher als Abstand je zweier Punkte die Länge des kürzesten sie auf ihr verbindenden Bogens erklärt ist.* Zum Beweise sei T ein Tetraeder im R_3, dessen 6 Kantenlängen $a_1, a_2, \ldots, a_6$ den Abständen der vier gegebenen Punkte gleich seien und von dem wir

[1]) Zusatz bei der Korrektur. Vgl. die inzwischen erschienene Arbeit von Wajsberg, Monatshefte f. Math. u. Phys., *40*, S. 113.

annehmen können, daß es nicht in einer Ebene liegt. R heiße der Radius der T umgeschriebenen Kugel. Wir setzen $a_{i,x} = \frac{2}{x} \sin \frac{a_i x}{2}$; dann ist also $a_{i,x}$ die Länge der Sehne eines Kreisbogens der Länge a_i auf einem Kreis vom Radius $\frac{1}{x}$; für $x = 0$ hat $a_{i,x}$ den Grenzwert a_i. Ein (ev. ebenes) Tetraeder im R_3, dessen 6 Kanten die Längen $a_{i,x}$ besitzen, wollen wir, falls ein solches Tetraeder existiert, mit $T(x)$ bezeichnen. Dabei setzen wir $T(0) = T$. Den reziproken Radius der $T(x)$ umgeschriebenen Kugel nennen wir $f(x)$; es ist dann also $f(0) = \frac{1}{R}$; liegt $T(x)$ in einer Ebene, so ist $f(x) = 0$. Zwischenbehauptung: Heißt a die größte der 6 Zahlen a_i, so existiert im offenen Intervall $(0, \frac{\pi}{a})$ eine Zahl x, sodaß $f(x) = x$. Beweisskizze: Es sei $\bar{x}$ die größte positive Zahl $\leqq \frac{\pi}{a}$, so daß für jedes x des abgeschlossenen Intervalles $[0, \bar{x}]$ das Tetraeder $T(x)$ existiert. (Die Existenz eines solchen $\bar{x}$ ergibt sich aus Stetigkeitsgründen). $f(x)$ ist eine im Intervall $[0, \bar{x}]$ stetige Funktion. Ist erstens $\bar{x} = \frac{\pi}{a}$, dann existiert also $T\left(\frac{\pi}{a}\right)$ und die größte Kante dieses Tetraeders hat die Länge $\frac{2a}{\pi}$, so daß der Radius der Umkugel $> \frac{a}{\pi}$ und $f\left(\frac{\pi}{a}\right) < \frac{\pi}{a}$ gilt. Da anderseits $f(0) = \frac{1}{R} > 0$, so muß für einen Zwischenwert $f(x) = x$ gelten. Ist zweitens $\bar{x} < \frac{\pi}{a}$, dann muß $T(\bar{x})$ in einer Ebene liegen, denn sonst würde für jedes hinreichend wenig von $\bar{x}$ verschiedene x auch $T(x)$ existieren entgegen der Definition von $\bar{x}$; also ist $0 = f(\bar{x}) < \bar{x}$, was zusammen mit $f(0) > 0$ die Existenz eines $x < \frac{\pi}{a}$ mit $f(x) = x$ ergibt, womit die Zwischenbehauptung bewiesen ist. Für die der Bedingung $f(x) = x$ gemäß bestimmte Zahl x hat das Tetraeder $T(x)$ die Kantenlängen $a_{i,x}$ und eine umgeschriebene Kugel vom Radius $\frac{1}{x}$. Also haben die Eckpunkte von $T(x)$ auf der Umkugel wegen $x < \frac{\pi}{a_i}$ die Bogenabstände a_i, so daß auf dieser Kugel ein mit den gegebenen vier Punkten kongruentes Quadrupel existiert.

Gödel: *Über die Waldsche Axiomatik des Zwischenbegriffes.*

In seiner Axiomatik des Zwischenbegriffes (diese Ergebnisse, *2*, S. 17 und Math. Ann., *104*, S. 476) beweist Wald, daß drei Punkte

2

a, b, c in einer gewissen Bedingungen genügenden Relation aller Punktetripel aller metrischen Räume dann und nur dann stehen, wenn für ihre Abstände die Beziehung $ab + bc = ac$ gilt, d. h. wenn b im Mengerschen Sinne zwischen a und c liegt. Unter den der Relation auferlegten Bedingungen tritt dabei die Kongruenzinvarianz auf, d. h. wenn drei Punkte a, b, c in der betreffenden Relation stehen, so sollen auch je drei mit ihnen kongruente Punkte a', b', c' in der betreffenden Relation zu einander stehen. Da die Gesamtheit aller zum Punktetripel a, b, c kongruenten Tripel durch die drei Abstandszahlen ab, bc, ac gekennzeichnet ist, so läßt sich die Waldsche Axiomatik des Zwischenbegriffes als ein Satz über Tripel reeller Zahlen, m. a. W. über den dreidimensionalen Cartesischen Raum aussprechen. Wir erhalten denselben, indem wir jedem metrischen Punktetripel a, b, c den Punkt

$$(+) \qquad x = ab, \ y = bc, \ z = ac$$

des R_3 zuordnen. Da für a, b, c die Dreiecksungleichung gilt und Abstände nicht negativ sind, treten dabei nur Punkte des R_3 auf, für die

$$(*) \quad x \geqq 0, \ y \geqq 0, \ z \geqq 0, \ (x+y-z)(x-y+z)(-x+y+z) \geqq 0$$

gilt. Es liegt b zwischen a und c dann und nur dann, wenn der dem Tripel a, b, c zugeordnete Punkt x, y, z in dem im Raumteil (*) enthaltenen Teil der Ebene $x + y = z$ liegt. Der Waldschen Axiomatik entspricht dann folgender Satz:

Es sei M eine Teilmenge des Raumteiles (*) des R_3 mit folgenden Eigenschaften:

1. Liegt der Punkt x, y, z in M, so auch der Punkt y, x, z.

2. Liegt der Punkt x, y, z in M, so liegt der Punkt z, y, x nicht in M.

3. Liegen die Punkte u, v, w und w, x, y in M, so liegen für jedes z die Punkte u, z, y und v, x, z, falls sie im Raumteil (*) liegen, auch in M.

4. Für jedes $\bar{z} > 0$ existiert mindestens ein Punkt x, y, $\bar{z}$ von M.

5. Für jedes $\bar{z} > 0$ ist die Menge bestehend aus den Punkten 0, $\bar{z}$, $\bar{z}$ und $\bar{z}$, 0, $\bar{z}$, sowie allen Punkten x, y, $\bar{z}$ von M abgeschlossen.

Dann ist M der den Ungleichungen (*) genügende Teil der Ebene $x + y = z$.

Es wird durch diesen Satz nahegelegt, allgemein für metrische Räume die durch (+) definierte Abbildung ihrer Punktetripel in den R_3 zu betrachten.

43. Kolloquium (25. II. 1932).

Nöbeling: *Ein Deformationssatz.*

Sind M und M' zwei höchstens eindimensionale kompakte Mengen des Euklidischen R_n ($n \geqq 3$), dann gibt es zu jedem $\varepsilon > 0$ ein zu M fremdes topologisches Bild $f(M')$ von M', sodaß für jeden Punkt *aus* M' *gilt* $\delta(p, f(p)) < 3\varepsilon$. M. a. W. die beiden Mengen M und M' können durch eine topologische 3ε-Deformation von M' allein von einander getrennt werden.

Beweis. Zunächst bilden wir die Menge M' durch eine ε-Deformation f_1 stetig auf einen eindimensionalen Komplex $K = f_1(M')$ ab[1]). Diesen Komplex K können wir darstellen als Summe endlich vieler Strecken, die zu je zwei höchstens einen Endpunkt gemein haben und deren Längen $< \varepsilon$ sind. Da M im R_n nirgends dicht ist, dürfen wir annehmen, daß die Endpunkte der Strecken nicht in M liegen (nötigenfalls können wir dies durch beliebig kleine Eckpunktsverrückungen erreichen). Es sei nun S eine der Strecken des Komplexes K; ihre Endpunkte seien e_1 und e_2. Da S eine Länge $< \varepsilon$ hat, ist S in einer offenen Kugel L von einem Durchmesser $< \varepsilon$ enthalten. Die offene Menge $L - L.M$ ist zusammenhängend, da M als höchstens eindimensionale Menge keinen Raumteil des $R_n (n \geqq 3)$ zerlegt; also ist $L - L.M$ auch bogenverknüpft. Daher kann man die Punkte e_1 und e_2 durch einen zu M fremden Teilbogen von L verbinden. Auf diesen Bogen bilden wir nun die Strecke S topologisch so ab, daß dabei jeder der Punkte e_1 und e_2 auf sich abgebildet wird; offenbar hat dann jeder Punkt der Strecke S von seinem Bildpunkt auf dem Bogen einen Abstand $< \varepsilon$. Führen wir dies für jede der endlichvielen Strecken von K durch, so entsteht also eine eindeutige, stetige Abbildung f_2 von M' auf eine zu M fremde Bogensumme $f_2(M')$, bei welcher jeder Punkt aus M' von seinem Bildpunkt in $f_2(M')$ einen Abstand $< 2\varepsilon$ hat. Die fremden Mengen M und $f_2(M')$ haben einen positiven Abstand. Daher kann man die Abbildung f_2 durch eine Modifikation $< \varepsilon$ in eine topologische Abbildung f von M' überführen[2]), sodaß $f(M')$ zu M fremd ist. Diese Abbildung f hat also die im Satze behaupteten Eigenschaften.

Nöbeling: *Beweis des Mengerschen Satzes* II_1^3.

Menger hat vermutungsweise folgenden Satz ausgesprochen[3]):

Satz II_k^n. *Eine* G_δ*-Menge* G *des* R_n *mit höchstens* $(n-k-1)$*-dimensionalem Komplement enthält zu jeder höchstens* k*-dimensionalen Menge des* R_n *einen homöomorphen Teil.*

Wir beweisen den Fall II_1^3. Wir stellen das Komplement $R_3 - G$ als Summe abzählbarvieler kompakter Mengen $K_i (i = 1, 2, \ldots)$ dar. Sei M ein kompakter eindimensionaler Raum. Die Menge $\mathfrak{E}$ aller eindeutigen stetigen Abbildungen φ von M auf Teilmengen des R_3 machen wir durch die Abstandsfestsetzung $D(\varphi, \varphi') = \operatorname{Max} \rho(\varphi(p), \varphi'(p))$, wo das Maximum für alle Punkte p von M zu nehmen ist, zu einem vollständigen metrischen Raum. Es sei $\mathfrak{E}_i$ die Menge aller φ aus $\mathfrak{E}$

---

1) Alexandroff, Ann. of Math., *30*, S. 101.

2) Pontrjagin und Tolstowa, Math. Ann., *105*, S. 744.

3) Kuratowski-Menger, Monatsh. f. Math. u. Phys., *37*, S. 173.

2*

mit der Eigenschaft, daß $\varphi(M)$ zu K_i fremd ist und für jeden Punkt aus $\varphi(M)$ die Urbildmenge in M einen Durchmesser $< \frac{1}{i}$ hat. $\mathfrak{C}_i$ ist in $\mathfrak{C}$ offen. Es ist aber $\mathfrak{C}_i$ auch dicht in $\mathfrak{C}$. Denn sei ein φ' von $\mathfrak{C}$ und ein $\varepsilon > 0$ gegeben. Dann existiert zunächst eine topologische Abbildung $\varphi''(M)$, sodaß $D(\varphi', \varphi'') < \varepsilon$. Da $\varphi''(M)$ und K_i höchstens eindimensional sind, existiert (vgl. dieses Heft, S. 18) eine topologische Abbildung $\varphi(M)$ mit $D(\varphi'', \varphi) < \varepsilon$ von M auf eine zu K_i fremde Menge. Dann ist $D(\varphi', \varphi) < 2\varepsilon$ und φ in $\mathfrak{C}_i$ enthalten. Also ist $\mathfrak{C}_i$ in $\mathfrak{C}$ offen und dicht. Mithin ist der Durchschnitt $\Pi\, \mathfrak{C}_i$ nichtleer. Jedes φ dieses Durchschnittes ist offenbar eine topologische Abbildung von M auf eine zu allen K_i fremde, also in G enthaltene Menge.

Nöbeling: *Eine rational n-dimensionale Universalmenge.*

In Beantwortung einer Frage von Menger (dieses Heft, S. 13) gilt der Satz: Die Menge aller Punkte des Q_ω mit höchstens $n-1$ rationalen oder nur rationalen Koordinaten, die fast alle verschwinden, ist eine rational n-dimensionale Universalmenge. Ebenso ist die Menge aller Punkte des R_{2n+1} mit höchstens $n-1$ oder genau $2n+1$ rationalen Koordinaten eine rational n-dimensionale Universalmenge. (Erscheint in den Math. Ann.)

44. Kolloquium (2. III. 1932).

Mimura: *Über die Bogenlänge.*

Ist E eine endliche Teilmenge eines euklidischen Raumes R, so bezeichnen wir für jede zusammenhängende E enthaltende und in R enthaltene Menge S, welche Summe von endlichvielen Strecken ist, mit $\iota(S)$ die Summe der Längen dieser Strecken und nennen $\varkappa(E)$ das Infimum aller dieser der Menge E zugeordneten Zahlen $\iota(S)$. Offenbar wird $\varkappa(E)$ vom Abstand keines Punktepaares von E übertroffen. Ist B ein Bogen der Länge $l(B)$ in R, so hat Menger die Frage aufgeworfen[1]), ob $l(B)$ gleich dem im Folgenden mit $\lambda(B)$ bezeichneten Supremum der Zahlen $\varkappa(E)$ für alle endlichen Teilmengen E von B sei. Im Folgenden wird bewiesen, daß diese Frage zu bejahen ist.

Ist E irgend eine endliche Teilmenge von B, deren Punkte in natürlicher Reihenfolge $p_1, p_2, \ldots, p_n$ sein mögen, wobei p_1 und p_n die Endpunkte von B und $p_i p_{i+1}$ den Abstand der Punkte p_i und p_{i+1} bezeichnen möge, so gilt offenbar $\varkappa(E) \leqq \sum_{i=1}^{n-1} p_i p_{i+1} \leqq l(B)$, also $\lambda(B) \leqq l(B)$. Um auch $\lambda(B) \geqq l(B)$ zu beweisen, genügt es, da $l(B)$ das Supremum der Zahlen $\sum_{i=1}^{n-1} p_i p_{i+1}$ für alle endlichen natürlich geordneten Teilmengen $p_1, p_2, \ldots, p_n$ von B ist, zu zeigen, daß, wenn wir mit B_i den zwischen p_i und p_{i+1} gelegenen Teilbogen von B bezeichnen, für den ja sicher $\lambda(B_i) \geqq p_i p_{i+1}$ gilt, die Ungleichung $\lambda(B) \geqq \sum_{i=1}^{n-1} \lambda(B_i)$

---

[1]) Ann. of Math., *32*, S. 741.

besteht. Diese aber ergibt sich durch vollständige Induktion unmittelbar aus der nun zu beweisenden $\lambda(B_1+B_2) \geqq \lambda(B_1)+\lambda(B_2)$. Für eine beliebige positive Zahl ε existieren zunächst zwei endliche Mengen $E_1' \subset B_1$ und $E_2' \subset B_2$, sodaß $\lambda(B_1)-\varepsilon \leqq \varkappa(E_1')$ und $\lambda(B_2)-\varepsilon \leqq \varkappa(E_2')$. Da sich die Zahl $\varkappa(E)$ bei hinreichend kleiner Verrückung eines Punktes von E um beliebig wenig ändert, können wir E_1' und E_2' so wählen, daß der gemeinsame Endpunkt p_2 von B_1 und B_2 weder in E_1' noch in E_2' liegt. Dann ist also der letzte Punkt q_1 von E_1' und der erste Punkt q_2 von E_2' verschieden von p_2. Sind U_1 und U_2 zwei offene Mengen, die den zwischen p_1 und q_1 bzw. zwischen q_2 und p_3 gelegenen Teilbogen B_1' bzw. B_2' von B enthalten und fremde abgeschlossene Hüllen besitzen, und bezeichnet d den Abstand der Menge $B_1'+B_2'$ von der Begrenzung von U_1+U_2, so ist $d>0$. Wir wählen nun für eine Zahl $\delta>0$, die $<d$ und $<\dfrac{\varepsilon d}{\lambda(B_1+B_2)+\varepsilon}$ ist, zwei Mengen E_1 und E_2, so daß $E_1' \subset E_1 \subset B_1'$ und $E_2' \subset E_2 \subset B_2'$ gilt und daß je zwei aufeinanderfolgende Punkte in E_1 und in E_2 einen Abstand $<\delta$ haben. Mit S bezeichnen wir eine $E=E_1+E_2$ enthaltende zusammenhängende Summe von endlichvielen Strecken, für die

$$(*) \qquad \varkappa(E) \geqq \iota(S)-\varepsilon.$$

Wir zerlegen nun die endliche Menge E_i $(i=1,2)$ in endlichviele Klassen, indem wir zwei Punkte von E_i dann und nur dann in eine Klasse aufnehmen, wenn die Menge $S.\overline{U}_i$ ein sie verbindendes Polygon enthält. Die Anzahl dieser Klassen sei n_i. Da wegen des Zusammenhanges von S je zwei Klassen von E_i durch ein Teilpolygon von S verbunden sind, dasselbe jedoch nicht $\subset \overline{U}_i$ ist, ist also jede Klasse mit der Begrenzung von U_i durch ein Polygon $\subset S.\overline{U}_i$ verbunden. Je zwei dieser n_1+n_2 Polygone sind nach Definition der Klassen und wegen der Fremdheit von $\overline{U}_1$ und $\overline{U}_2$ fremd; anderseits hat jedes Polygon eine Länge $\geqq d$, da E_1+E_2 als Teilmenge von $B_1'+B_2'$ von der Begrenzung von U_1+U_2 einen Abstand $\geqq d$ hat. Also ist $(n_1+n_2)\,d \leqq \iota(S)$. Hieraus folgt wegen (*), $\delta \leqq \varepsilon d.(\lambda(B_1+B_2)+\varepsilon)^{-1}$ und $\varkappa(E) \leqq \lambda(B_1+B_2)$ die Ungleichung $(n_1+n_2)\,\delta \leqq \varepsilon$.

Wir verbinden jetzt für jede Klasse, außer den beiden, die den letzten Punkt von E_1 bzw. E_2 enthalten, den letzten in ihr enthaltenen Punkt mit dem folgenden Punkt von E_1 bzw. E_2 durch eine Strecke. Da je zwei aufeinanderfolgende Punkte von E_1 bzw. E_2 einen Abstand $\leqq \delta$ haben, hat jede dieser n_1+n_2-2 Strecken eine Länge $\leqq \delta$ und ist daher wegen $\delta<d$ und $E_1+E_2 \subset B_1'+B_2'$ in $\overline{U}_1$ bzw. $\overline{U}_2$ enthalten. Fügen wir diese n_1+n_2-2 Strecken zu S hinzu, so entsteht eine zusammenhängende Streckensumme S', für welche

$$(**) \qquad \iota(S') \leqq \iota(S)+(n_1+n_2)\,\delta \leqq \iota(S)+\varepsilon$$

gilt und die zu je zwei Punkten von E_i ein sie verbindendes Polygon $\subset \overline{U}_i$ enthält. Wegen der Fremdheit von $\overline{U}_1$ und $\overline{U}_2$ enthält also S' zwei fremde zusammenhängende Streckensummen $S_1 \supset E_1$ und $S_2 \supset E_2$. Dann ist einerseits $\iota(S_1)+\iota(S_2) \leqq \iota(S')$, also wegen (**) und (*) $\leqq \varkappa(E)+2\varepsilon$. Anderseits ist $\varkappa(E_i) \leqq \iota(S_i)$; da aber $E_i' \subset E_i$, also $\varkappa(E_i') \leqq \varkappa(E_i)$ gilt und E_i' so gewählt war, daß $\lambda(B_i)-\varepsilon \leqq \varkappa(E_i')$ ist, folgt $\lambda(B_i)-\varepsilon \leqq \iota(S_i)$. Mithin ist $\lambda(B_1)+\lambda(B_2) \leqq \varkappa(E)+4\varepsilon$, woraus nach Definition von $\lambda(B_1+B_2)$ folgt $\lambda(B_1)+\lambda(B_2) \leqq \lambda(B_1+B_2)+4\varepsilon$. Dies gilt für jedes $\varepsilon > 0$. Also ist $\lambda(B_1)+\lambda(B_2) \leqq \lambda(B_1+B_2)$, w.z.b.w.

In anschließender Diskussion:

Anderer Beweis des vorhergehenden Satzes über die Bogenlänge.

Es sei E_i ($i=1, 2, \ldots$ ad inf.) eine ε_i-Kette im Bogen B, wobei $\lim_{i=\infty} \varepsilon_i = 0$, ferner S_i eine E_i enthaltende zusammenhängende Menge, welche Summe von endlichvielen Strecken ist und für welche $\iota(S_i)$ möglichst klein, also $=\varkappa(E_i)$ ist. Zum Beweise des Satzes genügt es offenbar, daß wir zeigen $\overline{\lim}\, \varkappa(E_i) \geqq l(B)$; wegen der Unterhalbstetigkeit und der Monotonie des linearen Inhalts (vgl. Menger, Math. Ann., *103*, p. 477) folgt dies daraus, daß der Limes einer konvergenten Teilfolge von $\{S_n\}$ den Bogen B als Teil enthält.

Wir wollen übrigens auch ohne Verwendung des eben erwähnten Mengerschen Satzes sogar zeigen, daß die Folge $\{S_i\}$ gegen B konvergiert. Daraus, daß S_i eine *kürzeste* E_i enthaltende zusammenhängende Streckensumme ist, folgt:

1. S_i ist ein Baum, dessen sämtliche Endpunkte in E_i liegen.

2. Jede Teilstrecke von S_i, die im Innern weder einen End- noch einen Verzweigungspunkt von S_i enthält, ist $\leqq \varepsilon_i$; denn wäre eine Teilstrecke $> \varepsilon_i$, so könnte man sie weglassen und den dadurch zerstörten Zusammenhang von S_i durch Verbindung zweier aufeinanderfolgender Punkte von E_i wieder herstellen, wodurch eine Verkürzung von S_i eintreten würde. Angenommen nun, S_i konvergiere nicht gegen B, dann gäbe es ein $\delta > 0$ und für unendlichviele i einen Punkt p_i von S_i, der von B einen Abstand $> \delta$ hätte. Die Anzahl der Punkte von S_i, die mit p_i durch einen Teilbogen der Länge $\frac{\delta}{2}$ von S_i verbindbar sind, ist $\geqq 2^{\left[\frac{\delta}{2\varepsilon_i}\right]}$, da sich S_i wegen 2. längs eines Polygones der Länge $\frac{\delta}{2}$ mindestens $\left[\frac{\delta}{2\varepsilon_i}\right]$ Mal verzweigt. Von jedem dieser Punkte führt innerhalb S_i ein Bogen von einer Länge $> \frac{\delta}{2}$ nach B und dieselben sind wegen 1. paarweise fremd. Also wäre $\varkappa(E_i) > \frac{\delta}{2} \cdot 2^{\left[\frac{\delta}{2\varepsilon_i}\right]}$ für unendlichviele i und daher $\overline{\lim}\, \varkappa(E_i) = \infty$, während doch alle $\varkappa(E_i) \leqq l(B)$.

Menger: *Zur mengentheoretischen Behandlung des Tangentenbegriffes und verwandter Begriffe.*

Als Grundlage würden folgende Voraussetzungen ausreichen: Ein regulärer topologischer Raum heiße ein Kompaßraum, wenn 1. *jedem Punkt p eine Umgebung U_p zugeordnet ist, für welche die abgeschlossene Hülle $\overline{U}_p$ kompakt ist und $\overline{U}_p-(p)$ Produkt einer halboffenen Strecke $(0, 1]$ und der Begrenzung $B(U_p)$ ist;* es ist dann also $\overline{U}_p$ als Summe gewisser in p endender und sonst paarweise fremder Bogen darstellbar, welche wir als die *Kompaßbogen* aus p bezeichnen wollen, während andere in B endende Bogen (auch solche, welche als echte Teilmengen Kompaßbogen enthalten oder in Kompaßbogen enthalten sind,) nicht Kompaßbogen heißen sollen; jeder Punkt von $\overline{U}_q-(p)$ ist durch zwei „Koordinaten", eine Zahl aus $(0, 1]$ und einen Punkt von $B(U_p)$, kennzeichenbar. Gefordert wird 2. *Ist q ein innerer Punkt eines Kompaßbogens C aus p, so ist jeder der beiden Teilbogen, in welche C durch q zerlegt wird, Teilmenge eines Kompaßbogens aus q.* Man sieht dann leicht ein, daß, wenn C ein Kompaßbogen aus p, und zwar mit dem Endpunkt c, und D ein Kompaßbogen aus q, und zwar mit dem Endpunkt d ist, einer der 5 folgenden Fälle vorliegt: 1) C und D sind *identisch*; 2) C und D sind *fremd*; 3) C und D haben *genau einen Punkt gemein*; 4) D ist *finale Teilmenge* von C (d. h. $D \subset C$ und $d=c$ oder $d=p$) oder C ist finale Teilmenge von D; 5) D ist *Fortsetzung* von C (d. h. q ist innerer Punkt von C und der in q und c endende Teilbogen von C ist $\subset D$) oder q ist von p ist verschieden und der in q und p endende Teilbogen von C ist $\subset D$) oder C ist Fortsetzung von D.

Es ist klar, wie in Kompaßräumen ein Tangentenbegriff erklärt werden kann. Um zu tangentenerhaltenden Abbildungen zweier Kompaßräume aufeinander zu gelangen, gehen wir von folgender (vermutlich auch in anderem Zusammenhange verwendbaren) Begriffsbildung aus: Eine Abbildung eines topologischen Raumes R auf einen topologischen Raum R' heiße im Punkt p von R *isoliert*, falls eine Umgebung V_p existiert, deren abgeschlossene Hülle keinen von p verschiedenen Punkt enthält, der denselben Bildpunkt in R' besitzt wie p. Eine in jedem Punkt isolierte Abbildung heiße isoliert schlechthin. *Kompaßabbildung* des Kompaßraumes R auf den Kompaßraum R' im Punkte p wollen wir eine im Punkt p isolierte stetige Abbildung nennen, die folgender Bedingung genügt: Ist $\{p_n\}$ eine gegen p konvergente Folge von Punkten aus U_p derart, daß die Folge der p_n enthaltenden Kompaßbogen aus p konvergiert, so konvergiert auch, falls p' und p_n' die Bildpunkte von p bzw. p_n sind, die Folge der p_n' enthaltenden Kompaßbogen aus p', soweit solche existieren, was, wie man leicht zeigt, für fast alle n der Fall ist.

45. Kolloquium (20. IV. 1932).

Wald: *Zur Axiomatik des Zwischenbegriffes.*

Ist G eine geordnete Abelsche Gruppe, so wollen wir als *G-Raum* eine Menge M dann bezeichnen, wenn je zwei Elementen

p und q von M ein mit pq bezeichnetes und der Abstand von p und q genanntes Element von G zugeordnet ist gemäß folgenden Bedingungen: Es gilt $pq = qp \begin{cases} > 0, \text{ wenn } p \neq q \\ = 0, \text{ wenn } p = q \end{cases}$ für je zwei Elemente und die Dreiecksungleichung $pq + qr \geqq pr$ für je drei Elemente von M. (Von der G-metrischen Menge im Sinne von Menger [diese Ergebnisse, *2*, S. 35] unterscheidet sich der G-Raum inhaltlich also nur durch die Forderung nach Giltigkeit der Dreiecksungleichung.) Es sei für die geordneten Punktetripel aller G-Räume eine Relation abc erklärt gemäß folgenden Bedingungen

1. *Kongruenzinvarianz.* Aus $ab = a'b'$, $bc = b'c'$, $ca = c'a'$ und abc folgt $a'b'c'$.
2. *Auszeichnung des Innengliedes.* Aus abc folgt, daß acb und bac nicht gelten.
3. *Transitivität.* Aus abc und acd folgt abd und bcd.
4. *Streckbarkeit.* Zu jedem aus drei Punkten a, b, c bestehenden G-Raum existiert ein aus drei Punkten a', b', c' bestehender G-Raum derart, daß $ab = a'b'$, $bc = b'c'$ und $a'b'c'$ gilt.
5. *Verschiedenheit der Außenglieder.* Aus abc folgt $a \neq c$.

Dann ist diese Relation mit der Mengerschen Zwischenrelation identisch, d. h. es gilt abc dann und nur dann, wenn $ab + bc = ac$ und $a \neq b \neq c$. Ist G speziell die Gruppe der reellen Zahlen, so ist 5. eine Folge der vier ersten Voraussetzungen. Der auf diese Gruppe angewendete Satz gestattet nach dem Verfahren von Gödel (dieses Heft, S. 17f.) eine Übersetzung in eine Aussage über den R_3.

46. Kolloquium (28. IV. 1932).

Haupt (Erlangen): Zur Theorie der Ordnung reeller Kurven in der Ebene bezüglich vorgegebener Kurvenscharen (erscheint in Monatshefte für Math. u. Phys., *40*).

47. Kolloquium (4. V. 1932).

Nöbeling: *Die Projektionen einer kompakten, n-dimensionalen Menge im R_k.*

Ist M eine n-dimensionale, kompakte Teilmenge des k-dimensionalen Cartesischen Raumes R_k, dann ist die Projektion von M auf mindestens eine n-dimensionale Koordinatenhyperebene des R_k ebenfalls n-dimensional.

Zum Beweise nehmen wir an, M sei eine kompakte Menge des R_k, für welche die Projektion auf jede n-dimensionale Koordinatenhyperebene des R_k höchstens $(n-1)$-dimensional ist, also in ihr nirgends dicht liegt[1]) und zeigen, daß M höchstens $(n-1)$-dimensional ist. Wir bezeichnen mit E^i ($i = 1, 2, \ldots$ ad inf.) die abzählbarvielen $(k-n)$-dimensionalen Hyperebenen der Gestalt $x_{j_1} = r_1, \ldots, x_{j_n} = r_n$, wobei die $j_1, \ldots, j_n$ alle n-tupel natürlicher Zahlen $\leqq k$ und die

[1]) Menger, Dimensionstheorie, S. 244.

$r_1, \ldots, r_n$ alle n-tupel rationaler Zahlen durchlaufen. Dann besteht die Menge $U = R_k - \sum_{i=1}^{\infty} E^i$ aus allen Punkten des R_k mit höchstens $n-1$ rationalen Koordinaten. Bezeichnen wir mit U^i die Menge aller Punkte des R_k mit genau i rationalen Koordinaten, so gilt $U = U^0 + \ldots + U^{n-1}$, und folglich, da jede Menge U^i nulldimensional ist[1]), dim $U \leqq n-1$. Da die Menge M kompakt, also beschränkt ist, können wir die Numerierung der E^i so gewählt annehmen, daß M zu E^1 fremd ist. Wir setzen $M = M^1$ und machen die Induktionsvoraussetzung, daß M^i eine zu $E^1 + \ldots + E^i$ fremde, aus M durch eine Parallelverschiebung hervorgegangene Menge ist. Da die Projektion von M^i auf den zu E^{i+1} orthogonalen R_n höchstens $(n-1)$-dimensional, also im R_n nirgends dicht ist, können wir M^i durch eine beliebig kleine Parallelverschiebung in eine zu E^{i+1} fremde Menge überführen; diese Parallelverschiebung wählen wir kleiner als das 2^{-i}-fache des Abstandes der Mengen M^i und $E^1 + \ldots + E^i$. Die so definierten Mengen M^i ($i = 1, 2, \ldots$ ad inf.) konvergieren offenbar gegen eine zu M homöomorphe Menge, die zu allen E^i fremd, also Teilmenge der höchstens $(n-1)$-dimensionalen Menge U ist.

48. Kolloquium (8. V. 1932).

Wald: *Über die Volumsdeterminante.*

In einem euklidischen Raum ist das Volumen $V(p_1, \ldots, p_n)$ des durch die Punkte $p_1, \ldots, p_n$ bestimmten Simplexes, wenn r_{ij} den Abstand der Punkte p_i und p_j bezeichnet, die positive Quadratwurzel aus einer $(n+1)$-reihigen Derminante, die wir (vgl. diese Ergebnisse, *1*, S. 21) kurz mit

$$V^2(p_1, \ldots, p_n) = \frac{(-1)^n}{2^{n-1}(n-1)!} \begin{vmatrix} 0 & 1 \\ 1 & r_{ij}^2 \end{vmatrix}_{i,j=1,\ldots,n}$$

bezeichnen. Menger hat das Problem gestellt, diese Funktion unter den Funktionen der n-Tupel von Punkten euklidischer Räume zu kennzeichnen. Im folgenden wird bewiesen: *Damit eine nicht negative Funktion $v(p_1, \ldots, p_n)$ der n-Tupel von Punkten euklidischer Räume ein Multiplum der Funktion $V(p_1, \ldots, p_n)$ sei*, d. h. damit eine reelle Zahl C existiert, so daß für je n Punkte $v(p_1, \ldots, p_n) = C . V(p_1, \ldots, p_n)$ gilt, *sind folgende Bedingungen notwendig und hinreichend:*

I. $v(p_1, p_2, \ldots, p_n) = v(p_{i_1}, p_{i_2}, \ldots, p_{i_n})$ für jede Permutation $i_1, i_2, \ldots, i_n$ der Zahlen 1, 2, ..., n.

II. Wenn $p_i = p_j$, also $r_{ij} = 0$, $r_{ik} = r_{jk}$ $(k = 1, 2, \ldots, n)$, so gilt $v(p_1, \ldots, p_n) = 0$.

III. $v^2(p_1, \ldots, p_n)$ ist ein Polynom $(n-1)$-ten Grades in den $\frac{n(n-1)}{2}$ Variablen r_{ij}^2 $(i, j = 1, \ldots, n,\ r_{ij} = r_{ji},\ r_{ii} = 0)$.

Es ist also $C = 1$ und $v(p_1, \ldots, p_n) = V(p_1, \ldots, p_n)$ unter der weiteren Bedingung

---

[1]) Menger, Dimensionstheorie, S. 147.

IV. Es existieren n Punkte $p_1, \ldots, p_n$, so daß $v(p_1, \ldots, p_n) = V(p_1, \ldots, p_n) \neq 0$.

Die Notwendigkeit der Bedingungen ist klar. Wir wollen sie nun, um sie als hinreichend zu erweisen, für die Funktion v als erfüllt annehmen. Wegen III. ist dann also v^2 ein Polynom in den Variablen $a_{ij} = r_{ij}^2$. Jeder Summand desselben ist ein mit einem von 0 verschiedenen Koeffizienten multipliziertes Produkt von Potenzen gewisser a_{ij}. Zwei Summanden mögen *von demselben Typus* heißen, wenn ihre Potenzprodukte durch eine Permutation der Indizes $1, 2, \ldots, n$ aus einander hervorgehen. *Glieder desselben Typus haben wegen I. gleiche Koeffizienten.*

Aus II ergibt sich: *II.′ Für $a_{ij} = 0$ und $a_{ik} = a_{jk}$ $(k = 1, 2, \ldots, n)$ verschwindet das Polynom v^2 in den $\frac{(n-1)(n-2)}{2}$ Variablen a_{rs} $(r, s = 1, 2, \ldots, i, \ldots, j-1, j+1, \ldots, n;\ r < s)$ identisch.*

Es sei nun S irgend ein Summand von v^2. Stellen wir S als Produkt der a_{ij} dar, wobei jedes in k-ter Potenz auftretende a_{ij} genau k Mal angeschrieben wird, so sei l_i die Zahl, welche angibt, wie oft in den Indizespaaren dieser Faktoren von S die Zahl i auftritt $(i = 1, 2, \ldots, n)$. Wegen III gilt offenbar

$$\textbf{1.} \qquad \frac{l_1 + l_2 + \ldots + l_n}{2} \leq n - 1.$$

2. *Höchstens für eine der Zahlen $i = 1, 2, \ldots, n$ ist $l_i = 0$.* Wäre nämlich etwa $l_1 = l_2 = 0$, so setzen wir $a_{12} = 0$, $a_{1i} = a_{2i}$ $(i = 3, 4, \ldots, n)$. Mit Rücksicht auf II′. muß dann v^2 identisch verschwinden, was nicht der Fall ist, da S wegen $l_1 = l_2 = 0$ die Variablen a_{12}, a_{1i}, a_{2i} nicht enthält.

3. *Wenn $l_i = 0$ ist, so gilt $l_k = 2$ für $k = 1, \ldots, i-1, i+1, \ldots, n$.* Sei etwa $i = 1$, also $l_1 = 0$. Wegen **2** gilt dann $l_k \geq 1$ $(k = 2, \ldots, n)$. Wir zeigen vor allem $l_k \geq 2$ $(k = 2, \ldots, n)$. Wäre nämlich etwa $l_2 = 1$, so setzen wir $a_{12} = 0$, $a_{1i} = a_{2i}$ $(i = 3, 4, \ldots, n)$. Wegen II′. muß dann v^2 identisch verschwinden. Also muß schon die Summe von S und allen jenen Summanden, die dadurch entstehen, daß man in S jeden auftretenden Index 1 und 2 beliebig durch 1 oder 2 ersetzt, verschwinden. Wegen $l_1 = 0$, $l_2 = 1$ gibt es nur einen Summanden S', der auf diese Weise entsteht. S und S' haben denselben Typus, also den gleichen Koeffizienten und daher eine nicht verschwindende Summe, womit aus $l_2 = 1$ ein Widerspruch hergeleitet ist. Aus $l_k \geq 2$ und **1** folgt $l_k = 2$ $(k = 2, \ldots, n)$.

4. *Sind alle $l_i \neq 0$ $(i = 1, 2, \ldots, n)$, so sind genau zwei Zahlen l_i und l_j gleich 1, alle anderen $l_k = 2$.* Daß nur ein l_i gleich 1 ist, ist wegen **1** unmöglich. Daß drei l_i gleich 1 sind, ist auch unmöglich. Sei nämlich etwa $l_1 = l_2 = l_3 = 1$. Von den Variablen a_{12} und a_{13} kann S wegen $l_1 = 1$ höchstens eine enthalten. Ist etwa a_{12} in S nicht enthalten, so setzen wir $a_{12} = 0$, $a_{1i} = a_{2i}$ $(i = 3, \ldots, n)$. Es gibt dann wegen $l_1 = l_2 = 1$ drei weitere Summanden, die aus S

entstehen, wenn man jeden in S auftretenden Index 1 und 2 durch 1 oder 2 ersetzt. Dasjenige S', das entsteht, wenn man 1 durch 2 und 2 durch 1 ersetzt, hat den selben Typus wie S. Die beiden anderen S'' und S''' haben untereinander den gleichen Typus, der von dem von S verschieden ist. Da v^2 wegen II'. identisch verschwinden muß, muß $S+S'+S''+S'''=2\,(S+S'')=0$ sein. Das ist aber unmöglich, denn da S'' aus S dadurch entsteht, daß man die Indizes 1 und 2 beide durch 1 oder beide durch 2 ersetzt, kommt entweder 1 oder 2 nicht als Index vor, also muß nach **3** in S'' der Index 3 zweimal auftreten, während $l_3=1$. Also sind genau zwei l_i gleich 1. Dann folgt aus **1**, daß alle übrigen $=2$ sind, w. z. b. w.

Ordnen wir jeder Variablen a_{ij} die p_i und p_j verbindende Strecke zu, so ist auf Grund von **2**, **3**, **4** für jeden Summanden S von v^2 die Summe der den Faktoren von S entsprechenden Strecken (wenn wir eine Doppelstrecke als geschlossenes Polygon bezeichnen) offenbar *entweder eine Summe von paarweise fremden geschlossenen Polygonen, welche $n-1$ Punkte verbinden, oder eine Summe eines einfachen Streckenzuges, der k ($\leq n$) Punkte verbindet, und von geschlossenen Polygonen, die paarweise und zum Streckenzug fremd sind und die übrigen $n-k$ Punkte verbinden.*

Es sei nun λ der Koeffizient eines Summanden S von v^2, dem ein einziges $n-1$ Punkte verbindendes geschlossenes Polygon P entspricht, bzw. $\lambda=0$, falls ein solcher Summand in v^2 nicht auftritt.

5. *Jeder Summand S' von v^2, dem zwei $n-1$ Punkte verbindende geschlossene Polygone P_1 und P_2 entsprechen, hat den Koeffizienten -2λ.* Es bestehe P_1 aus k und P_2 aus l Strecken $(k+l=n-1)$. Wir wählen auf P zwei Punkte p_1 und p_2, durch welche P in zwei einfache Streckenzüge, bestehend aus k bzw. l Strecken zerlegt wird. Da in S die Indizes 1 und 2 genau je zweimal auftreten, erhält man, wenn man in S jeden auftretenden Index 1 und 2 beliebig durch 1 oder 2 ersetzt, wie man leicht sieht, 15 weitere Summanden, von denen 10 einen in v^2 nicht auftretenden Typus haben, während von den 5 übrigen 3 den Typus von S und 2 den Typus von S' haben. Daher ist, wie man erkennt, wenn man $a_{12}=0$, $a_{1i}-a_{2i}$ setzt, $4\,S+2\,S'=0$, woraus unmittelbar die Behauptung über die Koeffizienten folgt.

6. *Jeder Summand S'', dem r paarweise fremde geschlossene Polygone entsprechen, hat den Koeffizienten $c''.\lambda$* (für eine analog wie $c'=-2$ bestimmbare rationale Zahl c''). Der Satz folgt aus **5** durch vollständige Induktion.

7. *Jeder Summand S''', dem ein die n Punkte $p_1,\ldots,p_n$ verbindender einfacher Streckenzug entspricht, hat den Koeffizienten $c'''.\lambda$* (wo c''' rational). Es sei S ein Summand von v^2, dem ein $n-1$ Punkte verbindendes geschlossenes Polygon P entspricht, so daß sein Koeffizient λ ist. p_1 sei der in P nicht enthaltene Punkt. Ersetzen wir die in S auftretenden Indizes 1 und 2 durch 1 oder 2, so erhalten wir nur Summanden vom Typus S und S'''

8. *Jeder Summand S^*, dem ein aus k ($< n-1$) Strecken bestehender einfacher Streckenzug T und geschlossene Polygone entsprechen, hat den Koeffizienten $c^* . \lambda$* (c^* rational). Sei S''' von dem in **7** betrachteten Typus und p_1 ein Eckpunkt von T, durch den T in zwei Streckenzüge T_1 und T_2 zerlegt wird, wobei T_1 aus k Strecken besteht. Der von p_1 verschiedene Endpunkt von T_2 heiße p_2. Durch die Ersetzung der Indizes 1 und 2 durch 1 oder 2 entstehen aus S''' Summanden vom Typus S''' sowie solche, denen ein aus k Strecken bestehender einfacher Streckenzug und ein einfaches, geschlossenes Polygon entspricht. Durch die mehrmals verwendeten Schlüsse drückt man den Koeffizienten der Summanden vom letzten Typus und dann durch vollständige Induktion den von S^* durch den Koeffizienten von S''' und damit durch λ aus.

Da damit die Koeffizienten aller Summanden von v^2 als Multipla von λ erwiesen sind, unterscheiden sich je zwei den Bedingungen I, II, III genügende Funktionen höchstens durch einen konstanten Faktor, womit unser Satz bewiesen ist.

49. Kolloquium (11. V. 1932).

Borsuk (Warschau): *Über Rand- und Kernpunkte kompakter metrischer Räume.*

Zum Ausgangspunkt unserer Definition dient folgende geometrische Tatsache: Ist B ein n-dimensionaler Würfel mit der Oberfläche A und p ein beliebiger Punkt von $B—A$, so wird p in gewissem Sinne von A in B umschlossen: Nach Tilgung jeder Umgebung $U \subset B—A$ von p entsteht in B eine Aushöhlung; präzise gesprochen: Es gibt eine stetige Abbildung von A auf eine $(n-1)$-dimensionale Sphäre S_n mit der Gleichung $x_1^2+x_2^2+\ldots+x_n^2=1$, welche eine Erweiterung relativ zu S_n auf die Menge $B—U$, nicht aber auf die ganze Menge B besitzt.

1. Im Folgenden bezeichnen stets: A und B insichkompakte metrische Mengen, die die Bedingung $A \subset B$ erfüllen, T einen beliebigen topologischen, M einen beliebigen metrischen Raum, M^T die Menge aller stetigen Abbildungen f, für welche $f(T) \subset M$ gilt und die dadurch, daß je zwei Elementen f_1 und f_2 von M^T der Abstand $\rho(f_1, f_2) = \text{Supr}\, \rho\, [f_1(x), f_2(x)]$ für alle x aus T zugeordnet wird, zu einem metrischen Raum wird.

Definition. Der Punkt p von $B-A$ heißt *in B durch A n-umschlossen, wenn es eine Funktion f aus S_n^A gibt, so daß f für jede Umgebung $U \subset B-A$ von p eine Erweiterung auf $B-U$ relativ zu S_n, aber keine Erweiterung auf B relativ zu S_n besitzt.*

2. Beispiel: *Wird ein Kontinuum B durch p zwischen den Punkten a und b zerschnitten, so wird p in B durch die Menge $A=(a)+(b)$ 1-umschlossen.* S_1 besteht aus zwei Punkten -1 und $+1$. Setzen wir $f(a)=-1$ und $f(b)=+1$, so hat f, da B zusammenhängend ist, keine Erweiterung auf B relativ zu S_1. Ist nun U eine Umgebung $\subset B-A$ von p und E die a enthaltende Komponente von $B-U$, so liegt, da B von p zwischen a und b zerschnitten wird, der Punkt b in $B-E$. Um nun eine Erweiterung von f auf $B-U$ relativ zu S_1 anzugeben, hat man also bloß $f(x)=-1$ bzw. $=+1$ zu setzen, je nachdem x in E oder in $B-U-E$ liegt.

3. *Ist* $B \subset R_n$ *und* $A = B.\overline{R_n - B}$, *so ist jeder Punkt* p *von* $B-A$ *in* B *durch* A *n-umschlossen.* Wählt man p als den Ursprung 0 des R_n, so ist $f(x) = \overrightarrow{0x} . S_n$ eine Funktion auf A, die keine Erweiterung auf B relativ zu S_n besitzt[1]), während eine Erweiterung von $f(x)$ auf $B-U$ relativ zu S_n für jede Umgebung U von p durch die Definitionsformel von $f(x)$ gegeben ist.

4. *Ist* $A \subset M$ *und* $\dim A < n-1$, *so hat jede Funktion* φ *aus* S_n^A *eine Erweiterung auf* M *relativ zu* S_n. Wir verwandeln die Abbildung φ für jede Zahl $k=1, 2, \ldots$ durch eine $\frac{1}{4k}$-Modifikation in eine topologische Abbildung φ'_k auf eine Teilmenge des R_{2n-1}, bilden die Menge $\varphi'_k(A)$ durch eine $\frac{1}{4k}$-Deformation auf einen Komplex $\varphi''_k(A)$ ab und projizieren $\varphi''_k(A)$ (evt. nach kleiner Eckpunktverrückung) so auf S_n, daß die entstehende Abbildung φ_k aus S_n^A nicht ganz S_n bedeckt. Dabei hat dann φ von φ_k einen Abstand $\rho(\varphi, \varphi_k) < \frac{1}{k}$. Es sei p_k Punkt von $S_n - \varphi_k(A)$. Da $S_n - (p_k)$ mit dem euklidischen R_{n-1} homöomorph und A insichkompakt, also in M abgeschlossen ist, gibt es eine Erweiterung[2]) von φ_k auf M relativ zu $S_n - (p_k)$ und somit auch relativ zu S_n. Da die Funktionen, welche eine Erweiterung auf M relativ zu S_n besitzen, im Raum S_n^A eine abgeschlossene Menge bilden und die Funktionen φ_k gegen φ konvergieren, hat auch φ eine Erweiterung auf M relativ zu S_n, w. z. b. w.

5. Hilfssatz: Ist U eine offene Teilmenge von T derart, daß jede stetige Funktion φ aus $A^{\overline{U}.(T-U)}$ eine Erweiterung auf $\overline{U}$ relativ zu A hat und ist $E \subset T-U$, so hat eine Funktion f aus A^E eine Erweiterung auf T relativ zu A dann und nur dann, wenn sie eine Erweiterung auf $T-U$ relativ zu A besitzt. Die Notwendigkeit der Bedingung ist klar. Ist nun φ eine Erweiterung von f auf $T-U$ relativ zu A und ψ eine Erweiterung der nur auf der Menge $\overline{U}.(T-U)$ betrachteten Funktion φ auf U relativ zu A, so erhalten wir eine Erweiterung f_0 von f auf T relativ zu A, wenn wir $f_0(x) = \varphi(x)$, bzw. $= \psi(x)$ setzen, je nachdem x in $T-U$ oder in U liegt.

6. *Wird* p *in* B *durch* A *n-umschlossen, so gilt für jede Umgebung* $U \subset B-A$ *von* p *die Ungleichung* $\dim \overline{U}.(B-U) \geqq n-1$. Wäre nämlich $\dim \overline{U}.(B-U) < n-1$, so hätte nach **5** und **4** jede

---

[1]) Borsuk, Fund. Math., *17*, S. 162.

[2]) Vgl. z. B. Tietze, Journ. f. reine u. angew. Math., *145*, S. 10; Caratheodory (nach Bohr), Vorl. über reelle Funktionen, S. 617; Brouwer, Math. Ann., *79*, S. 139; Hahn, Reelle Funktionen, S. 137 f.

Funktion f aus S_n^A, welche eine Erweiterung auf $B—U$ relativ zu S_n hat, auch eine Erweiterung auf B relativ zu S_n, während p in B durch A n-umschlossen sein soll.

7. Definition: p heiße ein *n-Kernpunkt* des Raumes M, *wenn M zwei Teilmengen A und B enthält, so daß p in B durch A n-umschlossen wird.* Jeder Punkt von M, der nicht n-Kernpunkt von M ist, möge ein n-Randpunkt von M heißen. Die Menge aller n-Randpunkte von M heiße der *n-Rand* von M[3]). Aus diesen Definitionen bzw. aus **6** folgt unmittelbar: *Jeder n-Kernpunkt einer Menge M ist auch n-Kernpunkt jeder Obermenge von M. Ein Raum M ist in jedem seiner n-Kernpunkte mindestens n-dimensional.*

8. *Die 1-Randpunkte eines im kleinen zusammenhängenden Kontinuums C sind mit den Endpunkten von C identisch.* Ist einerseits p ein Punkt einer Ordnung $\geqq 2$ von C, so existieren zwei in p endende, sonst fremde Teilbogen B_1 und B_2 von C. Nennen wir a und b die Endpunkte des Bogens $B = B_1 + B_2$ und $A = (a) + (b)$, so folgt aus **2**, daß p in B durch A 1-umschlossen wird, also 1-Kernpunkt von C ist. Nehmen wir anderseits an, daß p Endpunkt von C ist und A und B Teilmengen von C sind, so daß p in $B—A$ liegt, dann existiert eine Umgebung $U \subset C—A$ von p, deren Begrenzung $\overline{U}.(C—U)$ genau einen Punkt enthält, so daß also $U_0 = U.B$ eine Umgebung von p in B ist, deren Relativbegrenzung $\overline{U_0}.(B—U_0)$ höchstens einen Punkt enthält. Also hat jede Funktion φ aus $S_1^{\overline{U_0}.(B-U_0)}$ eine Erweiterung auf $\overline{U_0}$ relativ zu S_1, so daß mit Rücksicht auf **5** jede Funktion f aus S_1^A, welche eine Erweiterung auf $B—U$ hat, auch auf B erweiterbar ist; p ist also in B durch A nicht 1-umschlossen.

9. Hilfssatz: Ist $A \subset R_n$ und p ein Punkt der Komponente K von $R_n—A$, so ist jede Funktion f aus S_n^A einer Erweiterung auf $A + K—(p)$ relativ zu S_n fähig. Wenn nämlich K unbeschränkt ist, so besitzt f sogar[4]) eine Erweiterung auf $A + K$. Der Fall, daß K beschränkt ist, läßt sich auf den der Unbeschränktheit zurückführen, indem man den R_n an einer Kugel mit dem Zentrum p spiegelt.

10. *Der n-Rand einer Teilmenge M des R_n ist die Menge $M.\overline{R_n—M}$.* Liegt nämlich p nicht in $M.\overline{R_n—M}$, so gibt es eine p im Innern enthaltende Vollkugel $B \subset M$, von deren Begrenzung p wegen **3** in B n-umschlossen ist. Liegt dagegen p in $M.\overline{R_n—M}$ und sind A und B zwei Teilmengen von M, für welche p in $B—A$ liegt, so sei U die p enthaltende Komponente von $R_n—A$ und q irgend ein Punkt von $U—M$. Nach **9** hat jede Funktion aus $S_n^{\overline{U}.(R_n-U)}$ eine Erweiterung auf $\overline{U}.(R_n—U) + U—(q)$. Setzt man

---

[3]) Von den mannigfachen Versuchen, den Begriff des Kernes und Randes zu präzisieren, sei hier nur der von Alexandroff, Ann. of Math., *30*, S. 185, erwähnt.

[4]) Borsuk, Math. Ann., *106*, S.246 und Fund. Math., *20*, S. 182.

$T=B+U-(q)\supset A$, so folgt aus 5, daß jede Funktion φ aus S_n^A, die eine Erweiterung auf $T-U=B-U$ relativ zu S_n besitzt, auch eine auf T, also erst recht eine Erweiterung auf $B\subset T$ relativ zu S_n hat, so daß also p in B durch A nicht n-umschlossen und daher n-Randpunkt von M ist.

11. Ist $A\subset M$ und f eine Abbildung von A auf eine Teilmenge der n-dimensionalen Vollkugel H_n, so bezeichnen wir mit P_f die aus allen Erweiterungen von f auf M relativ zu H_n bestehende Teilmenge des Raumes H_n^M. Dann sind je zwei Elemente φ_0 und φ_1 aus P_f durch einen Teilbogen von P_f verbunden. Ein solcher ist, wenn wir mit φ_t den Punkt bezeichnen, der die Strecke zwischen φ_0 und φ_1 im Verhältnis $t:1-t$ teilt, die Menge der Abbildungen φ_t $(0\leqq t\leqq 1)$.

12. Hilfssatz: Existiert zu der Teilmenge A des insichkompakten metrischen Raumes C eine Funktion aus S_n^A, welche keine Erweiterung auf C relativ zu S_n besitzt, so existiert eine A als echten Teil enthaltende Teilmenge B von C, so daß jeder Punkt von $B-A$ in B durch A n-umschlossen wird, wobei, falls ein in C durch A n-umschlossener Punkt p_0 existiert, B so gewählt werden kann, daß p_0 in B liegt. Beweis: Von einer A enthaltenden abgeschlossenen Teilmenge E von C, auf welche f relativ zu S_n nicht erweiterbar ist, wollen wir sagen, daß sie die Eigenschaft W besitzt. Ist E_0 Durchschnitt einer monoton abnehmenden Folge $\{E_j\}$ $(j=1,2,\ldots)$ von Mengen mit der Eigenschaft W, so besitzt E_0 die Eigenschaft W, denn E_0 ist abgeschlossen und enthält A und es kann keine Erweiterung φ von f auf E_0 relativ zu S_n existieren, da sonst eine Umgebung U von E_0 im Raume C und eine Erweiterung ψ von φ auf U relativ zu S_n existieren würde [5]), was unmöglich ist, da E_j für hinreichend großes j Teil von U ist, so daß ψ (nur auf E_j betrachtet) eine Erweiterung von f auf E_j relativ zu S_n wäre, während E_j die Eigenschaft W besitzt.

Da die dem Raume C zukommende Eigenschaft W somit induzibel [6]) ist, existiert [6]) eine hinsichtlich der Eigenschaft W irreduzible Teilmenge B von C. Dieselbe enthält A als echten Teil, da A die Eigenschaft W nicht besitzt. Um zu zeigen, daß B der Behauptung genügt, d. h. daß jeder Punkt p von $B-A$ in B durch A n-umschlossen wird, braucht man, da f auf B relativ zu S_n nicht erweiterbar ist, wegen 7 nur zu beweisen, daß f für jede Umgebung $U\subset B-A$ von p eine Erweiterung auf $B-U$ relativ zu S_n besitzt.

$B-U$ ist abgeschlossen und $\supset A$. Da B hinsichtlich der Eigenschaft W irreduzibel ist, kann also f auf $B-U$ relativ zu S_n erweitert werden. Da in der Behauptung des Hilfssatzes von f keine Rede ist, kann man f durch eine beliebige Funktion aus S_n^A, die keine Erweiterung auf C relativ zu S_n besitzt, ersetzen und daher, wenn p_0

[5]) Borsuk, Math. Ann., *106*, S. 241.

[6]) Brouwer, Proc. Acad. Amsterdam, *14*, S. 138; vgl. auch Menger, Dimensionstheorie, 1928, S. 69.

ein durch A in C n-umschlossener Punkt ist, annehmen, daß f sich für jede Umgebung U von p_0 auf $C-U$ erweitern läßt. Läge dann p_0 nicht in der wie oben mit Hilfe von f konstruierten Menge B, so würde f, wenn man $U=C-B$ setzt, eine Erweiterung auf $C-(C-B)=B$ besitzen, während doch f auf B relativ zu S_n nicht erweiterbar ist. Damit ist der Hilfssatz in allen Stücken bewiesen.

13. *Jede insichkompakte Teilmenge des n-Randes eines metrischen Raumes M ist höchstens (n—1)-dimensional.* Ist die insichkompakte Menge C mindestens n-dimensional, so existiert eine wesentliche[7]) Abbildung f von C auf die n-dimensionale Vollkugel H_n. Daraus und aus **11** folgt, daß die nur auf $A=f^{-1}(S_n)$ betrachtete Funktion f keine Erweiterung auf C relativ zu S_n besitzt. Also existiert nach **12** eine abgeschlossene Menge B, die A als echten Teil enthält und so, daß jeder Punkt von $B-A$ durch A in B n-umschlossen wird. Nach **7** sind alle diese Punkte n-Kernpunkte von C, d. h. wenn dim $C \geqq n$, so ist C nicht im n-Rand von M enthalten. Offen bleibt die Frage, ob der n-Rand jedes metrischen Raumes, der ja selbst nicht notwendig insichkompakt ist, höchstens $(n-1)$-dimensional ist.

14. *Jeder n-Kernpunkt p_0 von M liegt in einer in p_0 mindestens n-dimensionalen insichkompakten Teilmenge von M, die nur aus n-Kernpunkten von M besteht.* Denn zum n-Kernpunkt p_0 existieren zwei Teilmengen A und C von M, so daß p_0 in C durch A n-umschlossen ist, so daß also eine Funktion von S_n^A keiner Erweiterung auf C relativ zu S_n fähig ist. Es sei B eine p_0 enthaltende dem Hilfssatz **12** genügende Menge. Ist dann U eine in B offene, p_0 enthaltende Menge, so daß $\overline{U}$ und A fremd sind, so ist nach **12** jeder Punkt von $\overline{U} \subset B-A$ in B durch A n-umschlossen, also ein n-Kernpunkt von M, und nach **6** ist $B-A$, also erst recht B in p_0 mindestens n-dimensional.

Wald: *Über das Volumen der euklidischen Simplexe.*

Das Volumen eines n-dimensionalen Simplexes S ist bekanntlich durch die Längen der $\binom{n+1}{2}$ Kanten von S (d. h. durch die Volumina der eindimensionalen Randsimplexe von S) eindeutig bestimmt. Menger hat (diese Erg., *2*, S. 21) die Vermutung ausgesprochen, daß das Volumen von S auch durch die Volumina der $\binom{n+1}{n-1}=\binom{n+1}{2}$ Randsimplexe der Dimension $n-2$ von S eindeutig bestimmt ist. Im Folgenden zeigen wir, daß dies nicht der Fall ist, indem wir *zwei vierdimensionale Simplexe S und S' mit verschiedenem Volumen* angeben, *deren 10 zweidimensionale Randsimplexe beziehungsweise gleich sind.* Die 5 Eckpunkte von S seien im R_4 durch ihre Koordinaten gegeben:

[7]) Alexandroff, C. R., *191*, S. 475.

$$p_1=(0,0,0,0),\ p_2=(\sqrt{3},0,0,0),\ p_3=\left(\frac{\sqrt{3}}{2},\frac{1}{2},0,0\right),$$

$$p_4=\left(\frac{\sqrt{3}}{2},\frac{1}{4},\frac{\sqrt{3}}{4},0\right),\ p_5=\left(\frac{\sqrt{3}}{2},-\frac{1}{4},\frac{\sqrt{3}}{4},0\right).$$

Die 5 Eckpunkte von S' seien

$$p_1'=(0,0,0,0),\ p_2'=(1,0,0,0),\ p_3'=\left(\frac{1}{2},\frac{\sqrt{3}}{2},0,0\right),$$

$$p_4'=\left(\frac{1}{2},\frac{5}{4\sqrt{3}},\frac{\sqrt{11}}{4\sqrt{3}},0\right),\ p_5'=\left(\frac{1}{2},\frac{\sqrt{3}}{4},\frac{5\sqrt{3}}{4\sqrt{11}},\frac{\sqrt{3}}{\sqrt{22}}\right).$$

Die Kanten von S haben die Längen

$$p_1p_2=\sqrt{3},\ p_1p_3=p_1p_4=p_1p_5=p_2p_3=p_2p_4=p_2p_5=1,$$

$$p_3p_4=p_4p_5=\frac{1}{2},\ p_3p_5=\frac{\sqrt{3}}{2}.$$

Die Kanten von S' haben die Längen

$$p_1'p_2'=p_1'p_3'=p_1'p_4'=p_1'p_5'=p_2'p_3'=p_2'p_4'=p_2'p_5'=1,$$

$$p_3'p_4'=p_4'p_5=\frac{1}{2},\ p_3p_5=\frac{\sqrt{3}}{2}.$$

Sämtliche entsprechende Flächen von S und S' haben denselben Inhalt. Da abgesehen von p_1p_2 und $p_1'p_2'$ alle entsprechenden Kanten von S und S' dieselbe Länge haben, so ist nur zu verifizieren, daß die Dreiecke $p_1p_2p_i$ und $p_1'p_2'p_i'$ ($i=3, 4, 5$) flächengleich sind. Da die drei ersteren die Kantenlängen $\sqrt{3}$, 1, 1, die drei letzteren die Kantenlängen 1, 1, 1 haben, ist dies klar. Anderseits haben S und S' verschiedene Volumina. Da alle Punkte p_1, p_2, p_3, p_4, p_5 die vierte Koordinate 0 haben, liegt S in einem R_3 und hat das Volumen 0. Hingegen sind die Punkte $p_1', p_2', p_3', p_4', p_5'$ linear unabhängig, wie man an der Matrix ihrer Koordinaten sofort erkennt, und bestimmen daher ein Simplex, dessen Volumen $\neq 0$ ist.

Es sei noch erwähnt, daß, wenn wir die 10 Gleichungen betrachten, welche die Quadrate der 10 Flächen von S als biquadratische Funktionen der 10 Kantenlängen ausdrücken, die Funktionaldeterminante für den Fall der Simplexe S und S', wie Herr Hiess ausgerechnet hat, nicht verschwindet, so daß also die Kanten algebraische Funktionen der Flächen und mithin das Volumen eine algebraische Funktion der Flächen ist.

50. Kolloquium (15. V. 1932).

Menger: Über die Bedeutung der Begriffe Kongruenzordnung und Quasikongruenzordnung (vgl. dieses Heft, S. 41) für die Axiomatik des R_n. — Neuer Beweis der Theoreme über pseudoeuklidische Mengen.

51. Kolloquium (25. V. 1932).

Zur Axiomatik der elementargeometrischen Verknüpfungsrelationen.

Schreiber: In einem Feld, dessen sämtliche Teilerketten endlich sind (vgl. diese Ergebnisse, *1*, S. 28—30) ist die Existenz von Einheitselementen eine Folge der übrigen Axiome.

Gödel: Man sollte das System aller jener Aussagen über Felder untersuchen, die in der Normalform keine Existenzpräfixe enthalten. Die Begriffe Punkt und Gerade, die unter Verwendung von Existenzpräfixen definierbar sind (z. B. Punkt ist ein Element, zu dem kein nicht leeres Element existiert, das sein echter Teil ist), sind in diesem engeren System undefinierbar. Untersuchungen in dieser Richtung finden sich bei Wernick (Crelles Journal, *161*).

52. Kolloquium (28. VI. 1932).

Gödel: *Zur intuitionistischen Arithmetik und Zahlentheorie.*

Läßt man den Grundbegriffen des Heytingschen Aussagenkalküls [1]) die gleich bezeichneten des klassischen und der „Absurdität" ($\neg$) die Negation ($\sim$) entsprechen, so erscheint der intuitionistische Aussagenkalkül H als echtes Teilsystem des gewöhnlichen A. Bei anderer Zuordnung (Übersetzung) der Begriffe ist jedoch umgekehrt *der klassische ein Teilsystem des intuitionistischen.* Denn es gilt: Jede nur aus Konjunktionen ($\wedge$) und Negationen ($\neg$) aufgebaute in A gültige Formel ist auch in H beweisbar. Jede solche Formel muß nämlich die Gestalt haben: $\neg A_1 \wedge \neg A_2 \wedge \ldots \wedge \neg A_n$, und ist sie in A gültig, so auch jedes einzelne $\neg A_i$; dann ist nach Glivenko [2]) $\neg A_i$ auch in H beweisbar und daher auch die Konjunktion der $\neg A_i$. Daraus folgt: Übersetzt man die klassischen Begriffe:

$$\sim p,\ p \rightarrow q,\ p \vee q,\ p \cdot q$$

durch die folgenden intuitionistischen:

$$\neg p,\ \neg(p \wedge \neg q),\ \neg(\neg p \wedge \neg q),\ p \wedge q$$

so ist jede klassische Formel auch in H gültig.

Das Ziel der vorliegenden Untersuchung ist, zu zeigen, daß etwas Ähnliches auch *für die ganze Arithmetik und Zahlentheorie* in dem Umfang, wie sie z B. durch die Axiome von Herbrand [3])

---

[1]) Die formalen Regeln der intuitionistischen Logik. Sitz.-Ber. preuß. Akad. Wiss. (phys.-math. Kl.), 1930, II. (zitiert als H_1).

[2]) Sur quelques points de la Logique de M. Brouwer. Acad. roy. de Belgique, Bull. de la Cl. des Sciences, Série 5, Tome *15*, 1929.

[3]) Sur la non-contradiction de l'Arithmétique, Crelles Journ., *166*, 1931.

gegeben ist, gilt[4]). Auch hier kann man eine Interpretation der klassischen Begriffe durch die intuitionistischen geben, *so daß sämtliche aus den klassischen Axiomen beweisbaren Sätze auch für den Intuitionismus gelten.*

Als *Grundzeichen* des Herbrandschen Systems betrachten wir: 1. die Operationen des Aussagenkalküls $\sim$, $\rightarrow$, $\vee$..; 2. Zahlvariable: $x, y, z \ldots$; 3. die Allzeichen: $(x), (y), \ldots$; 4. $=$; 5. 0 und $+1$; 6. Die abzählbare Menge der auf Grund von Axiomgruppe C eingeführten Funktionszeichen f_i, deren jedem eine Zahl n_i (Anzahl der Leerstellen) zugeordnet ist. Um zu präzisieren, wie sich aus diesen Grundzeichen die Formeln aufbauen, definieren wir zunächst den Terminus „*Zahlausdruck*" durch folgende rekursive Vorschrift: 1. 0 und sämtliche Variable $x, y \ldots$ sind Zahlausdrücke; 2. ist Z ein Zahlausdruck, so auch $Z+1$; 3. sind $Z_1, Z_2, \ldots Z_{n_i}$ Zahlausdrücke, so auch $f_i\,(Z_1\, Z_2 \ldots Z_{n_i})$. Unter einer *Elementarformel* verstehen wir einen Ausdruck der Gestalt $Z_1 = Z_2$, wobei Z_1, Z_2 Zahlausdrücke sind. „Sinnvolle Formeln der Zahlentheorie" (im folgenden *Z-Formeln* genannt) sind entweder Elementarformeln oder Ausdrücke, die sich aus Elementarformeln mittels der Operationen des Aussagenkalküls und der Allzeichen $(x), (y) \ldots$ aufbauen.

Zu den Herbrandschen Axiomgruppen A—D fügen wir folgende dort nicht explizit angeführte *logische Axiomgruppen* E—G hinzu: E. Jeder Ausdruck, der aus einer richtigen Formel des Aussagenkalküls durch Einsetzung von Z-Formeln für die Variablen entsteht, gilt als Axiom. F. Alle Formeln der Gestalt: $(x)\, F(x) \rightarrow F(Z)$, wobei $F(x)$ eine beliebige Z-Formel und Z ein Zahlausdruck ist, sind Axiome (mit der selbstverständlichen Einschränkung, daß die Variablen aus Z in F nicht gebunden sein dürfen). G. Alle Formeln der Gestalt: $x = y\,.\rightarrow.\, F(x) \rightarrow F(y)$ (wobei $F(x)$ eine beliebige Z-Formel ist), sind Axiome. Als *Schlußregeln* nehmen wir: I. Aus A und $A \rightarrow B$ folgt B. II. Aus $A \rightarrow B$ folgt $A \rightarrow (x) B$, wenn x in A nicht frei vorkommt. In F, G und Regel II bedeuten x, y beliebige Zahlvariable.

Im Gegensatz zum Herbrandschen System gibt es im Heytingschen[5]) keine Zahlvariable, sondern nur Variable $x, y \ldots$ für beliebige Gegenstände. Dadurch ergeben sich für das folgende einige Komplikationen, die man dadurch vermeiden kann, daß man zunächst im Heytingschen System Variable für natürliche Zahlen $x', y', z', \ldots$ einführt in der Weise, daß jeder die neue Variablenart enthaltende Satz mit einem ohne diese äquivalent ist, gemäß der Vorschrift: $(x')\, F(x')$ soll gleichbedeutend sein mit $(x)\,.\,x \varepsilon N \supset F(x)$ und eine Formel $A\,(x'\, y' \ldots)$, welche $x', y', \ldots$ als freie Variable enthält, soll gleichbedeutend sein mit: $x, y, \ldots \varepsilon N \supset A\,(x\, y \ldots)$[5a]).

[4]) Das von Glivenko l. c.[2]) für den Aussagenkalkül abgeleitete Resultat läßt sich auf die Zahlentheorie nicht ausdehnen.

[5]) Die formalen Regeln der intuitionistischen Mathematik. Sitz.-Ber. preuß. Akad. Wiss. (phys.-math. Kl.), 1930, II. (zitiert als H_2).

[5a]) Es ist klar, wie (Ex') zu definieren wäre, doch wird dies im Folgenden nicht verwendet.

3*

Es kann aus der eben gegebenen Übersetzungsvorschrift rein formal bewiesen werden, daß auch für die neue Variablenart fast alle Sätze aus H_2 § 5, § 6 bestehen bleiben. Insbesondere gilt dies von den in der vorliegenden Arbeit verwendeten: H_2 5·4, 5·5, 5·8, 6·26, 6·3, 6·4, 6·78, wobei nur in 5·4 $p = p$ durch $p' \varepsilon N$ zu ersetzen ist. In den Sätzen H_2 § 10 fallen die Voraussetzungen: $p \varepsilon N$, $q \varepsilon N$ etc. weg, wenn man sie unter Verwendung der Zahlenvariablen ausspricht.

Die Definition zahlentheoretischer Funktionen durch Rekursion ist auch intuitionistisch einwandfrei (vgl. H_2 10·03, 10·04). Die Funktionen f_i (Axiomgruppe C) kommen daher sämtlich auch in der intuitionistischen Mathematik vor und die sie definierenden Formeln denken wir zu den Heytingschen Axiomen hinzugefügt, ebenso die für den Intuitionismus selbstverständlich gültigen Formeln der Gruppe D. Das so erweiterte System heiße H'. Nun ordnen wir jeder Z-Formel A eine Formel A' aus H' (ihre „Übersetzung") zu durch die Festsetzungen: Die Variablen $x, y, \ldots$ sind zu übersetzen durch $x', y', \ldots$; jedes f_i durch das gleich bezeichnete f_i aus H'; $=$ durch $=$; 0 durch 1[6]); $+1$ durch seq'; die Operationen des Aussagenkalküls in der oben angegebenen Weise. $(x) A$ ist zu übersetzen durch $(x') A'$, wenn A' die Übersetzung von A ist. Eine Formel, die Übersetzung einer Z-Formel ist, soll Z'-Formel heißen.

Der zu beweisende Satz I lautet: *Wenn die Formel A im Herbrandschen System beweisbar ist, so ihre Übersetzung A' in H'.*

Hilfssatz 1. *Für jede Z'-Formel A' ist*

$$\neg\neg A' \supset A' \qquad (1)$$

in H' beweisbar. Dies zeigen wir durch vollständige Induktion.

α) (1) gilt, wenn A eine Elementarformel ist, denn für Zahlausdrücke Z gilt $Z' \varepsilon N$, was man analog wie H_2 10·4 beweist. Daher gilt wegen H_2 10·25 für Elementarformeln sogar $A' \vee \neg A'$, um so mehr (1) (wegen H_1 4·45).

β) Wenn (1) für zwei Z'-Formeln A' und B' gilt, so auch für $A' \wedge B'$. Denn nach H_1 4·61 gilt $\neg\neg(A' \wedge B') . \supset . \neg\neg A' \wedge \neg\neg B'$ und daher wegen der induktiven Annahme und H_1 2·23:

$$\neg\neg(A' \wedge B') . \supset . A' \wedge B'.$$

γ) Wenn (1) für A' gilt, so auch für $\neg A'$. Denn es gilt ganz allgemein $\neg\neg\neg A' \supset \neg A'$ (nach H_1 4·32).

δ) Wenn (1) für A' gilt, so auch für $(x') A'$. Beweis: Nach Annahme gilt $\neg\neg A' \supset A'$, daher nach H_2 5·8 $(x') . \neg\neg A' \supset A'$ und wegen H_2 6·4 $(x') \neg\neg A' . \supset . (x') A'$. Nach H_2 6·78 hat man weiter $\neg\neg (x') A' . \supset . (x') \neg\neg A'$ und die beiden letzten Formeln ergeben $\neg\neg (x') A' . \supset . (x') A'$, w. z. b. w.

Aus α)—δ) folgt Hilfssatz 1, da jede Z'-Formel sich aus Elementarformeln mittels der Operationen $\wedge$, $\neg$, (x') aufbaut.

[6]) Im Heytingschen System beginnt die Zahlenreihe mit 1.

Hilfssatz 2. *Für beliebige Z'-Formeln A', B' gilt*

$$A' \supset B' . \supset\subset . \neg(A' \wedge \neg B')$$

im System H'.

Beweis: $$A' \supset B' . \supset . \neg(A' \wedge \neg B') \tag{2}$$

gilt nach H_1 4·9, ferner $\neg(A' \wedge \neg B') . \supset . A' \supset \neg\neg B'$ nach H_1 4·52 und weil B' eine Z'-Formel ist, so folgt nach Hilfssatz 1 $\neg(A' \wedge \neg B') . \supset . A' \supset B'$, was zusammen mit (2) die Behauptung ergibt.

Wir zeigen jetzt: Für jedes Axiom A des Herbrandschen Systems ist die Übersetzung A' in H' beweisbar.

1. Die Axiome der Gruppe A gehen durch die Übersetzung in Sätze über, die nach Hilfssatz 2 mit H_2 10·2, 10·22, 10·221, 10·24, 10·25 äquivalent sind.

2. Die Übersetzung eines Axioms der Gruppe B lautet:

$$\neg[\Phi(1) \wedge (x')\neg(\Phi(x') \wedge \neg\Phi(\text{seq}`x')) \wedge \neg(x')\Phi(x')] \tag{3}$$

und dies folgt nach Hilfssatz 2 aus:

$$\Phi(1) \wedge (x') . \Phi(x') \supset \Phi(\text{seq}`x') : \supset (x')\Phi(x'), \tag{4}$$

indem man $p \supset q$ überall durch $\neg(p \wedge \neg q)$ ersetzt. (4) ist aber nichts anderes als H_2 10·14 in unserer Schreibweise.

3. Die Axiome der Gruppen C und D wurden ausdrücklich zum Heytingschen System adjungiert. Gruppe E folgt aus dem oben über den Heytingschen Aussagenkalkül Bewiesenen und F ergibt sich bei Beachtung von Hilfssatz 2 unmittelbar aus H_2 6·3 und 5·4; ebenso G aus H_2 6·26 und 10·01.

Es bleibt noch zu zeigen, daß die Anwendung der Schlußregeln I, II aus Formeln, deren Übersetzung in H' beweisbar ist, wieder solche Formel erzeugt. Für Regel I bedeutet das: Ist A' und $(A \rightarrow B)'$ in H' beweisbar, so auch B', d. h. aus A' und $\neg(A' \wedge \neg B')$ folgt B'. Dies ergibt sich aber unmittelbar aus Hilfssatz 2 und H_1 1·3. Die entsprechende Tatsache für Regel II folgt ebenso aus H_2 5·5 und Hilfssatz 2.

Der damit bewiesene Satz I zeigt, *daß die intuitionistische Arithmetik und Zahlentheorie nur scheinbar enger ist als die klassische, in Wahrheit aber die ganze klassische, bloß mit einer etwas abweichenden Interpretation, enthält.* Der Grund dafür liegt darin, daß das intuitionistische Verbot, Allsätze zu negieren und reine Existenzialsätze auszusprechen, in seiner Wirkung dadurch wieder aufgehoben wird, daß das Prädikat der Absurdität auf Allsätze angewendet werden kann, was zu formal genau den gleichen Sätzen führt, wie sie in der klassischen Mathematik behauptet werden. Wirkliche Einschränkungen scheint der Intuitionismus erst für die Analysis und Mengenlehre zu bringen, doch sind diese nicht durch Ablehnung des Tertium non datur, sondern der imprädikativen Begriffsbildungen bedingt. Durch die obigen Betrachtungen ist selbstverständlich ein intuitionistischer

Widerspruchsfreiheitsbeweis für die klassische Arithmetik und Zahlentheorie gegeben. Er ist allerdings nicht „finit" in dem Sinn wie ihn Herbrand[7]) im Anschluß an Hilbert angegeben hat.

Diskussion über die Axiomatik der Elementargeometrie.

Menger gibt eine Zusammenstellung der Ergebnisse dieses Kolloquiums über eine Axiomatik der Verknüpfungsbeziehungen (*1*, S. 28 ff.; dieses Heft, S. 6, 14, 34) durch welche die Theorie einerseits den Klassenkalkül und die Algebra der Logik und Wahrscheinlichkeitsrechnung mitumfaßt (vgl. Menger, Jahresber. d. D. M. V., *37*, S. 309 und Radaković, Monatshefte f. Math. u. Phys., *36*, S. 285), anderseits in Analogie zur abstrakten Körpertheorie gesetzt wird.

Wald gibt eine Axiomatik der Geometrie des R_n an, die allein auf den Kongruenzbegriff sich stützt.

Veblen (Princeton) hält den Kongruenzbegriff eher für die sphärische Geometrie als Grundbegriff geeignet, für die Geometrie des euklidischen Raumes dagegen zwei Grundbegriffe für zweckmäßig, den Zwischenbegriff und den Kongruenzbegriff, entsprechend der Tatsache, daß die Gruppe der Ähnlichkeitstransformationen des euklidischen Raumes Durchschnitt zweier Gruppen, nämlich der projektiven Abbildungen und der Kreisverwandtschaften ist. Es wäre ein Axiomensystem wünschenswert, in dem die Anteile dieser beiden Begriffe möglichst zweckmäßig verteilt wären und das einen Ausgleich zwischen den älteren Axiomatiken von Veblen und R. L. Moore darstellen würde, wobei die Arbeit von Dorroh (Ann. of Math., *29*, S. 229) Verwendung fände.

---

[7]) Les bases de la logique Hilbertienne. Rev. de Mét. et de Mor., *37*, 1930, p. 248 und Thèses présentées à la Faculté des Sciences de Paris (1930), p. 3 ff.

Gesammelte Mitteilungen des Jahres 1931/32.

Eine Interpretation des intuitionistischen Aussagenkalküls.

Von Kurt Gödel.

Man kann den Heytingschen Aussagenkalkül mittels der Begriffe des gewöhnlichen Aussagenkalküls und des Begriffes „p ist beweisbar" (bezeichnet mit Bp) interpretieren[1]), wenn man für den letzteren das folgende Axiomensystem $\mathfrak{S}$ annimmt:

1. $Bp \rightarrow p$
2. $Bp \rightarrow . B(p \rightarrow q) \rightarrow Bq$.
3. $Bp \rightarrow BBp$

Außerdem sind für die Begriffe $\rightarrow$, $\sim$, ., $\vee$ die Axiome und Schlußregeln des gewöhnlichen Aussagenkalküls anzunehmen, ferner die neue Schlußregel: Aus A darf auf BA geschlossen werden.

Die Heytingschen Grundbegriffe sind folgendermaßen zu übersetzen:

$\neg p$	$\sim Bp$
$p \supset q$	$Bp \rightarrow Bq$
$p \vee q$	$Bp \vee Bp$
$p \wedge q$	$p . q$.

Mit dem selben Erfolg könnte man auch $\neg p$ durch $B \sim Bp$ und $p \wedge q$ durch $Bp . Bq$ übersetzen. Die Übersetzung einer beliebigen im Heytingschen System gültigen Formel folgt aus $\mathfrak{S}$, dagegen folgt aus $\mathfrak{S}$ nicht die Übersetzung von $p \vee \neg p$ und allgemein keine Formel der Gestalt $BP \vee BQ$, für die nicht schon entweder BP oder BQ aus $\mathfrak{S}$ beweisbar ist. Vermutlich gilt eine Formel im Heytingschen Kalkül dann und nur dann, wenn ihre Übersetzung aus $\mathfrak{S}$ beweisbar ist.

Das System $\mathfrak{S}$ ist mit dem Lewisschen System of Strict Implication äquivalent, wenn Bp durch Np übersetzt wird (vgl. S. 15 dieses Heftes), und wenn man das Lewissche System durch das folgende Beckersche Zusatzaxiom $Np < NNp$ ergänzt[2]).

Es ist zu bemerken, daß für den Begriff „beweisbar in einem bestimmten formalen System S" die aus $\mathfrak{S}$ beweisbaren Formeln nicht alle gelten. Es gilt z. B. für ihn $B(Bp \rightarrow p)$ niemals, d. h. für kein System S, das die Arithmetik enthält. Denn andernfalls wäre

---

[1]) Eine etwas andere Interpretation der intuitionistischen Aussagenkalküls gab Kolmogoroff (Math. Zeitschr., *35*, S. 58) ohne allerdings einen präzisen Formalismus anzugeben.

[2]) Zur Logik der Modalitäten, Jahrb. f. Phil. u. phänomenol. Forsch., *11*, 1930, S. 497.

beispielsweise $B(0 \neq 0) \rightarrow 0 \neq 0$ und daher auch $\sim B(0 \neq 0)$ in S beweisbar d. h. die Widerspruchsfreiheit von S wäre in S beweisbar.

Zum intuitionistischen Aussagenkalkül.[1]) Von Kurt Gödel.

In Beantwortung einer von Hahn aufgeworfenen Frage gelten für das von Heyting[2]) aufgestellte System H des intuitionistischen Aussagenkalküls folgende Sätze:

I. *Es gibt keine Realisierung mit endlich vielen Elementen (Wahrheitswerten), für welche die und nur die in H beweisbaren Formeln erfüllt sind (d. h. bei beliebiger Einsetzung ausgezeichnete Werte ergeben).*

II. *Zwischen H und dem System A des gewöhnlichen Aussagenkalküls liegen unendlich viele Systeme, d. h. es gibt eine monoton abnehmende Folge von Systemen, welche sämtlich H umfassen und in A enthalten sind.*

Der Beweis ergibt sich aus folgenden Tatsachen: Sei F_n die Formel:

$$\sum_{1 \leq i < k \leq n} (a_i \supset \subset a_k),$$

wobei Σ die iterierte $\vee$-Verknüpfung bedeutet und die a_i Aussagevariable sind. F_n ist erfüllt für jede Realisierung mit weniger als n Elementen, für welche alle in H beweisbaren Formeln erfüllt sind. Denn bei jeder Einsetzung wird in mindestens einem Summanden von F_n a_i und a_k durch dasselbe Element e ersetzt und $e \supset \subset e \,.\vee\, b$ ergibt bei beliebigem b einen ausgezeichneten Wert, weil die Formel $a \supset \subset a \,.\vee\, b$ in H beweisbar ist. Sei ferner S_n die folgende Realisierung:

Elemente: $\{1, 2 \ldots n\}$, ausgezeichnetes Element: 1;

$$a \vee b = \min(a, b);\ a \wedge b = \max(a, b);\ a \supset b = 1 \text{ für } a \geqq b;$$
$$a \supset b = b \text{ für } a < b;\ \neg\, a = n \text{ für } a \neq n,\ \neg\, n = 1.$$

Dann sind für S_n sämtliche Formeln aus H und die Formel F_{n+1} sowie alle F_i mit größerem Index erfüllt, dagegen F_n sowie alle F_i mit kleinerem Index nicht erfüllt. Insbesondere ergibt sich daraus, daß kein F_n in H beweisbar ist. Es gilt übrigens ganz allgemein, daß eine Formel der Gestalt $A \vee B$ in H nur dann beweisbar sein kann, wenn entweder A oder B in H beweisbar ist.

Ein Überdeckungssatz für F_σ. Von Karl Menger.

Ist F ein F_σ eines kompakten metrischen Raumes, so enthält jedes unbegrenzt feine Überdeckungssystem von F eine F überdeckende Nullfolge $\{U_k\}$ von offenen Mengen derart, daß für jedes $\varepsilon > 0$ fast alle Mengen der Folge $\subset U(\overline{F} - F;\ \varepsilon)$ sind.

Wegen $U(\overline{F} - F;\ \varepsilon) \subset U(F;\ \varepsilon)$ ist in diesem Satz der in Dimensions- und Kurventheorie oft verwendete Überdeckungssatz für halb-

---

[1]) Abdruck aus dem Wiener akad. Anzeiger, 25. II. 1932.

[2]) Vgl. Sitz.-Ber. der Preuß. Akad. d. Wiss., phys.-math. Kl., 1930, II.

kompakte Räume (vgl. Kurventheorie, bei Teubner 1932. S. 134, sowie diese Ergebnisse, 2, S. 24) enthalten. Ist ferner F ein F_ϱ, d. h. gilt $F = A - B$, wo A und B abgeschlossene Mengen des kompakten Raumes sind, so ist $\overline{F} - F \subset B$ und aus dem eingangs ausgesprochenen Satze ergibt sich, daß jedes unbegrenzt feine Überdeckungssystem von F eine F überdeckende Nullfolge $\{U_k\}$ enthält, von der für jedes $\varepsilon > 0$ fast alle Mengen $\subset U(B; \varepsilon)$ sind, d. i. der Überdeckungssatz für F_ϱ (Kurventheorie S. 298).

Zum Beweise der allgemeinen Behauptung setzen wir, wenn F Summe der abgeschlossenen Mengen $\{F_n\}$ ($n = 1, 2, \ldots$ ad inf.) ist, indem wir mit $U(M; \varepsilon)$ bzw. $\overline{U}(M; \varepsilon)$ die Menge aller Punkte des Raumes bezeichnen, die von M einen Abstand $< \varepsilon$ bzw. $\leqq \varepsilon$ haben:

$$F = (F - F . \overline{\overline{F} - F}) + F . \overline{\overline{F} - F} =$$

$$\left[F - F . U(\overline{F} - F; \tfrac{1}{2})\right] + \sum_{n=1}^{\infty} F . \left[\overline{U}(\overline{F} - F; \frac{1}{n+1}) - U(\overline{F} - F; \frac{1}{n+2})\right] +$$

$$+ \sum_{n=1}^{\infty} F_n . \overline{\overline{F} - F},$$

wobei sämtliche auftretenden Summanden offenbar abgeschlossene, also kompakte Mengen sind. Ein unbegrenzt feines Überdeckungssystem $\mathfrak{U}$ von F enthält daher endlich viele offene Mengen $U_1^1, U_2^1, \ldots U_{m_1}^1$, welche $F - F . U(\overline{F} - F; \frac{1}{2})$ überdecken, ferner für jedes natürliche n endlichviele offene Mengen $U_1^n, U_2^n, \ldots, U_{k_n}^n$, welche $F . \left[\overline{U}(\overline{F} - F; \frac{1}{n+1}) - U(\overline{F} - F; \frac{1}{n+2})\right]$ überdecken, Durchmesser $< \frac{1}{n}$ besitzen und $\subset U(\overline{F} - F; \frac{1}{n})$ sind, sowie endlichviele offene Mengen $U_{k_n+1}^n, U_{k_n+2}^n, \ldots, U_{m_n}^n$, welche die Menge $F_n . \overline{\overline{F} - F}$ überdecken, Durchmesser $< \frac{1}{n}$ besitzen und $\subset U(\overline{F} - F; \frac{1}{n})$ sind. Die im Überdeckungssystem $\mathfrak{U}$ enthaltene Nullfolge

$$U_1^1, \ldots, U_{m_1}^1, U_1^2, \ldots, U_{m_2}^2, \ldots, U_1^n, \ldots, U_{m_n}^n, \ldots$$

genügt offenbar der Behauptung des Satzes.

Eine neue Kennzeichnung der Geraden. Von Karl Menger[1]).

Wir sagen, der halbmetrische Raum[2]) R habe die *Quasikongruenzordnung* k, wenn jeder halbmetrische Raum, der mehr als $k + 1$ Punkte enthält und von dem je k Punkte mit k Punkten von R

[1]) Vgl. Wiener akademischen Anzeiger, 21. V. 1931.

[2]) Dieser Begriff — eine Menge, in der je zwei Elementen p und q eine reelle Zahl $pq = qp$ zugeordnet ist, die > 0 für $p \neq q$ und $= 0$ für $p = q$ ist, — stammt von Fréchet, der ihn unter dem Namen *classe E* eingeführt hat.

kongruent sind, mit einer Teilmenge von R kongruent ist[2a]). In dieser Ausdrucksweise besagt ein Fundamentalsatz der metrischen Geometrie, daß der n-dimensionale euklidische Raum die Quasikongruenzordnung $n+2$ besitzt. Insbesondere hat also der R_1 die Quasikongruenzordnung 3. Diese Eigenschaft ist, wie wir nun im folgenden beweisen, unter den konvexen und nach außen konvexen[3]) metrischen Räumen für den R_1 charakteristisch.

Hilfssatz 1. Ein Raum R, der drei Punkte enthält, die paarweise denselben Abstand δ haben, besitzt nicht die Quasikongruenzordnung 3. Denn eine Menge, deren Mächtigkeit die von R übertrifft und in der je zwei Punkte den Abstand δ haben, ist nicht mit einer Teilmenge von R kongruent, obwohl je drei ihrer Punkte mit drei Punkten von R kongruent sind.

Hilfssatz 2. Ein konvexer, vollständiger metrischer Raum, der vier Punkte p_1, p_2, p_3, p_4 enthält, für welche $p_1 p_2 = p_2 p_3 = p_3 p_4 = p_4 p_1 = \frac{1}{2} p_1 p_3 = \frac{1}{2} p_2 p_4$ gilt, enthält ein gleichseitiges Dreieck. Betrachten wir nämlich[4]) auf einer p_2 und p_3 verbindenden Teilstrecke des Raumes den Punkt q im Abstand $\frac{1}{3} p_2 p_3$ von p_2 und auf einer p_3 und p_4 verbindenden Teilstrecke des Raumes den Punkt r im Abstand $\frac{2}{3} p_3 p_4$ von p_3, so haben je zwei von den drei Punkten p_1, q, r den Abstand $\frac{4}{3} p_1 p_2$.

Es sei nun R ein konvexer und nach außen konvexer vollständiger Raum der Quasikongruenzordnung 3. Sicher enthält R eine mit einer Geraden kongruente Teilmenge G. Wir behaupten, daß R mit G identisch ist; wir machen zu diesem Zweck die Annahme, R enthalte einen nicht in G liegenden Punkt s, und leiten aus dieser Annahme einen Widerspruch her. Es sei q irgend ein Punkt von G. Seinen Abstand von s bezeichnen wir mit d. Es existieren auf G genau zwei Punkte p und r, die beide von q den Abstand d haben. Ihre Abstände von s seien e, beziehungsweise f. Wir definieren nun einen aus fünf Punkten o', p', q', r', s' bestehenden Raum R' durch folgende Abstandsfestsetzungen:

[2a]) Offenbar besitzt jeder Raum R der Quasikongruenzordnung k die *Kongruenzordnung* $k+1$, d. h. jeder Raum, von dem je $k+1$ Punkte mit $k+1$ Punkten von R kongruent sind, ist mit einer Teilmenge von R kongruent.

[3]) Ein Raum R heißt *konvex*, bzw. *konvex nach außen*, wenn zu je zwei Punkten p und r von R ein Punkt q zwischen p und r bzw. ein Punkt s, so daß r zwischen p und s liegt, existiert.

[4]) Dieser Beweis wurde mir von Herrn Wald mit der Bemerkung mitgeteilt, daß aus den Voraussetzungen von Hilfssatz 2 nicht folgt, daß R eine mit einer konvex-metrisierten Kreislinie kongruente Teilmenge enthält. Bezeichnet nämlich R die Menge aller Paare reeller Zahlen und ist den Paaren x, y und x', y' der Abstand $|x-x'| + |y-y'|$ zugeordnet, so genügen die Punkte $p_1 = 0{,}0$, $p_2 = 0{,}1$, $p_3 = 1{,}1$, $p_4 = 1{,}0$ den Voraussetzungen von Hilfssatz 2, ohne daß doch R irgend eine mit einer konvex-metrisierten Kreislinie kongruente Teilmenge enthielte.

$$o'p' = o'r' = p'q' = q'r' = o's' = q's' = d, \quad o'q' = p'r' = 2d,$$
$$p's' = e, \quad r's' = f.$$

Je drei Punkte von R' sind dann offenbar mit drei Punkten von R kongruent (o', q', s' mit r, q, p). So wie R ist also auch R' ein metrischer Raum. Die Punkte o', p', q', r' genügen offenbar den Voraussetzuugen von Hilfssatz 2, demzufolge R' ein gleichseitiges Dreieck enthält. Da R die Quasikongruenzordnung 3 besitzt, ist also R' nach Hilfssatz 1 nicht mit einer Teilmenge von R kongruent. Damit ist, da R' ein mehr als vier Punkte enthaltender metrischer Raum ist, von dem je drei Punkte mit drei Punkten von R kongruent sind, der gesuchte Widerspruch, nämlich gegen die Voraussetzung, R habe die Quasikongruenzordnung 3, hergestellt. Den bewiesenen Satz können wir auch so aussprechen:

Damit ein konvexer und nach außen konvexer vollständiger Raum mit der Geraden kongruent sei, ist notwendig und hinreichend, daß er die Quasikongruenzordnung 3 besitzt.

Dieses Resultat legt mannigfache Fragen nahe. Vor allem, ob nicht auch dieser Satz einen allgemein-gruppentheoretischen Kern hat, so wie der Satz, daß die Gerade die Quasikongruenzordnung 3 besitzt[1]). Ferner wäre eine Aufzählung aller Räume, die die Quasikongruenzordnung 3 besitzen, wünschenswert. Es ist klar, daß diese Eigenschaft mit der Existenz irgend welcher einparametriger Bewegungsgruppen zusammenhängt[2]). Schließlich fragt es sich, ob auch für $n > 1$ der R_n unter den konvexen und nach außen konvexen vollständigen Räumen durch die Eigenschaft, die Quasikongruenzordnung $n+2$ zu besitzen, gekennzeichnet ist, bzw. welche weiteren Eigenschaften zu seiner Kennzeichnung erforderlich sind.

Metrische Charakterisierung der Kugel[3]). Von Laura Klanfer.

Bei der Untersuchung der n-dimensionalen Sphäre K_n^r vom Radius r, in der je zwei Punkten die Länge des kürzesten sie verbindenden Kreisbogens als Abstand zugeordnet ist, tritt an Stelle der von Menger zur Kennzeichnung der euklidischen Räume betrachteten Volumdeterminante die folgende Determinante: Wir setzen für je k Punkte $p_1, \ldots, p_k$ von K_n^r, falls p_{ij} den Abstand der Punkte p_i und p_j bezeichnet,

$$\Delta\,(p_1, \ldots, p_k;\, r) = \left|\cos\frac{p_{ij}}{r}\right| \quad (i, j = 1, \ldots, k).$$

---

[1]) Vgl. bezüglich dieser Tatsache diese Ergebnisse, *2*, S. 35 und Mathem. Zeitschrift, *33*, p. 415 ff.

[2]) Vgl. in diesem Zusammenhang auch meine Note Proc. Acad. Amst., *30*, S. 710. Ein aus vier Punkten bestehender Raum hat dann und nur dann die Quasikongruenzordnung 3, wenn je zwei gegenüberliegende Punktepaare gleichen, aber nicht alle sechs Paare gleichen Abstand haben. Punktequintupel, welche die Quasikongruenzordnung 3 haben, gibt es nicht.

[3]) Auszug aus der am 17. III. 1932 eingereichten Dissertation.

Wegen $p_{ij} = p_{ji}$ und $p_{ii} = 0$ ist diese Determinante symmetrisch in den k Punkten $p_1, \ldots, p_k$ und alle Glieder ihrer Hauptdiagonale sind 1. Sie ist ferner kongruenzinvariant.

Theorem. *Damit $n+2$ Punkte $p_1, \ldots, p_{n+2}$ eines halbmetrischen Raumes, von denen je $n+1$ mit $n+1$ Punkten der K_n^r kongruent sind, mit $n+2$ für den Radius r unabhängigen*[1]) *Punkten der K_{n+1}^r bzw. mit $n+2$ Punkten der K_n^r kongruent seien, ist notwendig und hinreichend, daß*

$\Delta\ (p_1, p_2, \ldots, p_{n+2};\ r) > 0$ bzw. $= 0$ und daß kein Abstand größer ist als $r\pi$.

Die Notwendigkeit ergibt sich daraus, daß für $n+2$ Punkte einer im $(n+2)$-dimensionalen euklidischen Raum gelegenen Oberfläche einer $(n+1)$-dimensionalen Sphäre mit dem Radius r und dem Ursprung als Zentrum die Determinante Δ bis auf den konstanten Faktor $\frac{r^{2k}}{k!^2}$ gleich dem Quadrat des euklidischen Volumens des $n+2$ dimensionalen Simplexes ist, welches durch die Punkte $p_1, \ldots, p_{n+2}$ und den Ursprung bestimmt wird. Als hinreichend läßt sich die Bedingung in leichter Abänderung[2]) des Mengerschen Beweisganges für den euklidischen Raum (Amer. Journ. of Math., *53*, S. 721) folgendermaßen erweisen: Wir können induktiv das für $n=0$ triviale Theorem für $n-1$ (d. h. für $n+1$ Punkte) als gültig annehmen. Wir nennen sodann ein $(n+3)$-Tupel von Punkten *pseudo-r-sphärisch,* wenn es nicht mit $n+3$ Punkten einer K_n^r kongruent ist, obwohl je $n+2$ Punkte mit $n+2$ Punkten der K_n^r kongruent sind, — bezeichnen jene quadratische Funktion von x, welche entsteht, indem man in der Determinante $\Delta\ (p_1, \ldots, p_k;\ r)$ die Terme $\cos \frac{p_{k-1\,k}}{r}$ und $\cos \frac{p_{k\,k-1}}{r}$ durch x ersetzt, mit $\Delta\ (p_1, \ldots, p_k;\ r,\ x)$ und beweisen der Reihe nach:

1. *Je $n+1$ Punkte eines pseudo-r-sphärischen $(n+3)$-Tupels sind für den Radius r unabhängig.*

2. *Sind $n+2$ Punkte $p_1, \ldots, p_{n+2}$ für den Radius r abhängig, während die n Punkte $p_1, \ldots, p_n$ für den Radius r unabhängig sind, so hat die Gleichung $\Delta\ (p_1, \ldots, p_{n+2};\ r,\ x) = 0$ die Doppelwurzel $x = \cos \frac{p_{n+1,\,n+2}}{r}$.*

3. *Für die Punkte $p_1, \ldots, p_{n+2}$ eines pseudo-r-sphärischen $(n+2)$-Tupels gilt $\Delta\ (p_1, \ldots, p_{n+2};\ r) < 0$.*

---

[1]) Wir nennen $n+2$ Punkte *für den Radius r unabhängig,* wenn sie nicht in einer n-dimensionalen Großkugel der K_{n+1}^r liegen.

[2]) Der Beweis des folgenden Satzes 1. verläuft, da die K_n^r im Gegensatz zum R_n nicht die Quasikongruenzordnung $n+2$ hat, anders als Mengers Beweis des entsprechenden Satzes über den R_n, welch letzterer sich jedoch ebenfalls ohne Bezugnahme auf die Quasikongruenzordnung $n+2$ des R_n beweisen läßt.

Hieraus ergibt sich die Bedingung des Theorems als hinreichend, womit die Sätze 1., 2., 3. für jedes n bewiesen sind. Aus 3. folgt, daß *ein pseudo-r-sphärisches $(n+2)$-Tupel auch nicht mit $n+2$ Punkten einer mehr als $(n+1)$-dimensionalen Sphäre vom Radius r kongruent ist.* Hingegen kann es kongruent sein mit $n+2$ Punkten einer mehr als $(n+1)$-dimensionalen Kugel eines anderen Radius, in welchem Falle es ein *unechtes* pseudo-r-sphärisches $(n+2)$-Tupel heißen möge. Überhaupt ist eine nicht mit einer Teilmenge der K_n^r kongruente Menge, von der je $n+2$ Punkte mit $n+2$ Punkten der K_n^r kongruent sind, für kein natürliches m mit einer Teilmenge der K_m^r kongruent.

Aus unserem Theorem folgt: *Damit ein halbmetrischer Raum R mit einer Teilmenge der K_n^r kongruent sei, ist notwendig und hinreichend, daß für je $n+3$ Punkte $p_1, \ldots, p_{n+3}$ die Beziehungen gelten: $\Delta(p_1, \ldots, p_{n+3}; r) = 0$, für je $n+2$ Punkte $\Delta(p_1 \ldots, p_{n+2}; r) = 0$ und für jedes $k \leq n+1$ und je k Punkte $p_1, \ldots, p_k$ die Ungleichung $\Delta(p_1, \ldots, p_k; r) \geqq 0$, während $p_{12} \leq r\pi$ für je zwei Punkte p_1 und p_2.*

Ein mit einer Teilmenge von K_n^r, aber nicht mit einer Teilmenge von K_{n-1}^r kongruenter halbmetrischer Raum ist mit K_n^r *selbst* dann und nur dann kongruent, *wenn R vollständig und konvex ist und zu jedem Punkt einen Gegenpunkt* (d. h. einen Punkt im Abstand $r\pi$) *enthält.*

Angelegenheiten des Kolloquiums und Nachrichten.

Wiener Dissertationen im Zusammenhange mit dem Kolloquium 1931/32:

F. Alt: *„Über eine metrische Definition der Krümmung einer Kurve“* (vgl. diese Ergebnisse *3*, S. 5 und *4*, S. 4).

L. Klanfer: *„Metrische Charakterisierung der Kugel“* (vgl. S. 43 dieses Heftes).

Stipendien.

Es arbeiteten 1931/32 im Kolloquium Dr. G. Nöbeling mit einem Stipendium der österreichisch-deutschen Wissenschaftshilfe, Dr. Y. Mimura (Tokio), ferner im Wintersemester Dr. K. Zarankiewicz (Dozent der Universität Warschau) mit einem Stipendium des polnischen Kulturfonds, Dr. W. T. Parry (Cambridge, Mass.) mit einer Harvard travelling fellowship, im Sommersemester Dr. K. Borsuk (Warschau) mit einem Stipendium des polnischen Kulturfonds.

Druck von Gottlieb Gistel & Cie., Wien, III., Münzgasse 6.

Bericht über das Kolloquium 1932/33.

53. Kolloquium (10. XI. 1932).

Gödel: *Bemerkung über projektive Abbildungen.*

Jede eineindeutige Abbildung φ der reellen projektiven Ebene E in sich, welche Gerade in Gerade überführt, ist eine Kollineation. Bezeichnet man für jeden Kegelschnitt der Ebene als die zugehörige *Kegelschnittumgebung* die Menge aller Punkte p von E, für welche jede p enthaltende Gerade genau zwei Punkte mit dem Kegelschnitt gemein hat, so ist das System $\mathfrak{K}$ aller Kegelschnittumgebungen ein unbegrenzt feines Überdeckungssystem [1]) von E. Da jeder Kegelschnitt durch zwei Geradenbüschel erzeugt werden kann, welche projektiv (d. h. durch eine endliche Kette von Perspektivitäten) auf einander bezogen sind, und da perspektiv auf einander bezogene Geradenbüschel bei jeder eineindeutigen Abbildung, die Gerade in Gerade überführt, in perspektiv auf einander bezogene Geradenbüschel übergehen, so wird durch die Abbildung φ jeder Kegelschnitt auf einen Kegelschnitt abgebildet. Ferner wird wegen der Eineindeutigkeit von φ und der obigen Definition einer Kegelschnittumgebung jede Kegelschnittumgebung durch φ in eine Kegelschnittumgebung übergeführt. Es wird demnach $\mathfrak{K}$ durch φ in sich übergeführt. Da $\mathfrak{K}$ ein unbegrenzt feines Überdeckungssystem von E ist, so ist also φ eine *stetige* Abbildung und daher nach einem Fundamentalsatz der projektiven Geometrie eine Kollineation.

Flores (Madrid): Kurzer Beweis des Satzes, daß der n-dimensionale projektive Raum für gerades n nicht orientierbar, für ungerades n orientierbar ist (erscheint in Revista matem. Hispano-Americ., 1933).

Menger: *Zur Begründung einer Theorie der Bogenlänge in Gruppen.*

In einer additiv symbolisierten Abelschen Gruppe G bezeichnen wir für jedes Element a als Betrag $|a|$ von a das ungeordnete Elementepaar $(a, -a)$ und mit $|G|$ die Menge der Beträge aller Elemente von G. Sind A und B zwei Teilmengen von G, so heiße $A+B$ die Menge aller Elemente s von G, für welche zwei Elemente a und b von A bzw. B existieren, so daß $s=a+b$ gilt. Diese Addition von Teilmengen ist offenbar assoziativ und kommutativ und hat

[1]) So heißt nach Menger ein System von offenen Mengen, in welchem zu jedem Punkt p des Raumes und zu jeder Umgebung U von p eine p enthaltende offene Menge $\subset U$ existiert.

die aus dem Element 0 bestehende Menge als Einheitselement. Insbesondere bezeichnen wir, wenn $|a_1|, |a_2|, \ldots, |a_n|$ irgend welche Elemente von $|G|$ sind, mit $\sum_{i=1}^{n} |a_i|$ die Summe der (aus je 2 Elementen bestehenden) Teilmengen $|a_1|, |a_2|, \ldots, |a_n|$ von G. Beispielsweise ist also $|a_1| + |a_2|$ das ungeordnete Quadrupel a_1+a_2, a_1-a_2, $-a_1+a_2$, $-a_1-a_2$ und $\sum_{i=1}^{n} |a_i|$ besteht aus den 2^n (nicht notwendig paarweise verschiedenen) Elementen $\sum_{i=1}^{n} \varepsilon_i a_i$ für alle aus $+1$ und -1 bestehenden n-Tupel $\varepsilon_1, \varepsilon_2, \ldots, \varepsilon_n$. Diese Menge ist bezüglich 0 symmetrisch.

Als *Abstand* ab der Elemente a und b von G wurde [1]) das ungeordnete Paar $(a-b, b-a) = |a-b|$ bezeichnet. Ist nun $E = \{p_1, p_2, \ldots, p_n\}$ eine geordnete endliche Teilmenge von G [oder von einer G-metrischen Menge [1])], so wollen wir die Menge $\sum_{i=1}^{n-1} p_i p_{i+1}$ mit $L(E)$ bezeichnen. Durch vollständige Induktion ergibt sich, falls E Teilmenge von G ist, unmittelbar, daß $L(E)$ aus den 2^{n-1} (nicht notwendig paarweise verschiedenen) Elementen

$$\pm \left[p_1 + 2 \sum_{r=1}^{m} (-1)^r p_{k_r} + (-1)^{m+1} p_n\right] \text{ für } 2 \leq k_1 < k_2 < \ldots < k_m \leq n-1,$$
$$m = 0, 1, \ldots, n-2$$

besteht. $L(E)$ ist bezüglich 0 symmetrisch. Sind $E = \{p_1, \ldots, p_n\}$ und $E' = \{p_1', \ldots, p_n'\}$ kongruent, d. h. gilt $p_i p_{i+1} = p_i' p'_{i+1}$ $(i = 1, \ldots, n-1)$, so ist $L(E) = L(E')$. Ist $E \subset G$ und $E' \subset E$ und sind E' und E koextremal (womit, wenn E die geordnete Menge $p_1, p_2, \ldots, p_n$ ist, gemeint ist, daß E' eine Menge $p_{i_1}, p_{i_2}, \ldots, p_{i_m}$ mit $1 = i_1 < i_2 < \ldots < i_m = n$ ist), so gilt $L(E') \subset L(E)$. Ist G eine Abelsche Limesgruppe [2]) und $\{E_k\}$ $(k = 1, 2, \ldots$ ad inf.) eine gegen E konvergente Folge von geordneten endlichen Mengen (womit, wenn E_k die Menge $p_1^k, \ldots, p_n^k$ ist, gemeint ist, daß $\lim_{k=\infty} p_i^k = p_i$ für $i = 1, \ldots, n$), so gilt $\lim_{k=\infty} L(E_k) = L(E)$.

Ist B Teilbogen der Abelschen Limesgruppe G, so bilden wir für jede die Endpunkte von B enthaltende endliche Teilmenge E von B in der B gemäßen Anordnung (d. h. in jener Reihenfolge, in welcher sie bei der Durchlaufung von B angetroffen wird), die Menge $L(E)$ und nennen $L'(B)$ die Vereinigungsmenge aller [3]) dieser Mengen $L(E)$. Ist G die additive Gruppe aller reellen Zahlen, so ist jeder Teilbogen B von G eine abgeschlossene Strecke und $L'(B)$ ist, wenn l die Länge von B be-

[1]) Vgl. diese Ergebn., *2*, S. 35 und Math. Ztschr., *33*, S. 126.

[2]) Vgl. Schreier, Abh. Hamb. math. Sem. *4* und Leja, Fund. Math., *9*, S. 37.

[3]) Offen ist die Frage, ob die folgenden Begriffe inhaltlich unverändert bleiben, falls man auch *die nicht beide Endpunkte von B enthaltenden endlichen Teilmengen von B mitberücksichtigt.*

zeichnet, das abgeschlossene Intervall $[-l, l]$ der Geraden G, also eine abgeschlossene Menge. Für die Gruppe G der Vektoren der Ebene ist, wenn B ein Teilbogen von G ist, der durch die Gleichung $y=f(x)$, $a \leqq x \leqq b$ gegeben ist, wo $f(x)$ die endliche Variation v in $[a, b]$ besitzt, $L'(B)$ offenbar eine Teilmenge des Rechteckes $|x| \leqq b-a$, $|y| \leqq v$ und es befinden sich in jeder Nachbarschaft der Geraden $y=v$ Punkte von $L'(B)$. Im Falle, daß für keine endliche Teilmenge von $[a, b]$ die Variation v tatsächlich erreicht wird, ist $L'(B)$ zur Geraden $y=v$ fremd und die Menge $L'(B)$ nicht abgeschlossen[4]). Als *Länge* $L(B)$ *des Teilbogens* B einer Abelschen Limesgruppe bezeichnen wir die abgeschlossene Hülle der Menge $L'(B)$.

Auf diese Weise ist jedem Teilbogen einer Gruppe eine bezüglich 0 symmetrische, abgeschlossene[5]) Teilmenge der Gruppe als Länge zugeordnet[6]). Die so definierte Länge ist *kongruenzinvariant*, d. h. kongruente (= eineindeutig und abstandstreu auf einander abbildbare) Bogen haben dieselbe Länge; sie ist *halbstetig*, d. h. aus $\lim_{n=\infty} B_n = B$ folgt $\underline{\lim}_{n=\infty} L(B_n) \supset L(B)$.[6a]) Ist F eine in B dichte Teil-

[4]) Offenbar kann die Definition von $L'(B)$, ebenso wie die gleich folgende von $L(B)$ auch *für Teilbogen G-metrischer Mengen* (vgl. diese Ergebn., *2*, S. 35) ausgesprochen werden. Ein metrischer Raum ist eine der Dreiecksungleichung genügende G-metrische Menge hinsichtlich der additiven Gruppe G der reellen Zahlen. Für Teilbogen eines metrischen Raumes ist die Menge $L'(B)$ offenbar dann und nur dann abgeschlossen, wenn B als Summe von endlichvielen Teilbogen dargestellt werden kann, von denen jeder mit einer Strecke kongruent ist. — Übrigens folgt für koextremale endliche Teilmengen E und E' eines metrischen Raumes M aus $E' \subset E$ nicht notwendig $L(E') \subset L(E)$. Wenn M die euklidische Ebene ist und E die geordnete Menge $p_1=(0,0)$, $p_2=(1,1)$, $p_3=(2,0)$ bezeichnet, so besteht $L(E)$ aus den Zahlen 0 und $\pm 2\sqrt{2}$; für die aus p_1 und p_3 bestehende Teilmenge E' von E besteht $L(E')$ aus den Zahlen ± 2.

[5]) Hiezu muß über G nur vorausgesetzt werden, daß jeder Häufungspunkt von Häufungspunkten einer beliebigen Teilmenge H von G selbst Häufungspunkt von H ist.

[6]) Offen bleibt die Frage, ob bzw. unter welchen Voraussetzungen über G für jeden Teilbogen B von G die Menge $L(B)$ das Element 0 enthält. Für die Gruppe der ebenen Vektoren ist die Frage zu bejahen (s. u.)

[6a]) [Zusatz bei der Korrektur]: Ist G eine Gruppe mit der Eigenschaft (0), *daß* $L(B)$ *für jeden Bogen* $B \subset G$ *das Element* 0 *enthält,* dann folgt ferner aus $B' \subset B \subset G$ stets $L(B') \subset L(B)$; die Länge ist dann m. a. W. *monoton.* Ist nämlich p ein Punkt von $L(B')$ und U irgend eine Umgebung von p, so ist zu zeigen, daß eine die Endpunkte von B enthaltende endliche Teilmenge E von B existiert, für welche $L(E)$ mit U einen Punkt q gemein hat. Nun existiert, da p in $L(B')$ liegt, sicher eine die Endpunkte von B' enthaltende endliche Teilmenge E' von B', so daß ein Punkt p' von $L(E')$ in U liegt. Sind B_1' und B_2' jene beiden (event. leeren) Bogen, in welche B nach Tilgung des Innern von B' zerfällt, so existieren wegen der Eigenschaft (0) von G zwei die Endpunkte von B_1' bzw. B_2' enthaltende endliche Teilmengen E_1' bzw. E_2' von B_1' bzw. B_2', für welche $L(E_1')$ und $L(E_2')$ Elemente p_1' bzw. p_2' enthalten, die 0 beliebig nahe, insbesondere so nahe liegen, daß $q = p' + p_1' + p_2'$ in U enthalten ist. Setzen wir $E = E_1' + E' + E_2'$, so ist E eine die Endpunkte von B enthaltende endliche Menge $\subset B$, für die q in $L(E)$ und in U liegt, w. z. b. w. Ebenso bestätigt man, wenn G die Eigenschaft (0) hat, daß für jeden Bogen $B \subset G$ der *„Durchmesser"* $D(B)$ von B, d. h. die Menge aller Elemente von G, welche Differenz irgend

1*

menge, so ist $L(B)$ die abgeschlossene Hülle der Summe der Mengen $L(E)$ für alle endlichen Teilmengen E von F. Ist B Teilbogen der Gruppe G und $\{E_n\}$ $(n = 1, 2, \ldots$ ad inf.) eine *ausgezeichnete Folge* von die Endpunkte von B enthaltenden endlichen Teilmengen von B, d. h. gilt $E_n \subset E_{n+1}$ für jedes n und ist $\sum_{n=1}^{\infty} E_n$ in B dicht, so ist $L(B)$ die abgeschlossene Hülle von $\sum_{n=1}^{\infty} L(E_n)$.

Diese Sätze lassen es wohl als berechtigt erscheinen, daß wir, während bisher nur für Teilbogen metrischer Räume eine Länge, und zwar eine Längen*zahl* definiert war, den Teilbogen beliebiger Gruppen eine Länge, und zwar eine *Menge*, nämlich die Längenmenge $L(B)$, zuordnen. Ist G die additive Gruppe der reellen Zahlen, so existiert, wenn B ein Teilbogen von G (oder auch nur von einer der Dreiecksungleichung genügenden G-metrischen Menge, d. h. von einem metrischen Raum) ist, eine reelle Zahl l (im Falle des metrischen Raumes eventuell ∞) derart, daß $L(B)$ das Intervall $[-l, l]$ ist, welches durch seine größte Zahl l eindeutig kennzeichenbar ist und diese Zahl l ist denn auch die Längenzahl $l(B)$ von B im gewöhnlichen Sinn [7]). Überhaupt kann die Länge der

---

zweier Elemente von B sind, Teilmenge von $L(B)$ ist. Ob es Gruppen ohne die Eigenschaft (0) gibt, ist unentschieden, vgl. [6]). Für etwaige solche Gruppen ist die Monotonie der Länge bisher unbewiesen. Jedenfalls monoton ist aber, wie O. Taussky bemerkt hat, die abgeschlossene Hülle $L^*(B)$ von der Vereinigungsmenge der Mengen $L(E)$ für *alle* endlichen Teilmengen E von B (auch die nicht die Endpunkte von B enthaltenden), oder, was gleichbedeutend ist, die abgeschlossene Hülle der Vereinigungsmenge der Mengen $L^*(E)$ für alle die Endpunkte von B enthaltenden endlichen Teilmengen E von B, wobei $L^*(E)$ die Vereinigung der Mengen $L(F)$ für alle Teilmengen F von E ist. Für eine ausgezeichnete Folge $\{E_n\}$ von Teilmengen von B ist $L^*(B)$ die abgeschlossene Hülle von $\sum_{n=1}^{\infty} L^*(E_n)$.

[7]) Ist G eine Gruppe und ein metrischer Raum, so hat jeder Teilbogen B von G eine Längenmenge $L(B)$ und eine Längenzahl $l(B)$. Bogen mit gleicher Längenzahl können natürlich durchaus verschiedene Längenmengen besitzen. Offen bleibt die Frage, unter welchen Voraussetzungen über G Bogen mit gleicher Längenmenge stets gleiche Längenzahl besitzen. Für die Gruppe der Vektoren der Ebene ist dies (s. u.) stets der Fall. Von besonderem Interesse ist die Übereinstimmung zweier Teilbogen einer Gruppe hinsichtlich ihrer Längenmenge für den Fall der Gruppe der Vektoren des n-dimensionalen Raumes, wo sie eine neue geometrische Verwandtschaft zweier Bogen darstellt, welche darin besteht, daß dieselben gleichsam durch Permutation ihrer Richtungselemente auseinander hervorgehen. Da die Längenmenge gegenüber Translationen, aber im allgemeinen nicht gegenüber Rotationen des Bogens invariant ist, könnte man daneben noch die Verwandtschaft, kongruente Längenmengen zu besitzen, untersuchen.

Auch für verschiedene Gruppen ergibt sich ein neues Vergleichsmittel. Als *ähnlich* hatte ich (Ann. of Math., *32*, S. 753) zwei Gruppen bezeichnet, die so eineindeutig aufeinander abgebildet werden können, daß je zwei kongruenten Punktepaaren (d. h. Punktepaaren mit gleichen Abständen) der einen Gruppe kongruente Punktepaare der anderen entsprechen. O. Taussky hat (Math. Ann., *108*, S. 615) bewiesen, daß zwei Gruppen dann und nur dann ähnlich sind, wenn sie isomorph sind. Nun könnte man zwei Gruppen als *verbiegungsähnlich* bezeichnen, wenn sie eineindeutig derart aufeinander abbildbar sind, daß je zwei

Teilbogen einer *geordneten* Gruppe als ein gewisses einzelnes Gruppen*element* erklärt werden, was für allgemeine Gruppen nicht in zwangloser Weise möglich zu sein scheint.

Unsere Längenmenge $L(B)$ dagegen gestattet einen in gewissem Sinne quantitativen Vergleich von je zwei Teilbogen einer beliebigen Abelschen Gruppe und insbesondere die Kennzeichnung kürzester Bogen. Wir wollen einen a und b verbindenden Teilbogen B einer Gruppe einen *Minimalbogen* nennen, wenn $L(B)$ ein Bogen, und zwar Summe zweier mit B kongruenter Bogen ist. Dieser für die Theorie der allgemeinen kontinuierlichen Gruppen wohl außerordentlich wichtige Begriff soll in späteren Mitteilungen näher untersucht werden.

Hier sei nur noch als Beispiel die vollständige Theorie der *Gruppe der ebenen Vektoren* entwickelt. Ein Bogen B der Ebene heiße ein *Elementar*bogen, wenn er erstens konvex ist und wenn zweitens, falls a und b die Endpunkte von B sind, für je zwei von a und b verschiedene Punkte p und q von B gilt: $\sphericalangle pab + \sphericalangle qba \leqq 180^0$. Dann ist $L(B)$ das Flächenstück F, dessen Rand $R(B)$ die Menge aller Punkte $\pm(2p-a-b)$ ist, wo p irgend ein Punkt von B ist[8]). Ist nämlich erstens E irgend eine B gemäß geordnete Teilmenge $p_1, p_2, \ldots, p_n$ von B, so bestätigt man durch Induktion nach n, daß die Punkte

$$p_1-p_n,\ 2p_2-p_1-p_n, \ldots, 2p_{n-1}-p_1-p_n,\ p_n-p_1,\ p_n+p_1-2p_2, \ldots,\ p_n+p_1-2p_{n-1}$$

in dieser Reihenfolge die Ecken eines geschlossenen konvexen Polygons bilden, welches $L(E)$ als Teil enthält und im Flächenstück F als Teil enthalten ist. Also[9]) gilt $L(B) \subset F$. Ist zweitens f irgend ein innerer Punkt von F, so existiert eine aus 4 Punkten bestehende Teilmenge E von B, so daß f in $L(E)$ liegt[10]). Um dieselbe zu konstruieren, bilden wir auf jedem von f ausgehenden Halbstrahl jenen Punkt, der gleichen Abstand von f und vom Schnittpunkt des Halbstrahles mit $R(B)$ besitzt. Die Menge dieser Halbierungspunkte ist eine geschlossene Kurve $H(f)$. Bezeichnen wir mit $B-a$ den Bogen, der aus B durch jene Translation entsteht, die a in 0 und b in $b-a$ überführt, so hat $B-a$ mit $H(f)$ genau einen Punkt q gemein. Ist q' jener Punkt von $R(B)$, der auf dem von f ausgehenden Halbstrahl durch q liegt, und r jener Punkt von $B-a$, der auf dem von $b-a$ ausgehenden Halbstrahl durch q' liegt, so

---

gleichlangen Bogen (d. h. Bogen mit gleicher Längenmenge) der einen Gruppe zwei gleichlange Bogen der andern entsprechen. Es ergibt sich dann die schwierige Frage, welche Gruppen mit allen zu ihnen verbiegungsgleichen Gruppen isomorph, also gleichsam *starr* sind, und welche nicht isomorphen Gruppen durch diese neue Verwandtschaft zusammengefaßt werden.

[8]) Dann und nur dann, wenn B eine Strecke ist, degeneriert F zu einem Bogen, ist also B Minimalbogen.

[9]) Diese Hälfte des Beweises stammt von Herrn F. Alt.

[10]) Liegt p in $R(B)$, so existiert offenbar eine aus höchstens 3 Punkten bestehende Menge E, für die p in $L(E)$ liegt.

bestätigt man leicht, daß $f = -a + 2q - 2r + b$ gilt. Setzen wir daher $\bar{q} = q + a$, $\bar{r} = r + a$, so liegt f für die Teilmenge a, $\bar{q}$, $\bar{r}$, b von B in $L(E)$. Also gilt $F \subset L(B)$ und folglich $L(B) = F$, wie behauptet. $R(B)$ besteht aus dem Elementarbogen $2p - a - b$, der entsteht, wenn man $B - a$ von $b - a$ aus im Verhältnis $1:2$ streckt, und aus dessen Spiegelbild am Punkt 0. Die Länge von $R(B)$ ist also die vierfache Länge von B. Elementarbogen mit gleicher Längenmenge haben somit gleiche Längenzahl.

Ist nun B irgend ein Bogen der Ebene, so bilden wir eine ausgezeichnete Folge $\{E_n\}$ von endlichen Teilmengen von B und die Summe der konvexen Hüllen aller Mengen $L(E_n)$ ($n = 1, 2, \ldots$ ad inf.). Die abgeschlossene Hülle F dieser Menge ist konvex und bezüglich 0 symmetrisch, also entweder eine konvexe Fläche oder ein bezüglich 0 symmetrischer Streifen zwischen zwei Parallelen oder die ganze Ebene oder sie degeneriert zu einer Strecke, was dann und nur dann der Fall ist, wenn B eine Strecke (Minimalbogen) ist. Offenbar gilt $L(B) \subset F$ und man kann umgekehrt auch zeigen, daß jeder Punkt von F Häufungspunkt der Menge $L'(B)$ ist. Fassen wir dieses Resultat mit seiner evidenten Umkehrung zusammen, so ergibt sich: *Damit eine Teilmenge der Ebene Längenmenge eines Bogens sei, ist notwendig und hinreichend, daß sie abgeschlossen, bezüglich* 0 *symmetrisch und konvex sei. Dann und nur dann, wenn sie beschränkt ist, sind die zugehörigen Bogen rektifizierbar und mit Elementarbogen in Längenmenge und Längenzahl identisch, wobei die Längenzahl des Elementarbogens gleich dem vierten Teil des Umfanges seiner Längenmenge ist. Bogen mit gleicher Längenmenge haben gleiche Längenzahl.*

Zum Schluß sei darauf hingewiesen, daß alle hier entwickelten Begriffe und Theoreme auch für nicht-Abelsche Gruppen verallgemeinert werden können, wenn man zwischen rechts- und linksseitigen Abständen unterscheidet.

54. Kolloquium (17. XI. 1932).

Diskussion über die Mengerschen Probleme, ob das Quadrat der Tetraederkonfiguration (vgl. diese Ergebnisse, 2, S. 30) mit einer Teilmenge des R_4 homöomorph sei und welche eindimensionalen Komplexe in Flächen vom Geschlecht p einbettbar sind.

Flores: *Das Produkt eines ebenen und eines beliebigen eindimensionalen Komplexes ist in den R_4 einbettbar.* G sei der in der x, y-Ebene E liegende, G' der im r, s, t-Raum R_3 liegende eindimensionale Komplex. Bei geeigneter Lage von G' im R_3 existiert ein $\varepsilon > 0$ derart, daß für je zwei Punkte $p = r, s, t$ und $\bar{p} = \bar{r}, \bar{s}, \bar{t}$ aus $r = \bar{r}$ und $s = \bar{s}$ stets folgt $|t - \bar{t}| > 3\varepsilon$. Nehmen wir dann G im Streifen $-\varepsilon < y < \varepsilon$ von E an, so ist die Menge $x, r, s, t + y$ im R_4 mit $G \times G'$ homöomorph. Seien nämlich $p \times p'$ und $q \times q'$ irgend zwei Punkte des Produktes. Wenn dann keine der Ungleichungen $x \neq \bar{x}$, $r \neq \bar{r}$, $s \neq \bar{s}$ gilt, so ist $|t - \bar{t}| > 3\varepsilon$, also, da $|y - \bar{y}| < 2\varepsilon$ gilt, $|(t + y) - (\bar{t} + \bar{y})| > 0$.

55. Kolloquium (24. XI. 1932).

Nöbeling: Eine kompakte Menge des R_k ist genau dann höchstens rational n-dimensional, wenn sie für jedes $\varepsilon > 0$ so in einen n-dimensionalen Komplex K ε-deformierbar ist, daß die Menge aller Punkte von K mit höchstens abzählbarvielen Urbildpunkten in K dicht liegt. (Erscheint in den Mathem. Annalen.) Menger: Der Beweis liefert auch: Eine kompakte n-dimensionale Menge kann für jedes $\varepsilon > 0$ so in einen n-dimensionalen Komplex K ε-deformiert werden, daß die Menge aller Punkte mit höchstens nulldimensionalen Urbildmengen in K dicht liegt. Nachträgliche Bemerkung: Von Warschauer Mathematikern wurde nach Kenntnisnahme dieses Satzes die Frage aufgeworfen, ob bzw. unter welchen Bedingungen eine n-dimensionale kompakte Menge für jedes ε so in einen n-dimensionalen Komplex K ε-deformierbar sei, *daß jeder Punkt von K eine höchstens nulldimensionale Urbildmenge besitzt*?

56. Kolloquium (8. XII. 1932).

Wald: *Halbmetrische Räume und Konvexifizierbarkeit.*

Ein halbmetrischer Raum R heißt nach Menger konvex, wenn zu je zwei verschiedenen Punkten p und q von R ein Punkt r existiert, so daß $p \neq r \neq q$ und $pr + rq = pq$. Wir sagen mit Menger (Dritte Unters. üb. allgem. Metrik, Math. Ann., *100*), daß R eine *stetige Metrik* besitzt, wenn für jede Folge $\{p_k, q_k\}$ $(k = 1, 2, \ldots$ ad inf.) von Punktepaaren aus $\lim_{k=\infty} p_k = p$ und $\lim_{k=\infty} q_k = q$ stets $\lim_{k=\infty} p_k q_k = pq$ folgt. Es gilt nun der Satz: *Jedes im kleinen zusammenhängende Kontinuum ist mit einem kompakten konvexen halbmetrischen Raum mit stetiger Metrik homöomorph.* Hinsichtlich der *halbmetrischen* Kontinua mit stetiger Metrik ist also Zusammenhang im Kleinen *hinreichend* für Konvexifizierbarkeit, aber, wie sich an Beispielen zeigen läßt, *nicht notwendig*, während hinsichtlich *metrischer* Kontinua Zusammenhang im Kleinen für die Konvexifizierbarkeit nach Menger (Erste Unters. üb. allg. Metrik, Math. Ann., *100*) notwendig ist und die Frage, ob er hinreichend ist, das bekannte schwierige Konvexifizierungsproblem bildet, das bisher ungelöst ist.

Ein halbmetrischer Raum heiße *schwachmetrisch*, wenn eine Zahl $\lambda > 0$ existiert, so daß für je drei Punkte p, q, r von R in jeder Anordnung $pq + \lambda qr \geqq pr$ gilt. Dann gilt der Satz: *jeder kompakte konvexe schwachmetrische Raum ist mit einem kompakten konvexen metrischen Raum homöomorph* (und daher insbesondere zusammenhängend im Kleinen).

57. Kolloquium (26. I. 1933).

Eine merkwürdige Kurve von Knaster und Mazurkiewicz. (Nach mündlichen Mitteilungen vorgetragen von Menger.) Die von Menger (Kurventheorie, 1932, S. 112) aufgeworfene Frage, ob ein Kontinuum K nach Weglassung aller Endpunkte von K in jedem irrationalen Punkt von K irrational bleibt, ist zu verneinen. Es gibt in der Ebene sogar *ein Kontinuum, welches bloß Punkte der Ordnung 1, $\aleph_0$ und $\mathfrak{c}$ enthält und* *in welchem die Menge aller Punkte von höherer als erster Ordnung Summe von abzählbarvielen Strecken ist,* so daß also nach Tilgung der Endpunkte von K eine rational-eindimensionale Menge verbleibt. (Aus diesen Eigenschaften folgt natürlich auch, daß K eine schwach irrationale Kurve ist. Zudem enthält K keinen topologischen Kreis und kann in gewissem Sinne als Baumkurve bezeichnet werden.) Die Konstruktion von K erfolgt so: Es seien P ein Parallelogramm, p und q zwei gegenüberliegende Ecken von P und D die sie verbindende Diagonale. Wir wählen in einem der beiden offenen Dreiecke, in welche das Innere von P durch D zerfällt, eine Punktefolge $\{p_i\}$, welche gegen den Punkt r von D konvergiert, dessen Abstand von p doppelt so groß ist wie sein Abstand von q, wobei keine zwei Punkte p_i und p_j mit p auf einer Geraden liegen. $D_{2i}(D, P)$ oder kurz D_{2i} heiße die p und p_i verbindende Strecke, $D_{2i+1}(D, P)$ oder D_{2i+1} ihr Spiegelbild am Zentrum von P. Wir setzen $\sum_{i=1}^{\infty} D_i = D^1$ und bilden für jede der Strecken D_i ($i = 1, 2, \ldots$ ad inf.) ein Parallelogramm P_i, welches $\subset P$ ist und D_i als Diagonale besitzt, wobei die abgeschlossenen Parallelogramme P_i abgesehen von dem einen Eckpunkt paarweise fremd gewählt werden. Für jede Strecke D_i ($i = 1, 2, \ldots$ ad inf.) und das entsprechende Parallelogramm P_i wiederholen wir die Konstruktion, d. h. wir bilden eine Folge von Strecken $D_j(D_i, P_i)$ oder kurz D_{ij} ($j = 1, 2, \ldots$ ad inf.), setzen $D^2 = \sum_{i_1, i_2 = 1}^{\infty} D_{i_1 i_2}$, bilden zu jeder Strecke $D_{i_1 i_2}$ ein Parallelogramm $P_{i_1 i_2}$ und fahren in dieser Weise fort. Setzen wir noch $D = D^0$, so ist die erwähnte Kurve K als die abgeschlossene Hülle der Menge $\sum_{i=0}^{\infty} D^i$ definiert. Nach der Methode von Mazurkiewicz (Fund. Math., *16*, S. 337, vgl. auch Menger, Kurventheorie, 1932, S. 143) zeigt man, daß K irrational ist, u. zw. in denjenigen Punkten, die dem mittleren Drittel von D oder von einer der Strecken $D_{i_1, i_2, \ldots, i_n}$ ($i_1, i_2, \ldots i_n$, $n = 1, 2, \ldots$ ad inf.) angehören. In jedem sonstigen Punkt der Menge $\sum_{i=0}^{\infty} D^i$ hat K offenbar die Ordnung $\aleph_0$. Jeder nicht in $\sum_{i=0}^{\infty} D^i$ liegende Punkt p von K ist Endpunkt von K, da er in einer auf p sich zusammenziehenden Folge $P_{i_1}, P_{i_1 i_2}, \ldots, P_{i_1 i_2 \ldots i_n}, \ldots$ von Parallelogrammen liegt, deren Begrenzungen mit K genau einen Eckpunkt gemein haben. K enthält also nur Punkte der Ordnungen 1, $\aleph_0$ und $\mathfrak{c}$. Die nach Weglassung der Endpunkte aus K verbleibende Menge ist $\sum_{i=0}^{\infty} D^i$, also Summe von abzählbarvielen Strecken.

Otto (Lemberg, nach mündlicher Mitteilung): *Eine abgeschlossene nulldimensionale Menge der Ebene, die mit jeder Geraden, die ein gewisses konvexes Gebiet trifft, Punkte gemein hat,* wurde schon von Mazurkiewicz angegeben. Man kann eine Menge M

dieser Art auch folgendermaßen erhalten: Man wähle in einem regulären Fünfeck F sechs seitenparallele Fünfecke, $F_1, \ldots, F_6$, deren Seitenlängen gleich einem Drittel der Seitenlänge von F sind, wobei die fünf ersten Fünfecke je eine Ecke mit F gemein haben, während das sechste Fünfeck denselben Mittelpunkt habe wie F. Jede Gerade, die F trifft, trifft dann auch mindestens eines der sechs Fünfecke F_i. Iteriert man daher die Konstruktion, indem man in jedem der F_{i_1} sechs Fünfecke $F_{i_1 i_2}$ bestimmt, in jedem von diesen wieder sechs Fünfecke $F_{i_1 i_2 i_3}$ usw., so ist, wenn wir für jedes natürliche n mit F^n die Summe der 6^n Fünfecke mit n Indizes bezeichnen, die abgeschlossene Menge $M = \prod_{n=1}^{\infty} F^n$ offenbar nulldimensional und hat mit jeder Geraden, die F trifft, mindestens einen Punkt gemein. Flores: Man kann dasselbe mit Quadraten erreichen, wenn man in jedem Quadrat Q fünf Quadrate gedrittelter Seitenlänge wählt, von denen vier je eine Ecke mit Q gemein haben und zu Q kantenparallel sind, während das fünfte das gleiche Zentrum wie Q hat und Seiten besitzt, die gegenüber denen von Q um 45^0 gedreht sind. Nöbeling: Man kann durch Zusammensetzung einiger solcher Quadrate leicht erreichen, daß jede Gerade, die das Ausgangsquadrat trifft, für ein gewisses n mindestens zwei Quadrate mit n Indizes trifft und daher mit der nulldimensionalen Menge einen Durchschnitt von der Mächtigkeit des Kontinuums besitzt.

58. Kolloquium (2. II. 1933).

Menger: *Eine Bemerkung über die Potenzen schwach eindimensionaler Mengen.*

Ist S eine Teilmenge der X, Y-Ebene derart, daß jede zur Y-Achse parallele Gerade mit S höchstens einen Punkt gemein hat und die Orthogonalprojektion von S auf die X-Achse nulldimensional ist, so ist S^n (die Menge aller n-Tupel von Punkten von S) *mit einer Teilmenge des* R_{n+1} (des $(n+1)$-dimensionalen euklidischen Raumes) *homöomorph.* Auf Grund der Voraussetzungen über S ist jeder Punkt von S durch seine Abszisse d, welche Element der nulldimensionalen Projektion D von S auf die X-Achse ist, und seine Ordinate $f(d)$, wobei f eine auf D definierte eindeutige Funktion ist, gekennzeichnet. Jeder Punkt von S^n ist daher durch einen Punkt im Raume R_{2n}, d. h. durch $2n$ Zahlen $d_1, f(d_1), d_2, f(d_2), \ldots, d_n, f(d_n)$ gekennzeichnet. Da die n-te Potenz einer nulldimensionalen Menge nulldimensional und daher mit einer linearen Menge homöomorph ist, so existiert eine topologische Abbildung von D^n auf eine nulldimensionale Teilmenge E der X_1-Achse des R_{n+1}. Bezeichnet dann F die Menge aller Punkte des R_{n+1}, deren x_1-Koordinate ein Element von E ist und deren n weitere Koordinaten, falls $d_1, d_2, \ldots, d_n$ der bei der Abbildung von D^n auf E dem Punkt e von E entsprechende Punkt von D^n ist, $f(d_1), f(d_2), \ldots, f(d_n)$ lauten, so ist F offenbar eine mit S^n homöomorphe Teilmenge des R_{n+1}.

Die äußerst schwach eindimensionalen Mengen von Sierpiński (Fund. Math., *2*, S. 81, vgl. auch Dimensionstheorie, 1928, S. 139) und Kuratowski (Mathematica, *6*, S. 120) besitzen die eben vorausgesetzten Eigenschaften von S; ihre n-ten Potenzen S^n sind also mit Teilmengen F des R_{n+1} homöomorph, was wohl das Studium ihrer Dimension erleichtert. Da für diese Mengen S die Funktion $f(d)$, wenn d die Menge D durchläuft, alle Werte eines Intervalles annimmt, bestätigt man leicht, daß F im R_{n+1} als Projektion auf die zur X_1-Achse senkrechte Hyperebene einen n-dimensionalen Würfel besitzt, während die Projektionen von F auf sämtliche anderen Koordinatenhyperebenen des R_{n+1} weniger als n-dimensional sein dürften. Überdies ist F eineindeutig und stetig auf die nulldimensionale Menge D abgebildet, welche im Falle der Kuratowskischen Menge das Cantorsche Diskontinuum ist. Wenn also S^n n-dimensional ist, so existiert *eine n-dimensionale Menge, die eineindeutig und stetig auf das Diskontinuum abgebildet werden kann.* Überdies existieren dann, wie ich (Prace Mat.-Fiz., *37*, S. 89) bewiesen habe, abzählbarviele schwach n-dimensionale Mengen, welche eine stark n-dimensionale Summe besitzen und in derselben abgeschlossen sind.

Nöbeling: Die Anwendung der Dimensionstheorie auf Riemannsche Flächen durch Mazurkiewicz (Fund. Math., *17*, S. 26).

Wald: $n+2$ Punkte eines komplexmetrischen Raumes sind dann und nur dann in den K_n einbettbar, wenn ihre Determinante $=0$ ist. (Siehe dieses Heft S. 35 f.)

59. Kolloquium (22. II. 1933).

Flexer: *Der K_2 hat nicht die Quasikongruenzordnung*[1]) *4*, d. h. es existiert ein aus 6 Punkten $p_1, \ldots, p_6$ bestehender komplexmetrischer Raum R, von dem je vier Punkte mit vier Punkten des K_2 abstandsgleich sind, während R nicht kongruent mit einer Teilmenge des K_2 ist. Die Abstände $p_i p_j$ in R sind die folgenden:

$$p_1 p_2 = a \neq 0; \qquad p_2 p_4 = a - b;$$
$$p_1 p_3 = b \neq 0, \neq a, \neq -a; \qquad p_2 p_6 = a + c - 2b;$$
$$p_1 p_6 = c = \frac{b^2}{a}; \qquad p_5 p_6 = c - b;$$

$$p_1 p_4 = p_1 p_5 = p_2 p_3 = p_2 p_5 = p_3 p_4 = p_3 p_6 = p_3 p_5 = p_4 p_5 = p_4 p_6 = 0.$$

Wald: *Vereinfachter Beweis des Steinitzschen Satzes über Vektorenreihen im R_n.*

Für eine bedingt konvergente Reihe $\sum_{i=1}^{\infty} \mathfrak{v}_i$ von Vektoren des R_n ist die Konvergenzmenge, d. h. die Menge K derjenigen Vektoren, die man durch Umordnen der Glieder von $\{\mathfrak{v}_i\}$ als Reihensumme erhalten kann, eine lineare Mannigfaltigkeit des R_n.[2])

[1]) Vgl. über diesen Begriff Menger, diese Ergeb., *4*, S. 41.
[2]) Vgl. R. Levi, Nouv. Ann. Math., 1905, S. 506, und E. Steinitz, Crelles Journ., *144*.

Die Richtung h des R_n heiße *Hauptrichtung*, wenn für jedes $\varepsilon > 0$ die Summe der absoluten Beträge der Vektoren aus $\{\mathfrak{v}_i\}$, welche mit h einen Winkel $< \varepsilon$ einschließen, unendlich ist. Die Richtung h heiße *Konvergenzrichtung*, wenn die Summe der Projektionen auf h von allen Vektoren aus $\{\mathfrak{v}_i\}$, die mit h einen Winkel $\leqq \frac{\pi}{2}$ einschließen, endlich ist.

Wir machen nun die induktive Voraussetzung, der Satz sei bewiesen für den R_k $(k < n)$ und unterscheiden bei der Betrachtung des R_n zwei Fälle.

I. Fall. *Es existiert eine Konvergenzrichtung h.* Sind dann $\mathfrak{v}'_i$ und $\mathfrak{v}''_i$ für jeden Vektor $\mathfrak{v}_i$ die Projektionen auf die h enthaltende Gerade bzw. auf die zu ihr senkrechte $(n-1)$-dimensionale Hyperebene H durch den Punkt 0, so ist $\Sigma\mathfrak{v}'_i$, da wegen der bedingten Konvergenz von $\Sigma\mathfrak{v}_i$ zugleich mit h auch die mit h den Winkel π einschließende Richtung eine Konvergenzrichtung ist, absolut konvergent, etwa gegen $\mathfrak{v}'$, während $\Sigma\mathfrak{v}''_i$ bedingt konvergiert, also laut induktiver Voraussetzung eine lineare Mannigfaltigkeit L als Konvergenzmenge hat. Dann ist K die lineare Mannigfaltigkeit $L + \mathfrak{v}'$.

II. Fall. *Es existiert keine Konvergenzrichtung.* Dann gilt *a). Es gibt eine 0 enthaltende, mindestens eindimensionale Hyperebene H, die zu jeder Richtung h von H eine mit h einen Winkel* $< \frac{\pi}{2}$ *einschließende Hauptrichtung enthält.* Die Annahme, es gebe keine solche H, impliziert nämlich, daß alle Hauptrichtungen in einem Kegel K mit der Spitze 0 und einem Öffnungswinkel $< \pi$ liegen[3]). Da dann mit dem zur Achse von K entgegengesetzten Halbstrahl k alle Hauptrichtungen Winkel $> \frac{\pi}{2}$ einschließen, folgt wegen der Kompaktheit der Einheitskugel, daß k eine Konvergenzrichtung ist, im Widerspruch dazu, daß keine Konvergenzrichtung existiert, womit *a)* bewiesen ist. Aus *a)* folgt wegen der Kompaktheit der Einheitskugel *b). Es existiert ein Winkel* $\alpha < \frac{\pi}{2}$, *so daß H zu jeder Richtung h von H eine mit h einen Winkel* $\leqq \alpha$ *einschließende Hauptrichtung enthält.*

Sei nun H' die zu H senkrechte Hyperebene durch 0 (falls $H = R_n$, reduziert sich H' auf den Punkt 0). Seien $\mathfrak{v}'_i$ und $\mathfrak{v}''_i$ die Projektionen von $\mathfrak{v}_i$ auf H' bzw. H. Da keine Konvergenzrichtung

[3]) Setzen wir nämlich diese für den R_2 evidente Implikation als für den R_{n-1} giltig voraus und betrachten wir den R_n! Sicher enthält der R_n eine Richtung h, mit der jede Hauptrichtung einen Winkel $\geqq \frac{\pi}{2}$ einschließt, da wir sonst entgegen der Annahme $H = R_n$ setzen könnten. In der auf h senkrechten $(n-1)$-dimensionalen Hyperebene H_{n-1} durch 0 liegen nach der induktiven Voraussetzung alle Hauptrichtungen in einem Kegel der H_{n-1} mit der Spitze in 0 und einem Öffnungswinkel $< \pi$, woraus für die Hauptrichtungen des R_n, da sie alle auf einer Seite von H_{n-1} liegen, folgt, daß sie in einem Kegel K mit der Spitze 0 und einem Öffnungswinkel $< \pi$ liegen, wie behauptet.

existiert, ist $\sum_{i=1}^{\infty} \mathfrak{v}'_i$ bedingt konvergent, hat also nach Induktionsvoraussetzung als Konvergenzmenge eine lineare Mannigfaltigkeit L (die im Falle $H = R_n$ aus dem Punkt 0 besteht). Um zu zeigen, daß die Konvergenzmenge K von $\Sigma \mathfrak{v}_i$ eine lineare Mannigfaltigkeit ist, genügt es zu zeigen: *Ist* $\mathfrak{v}'$ *irgend ein Vektor von* L *und* $\mathfrak{v}''$ *irgend einer von* H, *so liegt* $\mathfrak{v}' + \mathfrak{v}''$ *in* K. Sei etwa $\mathfrak{v}' = \sum_{i=1}^{\infty} \mathfrak{v}'_i$ und $\mathfrak{v}''$ ein gegebener Vektor aus H. Wir wollen eine Anordnung $\{\mathfrak{v}_{i_n}\}$ der Vektoren $\{\mathfrak{v}_i\}$ angeben, für welche $\sum_{n=1}^{\infty} \mathfrak{v}_{i_n} = \mathfrak{v}' + \mathfrak{v}''$ gilt. Dazu geben wir eine Folge $\{\varepsilon_i\}$ von Zahlen > 0 mit konvergenter Summe vor. Wir setzen $\mathfrak{v}_{i_1} = \mathfrak{v}_1$. Der Vektor $\mathfrak{w}'' = \mathfrak{v}'' - \mathfrak{v}''_{i_1}$ liegt in H. Es gibt dann eine Hauptrichtung l in H, die mit $\mathfrak{w}''$ einen Winkel $\sphericalangle \alpha < \frac{\pi}{2}$ einschließt. Sei nun φ ein Winkel, so daß $\alpha + |\varphi| < \frac{\pi}{2}$. Wir nennen $\mathfrak{v}_{i_2}$ den ersten Vektor $\neq \mathfrak{v}_{i_1}$ der Folge $\{\mathfrak{v}_i\}$, der mit l einen Winkel $\leqq \varphi$ einschließt. Falls die Projektion von $\mathfrak{v}_{i_2}$ auf l kleiner als die Projektion von $\mathfrak{w}''$ auf l ist, so nennen wir $\mathfrak{v}_{i_3}$ den ersten Vektor $\neq \mathfrak{v}_{i_1}$ und $\neq \mathfrak{v}_{i_2}$ der Folge $\{\mathfrak{v}_i\}$, der mit l einen Winkel $\leqq \varphi$ einschließt. Falls die Projektion von $\mathfrak{v}_{i_2} + \mathfrak{v}_{i_3}$ auf l kleiner ist als die von $\mathfrak{w}''$, so nennen wir $\mathfrak{v}_{i_4}$ den ersten Vektor $\neq \mathfrak{v}_{i_j}$ $(j = 1, 2, 3)$ der Folge $\{\mathfrak{v}_i\}$, der mit l einen Winkel $\leqq \varphi$ einschließt. So fahren wir fort, bis wir zu einem $\mathfrak{v}_{i_r}$ kommen, so daß die Projektion von $\sum_{j=2}^{r} \mathfrak{v}_{i_j}$ auf l nicht kleiner als die von $\mathfrak{w}''$ ist. Es sei $\mathfrak{w}''_1 = \mathfrak{w}'' - \sum_{j=2}^{r} \mathfrak{v}''_{i_j} = \mathfrak{v}'' - \sum_{j=1}^{r} \mathfrak{v}''_{i_j}$. Dann gibt es eine Hauptrichtung l_1 in H, die mit $\mathfrak{w}_1''$ einen Winkel $\leqq \alpha$ einschließt. Dann sei $\mathfrak{v}_{i_{r+1}}$ der erste Vektor $\neq \mathfrak{v}_{i_j}$ $(j = 1, 2, \ldots, r)$ der Folge $\{\mathfrak{v}_i\}$, der mit l_1 einen Winkel $\leqq \varphi$ einschließt u. s. f., bis wir zu einem $\mathfrak{v}_{i_s}$ gelangen, so daß die Projektion von $\sum_{j=r+1}^{s} \mathfrak{v}_{i_j}$ auf l_1 nicht kleiner als die Projektion von $\mathfrak{w}''_1$ auf l_1 ist. Es sei

$$\mathfrak{w}''_2 = \mathfrak{w}''_1 - \sum_{j=r+1}^{s} \mathfrak{w}''_{i_j} = \mathfrak{v}'' - \sum_{j=1}^{s} \mathfrak{v}''_{i_j}.$$

Durch Fortsetzung dieses Verfahrens gelangen wir bei hinlänglich kleinem φ offenbar nach endlichvielen Schritten zu einem $\mathfrak{w}''_n = \mathfrak{v}'' - \sum_{j=1}^{k} \mathfrak{v}_{i_j}$, so daß

$$|\mathfrak{w}''_n| = |\mathfrak{v}''_{i_1} + \ldots + \mathfrak{v}''_{i_k} - \mathfrak{v}''| \leqq \varepsilon_1 \text{ und } |\mathfrak{v}'_{i_1}| + \ldots + |\mathfrak{v}'_{i_k}| \leqq \varepsilon_1$$

gilt.

Wir nennen $\mathfrak{v}_{i_{k+1}}$ den ersten Vektor $\neq \mathfrak{v}_{i_j}$ $(j = 1, 2, \ldots, k)$ der Folge $\{\mathfrak{v}_i\}$. Analog bestimmen wir nun Vektoren $\mathfrak{v}_{i_{j+2}} \ldots, \mathfrak{v}_{i_l}$, so daß

$$|\mathfrak{v}''_{i_1}+\ldots+\mathfrak{v}''_{i_l}-\mathfrak{v}''|\leqq\varepsilon_2 \text{ und } |\mathfrak{v}'_{i_{k+2}}|+\ldots+|\mathfrak{v}'_{i_l}|\leqq\varepsilon_2$$

gilt und nennen $\mathfrak{v}_i$ den ersten Vektor $\neq \mathfrak{v}_{i_j}$ $(j=1, 2, \ldots l)$ der Folge $\{\mathfrak{v}_i\}$. Für die durch Fortsetzung dieses $l+1$-Verfahrens definierte Anordnung $\{\mathfrak{v}_{i_j}\}$ $(j=1, 2, \ldots$ ad inf.) bestätigt man mit Rücksicht auf $\mathfrak{v}'=\sum\limits_{i=1}^{\infty}\mathfrak{v}'_i$ und die Konvergenz der Reihe $\sum\limits_{i=1}^{\infty}\varepsilon_i$ mühelos, daß $\sum\limits_{j=1}^{\infty}\mathfrak{v}_{i_j}=\mathfrak{v}'+\mathfrak{v}''$, w. z. b. w.

60. Kolloquium (8. III. 1933).

Wald: *Bedingt konvergente Reihen von Vektoren im R_ω.*

Unter dem R_ω verstehen wir die Menge aller unendlichdimensionalen Vektoren (= Folgen reeller Zahlen) mit der Festsetzung, daß die Vektorenfolge $\{\mathfrak{v}_i\}$ $(i=1, 2, \ldots$ ad inf.), wo $\mathfrak{v}_i$ die Zahlenfolge $\{x_{i,k}\}$ $(k=, 1\ 2 \ldots$ ad inf.) ist, gegen den Vektor $\mathfrak{v}=\{x_k\}$ $(k=1, 2\ldots$ ad inf.), dann und nur dann konvergiert, wenn $\lim\limits_{i=\infty} x_{i,k}=x_k$ $(k=1, 2, \ldots$ ad inf.) gilt.

Theorem: *Die Konvergenzmenge K einer bedingt konvergenten Reihe $\sum\limits_{i=1}^{\infty}\mathfrak{v}_i$ von Vektoren des R_ω ist eine lineare Mannigfaltigkeit des R_ω.*

Zum Beweis verwenden wir den entsprechenden Steinitzschen Satz für den R_n (vgl. 59. Kolloquium) sowie folgendes Lemma von Gross (Monatshefte f. Math. u. Phys., *28*): *Endlichviele Vektoren $\mathfrak{u}_1, \ldots, \mathfrak{u}_k$ des R_n, für die $|\mathfrak{u}_i|\leqq\delta$ $(i=1, \ldots, k)$ und $|\mathfrak{u}_1+\ldots+\mathfrak{u}_k|\leqq\delta$ gilt, können so angeordnet werden: $\mathfrak{u}_{i_1}, \ldots, \mathfrak{u}_{i_k}$, daß für jedes $r=1, \ldots, k$ gilt: $|\mathfrak{u}_{i_1}+\ldots+\mathfrak{u}_{i_r}|\leqq(2^n-1)\,\delta$.*

Ist $\mathfrak{u}$ irgend ein Vektor des R_ω, so bezeichnen wir jenen Vektor des R_n, dessen n Koordinaten der Reihe nach mit den n ersten von $\mathfrak{u}$ übereinstimmen, mit $\mathfrak{u}^{(n)}$. Wir geben ferner eine monoton abnehmende Folge $\{\varepsilon_i\}$ von Zahlen >0 vor, für die $\lim\limits_{i=\infty} 2^i\varepsilon_i=0$ gilt. Zum Beweis unseres Theorems genügt es, daß wir zeigen: *Sind $\mathfrak{x}$ und $\mathfrak{y}$ irgend zwei gegebene Vektoren von K und liegt der Endpunkt von $\mathfrak{w}$ auf der durch die Endpunkte von $\mathfrak{x}$ und $\mathfrak{y}$ bestimmten Geraden, so gehört auch $\mathfrak{w}$ zu K.* Dann liegen $\mathfrak{x}^{(n)}$ und $\mathfrak{y}^{(n)}$ für jedes natürliche n in der Konvergenzmenge K_n der Reihe $\sum\limits_{i=1}^{\infty}\mathfrak{v}_i^{(n)}$ und aus dem Steinitzschen Satz folgt

$$(I_n)\quad \mathfrak{w}^{(n)}=\mathfrak{v}^{(n)}_{n_1}+\mathfrak{v}^{(n)}_{n_2}+\ldots+\mathfrak{v}^{(n)}_{n_k}+\ldots\quad (n=1,\ 2, \ldots \text{ ad inf.}),$$

wobei $n_1, n_2, \ldots, n_k$ nicht eine Zahlenfolge, sondern eine Folge von Zahlenfolgen symbolisiert, nämlich für $n=1$ die Folge $1_1, 1_2, \ldots, 1_k \ldots$, für $n=2$ die Folge $2_1, 2_2, \ldots, 2_k, \ldots$ usw.

Die Menge der r ersten Vektoren der Reihen (I_n) nennen wir (I_n^r), ihre Summe S_n^r. Wir ordnen nun die Reihe $\sum_{i=1}^{\infty} \mathfrak{v}_i$ in eine Reihe $\sum_{n=1}^{\infty} \mathfrak{v}_{i_n}$ mit der Summe $\mathfrak{w}$ um. Dazu bestimmen wir j_1 so groß, daß $|\mathfrak{w}^{(1)} - S_1^{j_1}| \leqq \varepsilon_1$ und $|\mathfrak{v}_j^{(1)}| \leqq \varepsilon_1$ für alle $j \neq 1_1, 1_2, \ldots, 1_{j_1}$. Die j_1 Vektoren $\mathfrak{v}_i$ $(i = 1_1, 1_2, \ldots, 1_{j_1})$ nehmen wir in die zu bildende Reihe als erste j_1 Vektoren $\mathfrak{v}_{i_1}, \mathfrak{v}_{i_2}, \ldots, \mathfrak{v}_{i_{j_1}}$ auf. Sodann bestimmen wir j_2 so, daß die Vektoren $\mathfrak{v}_{i_1}^{(2)}, \mathfrak{v}_{i_2}^{(2)}, \ldots, \mathfrak{v}_{i_{j_1}}^{(2)}$ in $(I_2^{j_2})$ vorkommen (Vektoren mit verschiedenen Indizes als verschieden betrachtet), daß ferner $|\mathfrak{w}^{(2)} - S_2^{j_2}| \leqq \varepsilon_2$ und daß schließlich $|\mathfrak{v}_j^{(2)}| \leqq \varepsilon_2$ für alle $j \neq 2_1, 2_2, \ldots, 2_{j_2}$. Sind $\mathfrak{v}_{l_1}^{(2)}, \ldots, \mathfrak{v}_{l_{j_2-j_1}}^{(2)}$ die von $\mathfrak{v}_{i_1}^{(2)}, \ldots, \mathfrak{v}_{i_{j_1}}^{(2)}$ verschiedenen Vektoren von $(I_2^{j_2})$, so gilt

$$|\mathfrak{v}_{l_i}^{(1)}| \leqq \varepsilon_1 \; (i = 1, 2, \ldots, j_2 - j_1) \text{ und } |\mathfrak{v}_{l_1}^{(1)} + \ldots + \mathfrak{v}_{l_{j_2-j_1}}^{(1)}| \leqq \varepsilon_2 + \varepsilon_1.$$

Also gibt es nach dem Grossschen Lemma eine Anordnung $\mathfrak{v}_{m_1}^{(1)}, \ldots, \mathfrak{v}_{m_{j_2-j_1}}^{(1)}$, so daß $|\mathfrak{v}_{m_1}^{(1)} + \ldots + \mathfrak{v}_{m_r}^{(1)}| \leqq (2^1 - 1)(\varepsilon_2 + \varepsilon_1)$ für $r = 1, 2, \ldots, j_2 - j_1$. Diese so geordneten $j_2 - j_1$ Vektoren nehmen wir nun als die $j_2 - j_1$ nächsten Glieder $\mathfrak{v}_{i_{j_1+1}}, \ldots, \mathfrak{v}_{i_{j_2}}$ in die zu bildende Reihe auf. Wir bestimmen nun j_3 so groß, daß die Vektoren $\mathfrak{v}_{i_1}^{(3)}, \mathfrak{v}_{i_2}^{(3)}, \ldots, \mathfrak{v}_{i_{j_2}}^{(3)}$ in $(I_3^{j_3})$ vorkommen, daß $|\mathfrak{w}^{(3)} - S_3^{j_3}| \leqq \varepsilon_3$ und daß $|\mathfrak{v}_j^{(3)}| \leqq \varepsilon_3$ für alle $j \neq 3_1, 3_2, \ldots, 3_{j_3}$. Sind $\mathfrak{v}_{k_1}^{(3)}, \ldots, \mathfrak{v}_{k_{j_3-j_2}}^{(3)}$ die von $\mathfrak{v}_{i_1}^{(3)}, \ldots, \mathfrak{v}_{i_{j_2}}^{(3)}$ verschiedenen Vektoren von $(I_3^{j_3})$, so gilt $|\mathfrak{v}_{k_i}^{(2)}| \leqq \varepsilon_2$ $(i = 1, \ldots, j_3 - j_2)$ und $|\mathfrak{v}_{k_1}^{(2)} + \ldots + \mathfrak{v}_{k_{j_3-j_2}}^{(2)}| \leqq \varepsilon_3 + \varepsilon_2$. Also gibt es nach dem Grossschen Lemma eine Anordnung $\mathfrak{v}_{n_1}^{(2)}, \ldots, \mathfrak{v}_{n_{j_3-j_2}}^{(2)}$, so daß $|\mathfrak{v}_{n_1}^{(2)} + \ldots + \mathfrak{v}_{n_r}^{(2)}| \leqq (2^2 - 1)(\varepsilon_3 + \varepsilon_2)$ für $r = 1, 2, \ldots, j_3 - j_2$. Diese so geordneten $j_3 - j_2$ Vektoren nehmen wir als die $j_3 - j_2$ nächsten Glieder in die zu bildende Reihe auf. Durch Fortsetzung dieses Verfahrens erhalten wir eine Reihe $\sum_{n=1}^{\infty} \mathfrak{v}_{i_n}$, als deren Summe man wegen $\lim_{n=\infty} 2^n \varepsilon_n = \lim_{n=\infty} (2^n - 1)(\varepsilon_{n+1} + \varepsilon_n) = 0$ mühelos $\mathfrak{w}$ erhält.

Wald: *Reihen in topologischen Gruppen.*

G sei eine *topologische Gruppe*, d. h. ein topologischer Raum, in dem je zwei Elementen x und y ein Element $x + y$ derart zugeordnet ist, daß G hinsichtlich dieser Operation eine Gruppe bildet und aus $\lim_{n=\infty} x_n = x$ und $\lim_{n=\infty} y_n = y$ stets $\lim_{n=\infty} (x_n + y_n) = x + y$ und $\lim_{n=\infty} (-x_n) = -x$ folgt. Ist M eine Teilmenge von G, so heißt der Punkt x *Häufungspunkt* von M, wenn eine gegen x konvergente Folge von Punkten von M existiert. Ist F eine Folge (d. h. eine geordnete abzählbare Menge) von Punkten $\{x_i\}$ $(i = 1, 2, \ldots$ ad inf.)

von G, so heißt x *Summenhäufungspunkt* von F oder *Häufungspunkt der Reihe* $\sum_{i=1}^{\infty} x_i$, wenn x Häufungspunkt der Menge der Punkte $\sum_{i=1}^{n} x_i$ ($n=1, 2, \ldots$ ad inf.) ist. Ist M eine abzählbare Teilmenge von G, so heißt x *Summenhäufungspunkt* von M, wenn x für mindestens eine Anordnung von M Summenhäufungspunkt der so geordneten Menge ist. Die Menge aller Summenhäufungspunkte von M heiße die *Summenhäufungsmenge* von M.

Theorem: *Ist M eine abzählbare Teilmenge einer Abelschen topologischen Gruppe G, so ist die Summenhäufungsmenge S von M entweder leer oder eine Nebengruppe einer Untergruppe von G.*

Zum Beweise genügt es offenbar zu zeigen: Wenn x, y und z in S liegen, so auch $x+y-z$. Es seien $\{m_n\}$ ($n=1, 2 \ldots$ ad inf.) die Elemente von M in eine Folge geordnet und $\{i_n\}$, $\{j_n\}$, $\{k_n\}$ Permutationen der Folge der natürlichen Zahlen derart, daß x, y, z Summenhäufungspunkte bzw. der Reihen

$$I=\sum_{n=1}^{\infty} m_{i_n}, \qquad J=\sum_{n=1}^{\infty} m_{j_n}, \qquad K=\sum_{n=1}^{\infty} m_{k_n}$$

sind. Wir nennen I_r, J_r, K_r die Menge der r ersten Summanden von I, J, K, wobei die Summanden mit verschiedenen Indizes als verschieden gelten. Die Summen der Elemente von I_r, J_r, K_r mögen I^*_r, J^*_r, K^*_r heißen. Ferner bezeichnen wir mit $\{V^n_p\}$ ($n=1, 2, \ldots$ ad inf., $p=x, y, z$) eine auf p sich zusammenziehende Folge von Umgebungen. Wir wollen nun eine Reihe $H=\sum_{n=1}^{\infty} m_{g_n}$ konstruieren, für die $u=x+y-z$ Häufungspunkt ist.

Dazu bestimmen wir j_1 derart, daß $J^*_{j_1}$ in V^1_y liegt, und k_1 so, daß $K_{k_1} \supset J_{j_1}$ gilt, d. h. daß die j_1 ersten Summanden von J unter den k_1 ersten Summanden von K vorkommen, und daß $K^*_{k_1}$ in V^1_z liegt. Die Menge $K_{k_1}-J_{j_1}$ der in J_{j_1} nicht vorkommenden Elemente von K_{k_1} (solche mit verschiedenen Indizes als verschieden betrachtet) heiße K'_{k_1}. Wir bestimmen nun i_1 so, daß $I_{i_1} \supset K'_{k_1}$ gilt und $I^*_{i_1}$ in V^1_x liegt. Ist h_1 die Anzahl der Elemente von $I_{i_1}-K'_{k_1}$, so nennen wir diese Menge H_{h_1} und wählen ihre Elemente in irgend einer Reihenfolge als die h_1 ersten Summanden der zu bildenden Reihe H. Sodann bestimmen wir $j_2 > j_1$ derart, daß $J_{j_2} \supset H_{h_1}$ gilt und $J^*_{j_2}$ in V^2_y liegt, k_2 so, daß $K_{k_2} \supset J_{j_2}$ gilt und $K^*_{k_2}$ in V^2_z liegt, setzen $K'_{k_2}=K_{k_2}-J_{j_2}$ und bestimmen i_2 so, daß $I_{i_2} \supset K'_{k_2}$ und $I_{i_2} \supset H_{h_1}$. Ist h_2 die Zahl der Summanden von $I_{i_2}-K'_{k_2}$, so nennen wir diese Menge, die offenbar $\supset H_{h_1}$ ist, H_{h_2} und wählen ihre Glieder, indem wir die in H_{h_1} enthaltenen als die h_1 ersten nehmen, in irgend einer Reihenfolge als die h_2 ersten der zu bildenden Reihe H. So fortfahrend, definieren wir H. Jeder Punkt m_i von M kommt offenbar für fast alle r in

I_{i_r}, J_{j_r}, K_{k_r}, daher in fast allen H_{h_r} vor und man bestätigt weiter mühelos, daß $\lim_{r=\infty} H^*_{h_r} = x + y - z$ gilt, w. z. b. w.

Menger: *Rein imaginäre Räume, indefinite Metriken und verwandte Probleme.*

Nennen wir *imaginären n-dimensionalen euklidischen Raum* die Menge I_n aller n-Tupel von rein imaginären Zahlen, wenn $x_1, \ldots, x_n$ und $y_1, \ldots, y_n$ die Zahl $\sum_{i=1}^{\infty}(x_i - y_i)^2$ als Abstandsquadrat zugeordnet ist, so ergibt sich aus meiner Kennzeichnung der euklidischen Räume (diese Ergebnisse, *1*, S. 21 f.): *Damit eine komplex-metrische Menge M* (vgl. diese Ergebnisse, *2*, S. 36) *mit einer Teilmenge von I_n kongruent sei, ist notwendig und hinreichend, daß, falls M mehr als $n + 3$ Elemente enthält,*

$D(p_1, \ldots, p_{n+2}) = 0$ für je $n + 2$ Elemente und $D(p_1, \ldots, p_k) \leqq 0$ für jedes k $(2 \leqq k \leqq n + 1)$ und je k Elemente von M gilt, und daß, falls M aus genau $n + 3$ Elementen $p_1, \ldots, p_{n+3}$ besteht, überdies $D(p_1, \ldots, p_{n+3}) = 0$ gilt. (Erscheint im Tohôku Journal.)

Wann sind drei Punkte p_1, p_2, p_3 mit den Abständen a, b, c zwischen p_1 und p_2, bzw. p_1 und p_3, bzw. p_2 und p_3, wobei a^2, b^2, c^2 reelle Zahlen sind, einbettbar in den Raum, dessen Punkte Paare reeller Zahlen sind und in dem den Punkten x_1, x_2 und y_1, y_2 das (reelle) Abstandsquadrat $(x_1 - y_1)^2 - (x_2 - y_2)^2$ zugeordnet ist? Sicher ist dies der Fall, wenn $a = b = c = 0$. Andernfalls sei etwa $a \neq 0$. Ist $a^2 > 0$, so bilden wir p_1 auf 0, 0 und p_2 auf a, 0 ab. Ordnen wir dann p_3 den Punkt x, y zu, so müssen, damit die Abbildung kongruent sei, die Beziehungen gelten $(x - a)^2 - y^2 = b^2$ und $x^2 - y^2 = c^2$, also $x = \frac{a^2 - b^2 + c^2}{2a}$. Damit y reell sei, ist wegen $y^2 = x^2 - c^2$ erforderlich, daß $D(p_1, p_2, p_3) = -(a + b + c)(a + b - c)(a - b + c)(-a + b + c) \geqq 0$. Dieselbe Bedingung erweist sich für $a^2 < 0$ als notwendig, wenn wir p_1 auf 0, 0 und p_2 auf 0, ia abbilden. Offenbar ist die Bedingung $D(p_1, p_2, p_3) \geqq 0$ auch hinreichend[1]). Diese Bemerkung legt die Frage nahe nach der metrischen Kennzeichnung des $R_{n,s}$, d. i. die Menge aller n-Tupel reeller Zahlen, für die den Elementen $x_1, \ldots x_n$ und $y_1, \ldots, y_n$ die Zahl $\sum_{i=1}^{n} \varepsilon_i (x_i - y_i)^2$ als Abstandsquadrat zugeordnet ist, wobei $\frac{n+s}{2}$ von den n Zahlen ε_i gleich 1 und $\frac{n-s}{2}$ gleich -1 sind, so daß s die Signatur der quadratischen Maßform ist. $R_{n,n}$ bzw. $R_{n,-n}$ ist offenbar der gewöhnliche, bzw. der imaginäre n-dimensionale euklidische Raum.

[1]) Die Bedingung $D \geqq 0$ ist, wenn die 4 Faktoren von D reell sind, offenbar damit äquivalent, daß entweder mindestens einer $= 0$ ist oder $a + b + c$ dasselbe Vorzeichen hat wie zwei von den drei übrigen Faktoren, während der restliche entgegengesetztes Vorzeichen hat. Ist $a + b + c > 0$, so ist dies gleichbedeutend mit der „Dreieckskontraungleichung“: Eine der Zahlen a, b, c ist größer als die Summe der beiden anderen (vgl. diese Ergebn., *2*, S. 17).

Der n-dimensionale *unitäre Raum*, d. i. die Menge aller n-Tupel von komplexen Zahlen, für welche $x_1, \ldots, x_n$ und $y_1, \ldots, y_n$ die reelle Zahl $(x_1 - y_1)(\bar{x}_1 - \bar{y}_1) + \ldots + (x_n - y_n)(\bar{x}_n - \bar{y}_n)$ als Abstandsquadrat zugeordnet ist, wobei $\bar{z}$ die zu z konjugierte Zahl bezeichnet, ist offenbar mit dem $2n$-dimensionalen euklidischen Raum kongruent.

Problem: Ist der *komplexe Hilbertsche Raum*, d. i. die Menge aller Folgen komplexer Zahlen mit konvergenter Quadratsumme, für welche den Folgen $\{x_i\}$ und $\{y_i\}$ die Zahl $\sum_{i=1}^{\infty} (x_i - y_i)^2$ als Abstandsquadrat zugeordnet ist, ein separabler komplexmetrischer Universalraum (d. h. enthält er zu jedem separabeln komplexmetrischen Raum eine kongruente Teilmenge)?

61. Kolloquium (15. III. 1933).

Wald: Die Räume $R_{n,s}$ (vgl. Menger, 60. Kolloquium) haben die Kongruenzordnung $n+3$. (Vgl. dieses Heft S. 37.)

Alt: Die Funktion von Gross, die über jedem Punkt der Ebene eine Singularität hat.

62. Kolloquium (10. V. 1933).

Flores: *Über die Existenz n dimensionaler Komplexe, die nicht in den R_{2n} topologisch einbettbar sind.*

Dem Menger-Nöbelingschen Einbettungssatz zufolge ist jeder n-dimensionale separable metrische Raum mit einer Teilmenge des R_{2n+1}, d. h. des $(2n+1)$-dimensionalen Cartesischen Raumes, homöomorph. Menger hat die Frage aufgeworfen, ob die Schranke $2n+1$ seines Fundamentalsatzes scharf sei[1]). Für $n=1$ ist die Frage, da es Kurven gibt, die nicht in die Ebene einbettbar sind, wie z. B. Mengers Tetraederkonfiguration[2]), offenbar zu bejahen. Doch war sie bisher schon für $n=2$ unentschieden[3]). Im Folgenden wird eine positive Lösung des Mengerschen Problems für beliebiges n gegeben: *Für jedes n existiert eine n-dimensionale Menge* (sogar ein Komplex), *die nicht mit einer Teilmenge des R_{2n} homöomorph ist.*

1. Eine Teilmenge A des Raumes R heiße *selbstverschlungen* in R, wenn keine stetige Schar stetiger Abbildungen φ_t $(0 \leqq t \leqq 1)$ von A auf Teilmengen $\varphi_t(A)$ von R existiert derart, daß φ_0 die Identität ist, daß $\varphi_t(A)$ für $t \neq 0$ zu A fremd ist und daß $\varphi_1(A)$ aus einem einzigen Punkt besteht. Es gilt: *Eine Teilmenge A einer Hyperebene des R_k ist nicht selbstverschlungen im R_k.* Es sei nämlich $x_k = 0$ die betreffende Hyperebene. Bezeichnet man mit p einen Punkt, für den $x_k = 1$, und für $0 \leqq t \leqq 1$ mit $\varphi_t(A)$ die Projektion von A aus p auf die Hyperebene $x_k = t$, so hat man offenbar eine stetige Schar stetiger Abbildungen von A derart, wie sie im Falle der Selbstverschlungenheit von A im R_k nicht existiert.

---

1) Dimensionstheorie, 1928. Vgl. auch diese Ergebn., *1*, S. 3.

2) Vgl. diese Ergebn., *2*, S. 30f.

3) Menger fragt l. c. 2) speziell, ob das Produkt zweier Tetraederkonfigurationen in den R_4 einbettbar sei.

2. Der Raum A heiße *absolut selbstverschlungen im Raume R*, wenn jede mit A homöomorphe Teilmenge von R selbstverschlungen in R ist. Aus 1. folgt unmittelbar: *Ein im R_k absolut selbstverschlungener Raum A ist nicht mit einer Teilmenge des R_{k-1} homöomorph.*

3. Ist A ein Komplex, d. h. Summe von endlichvielen Simplexen, die zu je zwei als Durchschnitt ein ihnen gemeinsames Randsimplex haben (wobei die leere Menge als das -1-dimensionale Simplex gilt), so bezeichnen wir als *reduziertes Quadrat von A* die Menge B aller derjenigen geordneten Paare von Punkten von A, die in zu einander fremden Simplexen von A liegen. B ist also ein Teilkomplex des Quadrates von A. Ist $b = a_1 \times a_2$ ein Punkt von B, so sind die Faktorpunkte a_1 und a_2, da sie in fremden Simplexen liegen, verschieden.

4. Ist φ eine topologische Abbildung des Komplexes A auf die Teilmenge $\varphi(A)$ des R_k, so bilden wir zu jedem Punkt $b = a_1 \times a_2$ des reduzierten Quadrates B von A den Vektor $\mathfrak{v}_b$, der vom Punkt $\varphi(a_1)$ zum Punkt $\varphi(a_2)$ führt. Da $a_1 \neq a_2$ und φ topologisch ist, gilt $\varphi(a_1) \neq \varphi(a_2)$; also ist $\mathfrak{v}_b$ für keinen Punkt b von B der Nullvektor. Wir tragen $\mathfrak{v}_b$ vom Ursprung des R_k aus ab, bringen die $\mathfrak{v}_b$ tragende Halbgerade mit der Oberfläche der Einheitskugel S_{k-1} des R_k zum Schnitt und nennen den Schnittpunkt $\bar{\varphi}(b)$. Auf diese Weise entspricht also jedem Punkt b von B ein Punkt $\bar{\varphi}(b)$ von S_{k-1}. Die so definierte Abbildung $\bar{\varphi}$ von B auf eine Teilmenge von S_{k-1}, welche offenbar stetig ist, heiße die *zu φ assoziierte Abbildung* des reduzierten Quadrates B von A.

5. Voraussetzungen: 1. *Das Komplement von $\varphi(A)$ zum R_k sei bogenverknüpft* (was z. B. sicher der Fall ist, wenn A ein höchstens $(k-2)$-dimensionaler Komplex ist). 2. *Das reduzierte Quadrat B von A sei ein Zyklus mod. 2.* Behauptung: *Wenn $\varphi(A)$ nicht selbstverschlungen ist, so ist der Abbildungsgrad mod. 2 der zu φ assoziierten Abbildung $\bar{\varphi}$ von B auf S_{k-1} gleich* 0.

Beweis: Da $\varphi(A)$ nicht selbstverschlungen im R_k ist, existiert eine stetige Schar φ_t $(0 \leq t \leq 1)$ von stetigen Abbildungen von A auf Teilmengen $\varphi_t(A)$ des R_k, so daß $\varphi_0(A) = \varphi(A)$, daß $\varphi_t(A)$ für $t \neq 0$ zu $\varphi_0(A)$ fremd ist und daß $\varphi_1(A)$ ein Punkt p_1 ist. $\varphi_0(A)$ ist als topologisches Bild eines Komplexes beschränkt, also in einer Kugel S des R_k enthalten. Es sei p_2 irgend ein Punkt außerhalb von S. Da $R_k - \varphi_0(A)$ bogenverknüpft ist, existiert ein Bogen C zwischen p_1 und p_2, der zu $\varphi_0(A)$ fremd ist. Wir bilden C topologisch auf das Intervall $1 \leq t \leq 2$ ab, und zwar $t = 1$ auf p_1 und bezeichnen mit p_t den Bildpunkt von t auf C. Es sei ferner für jede reelle Zahl t, die ≥ 1 und ≤ 2 ist, φ_t diejenige stetige Abbildung von A, welche A auf den Punkt p_t abbildet. Dadurch liegt insgesamt eine stetige Schar φ_t $(0 \leq t \leq 2)$ von stetigen Abbildungen von A mit folgenden Eigenschaften vor: Es ist $\varphi_0(A) = \varphi(A)$; für $t \neq 0$ ist $\varphi_t(A)$ zu A fremd; $\varphi_2(A)$ ist ein Punkt p_2 außerhalb der $\varphi(A)$

enthaltenden Kugel S. Wir definieren nun für jede Zahl t, für die $0 \leq t \leq 2$ ist, eine stetige Abbildung $\bar{\varphi}_t$ von B auf eine Teilmenge von S_{k-1}, indem wir für jeden Punkt $b = a_1 \times a_2$ von B den Vektor bilden, der vom Punkt $\varphi_0(a_1)$ zu $\varphi_t(a_2)$ führt, sodann diesen Vektor, der für $t=0$ mit $\mathfrak{v}_b$ identisch ist und für $t \neq 0$ wegen der Fremdheit der Mengen $\varphi_0(A)$ und $\varphi_t(A)$ ebenfalls vom Nullvektor verschieden ist, vom Ursprung aus abtragen und die ihn tragende Halbgerade mit S_{k-1} zum Schnitt bringen; der Schnittpunkt sei $\bar{\varphi}_t(b)$. Offenbar ist $\bar{\varphi}_0$ die zu φ_0 assoziierte Abbildung und die $\bar{\varphi}_t$ $(0 \leq t \leq 2)$ bilden eine stetige Schar von stetigen Abbildungen des Zyklus B (Voraussetzung 2) auf Teilmengen von S_{k-1}. Sie alle haben also den gleichen Abbildungsgrad mod. 2 wie die zu φ assoziierte Abbildung $\bar{\varphi}_0$. Insbesondere gilt dies von $\bar{\varphi}_2$. Der Abbildungsgrad mod. 2 von $\bar{\varphi}_2$ ist aber 0. Denn wenn $\bar{\varphi}_2(b)$ der irgend einem Punkt b von B entsprechende Punkte von S_{k-1} ist, so ist der $\bar{\varphi}_2(b)$ diametral gegenüberliegende Punkt von S_{k-1} nicht Bild eines Punktes von B bei der Abbildung $\bar{\varphi}_2$. Sonst gäbe es ja zwei entgegengesetzt gerichtete Vektoren, deren Anfangspunkte in $\varphi_0(A)$ liegen und deren Endpunkte mit p_2 zusammenfallen, was unmöglich ist, da $\varphi_0(A)$ in S und p_2 außerhalb von S liegt.

6. Es sei T eine aus $3(n+1)$ Punkten bestehende Teilmenge des R_{2n+1} mit der Eigenschaft, daß je zwei höchstens n-dimensionale Simplexe, deren Ecken in T liegen, die aber keine gemeinsame Ecke haben, fremd sind. Wir stellen T dar als Summe von $n+1$ Tripeln T_ν von Punkten $1_\nu, 2_\nu, 3_\nu$ $(\nu = 0, 1, \ldots, n)$ und bezeichnen[4]) mit M_n die Summe aller jener n-dimensionalen Simplexe, die mit jedem der $n+1$ Tripel genau einen Eckpunkt gemein haben, von denen jedes also in der Gestalt $(i_0, i_1, \ldots, i_n)$ dargestellt werden kann, wo jedes i_ν eine der Zahlen $1_\nu, 2_\nu, 3_\nu$ ist $(\nu = 0, 1, \ldots, n)$. Es ist M_0 das Punktetripel $1_0, 2_0, 3_0$ und M_1 homöomorph mit der Mengerschen Tetraederkonfiguration (diese Ergebn., 2, S. 30). Wir behaupten: *M_n ist nicht homöomorph mit einer Teilmenge des R_{2n}.* Zum Beweise genügt es nach **2** und **5** zu zeigen, 1. daß das reduzierte Quadrat A_{2n} von M_n ein Zyklus mod. 2 ist und 2., daß für jede topologische Abbildung φ von M_n in den R_{2n+1} der Abbildungsgrad mod. 2 der assoziierten Abbildung $\bar{\varphi}$ von A_{2n} auf S_{2n} gleich 1 ist, was in **7**, bzw. **9—11** geschehen wird.

7. Daß A_{2n} ein Zyklus mod. 2 ist, folgt daraus, daß an jeder $(2n—1)$-dimensionalen Randzelle von A_{2n} genau zwei 2_n-dimensionale Zellen zusammenstoßen; z. B. stoßen in der Randzelle $(i_1, i_2, \ldots, i_n)$

---

[4]) Nach Fertigstellung der obigen Arbeit fand ich bei van Kampen (Abh. d. Hamb. math. Sem., *9*, S. 72) den obigen Komplex M_n definiert und bewiesen, daß er nicht *simplizial* in R_{2n} eingebettet werden kann. Übrigens ist mir seither gelungen, die Mengersche Vermutung (l. c. [2]), daß das Quadrat der Tetraederkonfiguration nicht mit einer Teilmenge des R_4 homöomorph sei, zu beweisen.

2*

$\times(j_0, j_1, \ldots, j_n)$ nur $(i_0, i_1, \ldots, i_n)\times(j_0, j_1, \ldots, j_n)$ zusammen, wo i_0 die beiden von j_0 verschiedenen Punkte des Tripels T_0 durchläuft.

8. Hilfssatz. Voraussetzungen: $I=(i_0, i_1, \ldots, i_n)$ *sei ein Simplex von* M_n; φ *und* ψ *seien zwei bzgl. einer Unterteilung simpliziale topologische Abbildungen von* M_n *in den* R_{2n+1}, *die auf* $M_n - I$ *identisch sind und das Innere von I auf zwei fremde Mengen abbilden.* Behauptung: *Die Abbildungsgrade mod. 2 der assoziierten Abbildungen* $\bar{\varphi}$ (A_{2n}) *und* $\bar{\psi}$ (A_{2n}) *von* A_{2n} *auf* S_{2n} *sind gleich.*

Beweis: Verbindet man jeden Punkt m einer zu S_k homöomorphen Teilmenge M eines euklidischen Raumes mit zwei festen nicht in M liegenden Punkten p und q durch eine Strecke mp bzw. mq und sind dabei mp und mq bis auf den Punkt m fremd und für zwei verschiedene Punkte m und $\bar{m}$ von M die Bogen $mp+mq$ und $\bar{m}p+\bar{m}q$ bis auf die Punkte p und q fremd, so ist die Summe aller dieser Strecken zu S_{k+1} homöomorph. Hieraus folgt leicht durch vollständige Induktion nach n, daß die Summe S aller zu I fremder Simplexe $(j_0, j_1, \ldots, j_n)$ von M_n zu S_n homöomorph ist. Offenbar enthält das Produkt $I\times S$ bzw. $S\times I$ alle und nur diejenigen Produkte von Paaren fremder Simplexe von M_n, deren erster bzw. zweiter Faktor mit I übereinstimmt. Bezeichnen wir mit α die Differenz der Abbildungsgrade von $\bar{\varphi}$ (A_{2n}) und $\bar{\psi}$ (A_{2n}) auf S_{2n} und für einen Punkt p von S_{2n}, welcher weder im $\bar{\varphi}$-Bild noch im $\bar{\psi}$-Bild des Randes von $I\times S+S\times I$ liegt, mit $\alpha_{\bar{\varphi}(I\times S)}$, $\alpha_{\bar{\varphi}(S\times I)}$, $\alpha_{\bar{\psi}(I\times S)}$, $\alpha_{\bar{\psi}(S\times I)}$ den Abbildungsgrad mod. 2 von $I\times S$ bzw. $S\times I$ in p bzgl. der Abbildungen $\bar{\varphi}$ und $\bar{\psi}$, so gilt mit Rücksicht auf die Voraussetzung, daß $\bar{\varphi}$ und $\bar{\psi}$ außerhalb von $I\times S+S\times I$ identisch sind,

$$\alpha \equiv \alpha_{\bar{\varphi}(I\times S)} + \alpha_{\bar{\varphi}(S\times I)} + \alpha_{\bar{\psi}(I\times S)} + \alpha_{\bar{\psi}(S\times I)} \pmod{2}.$$

Bezeichnen wir nun mit Φ bzw. Ψ die Komplexe $\varphi(I)$ bzw. $\psi(I)$, dann erzeugen die Umkehrungen φ^{-1} und ψ^{-1} topologische simpliziale Abbildungen

$$\Phi\times S \rightarrow I\times S,\quad \Psi\times S \rightarrow I\times S,\quad S\times\Phi \rightarrow S\times I,\quad S\times\Psi \rightarrow S\times I.$$

Schalten wir nun $\bar{\varphi}$ hinter die erste und dritte von diesen Abbildungen und $\bar{\psi}$ hinter die zweite und vierte, so entstehen vier Abbildungen[5]) $\chi(\Phi\times S)$, $\chi(\Psi\times S)$, $\omega(S+\Phi)$, $\omega(S\times\Psi)$ auf Teilmengen von S_{2n}, für welche

$$\alpha_{\bar{\varphi}(I\times S)} = \alpha_{\chi(\Phi\times S)},\quad \alpha_{\bar{\varphi}(S\times I)} = \alpha_{\omega(S\times\Phi)},\quad \alpha_{\bar{\psi}(I\times S)} = \alpha_{\chi(\Psi\times S)},$$
$$\alpha_{\bar{\psi}(S\times I)} = \alpha_{\omega(S\times\Psi)}.$$

Also gilt

$$\alpha \equiv \alpha_{\chi(\Phi\times S)} + \alpha_{\chi(\Psi\times S)} + \alpha_{\omega(S\times\Phi)} + \alpha_{\omega(S\times\Psi)} \pmod{2}.$$

[5]) Die Bezeichnung der ersten und letzten beiden Abbildungen mit gleichen Buchstaben ist deshalb gerechtfertigt, da Φ und Ψ nur Randpunkte gemein haben und umgekehrt für jeden Randpunkt von I die Bilder $\varphi(p)$ und $\psi(p)$ gleich sind.

Nun ist $\Phi + \Psi$ ein Zyklus Z (eine topologische Sphäre) und die Abbildungen $\chi\,(\Phi \times S)$ und $\chi\,(\Psi \times S)$, bzw. $\omega\,(S \times \Phi)$ und $\omega\,(S \times \Psi)$ schließen sich zu Abbildungen $\chi\,(Z \times S)$ bzw. $\omega\,(S \times Z)$ zusammen[5]), für welche

$$\alpha_{\chi\,(Z \times S)} = \alpha_{\chi\,(\Phi \times S)} + \alpha_{\chi\,(\Psi \times S)} \text{ und } \alpha_{\omega\,(S \times Z)} = \alpha_{\omega\,(S \times \Phi)} + \alpha_{\omega\,(S \times \Psi)}, \text{ also}$$

$$\alpha \equiv \alpha_{\chi\,(Z \times S)} + \alpha_{\omega\,(S \times Z)} \pmod{2}$$

gilt. Nach Definition der zu einer Abbildung φ von M_n^* assoziierten Abbildung $\bar{\varphi}$ von A_{2n} auf S_{2n} geht $\bar{\varphi}\,(I \times S)$ in $\bar{\varphi}\,(S \times I)$ und $\bar{\psi}\,(I \times S)$ in $\bar{\psi}\,(S \times I)$ über durch Spiegelung am Mittelpunkt, d. h. für jeden Punkt $i \times s$ von $I \times S$ sind $\bar{\varphi}\,(i \times s)$ und $\bar{\varphi}\,(s \times i)$ diametral entgegengesetzt und ebenso $\bar{\psi}\,(i \times s)$ und $\bar{\psi}\,(s \times i)$. Folglich gehen auch $\chi\,(Z \times S)$ und $\omega\,(S \times Z)$ durch Spiegelung am Mittelpunkt von S_{2n} auseinander hervor und daher ist der Grad $\alpha_{\chi\,(Z \times S)}$ von $\chi\,(Z \times S)$ im Punkt p gleich dem Grad $\alpha'_{\omega\,(S \times Z)}$ von $\omega\,(S \times Z)$ in dem zu p diametral entgegengesetzten Punkt p'. Nun ist $S \times Z$ ein Zyklus mod. 2. Also hat die Abbildung $\omega\,(S \times Z)$ in allen Punkten von S_{2n} denselben Grad. Mithin gilt

$$\alpha_{\omega\,(S \times Z)} = \alpha'_{\omega\,(S \times Z)} \text{ und } \alpha \equiv 2\alpha_{\omega\,(S \times Z)} \equiv 0 \pmod{2}, \text{ w. z. b. w.}$$

9. Nun sei φ irgend eine topologische Abbildung von M_n in den R_{2n+1}. Die Bilder je zweier fremder Simplexe von M_n haben einen positiven Abstand. Daher existiert eine positive untere Schranke für die Längen der Vektoren, die in der Definition der zu φ assoziierten Abbildung $\bar{\varphi}$ auftreten, so daß bei hinreichend kleiner Modifikation von φ deren Richtungen sich beliebig wenig ändern und insbesondere kein Vektor in einen entgegengesetzt gerichteten übergeht. Die zu einer aus φ durch hinreichend kleine Modifikation hervorgehenden Abbildung assoziierte Abbildung hat also denselben Abbildungsgrad mod. 2 wie $\bar{\varphi}$. Da M_n ein n-dimensionaler Komplex ist und jede topologische Abbildung von M_n in den R_{2n+1} daher durch beliebig kleine Modifikation in eine topologische simpliziale Abbildung übergeführt werden kann, so dürfen wir, um zu zeigen, daß $\bar{\psi}$ den Abbildungsgrad 1 mod. 2 hat, ohne Beschränkung der Allgemeinheit annehmen, daß φ eine topologische *simpliziale* Abbildung ist.

10. Sind φ_1 und φ_2 zwei topologische simpliziale Abbildungen von M_n in den R_{2n+1}, so kann man $\varphi_2\,(M_n)$ im R_{2n+1} so deformieren, daß die entstehende topologische simpliziale Abbildung φ'_2 auf jedem Randsimplex $(i_0, \ldots, i_{k-1}, i_{k+1}, \ldots, i_n)$ von M_n mit φ_1 identisch ist und daß die Überführung von φ_2 in φ'_2 stückweise simplizial und ohne Singularitäten vor sich geht[6]), so daß gleichzeitig die assoziierten

[6]) Ursprünglich habe ich φ_1 und φ_2 ohne Bezugnahme auf diesen von van Kampen l. c. [2]) stammenden Satz verglichen. Übrigens kann man für jede Einbettung von M_n in den R_{2n+1} den Abbildungsgrad mod. 2 von $\bar{\varphi}\,(A_{2n})$ direkt, d. h. ohne Berufung auf die spezielle Einbettung von **11** berechnen, wie ich demnächst zeigen werde.

Abbildungen $\bar{\varphi}_2$ und $\bar{\varphi}'_2$ stetig ineinander übergehen und mithin denselben Abbildungsgrad mod. 2 besitzen. Durch mehrfache Anwendung des Hilfssatzes 8 folgt, daß $\bar{\varphi}_1$ und $\bar{\varphi}'_2$, also $\bar{\varphi}_1$ und $\bar{\varphi}_2$ denselben Abbildungsgrad mod. 2 haben. Um zu zeigen, daß für jede topologische Abbildung von M_n in den R_{2n+1} der Abbildungsgrad mod. 2 der assoziierten Abbildung 1 ist, genügt es, *eine* topologische simpliziale Abbildung φ von M_n in den R_{2n+1} anzugeben, für deren assoziierte Abbildung $\bar{\varphi}$ der Abbildungsgrad mod. 2 gleich 1 ist.

11. Dies geschieht durch vollständige Induktion. Man bestätigt leicht, daß für jede eineindeutige Abbildung der dreipunktigen Menge M_0 in den R_1 der Abbildungsgrad mod. 2 der assoziierten Abbildung gleich 1 ist. Wir machen die Induktionsvoraussetzung, daß φ_0 eine topologische simpliziale Abbildung[7]) des Komplexes M_{n-1} in den R_{2n-1} ist, so daß der Abbildungsgrad mod. 2 der assoziierten Abbildung $\bar{\varphi}_0$ gleich 1 ist. Es sei etwa R_{2n-1} der durch die Gleichungen $x_{2n} = x_{2n+1} = 0$ bestimmte lineare Teilraum des R_{2n+1}. Aus der Induktionsvoraussetzung folgt die Existenz eines Punktes S der Sphäre $x_1^2 + \ldots + x_{2n-1}^2 = 1$ des R_{2n-1}, welche $\bar{\varphi}_0$-Bild einer ungeraden Anzahl $2k+1$ von Punkten $p_1^* \times q_1^*, \ldots, p_{2k+1}^* \times q_{2k+1}^*$ des reduzierten Quadrates A_{2n-2} von M_{n-1} ist, und zwar von solchen Punkten, die nicht auf einer Randzelle von A_{2n-2} liegen, deren Faktoren p_i^*, q_i^* also innere Punkte fremder $(n-1)$-dimensionaler Simplexe von F_{n-1} sind. Nach Definition der assoziierten Abbildungen $\bar{\varphi}_0$ heißt dies Folgendes: Ist G die gerichtete Gerade durch den Ursprung des Koordinatensystems und den Punkt S, so sind die Paare p_i^*, q_i^* $(i = 1, 2, \ldots, 2k+1)$ die einzigen Paare p, q von Punkten aus M_{n-1} derart, daß die von $\varphi_0(p)$ nach $\varphi_0(q)$ führende gerichtete Strecke zu G parallel und gleichgerichtet ist. Daraus folgt: verschiebt man den Komplex $\varphi_0(M_{n-1})$ in beiden Richtungen parallel zu G, so durchdringen sich die beiden laufenden Komplexe $2k+1$mal, und zwar in den Punkten $\varphi_0(p_i^*)$, $\varphi_0(q_i^*)$. Genauer: es seien: $\varphi^*_{\frac{1}{2}}(M_{n-1})$ und $\varphi^*_{-\frac{1}{2}}(M_{n-1})$ zwei aus $\varphi_0(M_{n-1})$ durch Parallelverschiebung parallel zu G hervorgehende Komplexe, die durch eine Hyperebene H des Raumes R_{2n-1} getrennt werden; bezeichnet man nun für einen beliebigen Punkt p von M_{n-1} und beliebiges der Ungleichung $0 \leqq t \leqq \frac{1}{2}$ genügendes t mit $\varphi_t^*(p)$ bzw. $\varphi_{-t}^*(p)$ den Punkt, der die Strecke mit den Endpunkten $\varphi_0(p)$ und $\varphi^*_{\frac{1}{2}}(p)$ bzw. $\varphi_0(p)$ und $\varphi^*_{-\frac{1}{2}}(p)$ im Verhältnis $t:(\frac{1}{2}-t)$ teilt, so hat die Gleichung $\varphi_t^*(p) = \varphi_{-t}^*(q)$ mit $t > 0$ genau $2k+1$ Lösungen t, p, q, und zwar sind die Punktepaare p, q gerade die Paare p_i^*, q_i^*. Nun zeigt man leicht die Existenz eines $\varepsilon > 0$ mit folgender Eigenschaft:

[7]) Gemeint ist von nun an stets, daß jedes Simplex des Komplexes selbst, nicht etwa nur einer Unterteilung auf ein Simplex abgebildet wird.

bezeichnet man mit $\varphi_{\frac{1}{2}}(M_{n-1})$ und $\varphi_{-\frac{1}{2}}(M_{n-1})$ zwei aus $\varphi^*_{\frac{1}{2}}(M_{n-1})$ bzw. $\varphi^*_{-\frac{1}{2}}(M_{n-1})$ durch Eckpunktsverrückungen $< \varepsilon$ hervorgehende Komplexe und definiert man die topologischen simplizialen Abbildungen φ_t analog wie oben die φ^*_t, so hat die Gleichung $\varphi_t(p) = \varphi_{-t}(q)$ mit $t > 0$ genau $2k+1$ Lösungen t_i, p_i, q_i $(i = 1, 2, \ldots, 2k+1)$, und zwar sind die Punkte p_i und q_i, falls ε hinreichend klein ist, den Punkten p^*_i bzw. q^*_i beliebig benachbart, also ebenfalls innere Punkte fremder $(n-1)$-dimensionaler Simplexe von M_{n-1}.

Auf der Geraden G wählen wir zwei durch die Hyperebene H getrennte Punkte P und Q (es möge etwa P mit $\varphi_{\frac{1}{2}}(M_{n-1})$ und Q mit $\varphi_{-\frac{1}{2}}(M_{n-1})$ auf derselben Seite von H liegen) so weit von H entfernt, daß für jeden Eckpunkt und damit für jeden Punkt von M_{n-1} das $\varphi^*_{\frac{1}{2}}$- bzw. $\varphi^*_{-\frac{1}{2}}$-Bild einen Abstand $< \varepsilon$ hat von der Verbindungsstrecke seines φ_0-Bildes mit P bzw. Q. Bezeichnet man für jeden Punkt p von M_{n-1} und jedes der Beziehung $0 \leq t \leq 1$ genügende t mit $\varphi_t(p)$ bzw. $\varphi_{-t}(p)$ den Punkt, der die Strecke mit den Endpunkten $\varphi_0(p)$ und P bzw. $\varphi_0(p)$ und Q im Verhältnis $t:(1-t)$ teilt, so ist φ_t für jedes t mit $-1 < t < 1$ eine topologische simpliziale Abbildung von M_{n-1}, und da für $\frac{1}{2} \leq t \leq 1$ die Komplexe $\varphi_t(M_{n-1})$ und $\varphi_{-t}(M_{n-1})$ durch H getrennt, also fremd sind, so folgt aus der Definition von ε, daß die Gleichung $\varphi_t(p) = \varphi_{-t}(q)$, wo $t > 0$, genau 2_{k+1} Lösungen t_i, p_i, q_i $(i = 1, 2, \ldots, 2k+1)$ hat, wobei für jedes i die Punkte p_i und q_i innere Punkte zweier fremder $(n-1)$-dimensionaler Simplexe von M_{n-1} sind und $t_i > 0$ ist.

Wir definieren nun die gesuchte Abbildung $\varphi(M_n)$ folgendermaßen: Für alle Punkte p, die in dem durch die Eckpunkte $1_0, 2_0, 3_0, \ldots; 1_{n-1}, 2_{n-1}, 3_{n-1}$ aufgespannten Komplex M_{n-1} liegen, setzen wir $\varphi(p) = \varphi_0(p)$. Dann ist also $\varphi(M_{n-1}) \subset R_{2n-1} \subset R_{2n+1}$. Wir ordnen dem Punkte 1_n als Bild $\varphi(1_n)$ einen Punkt des R_{2n+1} zu, dessen letzte Koordinate $= 0$, dessen vorletzte Koordinate < 0 ist. Haben die Punkte P und $Q \subset R_{2n-1}$ die Koordinaten $x^P_1, \ldots, x^P_{2n-1}$ bzw. $x^Q_1, \ldots, x^Q_{2n-1}$, so ordnen wir den Punkten 2_n und 3_n die Punkte mit den Koordinaten $x^P_1, \ldots, x^P_{2n-1}, 1, 0$ bzw. $x^Q_1, \ldots, x^Q_{2n-1}, 1, 1$ zu. Man bestätigt leicht, daß die dadurch definierte simpliziale Abbildung $\varphi(M_n)$ topologisch ist. Um zu beweisen, daß der Abbildungsgrad mod. 2 der assoziierten Abbildung $\overline{\varphi}$ gleich 1 ist, genügt es, zu zeigen, daß die Anzahl der Punkte p, q, deren Bilder $\varphi(p)$ und $\varphi(q)$ eine zur x_{2n+1}-Achse parallele Verbindungsstrecke haben, ungerade ist und je zwei solche Punkte p, q innere Punkte zweier fremder n-dimensionaler Simplexe von M_n sind. Da die $2n$-te Koordinate von $\varphi(1_n)$ negativ ist, dagegen die $2n$-ten Koordinaten von $\varphi(2_n)$ und $\varphi(3_n)$ positiv sind, können die gesuchten Punkte p und q nur in Simplexen von M_n liegen, die den Punkt 2_n oder 3_n zur Ecke haben. Die Abbildung φ läßt sich nun auch folgendermaßen definieren. Ist p ein beliebiger Punkt eines Simplexes $(i_0, \ldots, i_{n-1}, 2_n)$, so bestimmt p eindeutig einen Punkt p' von $(i_0, \ldots, i_{n-1}) \subset M_{n-1}$ und eine Zahl s mit $0 \leqq s \leqq 1$, so daß p die Strecke mit den Endpunkten p' und 2_n im Verhältnis $s:(1-s)$ teilt; sind nun $x_1, \ldots, x_{2n-1}$ die Koordinaten des Punktes $\varphi_s(p') \subset R_{2n-1}$, so hat $\varphi(p)$ die Koordinaten $x_1, \ldots, x_{2n-1}, s, 0$; analog hat, wenn der Punkt q des Simplexes $(i_0, \ldots, i_{n-1}, 3_n)$ die Strecke mit den Endpunkten $q' \subset (i_0, \ldots, i_{n-1})$ und 3_n im Verhältnis $t:(1-t)$ teilt und $\varphi_{-t}(q')$ die Koordinaten $x_1, \ldots, x_{2n-1}$ hat, der Punkt $\varphi(q)$ die Koordinaten $x_1, \ldots, x_{2n-1}, t, t$. Damit also für zwei Punkte p und q von M_n die Verbindungsstrecke von $\varphi(p)$ und $\varphi(q)$ parallel zur x_{2n+1}-Achse sei, ist notwendig und hinreichend, daß $s=t>0$ und $\varphi_s(p')=\varphi_{-t}(q')$ sei. Nun hat die Gleichung $\varphi_t(p')=\varphi_{-t}(q')$ mit $t>0$ genau $2k+1$ Lösungen t, p', q'; dabei sind p' und q' innere Punkte zweier fremder $(n-1)$-dimensionaler Simplexe von M_{n-1} und t ist <1, so daß die zugehörigen $2k+1$ Punkte p und q innere Punkte fremder n-dimensionaler Simplexe von M_n sind. Damit ist alles bewiesen.

12. Im Vorangehenden wurde der Kürze halber der Abbildungsgrad von $\varphi(A_{2n})$ bloß mod. 2 betrachtet. Da A_{2n}, wie man leicht beweisen kann, nicht nur ein Zyklus mod. 2, sondern ein gewöhnlicher Zyklus (übrigens sogar eine Mannigfaltigkeit) ist, so kann man auch mit Orientierung arbeiten. Weitere Ergebnisse aus dem Problemenkreis der topologischen Einbettungen werden den Gegenstand weiterer Mitteilungen bilden.

63. Kolloquium (17. V. 1933).

Čech (Brünn): *Über einen kurventheoretischen Satz von Ayres.* (Nach brieflicher Mitteilung.)

Für einen metrischen Raum K sei K_ν^n die Menge derjenigen Punkte von K, die eine Umgebung U mit $\delta(U)<\frac{1}{\nu}$ und mit höchstens n-punktigem $K.B(U)$ besitzen[1]). Der Durchschnitt $\prod_{\nu=1}^{\infty} K_\nu^n$ ist die Menge K^n aller Punkte höchstens n-ter Ordnung von K. Ayres bewies folgenden Satz[2]): *Für jeden metrischen Raum K und für $n=3, 4, \ldots$ ist die Menge $K^{2n-3}-K^{n-1}$ nulldimensional*[3]) *oder leer.* Der Ayressche Beweis wurde von Menger in seinem Buche „*Kurventheorie*" (1932), S. 115, ausführlich dargestellt. Im Folgenden möchte ich darauf hinweisen, daß im Falle eines separablen Raumes K unter Verwendung des Summensatzes der Dimensionstheorie für separable Räume der ganze Beweis schon mit Hilfe der von Menger zum Beweise eines Hilfssatzes (Kurventheorie, S. 117), verwendeten Methode ohne die bei Menger folgende rekursive Konstruktion geführt

[1]) $\delta(U)$ bedeutet den Durchmesser, $B(U)$ die Begrenzung von U.
[2]) Trans. Amer. Math. Soc., *33* (1931), S. 253—256.
[3]) Im Menger-Urysohnschen Sinn.

werden kann. Da die Menge K_ν^{n-1} in K offen ist, ist $K^{2n-3} - K_\nu^{n-1}$ ($\nu = 1, 2, \ldots$ ad inf.) in $K^{2n-3} - K^{n-1} = \sum_{\nu=1}^{\infty} (K^{2n-3} - K_\nu^{n-1})$ abgeschlossen. Nach dem Summensatz der Dimensionstheorie[4]) hat man also nur zu beweisen, daß $K^{2n-3} - K_\nu^{n-1}$ für jedes ν nulldimensional oder leer ist. Sei also p ein Punkt von K^{2n-3} und sei Z_0 eine vorgelegte Umgebung von p. Man hat eine Umgebung Z' von p mit den Eigenschaften $Z' \subset Z_0$ und $K^{2n-3} \cdot B(Z') \subset K_\nu^{n-1}$ anzugeben. Es gibt eine Umgebung $Z \Subset Z_0$ von p, deren Begrenzung mit K einen endlichen (sogar höchstens $(2n-3)$-punktigen) Durchschnitt hat. Falls $K^{2n-3} \cdot B(Z)$ leer ist, setzen wir $Z' = Z$. Sonst seien q_j $(1 \leqq j \leqq c)$ die endlich vielen Punkte von $K^{2n-3} \cdot B(Z)$. Wie bei Menger (l. c. S. 117) bestimmen wir Umgebungen Q_j von q_j mit den Eigenschaften 1*), 2*), 3*), 4*), wobei jedoch 1*) jetzt aussagt, daß p nicht in $\overline{Q}_j$ liegt und in 2*) $\varepsilon = \frac{1}{\nu}$ zu setzen ist. Man hat dann (l. c., S. 118)

$$\delta\,(V_{h_j}) < \frac{1}{\nu},\quad \delta\,(W_{i_j}) < \frac{1}{\nu}$$ und die Mengen $K \cdot B(V_{h_j})$, $K \cdot B(W_{i_j})$ sind höchstens $(n-1)$-punktig; folglich ist $K \cdot V_{h_j} \subset K_\nu^{n-1}$, $K \cdot W_{i_j} \subset K_\nu^{n-1}$. Wir definieren Z' wie l. c., S. 118 (**). Dann ist $(p) \subset Z' \subset Z_0$ und (l. c., S. 119, Zeile 17) $K^{2n-3} \cdot B(Z') \subset \sum_{j=1}^{a} K \cdot V_{h_j} + \sum_{j=1}^{b} K \cdot W_{i_j} \subset K_\nu^{n-1}$.

Gödel, Menger, Wald: *Diskussion über koordinatenlose Differentialgeometrie.*

Um die Aussage, daß die Riemannschen Räume sich im Kleinen wie euklidische Räume verhalten (daß „für unendlich benachbarte Punkte Riemannscher Räume die euklidische Geometrie gelte") im Sinne des von Menger (Math. Ann., *103*, vgl. auch diese Ergebn., *2*, S. 22) entwickelten Programmes einer koordinatenlosen Differentialgeometrie zu präzisieren, wären die zur metrischen Kennzeichnung der euklidischen Räume dienenden Determinanten („Volumsdeterminanten") geeignet. Setzen wir für vier Punkte p_1, p_2, p_3, p_4 eines metrischen Raumes

$$D\,(p_1, p_2, p_3, p_4) = \begin{vmatrix} 0 & 1 \\ 1 & (p_i\,p_j)^2 \end{vmatrix} \quad (i, j = 1, 2, 3, 4)$$

wo $p_i p_j$ den Abstand von p_i und p_j bezeichnet, so ist die Ebene unter den metrischen Räumen nach Menger (Math. Ann., *100*, S. 113 und diese Ergebn., *1*, S. 20 f.) dadurch gekennzeichnet, daß sie vollständig, konvex und konvex nach außen ist und daß für je vier ihrer Punkte $D\,(p_1, p_2, p_3, p_4) = 0$ gilt. Zur Kennzeichnung Gaußscher Flächen müßte statt dessen die Tatsache herangezogen werden, daß für Punktequadrupel, die gegen einen Punkt konvergieren, der Wert

[4]) Die Voraussetzung der Separabilität brauche ich, da sonst die Giltigkeit des Summensatzes nicht feststeht.

von D *stark* (d. h. so wie gewisse Potenzen der Abstände) *gegen 0 konvergiert.* Bezeichnen wir mit $S_n(p_1, p_2, p_3, p_4)$ die Summe der n-ten Potenzen der 6 Abstände $p_i p_j$, so gilt folgender Satz: Ist F eine Fläche des R_3, die in Parameterdarstellung durch zweimal stetig differenzierbare Funktionen $x = x(u, v)$, $y = y(u, v)$, $z = z(u, v)$ gegeben und entspricht der Punkt p_0 Parameterwerten u_0, v_0, für welche eine der Funktionaldeterminanten, etwa $\frac{\partial(x, y)}{\partial(u, v)}$, nicht verschwindet, so besteht, wenn $\{p_1^n\}$, $\{p_2^n\}$, $\{p_3^n\}$, $\{p_4^n\}$ $(n = 1, 2, \ldots$ ad inf.) irgend vier gegen p_0 konvergente Folgen von Punkten von F sind, die Beziehung

$$\lim_{n=\infty} \frac{D(p_1^n, p_2^n, p_3^n, p_4^n)}{S_6(p_1^n, p_2^n, p_3^n, p_4^n)} = 0.$$

64. Kolloquium (26. V. 1933).

Borsuk (Warschau): *Zur topologischen Charakterisierung der euklidischen Sphären.*

Ist E irgend eine topologische Eigenschaft, so wollen wir sagen, der metrische Raum M habe die Eigenschaft $J(E)$, wenn jeder Punkt von M in beliebig kleinen Umgebungen enthalten ist, deren Komplemente zu M die Eigenschaft E besitzen. Ist E beispielsweise die Kompaktheit, so ist $J(E)$ ebenfalls die Kompaktheit. Ist E die Eigenschaft Streckenbild zu sein, so hat ein Raum R die Eigenschaft $J(E)$ dann und nur dann, wenn R ein keinen Zerschneidungspunkt enthaltendes Streckenbild oder zweipunktig ist.

Die Menge P des metrischen Raumes M heißt in M *zusammenziehbar,* wenn eine stetige Funktion $\varphi(x, t)$ existiert, so daß $\varphi(x, 0) = x$, $\varphi(x, t) \varepsilon M$ und $\varphi(x, 1) =$ const. für jedes $x \varepsilon P$ und $0 \leqq t \leqq 1$ gilt. Der Raum M heißt *lokal zusammenziehbar*, wenn zu jeder Umgebung U jedes Punktes p von M eine in U zusammenziehbare Umgebung V von p existiert. Die kompakten, lokal zusammenziehbaren und in sich zusammenziehbaren Räume heißen *absolute Retrakte* (vgl. Borsuk, Fund. Math., *17*, S. 159 f.). Solche Mengen haben einen ziemlich einfachen Bau im Großen und im Kleinen. Alle ihre Bettischen Gruppen (außer der nulldimensionalen) verschwinden, ebenso ihre Fundamentalgruppe. Doch gibt es unter den absoluten Retrakten immer noch Mengen mit ziemlich paradoxen Eigenschaften: Z. B. läßt sich leicht ein zweidimensionaler Retrakt konstruieren, welcher kein topologisches Bild der Kreisscheibe enthält. Die Eigenschaft $J(R)$ dagegen, wo R die Eigenschaft, ein absoluter Retrakt zu sein, bedeutet, führt vermutlich auf Räume von ganz regelmäßigem Bau. Ich halte es selbst nicht für ausgeschlossen, daß diese Räume mit den euklidischen Sphären S_n homöomorph sind. Jedenfalls haben die S_n die Eigenschaft $J(R)$ und sind für $n = 2$ durch diese Eigenschaft unter den Flächen (= zweidimensionalen Kontinua) gekennzeichnet. Nennen wir Σ_n die n-dimensionalen metrischen Räume mit der Eigenschaft $J(R)$, so wird ein Σ_n von einer abgeschlossenen echten Teilmenge B dann und nur dann zerschnitten, wenn B einen $(n-1)$-dimensionalen wahren Zyklus enthält. Ein Polyeder, das ein Σ_n ist, ist eine kombinatorische Sphäre im Sinne von Alexander und van Kampen, aber nicht umgekehrt. Denn Poincarés dreidimensionale Mannigfaltigkeit mit nicht verschwindender Fundamentalgruppe, die in allen Homologieeigenschaften mit der S_3 übereinstimmt, ist keine Σ_3, da die Fundamentalgruppe jeder Σ_n verschwindet. Für Polyeder ist also die Eigenschaft, Σ_n zu sein, stärker, als eine kombinatorische Sphäre zu sein.

65. Kolloquium (1. VI. 1933).

Menger: *Neuer Aufbau der Vektoralgebra.*

Damit ein halbmetrischer Raum R, d. h. eine Menge, in der je zwei Elementen („Punkten") p_i und p_j eine Zahl $p_{ij}=p_{ji}>0$ für $p_i \neq p_j$ und $p_{ij}=0$ zugeordnet ist, mit einer Teilmenge des n-dimensionalen euklidischen Raumes R_n kongruent sei, ist bekanntlich (vgl. Math. Ann., *100* und Amer. Journ., *53*, S. 721) notwendig und hinreichend, daß, wenn wir für k Punkte $p_1, \ldots, p_k$ von R mit $D(p_1, \ldots, p_k)$ die $(k+1)$-reihige Determinante $\begin{vmatrix} 0 & 1 \\ 1 & p_{ij}^2 \end{vmatrix}_{(i,j=1,\ldots,k)}$ bezeichnen, folgende Bedingungen erfüllt sind: Für je $n+2$ Punkte $p_1, \ldots, p_{n+2}$ von R ist $D(p_1, \ldots, p_{n+2})=0$ und für jedes $k \leq n+1$ und für je k Punkte $p_1, \ldots, p_k$ von R ist $\operatorname{sgn} D(p_1, \ldots, p_k)= \operatorname{sgn}(-1)^k$ oder 0. Im Falle, daß R aus genau $n+3$ Punkten $p_1, \ldots, p_{n+3}$ besteht, tritt hierzu noch die Bedingung $D(p_1, \ldots, p_{n+3})=0$. Damit R mit dem R_n selbst kongruent sei, ist notwendig und hinreichend, daß außerdem $n+1$ Punkte $p_1, \ldots, p_{n+1}$ in R enthalten sind, für welche $D(p_1, \ldots, p_{n+1}) \neq 0$ und daß R vollständig, konvex und nach außen konvex ist.

Von mehreren Seiten (so von Ph. Frank in Prag und N. Wiener in Cambridge Mass.) wurde nach Vorträgen über dieses Theorem darauf hingewiesen, daß die in ihm auftretende Determinante von Punkten eines halbmetrischen Raumes offenbar eng verwandt mit der *Gramschen Determinante* von Vektoren sei. Tatsächlich ist, wenn man für $n+1$ Punkte $p_0, p_1, \ldots, p_n$ eines halbmetrischen Raumes $p_{0i}=p_{i0}=v_i$ setzt, v_i^2 das Betragsquadrat des von p_0 nach p_i führenden Vektors v_i und p_{ij}^2 das Betragsquadrat des Vektors v_i-v_j. Wegen $p_{ij}^2-p_{0i}^2-p_{0j}^2=(v_i-v_j)^2-v_i^2-v_j^2=-2\,(v_i v_j)$, wo der Klammerausdruck das innere Produkt bedeutet, gilt

$$D(p_0, p_1, \ldots, p_n)=\begin{vmatrix} 0 & 1 \\ 1 & p_{ij}^2 \end{vmatrix}_{(i,j=0,1,\ldots,n)}=\left|p_{ij}^2-p_{0i}^2-p_{0j}^2\right|_{(i,j=1,\ldots,n)}=$$
$$=\left|-2\,(v_i v_j)\right|_{(i,j=1,\ldots,n)}=(-2)^n\,\Gamma(v_1, \ldots, v_n),$$

wobei $\Gamma(v_1, \ldots, v_n)$ die Gramsche Determinante der Vektoren $v_1, \ldots, v_n$ bezeichnet. Die einfache Tatsache, daß die Bedingungen des eingangs erwähnten Theorems für die Kongruenz von R mit einer Teilmenge des R_n notwendig sind, läuft also auf den bekannten Satz hinaus, daß die Gramsche Determinante für n Vektoren des euklidischen Raumes $\geqq 0$ u. zw. $= 0$ dann und nur dann ist, wenn die Vektoren abhängig sind. Wir übersetzen nun die viel tiefer liegende Tatsache, daß die Bedingungen unseres Theorems hinreichend sind, in die Sprache der Vektoren.

Eine Menge V, deren Elemente wir *Vektoren* nennen, heiße eine *Vektormenge*, wenn je zwei Elementen v und w von V eine reelle Zahl $(v\,w) = (w\,v)$ zugeordnet ist, welche das *innere Produkt* der Vektoren v und w heißen möge und für je zwei *verschiedene* Vektoren v und w der Bedingung

$$(\beta) \qquad (v\,v) + (w\,w) \neq 2\,(v\,w)$$

genügen möge. *Damit V mit einer Menge von Vektoren des R_n isomorph sei,* d. h. auf eine Teilmenge V' des n-dimensionalen euklidischen Vektorraumes eineindeutig derart abgebildet werden könne, daß für je zwei Elemente v und w von V die ihnen zugeordnete Zahl $(v\,w)$ gleich dem inneren Produkt ihrer Bildvektoren v' und w' in V' sei, *sind folgende Bedingungen notwendig und hinreichend:*

1_{n+1}) $\Gamma(v_1, \ldots, v_{n+1}) = 0$ *für je $n+1$ Elemente $v_1, \ldots, v_{n+1}$ von V.*

2_k) $\Gamma(v_1, \ldots, v_k) \geqq 0$ *für je k Elemente $v_1, \ldots, v_k$ von V für $k = 1, 2, \ldots, n$.*

Dazu tritt im Falle, daß V aus genau $n+2$ Elementen $v_1, \ldots, v_{n+2}$ besteht, die Bedingung

1_{n+2}) $\Gamma(v_1, \ldots, v_{n+2}) = 0$.

Sind diese (offenbar notwendigen) Bedingungen erfüllt und ist v_0 ein Element von V, so daß $(v_0\,v_0) = 0$, so gilt für jedes v von V wegen $\Gamma(v_0, v) \geqq 0$ offenbar $(v_0\,v) = 0$, also $(v\,v) \neq 0$, wenn $v \neq v_0$ ist, wegen (β). M. a. W. es gibt höchstens *ein* v_0 in V, welches „*Nullvektor*" ist, d. h. $v_0^2 = 0$ erfüllt. Enthält V keinen Nullvektor, so sei V_0 jene Vektormenge, welche außer den Elementen von V genau ein Element v_0 enthält, für welches $(v_0\,v_0) = (v_0\,v) = 0$ für alle Elemente v von V gilt. Zugleich mit V erfüllt offenbar auch V_0 die Bedingungen 1_{n+1}), 2_k) bzw. evt. 1_{n+2}). Ordnen wir je zwei Elementen v und w von V_0 als „*Abstand*" die Zahl $v\,w = +\sqrt{(v\,v) + (w\,w) - 2\,(v\,w)}$ zu — wobei $v\,w$ nicht mit $(v\,w)$ zu verwechseln ist! —, so gilt offenbar $v\,w = w\,v$, ferner $v\,v = 0$ für jedes v und wegen (β) folgt $v\,w \neq 0$ aus $v \neq w$. Also ist V_0 vermöge der erwähnten Abstandsdefinition ein *halbmetrischer Raum.* Derselbe ist auf Grund der eingangs erwähnten Kennzeichnung der Teilmengen des R_n mit einer Teilmenge V_0' des R_n kongruent. Ist V_0'' irgend eine Teilmenge des R_n, die aus V_0' durch eine Bewegung hervorgeht, welche den Bildpunkt v_0' des Nullvektors v_0 in den Ursprung des R_n überführt, so ist die Menge der vom Ursprung zu den Punkten V_0'' führenden Vektoren offenbar mit V_0 isomorph. Damit sind die Bedingungen als hinreichend erwiesen.

Es ist klar, welche Eigenschaften die Vektormenge V besitzen muß, damit der aus ihr eben abgeleitete halbmetrische Raum ein *vollständiger* metrischer Raum sei [die Giltigkeit der Dreiecksungleichung ist durch die Bedingung 2_2) gewährleistet]. Daß dieser metrische Raum *konvex und nach außen konvex* sei, ist damit gleichbedeutend, daß in der gegebenen Vektormenge zu je zwei Vektoren u und w ein Vektor v zwischen u und w existiert und ein Vektor x, so daß w zwischen u und x liegt, wobei wir sagen, der Vektor v liege *zwischen* u und w, wenn entweder $\Gamma(u, w) \neq 0$ und $\Gamma(u, v, w) = 0$, $\Gamma(u, v) + \Gamma(v, w) = \Gamma(u, w)$ gilt oder $\Gamma(u, w) = 0$ und $uv + vw = uw$. Dieser bloß auf den Begriff des inneren Produktes gestützte Aufbau der Vektoralgebra zeigt, *daß die Multiplikation mit Skalaren und die Vektoraddition durch das innere Produkt definierbar sind.* Sind v und w irgend zwei Elemente von V und λ eine reelle Zahl, so nennen wir λv jenen Vektor u, für den $\Gamma(u, v) = 0$ und $(u\,v) = \lambda . (v\,v)$ gilt und $v + w$ jenen Vektor z, für welchen $\Gamma(v, w, z) = 0$ und $\Gamma\left(v, \frac{z}{2}\right) = \Gamma\left(w, \frac{z}{2}\right) = \frac{1}{2}\,\Gamma(v, w)$, bzw. $(z\,z) = (v\,z) + (w\,z)$ gilt, je nachdem ob $\Gamma(v, w) \neq 0$ oder $= 0$ ist. Dabei kann nämlich bewiesen werden, daß genau *ein* derartiger Vektor u bzw. z existiert. Bemerkenswert ist auch, daß durch die Existentialforderung der Konvexität die Bedingungen 2_k) für großes k entbehrlich werden.

Es sei noch erwähnt, daß in analoger Weise die Vektoralgebra in *unitären* Räumen aufgebaut werden kann, wobei die Symmetriebedingung $(u\,v) = (v\,u)$ durch die Hermitesche Symmetrie $(u\,v) = \overline{(v\,u)}$ zu ersetzen ist, sowie eine Vektoralgebra in *indefiniten* euklidischen Räumen (vgl. 60. Kolloquium) auf Grund der Resultate von Wald (dieses Heft S. 36 f.).

Wald: *Eine koordinatenlose Definition der Flächenkrümmung* wird durch Betrachtung der Abstände von gegen einen Punkt konvergenten Punktequadrupeln gewisser Art eingeführt und es wird für Flächen des R_3, die in Parameterdarstellung durch dreimal stetig differenzierbare Funktionen gegeben sind, in Punkten, in denen nicht alle Funktionaldeterminanten verschwinden, die Identität dieser metrisch-koordinatenlos eingeführten Krümmung mit der Gaußschen Krümmung bewiesen. (Erscheint in Heft 6 dieser Ergebnisse.)

66. Kolloquium (21. VI. 1933).

Wald: *Der $R_{n.s.}$* (Vgl. dieses Heft S. 36 f.)

67. Kolloquium (4. VII. 1933).

Čech: *Eine Verallgemeinerung des Jordan-Brouwerschen Satzes.*

Sei R eine n-dimensionale Mannigfaltigkeit, deren $(n-1)^{\text{te}}$ Bettische Zahl verschwindet und S eine abgeschlossene Teilmenge von R $(0 \neq S \neq R)$. Dann besagt der einfachste und wichtigste Fall des Dualitätssatzes, daß die um eins verminderte Komponentenzahl von $R - S$ gleich der $(n-1)^{\text{ten}}$ Bettischen Zahl von S ist.

Es ist bemerkenswert, daß in diesem Satze, der bekanntlich den Jordan-Brouwerschen Satz als Spezialfall enthält, die Voraussetzung, R sei eine Mannigfaltigkeit, wesentlich abgeschwächt werden darf. Es genügt nämlich, Folgendes über R anzunehmen (in der Theorie mod. 2, auf die ich mich hier beschränke; sonst kommt noch die Orientierbarkeitsannahme hinzu): R ist ein zusammenhängendes Polyeder, das Summe ist von einer endlichen Anzahl abgeschlossener (simplizialer) n-Zellen. Jede $(n-1)$-Zelle ist Randzelle von genau zwei n-Zellen. Sind σ_1 und σ_2 zwei beliebige n-Zellen mit gemeinsamem Eckpunkt p, so kann man σ_1 und σ_2 durch eine Kette von n-Zellen mit dem Eckpunkt p verbinden so, daß je zwei benachbarte Glieder der Kette eine gemeinsame $(n-1)$-Zelle haben. Außerdem wird, wie gesagt, vorausgesetzt, daß die $(n-1)^{\text{te}}$ Bettische Zahl von R gleich Null sei.

Der Beweis des angeführten Satzes möge in dem Falle hier durchgeführt werden, daß S ein Unterkomplex der gegebenen Zelleneinteilung von R ist. Sei Z^n die Summe sämtlicher n-Zellen von R. Dann ist Z^n der *einzige* n-Zyklus in R. Seien K_i $(i=1, 2, \ldots, r)$ die Komponenten von $R-S$. Sei C_i^n die Summe derjenigen n-Zellen von R, deren Inneres in K_i liegt. Sei $C_i^n \rightarrow \Gamma_i^{n-1}$. Dann sind die Γ_i^{n-1} $(n-1)$-Zykeln in S. Da jeder $(n-1)$-Zyklus in $R \sim 0$ ist, schließt man leicht, daß jeder $(n-1)$-Zyklus in S von den Γ_i^{n-1} linear abhängt. Zu beweisen bleibt, daß zwischen den Γ_i^{n-1} genau eine lineare Abhängigkeit besteht. Sei $\Sigma a_i \Gamma_i^{n-1} \sim 0$ in S $(a_i=0$ oder $1)$. Es gibt also einen n-Komplex D^n in S so, daß $D^n \rightarrow \Sigma a_i \Gamma_i^{n-1}$ ist. Dann ist $D^n + \Sigma a_i \Gamma_i^{n-1} \rightarrow 0$, folglich $D^n = \Sigma a_i C_i^n + a Z^n$ $(a=0$ oder $1)$. Der n-Komplex $\Sigma a_i C_i^n$ unterscheidet sich daher von $a Z^n$ nur in den Zellen $\subset S$. Dasselbe gilt aber auch für den n-Komplex $a \,.\, \Sigma C_i^n$; also ist $a_1 = a_2 = \ldots = a_r = a$. Umgekehrt ist offenbar $\Sigma \Gamma_i^{n-1} \sim 0$ in S.

Der Beweis für den Fall, daß S eine beliebige kompakte Teilmenge von R ist, ist gegenüber dem durchgeführten Fall nur insofern komplizierter, als es schon für den *Wortlaut* des Satzes (Begriff der $(n-1)^{\text{ten}}$ Bettischen Zahl von S!) der Fall ist. Indem man sich auf die von mir angegebene (Fund. Math., *19*) Definition der Bettischen Zahl stützt, führt man für beliebiges S den Beweis völlig analog und direkt (es wird also nicht etwa S polyedral approximiert!) durch. Der einzige Unterschied besteht darin, daß man nicht mit der gegebenen Zelleneinteilung von R auskommt, sondern gleichzeitig deren sukzessive baryzentrische Unterteilungen (die eine „famille complète de réseaux dans R“ definieren) in Betracht ziehen muß.

Wichtig ist nun, daß auch der polyedrale Charakter von R für den Beweis gar nicht wesentlich ist und leicht durch mengentheoretische (den Homologiecharakter von R betreffende) Annahmen ersetzt werden kann, wobei der Beweis noch ganz durchsichtig bleibt.

Die naheliegende Frage, ob auch in den höheren Fällen des Dualitätssatzes die Annahmen über *R* analog abgeschwächt werden dürfen, ist zu bejahen. Doch ist der allgemeine Fall ungemein komplizierter (Vgl. meine ausführliche Darstellung in den Annals of Math., Oktober 1932).

Čech: Neue Beweisanordnung für den Nöbelingschen Beweis des n-Bogensatzes.

68. Kolloquium (5. VII. 1933).

Čech: Definition lokaler Bettischer Zahlen. Lokalisierung der Dualitätssätze und der Additionssätze der Homologietheorie. Invarianz der lokalen Zerschneidungszahl. Externe Charakterisierung des Zusammenhanges im Kleinen.

Gesammelte Mitteilungen des Jahres 1932/33.

Komplexe und indefinite Räume. Von A. Wald.

I. Der komplexe euklidische Raum.

Eine Menge R heiße ein *komplexmetrischer Raum*[1]), wenn je zwei Elementen p und q von R eine komplexe Zahl $pq = qp$ zugeordnet ist, so daß erstens $pp = 0$ für jedes p gilt und zweitens[2]), falls $p \neq q$ ist, entweder $pq \neq 0$ gilt oder mindestens ein Element r von R existiert, so daß $pr \neq qr$. Eine Teilmenge M eines komplexmetrischen Raumes ist (mit Rücksicht auf diese zweite Bedingung) nicht notwendig ein komplexmetrischer Raum, doch entspricht ihr ein bestimmter komplexmetrischer Raum $\hat{M}$, dessen Punkte die maximalen Teilmengen N von M sind, welche die Eigenschaft haben, daß, wenn n_1 und n_2 irgend zwei Punkte von N sind und m irgend ein Punkt von M ist, $n_1 n_2 = 0$ und $n_1 m = n_2 m$ gilt[2a]). Die Teilmenge M eines komplexmetrischen Raumes heißt *abstandsgleich* bzw. *kongruent* mit der Teilmenge M' eines komplexmetrischen Raumes, wenn sie eindeutig und abstandstreu bzw. eineindeutig und abstandstreu auf M' abgebildet werden kann. Ist M kongruent mit M', so ist M' kongruent mit M. *Bemerkung:* Ist M abstandsgleich mit M', so sind die entsprechenden komplexmetrischen Räume $\hat{M}$ und $\hat{M}'$ kongruent. M' ist dann nicht notwendig abstandsgleich mit M, wohl aber mit einer Teilmenge von M. Als K_n bezeichnen wir die Menge aller n-Tupel komplexer Zahlen, wenn je zwei Elementen $x_1, \ldots, x_n$ und $y_1, \ldots, y_n$ die Zahl $\sum_{i=1}^{n} (x_i - y_i)^2$ als Abstand (nicht Abstandsquadrat!) zugeordnet ist. Eine Teilmenge M des K_n heiße *linear abhängig,* wenn ein $(n+1)$-Tupel $\lambda_0, \ldots, \lambda_n$ von komplexen Zahlen, die nicht alle 0 sind, existiert, so daß für alle Punkte $x_1, \ldots, x_n$ von M die Gleichung gilt

$$(*) \qquad \lambda_0 + \lambda_1 x_1 + \ldots + \lambda_n x_n = 0.$$

1. *Der einer linear abhängigen Teilmenge M des K_n entsprechende komplexmetrische Raum $\hat{M}$ ist mit dem einer Teilmenge des K_{n-1} entsprechenden komplexmetrischen Raum kongruent*[3]). Es sei

---

[1]) Vgl. Menger, diese Ergebn., *1*, S. 36.

[2]) Diese Bedingung stammt aus der Dissertation von Flexer, Wien 1933.

[2a]) Als Abstand der Punkte N und N' von $\hat{M}$ gilt dabei der Abstand irgend eines Punktes n von N von irgend einem Punkt n' von N'.

[3]) Dieser Satz wird auch in der Dissertation von Flexer bewiesen.

M' eine Menge, die aus M durch Translation hervorgeht und den Punkt $(0, \ldots, 0)$ enthält. Da $\hat{M}'$ und $\hat{M}$ kongruent sind, genügt es zu zeigen, daß M' abstandsgleich einer Teilmenge von K_{n-1} ist. Es sei (*) die Gleichung, der die Punkte von M' genügen, und es sei H die Menge aller Punkte des K_n, die (*) genügen. Nach einem Satz über quadratische Formen existieren r lineare Formen $L_i(x_1, \ldots, x_n)$ $(i=1, \ldots, r)$ $(r \leqq n-1)$, so daß für jeden Punkt $x_1, \ldots, x_n$ von H gilt: (**) $x_1^2 + \ldots + x_n^2 = L_1^2 + \ldots + L_r^2$. Ordnen wir nun jedem Punkt $x_1, \ldots, x_n$ von H den Punkt $x'_1 \ldots, x'_r$ des K_r zu, wo $x'_i = L_i(x_1, \ldots, x_n)$ $(i=1, \ldots, r)$, so ist diese eindeutige Abbildung abstandstreu. Denn sind $x_1, \ldots, x_n$ und $y_1, \ldots, y_n$ irgend zwei Punkte von H mit den Bildpunkten $x'_1, \ldots, x'_r$ und $y'_1, \ldots, y'_r$, so gilt, da auch der Punkt $(x_1-y_1, \ldots, x_n-y_n)$ zu H gehört und sein Bildpunkt die Koordinaten $x'_1-y'_1, \ldots, x'_r-y'_r$ hat, wegen (**): $(x'_1-y'_1)^2 + \ldots + (x'_r-y'_r)^2 = (x_1-y_1)^2 + \ldots + (x_n-y_n)^2$. Da durch diese Abbildung M' eindeutig und abstandstreu auf eine Teilmenge von K_r abgebildet wird, ist **1** auf Grund der einleitenden Bemerkung bewiesen.

2. *Es gibt im K_n höchstens einen Punkt p, der von einem linear unabhängigen $(n+1)$-Tupel $p_0, \ldots, p_n$ von Punkten des K_n vorgeschriebene Abstände hat.* Wir können $p_0 = (0, \ldots, 0)$ annehmen. Ferner sei $p_i = (\alpha_{i1}, \ldots, \alpha_{in})$ und $p = (x_1, \ldots, x_n)$. Dann gilt
1) $x_1^2 + \ldots + x_n^2 = pp_0$ 2) $(x_1-\alpha_{i1})^2 + \ldots + (x_n-\alpha_{in})^2 = pp_i$ $(i=1, \ldots, n)$.

Aus 2) folgt wegen 1), daß $-2\sum_{k=1}^{n} x_k \alpha_{ik} = pp_i - pp_0 - \sum_{k=1}^{n} \alpha_{ik}^2$ $(i=1, \ldots, n)$ und diese n linearen Gleichungen in $x_1 \ldots, x_n$ haben wegen der linearen Unabhängigkeit von $p_0, \ldots, p_n$ eine nicht verschwindende Determinante, also genau eine Lösung.

3. *Eine lineare Transformation, die ein linear unabhängiges $(n+1)$-Tupel $p_0, \ldots, p_n$ des K_n in ein abstandsgleiches $(n+1)$-Tupel $p'_0, \ldots, p'_n$ überführt, ist eine kongruente Selbstabbildung des K_n.* Wir können $p_0 = p'_0 = (0, \ldots, 0)$ annehmen. Ferner sei $p_i = (x_{i1}, \ldots, x_{in})$, $p'_i = (x'_{i1}, \ldots, x'_{in})$ $(i=1, \ldots, n)$. Wegen der linearen Unabhängigkeit von $p_0, \ldots, p_n$ ist $|x_{ik}|_{(i,k=1,\ldots,n)} \neq 0$. Setzen wir $x'_i = \lambda_{i1} x_1 + \ldots + \lambda_{in} x_n$ $(i=1, \ldots, n)$, so gilt für $i=1, \ldots, n$ das Gleichungssystem $x'_{ki} = \lambda_{i1} x_{k1} + \ldots + \lambda_{in} x_{kn}$ $(k=1, \ldots, n)$. Also sind wegen $|x_{ik}| \neq 0$ die λ_{ik} eindeutig bestimmt. Es gibt daher genau eine homogene lineare Transformation, welche $p_1, \ldots, p_n$ in $p'_1, \ldots, p'_n$ überführt. Sind nun p und q irgend zwei Punkte des K_n, so kann man sie wegen der linearen Unabhängigkeit von $p_0, \ldots, p_n$ folgendermaßen darstellen:

$$p = \lambda_1 p_1 + \ldots + \lambda_n p_n, \quad q = \mu_1 p_1 + \ldots + \mu_n p_n. \text{ Dann ist}$$
$$p' = \lambda_1 p'_1 + \ldots + \lambda_n p'_n, \quad q' = \mu_1 p'_1 + \ldots + \mu_n p'_n.$$

Aus $p_i p_0 = p'_i p'_0$ folgt

$$(1) \quad \sum_{r=1}^{n} x_{ir}^2 = \sum_{r=1}^{n} x_{ir}'^2 \quad (i=1, \ldots, n).$$

3

Aus $p_i p_k = p'_i p'_k$ folgt unter Berücksichtigung von (1):

(2) $\sum_{r=1}^{n} x_{ir} x_{kr} = \sum_{r=1}^{n} x'_{ir} x'_{kr} \quad (i \neq k)$.

Nun gilt $pq = [(\lambda_1 - \mu_1) x_{11} + \ldots + (\lambda_n - \mu_n) x_{n1}]^2 + \ldots + [(\lambda_1 - \mu_1) x_{1n} + \ldots + (\lambda_n - \mu_n) x_{nn}]^2 = \sum_{i=1}^{n} \sum_{k=1}^{n} (\lambda_i - \mu_i)(\lambda_k - \mu_k) \sum_{r=1}^{n} x_{ir} x_{kr}$.

Ersetzen wir x_{ik} durch x'_{ik}, so entsteht der Ausdruck, den wir für $p'q'$ erhalten, woraus wegen (1) und (2) folgt $pq = p'q'$ w. z. b. w.

4. *Ein linear unabhängiges $(n+3)$-Tupel von Punkten des K_n enthält ein linear unabhängiges $(n+1)$-Tupel.* Sind nämlich je $n+1$ von den $n+3$ Punkten $p_i = (\alpha_{i1}, \ldots, \alpha_{in})$ $(i = 0, 1, \ldots, n+2)$ linear abhängig, dann hat die Matrix des Gleichungssystems $\lambda_0 + \lambda_1 \alpha_{i1} + \ldots + \lambda_n \alpha_{in} = 0$ $(i = 0, 1, \ldots, n+2)$ einen Rang $\leqq n$ und das System besitzt auch eine Lösung, wobei mindestens eine der $n+1$ Unbekannten λ_k $(k = 0, 1, \ldots, n)$ nicht 0 ist, d. h. es sind $p_0, p_1, \ldots, p_{n+2}$ linear abhängig.

5. *Der K_n hat die Kongruenzordnung $n+3$, d. h. ist R ein komplexmetrischer Raum, von dem je $n+3$ Punkte abstandsgleich einer Teilmenge des K_n sind, so ist auch R abstandsgleich einer Teilmenge des K_n.*[4]) Enthält R erstens $n+1$ Punkte $p_0, \ldots, p_n$, die mit $n+1$ linear unabhängigen Punkten $p'_0, \ldots p'_n$ des K_n abstandsgleich sind, so sind für je zwei Punkte p und q von R die Punkte $p_0, \ldots, p_n, p, q$ mit $n+3$ Punkten $p''_0, \ldots, p''_n, p'', q''$ von K_n abstandsgleich. Die lineare Transformation, die $p''_0, \ldots, p''_n$ in $p'_0, \ldots, p'_n$ überführt, ist nach **3** eine kongruente Selbstabbildung des K_n. Die Punkte, in welche p'' und q'' durch sie überführt werden, mögen p' und q' heißen und p bzw. q zugeordnet werden. Nach **2** ist dadurch eine eindeutige Abbildung von R auf eine Teilmenge von K_n definiert, welche offenbar abstandstreu ist. Für den K_1 ist bereits durch diese Überlegung die Kongruenzordnung 4 erwiesen, da zwei Punkte eines komplexmetrischen Raumes R, deren Abstand $\neq 0$ ist, mit zwei linear unabhängigen Punkten des K_1 kongruent sind. Sind zweitens je $n+1$ Punkte von R mit $n+1$ linear abhängigen Punkten von K_n abstandsgleich, so sind nach **4** und **1** je $n+3$ Punkte von R einer Teilmenge des K_{n-1} abstandsgleich. Machen wir die Annahme, daß für den K_{n-1} die Kongruenzordnung $n+2$ bereits bewiesen ist, so ist also auch der zweite Fall erledigt. Da diese Annahme, wie wir eben sahen, für $n=2$ zutrifft, ist **5** durch vollständige Induktion bewiesen.

6. *Ist R ein aus $n+1$ Punkten bestehender komplexmetrischer Raum, so existiert eine ganze Zahl m, so daß R mit $n+1$ Punkten*

---

[4]) Über den Begriff der Kongruenzordnung vgl. Menger, Amer. Journ. Math., *53*, S. 722. Daß der K_1 die Kongruenzordnung 4 hat, ist ein Spezialfall eines gruppentheoretischen Satzes von Menger, Math. Zschr., *33*, S. 126. Einen ersten Beweis dafür, daß der K_n die Kongruenzordnung $n+3$ hat, gab Flexer in seiner Dissertation. Der oben angegebene Beweis läuft dem von Menger (Amer. Journ. Math., *53*, S. 722) für die Kongruenzordnung $n+3$ des R_n parallel.

des K_m kongruent ist. Da der Satz für $n=1$ trivial ist, haben wir ihn unter der induktiven Annahme, daß er für aus n Punkten bestehende Räume gelte, für den aus den $n+1$ Punkten $p_0, \ldots, p_n$ bestehenden Raum zu beweisen. Wir betrachten zu diesem Zweck eine Menge Q von n Elementen $q_0, \ldots, q_{n-1}$, für welche wir folgende Abstände definieren:

$$q_k q_i = q_i q_k = p_i p_k - 2 \quad \text{für } i \neq 0 \text{ und } k \neq 0$$
$$q_0 q_i = q_i q_0 = p_i p_0 - 1 \quad \text{für } i = 1, \ldots, n-1.$$

Der den Punkten $q_0, \ldots, q_{n-1}$ entsprechende komplexmetrische Raum $\hat{Q}$ besteht aus höchstens n Punkten. Nach der induktiven Voraussetzung existiert eine ganze Zahl s, so daß $\hat{Q}$ mit einer Teilmenge des K_s kongruent ist. Daraus folgt, daß die Punkte $q_0, \ldots, q_{n-1}$ abstandsgleich einer Teilmenge des K_s sind. Dem Punkt q_k entspreche dabei der Punkt $q_k = (\alpha_{k1}, \alpha_{k2}, \ldots, \alpha_{ks})$ des K_s $(k = 1, \ldots, n-1)$, wobei wir $q_0 = (0, 0, \ldots, 0)$ annehmen können. Wir betrachten nun im K_{s+n} die n Punkte

$$\begin{array}{lllllll} & 1, & 2, & n-1, & n, & s+n-1, & s+n \\ p'_0 & = (0, & 0, \ldots, & 0, & 0 \quad, \ldots, & 0 & , 0) \\ p'_1 & = (1, & 0, \ldots, & 0, & \alpha_{11} \quad, \ldots, & \alpha_{1s} & , 0) \\ p'_2 & = (0, & 1, \ldots, & 0, & \alpha_{21} \quad, \ldots, & \alpha_{2s} & , 0) \\ \cdot & \cdot & \cdot \quad \cdot & \cdot & \cdot & \cdot & \cdot \\ p'_{n-1} & = (0, & 0, \ldots, & 1, & \alpha_{n-11}, \ldots, & \alpha_{n-1s}, & 0). \end{array}$$

Sie sind abstandsgleich mit $p_0, p_1, p_2, \ldots, p_{n-1}$. Setzen wir

$$p'_n = (\alpha_1, \alpha_2, \ldots, \alpha_{n-1}, \quad 0, \ldots 0, \quad \alpha),$$

so müssen, damit $p'_0, p'_1, \ldots, p'_n$ mit $p_0, p_1, \ldots, p_n$ kongruent seien, die Zahlen $\alpha_1, \alpha_2, \ldots, \alpha_{n-1}, \alpha$ folgenden Bedingungen genügen:

$$\begin{aligned} \alpha_1^2 + \ldots + \alpha_{n-1}^2 + \alpha^2 &= p_0 p_n, \\ -2\alpha_1 &= p_n p_1 - p_n p_0 - p_1 p_0, \\ -2\alpha_2 &= p_n p_2 - p_n p_0 - p_2 p_0, \\ &\cdots \\ -2\alpha_{n-1} &= p_n p_{n-1} - p_n p_0 - p_{n-1} p_0. \end{aligned}$$

Diese Gleichungen sind aber offenkundig lösbar; also sind $p_0, \ldots, p_n$ mit $n+1$ Punkten des K_{s+n} kongruent.

7. *Ein aus $n+1$ Punkten bestehender komplexmetrischer Raum R ist mit $n+1$ Punkten des K_n kongruent*[5]). Nach **6** existiert eine ganze Zahl m, so daß R mit einer Teilmenge des K_m kongruent ist. Es sei m die kleinste Zahl dieser Art. Wäre $m > n$, so wären die mit R kongruenten $n+1$ Punkte des K_m linear abbhängig; also wären sie (und daher auch R) nach **1** kongruent mit einer Teilmenge des K_{m-1} im Widerspruch zur Definition von m. Daher gilt $m = n$.

8. *Damit $n+2$ Punkte $p_0, \ldots, p_{n+1}$ eines komplexmetrischen Raumes mit $n+2$ Punkten des K_n abstandsgleich seien, ist, wenn wir*

[5]) Über den Fall $n=3$ dieses Satzes vgl. diese Ergebn., *4*, S. 13.

mit $\Delta(p_0,\ldots,p_{n+1})$ *die geränderte Determinante* $\begin{vmatrix} 0 & 1 \\ 1 & p_i p_j \end{vmatrix}_{(i,j=0,\ldots,n+1)}$ *bezeichnen, notwendig und hinreichend, daß* $\Delta(p_0,\ldots,p_{n+1})=0$ *gilt*[6]). Nach **7** existieren sicher $n+2$ mit $p_0,\ldots,p_{n+1}$ abstandsgleiche Punkte $p'_i=(\alpha_{i1},\ldots,\alpha_{in+1})$ $(i=0,\ldots,n+1)$ im K_{n+1}. Nun gilt bekanntlich

$$\Delta(p_0,\ldots,p_{n+1})=\begin{vmatrix} 0 & 1 \\ 1 & p_i p_j \end{vmatrix}_{(i,j=0,\ldots,n+1)}=\begin{vmatrix} 0 & 1 \\ 1 & p'_i p'_j \end{vmatrix}_{(i,j=0,\ldots,n+1)}=$$

$$C.\begin{vmatrix} 1 & \alpha_{0j} \\ 1 & \alpha_{ij} \end{vmatrix}^2_{(i,j=1,\ldots,n+1)} \quad (C\neq 0).$$

Ist $\Delta(p_0,\ldots,p_{n+1})\neq 0$, dann sind $p'_0,\ldots,p'_{n+1}$, linear unabhängig und daher nicht mit $n+2$ Punkten eines K_n kongruent. Ist $\Delta(p_0,\ldots,p_{n+1})=0$, so sind sie linear abhängig und daher nach **1** mit $n+2$ Punkten des K_n abstandsgleich.

9. *Damit* $n+3$ *Punkte* $p_0,\ldots,p_{n+2}$ *eines komplexmetrischen Raumes mit* $n+3$ *Punkten des* K_n *abstandsgleich seien, ist notwendig und hinreichend, daß* $\Delta(p_0,\ldots,p_{n+2})=0$ *und* $\Delta(p_{i_0},\ldots,p_{i_{n+1}})=0$ *für je* $n+2$ *Punkte* $p_{i_0},\ldots,p_{i_{n+1}}$. Nach **8** sind $p_0,\ldots,p_{n+2}$ mit $n+3$ Punkten $p'_0,\ldots,p'_{n+2}$ des K_{n+1} abstandsgleich und je $n+2$ Punkte $p'_{i_0},\ldots,p'_{i_{n+1}}$ sind der Voraussetzung zufolge linear abhängig. Daher sind nach **4** auch $p'_0,\ldots,p'_{n+2}$ linear abhängig, also mit $n+3$ Punkten des K_n abstandsgleich. Durch diese Sätze ist das Problem von Menger, die Teilmengen des K_n zu kennzeichnen, vollkommen gelöst.

II. Indefinit-euklidische Räume.

Als $R_{n,s}$ bezeichnen wir ähnlich[7]) Menger (dieses Heft, S. 16) die Menge aller n-Tupel reeller Zahlen, in der den Elementen $x=(x_1,\ldots,x_n)$ und $y=(y_1,\ldots,y_n)$ als Abstand die Zahl $\sum_{i=1}^{n}\varepsilon_i(x_i-y_i)^2$ mit $\varepsilon_i=1$ für $1\leqq i\leqq\frac{n+s}{2}$ und $\varepsilon_i=-1$ für $\frac{n+s}{2}<i\leqq n$ zugeordnet ist, so daß also s die Signatur der quadratischen Maßform ist. $R_{n,n}$ ist der reelle, $R_{n,-n}$ der rein-imaginäre euklidische Raum (vgl. Menger, dieses Heft, S. 16). Wir lösen im folgenden das von Menger (dieses Heft, S. 16) formulierte und für den $R_{n,n}$ und $R_{n,-n}$ gelöste Problem der Kennzeichnung des $R_{n,s}$. Kongruent mit dem $R_{n,s}$ ist offenbar die Menge $R^*_{n,s}$ aller Punkte des K_n, deren $\frac{n+s}{2}$ erste Koordinaten reell und deren sämliche übrigen Koordinaten rein-imaginär sind. Dieser für die Behandlung des $R_{n,s}$ wichtige Gedanke stammt von O. Taussky.

---

[6]) Der Fall $n=3$ findet sich in der Dissertation von Flexer.

[7]) Was wir als Abstand bezeichnen ist das Quadrat des Mengerschen Abstandes!

10. *Der $R_{n,s}$ hat die Kongruenzordnung $n+3$ für alle s.* Es sei R ein komplexmetrischer Raum, von dem je $n+3$ Punkte abstandsgleich mit $n+3$ Punkten von $R_{n,s}$ und daher von $R^*_{n,s}$ sind. Nach **9** gilt wegen $R^*_{n,s} \subset K_n$ für je $n+2$ und je $n+3$ Punkte von R die Beziehung $\Delta(p_0, \ldots, p_{n+1}) = 0$, $\Delta(p_0, \ldots, p_{n+2}) = 0$. Es sei r die größte ganze Zahl $\leqq n$ mit der Eigenschaft, daß mindestens ein $(r+1)$-Tupel $p_0, \ldots, p_r$ von Punkten von R existiert, für welches $\Delta(p_0, \ldots, p_r) \neq 0$ gilt. R ist dann wegen **5** und **9** mit einer Teilmenge von K_r kongruent. Da $p_0, \ldots, p_r$ wegen $r \leqq n$ mit $r+1$ Punkten von $R^*_{n,s}$ abstandsgleich sind, so ist R wegen **3** mit einer solchen Teilmenge von K_n kongruent, daß die Bildpunkte $p'_0, \ldots, p'_r$ von $p_0, \ldots, p_r$ in einem linearen Teilraum $R^*_{r,t}$ von $R^*_{n,s}$ liegen, wobei durch eine Translation $p'_0 = (0, \ldots, 0)$ erreicht werden kann. Es sei $p'_i = (\alpha_{i1}, \ldots, \alpha_{ir})$ $(i = 1, 2, \ldots, r)$. Wir betrachten nun irgend einen von $p_0, \ldots, p_r$ verschiedenen Punkt p, dessen Bildpunkt p' in K_r die Koordinaten $(x_1, \ldots, x_r)$ habe. Da $\{p'_0, \ldots, p'_r, p'\}$ abstandsgleich zu $\{p_0, \ldots, p_r, p\}$ ist, gilt

$$(+) \quad -2(\alpha_{i1} x_1 + \ldots + \alpha_{ir} x_r) = pp_i - pp_0 - p_i p_0 = A_i \quad (i = 1, 2, \ldots, r).$$

Aus $\Delta(p_0, \ldots, p_r) \neq 0$ folgt $|\alpha_{ik}|_{(i,k=1,\ldots,r)} \neq 0$. Also hat das Gleichungssystem (+) genau eine Lösung in den Unbekannten x_i $(i = 1, \ldots, r)$. Da $p'_1, \ldots, p'_r$ in $R^*_{r,t}$ liegen, so gilt für $i = 1, \ldots, r$ daß α_{ik} reell ist für $1 \leqq k \leqq \frac{r+t}{2}$ und rein-imaginär für $\frac{r+t}{2} < k \leqq r$. Ferner sind die Zahlen A_i $(i = 1, 2, \ldots, r)$ reell, da sie $= pp_i - pp_0 - p_i p_0$ sind, sich also additiv aus Abständen des Raumes R zusammensetzen, welch letztere reell sein müssen, da je $n+3$ Punkte von R mit $n+3$ Punkten von $R_{n,s}$ abstandsgleich sind. Also ist x_i reell für $1 \leqq i \leqq \frac{r+t}{2}$ und rein-imaginär für $\frac{r+t}{2} < i \leqq r$. Daher liegt p' in $R^*_{r,t}$ und R ist mithin kongruent abgebildet auf eine Teilmenge von $R_{n,s}$.

11. *Ein aus $n+1$ Punkten $p_0, \ldots, p_n$ bestehender komplexmetrischer Raum, für den $\Delta(p_0, \ldots, p_n) \neq 0$ gilt und der mit $n+1$ Punkten des $R_{n,s}$ kongruent ist, ist für $s \neq t$ mit keinen $n+1$ Punkten des $R_{n,t}$ kongruent.*

Wenn es zu $p_0, \ldots, p_n$ kongruente $(n+1)$-Tupel $q_0, \ldots, q_n$ bzw. $r_0, \ldots, r_n$ im $R^*_{n,s}$ bzw. $R^*_{n,t}$ gibt, wobei durch eine Translation $q_0 = r_0 = (0, \ldots, 0)$ erreicht werden kann, so gibt es, da wegen $\Delta(p_0, \ldots, p_n) \neq 0$ beide $(n+1)$-Tupel linear unabhängig sind, genau eine lineare homogene Transformation $y_i = \alpha_{i1} x_1 + \ldots + \alpha_{in} x_n = L_i(x_1, \ldots, x_n)$ $(i = 1, 2, \ldots, n)$, welche $q_0, \ldots, q_n$ in $r_0, \ldots, r_n$ überführt. Sie führt jeden Punkt q von $R^*_{n,s}$, falls $\lambda_1, \ldots, \lambda_n$ jene reelle Zahlen sind, für die $q = \lambda_1 q_1 + \ldots + \lambda_n q_n$ gilt, in den Punkt $r = \lambda_1 r_1 + \ldots + \lambda_n r_n$, der also in $R^*_{n,t}$ liegt, über, bildet also $R^*_{n,s}$ auf $R^*_{n,t}$ ab und muß wegen **3** eine kongruente Selbstabbildung des K_n sein. Dieser Abbildung entspricht eine kongruente reelle lineare Abbildung $y_i = L'_i(x_1, \ldots, x_n)$ $(i = 1, 2, \ldots, n)$ des $R_{n,s}$ auf den $R_{n,t}$, die den Punkt $(0, \ldots, 0)$ fest läßt, so daß

$$\varepsilon_1 x_1^2 + \ldots + \varepsilon_n x_n^2 = \varepsilon'_1 L_1'^2 + \ldots + \varepsilon'_n L_n'^2,$$

wobei $\varepsilon_i = 1$ für $1 \leqq i \leqq \frac{n+s}{2}$, $\varepsilon_i = -1$ für $\frac{n+s}{2} < i \leqq n$ und

$$\varepsilon'_i = 1 \text{ für } 1 \leqq i \leqq \frac{n+t}{2},\ \varepsilon'_i = -1 \text{ für } \frac{n+t}{2} < i \leqq n.$$

Dem Trägheitsgesetz für quadratische Formen zufolge muß dann $s = t$ sein, w. z. b. w.

12. *Ist* $\{p_0, p_1, \ldots, p_n\}$ *ein komplexmetrischer Raum, so daß* $\Delta(p_0, \ldots, p_n) \neq 0$, *so gibt es eine ausgezeichnete Anordnung* p_{i_0}, $p_{i_1}, \ldots, p_{i_n}$, *so daß in der Zahlenreihe*

$$\Delta(p_{i_0}),\ \Delta(p_{i_0}, p_{i_1}), \ldots, \Delta(p_{i_0}, p_{i_1}, \ldots, p_{i_n})$$

keine zwei benachbarten Nullen auftreten. Gilt für jedes n-Tupel $\Delta(p_{i_0}, \ldots, p_{i_{n-1}}) = 0$, so muß ein $(n-1)$-Tupel existieren, so daß $\Delta(p_{i_0}, \ldots, p_{i_{n-2}}) \neq 0$. Andernfalls wären nach **9** je n Punkte mit n Punkten des K_{n-3} abstandsgleich und folglich wären wegen **5** auch $p_0, \ldots, p_n$ mit Punkten des K_{n-3} kongruent, was wegen $\Delta(p_0, \ldots, p_n) \neq 0$ unmöglich ist. Behandeln wir eines der n-Tupel oder $(n-1)$-Tupel, für welche $\Delta \neq 0$ ist, in derselben Weise usw., so gelangen wir schließlich zur gewünschten ausgezeichneten Anordnung.

13. *Voraussetzungen:*

1) $\{p_0, \ldots, p_n\}$ *sei ein komplexmetrischer Raum R mit lauter reellen Abständen.*

2) $\Delta(p_0, \ldots, p_n) \neq 0$.

3) $p_0, p_1, \ldots, p_n$ *sei eine ausgezeichnete Anordnung im Sinne von* **12.**

Behauptung: R ist kongruent mit einer Teilmenge des $R_{n,s}$, *wo* $\frac{n+s}{2}$ *die Anzahl der Zeichenwechsel in der Zahlenreihe*

$$(\Delta) \qquad \Delta(p_0) = -1,\ \Delta(p_0, p_1), \ldots, \Delta(p_0, p_1, \ldots, p_n)$$

nach Streichung der in der Reihe (Δ) *etwa auftretenden Nullen ist.*

Wir setzen $p_0' = (0, \ldots, 0)$ im K_n und machen für die ganze Zahl $k \leqq n-1$ die Annahme (α_k), es gelte $\Delta(p_0, p_1, \ldots, p_k) \neq 0$ und es seien bereits die mit $p_0, p_1, \ldots, p_k$ kongruenten Punkte $p'_0, p'_1, \ldots, p'_k$ so definiert, daß

$$p'_i = \left\{ \begin{array}{l} (\alpha_{i1}, \alpha_{i2}, \ldots, \alpha_{ii}, 0, \ldots, 0), \text{ falls} \\ \qquad\qquad \Delta(p_0, p_1, \ldots, p_i) \neq 0 \\ (\alpha_{i1}, \alpha_{i2}, \ldots, \alpha_{i\,i-1}, 1, \sqrt{-1}, 0, \ldots, 0), \text{ falls} \\ \qquad\qquad \Delta(p_0, p_1, \ldots, p_i) = 0 \end{array} \right\} \ (i = 1, 2, \ldots, k),$$

wobei jede einzelne Zahlenspalte $\alpha_{1j}, \alpha_{2j}, \ldots, \alpha_{kj}$ entweder aus lauter reellen oder aus 0 und lauter rein imaginären Zahlen besteht ($j=1, 2, \ldots, k$) und wobei genau so viele Spalten aus lauter reellen Zahlen bestehen, als Zeichenwechsel auftreten in der Zahlenreihe

$$(\Delta_k) \qquad \Delta(p_0), \Delta(p_0, p_1), \ldots, \Delta(p_0, p_1, \ldots, p_k).$$

Wir bezeichnen nun für zwei Punkte $x_1=(x_{11}, \ldots, x_{1n})$ und $x_2=$ $=(x_{21}, \ldots, x_{2n})$ des K_n mit $(x_1 x_2)$ das innere Produkt $\sum_{i=1}^{n} x_{1i} x_{2i}$, setzen für n Punkte $x_i=(x_{i1}, \ldots, x_{in})$ des K_n $|x_1, \ldots, x_n|=|x_{ij}|_{(i,j=1,\ldots,n)}$ und unterscheiden zwei Fälle:

I. Es gelte $\Delta(p_0, p_1, \ldots, p_{k+1}) \neq 0$. Setzen wir $p'_{k+1}=(\alpha_{k+1\,1}, \alpha_{k+1\,2}, \ldots, \alpha_{k+1\,k+1}, 0, \ldots, 0)$, so haben wir, damit $p'_0, \ldots, p'_{k+1}$ mit $p_0, \ldots, p_{k+1}$ kongruent seien, die $\alpha_{k+1\,j}$ so zu bestimmen, daß die Gleichungen erfüllt sind:

$$(\mathrm{I}_{k+1}) \qquad \begin{aligned} -2\,(p'_{k+1} p'_j) &= p_{k+1} p_j - p_{k+1} p_0 - p_j p_0 \quad (j=1, 2, \ldots, k) \\ (p'_{k+1} p'_{k+1}) &= \alpha^2_{k+1\,1} + \alpha^2_{k+1\,2} + \ldots + \alpha^2_{k+1\,k+1} = p_{k+1} p_0. \end{aligned}$$

Da bekanntlich

$$(+) \qquad \Delta(p_0, \ldots, p_j) = (-1)^{j+1} C_j (|p_1', \ldots, p_j'|)^2 \quad (C_j > 0),$$

gilt, falls $p'_1, \ldots, p_j'$ in K_j liegen, und $\Delta(p_0, \ldots, p_k) \neq 0$ vorausgesetzt ist, so gilt $|p'_1, \ldots, p'_k| \neq 0$. Also ist die Determinante der k ersten Gleichungen (I_{k+1}), die $=-2^k |p'_1, \ldots, p'_k|$ ist, $\neq 0$. Daher sind die Zahlen $\alpha_{k+1\,1}, \ldots, \alpha_{k+1\,k}$ eindeutig bestimmt. Da die rechten Seiten dieser Gleichungen reell sind, folgert man mühelos, daß für jedes $j=1, \ldots, k$ die Zahl $\alpha_{k+1\,j}$ reell bzw. 0 oder rein imaginär ist, je nachdem die Zahlenspalte $\alpha_{1j}, \alpha_{2j}, \ldots, \alpha_{kj}$ reell oder rein imaginär ist. Die $(k+1)$-te Gleichung (I_{k+1}) ergibt $\alpha_{k+1\,k+1}$. Da $|p'_1, \ldots, p'_{k+1}| =$ $= \alpha_{k+1\,k+1}\, |p'_1, \ldots, p'_k|$ gilt, folgt aus (+)

$$(++) \qquad \alpha^2_{k+1\,k+1} = -\frac{C_k}{C_{k+1}} \frac{\Delta(p_0, \ldots, p_{k+1})}{\Delta(p_0, \ldots, p_k)}.$$

Es ist $\alpha_{k+1\,k+1}$ daher $\neq 0$, und zwar reell oder rein imaginär, je nachdem *sgn* $\Delta(p_0, \ldots, p_{k+1}) \neq$ oder $=$ sgn $\Delta(p_0, \ldots, p_k)$ gilt. Damit ist die Annahme (α_{k+1}) realisiert.

II Es ist $\Delta(p_0, \ldots, p_{k+1})=0$. Ist $\{\alpha_{k+1\,1}, \ldots, \alpha_{k+1\,k}\}$ das Lösungssystem der ersten k Gleichungen von (I_{k+1}), dann sind $p_0, \ldots, p_k, p_{k+1}$, mit Rücksicht auf (++) abstandsgleich mit $p'_0, \ldots,$ p'_k, p^*_{k+1}, wenn wir $p^*_{k+1}=(\alpha_{k+1\,1}, \ldots, \alpha_{k+1\,k}, 0, \ldots, 0)$ setzen. Daraus folgt, daß $p_0, \ldots, p_{k+1}$ auch mit $p'_0, \ldots, p'_{k+1}$ abstandsgleich sind, wobei $p'_{k+1}=(\alpha_{k+1\,1}, \ldots, \alpha_{k+1\,k}, 1, \sqrt{-1}, 0, \ldots 0)$ ist.

Wegen $\Delta(p_0, \ldots, p_{k+1})=0$ folgt aus der Voraussetzung 3), daß $\Delta(p_0, \ldots, p_{k+2}) \neq 0$ ist. Um die Annahme (α_{k+2}) zu realisieren, haben wir die Koordinaten $\alpha_{k+2\,1}, \ldots, \alpha_{k+2\,k+2}$ von p'_{k+2} so zu wählen, daß $p_0, \ldots, p_{k+2}$ mit $p_0', \ldots, p'_{k+2}$ abstandsgleich sind, d. h. folgendes Gleichungssystem befriedigt ist:

$$(\mathrm{II}_{k+2}) \quad -2\,(p'_{k+2}\,p_j) = p_{k+2}\,p_j - p_{k+2}\,p_0 - p_j\,p_0 \quad (j = 1, \ldots, k)$$

$$-2\,(p'_{k+2}\,p'_{k+1}) = -2\,(\alpha_{k+2\,1}\,\alpha_{k+1\,1} + \ldots + \alpha_{k+2\,k}\,\alpha_{k+1\,k} + {}$$

$$+ \alpha_{k+2\,k+1} + \sqrt{-1}\,\alpha_{k+2\,k+2}) = p_{k+2}\,p_{k+1} - p_{k+2}\,p_0 - p_{k+1}\,p_0$$

$$(p'_{k+2}\,p'_{k+2}) = \alpha^2_{k+2\,1} + \ldots + \alpha^2_{k+2\,k} + \alpha^2_{k+2\,k+1} + \alpha^2_{k+2\,k+2} =$$

$= p_{k+2}\,p_0$. Die Determinante der ersten k Gleichungen von (II_{k+2}) ist $-2^k\,|p'_1, \ldots, p'_k| \neq 0$. Durch die ersten k Gleichungen von (II_{k+2}) sind also die Zahlen $\alpha_{k+2\,1}, \ldots, \alpha_{k+2\,k}$ bestimmt. Da die rechten Seiten von (II_{k+2}) reell sind, so ist $\alpha_{k+2\,j}$ $(j = 1, \ldots, k)$ reell bzw. 0 oder rein imaginär, je nachdem die Zahlenspalte $\alpha_{1j}, \alpha_{2j}, \ldots, \alpha_{kj}$ reell oder rein imaginär ist. Substituiert man die Werte $\alpha_{k+2\,j}$ $(j = 1, \ldots, k)$ in die letzten zwei Gleichungen von (II_{k+2}), so folgt

$$(\mathrm{III}) \qquad \begin{aligned} \alpha_{k+2\,k+1} + \sqrt{-1}\;\alpha_{k+2\,k+2} &= A \\ \alpha^2_{k+2\,k+1} \quad + \quad \alpha^2_{k+2\,k+2} &= B, \end{aligned}$$

wobei A und B reell sind. Da $\Delta\,(p_0, \ldots, p_{k+2}) \neq 0$ und $|p'_1, \ldots, p'_{k+2}| = (\alpha_{k+2\,k+2} - \sqrt{-1}\,\alpha_{k+2\,k+1})\,|p'_1, \ldots, p'_k|$ ist, so ergibt sich aus (+), daß $\alpha_{k+2\,k+2} - \sqrt{-1}\;\alpha_{k+2\,k+1} = -\sqrt{-1}\;A \neq 0$, also auch $A \neq 0$. Die Gleichungen (III) kann man dann auch so schreiben:

$$(\mathrm{IV}) \qquad \begin{aligned} \alpha_{k+2\,k+1} + \sqrt{-1}\;\alpha_{k+2\,k+2} &= A \\ \alpha_{k+2\,k+1} - \sqrt{-1}\;\alpha_{k+k\,k+2} &= \frac{B}{A}. \end{aligned}$$

Die Gleichungen (IV) haben genau eine Lösung in $\alpha_{k+2\,k+1}$ und $\alpha_{k+2\,k+2}$ und zwar ist $\alpha_{k+2\,k+1}$ reell und $\alpha_{k+2\,k+2}$ rein imaginär (oder 0). Aus der Beziehung

$$|p'_1, \ldots, p'_{k+2}| = (\alpha_{k+2\,k+2} - \sqrt{-1}\;\alpha_{k+2\,k+1}) \cdot |p'_1, \ldots, p'_k|$$

folgt dann mit Rücksicht auf (+), daß $\operatorname{sgn} \Delta\,(p_0, \ldots, p_{k+2}) \neq \operatorname{sgn} \Delta\,(p_0, \ldots, p_k)$. Damit ist die Annahme (α_{k+2}) realisiert. Da die Annahme (α_1) bzw. (α_2) (je nachdem $\Delta\,(p_0, p_1) \neq$ oder $= 0$ ist) offenbar realisierbar ist, so kann wegen $\Delta\,(p_0, p_1, \ldots, p_n) \neq 0$ sukzessive die Annahme (α_n) realisiert werden, wodurch **13** bewiesen ist.

14. *Voraussetzungen:*

1) R sei ein aus $n+1$ Punkten $p_0, \ldots, p_n$ bestehender komplexmetrischer Raum mit reellen Abständen.

2) Es sei $\Delta\,(p_0, \ldots, p_r) \neq 0$ für eine Zahl $r \leq n-1$.

3) Für je $r + 2$ Punkte $p_{i_0}, \ldots, p_{i_{r+1}}$ sei $\Delta\,(p_{i_0}, \ldots, p_{i_{r+1}}) = 0$.

Behauptung: $p_0, \ldots, p_n$ sind kongruent mit $n+1$ Punkten jenes $R_{r,s}$, welcher $r+1$ mit $p_0, \ldots, p_r$ kongruente Punkte enthält.

Zunächst ist klar, daß $p_0, \ldots, p_n$ mit $n+1$ Punkten $p'_0, \ldots, p'_n$ von K_r kongruent wird, wobei nach **3** und **13** $p_0' = (0, \ldots, 0)$ und $p'_1, \ldots, p'_r$ in $R^*_{r,s}$ gewählt werden können. Ist $p'_i = (\alpha_{i1}, \ldots, \alpha_{ir})$ $(i = 1, \ldots, r)$ und $p'_i = (x_{i1}, \ldots, x_{ir})$ $(i = r+1, \ldots, n)$, so gilt für $i = r+1, \ldots, n$:

$$-2\,(x_{i1}\alpha_{11}+\ldots+x_{ir}\alpha_{1r})=p_ip_1-p_ip_0-p_1p_0$$
$$-2\,(x_{i1}\alpha_{21}+\ldots+x_{ir}\alpha_{2r})=p_ip_2-p_ip_0-p_2p_0$$
$$\ldots\ldots\ldots\ldots\ldots\ldots\ldots\ldots$$
$$-2\,(x_{i1}\alpha_{r1}+\ldots+x_{ir}\alpha_{ir})=p_ip_r-p_ip_0-p_rp_0.$$

Da die Determinante $-2\,|\alpha_{ik}|$ dieses Gleichungssystemes $\neq 0$ ist, sind die x_{ik} eindeutig bestimmt und da die rechten Seiten der Gleichungen reell sind, so müssen die x_{ik} für $K=1,\ldots,\frac{r+s}{2}$ reell und für $K>\frac{r+s}{2}$ rein-imaginär oder 0 sein ($i=r+1,\ldots,n$); d. h. $p'_0,\ldots,p'_n$ liegt in $R^*_{r,s}$.

III. Pseudo-$R_{n,s}$-Systeme.

Ein aus $n+3$ Punkten bestehender komplexmetrischer Raum heiße ein $P_{n,s}$, wenn er nicht mit $n+3$ Punkten des $R_{n,s}$ kongruent ist, aber je $n+2$ seiner Punkte mit $n+2$ Punkten des $R_{n,s}$ abstandsgleich sind. $P_{n,n}$ sind offenbar Mengers pseudoeuklidische $(n+3)$-Tupel.

15. $\Delta\,(P_{n,s})$ *ist* $\neq 0$. Denn wäre $\Delta\,(P_{n,s})=0$, so müßte wegen **9** $P_{n,s}$ mit einer Teilmenge des K_n kongruent sein gegen unsere Voraussetzung.

16. *Ein* $P_{n,s}$ *ist mit einer Teilmenge des* $R_{n+2,s}$ *kongruent.* Es sei $p_0, p_1,\ldots,p_{n+2}$ eine ausgezeichnete Anordnung von $P_{n,s}$ im Sinne von **12**. Da nach Voraussetzung $\Delta\,(p_0,\ldots,p_{n+1})=0$ ist, so ist $\Delta\,(p_0,\ldots,p_n)\neq 0$. Die Punkte $p_0,\ldots,p_n$ sind nach Voraussetzung mit $n+1$ Punkten des $R_{n,s}$ kongruent. Wegen **13** ist dann die Anzahl der Zeichenwechsel in der Zahlenreihe:

$$(\Delta_n)\quad \Delta\,(p_0)=-1,\quad \Delta\,(p_0,p_1),\;\ldots,\quad \Delta\,(p_0,\;\ldots,\;p_n),$$

gleich $\frac{n+s}{2}$. Da $\Delta\,(p_0,\ldots,p_{n+1})=0$ ist, so ist $\operatorname{sgn}\Delta\,(p_0,\ldots,p_{n+2})\neq$ $\neq\operatorname{sgn}\Delta\,(p_0,\ldots,p_n)$, folglich ist die Anzahl der Zeichenwechsel in der Zahlenreihe (Δ_{n+2}) gleich $\frac{n+s}{2}+1$. Wegen **13** ist dann $P_{n,s}$ mit einer Teilmenge des $R_{n+2,s}$ kongruent.

17. $\operatorname{sgn}\;\Delta\;(P_{n,s})=(-1)^{\frac{n+s}{2}}$. Dies folgt unmittelbar aus der Tatsache, daß $\Delta\,(p_0)=-1$ und die Anzahl der Zeichenwechsel in der Zahlenreihe (Δ_{n+2}) gleich $\frac{n+s}{2}+1$ ist.

Bemerkung: Ein $P_{n,n}$ ist nach **16** mit einer Teilmenge des $R_{n+2,n}$ kongruent. $R_{n+2,n}$ hat die Maßform: $x_1^2+\ldots+x_{n+1}^2-$ $-x_{n+2}^2$. Man kann nun mühelos zeigen, daß ein $n+3$ Tupel des $R_{n+2,n}$ dann und nur dann ein $P_{n,n}$ ist, falls es linear unabhängig

4

ist und je $n+2$ Punkte eine Hyperebene bestimmen, die mit der positiven oder negativen Richtung der x_n Achse den Winkel $\frac{\pi}{4}$ einschließt.

Angelegenheiten des Kolloquiums und Nachrichten.

Wiener Dissertationen im Zusammenhange mit dem Kolloquium 1932/33:

O. Schreiber: *„Zur Axiomatik der geometrischen Verknüpfungsbeziehungen“* (vgl. diese Ergebn., *3*, S. 14 u. 34, und einen in den Monatsheften f. Math. u. Phys. erscheinenden Aufsatz).

A. Flexer: *„Einige Sätze über komplexmetrische Räume“* (vgl. dieses Heft, S. 10, 32, 34, 36).

Stipendien:

1932/33 arbeitete im Kolloquium Herr A. Flores (Madrid) mit einem Stipendium vom Laboratorio Matematico de Madrid im Wintersemester und von der Fundación Conde de Cartagena der Madrider Akademie der Wissenschaften im Sommersemester.

Begründet wurde aus dem Gesamterträgnis eines im Winter 1932 abgehaltenen Wiener Vortragszyklus „Krise und Neuaufbau in den exakten Wissenschaften“ [1]) ein Fonds zur Subventionierung wissenschaftlicher Arbeiten des akademischen Nachwuchses, aus welchem u. a. Forschungen von Teilnehmern des mathematischen Kolloquiums subventioniert wurden.

[1]) Erschienen bei F. Deuticke, Leipzig und Wien. Von den fünf Vorträgen sind drei mathematischen Inhaltes: H. Hahn, „Die Krise der Anschauung“, K. Menger, „Die neue Logik“, G. Nöbeling, „Die vierte Dimension und der krumme Raum“.

Druck von Gottlieb Gistel & Cie., Wien.

Bericht über das Kolloquium 1933/34.

69. Kolloquium (17. XI. 1933).

Menger: *Eine Bemerkung über Längenmengen.*

Es sei B ein Teilbogen eines euklidischen Raumes. Ist $p_0, p_1, \ldots, p_n$ irgend eine bogengemäß geordnete endliche Teilmenge von B, welche die Endpunkte von B enthält, so bilden wir die Menge der 2^n Vektoren $\sum_{i=1}^{n} \varepsilon_i \overrightarrow{p_{i-1}p_i}$ für alle $\varepsilon_i = \pm 1$ $(i = 1, 2, \ldots, n)$. Die abgeschlossene Hülle der Summe dieser Vektorenmengen für alle bogengemäß geordneten endlichen Teilmengen von B nennen wir[1]) die Längenmenge $L(B)$ von B und beweisen: *Die Längenmenge jedes rektifizierbaren Teilbogens B eines euklidischen Raumes enthält den Vektor* 0. Zum Beweise genügt es wegen der definitionsgemäßen Abgeschlossenheit der Menge $L(B)$ zu zeigen, daß $L(B)$ Vektoren beliebig kleiner Länge enthält.

Sei P die bogengemäß geordnete endliche Teilmenge $p_0, p_1, \ldots, p_n$ von B. Für jedes i ist der Durchschnitt von B mit der Symmetralen von p_{i-1} und p_i nicht leer. Es sei q_i ein Punkt von B, für den $p_{i-1} q_i = p_i q_i$ gilt. Nennen wir h_i die Höhe des gleichschenkeligen Dreiecks p_{i-1}, q_i, p_i $(i = 1, \ldots, n)$ und setzen wir $h(P) = \sum_{i=1}^{n} h_i$, so besitzt der in $L(B)$ enthaltene Vektor $\sum_{i=1}^{n} \overrightarrow{p_{i-1}\, q_i} - \sum_{i=1}^{n} \overrightarrow{q_i p_i}$ offenbar eine Länge $\leqq 2h(P)$. Wir sehen also:

(+) Zu jeder endlichen Teilmenge P von B existiert ein Vektor einer Länge $\leqq 2h(P)$ in $L(B)$.

Wir nennen nun in jedem der gleichschenkeligen Dreiecke p_{i-1}, q_i, p_i die Basislänge π_i, die Summe der Schenkellängen $2\rho_i$. Offenbar gilt

$$(++) \qquad 2\rho_i = \pi_i \sqrt{1 + \frac{4h_i^2}{\pi_i^2}} \geqq \pi_i.$$

Es sei nun $\bar{h}$ irgend eine positive Zahl, die < 1, $< h(P)$ und $< 2l$ ist, wo l die Länge des rektifizierbaren Bogens B bedeutet. Für jede der Zahlen $i = 1, \ldots, n$ gilt entweder $\frac{h_i}{\pi_i} > \frac{\bar{h}}{2l}$ (es gelte dies etwa

[1]) Vgl. diese Ergebnisse 5, S. 1 ff.

für die Zahlen $i=i_1, \ldots, i_k$), — oder $\frac{h_i}{\pi_i} \leqq \frac{\bar{h}}{2l}$ (dies gelte für $i=j_1, \ldots, j_{n-k}$). Offenbar gilt $\sum_{\alpha=1}^{n-k} h_{j_\alpha} \leqq \frac{\bar{h}}{2}$ und daher wegen $\bar{h} < h(P)$

$$\begin{pmatrix}+\,+\\+\end{pmatrix} \qquad \sum_{\alpha=1}^{k} h_{i_\alpha} > \frac{\bar{h}}{2}.$$

In $\left(1+\frac{h_i}{\pi_i}\frac{\bar{h}}{2l}\right)^2 = 1+\frac{2h_i}{\pi_i}\frac{\bar{h}}{2l}+\frac{h_i^2}{\pi_i^2}\left(\frac{\bar{h}}{2l}\right)^2$ ist rechts wegen $\bar{h} < 2l$ der letzte Summand $<\frac{h_i^2}{\pi_i^2}$ und der zweite, wenn i eine der Zahlen $i_1, \ldots, i_k$ ist, $<\frac{2h_i^2}{\pi_i^2}$. Für $i=i_1, \ldots, i_k$ gilt also $1+\frac{h_i}{\pi_i}\frac{\bar{h}}{2l} <$ $<\sqrt{1+\frac{4h_i^2}{\pi_i^2}}$. Wegen $(+\,+)$ und $\begin{pmatrix}+\,+\\+\end{pmatrix}$ folgt daraus $\sum_{\alpha=1}^{k} 2\rho_{i_\alpha} - \sum_{\alpha=1}^{k} \pi_{i_\alpha} >$ $> \sum_{\alpha=1}^{k} h_{i_\alpha}\frac{\bar{h}}{2l} > \frac{\bar{h}^2}{4l}$. Wegen $2\rho_i \geqq \pi_i$ für $i=1, \ldots, n$ gilt also $\sum_{i=1}^{n} 2\rho_i -$ $-\sum_{i=1}^{n} \pi_i > \frac{\bar{h}^2}{4l}$. Da $\sum_{i=1}^{n} 2\rho_i$ die Länge $l(P_1)$ des Polygons ist, das durch die Menge $P_1 = r(P)$ bestehend aus den Punkten $p_0, q_1, p_1, q_2, p_2, \ldots, p_n$ bestimmt ist, so gilt $l(P_1) - l(P) > \frac{\bar{h}^2}{4l}$.

Bilden wir nun die Mengen $r(P_1) = P_2$ mit der Höhensumme $h(P_1)$ und allgemein $r(P_{n-1}) = P_n$ mit der Höhensumme $h(P_n)$ für jedes n, so muß $\lim_{n=\infty} h(P_n) = 0$ gelten. Denn andernfalls existierte ein $\varepsilon > 0$, so daß $h(P_n) > \varepsilon$ für unendlich viele n wäre und wenn $\bar{h}$ eine positive Zahl < 1, $< 2l$ und $< \varepsilon$ bezeichnete, so würde $l(P_n) -$ $-l(P_{n-1}) > \frac{\varepsilon^2}{4l}$ für unendlichviele n gelten, also $\lim_{n=\infty} l(P_n) = \infty$, während das durch P_n bestimmte Polygon dem Bogen B der Länge l eingeschrieben ist, also $l(P_n) \leqq l$ für jedes n gilt. Wir haben somit gezeigt, daß es endliche Teilmengen P_n von B gibt, für die $h(P_n)$ beliebig klein ist, woraus wegen (+) der eingangs behauptete Satz folgt. Aus ihm ergibt sich unmittelbar, *daß die Längenmenge jedes rektifizierbaren Teilbogens eines euklidischen Raumes konvex ist.*

Vilim: *Damit ein Polytop eines euklidischen Raumes Längenmenge eines Polygons sei, ist notwendig und hinreichend, daß es zentrisch symmetrisch und konvex sei und seine Randpolytope Längenmengen seien.* (Ein Beweis ist in der Wiener Dissertation von Vilim enthalten.)

70. Kolloquium (24. XI. 1933).

Flores: *Über stetige Selbstabbildungen der S_n.*

Die drei folgenden Sätze, in denen R_n den n-dimensionalen euklidischen Raum, S_n die Oberfläche einer Kugel des R_{n+1} bezeichnet, sind äquivalent:

I_{n+1}). Für jede stetige Abbildung von S_{n+1} in R_{n+1} gibt es mindestens ein Paar diametral gegenüberliegender Punkte von S_{n+1}, die auf denselben Punkt von R_{n+1} abgebildet werden.

II_n). Jede antipodentreue Selbstabbildung von S_n ist wesentlich[1]).

III_n). Jede stetige Selbstabbildung $f(S_n)$ von S_n, für welche die Bilder von je zwei diametral gegenüberliegenden Punkten verschieden sind, ist wesentlich.

Mit Hilfe des Satzes I_{n+1}) kann man II_n) beweisen.

Aus II_n) kann man III_n) ableiten. Denn es sei S_n die Einheitssphäre des R_{n+1} mit dem Zentrum O. Es sei p ein Punkt der S_n. Mit $\Pi_t(p)$ bezeichnen wir den Durchschnittspunkt des Radius Op mit der mit S_n konzentrischen Sphäre von Radius t. Für ein festes t und für jeden Punkt a von S_n betrachten wir den vom Punkt $\Pi_t(f(a'))$ zu $f(a)$ führenden Vektor (a' bedeutet den zu a diametral entgegengesetzten Punkt von S_n) und den Durchschnitt $\varphi_t(a)$ von dem zu diesen Vektor durch O parallelen Halbstrahl mit S_n. Indem man nun t von 1 bis 0 variieren läßt, erhält man eine Schar von stetigen Selbstabbildungen $\varphi_t(a)$ von S_n, welche die mit $\varphi_0(S_n)$ offenbar identische Abbildung $f(S_n)$ in die (als antipodentreu nach Satz II_n wesentliche) Abbildung $\varphi_t(S_n)$ stetig überführt.

Aus III_n) ergibt sich unmittelbar I_{n+1}).

Aus I_n) kann man den Satz der Invarianz der Dimensionszahl bekommen. Denn wäre R_m in R_n $(m>n)$ topologisch abgebildet, dann müßte S_n auch in R_n topologisch enthalten sein und das ist wegen I_n) unmöglich.

Terasaka (Tokio): *Ein Satz über Längenmengen.*

Es sei $B_{ab}=B$ ein Bogen (d. h. topologisches Bild der Strecke $0\leq x\leq 1$) in einem euklidischen Raum R_k mit den Endpunkten a und b. Es seien $p_1, p_2, \ldots p_n$ Punkte von B die bzw. den n Werten $x_1, x_2, \ldots, x_n$ zwischen 0 und 1 entsprechen. Wir bezeichnen die Vektoren $ap_1, p_1p_2, \ldots, p_nb$ der Reihe nach mit $\mathfrak{v}_1, \mathfrak{v}_2, \ldots, \mathfrak{v}_{n+1}$ und betrachten den Vektor $0p=0a+\sum_{i=1}^{n+1}\varepsilon_i\mathfrak{v}_i$, wobei $\varepsilon_i=\pm 1$ und 0 der Koordinatenursprung ist. Die Menge der Endpunkte p aller solcher Vektoren $0p$ für alle x_i und $\varepsilon_i (i=1, 2, \ldots, n)$ bei festem n wollen wir die *Längenmenge n-ten Grades* von B nennen und mit $L_n(B)$ bezeichnen.

Satz. *Ist B_{0b} ein im Punkt 0 beginnender Bogen im R_n, dann enthält die Längenmenge n-ten Grades $L_n(B_{0b})$ den Punkt 0.*

$L_n(B_{0b})$ ist, wie man leicht sieht, gleich der Summe aller Mengen $L_{n-1}(B_{ab})$ für alle Punkte a auf B_{0b} und der dazu in bezug auf den Punkt 0 symmetrischen Mengen. Wir haben also die Existenz eines Punktes a_0 auf B_{0b} zu zeigen derart, daß $L_{n-1}(B_{a_0b})$ den Punkt 0 enthält. Es sei nun a ein beliebiger Punkt auf B_{0b} und sei der Teilbogen B_{ab} von B_{0b} das topologische Bild der Strecke $0\leq x\leq 1$.

[1]) Für I_n) und II_n) vgl. Borsuk, Fund. Math. *20*, S. 178 und dessen Hinweis auf Ulam bezüglich I_n).

Je $n-1$ Werten $x_1, x_2, \ldots, x_{n-1}$ mit $0 \leqq x_1 \leqq x_2 \leqq \cdot - \leqq x_{n-1} \leqq 1$ und jedem n-Tupel $\varepsilon_i = \pm 1$ $(i = 1, 2, \cdot -, n)$ entspricht einerseits der Vektor $0p = 0a + \sum_{i=1}^{n} \varepsilon_i \mathfrak{v}_i$, andererseits aber auch eineindeutig ein Punkt s der $(n-1)$-dimensionalen Einheitsphäre S_{n-1} mit den Koordinaten

$$(\varepsilon_1 \sqrt{x_1}, \quad \varepsilon_2 \sqrt{x_2 - x_1}, \quad \ldots \quad, \varepsilon_n \sqrt{1 - x_{n-1}}).$$

Wenn man also dem Punkt s den Schnittpunkt p_s des Vektors $0p$ mit der S_{n-1} entsprechen läßt, so ist diese Abbildung φ eine stetige Abbildung der S_{n-1} in sich selbst. Fällt a mit 0 zusammen, so ist $0p = \sum_{i=1}^{n} \varepsilon_i \mathfrak{v}_i$ und es entsprechen den antipodischen Punkten s' und s'' von S_{n-1} vermöge der Abbildung φ antipodische Punkte p'_s und p''_s. Also ist nach einem Satz von Borsuk[1]) der Abbildungsgrad von φ nicht gleich Null. Liegt hingegen a hinreichend nahe an b, ist der Abbildungsgrad von φ gleich Null. Aus Stetigkeitsgründen folgt hieraus, daß es einen Punkt a_0 auf B_{ob} gibt, für den 0 in $L_{n-1}(B_{a_0 b})$ enthalten ist, w. z. b. w.

71. Kolloquium (8. XII. 1938).

Flores: *Über n-dimensionale Komplexe, die im R_{2n+1} absolut selbstverschlungen sind.*

1. Es sei C ein n-dimensionaler Komplex. Zwei n-dimensionale Seiten von C mögen getrennt heißen, wenn sie keinen gemeinsamen Eckpunkt besitzen. C' sei ein mit C homöomorpher Komplex. Dann bezeichnen wir mit $U(C)$ den Komplex, welcher aus allen $(2n+1)$-dimensionalen Simplexen $V(a, b')$ und deren Seiten besteht, wo a irgend ein n-dimensionales Simplex von C und b' Bild in C' von irgend einem von a getrennten Simplex b von C ist und $V(A, B)$ hier und im Folgenden die Verbindung („join") von A und B bedeutet.

2. Es sei jetzt P der aus einem einzigen Punkt bestehende nulldimensionale Komplex. Man bilde $V(C, P)$. Die Menge aller geordneten Punktepaare von $V(C, P)$, für welche der eine Punkt in einem n-dimensionalen Simplex s oder dessen Seiten liegt und der andere in der Verbindung mit P von einem von s getrennten Simplex t oder dessen Seiten, macht man zu einem Raum $L(C)$, indem man sie in üblicher Art metrisiert. Von zwei Punktepaaren, die sich nur durch die Reihenfolge ihrer Komponenten unterscheiden, sagen wir, daß sie in $L(C)$ symmetrisch liegen.

3. *Es sind $L(C)$ und $U(C)$ homöomorph.* Sei nämlich (α, β) ein Punkt von $L(C)$, für den α und β zu C gehören. Die Strecke

[1]) K. Borsuk, Fund. Math. *20*, S. 178. Den Hinweis auf diesen Aufsatz verdanke ich Herrn Flores.

$(\alpha, V(\beta, P))$ von $L(C)$ bilde man affin auf die α enthaltende Hälfte der α und β' in $U(C)$ verbindenden Strecke $V(\alpha, \beta')$ derart ab, daß α dem Punkt (α, P), der Mittelpunkt von $V(\alpha, \beta')$ dem Punkt (α, β) entspricht. Analog bilde man $(V(\alpha, P), \beta)$ auf die andere Hälfte von $V(\alpha, \beta')$ ab, so daß der Mittelpunkt von $V(\alpha, \beta')$ dem Punkt (α, β), der Punkt (P, β) dem Punkt β' entspricht. Man sieht leicht, daß die so definierte Abbildung eine Homöomorphie ist und daß die Urbilder von zwei Punkten von $U(C)$, die einander vermöge des Homöomorphismus zwischen C und C' entsprechen, in $L(C)$ symmetrisch liegen. Wir sagen von zwei Punkten, daß sie in $U(C)$ symmetrisch liegen, wenn ihre Urbilder in $L(C)$ symmetrisch liegen.

4. Der Komplex C heißt im R_m *absolut selbstverschlungen*[1]), falls für keine stetige Abbildung $\varphi(V(C, P))$ von $V(C, P)$ in den R_m, für welche φ auf C topologisch ist, $\varphi(C)$ und $\varphi(V(C, P) - C)$ fremd sind. Offenbar ist keine Teilmenge des R_{m-1} absolut selbstverschlungen im R_m. Wird von einem Komplex C nachgewiesen, daß er im R_m absolut selbstverschlungen ist, so ist damit zugleich gezeigt, *daß C nicht topologisch in den R_{m-1} einbettbar ist.*

5. Für jede stetige Abbildung S von $V(C, P)$ in den R_m erhält man ein stetiges Bild $\psi(L(C))$ von $L(C)$ in denselben R_m, wenn man für jeden Punkt (α, β) von $L(C)$ den Vektor von $\varphi(\alpha)$ nach $\varphi(\beta)$ parallel verschiebt, bis sein Anfangspunkt mit dem Koordinatenursprung zusammenfällt, und dann den Endpunkt als Bild von (α, β) erklärt. Offenbar sind die Bilder je zweier in $L(C)$ symmetrisch liegender Punkte bezüglich des Koordinatenursprunges symmetrisch. Es ist klar, daß, wenn für jede $\varphi(V(C, P))$ der Ursprung zu $\psi(L(C))$ gehört, C im R_m absolut selbstverschlungen, also wegen **4** nicht mit einer Teilmenge des R_{m-1} homöomorph ist.

6. Dies ist immer der Fall, wenn $U(C)$ mit S_{2n+1} (der Oberfläche der $(2n+2)$-dimensionalen Vollkugel) homöomorph ist, die Bilder von in $U(C)$ symmetrisch liegenden Punkten in S_{2n+1} diametral gegenüberliegen und $m = 2n+1$ gilt; denn für jede stetige Abbildung von S_m in R_m existiert mindestens ein Paar diametral gegenüberliegender Punkte von S_m, die auf denselben Punkt von R_m abgebildet werden[2]).

7. Die Komplexe K_n bestehend aus allen höchstens n-dimensionalen Simplexen, deren Endpunkte zu $2n+3$ gegebenen Punkten gehören und M_n, induktiv definiert als $V(M_{n-1}, M_0)$, wo M_0 eine aus drei Punkten bestehende Menge ist[3]), genügen den Bedingungen

---

[1]) Vgl. diese Ergebnisse *5*, S. 18.

[2]) Dieser von Ulam vermutete Satz ist von Borsuk, Fund. Math. *20*, S. 178 bewiesen worden.

[3]) van Kampen, Abhandl. Hamb. Sem. *9*, S. 72. Dabei ist M_1 die Mengersche Tetraederkonfiguration (vgl. diese Ergebnisse *2*, S. 30). K_1 und M_1 finden sich bei Kuratowski, Fund. Math. *15*.

von **6**, sind also nicht in den R_{2n} topologisch einbettbar[1]). Um dies für K_n zu beweisen, wählt man die $2n+3$ Eckpunkte des S_{2n+1} eingeschriebenen regulären $(2n+2)$-dimensionalen Simplexes als Eckpunkte von K_n, ihre diametral gegenüberliegenden Punkte als Eckpunkte von K'_n. Bildet man alle diejenigen Gruppen von höchstens $2n+2$ Punkten aus diesen $4n+6$ Ecken von K_n und K'_n, unter denen keine zwei diametral gegenüberliegende Punkte vorkommen, und alle Simplexe, die als Ecken solche Punktgruppen besitzen, so erhält man $U(K_n)$. Die Projektion aus dem Mittelpunkt von S_{2n+1} ist eine Homöomorphie, die je zwei in $U(K_n)$ symmetrisch liegende Punkte in diametral gegenüberliegende Punkte von S_{2n+1} überführt. Was M_n betrifft, so ist klar, daß M_0 mit K_0 homöomorph ist. Machen wir die induktive Voraussetzung $U(M_{n-1}) \equiv S_{2n-1}$, so ist $U(M_n) \equiv U(V(M_{n-1}, M_0)) \equiv V(U(M_{n-1}), U(M_0)) \equiv V(S_{2n-1}, S_1) \equiv$ $\equiv S_{2n+1}$. Mithilfe dieser Formel beweist man auch, daß die Bilder von in $U(M_n)$ symmetrisch liegenden Punkten in S_{2n+1} diametral gegenüberliegen. Übrigens ist es leicht, zu beweisen, daß die Komplexe $V(K_m, K_{n-m-1})$ und $V(K_m, M_{n-m-1})$ den Bedingungen von **6** genügen.

8. Kuratowski hat bewiesen, daß ein eindimensionaler Komplex G dann und nur dann plättbar, d. h. mit einer Teilmenge der Ebene homöomorph ist, wenn G keine M_1 oder K_1 homöomorphe Teilmenge enthält[2]). Sagt man von einem Teilkomplex T von $U(G)$, er liege in $U(G)$ symmetrisch, wenn T zu jedem Punkt p auch den zu p in $U(G)$ symmetrisch liegenden Punkt enthält, so ergibt sich aus dem Vorangehenden: Notwendig und hinreichend für die Plättbarkeit eines eindimensionalen Komplexes G ist, daß $U(G)$ keine symmetrisch liegende S_3 enthält.

9. Aus der Existenz von im R_{2n+1} absolut selbstverschlungenen n-dimensionalen Komplexen ergibt sich unmittelbar der Lebesgue-Brouwersche Satz von der *Invarianz der Dimensionszahl*[3]). Wären nämlich der R_m mit einer Teilmenge des R_n homöomorph für $m > n$, dann wäre auch der R_{n+1} im R_n topologisch enthalten. Ist n gerade $= 2k$, so wäre ein im R_{2k+1} absolut selbstverschlungener Komplex im R_{2k} topologisch enthalten und das ist unmöglich. Ist $n = 2k+1$, so ergibt sich der Widerspruch daraus, daß wenn C ein im R_{2k+1} absolut selbstverschlungener Komplex und P ein einpunktiger Komplex, ist $V(C, P)$ in den R_{2k+2}, aber nicht in den R_{2k+1} einbettbar ist.

---

[1]) Im folgenden erbringe ich einen, soweit mir bekannt, neuen Beweis der Nichteinbettbarkeit der Komplexe K_n und M_n in den R_{2n}. Übrigens kann ich auch was meinen Beweis dieser Tatsache für M_n in diesen Ergebnissen *5*, S. 17ff. betrifft, der Bemerkung von Alexandroff im Mathem. Zentralblatt, daß derselbe im wesentlichen mit v. Kampens Ausführungen in den Abhandl. Hamb. Sem. *9* übereinstimmt, nicht beipflichten.

[2]) l. c. Fund. Math. *15*, S. 271.

[3]) Über dessen direkte Herleitung aus dem Satz von Ulam-Borsuk vgl. S. 3 dieses Heftes.

Alt und Beer: *Der n-Gittersatz in Bogen.*

In der Bogenmetrik[1]) verwendet Menger den Satz, *daß auf jedem Bogen B für jedes natürliche n ein n-Gitter existiere,* d. h. eine geordnete Teilmenge $b^0, b^1, \ldots, b^n$ von B, für die b^0 und b^n die Endpunkte von B und die Abstände $b^{i-1} b^i$ einander gleich sind für $i = 1, 2, \ldots, n$. Der Beweis ist jedoch lückenhaft, denn es werden die reellen Zahlen in zwei Klassen geteilt, wobei der ersten Klasse eine Zahl r dann und nur dann zugeteilt wird, wenn $n + 1$ Punkte $b^0, b^1, \ldots, b^n$ von B existieren, von denen b^0 der Anfangspunkt von B ist und b^i der erste auf b^{i-1} folgende Punkt, der von b^{i-1} den Abstand r hat $(i = 1, \ldots, n)$, worauf behauptet wird, daß diese Klasseneinteilung ein Dedekindscher Schnitt sei. Verbindet man aber z. B. in der Ebene die Punkte (0, 0), (0, 4), (3, 4), (0, 5), (—2, 4) in dieser Reihenfolge durch einen Streckenzug, so ist für den so entstehenden Streckenzug für $n = 2$ die erwähnte Klasseneinteilung kein Schnitt. da $r = 2$ und $r = 5$ zur ersten, $r = 4$ zur zweiten Klasse gehören.

Wir beweisen die Existenz von n-Gittern zunächst für den Fall, daß B ein *Streckenzug* mit den Endpunkten 0 und p ist. Wir wählen irgend eine Zahl r, und schlagen eine Kugel mit dem Radius r um 0. Sie schneidet jede Strecke in höchstens 2 Punkten, wobei hier und im Folgenden ein Schnittpunkt doppelt gezählt werden soll, wenn in ihm der Streckenzug die Kugel nicht durchsetzt. Ist für eine Zahl $k \leqq n-1$ eine endliche Punktmenge $P_{k-1}(r)$ auf B definiert, so schlagen wir um jeden Punkt q von $P_{k-1}(r)$ eine Kugel vom Radius r, schneiden sie mit dem Bogenstück qp und bezeichnen die Menge aller dieser Schnittpunkte für alle q von $P_{k-1}(r)$ mit $P_k(r)$. Gilt $0 < r < \frac{0p}{n}$, so zeigt man durch Induktion leicht, daß jede Menge $P_k(r)$ $(k = 1, 2, \ldots, n)$ bei der oben angegebenen Zählung eine ungerade Anzahl von Punkten enthält. Ferner kann man zeigen, daß zu jedem $\bar{r}$, für welches $P_n(\bar{r})$ nicht den Endpunkt p von B enthält, ein $\bar{r}$ enthaltendes offenes Intervall (r', r'') existiert, so daß für jedes r aus (r', r'') die Anzahl der Menge $P_n(r)$ sich von der von $P_n(\bar{r})$ entweder gar nicht oder um eine gerade Zahl unterscheidet. Da die Anzahl von $P_n(r)$ für kleines r ungerade, für ein r, das größer als der Durchmesser von B ist, aber 0 ist, so muß also ein r_0 existieren, für welches p in $P_n(r_0)$ liegt. Hieraus folgt die Existenz eines n-Gitters in B unmittelbar.

Ist nun B ein *beliebiger* Bogen eines euklidischen Raumes, so schreiben wir ihm eine Folge von Streckenzügen, die gegen B konvergieren, ein, konstruieren auf jedem ein n-Gitter und greifen aus dieser Folge von $(n + 1)$-Tupeln eine konvergente Teilfolge heraus. Ihr Limes ist offenbar ein n-Gitter von B.

[1]) Mathem. Annalen, *104*, S. 487.

72. Kolloquium (19. I. 1934).

Wald: *Über Quasikongruenzordnung.* Erscheint in verallgemeinerter Form in Heft 7 dieser Ergebnisse.

73. Kolloquium (10. II. 1934).

R. G. Putnam (New York): *Über Endpunkte höheren Geschlechtes.*

Menger hat gesprächsweise die Frage aufgeworfen, ob für jede Zahl $n \neq 2$ die Menge aller Punkte eines Raumes, welche vom Typus (α, n) sind, und vielleicht sogar bei festem $n \neq 2$ die Summe aller dieser Mengen über alle isolierten Zahlen α nulldimensional sei? Im Folgenden wird diese Frage für den Fall $n=1$ und $\alpha = 1, 2$ in positivem Sinne entschieden, d. h. wir zeigen: *In jedem metrischen Raum ist die Summe der Menge aller Endpunkte und der Menge aller Endpunkte vom Geschlecht 2 nulldimensional.* Dabei heißt p Endpunkt vom Geschlecht 2, wenn p irregulär ist und in beliebig kleinen Umgebungen enthalten ist, deren Begrenzungen einen einzigen Häufungspunkt besitzen. Solche Umgebungen eines Endpunktes vom Geschlecht 2 nennen wir *charakteristische* Umgebungen, während wir als charakteristische Umgebungen eines (regulären) Endpunktes solche mit einpunktigen Begrenzungen bezeichnen.

Hilfssatz: *Jeder Endpunkt vom Geschlecht 2 ist in beliebig kleinen charakteristischen Umgebungen enthalten, für die der einzige Häufungspunkt der Begrenzung von einem Typus* $>(2, 1)$, d. h. irregulär und nicht in beliebig kleinen Umgebungen enthalten ist, deren Begrenzungen nur einen Häufungspunkt besitzen. Sei nämlich Z eine vorgelegte Umgebung von p und U eine charakteristische Umgebung $\Subset Z$ von p. Wir wenden die Mengersche Methode der Modifikation der Umgebungen an. Ist der Häufungspunkt q ihrer Begrenzung von einem Typus $\leqq (2, 1)$, dann existiert eine Umgebung $\Subset Z$ von q, deren abgeschlossene Hülle p nicht enthält und deren Begrenzung entweder endlich ist oder nur einen Häufungspunkt r enthält. Im ersteren Fall oder wenn r in U liegt, ist $U + V$ eine Umgebung $\subset Z$ von p, deren Begrenzung endlich ist. Wenn r außerhalb $\overline{U}$ liegt, ist $U - U \cdot \overline{V}$ eine Umgebung $\subset Z$ von p mit endlicher Begrenzung. Liegt r in $B(U)$ und ist r von einem Typus $\leqq (2, 1)$, so existiert eine Umgebung $W \Subset Z$ von r, deren Begrenzung höchstens einen Häufungspunkt enthält und die so klein ist, daß ihre abgeschlossene Hülle weder p noch einen von r verschiedenen Punkt von $B(U)$ enthält. Das letztere kann erreicht werden, da ja q der einzige Häufungspunkt von $B(U)$ und daher r in $B(U)$ isoliert ist. Der Häufungspunkt von $B(W)$ liegt entweder in U oder außerhalb von $\overline{U}$. Im ersteren Falle ist $U + V + W$, im letzteren $U - U \cdot \overline{V + W}$ eine Umgebung $\subset Z$ von p mit endlicher Begrenzung. Da nun, wenn Z hinlänglich klein ist, wegen der Irregularität von p Umgebungen $\subset Z$ von p mit endlichen Begrenzungen nicht existieren, so muß entweder q oder wenn nicht q, so r von einem Typus $>(2, 1)$ sein und es ist entweder U oder wenn nicht U, so

$U + V$ eine charakteristische Umgebung $\subset Z$ von p, für die der einzige Häufungspunkt der Begrenzung von einem Typus $> (2, 1)$ ist, w. z. b. w.

Zum Beweise des Satzes nach der Methode von Ayres sei p ein Endpunkt vom Geschlecht 2. Wir zeigen, daß die Menge der Endpunkte vom Geschlecht 2 in p nulldimensional ist. Sei zu diesem Zweck Z eine Umgebung von p. Nach dem Hilfssatz existiert eine charakteristische Umgebung $\Subset Z$ von p, für die der einzige Häufungspunkt q der Begrenzung von einem Typus $> (2, 1)$ ist. Die übrigen Punkte von $B(U)$ bilden eine gegen q konvergente Folge $\{q_k\}$ ($k = 1, 2, \ldots$ ad inf.). Für jedes k, für welches q_k ein Endpunkt vom Geschlecht 2 ist, existiert eine charakteristische Umgebung $V_k \Subset Z$, so daß jedes V_k einen Durchmesser < 1 besitzt, daß $\overline{V_k}$ weder p noch einen von q_k verschiedenen Punkt von $B(U)$ enthält und daß die Mengen $\overline{V_k}$ paarweise fremd sind und eine Nullfolge bilden (d. h. gegen 0 konvergente Durchmesser besitzen). Der Häufungspunkt der Begrenzung von V_k liegt, da er nicht in $B(U)$ liegen kann, entweder in U oder außerhalb $\overline{U}$. Im zweiten Falle ist $V_k^i = V_k \cdot U$, im ersten $V_k^a = V_k - \overline{U} \cdot V_k$ eine offene Menge mit endlicher Begrenzung. Ferner wählen wir für jedes k eine charakteristische Umgebung $V_k' \Subset V_k$ von q_k, deren einziger Begrenzungshäufungspunkt von einem Typus $> (2, 1)$ ist. Wir fügen für jedes k, für welches V_k^a eine endliche Begrenzung besitzt, V_k' zu U hinzu und tilgen für jedes sonstige k, für welches also V_k^i eine endliche Begrenzung besitzt, $\overline{V_k'}$ aus U. Da die V_k eine Nullfolge bilden, ist die aus U so entstehende Menge U_1 eine Umgebung $\Subset Z$ von p, deren Begrenzung nur Häufungspunkte enthält, die einen Typus $> (2, 1)$ haben, während jeder etwa in ihr enthaltene Endpunkt vom Geschlecht 2 in einer offenen Menge mit einem Durchmesser < 1 und endlicher Begrenzung liegt. Indem wir auf diese letzteren Punkte dasselbe Verfahren mit offenen Mengen V_{kl}, deren Durchmesser $< \frac{1}{2}$ sind, wiederholen, gelangen wir zu einer Umgebung $U_2 \Subset Z$ von p, deren Begrenzung nur Häufungspunkte von einem Typus $> (2, 1)$ enthält, während jeder ihrer Endpunkte vom Geschlecht 1 oder 2 in zwei offenen Mengen von Durchmessern < 1 und $< \frac{1}{2}$ mit endlichen Begrenzungen liegt. So fortfahrend gelangen wir zu einer Umgebungsfolge $\{U_k\}$, deren Limes U^* eine Umgebung $\subset Z$ von p ist, deren Begrenzung außer Punkten, deren Typus $> (2, 1)$ ist, nur reguläre Punkte, also keine Endpunkte vom Geschlecht 2 enthält. Die Menge der Endpunkte vom Geschlecht 2 ist also in jedem ihrer Punkte 0-dimensional. Es bereitet keine wesentliche Schwierigkeit, dasselbe von der Summe dieser Menge und der Menge aller (regulären) Endpunkte zu beweisen.

Der Übergang zu Geschlechtern > 2 ist nicht ohne weiters möglich, denn die wesentlich verwendete Behauptung, daß entweder V_k^i oder V_k^a eine endliche Begrenzung besitzt, konnte nur deshalb bewiesen werden, weil der Begrenzungshäufungspunkt von V_k nicht in $B(U)$ liegt, was deshalb bewirkt werden konnte, weil q_k isoliert in $B(U)$ ist. Die Analoga für Geschlechter > 2 versagen. Die Außerachtlassung dieses Punktes bildet übrigens m. E. eine Lücke im Beweis des Satzes V von Reschovsky (Fund. Math. *14*, S. 32f). Die Behauptung, daß die Mengen $B^{\alpha-1}(V_1^a)$ und $B^{\alpha-1}(V_1^i)$ fremd seien (a. a. O. S. 33) trifft nicht notwendig zu.

74. Kolloquium. (19. III. 1934.)

Karl Schlesinger (Wien): *Über die Produktionsgleichungen der ökonomischen Wertlehre.*

Zur Bestimmung der Preise der Produktionsmittel und der herzustellenden Mengen von Produkten auf Grund der gegebenen Mengen der Produktionsmittel und bekannter Daten (betreffend erstens die Arten produktiver Verwendungen der Produktionsmittel und zweitens die Abhängigkeit der Preise der Produkte von den produzierten Produktmengen) hat Walras ein Gleichungssystem aufgestellt, welches durch Cassel in einer vereinfachten Form popularisiert worden ist. Sind $R_1, \ldots, R_m$ Produktionsmittel, durch deren verschiedene Kombination n Produkte $S_1, \ldots, S_n$ hergestellt werden können, und zwar derart, daß, um eine Einheit des Produktes S_j zu produzieren, a_{1j} Einheiten des Produktionsmittels R_1, a_{2j} Einheiten des Produktionsmittels $R_2, \ldots$ und a_{mj} Einheiten des Produktionsmittels R_m verwendet werden müssen (für $j = 1, 2, \ldots, n$), und weiß man, daß der Preis einer Einheit des Produktes S_j, falls s_1 Einheiten von $S_1, \ldots, s_n$ Einheiten von S_n produziert werden, $f_j(s_1, \ldots, s_n)$ beträgt, stehen ferner dem Produzenten r_i Einheiten des Produktionsmittels R_i zur Verfügung ($i = 1, \ldots, m$), so lauten die $m + 2n$ Casselschen Gleichungen für die m unbekannten Preise ρ_i der Einheit des Produktionsmittels R_i, für die n unbekannten Zahlen s_j der zu produzierenden Einheiten des Produktes S_j ($j = 1, \ldots, n$) und für die n unbekannten Preise σ_j der Einheit des Produktes S_j ($j = 1, \ldots, n$)

$$\begin{aligned}
r_1 &= a_{11} s_1 + a_{12} s_2 + \ldots + a_{1n} s_n \\
r_2 &= a_{21} s_1 + a_{22} s_2 + \ldots + a_{2n} s_n \\
&\ldots\ldots\ldots\ldots\ldots\ldots \\
r_m &= a_{m1} s_1 + a_{m2} s_2 + \ldots + a_{mn} s_n \\
\sigma_1 &= a_{11} \rho_1 + a_{21} \rho_2 + \ldots + a_{m1} \rho_m \\
\sigma_2 &= a_{12} \rho_1 + a_{22} \rho_2 + \ldots + a_{m2} \rho_m \\
&\ldots\ldots\ldots\ldots\ldots\ldots \\
\sigma_n &= a_{1n} \rho_1 + a_{2n} \rho_2 + \ldots + a_{mn} \rho_m \\
\sigma_1 &= f_1(s_1, s_2, \ldots, s_n) \\
\sigma_2 &= f_2(s_1, s_2, \ldots, s_n) \\
&\ldots\ldots\ldots\ldots \\
\sigma_n &= f_n(s_1, s_2, \ldots, s_n).
\end{aligned}$$

Daß diese Gleichungen nicht notwendig Lösungen besitzen und vor allem nicht notwendig positive Lösungen besitzen (und nur solche kommen mit Rücksicht auf die Bedeutung der Zahlen ρ_i, σ_j, s_j in Betracht), wurde von Stackelberg[1]) und Neisser[2]) bemerkt.

Nun sind bei Walras und Cassel unter den Produktionsmitteln $R_1, \ldots, R_m$ bloß die „seltenen" Produktionsmittel gemeint, d. h. diejenigen, deren gesamte verfügbare Menge in der Produktion Verwendung findet und die dementsprechend nicht unentgeltlich erworben werden können. Die Serie der seltenen Produktionsmittel wird von diesen Autoren als ein Datum der Wirtschaft angesehen. Diese Annahme ist jedoch unzuläßig, weil der Sachverhalt der Seltenheit oder Nicht-Seltenheit eines Produktionsmittels seinerseits von den Nachfragekurven, technischen Produktionsmöglichkeiten usw. abhängig ist.

Wenn wir nun unter $R_1, \ldots, R_m$ *sämtliche* (seltene *und* freie) Produktionsmittel verstehen, so gliedern sich die Produktionsmittel $R_1, \ldots, R_m$ in zwei Gruppen; für die Einen (d. h. für die seltenen Produktionsmittel) gilt

$$r_i = a_{i1} s_1 + a_{i2} s_2 + \ldots + a_{in} s_n \text{ und } \rho_i > 0,$$

für die Anderen (d. h. für die freien Produktionsmittel)

$$r_j > a_{j1} s_1 + a_{j2} s_2 + \ldots + a_{jn} s_n \text{ und } \rho_j = 0.$$

Bezeichnet man den nichtverwendeten Überschuß des Produktionsmittels R_j mit $\ddot{u}_j$, so ergeben sich an Stelle der ersten m Gleichungen des obigen Gleichungssystems folgende Gleichungen für sämtliche Produktionsmittel:

$$\begin{aligned} r_1 &= a_{11} s_1 + a_{12} s_2 + \ldots + a_{1n} s_n + \ddot{u}_1 \\ r_2 &= a_{21} s_1 + a_{22} s_2 + \ldots + a_{2n} s_n + \ddot{u}_2 \\ &\ldots\ldots\ldots\ldots\ldots\ldots\ldots \\ r_m &= a_{m1} s_1 + a_{m2} s_2 + \ldots + a_{mn} s_n + \ddot{u}_m, \end{aligned}$$

wobei diese m neuen Unbekannten $\ddot{u}_1, \ldots, \ddot{u}_m$, die nicht negativ sein können, durch folgende m Nebenbedingungen mit den übrigen verknüpft sind:

Wenn $\ddot{u}_i > 0$, *so gilt* $\rho_i = 0$ $(i = 1, \ldots, m)$

oder, was gleichbedeutend ist:

Für jedes $i = 1, \ldots, m$ *ist von den Unbekannten* ρ_i *und* $\ddot{u}_i$ *mindestens eine* $= 0$.

Von diesem modifizierten System von $m + 2n$ Gleichungen mit m Nebenbedingungen in $2m + 2n$ Unbekannten $\ddot{u}_i, \rho_i, s_j, \sigma_j$ $(i = 1, \ldots, m, j = 1, \ldots, n)$ hat nun auf meine Anregung Herr Wald zunächst unter der vereinfachenden Annahme, daß die n letzten Gleichungen $\sigma_j = f_j(s_j)$ $(j = 1, \ldots, n)$ lauten, bewiesen, daß es tatsächlich stets nicht negative Lösungen besitzt, und zwar nur eine *einzige* Lösung in den Unbekannten $\ddot{u}_i, s_j, \sigma_j$ $(i = 1, \ldots, m;\ j = 1, \ldots, n)$.[3])

[1]) Zeitschrift f. Nationalökonomie 1933.

[2]) Weltwirtschaftliches Archiv 1932.

[3]) *Zusatz bei der Korrektur:* Nach Abhaltung vorstehenden Vortrages wurde ich darauf aufmerksam, daß in der Arbeit von F. Zeuthen: „Das Prinzip der Knappheit, technische Kombination und ökonomische Qualität" im Band IV. der Zeitschrift für Nationalökonomie, 1933, schon auf Folgendes hingewiesen wurde: „Da man im voraus nicht weiß, welche produktiven Leistungen freie

Wald: *Über die eindeutige positive Lösbarkeit der neuen Produktionsgleichungen.*

Theorem: *Das Gleichungssystem* (*Sch*)

$$r_i = \sum_{j=1}^{n} a_{ij} s_j + ü_i \; (i=1,\ldots,m), \; \sigma_j = \sum_{i=1}^{m} a_{ij} \rho_i, \; \sigma_j = f_j(s_j), \; (j=1,\ldots,n),$$

in welchem die r_i *und* a_{ij} *gegebene Zahlen, die* f_j *bekannte Funktionen, die* $ü_i, \rho_i, s_j, \sigma_j$ *unbekannte Zahlen sind, besitzt, wenn folgende vier Annahmen gemacht werden:*

1. $r_i > 0$ $(i=1,\ldots,m)$.

2. $a_{ij} \geqq 0$ $(i=1,\ldots,m;\; j=1,\ldots,n)$.

3. Zu jedem j $(j=1,\ldots,n)$ *gibt es mindestens ein* i $(i=1,\ldots,m)$, *so daß* $a_{ij} \neq 0$.

4. Für jede der Zahlen $j=1,\ldots,n$ *ist die Funktion* $f_j(s_j)$ *für jeden positiven Wert von* s_j *definiert, nicht negativ, stetig und im schärferen Sinne monoton abnehmend,* d. h. aus $s'_j < s_j$ folgt stets $f_j(s'_j) > f_j(s_j)$; *ferner gilt* $\lim_{s_j=0} f_j(s_j) = \infty$.

ein in den Unbekannten $ü_i, s_j, \sigma_j$ *eindeutiges Lösungssystem, das folgenden Nebenbedingungen genügt:*

a) $s_j \geqq 0$ $(j=1,\ldots,n)$;

b) $\sigma_j \geqq 0$ $(j=1,\ldots,n)$;

c) $\rho_i \geqq 0$ $(i=1,\ldots,m)$;

d) $ü_i \geqq 0$ $(i=1,\ldots,m)$;

e) wenn $ü_i > 0$, *so ist* $\rho_i = 0$ $(i=1,\ldots,m)$.

Wir beweisen, indem wir unter *„Lösungssystem"* stets ein den Nebenbedingungen *a)—e)* genügendes Lösungssystem der Gleichungen (*Sch*) verstehen, zunächst drei Hilfssätze.

1. Ist $ü_1,\ldots,ü_m, \rho_1,\ldots,\rho_m, s_1,\ldots,s_n, \sigma_1,\ldots,\sigma_n$ ein Lösungssystem und $ü_1+\Delta u_1,\ldots,ü_m+\Delta ü_m, \rho_1+\Delta\rho_1,\ldots,\rho_m+\Delta\rho_m, s_1+\Delta s_1,\ldots, s_n+\Delta s_n, \sigma_1+\Delta\sigma_1,\ldots,\sigma_n+\Delta\sigma_n$ ein Lösungssystem der Gleichungen, die aus (*Sch*) entstehen, wenn man die $r_1,\ldots,r_m$ durch $r_1+\Delta r_1,\ldots, r_m+\Delta r_m$ ersetzt, so gilt

$$\Delta r_1 \Delta\rho_1 + \ldots + \Delta r_m \Delta\rho_m \leqq 0.$$

Beweis: Zunächst ist klar, daß folgende Beziehungen gelten:

$$\Delta r_i - \Delta ü_i = a_{i1}\Delta s_1 + a_{i2}\Delta s_2 + \ldots + a_{in}\Delta s_n \quad (i=1,\ldots,m)$$
$$\Delta\sigma_j = a_{1j}\Delta\rho_1 + a_{2j}\Delta\rho_2 + \ldots + a_{mj}\Delta\rho_m \quad (j=1,\ldots,n).$$

Multipliziert man für $i=1,\ldots,m$ die erste Beziehung mit $\Delta\rho_i$ und addiert diese m Beziehungen, so erhält man

$$(\Delta r_1 - \Delta ü_1)\Delta\rho_1 + \ldots + (\Delta r_m - \Delta ü_m)\Delta\rho_m =$$
$$(a_{11}\Delta s_1 + \ldots + a_{1n}\Delta s_n)\Delta\rho_1 + \ldots + (a_{m1}\Delta s_1 + \ldots + a_{mn}\Delta s_n)\Delta\rho_m =$$
$$(a_{11}\Delta\rho_1 + \ldots + a_{m1}\Delta\rho_m)\Delta s_1 + \ldots + (a_{1n}\Delta\rho_1 + \ldots + a_{mn}\Delta\rho_m)\Delta s_n =$$
$$\Delta\sigma_1\Delta s_1 + \ldots + \Delta\sigma_n\Delta s_n.$$

Güter sind, sollte man in diesen Gleichungen als letztes Glied einen etwaigen unbenützten Rest einschalten und gleichzeitig als Bedingung festsetzen, daß entweder dieser Rest oder der Preis der Produktionsleistung gleich 0 ist." Mit der Frage der allgemeinen positiven und eindeutigen Lösbarkeit des Gleichungssystems wurde diese Modifikation von **Zeuthen** nicht in Verbindung gebracht.

Aus Annahme 4 über $f_j(s_j)$ folgt unmittelbar $\Delta\sigma_1 \Delta s_1 + \ldots + \Delta\sigma_n \Delta s_n \leqq 0$, also gilt $(\Delta r_1 - \Delta ü_1)\Delta\rho_1 + \ldots + (\Delta r_m - \Delta ü_m)\Delta\rho_m \leqq 0$ und daher $\Delta r_1 \Delta\rho_1 + \ldots + \Delta r_m \Delta\rho_m \leqq \Delta ü_1 \Delta\rho_1 + \ldots + \Delta ü_m \Delta\rho_m$. Zum Beweise von Satz 1 genügt es also zu zeigen: $\Delta ü_1 \Delta\rho_1 + \ldots + \Delta ü_m \Delta\rho_m \leqq 0$. Wir beweisen sogar, daß jeder Summand $\Delta ü_i \Delta\rho_i \leqq 0$. Ist nämlich erstens $\Delta ü_i > 0$, so folgt aus Bedingung *d)*, daß $ü_i + \Delta ü_i > 0$ und daher aus *e)*, daß $\rho_i + \Delta\rho_i = 0$; da $\rho_i \geqq 0$ nach Bedingung *c)*, muß also $\Delta\rho_i \leqq 0$ und demnach $\Delta ü_i \Delta\rho_i \leqq 0$ gelten. Ist zweitens $\Delta ü_i = 0$, so ist $\Delta ü_i \Delta\rho_i = 0$. Ist drittens $\Delta ü_i < 0$, so folgt aus *d)*, daß $ü_i > 0$ und daher aus *e)*, daß $\rho_i = 0$; wegen *c)* muß dann $\Delta\rho_i \geqq 0$ sein, also gilt auch in diesem und daher in jedem Fall $\Delta ü_i \Delta\rho_i \leqq 0$, w. z. b. w.

2. Ist $\{r_1^{(k)}, \ldots, r_m^{(k)}\}$ $(k = 1, 2, \ldots$ ad inf.) eine konvergente Folge von m-Tupeln positiver Zahlen, die gegen das m-Tupel $r_1, \ldots, r_m$ konvergieren und ist $ü_1^{(k)}, \ldots, ü_m^{(k)}, \rho_1^{(k)}, \ldots, \rho_m^{(k)}, s_1^{(k)}, \ldots, s_n^{(k)}, \sigma_1^{(k)}, \ldots, \sigma_n^{(k)}$ ein Lösungssystem der Gleichungen $(Sch^{(k)})$, die entstehen, wenn in den Gleichungen (Sch) $r_1, \ldots, r_m$ durch $r_1^{(k)}, \ldots, r_m^{(k)}$ ersetzt werden, dann sind die Zahlenfolgen $\{ü_1^{(k)}\}, \ldots, \{ü_m^{(k)}\}, \{s_1^{(k)}\}, \ldots, \{s_n^{(k)}\}$ $(k = 1, 2, \ldots$ ad inf.) beschränkt und es ist für jedes i, für welches $r_i \neq 0$, auch die Zahlenfolge $\{\rho_i^{(k)}\}$ $(k = 1, 2, \ldots$ ad inf.) beschränkt.

Beweis: Nach Annahme 3 gibt es zu jedem j mindestens ein i, so daß $a_{ij} \neq 0$. Wegen $ü_i^{(k)} \geqq 0$ folgt aus der Gleichung $r_i^{(k)} = a_{i1} s_1^{(k)} + \ldots + a_{in} s_n^{(k)} + ü_i^{(k)}$, daß $r_i^{(k)} \geqq a_{i1} s_1^{(k)} + \ldots + a_{in} s_n^{(k)} \geqq \geqq a_{ij} s_j^{(k)} \geqq 0$. Da die Folge $\{r_i^{(k)}\}$ gegen r_i konvergiert, muß also vor allem die Folge $\{s_j^{(k)}\}$ beschränkt sein, wie behauptet. Aus der Beschränktheit der Folgen $\{s_1^{(k)}\}, \ldots, \{s_n^{(k)}\}$ folgt auf Grund der m ersten Gleichungen (Sch) die behauptete Beschränktheit der Folgen $\{ü_1^{(k)}\}, \ldots, \{ü_m^{(k)}\}$. Um noch die Beschränktheit der $\{\rho_i^{(k)}\}$ mit $r_i \neq 0$ zu beweisen, setzen wir voraus, es sei etwa $r_1 > 0$, und leiten aus der Annahme, die Folge $\{\rho_1^{(k)}\}$ sei unbeschränkt, einen Widerspruch her. Aus den Gleichungen (Sch) folgt dann, daß für jedes j, für das $a_{1j} \neq 0$, die Folge $\{\sigma_j^{(k)}\}$ unbeschränkt ist. Es gibt also eine Teilfolge $\{g_k\}$ der natürlichen Zahlen, so daß $\lim\limits_{k=\infty} \rho_1^{(g_k)} = \infty$ und $\lim\limits_{k=\infty} \sigma_j^{(g_k)} = \infty$ für jedes j, für das $a_{1j} \neq 0$. Aus Annahme 4 folgt dann $\lim\limits_{k=\infty} s_j^{(g_k)} = 0$ für jedes j, für das $a_{1j} \neq 0$, mithin $\lim\limits_{k=\infty} (a_{11} s_1^{g_k} + \ldots + a_{1n} s_n^{g_k}) = 0$. Also muß $\lim\limits_{k=\infty} ü_1^{(g_k)} = r_1 > 0$ gelten, woraus wegen Bedingung *e)* folgt, daß $\lim\limits_{k=\infty} \rho_1^{(g_k)} = 0$ gilt, womit der behauptete Widerspruch hergeleitet und alle Behauptungen von Satz 2 bewiesen sind.

Aus der Tatsache, daß die Bedingungen *a)*—*e)* beim Grenzübergang $k \to \infty$ erhalten bleiben, folgt:

3. Konvergiert eine Folge von Lösungssystemen der Gleichungen $(Sch^{(k)})$ mit $M \to \infty$ gegen ein Zahlensystem $ü_1, \ldots, ü_m, \rho_1, \ldots \rho_m$,

$s_1, \ldots, s_n, \sigma_1, \ldots, \sigma_n$, so ist dieses letztere ein Lösungssystem der Gleichungen (Sch).

Wir beweisen nun das Theorem für $n = 1$, in welchem Falle das Gleichungssystem lautet:

$$(Sch_1): \; r_1 = a_{11} s_1 + \ddot{u}_1, \ldots, r_m = a_{m1} s_1 + \ddot{u}_m, \sigma_1 = a_{11} \rho_1 + \ldots + a_{m1} \rho_m, \\ \sigma_1 = f_1(s_1).$$

Nach Annahme 3 gibt es mindestens ein i, so daß $a_{i1} \neq 0$. Für alle diese i bilden wir $\frac{r_i}{a_{i1}}$ und nennen α die kleinste dieser Zahlen; es sei etwa $\frac{r_1}{a_{11}} = \alpha$. Dann kann weder $s_1 > \alpha$ eine Lösung von (Sch) sein, da sonst $\ddot{u}_1 < 0$ wäre, was unmöglich ist, noch $s_1 < \alpha$, da sonst alle $\ddot{u}_i > 0$ wären und daher $\rho_1 = \ldots = \rho_m = 0$ gelten müßte, was der vorletzten Gleichung (Sch_1) widerspricht, da aus Annahme 4 folgt, daß $\sigma_1 > 0$. Wir setzen also $s_1 = \alpha$ und können dann aus den m ersten Gleichungen (Sch_1) eindeutig $\ddot{u}_1, \ldots, \ddot{u}_m$ bestimmen, wobei, da $\frac{r_1}{a_{11}} = \alpha$ die kleinste unter den Zahlen $\frac{r_i}{a_{i1}}$ ist, $\ddot{u}_1 = 0$ und $\ddot{u}_i \geqq 0 \; (i = 2, \ldots, m)$ gilt. Für alle i, für welche $\ddot{u}_i > 0$ gilt, setzen wir $\rho_i = 0$ in der letzten Gleichung (Sch_1). Wegen $\ddot{u}_1 = 0$ bleibt mindestens ρ_1 stehen und da $a_{11} \neq 0$ gilt, ist diese letzte Gleichung (im allgemeinen auf zahlreiche Arten) gemäß den Bedingungen *a)*—*e)* in den $\rho_i \; (i = 1, \ldots, m)$ lösbar, womit das Theorem für $n = 1$ bewiesen ist.

Nach dem Prinzip der vollständigen Induktion genügt es nun, das Theorem unter der induktiven Annahme, daß es für $n - 1$ gelte, für den Fall n zu beweisen.

Wir nennen $\mu(s_n)$ die kleinste der Zahlen $r_1 - a_{1n} s_n$, $r_2 - a_{2n} s_n, \ldots, r_m - a_{mn} s_n$. Wegen Annahme 3 gibt es eine Zahl $\bar{\lambda}$, so daß $\mu(\bar{\lambda}) = 0$. Wir bilden nun für jede Zahl λ, für die $0 \leqq \lambda < \bar{\lambda}$ gilt, das Gleichungssystem

$$(Sch^{\lambda}_{n-1}) \quad r'_i = r_i - a_{in} \lambda = a_{i1} s_1 + \ldots + a_{in-1} s_{n-1} \; (i = 1, \ldots, m) \\ \sigma_j = a_{1j} \rho_1 + \ldots + a_{mj} \rho_m, \; \sigma_j = f_j(s_j) \; (j = 1, \ldots, n-1).$$

Wegen $0 \leqq \lambda < \bar{\lambda}$ gilt $r'_i > 0 \; (i = 1, \ldots, m)$. Offenbar genügen die gegebenen Größen des Systems (Sch^{λ}_{n-1}) den Annahmen 1—4. Nach der induktiven Annahme ist dieses Gleichungssystem also lösbar.

Es sei nun λ irgend eine bestimmte Zahl des Intervalles $[0, \bar{\lambda})$. Wir bilden für jedes Lösungssystem $\rho_1, \ldots, \rho_m$ von (Sch^{λ}_{n-1}) die Zahl $a_{1n} \rho_1 + \ldots + a_{mn} \rho_m$ und nennen die so erhaltene Zahlenmenge $\Pi(\lambda)$. Sind λ_1 und λ_2 zwei Zahlen, so daß $0 \leqq \lambda_1 < \lambda_2 < \bar{\lambda}$ gilt, ist $\rho_1, \ldots, \rho_m$ irgend ein Lösungssystem von $(Sch^{\lambda_1}_{n-1})$ und $\rho_1 + \Delta\rho_1, \ldots, \rho_m + \Delta\rho_m$ irgend eines von $(Sch^{\lambda_2}_{n-1})$, so gilt Hilfssatz 1 zufolge

$$-a_{1n}(\lambda_2 - \lambda_1)\Delta\rho_1 - \ldots - a_{mn}(\lambda_2 - \lambda_1)\Delta\rho_m \leqq 0, \text{ also } a_{1n}\Delta\rho_1 + \ldots + \\ a_{mn}\Delta\rho_m \geqq 0.$$

Wir sehen also:

(+) *Ist* $0 \leqq \lambda_1 < \lambda_2 < \bar\lambda$, *so ist keine Zahl von* $\Pi(\lambda_2)$ *kleiner als irgend eine Zahl von* $\Pi(\lambda_1)$.

Da nach induktiver Annahme für $0 \leqq \lambda < \bar\lambda$ die Menge $\Pi(\lambda)$ *nichtleer* ist, folgt aus (+), daß für $0 \leqq \lambda < \bar\lambda$ die Menge $\Pi(\lambda)$ *beschränkt* ist. Nach Hilfssatz 3 ist $\Pi(\lambda)$ *abgeschlossen*. Sind $\rho_1', \ldots, \rho_m'$ und $\rho_1'', \ldots, \rho_m''$ zwei Lösungssysteme von (Sch_{n-1}^{λ}), so ist auch $\rho_1^{(t)} = \frac{\rho_1' + t\rho_1''}{1+t}, \ldots, \rho_m^{(t)} = \frac{\rho_m' + t\rho_m''}{1+t}$ für jedes positive t ein Lösungssystem von (Sch_{n-1}^{λ}). Ist also $a_{1n}\rho_1' + \ldots + a_{mn}\rho_m' = \alpha'$ und $a_{1n}\rho_1'' + \ldots + a_{mn}\rho_m'' = \alpha''$, so nimmt der Ausdruck $a_{1n}\rho_1^{(t)} + \ldots + a_{mn}\rho_m^{(t)}$ für geeignetes t offenbar jeden vorgelegten Wert zwischen α' und α'' an, d. h. zur Menge $\Pi(\lambda)$ gehören zugleich mit je zwei Zahlen α' und α'' auch alle Zahlen zwischen α' und α'', m. a. W. die Menge $\Pi(\lambda)$ ist *konvex*. Als nichtleere beschränkte abgeschlossene Menge ist $\Pi(\lambda)$ ein *abgeschlossenes Intervall*, dessen linker bzw. rechter Endpunkt $\underline{\pi}(\lambda)$ bzw. $\bar\pi(\lambda)$ heißen möge, das sich aber auch auf einen Punkt $\underline{\pi}(\lambda) = \bar\pi(\lambda)$ reduzieren kann. Hieraus ergibt sich auf Grund von (+):

(++) *Ist* $\lambda_1 < \lambda_2$, *so sind die Mengen* $\Pi(\lambda_1)$ *und* $\Pi(\lambda_2)$ *entweder fremd oder sie haben einen einzigen Punkt, nämlich* $\bar\pi(\lambda_1) = \underline{\pi}(\lambda_2)$ *gemein,* in welch letzterem Falle für jedes λ zwischen λ_1 und λ_2 die Menge $\Pi(\lambda)$ aus diesem Punkt besteht.

Es sei nun $\{\lambda_k\}$ $(k = 1, 2, \ldots$ ad inf.) irgend eine Zahlenfolge, für die $\lambda < \lambda_k < \bar\lambda$ für alle k und $\lim\limits_{k=\infty} \lambda_k = \lambda$ gilt. Offenbar ist dann die Zahlenfolge $\{\underline{\pi}(\lambda_k)\}$ $(k = 1, 2, \ldots$ ad inf.) beschränkt. Ist α irgend ein Häufungselement dieser Folge, so ist $\alpha \geqq \bar\pi(\lambda)$ wegen (+). Nach Hilfssatz 2 ist die Folge der Lösungssysteme der Gleichungen $(Sch_{n-1}^{\lambda_k})$ $(k = 1, 2, \ldots$ ad inf.) beschränkt. Aus Hilfssatz 3 folgert man leicht, daß α in der Menge $\Pi(\lambda)$ liegt, also wegen $\lambda_k > \lambda$ mit $\bar\pi(\lambda)$ identisch sein muß. Damit ist bewiesen:

$\binom{++}{+}$ Aus $\lambda < \lambda_k < \bar\lambda$ und $\lim\limits_{k=\infty} \lambda_k = \lambda$ folgt $\lim\limits_{k=\infty} \pi(\lambda_k) = \bar\pi(\lambda)$; und analog beweist man: aus $0 \leqq \lambda_k < \lambda$ und $\lim\limits_{k=\infty} \lambda_k = \lambda$ folgt $\lim\limits_{k=\infty} \bar\pi(\lambda_k) = \underline{\pi}(\lambda)$.

Ferner ist klar, daß mit $\lambda \to \bar\lambda$ die Folge $\{\underline{\pi}(\lambda)\}$ konvergiert (ev. gegen $+\infty$), d. h. es existiert $\lim\limits_{\lambda=\bar\lambda} \underline{\pi}(\lambda) = \pi^*$.

Wir setzen nun $\underline{\tau}(\lambda) = \underline{\pi}(\lambda) - f_n(\lambda)$, $\bar\tau(\lambda) = \bar\pi(\lambda) - f_n(\lambda)$ für jedes λ aus $[0, \bar\lambda]$ und nennen $T(\lambda)$ das abgeschlossene Intervall mit den Endpunkten $\underline{\tau}(\lambda)$ und $\bar\tau(\lambda)$. Ist $0 \leqq \lambda_1 < \lambda_2 < \bar\lambda$, so folgt aus Annahme 4, daß $f_n(\lambda_1) > f_n(\lambda_2)$ ist; wegen (++) gilt also $\bar\tau(\lambda_1) < \underline{\tau}(\lambda_2)$, d. h. die Intervalle $T(\lambda_1)$ und $T(\lambda_2)$ haben keinen einzigen Punkt gemein, also

(*) *aus* $\lambda_1 \neq \lambda_2$ *folgt, daß* $\mathrm{T}(\lambda_1)$ *und* $\mathrm{T}(\lambda_2)$ *fremd sind.*

Nach Annahme 4 ist f_n stetig. Auf Grund von $\binom{+\,+}{+}$ beweist man daher leicht:

(**) *Ist* τ *eine Zahl, zu der zwei Zahlen* λ_1 *und* λ_2 *existieren, für die* $0 \leqq \lambda_1 < \lambda_2 < \bar{\lambda}$ *und* $\overline{\tau}(\lambda_1) < \tau < \underline{\tau}(\lambda_2)$ *gilt, so gibt es ein* λ *zwischen* λ_1 *und* λ_2, *so daß* τ *in* $\mathrm{T}(\lambda)$ *liegt.*

Zum Beweis der Lösbarkeit des Systems (*Sch*) unterscheiden wir nun zwei Fälle.

1. Es gilt $\pi^* = \lim\limits_{\lambda=\bar{\lambda}} \underline{\pi}(\lambda) > f_n(\bar{\lambda})$. In diesem Fall gilt $\lim\limits_{\lambda=\bar{\lambda}} \underline{\tau}(\lambda) > 0$. Da $\lim\limits_{\lambda=0} \overline{\pi}(\lambda)$ endlich ist und $\lim\limits_{\lambda=0} f_n(\lambda) = \infty$ nach Voraussetzung 4 gilt, so ist $\lim\limits_{\lambda=0} \overline{\tau}(\lambda) = -\infty$. Nach (**) gibt es zur Zahl $\tau = 0$ zwischen $\lambda_1 = 0$ und $\lambda_2 = \bar{\lambda}$ ein λ_0, so daß 0 in $\mathrm{T}(\lambda_0)$ liegt, und nach (*) gibt es nur ein solches λ_0. Dann bildet $s_n = \lambda_0$ zusammen mit jedem Lösungssystem von $(Sch_{n-1}^{\lambda_0})$ ein Lösungssystem der Gleichungen (*Sch*) und dasselbe ist eindeutig in $\ddot{u}_1, \ldots, \ddot{u}_m, s_1, \ldots, s_n, \sigma_1, \ldots, \sigma_n$, da s_n nur $= \lambda_0$ sein kann und $(Sch_{n-1}^{\lambda_0})$ nach induktiver Annahme nur eine Lösung in den genannten Unbekannten besitzt.

2. Es gilt $\pi^* = \lim\limits_{\lambda=\bar{\lambda}} \underline{\pi}(\lambda) \leqq f_n(\bar{\lambda})$. Es sei dann $\rho_1^{(k)}, \ldots, \rho_m^{(k)}$ ein Lösungssystem von $(Sch_{n-1}^{\lambda_k})$, $(k = 1, 2, \ldots \text{ ad inf.})$, wobei $\lim\limits_{k=\infty} \lambda_k = \bar{\lambda}$ sei. Die Folgen $\{\rho_i^{(k)}\}$ $(k = 1, 2, \ldots$ ad inf.) sind beschränkt, denn wäre etwa $\{\rho_1^{(k)}\}$ unbeschränkt, so müßte wegen $\lim\limits_{k=\infty} \underline{\pi}(\lambda_k) \leqq f_n(\bar{\lambda})$ offenbar $a_{1n} = 0$ gelten, dann aber könnte Hilfssatz 2 auf (Sch_{n-1}^{λ}) angewendet werden und würde ergeben, daß $\{\rho_1^{(k)}\}$ beschränkt ist.

Es seien nun etwa $i = 1, 2, \ldots, m_0$ diejenigen der Zahlen $1, 2, \ldots, m$, für welche $r_i - a_{in}\bar{\lambda} = 0$ gilt. Auf Grund der Definition von $\bar{\lambda}$ gibt es mindestens eine Zahl i, für die $r_i - a_{in}\bar{\lambda} = 0$ gilt, also ist $1 \leqq m_0 \leqq m$. Würde nun $a_{ij} \neq 0$ gelten, wo $i = 1, \ldots, m_0$; $j = 1, \ldots, n-1$ ist, so würde, wenn $s_j^{(k)}$ die Lösung für s_j in $(Sch_{n-1}^{\lambda_k})$ bezeichnet, $\lim\limits_{k=\infty} s_j^{(k)} = 0$ und $\lim\limits_{k=\infty} \sigma_j^{(k)} = \infty$, also für mindestens ein h auch $\lim\limits_{k=\infty} \rho_h^{(k)} = \infty$ gelten, was unmöglich ist. Es gilt demnach $a_{ij} = 0$ für $i = 1, \ldots, m_0$; $j = 1, \ldots, n-1$; nach Annahme 3 ist dann $m_0 < m$ und (*Sch*) hat folgende Gestalt (*Sch**):

$$
\begin{aligned}
r_i &= \frac{r_i}{\bar{\lambda}} s_n + \ddot{u}_i \quad (i = 1, 2, \ldots, m_0)\\
r_i &= a_{i1} s_1 + \ldots + a_{in} s_n + \ddot{u}_i \quad (i = m_0 + 1, \ldots, m)\\
\sigma_j &= a_{m_0+1\,j}\, \rho_{m_0+1} + \ldots + a_{mj}\, \rho_m \quad (j = 1, \ldots, n-1)\\
\sigma_n &= a_{1n} \rho_1 + \ldots + a_{mn} \rho_m\\
\sigma_j &= f_j(s_j) \quad (j = 1, \ldots, n).
\end{aligned}
$$

Bezeichnen wir mit (Sch^{**}) das Gleichungssystem, das aus (Sch^*) entsteht, wenn man die m_0 ersten und σ_n enthaltenden Gleichungen streicht und $s_n = \bar{\lambda}$ setzt, also

$$r_i - a_{in}\bar{\lambda} = a_{i1}s_1 + \ldots + a_{in-1}s_{n-1} + \ddot{u}_i \quad (i = m_0 + 1, \ldots, m)$$
$$\sigma_j = a_{m_0+1j}\rho_{m_0+1} + \ldots + a_{mj}\rho_m, \quad \sigma_j = f_j(s_j) \quad (j = 1, \ldots, n-1)$$

so hat (Sch^{**}) nach induktiver Annahme Lösungen, die in $s_1, \ldots, s_{n-1}$ eindeutig sind; eine sei

$$\ddot{u}^0_{m_0+1}, \ldots, \ddot{u}^0_m, \rho^0_{m_0+1}, \ldots, \rho^0_m, s^0_1, \ldots, s^0_{n-1}, \sigma^0_1, \ldots, \sigma^0_{n-1}.$$

Wenn man in (Sch^*) setzt: $\ddot{u}_i = 0$ $(i = 1, \ldots, m_0)$, $\ddot{u}_i = u^0_i$ $(i = m_0 + 1, \ldots, m)$, $\rho_i = \rho^0_i$ $(i = m_0 + 1, \ldots, m)$, $s_j = s^0_j$, $\sigma_j = \sigma^0_j$ $(j = 1, \ldots, n-1)$, $s_n = \bar{\lambda}$, so sind alle Gleichungen befriedigt, abgesehen von denen, die σ_n enthalten:

$$\sigma_n = a_{1n}\rho_1 + \ldots + a_{mn}\rho_m, \quad \sigma_n = f_n(\bar{\lambda}).$$

Da wir $\rho_1, \ldots, \rho_{m_0}$ beliebig wählen dürfen, können auch diese Gleichungen befriedigt werden, wofern $f_n(\bar{\lambda}) \geqq a_{m_0+1n}\rho^0_{m_0+1} + \ldots + a_{mn}\rho^0_m$ gilt. Da in dem jetzt betrachteten Fall $f_n(\bar{\lambda}) \geqq \pi^*$ gilt, genügt es zu zeigen, daß eine Lösung $\rho^0_{m_0+1}, \ldots, \rho^0_m$ von (Sch^{**}) existiert, so daß $a_{m_0+1n}\rho^0_{m_0+1} + \ldots + a_{mn}\rho^0_m \leqq \pi^*$. Zu diesem Zweck betrachten wir das Gleichungssystem $(Sch^*_{\lambda_k})$, das aus (Sch^*) entsteht, wenn man in ihm die σ_n enthaltenden Gleichungen streicht und $s_n = \lambda_k$ setzt, wobei $\{\lambda_k\}$ $(k = 1, 2, \ldots$ ad inf.) eine Zahlenfolge mit $\lim\limits_{k=\infty} \lambda_k = \bar{\lambda}$ sei. Auf Grund der induktiven Annahme existiert ein Lösungssystem $\rho^{(k)}_i$ $(i = 1, \ldots, m)$ für jedes k. Es ist offenbar $\rho^{(k)}_1 = \ldots = \rho^{(k)}_{m_0} = 0$. Nach Hilfssatz 2 sind die Folgen $\{\rho^{(k)}_i\}$ $(i = m_0 + 1, \ldots, m)$ beschränkt. Ist $\rho^0_{m_0+1}, \ldots, \rho^0_m$ ein Häufungs-$(m-m_0)$-Tupel der $(m-m_0)$-Tupelfolge $\{\rho^{(k)}_{m_0-1}, \ldots, \rho^{(k)}_m\}$, so folgt aus $\lim\limits_{\lambda=\bar{\lambda}} \bar{\pi}(\lambda) = \pi^*$, daß

$a_{m_0+1n}\rho^0_{m_0+n} + \ldots + a_{mn}\rho^0_m \leqq \pi^*$ gilt und da $\rho^0_{m_0+1}, \ldots, \rho^0_m$ nach Hilfssatz 3 offenbar eine Lösung von (Sch^{**}) ist, sind wir am Ziel.

Zum Abschluß zeigen wir, *daß die Annahme 1 über die n Funktionen $f_j(s_j)$ für die Existenz einer einzigen positiven Lösung des Gleichungssystems (Sch) unentbehrlich ist.*

a) Um die Unentbehrlichkeit der Monotonie im schärferen Sinne von $f_j(s_j)$ zu beweisen, betrachten wir das Gleichungssystem

$$r_1 = a_{11}s_1 + a_{12}s_2 + \ddot{u}_1$$
$$\sigma_1 = a_{11}\rho_1, \quad \sigma_2 = a_{12}\rho_1$$
$$\sigma_1 = f_1(s_1), \quad \sigma_2 = f_2(s_2),$$

wobei $r_1 > 0$ und $a_{11} = a_{12} = a > 0$ gelten möge und f_1 und f_2 auf der Halbgeraden > 0 stetige positive Funktionen seien, für die $\lim\limits_{s_j=0} f_j(s_j) = \infty$ gelten möge, welche aber nicht im schärferen Sinne monoton abnehmend seien, sondern für welche $f_1(x) = f_2(x) = c$ für alle x des

2

Intervalles $\left[\frac{r_1}{4a}, \frac{3r_1}{4a}\right]$ sei. (Die Existenz solcher Funktionen f_1 und f_2 ist evident.) Ist dann λ irgend eine Zahl des Intervalles $\left[\frac{r_1}{4a}, \frac{r_1}{2a}\right]$, so sieht man unmittelbar, daß

$$s_1 = \lambda,\ s_2 = \frac{r_1}{a} - \lambda$$

eine Lösung des Gleichungssystems ist. Es gibt also in diesem Falle unendlich viele Lösungen in s_1, s_2.

b) Um die Unentbehrlichkeit der Annahme $\lim\limits_{s_j=0} f(s_j) = \infty$ zu beweisen, betrachten wir das Gleichungssystem

$$\begin{array}{ll} r_1 = a_{11}s_1 + a_{12}s_2 + \ddot{u}_1, & r_2 = a_{21}s_1 + a_{22}s_2 + \ddot{u}_2, \\ \sigma_1 = a_{11}\rho_1 + a_{21}\rho_2, & \sigma_2 = a_{12}\rho_1 + a_{22}\rho_2, \\ \sigma_1 = f_1(s_1), & \sigma_2 = f_2(s_2), \end{array}$$

wobei $r_1 = r_2 = r > 0$ und $a_{11} = a_{12} = a_{21} = a_{22} = a > 0$ ist und $f_1(s_1)$ eine für $s_1 > 0$ stetige, im schärferen Sinn monoton abnehmende positive Funktion sei, für welche aber $\lim\limits_{s_1=0} f_1(s_1) = \alpha$ sei, wo α endlich ist und $f_1\left(\frac{r}{a}\right) > \frac{\alpha}{2}$ gilt; es sei ferner $f_2(x) = \frac{1}{2}f_1(x)$ für jedes x. (Die Existenz solcher Funktionen f_1 und f_2 ist evident.) Wir zeigen nun, daß das Gleichungssystem keine positiven Lösungen besitzt. Aus den Gleichungen folgt nämlich unmittelbar $\sigma_1 = \sigma_2$, also $f_1(s_1) = f_2(s_2) = \frac{1}{2}f_1(s_2)$. Aus der Gleichung $r = a(s_1 + s_2) + \ddot{u}_1$ folgt wegen der Nichtnegativität von s_1, s_2, $\ddot{u}_1$, daß $s_1 \leqq \frac{r}{a}$. Wegen der Monotonie von f_1 ist also $f_1\left(\frac{r}{a}\right) \leqq f_1(s_1)$. Wegen $f_1(s_1) = \frac{1}{2}f_1(s_2)$ folgt daraus $f_1\left(\frac{r}{a}\right) \leqq \frac{1}{2}f_1(s_2)$, was aber wegen $\frac{\alpha}{2} < f_1\left(\frac{r}{a}\right)$ unmöglich ist, denn es würde $\alpha < f_1(s_2)$ ergeben, während $\alpha = \lim\limits_{x=0} f_1(x)$ und f_1 monoton abnehmend ist, so daß $\alpha \geqq f_1(x)$ für alle x gilt.

Schams (Wien): Durch die Annahme, daß die Funktionen $f_i(s_i)$ monoton abnehmen, daß also bei größerer Menge der Produktionsgüter die Preise der Einheit der Produktionsgüter geringer sind, werden werttheoretische Voraussetzungen in die Untersuchung hineingetragen, während Cassels Gleichungssystem unabhängig von der Wertlehre ist und diese von Cassel bestritten wird.

Menger: Was von Wald bewiesen wurde, ist der Satz: Modifiziert man die Walras-Casselschen Gleichungen, die im allgemeinen keine positiven Lösungen besitzen, nach dem neuen Vorschlag, so besitzen sie, wenn die von Wald präzisierten Annahmen über die Funktionen $f_j(s_j)$ gemacht werden, eindeutige positive Lösungen. Zugleich sind die Waldschen Annahmen *unentbehrliche* Bedingungen für die eindeutige Lösbarkeit der neuen Produktionsgleichungen, d. h. macht man diese Annahmen nicht, so besitzen diese Gleichungen nicht notwendig eindeutige positive Lösungen[1]). Allerdings sind diese Annahmen über die f_j in der Wirklichkeit nicht exakt erfüllt, insbesondere ist es nicht richtig, daß jeder Vergrößerung eines Gütervorrates eine Abnahme des Preises der betreffenden Gütereinheit entspricht. Aber die Waldschen Annahmen sind eine gute Annäherung an die Wirklichkeit und mehr als eine näherungsweise Beschreibung der Wirklichkeit liefert bekanntlich auch die theoretische Physik nicht. Beispielsweise sind die Funktionen $f_i(s_i)$ in Wirklichkeit zwar nicht immer im schärferen Sinne monoton abnehmend, aber sie sind im allgemeinen monoton abnehmend, d. h. es folgt in Wirklichkeit aus $s_j < s'_j$ nicht immer $f_j(s_j) > f_j(s'_j)$, aber im allgemeinen $f_j(s_j) \geqq f_j(s'_j)$. Das besagt nur, daß in der Nachbarschaft von s_j-Werten, für welche die Funktion nicht im schärferen Sinne monoton abnimmt, zahlreiche Lösungen der neuen Produktionsgleichungen existieren können, was aber der Bedeutung und Schönheit des im Vorangehenden dargelegten Ergebnisses, das ich zu den wichtigsten der mathematischen Ökonomie rechne, nicht den mindesten Abbruch tut. Wollte Cassel hingegen die von Wald präzisierten Annahmen *grundsätzlich* ablehnen, so würde er auf Grund des Waldschen Satzes damit zugleich behaupten, daß die von ihm selbst aufgestellten Gleichungen auch nach der scharfsinnigen Modifikation von Schlesinger[2]) nicht eindeutige positive Lösungen besitzen. Welcher Forscher würde aber, um von einem von ihm selbst aufgestellten Gleichungssystem nachzuweisen, daß es *un*lösbar ist, Voraussetzungen bestreiten, die eine gute Annäherung an die Wirklichkeit darstellen?

Richtig ist vielmehr (und wegen seines besonderen Interesses möchte ich diesen Umstand eigens hervorheben), daß durch die Waldsche Arbeit zwei der Grundgedanken der subjektiven Wertlehre, nämlich einerseits das Bedürfnissättigungsgesetz, bzw. der Satz, daß bei größerem Vorrat der Preis der Einheit geringer ist, und andrerseits der Satz, daß der Wert der Produktionsmittel sich nach dem voraussichtlichen Wert der Produkte richte, bzw. die Aussagen über die Verknüpfung der Preise der Produktionsmittel und der Preise der Produkte miteinander in Verbindung gebracht wurden, und zwar besteht die Verbindung darin, daß der Satz vom Grenznutzen in gewisser Hinsicht als unentbehrliche und (zusammen mit anderen Voraussetzungen) als hinreichende Bedingung für die Lösbarkeit des Zurechnungsproblemes dargestellt wird.

---

[1]) Ich möchte übrigens darauf hinweisen, daß die Bedingung $\lim_{s_j=0} f_j(s_j) = \infty$, die zwar (vgl. S. 18) nicht schlechthin weggelassen werden darf, doch *abgeschwächt* werden kann. Eine Analyse des Beweises von Wald zeigt nämlich, daß eine nur von den mn Koeffizienten a_{ij} abhängige und aus ihnen leicht berechenbare endliche Zahl A existiert derart, daß schon die Voraussetzung $\lim_{s_j=0} f_j(s_j) > A$ zusammen mit den übrigen Teilen der Annahme 4 die Existenz und Unizität der Lösungen gewährleistet.

[2]) *Zusatz bei der Korrektur:* Von Zeuthen und Schlesinger vgl. Fußnote zu S. 11.

Eine sehr wichtige Aufgabe wäre nun, hinreichende und zugleich unentbehrliche Bedingungen für Existenz und Eindeutigkeit der Lösungen der neuen Produktionsgleichungen zu formulieren unter der Voraussetzung, daß der Preis der Einheit jedes Produktes nicht nur von der hergestellten Menge dieses einen Produktes, sondern auch von der Menge der übrigen Produkte abhängt. Da unter der vereinfachenden Annahme, daß der Preis jedes Produktes bloß von der Menge dieses einen Produktes abhängt, als Bedingungen für die Lösbarkeit der neuen Produktionsgleichungen sich Annahmen ergeben, die eine gute Annäherung an die Wirklichkeit darstellen, so kann man mit großem Interesse den Bedingungen entgegensehen, welche für die Lösbarkeit der Gleichungen im allgemeinen Falle charakteristisch sind. Da über diese sogenannten Interdependenzerscheinungen bisher nur wenige allgemeine Regelmäßigkeiten formuliert worden sind, so kommt also der Untersuchung der allgemeinen neuen Produktionsgleichungen auch heuristische Bedeutung für prinzipielle preistheoretische Fragen zu.

Jedenfalls möchte ich zum Schluß bemerken, daß mir mit der Waldschen Arbeit die Periode abgeschlossen erscheint, in der die Ökonomen Gleichungen bloß *aufgestellt* haben, ohne sich um Existenz und Eindeutigkeit ihrer Lösungen zu kümmern, und bestenfalls darauf sahen, daß die Anzahl der Unbekannten mit der Anzahl der Gleichungen übereinstimme (was für Existenz und Unizität von Lösungen natürlich weder notwendig noch hinreichend ist). Künftig werden sich die Ökonomen, da in einem auch rein mathematisch tiefliegenden Fall der Existenz- und Eindeutigkeitsbeweis gelungen ist, wenn sie Gleichungen aufstellen, wohl auch (wie die Physiker seit jeher) um ihre Lösungen bemühen müssen.

75. Kolloquium (27. IV. 1934).

Taussky: *Abstrakte Körper und Metrik. 1. Mitteilung: Endliche Mengen und Körperpotenzen.*

P sei ein abstrakter Körper mit der Charakteristik 0. Eine Menge R heißt ein P-*Raum* oder P-*metrische Menge*, wenn je zwei Elementen p und q von R ein Element pq von P als Abstand zugeordnet ist, wobei $pq = qp$ und $pq = 0$ für $p = q$ gilt. Zwei P-Räume heißen *kongruent*, wenn sie eineindeutig und abstandstreu aufeinander abgebildet werden können. Ein P-Raum wird *abstandsgleich* mit dem P-Raum R' genannt, wenn R eindeutig und abstandstreu auf R' abgebildet werden kann. Ist R abstandsgleich mit einer Teilmenge von R', so sagen wir, daß R in R' *einbettbar* ist. Ein P-Raum R hat die *Kongruenzordnung* m, wenn folgendes gilt: Jeder P-Raum, von dem je m Punkte in R einbettbar sind, ist in R einbettbar.

Unter P_n verstehen wir die Menge aller n-Tupel von Elementen aus P, wenn je zwei Punkten $p = (x_1, \ldots, x_n)$ und $q = (y_1, \ldots, y_n)$

der Abstand $\sum_{i=1}^{n}(x_i - y_i)^2$ zugeordnet wird. P_n ist natürlich ein P-Raum. Der Begriff des P_n ist eine Verallgemeinerung des n-dimensionalen euklidischen Raumes R_n und des komplex-metrischen Raumes K_n[1]). Während jede endliche komplex-metrische Menge für hinreichend großes n in den K_n einbettbar ist[2]), gibt es endliche reellmetrische Mengen, welche für kein n in den R_n eingebettet werden können[3]).

Theorem I. *Damit für den Körper* P *jede endliche* P-*metrische Menge für hinreichend großes n in einen* P_n *eingebettet werden kann, ist notwendig und hinreichend, daß* P *nicht formal reell ist,* d. h. daß -1 Summe von endlichvielen Quadraten von Elementen aus P ist. *Ist* -1 *Summe von* μ *Quadraten aus* P, *so ist jeder n-punktige* P-*Raum einbettbar in den* P_{s_n}, *wo* $s_n = (n-1)\mu + \binom{n}{2}$.

Ist irgend ein Element a von P nicht als Summe von Quadraten in P darstellbar, so ist eine Menge, welche zwei Elemente mit dem Abstand a enthält, für kein n in den P_n einbettbar[4]). Also ist die Bedingung notwendig. Der Beweis dafür, daß sie hinreichend ist, läßt sich durch eine Abänderung des Beweises von Satz 6 von Wald[5]) erbringen. Wir setzen voraus, P sei nicht reell, und behaupten: Jede aus n Elementen bestehende P-metrische Menge kann in den P_{s_n} eingebettet werden. Diese Behauptung ist für $n = 2$ richtig, da jede Zahl von P Summe von $\mu + 1$ Quadraten ist[6]). Wir machen die Induktionsannahme, sie gelte für n und beweisen sie für $n+1$ Punkte $p_0, \ldots, p_n$ mit den Abständen $p_i p_k$. Der aus n Punkten bestehende Raum $q_0, \ldots, q_{n-1}$ mit den Abständen $q_k q_i = q_i q_k = p_i p_k - 2$ für $i, k \neq 0$ und $q_0 q_i = q_i q_0 = p_i p_0 - 1$ für $i = 1, \ldots, n-1$ ist laut Induktionsannahme abstandsgleich einer Teilmenge des P_{s_n}. Dabei werde q_0 auf $(0, \ldots, 0)$ abgebildet und dem Punkt q_k entspreche der

---

1) Dabei verstehen wir hier unter dem R_n bzw. K_n die Menge aller n-Tupel reeller bzw. komplexer Zahlen mit der Metrik $x_1^2 + \ldots + x_n^2$. Dieser R_n bzw. K_n ist zwar mit demjenigen Raum, dessen Metrik durch $(x_1^2 + \ldots + x_n^2)^{\frac{1}{2}}$ gegeben ist, nicht kongruent, aber die Bedingungen für die Einbettbarkeit sind analog. Es gilt allgemein: *Führt man in einem Raum R mit der Metrik* $\mathfrak{M}$ *eine neue Metrik* $\mathfrak{M}'$ *ein, welche eine umkehrbar eindeutige Funktion der Metrik* $\mathfrak{M}$ *ist, und sind* $B(p_i p_j)$ *die Bedingungen für die Einbettbarkeit einer Menge in den Raum* $R(\mathfrak{M})$, *so sind* $B(f^{-1}(p_i p_j))$ *die Bedingungen für die Einbettbarkeit derselben Menge in den Raum* $R(\mathfrak{M}')$, *wenn* f^{-1} *die inverse Funktion von* f *ist.*

2) Siehe A. Wald, diese Ergebnisse, *5*, S. 34, Satz 6.

3) Siehe K. Menger, Math. Annalen *100*, S. 134.

4) Dieselbe Bedingung ist übrigens auch dann notwendig, wenn in der P-metrischen Menge nur solche Abstände vorkommen, welche Summen von Quadraten sind.

5) l. c. S. 34 f.

6) Vgl. E. Artin, Über die Zerlegung definiter Funktionen in Quadrate Hamb. Sem. Abt. *5*, S. 101.

Punkt $(\alpha_{k1}, \ldots, \alpha_{ks_n})$ $(k=1, \ldots, n-1)$ des P_{s_n}. Wir betrachten nun im $P_{s_n+n+\mu}$ die folgenden n Punkte

$$\begin{array}{llllllll} & 1 & 2 & & n-1 & n & & s_n+n-1 \quad s_n+n \quad s_n+n+\mu \\ p'_0 = (0, & 0, & \ldots, & 0, & 0, & \ldots, & 0, & 0, \ldots, 0) \\ p'_1 = (1, & 0, & \ldots, & 0, & \alpha_{11}, & \ldots, & \alpha_{1s_n}, & 0, \ldots, 0) \\ \ldots & & & & & & & \\ p'_{n-1} = (0, & 0, & \ldots, & 1, & \alpha_{n-1,1}, & \ldots, & \alpha_{n-1,s_n}, & 0, \ldots, 0). \end{array}$$

Sie sind abstandsgleich mit $p_0, \ldots, p_{n-1}$. Setzen wir

$$p'_n = (\alpha_1, \alpha_2, \ldots, \alpha_{n-1}, 0, \ldots, 0, \beta_1, \ldots, \beta_{\mu+1}),$$

so müssen die Zahlen α_i, β_k folgenden Bedingungen genügen:

$$\alpha_1^2 + \ldots + \alpha_{n-1}^2 + \beta_1^2 + \ldots + \beta_{\mu-1}^2 = p_0 p_n, \quad -2\alpha_i = p_n p_i - p_n p_0 - p_i p_0$$

$(i=1, 2, \ldots, n-1)$. $\alpha_1, \ldots, \alpha_{n-1}$ lassen sich aus den letzten $n-1$ Gleichungen bestimmen. Da jede Zahl von P Summe von $\mu+1$ Quadraten ist, existieren $\mu+1$ Zahlen $\beta_1, \ldots, \beta_{\mu+1}$ aus P, so daß

$$p_0 p_n - \alpha_1^2 - \ldots - \alpha_{n-1}^2 = \beta_1^2 + \ldots + \beta_{\mu+1}^2.$$

Also sind $p_0, \ldots, p_n$ mit $n+1$ Punkten des $P_{s_n+n+\mu} = P_{s_{n+1}}$ kongruent, w. z. b. w.

Wir nennen eine Teilmenge M des P_n *linear abhängig*[1]), wenn $n+1$ nicht sämtlich verschwindende Elemente $\lambda_0, \ldots, \lambda_n$ existieren, so daß $\lambda_0 + \lambda_1 x_1 + \ldots + \lambda_n x_n = 0$ für jeden Punkt $x_1, \ldots, x_n$ von M gilt.

Hilfssatz. *Wenn in* P *jede Quadratsumme ein Quadrat ist, so ist jede linear abhängige Teilmenge M des P_n in den P_{n-1} einbettbar*[2]). Es sei M' eine Menge, welche aus M durch Translation hervorgeht und den Punkt $(0, \ldots, 0)$ des P_n enthält. Auch M' ist eine linear abhängige Menge, und zwar genügen ihre Elemente einer Gleichung $\mu_1 x_1 + \ldots + \mu_n x_n = 0$. Es sei etwa $\mu_n \neq 0$; wir können dann $\mu_n = -1$ annehmen, also $x_n = \mu_1 x_1 + \ldots + \mu_{n-1} x_{n-1}$. Um zu beweisen, daß M' und damit auch M in den P_{n-1} einbettbar ist, zeigen wir, daß $n-1$ Linearformen $L_i(x_1, \ldots, x_n)$ mit Koeffizienten aus P existieren, so daß $x_1^2 + \ldots + x_{n-1}^2 + (\mu_1 x_1 + \ldots + \mu_{n-1} x_{n-1})^2 = L_1^2 + \ldots + L_{n-1}^2$. Ordnet man dann jedem Punkt $(x_1, \ldots, x_n)$ von M' den Punkt $(L_1, \ldots, L_{n-1})$ zu, so ist dies eine eindeutige und abstandstreue Abbildung. Zunächst ist

$$x_1^2 + \ldots + x_{n-1}^2 + (\mu_1 x_1 + \ldots + \mu_{n-1} x_{n-1})^2 = \sum_{i=1}^{n-1} x_i^2 (1 + \mu_i^2) + 2 \sum_{i>k} x_i x_k \cdot \mu_i \mu_k.$$

Ist $1 + \mu_i^2 = 0$ für eine der Zahlen $i = 1, \ldots, n-1$, so liegt $\sqrt{-1}$ in P; bekanntlich ist dann jedes Element von P Summe von Quadraten, also nach Voraussetzung ein Quadrat und es existiert sicher eine Substitution mit Koeffizienten aus P, welche die quadratische

[1]) Dieser Begriff der linearen Abhängigkeit einer Menge deckt sich natürlich im allgemeinen nicht mit dem sonst üblichen.

[2]) Dieser Satz gilt also gleichzeitig für den R_n und K_n, aber, wie Herr Wald bemerkt, z. B. nicht für den P_n, wenn P der Körper der rationalen Zahlen ist.

Form in eine Summe von Quadraten überführt. Ist $1+\mu_i^2 \neq 0$ $(i=1,\ldots,n)$, so wenden wir zunächst die Substitution $x_1=\frac{x_1'}{\sqrt{1+\mu_1^2}}$, $x_i=x_i'$ $(i>1)$ an und erhalten, wenn wir statt x_i' wieder x_i schreiben:

$$x_1^2+x_2^2(1+\mu_2^2)+\ldots+x_{n-1}^2(1+\mu_{n-1}^2)+2\sum_{k=1}^{n-1} x_1 x_k \frac{\mu_1\mu_k}{\sqrt{1+\mu_1^2}}+2\sum_{\substack{i>k\\ i,k\neq 1}} x_i x_k\,.\,\mu_i\mu_k.$$

Durch die Substitution

$$x_1=x_1'-\frac{\mu_1\mu_2}{\sqrt{1+\mu_1^2}}x_2'-\ldots-\frac{\mu_1\mu_{n-1}}{\sqrt{1+\mu_1^2}}x_{n-1}',\ x_i=x_i'\ (i=2,\ldots,n-1)$$

geht dann die Form über in

$$x_1^2+x_2^2\,.\,\frac{1+\mu_1^2+\mu_2^2}{1+\mu_1^2}+\ldots+x_{n-1}^2\,.\,\frac{1+\mu_1^2+\mu_{n-1}^2}{1+\mu_1^2}+2\sum_{\substack{i>k\\ i,k\neq 1}} x_i x_k\,.\,\frac{\mu_i\mu_k}{1+\mu_1^2}.$$

Durch Fortsetzung dieses Verfahrens gelangen wir schließlich zu einer Summe von $n-1$ Quadraten, wobei die verwendeten linearen Substitutionen nur Quadratwurzeln von Quadratsummen als Koeffizienten besitzen.

Mit Verwendung dieses Hilfssatzes und des Theorems I beweist man völlig analog den entsprechenden Sätzen für den K_n (vgl. Wald, diese Ergebnisse 5, S. 35 ff., Satz 5, 7, 8, 9) das

Theorem II. *Ist* P *ein Körper, in dem jedes Element ein Quadrat ist, so ist jeder* $(n+1)$*-punktige* P*-Raum in den* P_n *einbettbar. Damit ein aus* $n+2$ *Punkten* $p_0,\ldots,p_{n+1}$ *bestehender* P*-Raum in* P_n *einbettbar sei, ist notwendig und hinreichend, daß*

$$\Delta(p_0,\ldots,p_{n+1})=\begin{vmatrix} 0 & 1 \\ 1 & p_i p_j \end{vmatrix}_{i,j=0,1,\ldots,n-1}=0$$

sei. Damit ein aus $n+3$ *Punkten bestehender* P*-Raum in* P_n *einbettbar sei, ist notwendig und hinreichend, daß je* $n+2$ *seiner Punkte in den* P_n *einbettbar seien und* $\Delta(p_0,\ldots,p_{n+2})=0$. *Damit ein mehr als* $n+3$ *Punkte enthaltender* P*-Raum in* P_n *einbettbar sei, ist notwendig und hinreichend, daß je* $n+3$ *seiner Punkte in den* P_n *einbettbar seien*[1]).

Wald: *Abstrakte Körper und Metrik. II. Mitteilung.* (Wird in Heft 7 dieser Ergebnisse erscheinen).

76. Kolloquium (15. V. 1934).

M. Morse (Cambridge, Mass.): *Bericht über die Theorie der kritischen Punkte und ihre Anwendung auf die Variationsrechnung im Großen.*

77. Kolloquium (30. V. 1934).

Menger: *Ein Satz über endliche Mengen mit Anwendungen auf die formale Ethik.*

---

[1]) Daß der P_n die Kongruenzordnung $n+3$ hat, läßt sich analog Satz 5 von Wald (l. c.) unter der bloßen Annahme beweisen, daß jede Summe von Quadraten aus P ein Quadrat ist.

Es sei N die aus den Zahlen $1, 2, \ldots, n$ bestehende Menge. $\mathfrak{N}_0$ sei die Menge aller $2^n - 1$ Teilmengen von N. Ist N' (echte oder unechte) Teilmenge von N'', so schreiben wir $N' \subset N''$. Wir betrachten die $2^{2^n-1} - 1$ Teilmengen von $\mathfrak{N}_0$. Ist $\mathfrak{N}$ irgend eine Teilmenge von $\mathfrak{N}_0$, d. h. eine Menge von Teilmengen von N, so sagen wir:

1. $\mathfrak{N}$ genüge dem *Zwischenprinzip*, wenn für je drei Mengen N_1, N_2, N_3 aus $N_1 \subset N_2 \subset N_3 \subset N$ und daraus, daß N_1 und N_3 Elemente von $\mathfrak{N}$ sind, stets folgt, daß N_2 Element von $\mathfrak{N}$ ist:

2. $\mathfrak{N}$ genüge dem *Konjunktionsprinzip*, wenn zugleich mit N_1 und N_2 stets auch $N_1 + N_2$ Element von $\mathfrak{N}$ ist;

3. $\mathfrak{N}$ genüge dem *Disjunktionsprinzip*, wenn zugleich mit N_1 und N_2 stets auch $N_1 . N_2$, wofern diese Menge nicht genau eine der Zahlen $1, 2, \ldots, n$ enthält, Element von $\mathfrak{N}$ ist;

4. $\mathfrak{N}$ sei *abgerundet*, wenn $\mathfrak{N}$ allen drei obigen Prinzipien genügt.

Wir setzen ferner $\mathfrak{N}_0(i) = \mathfrak{N}_0$ für jede der Zahlen $i = 1, 2, \ldots, n$ und bezeichnen mit $\mathfrak{N}_1(i)$ die Menge aller Teilmengen von N, die die Zahl i als Element enthalten, mit $\mathfrak{N}_2(i)$ die Menge aller Teilmengen von N, die die Zahl i nicht als Element enthalten, mit $\mathfrak{N}_3(i)$ die Menge, welche nur die aus der Zahl i allein bestehende Menge (i) als Element enthält, mit $\mathfrak{N}_4(i)$ die Menge aller Teilmengen von N ausgenommen die Menge (i), mit $\mathfrak{N}_5(i)$ den Durchschnitt $\mathfrak{N}_1(i) . \mathfrak{N}_4(i)$. Dann gilt der

Satz: *Damit die Menge $\mathfrak{N}$ sich als Durchschnitt der Form*

$$\mathfrak{N} = \mathfrak{N}_{i_1}(1) . \mathfrak{N}_{i_2}(2) \ldots \mathfrak{N}_{i_n}(n) \quad (i_k = 0, 1, 2, 3, 4, 5 \text{ für } k = 1, 2, \ldots, n)$$

darstellen lasse, ist notwendig und hinreichend, daß $\mathfrak{N}$ abgerundet ist.

Daß die Bedingung notwendig ist, ergibt sich daraus, daß jede Menge $\mathfrak{N}_k(i)$ $(k = 0, 1, \ldots, 5;\ i = 1, 2, \ldots, n)$ abgerundet ist und der Durchschnitt je zweier abgerundeter Menge abgerundet ist. Zum Beweise daß sie hinreicht, sei $\mathfrak{N}$ eine abgerundete Menge. Es seien $i_1, i_2, \ldots, i_r$ alle diejenigen von den Zahlen $1, 2, \ldots, n$, welche in mindestens einer der Mengen $\subset N$, die Element von $\mathfrak{N}$ sind, auftreten, $k_1, k_2, \ldots, k_{n-r}$ die übrigen Zahlen von den Zahlen $1, 2, \ldots, n$; es seien $j_1, j_2, \ldots, j_s$ alle diejenigen von den Zahlen $1, 2, \ldots, n$, welche in allen Mengen, die Element von $\mathfrak{N}$ sind, auftreten. Ist *erstens* $s \geqq 2$, so ist in $\mathfrak{N}$ nach dem Disjunktionsprinzip die Menge $(j_1, j_2, \ldots, j_s)$ enthalten, nach dem Konjunktionsprinzip die Menge $(i_1, i_2, \ldots, i_r)$ und nach dem Zwischenprinzip besteht $\mathfrak{N}$ aus allen und nur den Mengen zwischen den beiden genannten. Dann gilt offenbar

$$(1)\quad \mathfrak{N} = \mathfrak{N}_1(j_1) . \mathfrak{N}_1(j_2) \ldots \mathfrak{N}_1(j_s) . \mathfrak{N}_2(k_1) . \mathfrak{N}_2(k_2) \ldots \mathfrak{N}_2(k_{n-r}).$$

Ist *zweitens* $s = 1$ und ist $j_1 = i_1$ die in allen Elementmengen von $\mathfrak{N}$ auftretende Zahl, so ist, wenn $r = 1$ gilt

$$(2)\qquad \mathfrak{N} = \mathfrak{N}_3(j_1).$$

Wenn $r \geqq 2$, so behaupten wir: Für jede der Zahlen i_ϱ $(\varrho = 2, 3, \ldots, r)$ ist die Menge (i_1, i_ϱ) Element von $\mathfrak{N}$. Dies ist evident für $r = 2$, in welchem Falle $\mathfrak{N}$ offenbar aus den Mengen (i_1) und (i_1, i_2) besteht und

(3) $$\mathfrak{N} = \mathfrak{N}_1(i_1) \, . \, \mathfrak{N}_2(k_1) \, . \, \mathfrak{N}_2(k_2) \ldots \mathfrak{N}_2(k_{n-2})$$

gilt. Für $r > 2$ existiert zu jeder der Zahlen $\sigma = 2, \ldots, \rho - 1, \rho + 1, \ldots, r$, da i_σ nicht in allen Mengen, die Element von $\mathfrak{N}$ sind, enthalten ist, eine Menge N_σ, welche Element von $\mathfrak{N}$ ist und i_σ nicht enthält. Wenn i_ϱ in N_σ enthalten ist, setzen wir $N_\sigma = N'_\sigma$; wenn nicht, so ist, da nach dem Konjunktionsprinzip die Menge $(i_1, i_2, \ldots, i_r)$ in $\mathfrak{N}$ enthalten ist, nach dem Zwischenprinzip die aus den Zahlen von N_σ und i_ϱ bestehende Menge N'_σ Element von $\mathfrak{N}$. Der Durchschnitt $N'_2 \ldots N'_{\varrho-1} \, . \, N'_{\varrho+1} \ldots N'_r$ ist offenbar die Menge (i_1, i_ϱ) und ist, da diese Menge mehr als ein Element enthält, nach dem Disjunktionsprinzip ein Element von $\mathfrak{N}$, wie behauptet. Je nachdem ob die Menge (i_1) in $\mathfrak{N}$ enthalten ist oder nicht, gilt dann offenbar

(4) $$\mathfrak{N} = \mathfrak{N}_1(i_1) \, . \, \mathfrak{N}_2(k_1) \, . \, \mathfrak{N}_2(k_2) \ldots \mathfrak{N}_2(k_{n-r}) \text{ bzw.}$$
$$= \mathfrak{N}_5(i_1) \, . \, \mathfrak{N}_2(k_1) \, . \, \mathfrak{N}_2(k_2) \ldots \mathfrak{N}_2(k_{n-r}).$$

Ist *drittens* $s = 0$, so gilt, wenn die Mengen $(i_{h_1}), (i_{h_2}), \ldots, (i_{h_p})$ nicht Elemente von $\mathfrak{N}$ sind, während für die $r - p$ übrigen von den Zahlen $h = 1, 2, \ldots, r$ die Menge (i_h) Element von $\mathfrak{N}$ ist, wie man leicht bestätigt,

(5) $$\mathfrak{N} = \mathfrak{N}_4(i_{h_1}) \, . \, \mathfrak{N}_4(i_{h_2}) \ldots \mathfrak{N}_4(i_{h_p}) \, . \, \mathfrak{N}_2(k_1) \, . \, \mathfrak{N}_2(k_2) \ldots \mathfrak{N}_2(k_{n-r}).$$

Fügt man zu den Darstellungen (1)—(5) die fehlenden Mengen $\mathfrak{N}_0(i)$ hinzu, so ist in jedem Falle die gewünschte Durchschnittsdarstellung der abgerundeten Menge $\mathfrak{N}$ geliefert.

Bezeichnet[1]) jede der Zahlen $1, 2, \ldots, n$ eine kategorische Norm zur Regelung eines Verhaltens V und stellen diese n Zahlen die sämtlichen zur Regelung von V vorgeschlagenen kategorischen Normen dar, bezeichnet ferner jede Teilmenge $(i_1, i_2, \ldots, i_k)$ von N die Disjunktivnorm „i_1 oder i_2 oder ... oder i_k“, die Menge (i) die „kategorische“ Disjunktivnorm i, so wird durch jede Menge $\mathfrak{N}$ von Teilmengen von N ein Wunsch bezüglich der Regelung des Verhaltens V repräsentiert, nämlich der Wunsch, daß eine von den Disjunktivnormen, deren entsprechende Teilmenge von N Element von $\mathfrak{N}$ ist und keine andere zur Regelung von V festgesetzt werden möge. Die sämtlichen Teilmengen von $\mathfrak{N}_0$ stellen dann die sämtlichen möglichen Wünsche betreffend die Regelung des Verhaltens V dar. $\mathfrak{N}_1(i), \ldots, \mathfrak{N}_5(i)$ entsprechen bzw. den Wünschen, daß das der Norm i gemäße Verhalten erlaubt, verboten, Pflicht, unerzwingbar, freistehend (= erlaubt und unerzwingbar) sei. Die Wünsche, welche nur darauf hinzielen, daß für manche der n Normen eine dieser fünf Regelungen erfolge (während für die übrigens nichts, d. h. $\mathfrak{N}_0(i)$

[1]) Vgl. zum folgenden mein Buch „Moral, Wille und Weltgestaltung“ bei Springer 1934.

gefordert wird), sind dem bewiesenen Satz zufolge identisch mit den „abgerundeten" Wünschen. Das Zwischenprinzip des Wünschens kann hinsichtlich der Stellungnahme gegenüber Normen tatsächlich beobachtet werden, das Konjunktionsprinzip bis zu einem gewissen Grade. Minder entspricht für $n \geqq 4$ das Disjunktionsprinzip der Erfahrung (für $n \leqq 3$ ist es von selbst erfüllt). Es erheben sich im Anschluß an diese Untersuchungen mehrere Fragen, z. B. wie die bloß dem Zwischen- und dem Konjunktionsprinzip genügenden Wünsche charakterisiert sind, usw. Auch auf die Modalitätenlogik der Begriffe möglich, unmöglich, notwendig, nicht notwendig, möglich und nicht notwendig (entsprechend den fünf Begriffen erlaubt, . . ., freistehend) sind der bewiesene und verwandte Sätze anwendbar.

Wald: *Über die Anzahl der abgerundeten Mengen.*

Um dieselbe zu bestimmen, genügt es nach dem vorhergehenden Satz die Zahl derjenigen unter den Teilmengen von $\mathfrak{N}_0$ zu bestimmen, welche die angegebene Durchschnittsdarstellung gestatten. Nun ist die Anzahl der Mengen, welche Durchschnitt von gewissen Mengen $\mathfrak{N}_0(i), \mathfrak{N}_1(i), \mathfrak{N}_2(i)$ sind, $3^n - n - 1$. Dabei sind wegen $\mathfrak{N}_3(i) = \mathfrak{N}_2(1) \ldots \mathfrak{N}_2(i-1) \,.\, \mathfrak{N}_1(i) \,.\, \mathfrak{N}_2(i+1) \ldots \mathfrak{N}_2(n)$ diejenigen Mengen, welche Faktoren $\mathfrak{N}_3(i)$ enthalten, schon inbegriffen. Nimmt man noch die hinzu, welche durch $\mathfrak{N}_0(i), \mathfrak{N}_1(i), \mathfrak{N}_2(i), \mathfrak{N}_4(i)$, aber nicht ohne $\mathfrak{N}_4(i)$ darstellbar sind, so kommen noch $3^n - 2^n - n - 3\binom{n}{2}$ Mengen hinzu. Die Anzahl der Mengen, die nicht ohne Faktoren $\mathfrak{N}_5(i)$ darstellbar sind, ist $n(2^{n-1} - (n-1) - 1)$. Insgesamt befinden sich unter den $2^{2^n - 1} - 1$ Mengen von Teilmengen von N also $2 \,.\, 3^n + 2^{n-1}(n-2) - (n+1)^2 - 3\binom{n}{2}$ abgerundete.

Putnam: *Über Konvergenzkontinua höheren Geschlechtes.*

Ein Teilkontinuum C des Kontinuums K möge *Konvergenzkontinuum vom Geschlecht 2* von K heißen, wenn eine Folge $\{C_i\}$ ($i = 1, 2, \ldots$, ad inf.) von paarweise fremden und zu C fremden Teilkontinua von K existiert und jedes C_i ein Konvergenzkontinuum von K ist, d. h. zu jedem C_i eine Folge $\{C_{i,j}\}$ ($j = 1, 2, \ldots$, ad inf.) von paarweise fremden und zu C_i fremden Teilkontinua von K existiert, derart daß $\lim_{i=\infty} C_i = C$ und $\lim_{j=\infty} C_{ij} = C_i$ ($i = 1, 2, \ldots$, ad inf.) gilt. Dann gilt der Satz: *Damit ein Kontinuum M kein Konvergenzkontinuum vom Geschlecht 2 enthalte, ist notwendig und hinreichend, daß jede Komponente der Menge aller in einem Konvergenzkontinuum von K enthaltenen Punkte erblich im kleinen zusammenhängend ist und daß diese Komponenten eine Nullfolge bilden.*

78. Kolloquium (19. VI. 1934).

Menger: *Bernoullische Wertlehre und Petersburger Spiel.*

Eine Münze wird so lange geworfen, bis zum ersten Male Kopf fällt; tritt dies beim n-ten Wurf ein, so erhält B von A den Betrag von 2^{n-1} Dukaten. Für die mathematische Hoffnung des B bei diesem „Petersburger Spiel" ergibt sich bekanntlich $\sum_{n=1}^{\infty} \frac{2^{n-1}}{2^n} = \infty$.

D. Bernoulli hat die Tatsache, daß dennoch kein vernünftiger Mensch einen beträchtlichen Einsatz für dieses Spiel riskiert, darauf zurückgeführt, daß die möglichen Gewinne des B Zuwächse zum Vermögen v sind, mit dem er ans Spiel herantritt, und als solche von ihm geringer eingeschätzt werden als ihrem Betrag entspricht. In seiner berühmten werttheoretischen Abbhandlung nimmt Bernoulli an, daß der Zuwachs w zum Vermögen v bloß wie $C \log \frac{v+w}{v}$ eingeschätzt wird. Der Spieler B schätzt seine Hoffnung beim Petersburger Spiel $\sum_{n=1}^{\infty} \frac{1}{2^n} \log \frac{v+2^{n-1}}{v}$ und diese Summe ist, falls $v > 0$, *endlich.*

Während alle klassischen Lehrbücher der Wahrscheinlichkeitsrechnung Bernoullis Theorie des Petersburger Spieles gelten lassen oder aus inhaltlichen Gründen ablehnen, sei hier darauf hingewiesen, daß dieselbe schon aus formalen Gründen unhaltbar ist, da sie völligen ad-hoc-Charakter besitzt. Es gilt nämlich offenbar der Satz: *Ist der Wert $f(w)$, den B einem Gewinn w zuschreibt, irgend eine in ihrer Abhängigkeit von w unbeschränkte Funktion, so gibt es ein dem Petersburger analoges Spiel, bei dem B seine Hoffnung mit ∞ bewertet.* Zum Beweise hat man für jedes natürliche n mit w_n einen Betrag zu bezeichnen, für welchen beispielsweise $f(w_n) \geqq 2^{n-1}$ gilt, was wegen der Unbeschränktheit von f möglich ist, und als Spielregel festzusetzen, daß B, wenn Kopf zum ersten Male beim n-ten Wurf fällt, den Betrag w_n erhält. Um bei Bernoullischer Bewertung die Hoffnung ∞ zu erlangen, genügt es, etwa $w_n = 2^{2^n}$ zu setzen. Ausführlich behandelt im Artikel „Das Unsicherheitsmoment in der Wertlehre" in Zeitschr. f. Nationalökonomie 5, S. 459).

Wald: *Eine Charakterisierung des Lebesgueschen Maßes.*

Ist $\mu(M)$ eine für alle Teilmengen M des R_n definierte Mengenfunktion mit folgenden Eigenschaften[1])

1. *für jede Menge M gilt* $\mu(M) \geqq 0$,
2. *aus $M' \subset M$ folgt* $\mu(M') \leqq \mu(M)$,
3. *sind A und B zwei fremde abgeschlossene Mengen, so gilt* $\mu(A) + \mu(B) = \mu(A+B)$,
4. *ist* $\{A_i\}$ $(i = 1, 2, \ldots$ ad inf) *eine Folge von abgeschlossenen Mengen, so gilt* $\mu\left(\sum_{i=1}^{\infty} A_i\right) \leqq \sum_{i=1}^{\infty} \mu(A_i)$,
5. *für die abgeschlossenen n-dimensionalen Intervalle I ist $\mu(I)$ der elementargeometrische Inhalt,* —

dann gilt für jede im Lebesgueschen Sinne meßbare Menge M, wenn $\mu_L(M)$ ihr Lebesguesches Maß bezeichnet, $\mu(M) = \mu_L(M)$.

Wir beweisen unter den angegebenen Voraussetzungen über $\mu(M)$ zunächst vier Hilfssätze.

---

[1]) Eine die Forderungen 1.—5. realisierende Funktion aller Teilmengen des R_n ist z. B. das äußere Lebesguesche Maß.

1. *Ist U eine offene Menge mit endlichem Lebesgueschen Maß, so gilt* $\mu(U)=\mu_L(U)$. Bekanntlich ist $U=\sum_{k=1}^{\infty} I_k$, wobei die I_k abgeschlossene Intervalle sind, die paarweise keine inneren Punkte gemein haben, und es gilt $\sum_{k=1}^{\infty}\mu_L(I_k)=\mu_L(U)$. Wegen 5. können wir auch schreiben: $\sum_{k=1}^{\infty}\mu(I_k)=\mu_L(U)$. Wegen 4. gilt $\sum_{k=1}^{\infty}\mu(I_k)\geqq\mu(U)$, also

$$(+) \qquad \mu(U)\leqq\mu_L(U).$$

Es sei nun ε eine beliebige positive Zahl. Es gibt dann eine natürliche Zahl n, so daß $\sum_{k=1}^{n}\mu(I_k)>\mu_L(U)-\frac{\varepsilon}{2}$. Es sei nun $I_k^*(k=1, 2,\ldots, n)$ ein abgeschlossenes Intervall, das im Innern von I_k liegt und für das $\mu(I_k)-\mu(I_k^*)<\frac{\varepsilon}{2n}$ gilt, $(k=1, 2, \ldots, n)$. Dann gilt offenbar $\sum_{k=1}^{n}\mu(I_k^*)>\mu_L(U)-\varepsilon$. Aus 3. folgt, da die Intervalle I_k^* paarweise fremd sind,

$$\sum_{k=1}^{n}\mu(I_k^*)=\mu\left(\sum_{k=1}^{n}I_k^*\right), \text{ also } \mu\left(\sum_{k=1}^{n}I_k^*\right)>\mu_L(U)-\varepsilon.$$

Aus 2. folgt $\mu(U)>\mu_L(U)-\varepsilon$. Da dies für ein beliebiges $\varepsilon>0$ gilt, so muß

$$(++) \qquad \mu(U)\geqq\mu_L(U) \quad \text{sein.}$$

Aus $(+)$ und $(++)$ folgt $\mu(U)=\mu_L(U)$.

2. *Für jede offene Menge U gilt* $\mu(U)=\mu_L(U)$. Wir haben nur den Fall zu betrachten, daß $\mu_L(U)=\infty$ ist. Wir nehmen an, daß $\mu(U)$ endlich sei und leiten daraus einen Widerspruch her. Da $\mu_L(U)=\infty$ ist, so gibt es sicher eine offene Teilmenge U' von U, so daß $\mu_L(U')$ endlich und $>\mu(U)$ ist. Wegen Satz **1** ist $\mu(U')=\mu_L(U')$, also ist $\mu(U')>\mu(U)$, im Widerspruch zu 2.

3. *Ist A eine abgeschlossene Menge mit endlichem Lebesgueschen Maß, so gilt* $\mu(A)=\mu_L(A)$. Zu einer beliebigen positiven Zahl ε kann man eine A enthaltende offene Menge U so bestimmen, daß $\mu_L(U)<\mu_L(A)+\varepsilon$ ist. Wegen Satz **2** ist $\mu_L(U)=\mu(U)$. Wegen 2. gilt $\mu(U)\geqq\mu(A)$, also $\mu(A)\leqq\mu(U)<\mu_L(A)+\varepsilon$ und daher

$$(+) \qquad \mu(A)\leqq\mu_L(A).$$

Da $U-A$ offen ist, so ist $U-A=\sum_{k=1}^{\infty}I_k$, wobei die I_k abgeschlossene Intervalle sind, die paarweise keinen inneren Punkt gemein haben. Wegen 4. gilt $\sum_{k=1}^{\infty}\mu(I_k)+\mu(A)\geqq\mu(U)$. Wegen Satz **2** ist

$\sum_{k=1}^{\infty}\mu(I_k)=\mu_L(U-A)$ und $\mu(U)=\mu_L(U)$. Wir können also statt der obigen Ungleichung auch $\mu_L(U-A)+\mu(A)\geqq\mu_L(U)$ schreiben oder

$$(++)\qquad \mu(A)\geqq\mu_L(U)-\mu_L(U-A)=\mu_L(A).$$

Aus $(+)$ und $(++)$ folgt $\mu(A)=\mu_L(A)$.

4. *Für jede abgeschlossene Menge A gilt* $\mu(A)=\mu_L(A)$. Wir haben nur den Fall zu betrachten, daß $\mu_L(A)=\infty$ ist. Wir machen die Annahme, es sei $\mu(A)$ endlich. Sicher gibt es eine abgeschlossene Teilmenge A' von A, so daß $\mu_L(A')$ endlich und $>\mu(A)$ ist. Wegen Satz **3** ist $\mu(A')=\mu_L(A')$. Also müßte $\mu(A')>\mu(A)$ sein, was wegen 2. nicht möglich ist.

Da das äußere Lebesguesche Maß einer Menge M die untere Schranke der Maße aller M enthaltenden offenen Mengen ist und das innere Maß die obere Schranke der Maße aller in M enthaltenen abgeschlossenen Mengen ist, so muß $\mu(M)$ zwischen dem inneren Maß und dem äußeren Maß von M liegen. Für jede meßbare Menge M ist daher $\mu(M)=\mu_L(M)$, w. z. b. w.

Bemerkung. Ersetzt man die Forderungen 3 und 4 durch die folgende schwächere

3'. *ist* $\{A_i\}$ *eine Folge von paarweise fremden abgeschlossenen Mengen, so gilt* $\sum_{i=1}^{\infty}\mu(A_i)=\mu(\sum_{i=1}^{\infty}A_i)$, dann erfüllt auch das Peanosche innere Maß die Forderungen. Daß 1., 2. und 5. erfüllt sind, ist unmittelbar klar. Aus dem bekannten Satz, daß ein Kontinuum nicht in abzählbar viele paarweise fremde abgeschlossene Mengen zerlegt werden kann, folgt mühelos, daß auch 3'. erfüllt ist.

79. Kolloquium (27. VI. 1934).

Wald: *Zur Differentialgeometrie der Flächen. I. Mitteilung: Eine neue Definition der Flächenkrümmung.*

Im Folgenden wird das von Menger formulierte[1]) Problem einer Behandlung der Flächenkrümmung im Sinne seiner metrischen Geometrie gelöst. Eine metrische Theorie der Kurvenkrümmung gab bekanntlich Menger selbst[2]). Er ordnet zunächst drei Punkten eines metrischen Raumes, die ja stets mit drei Punkten der Ebene kongruent sind, den Radius des diesen ebenen Punkten umgeschriebenen Kreises als Krümmungsradius zu und bezeichnet sodann, wenn je drei Punkte eines Bogens B, die einem Punkt p hinlänglich benachbart sind, einen Krümmungsradius r besitzen, für den $\frac{1}{r}$ sich von einer Zahl ρ beliebig wenig unterscheidet, ρ als Krümmung von B in p. Entsprechend bilden wir im Folgenden für vier Punkte eines metrischen Raumes eine Kugel, in die sich die vier Punkte einbetten lassen u. zw. so, daß die 6 Abstände gleich den Längen der

---

[1]) Mathem. Ann. *103* und Jahresber. DMV *40*.

[2]) Mathem. Ann. *103*.

kürzesten die vier Kugelpunkte verbindenden Kugelbogen (nicht gleich den Sehnenabständen) werden. Es kann zu vier Punkten Kugeln verschiedener Radien geben, in die sie sich in dieser Art einbetten lassen (Mehrdeutigkeitssatz). Wenn aber zu einem Punkt p eines metrischen Raumes R eine Zahl $\varkappa$ existiert derart, daß sich für jede gegen p konvergente Folge von Punktequadrupeln gewisser Art (sog. p-Folgen) fast alle Quadrupel unter anderem auch in eine Kugel einbetten lassen, für deren Radius r sich $\frac{1}{r^2}$ beliebig wenig von $\varkappa$ unterscheidet, dann gibt es keine zweite Zahl mit dieser Eigenschaft *(Eindeutigkeitstheorem)* und wir nennen $\varkappa$ die Flächenkrümmung von R in p. Für dreimal stetig differenzierbare Flächen beweisen wir sodann in jedem Punkt p die *Existenz der Flächenkrümmung und ihre Übereinstimmung mit der Gaußschen Krümmung in p*, so daß für diese letztere eine neue einfache und anschauliche Einführungsart gewonnen ist, die im wesentlichen auf Folgendes hinausläuft: Man betrachtet Punktequadrupel, die hinlänglich nahe an p liegen und nicht (in einem durch den Begriff der *p-Folge* präzisierten Sinne) besonders verzerrt sind, und bildet das reziproke Quadrat vom Radius irgend einer Kugel, in welchen sich die vier Punkte einbetten lassen; diese Zahl wird sich von der Gaußschen Krümmung in p beliebig wenig unterscheiden. Während es sich bei den Bogen, da die innere Bogenmetrik trivial ist (je zwei gleichlange Bogen sind ja längentreu auf einander abbildbar), nur um Kongruenzinvarianz handeln kann, bezieht sich unsere Flächenkrümmung auf die innere Flächenmetrik und ist verbiegungsinvariant. Damit bahnen wir die von Menger konzipierte *„koordinatenlose Differentialgeometrie ohne Differentialrechnung“* [1]) (übrigens unter wiederholter Heranziehung von Mengers Determinantenmethoden und Ergebnissen über die euklidische Metrik) auch für Flächen an und werden dieselben in folgenden Mitteilungen durch Kennzeichnungen der Gaußschen Flächen unter den metrischen Räumen und durch strenge Behandlung flächentheoretischer Fragen ausbauen.

Ist ein metrisches n-Tupel gegeben (d. h. n Punkte $p_1, \ldots, p_n$, von denen je zweien p_i und p_j eine Zahl $p_i p_j = p_j p_i$ zugeordnet ist, die $=0$ ist, wenn $p_i = p_j$, und >0, wenn $p_i \neq p_j$, ferner für je drei Punkte p_i, p_j, p_k die Dreiecksungleichung $p_i p_j + p_j p_k \geqq p_i p_k$ gilt), so setzen wir

$$D(p_1, \ldots, p_n) = \begin{vmatrix} 0 & 1 \\ 1 & (p_i p_j)^2 \end{vmatrix} \quad (i, j = 1, \ldots, n)$$

$$E(p_1, \ldots, p_n) = |(p_i p_j)^2| \quad (i, j = 1, \ldots, n)$$

$$D_r(p_1, \ldots, p_n) = \begin{vmatrix} 0 & 1 \\ 1 & (2r \sin \frac{p_i p_j}{2r})^2 \end{vmatrix} \quad (i, j = 1, \ldots, n)$$

---

[1]) Mathem. Ann. *103* und diese Ergebnisse *2*, S. 22.

$$E_r(p_1, \ldots, p_n) = \left| (2r \sin \frac{p_i p_j}{2r})^2 \right|_{(i,j=1,\ldots,n)},$$

wobei die D und E Determinanten mit $n+1$ bzw. n Reihen sind, und nennen $d(p_1, \ldots, p_n)$ die größte der $\binom{n}{2}$ Zahlen $p_i p_j$, also den *Durchmesser* des metrischen n-Tupels. Sind $\binom{n}{2}$ Zahlen r_{ij} $(i,j=1, \ldots, n)$ gegeben, so sagen wir, das Zahlensystem sei im Raume R *realisierbar*, u. zw. als Abstandssystem der Punkte $p_1, \ldots, p_n$ von R, wenn für diese Punkte $p_i p_j = r_{ij}$ $(i,j=1, \ldots, n)$ gilt.

Als R_n bezeichnen wir den n-dimensionalen euklidischen Raum, als S^r die Oberfläche der dreidimensionalen Kugel mit dem (reellen oder imaginären) Radius r, in der je zwei Punkten die Länge des kürzeren sie verbindenden Großkreisbogens als Abstand zugeordnet ist. Sind p_1, p_2, p_3, p_4 Punkte von S^r oder mit Punkten von S^r kongruent, so gilt bekanntlich

$$(+) \qquad \frac{1}{r^2} = -\frac{2D_r(p_1, \ldots, p_n)}{E_r(p_1, \ldots, p_n)}.$$

Es gibt metrische Quadrupel, die für kein r mit vier Punkten von S^r kongruent sind, z. B. das Quadrupel mit den Abständen $p_1p_2 = p_1p_3 = p_1p_4 = 1$, $p_2p_3 = p_2p_4 = p_3p_4 = 2$. Anderseits *gibt es Quadrupel, die für verschiedene r mit vier Punkten von S^r kongruent sind.* Es gilt sogar folgender

Mehrdeutigkeitssatz. *Für jedes reelle $r>0$ gibt es in S^r in jeder Umgebung U jedes Punktes p Quadrupel, die nicht auf einem Großkreis liegen und mit vier Punkten der Ebene kongruent sind.*

Wir wählen in U drei paarweise verschiedene Punkte p_1, p_2, p_3 auf einem Großkreisbogen einer Länge $\leqq r\pi$ und einen Punkt p_4 außerhalb dieses Großkreises. Wenn p_1, p_2, p_3, p_4 mit vier Punkten der Ebene kongruent sind, so sind wir am Ziel. Wir nehmen also an, dies sei nicht der Fall. Dann sind, da p_1, p_2, p_3 mit drei Punkten einer Geraden kongruent sind, p_1, p_2, p_3, p_4 auch nicht kongruent mit vier Punkten des R_3, also[1]) gilt $D(p_1, \ldots, p_4) < 0$. Wir bezeichnen mit K_1 und K_2 zwei auf einander senkrechte Großkreisbogen $\subset U$ durch p, ferner mit q_1 und q_2 zwei Punkte von K_1, so daß $q_1 p = q_2 p \neq 0$, $q_1 p + p q_2 = q_1 q_2$ gilt, und mit q einen Punkt auf K_2, so daß $2\,qp < r\pi$. Es seien r_1, r_2, r die zu q_1, q_2, q kongruenten Eckpunkte eines (gleichschenkeligen) ebenen Dreiecks, dessen Höhe h sei. Ist *erstens* $h > qp$, dann wählen wir einen Punkt $q^* \neq q$ auf K_2, so daß $pq = pq^*$. Offenbar sind q_1, q_2, q, q^* mit vier Punkten des R_3 kongruent, also[2]) gilt $D(q_1, q_2, q, q^*) > 0$. Führen wir q_1, q_2, q, q^* innerhalb U stetig (ohne daß die vier Punkte während der Überführung auf einem Großkreis liegen) in p_1, p_2, p_3, p_4 über, so muß wegen $D(p_1, p_2, p_3, p_4) < 0$ ein Quadrupel passiert werden, für welches

[1]) Vgl. Menger, Math. Ann. *100* und diese Ergebnisse *1*, S. 21.
[2]) Vgl. Menger l. c. [4]).

$D=0$ gilt, welches also[1]) mit vier Punkten der Ebene kongruent ist. Ist *zweitens* $h=qp$, so sind q_1, q_2, q, q^* mit vier Punkten der Ebene kongruent. Ist *drittens* $h<pq$, so sei x ein variabler Punkt von K_2 zwischen p und q, ferner y ein Punkt der Ebene, so daß y, r_1, r_2 mit x, q_1, q_2 kongruent ist und daß y und r auf verschiedenen Seiten der Geraden durch r_1 und r_2 liegen. Für $x=p$ gilt $xq>yr$, für $x=q$ gilt $xq=0<yr$. Also existiert ein Punkt x_0 zwischen p und q, so daß $x_0 q=y_0 r$ gilt. Dann ist q_1, q_2, x_0, q mit r_1, r_2, y_0, r kongruent. In jedem der drei Fälle gibt es also ein mit einem ebenen kongruentes Punktequadrupel in S^r, w. z. b. w.

Ist in einem metrischen Raum R eine Punktefolge gegeben, so bezeichnen wir sie im Folgenden stets mit $p(t)$, wobei wir t eine gegen 0 konvergente Zahlenfolge, etwa $\frac{1}{n}$, durchlaufen lassen. Wenn die Punktefolge gegen Punkt p_0 konvergiert, so schreiben wir $\lim\limits_{t=0} p(t)=p_0$.

Ist p ein Punkt von R, so nennen wir eine Folge $\{p_1(t), p_2(t), p_3(t)\}$ von Punktetripeln eine *p-Folge*, wenn folgende Bedingungen erfüllt sind, wobei wir $p_0(t)=p$ für alle t setzen:

1. $\lim\limits_{t=0} p_1(t)=\lim\limits_{t=0} p_2(t)=\lim\limits_{t=0} p_3(t)=p$

2. Bezeichnet $d(t)$ den Durchmesser $d\big(p_0(t), p_1(t), p_2(t), p_3(t)\big)$, so häuft sich die Folge der Zahlensextupel $\left\{\frac{p_i(t)\, p_j(t)}{d(t)}\right\}$ $(i, j=0, 1, 2, 3)$ nicht gegen ein Zahlensextupel, das als Abstandssystem realisierbar ist von vier (nicht notwendig paarweise verschiedenen) Punkten p_0, p_1, p_2, p_3, die im R_1 oder auf einem Kreise des R_2 liegen.

3. Für keine Teilfolge von $\{p_1(t), p_2(t), p_3(t)\}$ gilt
$$\lim_{t=0} \frac{2\, D_r\big(p, p_1(t), p_2(t), p_3(t)\big)}{E_r\big(p, p_1(t), p_2(t), p_3(t)\big)}=\frac{1}{r^2} \text{ in } r \text{ identisch.}$$

Wir sagen nun, der metrische Raum R habe im Punkt p die *Flächenkrümmung* $\varkappa$, wenn

1. mindestens eine p-Folge existiert,

2. für jedes $\varepsilon>0$ und jede p-Folge $\{p_1(t), p_2(t), p_3(t)\}$ für fast alle t ein $r(t)$ existiert, so daß $\left|\frac{1}{r^2(t)}-\varkappa\right|<\varepsilon$ gilt und $p, p_1(t), p_2(t)$, $p_3(t)$ mit vier Punkten von $S^{r(t)}$ kongruent sind.

Wir beginnen mit einer Reihe von Hilfssätzen. Nach Taylor ist, wobei $0<\vartheta_{ij}<1$ gilt, und wenn wir $\eta_{ij}=\cos\left(\vartheta_{ij}\frac{p_i p_j}{r^2}\right)$ setzen,
$$2r \sin\frac{p_i p_j}{2r}=2r\left[\frac{p_i p_j}{2r}-\frac{1}{6}\left(\frac{p_i p_j}{r^2}\right)^3 \cos\left(\vartheta_{ij}\frac{p_i p_j}{2r}\right)\right]$$
$$=p_i p_j-\frac{\eta_{ij}}{24 r^2}\left(p_i p_j\right)^3 \qquad (-1\leqq\eta_{ij}\leqq 1).$$

---

[1]) Vgl. Menger l. c. [4]).

Führt man dies in $E_r(p_0, p_1, p_2, p_3)$ ein, so erhält man den Satz

H₁. $E_r(p_0, \ldots, p_3) = E(p_0, \ldots, p_3) + T$, wo T ein Polynom in den Abständen $p_i p_j$ ist, dessen sämtliche Glieder Grade $\geqq 10$ besitzen.

Schreibt man nach Taylor

$$2r \sin \frac{p_i p_j}{2r} = p_i p_j - \frac{1}{24 r^2}(p_i p_j)^3 + \zeta_{ij} \frac{1}{16 \cdot 5! \, r^4} \left(p_i p_j\right)^5 \qquad (-1 \leqq \zeta_{ij} \leqq 1)$$

und setzt dies in $D_r(p_0, \ldots, p_3) = D_r$ ein, so erhält man Satz

H₂. $D_r = D + \frac{S}{r^2} + Q$, wo S ein homogenes Polynom achten Grades in den $p_i p_j$ ist und Q ein Polynom, dessen sämtliche Glieder Grade > 8 besitzen.

Es sei nun $\{p_1(t), p_2(t), p_3(t)\}$ eine p-Folge. Da E ein homogener Ausdruck achten Grades in den $p_i p_j$ ist, so ist, wenn wir $d(p, p_1(t), p_2(t), p_3(t)) = d(t) = d$ setzen, die Zahlenfolge $\frac{E}{d^8}$ beschränkt. Ist überdies $\frac{1}{r^2(t)}$ eine beschränkte Funktion von t, so sind die Koeffizienten der Polynome T und Q der Sätze H_1 und H_2 beschränkte Funktionen von t. Da S ein homogenes Polynom achten Grades in den $p_i p_j$ ist, dessen Koeffizienten unabhängig von t und r sind, so ist dann die Zahlenfolge $\frac{S}{d^8}$ beschränkt; es gilt also:

H₃. Für eine p-Folge ist $\frac{E}{d^8}$ und $\frac{S}{d^8}$ beschränkt.

Da sämtliche Glieder von T und Q Grade > 8 besitzen, so gilt, wenn $\frac{1}{r^2}$ beschränkt ist, $\lim\limits_{t=0} \frac{T}{d^8} = 0$ und $\lim\limits_{t=0} \frac{Q}{d^8} = 0$; also folgt aus H_1 bzw. H_2

H₄. Ist $\frac{1}{r^2}$ beschränkt und ist für eine p-Folge die Zahlenfolge $\frac{E}{d^8}$ konvergent, so gilt $\frac{E_r}{d^8} = \lim \frac{E}{d^8} + \varepsilon_1(t)$, wo $\lim\limits_{t=0} \varepsilon_1(t) = 0$.

H₅. Ist $\frac{1}{r^2}$ beschränkt, so gilt $\frac{D_r}{d^8} = \left(\frac{D}{d^8} + \frac{S}{d^8 r^2}\right) + \varepsilon_2(t)$, wo $\lim\limits_{t=0} \varepsilon_2(t) = 0$.

Gilt für eine p-Folge $\lim\limits_{t=0} \frac{D}{d^8} = 0$ so gilt, wenn wir $\pi_{ij} = \frac{p_i p_j}{d}$ $(i, j = 0, 1, 2, 3)$ setzen, $\lim\limits_{t=0} \begin{vmatrix} 0 & 1 \\ 1 & \pi_{ij}^2 \end{vmatrix}_{(i,j=0,\ldots,3)} = 0$ und die Zahlensextupel π_{ij} $(i, j = 0, 1, 2, 3)$ häufen sich mit $t \to 0$ gegen ein Sextupel ρ_{ij}, für welches $D = 0$ gilt, welches also[1]) als Abstandssystem von vier Punkten der Ebene realisierbar ist. Diese vier Punkte lägen, wenn auch $\lim \frac{E}{d^8} = 0$ gälte, nach bekannten

[1]) Vgl. Menger, l. c.[4]).

3

Sätzen auf einer Geraden oder einem Kreis, was nach Bedingung 2) für p-Folgen nicht zutrifft. Wenn für eine p-Folge $\lim\limits_{t=0} \frac{D}{d^8} = 0$ gilt, so gilt für sie nicht $\lim\limits_{t=0} \frac{E}{d^8} = 0$. Wenn mithin für jede p-Folge $\lim\limits_{t=0} \frac{E}{d^8} = 0$ gilt, so gilt für keine $\lim \frac{D}{d^8} = 0$ und dann ist für jede p-Folge die Zahlenfolge $\frac{D}{d^8}$ unbeschränkt. Wir haben also bewiesen

H_6. Wenn für jede p-Folge $\lim \frac{E}{d^8} = 0$ gilt, dann ist für jede p-Folge $\frac{D}{d^8}$ unbeschränkt.

Wenn $\frac{1}{r^2}$ beschränkt ist, dann ist nach H_3 auch $\frac{S}{d^8}$ beschränkt, also ist dann, wenn für jede p-Folge $\lim \frac{E}{d^8} = 0$ gilt und dabei $\frac{D}{d^8}$ nach H_6 unbeschränkt ist, nach H_5 die Folge $\frac{D_r}{d^8}$ unbeschränkt, ferner, da $\frac{E}{d^8}$ nach H_3 beschränkt ist, nach H_4 die Folge $\frac{E_r}{d^8}$ beschränkt und demnach $\frac{\frac{-2D_r}{d^8}}{\frac{E_r}{d^8}} = \frac{-2D_r}{E_r}$ unbeschränkt. Die Gleichung $\frac{1}{r^2} = \frac{-2D_r}{E_r}$ kann dann unmöglich für fast alle t Lösungen $\frac{1}{r^2(t)}$ besitzen, für welche $\frac{1}{r^2(t)}$ gegen eine endliche Zahl konvergiert, d. h. R kann in p keine endliche Flächenkrümmung besitzen. Wenn R in p eine endliche Flächenkrümmung besitzt, so kann also nicht für jede p-Folge $\lim\limits_{t=0} \frac{E}{d^8} = 0$ gelten. Da anderseits nach H_3 für jede p-Folge $\frac{E}{d^8}$ beschränkt ist, also eine konvergente Teilfolge besitzt, gilt

H_7. Besitzt R in p eine endliche Flächenkrümmung, so existiert eine p-Folge, für die $\lim\limits_{t=0} \frac{E}{d^8} = \alpha \neq 0$.

Diese p-Folge betrachten wir nun. Wenn $\frac{1}{r^2}$ beschränkt ist, können wir H_4 und H_5 anwenden und sehen, wenn $\varkappa_0$ die Flächenkrümmung von R in p ist, daß die Gleichung

$$(+) \qquad \frac{1}{r^2} = \frac{-2D_r}{E_r} = \frac{-2\left[\frac{D}{d^8} + \frac{S}{d^8 r^2} + \varepsilon_2(t)\right]}{\alpha + \varepsilon_1(t)}$$

für fast alle t eine Lösung $\frac{1}{r^2}$ besitzt mit $\lim\limits_{t=0} \frac{1}{r^2} = \varkappa_0$. Daraus folgt offenbar, daß $\frac{D}{d^8}$ eine beschränkte Funktion von t ist. Da nach H_3 auch $\frac{S}{d^8}$ beschränkt ist, können wir also aus der p-Folge eine Teilfolge herausgreifen, für welche die Zahlenfolgen $\frac{D}{d^8}$ und $\frac{S}{d^8}$ konvergieren. Ist $\lim\limits_{t=0} \frac{-2D}{d^8} = \beta$ und $\lim\limits_{t=0} \frac{-2S}{d^8} = \gamma$, dann kann (+) offenbar in der Form geschrieben werden.

$$(++) \qquad \frac{1}{r^2} = \frac{\beta}{\alpha} + \frac{\gamma}{\alpha r^2} + \varepsilon(t), \text{ wobei } \lim_{t=0} \varepsilon(t) = 0.$$

Ist $\beta = 0$, so gilt wegen Forderung 3) an die p-Folgen $\gamma \neq \alpha$. Ist $\beta \neq 0$ und wäre $\gamma = \alpha$, so würde die Gleichung (++) lauten: $\frac{\beta}{\alpha} + \varepsilon(t) = 0$ (wo ε auch von r abhängen kann), die wegen $\beta \neq 0$ unmöglich eine Lösung $\frac{1}{r^2}$ mit $\lim\limits_{t=0} \frac{1}{r^2} = \varkappa_0$ haben kann, wie es für (+) der Fall ist. Also gilt in jedem Fall $\gamma \neq \alpha$ und wir können (++) in der Form schreiben

$$\frac{1}{r^2} = \frac{\beta}{\alpha\left(1 - \frac{\gamma}{\alpha}\right)} + \frac{\varepsilon(t)}{1 - \frac{\gamma}{\alpha}}.$$

Offenbar gilt für die Wurzeln dieser Gleichung $\lim\limits_{t=0} \frac{1}{r^2} = \frac{\beta}{\alpha - \gamma}$. Also muß $\varkappa_0 = \frac{\beta}{\alpha - \gamma}$ gelten und keine von $\varkappa_0$ verschiedene Zahl kann Flächenkrümmung von R in p sein, womit das Eindeutigkeitstheorem bewiesen ist.

Analog H_6 beweist man folgenden auch an sich bemerkenswerten Satz

H_8. *Zu jeder p-Folge existiert eine positive untere Schranke für die Verhältnisse von je zwei von den sechs Abständen der Quadrupel $p = p_0(t), p_1(t), p_2(t), p_3(t)$, d. h. ein $\eta > 0$, so daß $\frac{p_i p_j}{p_k p_l} > \eta$ für alle t und alle $i \neq j$ und $k \neq l$ gilt.* Würde sich nämlich etwa $\frac{p_0 p_1}{p_0 p_2}$ mit $t \to 0$ gegen 0 häufen, dann auch $\frac{p_0 p_1}{d}$. Also existierte eine Teilfolge der p-Folge, für welche die Zahlensextupel $\pi_{ij} = \frac{p_{ij}}{d}$ $(i, j = 0, 1, 2, 3)$ gegen ein Sextupel ρ_{ij} konvergieren würden, für welches $\rho_{01} = 0$ ist, welches also realisierbar wäre als Abstandssystem von vier ebenen Punkten q_0, q_1, q_2, q_3, von denen q_0 und q_1 identisch wären, welche also auf einer Geraden oder einem Kreis lägen, was nach Bedingung 2 für eine p-Folge nicht zutrifft.

Theorem. *Ist im R_3 die Fläche F durch die in einer Umgebung des Punktes u_0, v_0 dreimal stetig differenzierbaren Funktionen $x = x(u, v)$, $y = y(u, v)$, $z = z(u, v)$ gegeben und gilt in u_0, v_0 für die Funktionaldeterminante $\frac{\partial(x, y)}{\partial(u, v)} \neq 0$, dann existiert im Punkt p_0 mit den Koordinaten $x_0 = x(u_0, v_0)$, $y_0 = y(u_0, v_0)$, $z_0 = z(u_0, v_0)$,*

3*

die Flächenkrümmung $\varkappa(p_0)$ von F und stimmt mit der Gaußschen Krümmung überein.

Wegen $\frac{\partial(x, y)}{\partial(u, v)} \neq 0$ gilt in einer Umgebung von $x_0, y_0 : u = u(x,y)$, $v = v(x, y)$. Substituiert man dies in $z = z(u, v)$, so kann man F in einer Umgebung von x_0, y_0 durch eine dreimal stetig differenzierbare Funktion $z = f(x, y)$ definieren. Bewegen wir das Koordinatensystem so, daß x_0, y_0 in den Ursprung $p = (0, 0, 0)$ übergeht und die X- und Y-Achse mit den Hauptkrümmungsrichtungen zusammenfallen, so nimmt die Gleichung von F die Form an

$$z = a(x, y)\, x^2 + b(x, y)\, y^2 + c(x, y)\, xy,$$

wo $a(x, y)$, $b(x, y)$, $c(x, y)$ in einer Umgebung von $0, 0$ stetige Funktionen sind. Wir setzen $a(0, 0) = a$, $b(0, 0) = b$. Offenbar besitzt F in p die Gaußsche Krümmung $4\,a\,b$. Wir zeigen nun, daß F in p die Flächenkrümmung $\varkappa = 4\,a\,b$ besitzt.

Es sei $\{p_1(t), p_2(t), p_3(t)\}$ eine p-Folge in F, wobei t eine gegen 0 konvergente Zahlenfolge durchläuft. Wir setzen $p_0(t) = p$ für alle t. Die Projektion von $p_i(t)$ auf die XY-Ebene heiße $\bar{p}_i(t)$. Es gilt $\bar{p}_0(t) = p$ für jedes t. Es sei $\bar{d}(t)$ der Durchmesser $d(\bar{p}_0(t), \bar{p}_1(t), \bar{p}_2(t), \bar{p}_3(t))$. Die Koordinaten von $\bar{p}_i(t)$ seien $\lambda_i(t)\,\bar{d}(t)$, $\mu_i(t)\,\bar{d}(t)$, 0, wobei offenbar $\lambda_0(t) = \mu_0(t) = 0$ für alle t ist. p_{ij} sei die Länge des kürzesten p_i und p_j in F verbindenden Bogens (wobei wir das Argument t weglassen), p^*_{ij} der Abstand von p_i und p_j im R_3 (die Länge der Sehne zwischen p_i und p_j), $\bar{p}_{ij}$ der Abstand $\bar{p}_i\bar{p}_j$. Ferner setzen wir $\bar{\bar{p}}_{ij} = \frac{\bar{p}_{ij}}{\bar{d}}$. Wegen $\lambda_i^2 \bar{d}^2 + \mu_i^2 \bar{d}^2 = \bar{p}_{0i}^2 \leq \bar{d}^2$ gilt $\lambda_i^2 + \mu_i^2 \leq 1$, also sind λ_i und μ_i beschränkt und es gilt

$$\text{(1)} \qquad \bar{p}_{ij} = \bar{\bar{p}}_{ij}\, \bar{d} = \sqrt{(\lambda_i - \lambda_j)^2 + (\mu_i - \mu_j)^2}\, \bar{d}.$$

Wegen der Stetigkeit der Funktionen $a(x, y)$, $b(x, y)$ gilt $z = ax^2 + by^2 + \eta(x, y)$, wobei $\eta(x, y) \to 0$ mit $(x, y) \to (0,0)$. Wegen der Beschränktheit von λ_i und μ_i folgt

$$\text{(2)} \qquad p^{*2}_{ij} = \bar{p}^2_{ij} + \left\{[a(\lambda_i^2 - \lambda_j^2) + b(\mu_i^2 - \mu_j^2)]^2 + \eta_{ij}(t)\right\} \bar{d}^4, \quad \lim_{t=0} \eta_{ij}(t) = 0.$$

Wegen H_8 und $\lim\limits_{t=0} \frac{p_{ij}}{p^*_{ij}} = \lim\limits_{t=0} \frac{p^*_{ij}}{p_{ij}} = 1$ haben die $\frac{\bar{p}_{ij}}{\bar{p}_{kl}}$, insbesondere die $\frac{\bar{p}_{ij}}{\bar{d}}$ eine positive untere Schranke. Dividiert man (2) durch $\left(\frac{\bar{p}_{ij}}{\bar{d}}\right)^2$, zieht die Quadratwurzel und berücksichtigt (1) und die Beschränktheit von λ_i und μ_i, so erhält man

$$\text{(3)} \qquad p^*_{ij} = \sqrt{(\lambda_i - \lambda_j)^2 + (\mu_i - \mu_j)^2}\, \bar{d} + \frac{[a(\lambda_i^2 - \lambda_j^2) + b(\mu_i^2 - \mu_j^2)]^2\, \bar{d}^3}{2\sqrt{(\lambda_i - \lambda_j)^2 + (\mu_i - \mu_j)^2}} + {}$$
$$+ \vartheta_{ij}(t)\, \bar{d}^3, \quad \lim_{t=0} \vartheta_{ij}(t) = 0$$

Unter Benützung der Frénetschen Formeln beweist man mühelos

H$_9$. Ist im R_3 die Kurve K durch die in einer Umgebung von $s=0$ dreimal stetig differenzierbaren Funktionen $x=x(s)$, $y=y(s)$, $z=z(s)$ gegeben, wo s die Bogenlänge bedeutet, und ist h die zur Bogenlänge s gehörige Sehne, so gilt, wenn $\frac{1}{\rho_0}$ die Krümmung von K im Punkt $s=0$ bezeichnet, $s=h+\frac{h^3}{24\rho_0^2}+\zeta(h)h^3$, wobei $\lim\limits_{h=0}\zeta(h)=0$.

Ebenso beweist man leicht

H$_{10}$. Die Krümmung des geodätischen Bogens zwischen $p_i(t)$ und $p_j(t)$ konvergiert mit $t\to 0$ gegen die Krümmung $\frac{1}{\rho_{ij}}$ des Normalschnittes von F in p mit der Richtungstangente $\frac{\mu_i-\mu_j}{\lambda_i-\lambda_j}$.

Aus H$_9$ und H$_{10}$ folgt

$$p_{ij}=p_{ij}^*+\frac{p_{ij}^{*\,3}}{24\rho_{ij}^2}+\zeta_{ij}(t)\,\bar{d}^3,\ \text{wo}\ \lim_{t=0}\zeta_{ij}(t)=0,\ \frac{1}{\rho_{ij}}=\frac{2a(\lambda_i-\lambda_j)^2+2b(\mu_i-\mu_j)^2}{(\lambda_i-\lambda_j)^2+(\mu_i-\mu_j)^2}$$

also wegen (3)

$$(4)\qquad p_{ij}=p_{ij}^*+\frac{\bar{p}_{ij}^3}{24\rho_{ij}^2}+\varepsilon_{ij}(t)\,\bar{d}^3,\ \text{wobei}\ \lim_{t=0}\varepsilon_{ij}(t)=0.$$

Wir setzen nun

$$(5)\qquad p_{ij}^{**}=2r\sin\frac{p_ip_j}{2r}=p_{ij}-\frac{p_{ij}^3}{24r^2}+\gamma_{ij}(t)\,\bar{d}^3,\ \text{wo}\ \lim_{t=0}\gamma_{ij}(t)=0,$$

wenn $\frac{1}{r^2}$ eine beschränkte Funktion von t ist. Zur Berechnung der Flächenkrümmung von F in p haben wir die Gleichung aufzulösen

$$(6)\qquad \frac{1}{r^2}=\frac{-2D_r}{E_r},\ \text{wo}\ D_r=\begin{vmatrix}0 & 1\\ 1 & p_{ij}^{**2}\end{vmatrix}\ \text{und}\ E_r=\begin{vmatrix}p_{ij}^{**2}\end{vmatrix}.$$

Ersetzen wir in D_r die p_{ij}^{**}, p_{ij}, p_{ij}^* durch ihre Ausdrücke (5), (4), (3) (2) und ordnen wir nach Potenzen von $\bar{d}$, so erhalten wir offenbar

$$(7)\qquad D_r=\begin{vmatrix}0 & 1\\ 1 & \bar{\bar{p}}_{ij}^2\end{vmatrix}\bar{d}^6+\left(Pa^2+Qab+Rb^2+\frac{S}{r^2}+\nu(t)\right)\bar{d}^8+T,$$
$$\text{wo}\ \lim_{t=0}\nu(t)=0$$

und wobei P, Q, R, S nur Funktionen von λ_j und μ_j und beschränkte Funktionen in t sind und T ein Polynom in $\bar{d}$ ist, dessen sämtliche Glieder Grade >8 besitzen und dessen Koeffizienten, wenn $\frac{1}{r^2}$ beschränkt ist, beschränkte Funktionen von t sind. Da $\bar{p}_0, \bar{p}_1, \bar{p}_2, \bar{p}_3$ in einer Ebene liegen, ist $D(\bar{p}_0,\ldots,\bar{p}_3)=\begin{vmatrix}0 & 1\\ 1 & \bar{\bar{p}}_{ij}^2\end{vmatrix}\bar{d}^6=0$, also folgt, wenn $\frac{1}{r^2}$ beschränkt ist, aus (7)

$$(8)\qquad \frac{D_r}{\overline{d}^8}=Pa^2+Qab+Rb^2+\frac{S}{r^2}+\nu_1(t),\ \lim_{t-0}\nu_1(t)=0.$$

Ordnen wir E_r nach Potenzen von $\overline{d}$, so erhalten wir offenbar $E_r=|\overline{\overline{p}}_{ij}^2|\,\overline{d}^8+N$, wobei N ein Polynom ist, dessen sämtliche Glieder Grade >8 besitzen und dessen Koeffizienten, falls $\frac{1}{r^2}$ beschränkt ist, beschränkte Funktionen von t sind; es gilt also

$$(9)\qquad \frac{E_r}{\overline{d}^8}=\left|\overline{\overline{p}}_{ij}^2\right|+\nu_2(t),\ \lim_{t=0}\nu_2(t)=0.$$

Es kann sich $|\overline{\overline{p}}_{ij}^2|$ nicht gegen 0 häufen, sonst würden die Zahlensextupel $\overline{\overline{p}}_{ij}$ $(i,j=0,1,2,3)$ gegen ein Sextupel π_{ij} sich häufen, das als Abstandssystem von vier Punkten eines ebenen Kreises realisierbar wäre, was, da $\frac{p_{ij}}{d}$ offenbar ebenfalls gegen π_{ij} sich häuft, wegen Bedingung 2) für p-Folgen unmöglich ist. Wir dürfen also durch $|\overline{\overline{p}}_{ij}^2|$ dividieren und erhalten aus (6) wegen (7) und (9):

$$(10)\qquad \frac{1}{r^2}=-2\frac{Pa^2+Qab+Rb^2+\frac{1}{r^2}S}{\left|\overline{\overline{p}}_{ij}^2\right|}+\nu_3(t),\ \lim_{t=0}\nu_3(t)=0.$$

Nun muß für abwickelbare Flächen (10) durch $\frac{1}{r^2}=0$ in t identisch erfüllt sein. Daraus folgt, wenn etwa $b=0$, daß für beliebige beschränkte Funktionen λ_j und μ_j gelten muß $\lim\limits_{t=0}P=0$, also $P\equiv 0$ und analog $R\equiv 0$. Aus (10) wird also

$$(11)\qquad \frac{1}{r^2}=\frac{-2Qab-\frac{2S}{r^2}}{\left|\overline{\overline{p}}_{ij}^2\right|}+\nu_3(t),\ \lim_{t=0}\nu_3(t)=0.$$

Für die Kugel $b=a$ muß (11) durch $\frac{1}{r^2}=4a^2$ in t identisch erfüllt sein. Da $|\overline{\overline{p}}_{ij}^2|$ sich nicht gegen 0 häuft, gilt also $4a^2\{|\overline{\overline{p}}_{ij}^2|+2S\}=-2a^2Q+\nu_4(t)$, wo $\lim\limits_{t=0}\nu_4(t)=0$; für beliebige beschränkte Funktionen λ_i und μ_j gilt also $\lim\limits_{t=0}Q=\lim\limits_{t=0}-2\{|\overline{\overline{p}}_{ij}^2|+2S\}$, daher $Q\equiv-2\{|\overline{\overline{p}}_{ij}^2|+2S\}$.

Aus (11) wird

$$(12)\qquad \frac{1}{r^2}=\frac{4ab\{|\overline{\overline{p}}_{ij}^2|+2S\}-\frac{2S}{r^2}}{\left|\overline{\overline{p}}_{ij}^2\right|}+\nu_3(t),\ \lim_{t=0}\nu_3(t)=0.$$

$|\overline{\overline{p}}_{ij}^2|+2S$ kann sich nicht gegen 0 häufen, sonst würde die rechte Seite von (12) identisch in $\frac{1}{r^2}$ sich gegen $\frac{1}{r^2}$ häufen, was wegen Bedingung 3) für p-Folgen unmöglich ist. Also dürfen wir durch $|\overline{\overline{p}}{}_{ij}^2|+2S$ dividieren und erhalten aus (12)

$$(13)\qquad \frac{1}{r^2}=4ab+\nu_4(t),\ \lim_{t=0}\nu_4(t)=0.$$

Damit ist gezeigt, daß F in p die Flächenkrümmung $4ab$ besitzt, wenn p-Folgen in F überhaupt existieren. Dies letztere beweist man leicht. Es sei $\{p_1(t), p_2(t), p_3(t)\}$ irgend eine Folge von Punktetripeln in F, die gegen $p=p_0(t)$ konvergiert. Wenn $\overline{d}$ den Durchmesser $d(\overline{p}_0, \overline{p}_1, \overline{p}_2, \overline{p}_3)$ bezeichnet (unter Weglassung des Argumentes t), $\overline{p}_i$ die Projektion von p_i in die XY-Ebene, $\lambda_i\overline{d}$ und $\mu_i\overline{d}$ die Koordinaten von $\overline{p}_i$, wobei λ_i und μ_i von t unabhängige Zahlen seien, so ist, damit $\{p_1, p_2, p_3\}$ eine p-Folge sei, offenbar notwendig und hinreichend, daß

1. die vier Punkte $\overline{\overline{p}}_i$ der Ebene mit den Koordinaten λ_i, μ_i nicht auf einer Geraden oder einem Kreis liegen

2. $|(\overline{\overline{p_i p_j}})^2|+2S\neq 0$, wo S die in (12) auftretende Funktion ist.

Man konstruiert leicht ein ebenes Quadrupel, das 2) genügt und kann durch kleine Verschiebung der Punkte erreichen, daß auch 1) erfüllt ist. Es gibt also p-Folgen auf F, womit das Theorem über die Gaußsche Krümmung bewiesen ist.

Hans Hahn †

Von Karl Menger.

Am 24. Juli starb in seinem 55. Lebensjahre unerwartet Hans Hahn, der Lehrer von allen, die regelmäßig an diesem Kolloquium teilnehmen. Wir alle betrachten es als eine Dankespflicht, an dieser Stelle Lebenswerk, Lehrtätigkeit und Persönlichkeit des Verstorbenen — wenigstens mit der dem Stile dieser Ergebnisse gemäßen Konzision — zu würdigen.

Die erste Gruppe von Hahns Arbeiten war der *Variationsrechnung* gewidmet. Zunächst an Untersuchungen seines Lehrers v. Escherich anknüpfend (Monatsh. *14*, Math. Ann. *58*), wandte er sich dem Studium der Weierstraßschen Theorie und ihren Beziehungen zur zweiten Variation zu (Rend. Circ. Mat. Palermo *29*). Eingehend beschäftigte er sich mit dem Satz von Osgood, für den er einen einfachen Beweis, sowie Verallgemeinerungen angab (Monatsh. *17* u. *24*, Weber-Festschrift 1912). Weitere Arbeiten schwächen bekannte Bedingungen ab: bei der Lagrangeschen Multiplikatorenmethode gewährleistet die Voraussetzung stetiger Differenzierbarkeit der Lösungen im allgemeinen deren zweimalige Differenzierbarkeit (Monatsh. *14*, Math. Ann. *63*). Ein Extremalenbogen, der ein schwaches Extremum liefert, liefert, wenn die E-Funktion in seiner Umgebung einerlei Zeichen hat, ein starkes Extremum (Wien. Ak. Ber. *118* u. Monatsh. *22*). Für räumliche Variationsprobleme führt Hahn die Frage, ob ein Extremalenbogen, dessen Endpunkt zum Anfangspunkt konjugiert ist, ein Minimum liefere, auf die Frage zurück, ob eine Funktion von zwei Veränderlichen an einer gewissen Stelle ein Minimum hat (Math. Ann. *70*). In derselben Arbeit wird insbesondere auch das Verhalten von Extremalenbogen jenseits des zum Anfangspunkt konjugierten Punktes untersucht, eine Frage, die bekanntlich für die neueste „Variationsrechnung im Großen" von besonderer Bedeutung ist. Später hat Hahn die Lagrangesche Multiplikatorenmethode in die Theorie der Funktionaloperationen eingeordnet (Wien. Ak. Ber. *131*) und ein Existenztheorem für semidefinite Variationsprobleme bewiesen, welches mehrere Sätze aus Tonellis grundlegendem Buche als Spezialfälle enthält (Wien. Ak. Ber. *134*). Ein ausgezeichneter kurzer Artikel über Variationsrechnung aus Hahns Feder findet sich in Pascals Repertorium.

Über *Punktfunktionen* bewies Hahn in Beantwortung einer Frage von Lebesgue, daß bei Zulassung der Ableitung ∞ der Fundamentalsatz der Integralrechnung seine Giltigkeit verliert (Monatsh. *16*). Weiter fand er, daß zwischen eine oberhalb stetige und eine von ihr nirgends überschrittene unterhalb stetige Funktion immer eine stetige Funktion eingeschoben werden kann (Wien. Ak. Ber. *126*) und daß für eine Funktion $f(x_1, \ldots x_n)$, die in jeder Variablen stetig ist, die Menge der Stetigkeitpunkte dicht und sogar in jeder Menge $x_i = \text{const.}$ dicht liegt (Math. Zeitschr. *4*). Damit eine Menge aus den Konvergenzpunkten einer Folge stetiger Funktionen besteht, ist nicht nur notwendig, sondern auch hinreichend, daß sie ein $F_{\sigma\delta}$ sei (Arch. d. Math. u. Phys. III, *28*).

Aus der Lehre von den *Mengenfunktionen* hat Hahn besonders den Integralbegriff untersucht, so den Hellingerschen und Borelschen Begriff, sowie die Approximation Lebesguescher Integrale durch Riemannsche Summen (Monatsh. *23* u. *26*, Wien. Ak. Ber. *123*). Sehr elegant ist seine Einführung des Lebesgueschen Integrals durch den Satz, daß $\int_M f\,d\mu$ die einzige total additive Funktion aller meßbaren Mengen M sei, welche für jede Menge M und jede integrierbare Funktion f, deren Werte in den Punkten von M zwischen zwei Schranken c' und c'' liegen, zwischen dem c'-fachen und dem c''-fachen des Maßes von M liegt (Wien. Ak. Anz. 1928). Die Lehre von den mehrfachen Integralen ordnete er in eine Theorie des Produktes von Mengenfunktionen ein, welches er als eine Mengenfunktion des Produktraumes definierte, wobei das Reduktionstheorem für Integrale sich als assoziatives Gesetz der Multiplikation ergab (An. di Pisa *2*). In den letzten Jahren bewies Hahn einige Sätze über Mengenfunktionen in überraschend einfacher Weise durch Heranziehung von Sätzen über unendliche Reihen (Wien. Ak. Anz. 1928). Hahns Buch „Theorie der reellen Funktionen", von dem leider nur der erste Band erschienen ist (1. Aufl. 1921, 2. Aufl. 1932), ist eines der Standardwerke dieses Gebietes.

Fouriersche Reihen und Integrale hat Hahn (Acta Math. *49*) in einer von ihm wiederentdeckten, bereits von Weyl angegebenen Stieltjesintegralformel zusammengefaßt, welche erheblich weiteren Giltigkeitsbereich besitzt, als die Fouriersche Integralformel. Darstellungen durch *singuläre Integrale* untersuchte Hahn auch an gewissen Unstetigkeitsstellen, wobei er zugleich Differenzierbarkeitsbedingungen für die Darstellung angab, ferner Summationsverfahren, Bedingungen für gliedweise Integration von Entwicklungen nach orthogonalen Funktionen u. a. (Denkschr. Wien. Ak. *93*). Von besonderem Interesse ist Hahns Anwendung der für die singulären Integrale entwickelten Methoden auf das *Interpolationsproblem* (Math. Zeitschr. *1*). Er gab notwendige und hinreichende Bedingungen erstens dafür an, daß ein Interpolationsverfahren für alle stetigen Funktionen konvergiere, und zweitens dafür, daß es von endlicher Empfindlichkeit sei.

d. h. bei kleinen Änderungen der gegebenen Funktionswerte nicht allzu große Änderungen der interpolierten Funktion mit sich bringt. Es stellt sich dabei insbesondere heraus, daß Konvergenz die endliche Empfindlichkeit nach sich zieht.

In einer Theorie der *linearen Operationen* entwickelte Hahn den auch von Fréchet und Banach entdeckten Begriff des linearen Raumes (des abstrakten Vektorraumes) und faßte auf Grund der Bemerkung, daß die linearen Funktionen beschränkter Steigung der Punkte eines linearen Raumes selbst einen linearen Raum bilden, zahlreiche Sätze über Transformationen von Zahlenfolgen und von Funktionen mit Sätzen über singuläre Integrale zusammen (Monatsh. *32*). Anwendungen der Theorie machte Hahn auf lineare Gleichungssysteme in linearen Räumen (Crelles Journ. *157*) und auf Reihen (Monatsh. *33*). Schon aus früheren Jahren stammt ein vorzüglicher Bericht über die Theorie der linearen Integralgleichungen (Jahresber. d. D. M. V. *20*).

Einen wichtigen Beitrag lieferte Hahn zur *allgemeinen Arithmetik* durch seine Arbeit über nichtarchimedische Größensysteme (Wien. Ak. Ber. *116*). Seine Ergebnisse sind in der heutigen Terminologie als eine Charakterisierung der geordneten Abelschen Gruppen zu bezeichnen, u. zw. dadurch, daß sie isomorph mit Systemen hyperkomplexer Zahlen sind, deren Einheiten eine geordnete Menge (den „Klassentypus" Γ der Gruppe G) bilden und deren Koeffizienten reelle Zahlen sind, wobei zu jeder Gruppe G eine Mächtigkeit $\aleph$ gehört derart, daß jedem Elemente von G bei der Isomorphie eine hyperkomplexe Zahl entspricht, für welche nur die Einheiten einer absteigend wohlgeordneten Teilmenge von Γ, deren Mächtigkeit $<\aleph$ ist, Koeffizienten $\neq 0$ besitzen. Ist der Klassentypus Γ der Gruppe G selbst eine geordnete Abelsche Gruppe, so kann für die Elemente von G eine Multiplikation definiert werden, die G zu einem geordneten Körper macht. — Weiter hat Hahn (Wien. Ak. Ber. *122*) den Cantor-Bendixsonschen Satz und die Theorie der Kohärenzen auf geordnete Mengen übertragen.

Hahn war einer der ersten, welche die Bedeutung von Fréchets abstrakten Raumbegriffen erkannten. Er bewies (Monatsh. *19*), daß in jeder Klasse V nichtkonstante stetige Funktionen existieren (in Limesklassen nicht notwendig), bekanntlich der Kernsatz des Metrisationsproblemes — und zwar entwickelte Hahn zu seinem Beweise die später von Urysohn zum Beweis des Hauptlemmas der Metrisationstheorie benützte Methode der Umgebungsringe.

Hahns erster Beitrag zur eigentlichen *mengentheoretischen Geometrie* war der Beweis, daß bei jeder stetigen Abbildung der Strecke aufs Quadrat kontinuierlichviele Quadratpunkte mindestens je zwei Urbildpunkte besitzen und die Punkte einer im Quadrat dichten Menge mindestens je drei Urbilder (Ann. di mat. pur. ed appl. III. *21*). Die Fruchtbarkeit der damit angeschnittenen Frage geht aus den allgemeinen dimensionstheoretischen Sätzen von Hurewicz über die Multiplizität von Punkten bei dimensionserhöhenden stetigen Abbildungen hervor. — Hahns Kennzeichnung der stetigen Streckenbilder zunächst unter den ebenen Punktmengen (Jahresber. D. M. V. *24*), dann unter den metrischen Räumen (Wien. Ak. Ber. *123,* einen anderen Beweis in Atti Congr. Int. Bologna 1928) durch Kompaktheit, Zusammenhang und die (gleichzeitig und unabhängig auch von Mazurkiewicz entdeckte) Eigenschaft des Zusammenhanges im Kleinen gehört schon heute zu den klassischen Ergebnissen der mengentheoretischen Geometrie. Für den Fall ebener Punktmengen bewies Hahn auch direkt die Äquivalenz seiner Kennzeichnung mit der Schoenfliesschen durch Erreichbarkeitseigenschaften (Math. Zeitschr. *9*). Als kennzeichnend für den Zusammenhang im kleinen eines Raumes fand Hahn die Offenheit aller Komponenten aller offener Mengen (Fund. Math. *2*), was allerdings durch Kuratowski vorweggenommen war. — Hahn faßte ferner in jedem Kontinuum C die Punktepaare p, p', zu denen für jedes $\varepsilon > 0$ endlichviele Punkte $p_1 = p, p_2, \ldots p_{n-1}, p_n = p'$ existieren, so daß p_{i+1} und p_i Abstände $< \varepsilon$ haben und C in jedem Punkt p_i nicht zusammenhängend im Kleinen ist, in Primteile von C zusammen und bewies, daß jedes irreduzible Kontinuum entweder aus einem einzigen Primteil besteht oder ein Bogen von Primteilen ist (Wien. Ak. Ber. *130*; für einen sehr kurzen Beweis vgl. Hausdorff, Mengenlehre 2. Aufl.). Die Fruchtbarkeit des Primteilbegriffes geht auch aus einem Satz von R. L. Moore hervor (Vgl. Hausdorff a. a. O.).

Von sonstigen mathematischen Arbeiten Hahns seien insbesondere noch zwei erwähnt, die anderen als den bisher besprochenen Gebieten angehören: In der einen (Monatsh. *16*) übertrug er den Weierstraßschen Produktsatz auf Funktionen von zwei komplexen Veränderlichen. In der anderen (Monatsh. *19*) gab er den ersten lückenlosen Beweis des Jordanschen Kurvensatzes für Polygone mit elementaren Mitteln (Verknüpfungs- und Anordnungaxiomen). Diese Ergänzung zu Veblen's Arbeit über die Axiome der Geometrie (Ann. Trans. *5*) lieferte zugleich eine restlose Vervollständigung von Veblens Beweis des Jordanschen Satzes für beliebige einfache geschlossene Kurven aus dem Jahre 1905 (Ann. Trans. *6*). Zerlegungen des n-dimensionalen Raumes wurden im Anschluß an Hahns Arbeit von Lilly Hahn behandelt (Monatsh. *25*).

Neben der Mathematik beschäftigte sich Hahn in den letzten Jahren zunehmend mit Logik und Erkenntnistheorie. Nach Veröffentlichung einer noch heute lesenswerten Besprechung des bekannten Pringsheimschen Lehrbuches (Gött. gel. Anz. 1919) begann er in den ersten Nachkriegsjahren ein eingehendes Studium der Logistik, das ihn nicht nur zu einem besonderen Bewunderer, sondern sicher auch zu einem der besten Kenner von Russell machte. In der Folge schloß er sich dem bekannten Wiener Kreis an. Seine eigenen Auffassungen sind insbesondere in dem Artikel „Die Bedeutung der wissenschaftlichen Weltauffassung für Mathematik und Physik“ (Erkenntnis *1*) und in drei Vorträgen niedergelegt: „Logik, Mathematik und Naturerkennen“ (bei Gerold, Wien 1933) setzt sich mit der empiristischen

Auffassung auseinander, „Die Krise der Anschauung" (bei Deuticke, Wien 1933) mit der intuitionistischen, der Vortrag „Gibt es Unendliches?" (bei Deuticke, Wien 1934) skizziert Hahns eigenen logizistischen Standpunkt.

Als akademischer Lehrer hielt Hahn Vorlesungen über die verschiedensten Teile der Mathematik, wie überhaupt sein mathematisches Wissen sehr umfassend war. (Als auf einer deutschen Mathematikertagung an einem Tisch, an dem mehrere bekannte Kollegen saßen, darunter einer der hervorragendsten lebenden Zahlentheoretiker, ein Frage- und Antwort-Spiel zur Feststellung der mathematischen Allgemeinbildung gespielt wurde, blieb Hahn Sieger.) Alle Vorlesungen wurden von Hahn sorgfältig ausgearbeitet und waren vorbildlich an Klarheit. — Unübertrefflich war er als Seminarleiter und viele Anregungen, die er durch seine Seminare gab, waren außerordentlich wertvoll. Zahlreiche Schüler haben im Laufe der Jahre unter Hahns Anleitung gearbeitet und Dissertationen verfaßt, darunter einige Abhandlungen ersten Ranges, wie die von Hurewicz und Gödel. Ich selbst verdanke die Anregungen zu meiner gesamten wissenschaftlichen Tätigkeit der ersten Seminarstunde, die er im April 1921 nach seiner Berufung an die Wiener Universität abhielt und in der er nach einer unsere anschauliche Kurvenvorstellung wiedergebenden Definition fragte. — Auch Hahns Vorträge für größeres Publikum erfreuten sich großer Beliebtheit. Sie waren nicht nur stets von besonderer Klarheit — so wie auch die von Hahn gemeinsam mit Tietze verfaßte „Einführung in die Elemente der höheren Mathematik" — sondern enthielten stets auch irgendwelche neue Gedanken.

Grundzug von Hahns Persönlichkeit war eine Sachlichkeit, die ich nicht besser kennzeichnen kann, als mit Worten, die R. Meister bei seiner Bestattung sprach: „Hahn ist nie für eine Sache eingetreten, von der er nicht voll und ganz überzeugt war". Er wurde deshalb stets auch von seinen Gegnern hochgeachtet. Wenn er seine persönlichen Gefühle zeigte, so waren es stets solche der Güte und des Wohlwollens. Er scheute keine Mühe, um arme Studenten zu fördern oder einem allgemein nützlichen Zweck zu dienen. Wo er eine Gelegenheit fand, für Hilflose einzutreten, ließ er sie nie vorübergehen, mochte es sich auch nur um arme Tiere handeln. Als er einmal auf der Straße einen Kutscher sein Pferd mißhandeln sah und sein Einspruch erfolglos blieb, schleppte er den Rohling mit Gewalt vor einen Polizisten. Überhaupt hat Hahn, dessen normale Umgangsformen sich durch besondere Liebeswürdigkeit auszeichneten, wenn er beschlossen hatte, für etwas einzutreten, stets ohne Rücksicht auf die ihm selbst daraus erwachsenden Schwierigkeiten mit Energie und Konsequenz es zu Ende geführt.

In diesem Kolloquium wird seiner stets mit Verehrung und Dankbarkeit gedacht werden.

Neuer Beweis der Streckenverbundenheit vollständiger konvexer Räume. Von N. Aronszajn in Paris.

Im folgenden geben wir einen neuen, transfinite Induktion nicht verwendenden Beweis des Mengerschen Theorems[1]):

Jeder vollständige konvexe Raum R enthält zu je zwei Punkten p und q einen p und q verbindenden Bogen, der mit einer Strecke der Länge pq kongruent ist.

R sei ein metrischer Raum. Ist jedem Punkt p einer Teilmenge D von R eine reelle Zahl $f(p)$ zugeordnet, so sprechen wir von einer Funktion f in R mit dem Definitionsbereich $D(f) = D$ und nennen die Menge aller reellen Zahlen, die den Punkten von D zugeordnet sind, den Wertbereich $W(f)$ von f. Gelten für zwei Funktionen f und g in R die Beziehungen $D(f) \subset D(g)$ und $f(p) = g(p)$ für jeden Punkt p von $D(f)$, so schreiben wir $f \subset g$ und nennen g eine *Erweiterung* von f. Eine Funktion f in R heiße *isometrisch*, wenn $D(f)$ und $W(f)$ kongruent sind, d. h. wenn für je zwei Punkte p und q von $D(f)$ der Abstand pq gleich dem Abstand $|f(p) - f(q)|$ ihrer Bilder ist.

Es seien nun zwei Punkte p und q von R gegeben. Ihren Abstand pq bezeichnen wir mit ρ. Nennen wir D_1 die aus p und q bestehende Menge und setzen wir $f_1(p) = 0$, $f_1(q) = \rho$, so ist f_1 offenbar eine isometrische Funktion mit $D(f_1) = D_1$. Das abgeschlossene Zahlenintervall zwischen 0 und ρ bezeichnen wir mit $[0, \rho]$. Wir machen nun die (für $n = 1$ eben realisierte) Annahme, daß n isometrische Funktionen $f_1, f_2, \ldots, f_n$ in R vorliegen, so daß ihre Wertemengen $W(f_i) \subset [0, \rho]$ $(i = 1, 2, \ldots, n)$ und daß $f_1 \subset f_2 \subset \ldots \subset f_n$ gilt. Dann bezeichnen wir mit $\mathfrak{F}_n$ die Gesamtheit aller isometrischen Funktionen f in R, für welche $W(f) \subset [0, \rho]$ und $f_n \subset f$ gilt. Ist $\mathfrak{J}_n$ die Menge der 2^n abgeschlossenen Intervalle $\left[\frac{i-1}{2^n}\rho, \frac{i}{2^n}\rho\right]$ $(i = 1, 2, \ldots, 2^n)$, so bezeichnen wir für jede Funktion f von $\mathfrak{F}_n$ mit $z_n(f)$ die Anzahl derjenigen Intervalle von $\mathfrak{J}_n$, deren Durchschnitt mit $W(f)$ nichtleer ist. Da für jede Funktion f von $\mathfrak{F}_n$ offenbar $z_n(f) \leqq 2^n$ gilt, so können wir eine Funktion f_{n+1} aus $\mathfrak{F}_n$ wählen, für welche $z_n(f_{n+1})$ das Maximum erreicht, d. h. für welche $z_n(f_{n+1}) \geqq z_n(f)$ für jede Funktion f aus $\mathfrak{F}_n$ gilt.

Die auf diese Weise ausgehend von f_1 gewählte Funktionenfolge $\{f_n\}$ $(n = 1, 2, \ldots$ ad inf.) bestimmt offenbar eindeutig eine isometrische Funktion f, für welche

$$D(f) = \sum_{n=1}^{\infty} D(f_n) \subset R, \quad W(f) = \sum_{n=1}^{\infty} W(f_n) \subset [0, \rho]$$

und $f_n \subset f$ für $n = 1, 2, \ldots$ ad inf. gilt. Da das Intervall $[0, \rho]$ ein vollständiger Raum ist, so kann man, wenn auch R *vollständig* ist, nach einem bekannten Satz die Funktion f zu einer isometrischen Funktion g erweitern, für welche $D(g) = \overline{D(f)}$ und $W(g) = \overline{W(f)}$

[1]) Menger, Math. Annalen *100*, S. 87.

gilt, wobei wir für jede Menge M mit $\overline{M}$ die abgeschlossene Hülle von M bezeichnen.

Wir zeigen nun: Wenn R *konvex* ist, gilt $W(g) = [0, \rho]$. Wir machen zu diesem Zweck die Annahme, es gebe ein Element von $[0, \rho]$, das nicht in $W(g)$ enthalten ist, und leiten einen Widerspruch her. Da $W(g)$ als abgeschlossene Hülle von $W(f)$ eine abgeschlossene Teilmenge von $[0, \rho]$ ist und 0 und ρ enthält (denn g ist Erweiterung von f, also von f_1), so folgt aus der Annahme die Existenz eines zu $W(g)$ fremden offenen Teilintervalles (α, β) von $(0, \rho)$, dessen Endpunkte α und β $(\alpha < \beta)$ in $W(g)$ liegen. Da α und β in $W(g)$ liegen, existieren zwei Punkte a und b von $D(g) \subset R$, für welche $g(a) = \alpha$ und $g(b) = \beta$ gilt. Ihr Abstand ab ist, da g isometrisch ist, $= \beta - \alpha$. Wegen der Konvexität von R existiert ein Zwischenpunkt c von a und b, d. h. ein Punkt, für welchen $a \neq c \neq b$ und $ac + cb = ab$ gilt. Hieraus folgt, daß eine einzige Zahl γ zwischen α und β existiert, so daß $ac = \gamma - \alpha$ und $cb = \beta - \gamma$ gilt. Nennen wir D^* die um den Punkt c vermehrte Menge $D(g)$ und setzen wir $g^*(p) = g(p)$ für jeden Punkt p von $D(g)$ und $g^*(c) = \gamma$, so ist g^* offenbar eine isometrische Erweiterung von g (und daher von allen f_n) mit dem Definitionsbereich D^*. Bestimmen wir eine natürliche Zahl n so groß, daß $\frac{\rho}{2^n} < \text{Min}\,(\gamma - \alpha, \beta - \gamma)$, so hat $W(g^*)$ einen nichtleeren Durchschnitt mit wenigstens einem γ enthaltenden Intervall von $\mathfrak{I}_n$, zu welchem $W(f_{n+1})$ fremd ist. Da g^* Erweiterung von f_{n+1} ist, hat $W(g^*)$ mit allen Intervallen, mit denen $W(f_{n+1})$ Punkte gemein hat, einen nichtleeren Durchschnitt. Es gilt also $z_n(g^*) \geqq z_n(f_{n+1}) + 1$. Da g^* Erweiterung von f_n ist und $W(g^*) \subset [0, \rho]$ gilt, gehört g^* zu $\mathfrak{F}_n$, womit ein Widerspruch gegen die Definition von f_{n+1} hergeleitet und $W(g) = [0, \rho]$ bewiesen ist. Die Menge $D(g)$ ist, da g isometrisch ist, eine mit der Strecke $[0, \rho]$ kongruente Teilmenge von R, womit das eingangs erwähnte Theorem bewiesen ist.

Zur Topologie der Mannigfaltigkeiten. Von Georg Nöbeling.

Zu den bisher ungelösten Grundproblemen der Topologie zählen die Frage der Triangulierbarkeit einer separablen topologischen Mannigfaltigkeit und die Steinitzsche Vermutung. Dabei nennt man topologische Mannigfaltigkeit einen topologischen Raum M, in dem zu jedem Punkt eine Umgebung existiert, die homöomorph ist zum Innern der n-dimensionalen euklidischen Vollkugel. M heißt triangulierbar, wenn M dargestellt werden kann als Summe von endlich oder abzählbar unendlich vielen Mengen, von denen jede einem euklidischen Simplex homöomorph ist und die mit einander genau so zusammenhängen wie die Simplexe eines Komplexes. Die Steinitzsche Vermutung besagt, daß wenn zwei Triangulierungen von M vorliegen, dieselben so unterteilt werden können, daß die entstehenden Triangulierungen kombinatorisch äquivalent sind. Der Schlüssel[1]) zu

[1]) vgl. hierzu die Ausführungen von Alexander in den Berichten des Internationalen Mathematikerkongresses in Zürich, 1, S. 253.

den beiden genannten Problemen ist folgender Glättungssatz: Ist Φ eine topologische Abbildung der offenen Menge M des euklidischen E_n auf eine zweite offene Menge m des E_n und ist auf M eine stetige positive Funktion $\varkappa(p)$ definiert, so existiert eine topologische Abbildung Δ von M auf m, sodaß bei geeigneter Triangulierung von M jedes Simplex derselben affin abgebildet wird und für jeden Punkt p von M die Bilder $\Phi(p)$ und $\Delta(p)$ einen Abstand $< \varkappa(p)$ haben. Der Beweis dieses Glättungssatzes kann in drei Schritten geführt werden. Beim ersten wird aus der gegebenen Abbildung Φ zunächst eine simpliziale (nicht notwendig topologische) Abbildung von M auf m und aus dieser dann eine simpliziale Abbildung von M auf m hergeleitet, bei welcher jeder Punkt von m eine höchstens eindimensionale Urbildmenge und jeder Punkt jedes n-dimensionalen Simplexes einer gewissen Triangulierung von m genau einen Punkt zum Urbild hat. Der zweite und schwierigste Schritt leitet dann eine simpliziale Abbildung Γ von M auf m her, bei welcher jeder Punkt von m eine höchstens eindimensionale zusammenhängende Urbildmenge hat. Der dritte Schritt leitet dann aus Γ die gesuchte topologische simpliziale Abbildung Δ her. — Eine Ausführung dieser Beweisskizze und eine darauf fußende Lösung der genannten beiden Probleme werden demnächst in den Monatsheften für Mathematik und Physik erscheinen.

Angelegenheiten des Kolloquiums und Nachrichten.

Wiener Dissertationen im Zusammenhang mit dem Kolloquium 1933/34.

11. Dissertation: A. Vilim „Längenmengen".

Veränderungen des Teilnehmerkreises 1933/34.

Dozent K. Gödel hat während des Studienjahres 1933/34 einer Einladung an das Institute of Advanced Study, Princeton, Folge geleistet. Dr. G. Nöbeling hat einen Lehrauftrag an der Universität Erlangen erhalten und ist aus dem Kolloquium ausgeschieden. Prof. R. G. Putnam (New York University), Prof. K. Midutani (Handelshochschule Kobe), Prof. H. Terasaka (Universität Osaka). und A. Flores (Madrid), letzterer mit einem Stipendium der Fundación Conde de Cartagena der Madrider Akademie der Wissenschaften, nahmen im ganzen Studienjahr 1933/34, Prof. T. Hirano (Naturwissenschaftliche Hochschule Tokio) im Sommersemester 1934 am Kolloquium teil.

Druck von Gottlieb Gistel & Cie., Wien.

Bericht über des Kolloquium 1934/35.

80. Kollcquium (6. XI. 1934).

Über die Produktionsgleichungen der ökonomischen Wertlehre (II. Mitteilung). Von A. Wald.

In der ersten Mitteilung in Heft 6, S. 12, dieser Ergebnisse (im Folgenden mit P I zitiert) wurde von einem modifizierten System der Walras-Casselschen Produktionsgleichungen unter gewissen Voraussetzungen eindeutige nicht-negative Lösbarkeit nachgewiesen. In dieser Mitteilung wird die Lösbarkeit schon unter viel schwächeren Voraussetzungen bewiesen.

Es seien $R_1, R_2, \ldots, R_m$ sämtliche (wirtschaftliche und freie) Produktionsmittel, aus denen in verschiedenen Kombinationen n Produkte $S_1, S_2, \ldots, S_n$ herstellbar sind. Bekannt sei die Zahl r_i der dem Produzenten verfügbaren Einheiten von R_i für $i=1, 2, \ldots, m$; ferner, daß zur Herstellung einer Einheit von S_j die Aufwendung von a_{1j} Einheiten von R_1, von a_{2j} Einheiten von $R_2, \ldots$, von a_{mj} Einheiten von R_m erforderlich sei für $j=1, 2, \ldots, n$; schließlich, daß, wenn s_1 Einheiten von S_1, ferner s_2 Einheiten von $S_2, \ldots, s_n$ Einheiten von S_n produziert würden, eine Einheit von S_j den Preis $f_j(s_1, s_2, \ldots, s_n)$ erzielen würde, für $j=1, 2, \ldots, n$. *Unbekannt* sind: die Zahl s_j der herzustellenden Einheiten von S_j $(j=1, 2, \ldots, n)$, ferner die Zahl $\ddot{u}_i$ der eventuell bei der Produktion nicht verwendeten Einheiten von R_i $(i=1, 2, \ldots, m)$, schließlich der Preis ρ_i der Einheit von R_i $(i=1, 2, \ldots, m)$ und der Preis σ_j der Einheit von S_j $(j=1, 2, \ldots, n)$. Für diese $2m+2n$ Unbekannten nehmen wir in dieser zweiten Mitteilung die folgenden $2m+2n$ Gleichungen an:

$$
\text{(II)}\quad
\begin{aligned}
&r_i = \sum_{j=1}^{n} a_{ij} s_j + \ddot{u}_i \quad (i=1, 2, \ldots, m), \qquad \rho_i . \ddot{u}_i = 0 \quad (i=1, 2, \ldots, m)\\
&\sigma_j = \sum_{i=1}^{m} a_{ij} \rho_i \qquad (j=1, 2, \ldots, n),\\
&\qquad\qquad \sigma_j = f_j(s_1, s_2, \ldots, s_n) \quad (j=1, 2, \ldots, n)
\end{aligned}
$$

In P I wurde dieses Gleichungssystem unter der vereinfachenden Annahme untersucht, daß für $j=1, 2, \ldots, n$ die Funktion $f_j(s_1, s_2, \ldots, s_n)$ nur von s_j abhängt und mit wachsendem s_j monoton abnimmt. Diese den Tatsachen nicht entsprechende Annahme — der Preis eines Gutes hängt ja nicht bloß von der vorhandenen Menge dieses Gutes, sondern auch von den Quantitäten der übrigen Güter ab — lassen wir nun fallen und beweisen das.

Theorem: *Das Gleichungssystem (II) besitzt ein System von nicht negativen Lösungen in den $2m+2n$ Unbekannten, und zwar in den Unbekannten $s_1, \ldots, s_n, \ddot{u}_1, \ldots, \ddot{u}_m, \sigma_1, \ldots, \sigma_n$ eine einzige Lösung, wenn folgende Bedingungen erfüllt sind:*

1) $r_i > 0 (i=1, \ldots, m)$

2) $a_{ij} \geqq 0 (i=1, \ldots, m; j=1, \ldots, n)$

3) *Zu jedem j gibt es mindestens ein i, so daß $a_{ij} > 0$ ist.* (Daß die Bedingungen 1), 2), 3) erfüllt sind, ergibt sich unmittelbar aus der ökonomischen Bedeutung der Zahlen r_i und a_{ij}).

4) *Die Funktion $f_j(s_1, \ldots, s_n)$ ist für alle n-Tupel von nicht negativen Zahlen $s_1, \ldots, s_n$, für welche $s_j \neq 0$ ist, definiert, positiv und stetig für $j=1, 2, \ldots, n$.*

5) *Für jede der Zahlen $j=1, \ldots, n$ gilt*[1] $\lim_{k=\infty} f_j(s_1^k, \ldots, s_n^k) = \infty$, *wenn die n-Tupel nicht negativer Zahlen $s_1^k, \ldots, s_n^k$ ($k=1, 2, \ldots$ ad inf.), denen $s_j^k > 0$ für jedes k ist, gegen ein n-Tupel $s_1, \ldots, s_n$ konvergieren, in dem $s_j = 0$.*

6) *Sind $\Delta s_1, \ldots, \Delta s_n$ irgend n Zahlen, unter denen mindestens eine < 0 ist, und gilt $\sum_{j=1}^{n} \sigma_j \Delta s_j \leqq 0$, so gilt $\sum_{j=1}^{n} \sigma_j' \Delta s_j < 0$, wobei $\sigma_j' = f_j(s_1 + \Delta s_1, \ldots, s_n + \Delta s_n)$ ($j=1, 2, \ldots, n$).*

Es ist klar, daß 6) aus folgender schärferen Bedingung folgt:

6') *Ändern sich die Quantitäten $s_1, \ldots, s_n$ der $S_1, \ldots, S_n$ um $\Delta s_1, \ldots, \Delta s_n$, wobei mindestens eine dieser $\Delta s_j \neq 0$, so gilt für die dadurch bewirkte Änderung $\Delta \sigma_j = f_j(s_1 + \Delta s_1, \ldots, s_n + \Delta s_n) - f_j(s_1, \ldots, s_n)$ des Preises der Einheit von S_j ($j=1, \ldots, n$) stets:*

$$\sum_{j=1}^{n} \Delta \sigma_j \Delta s_j < 0.$$

6') ist eine wesentliche Abschwächung der Voraussetzung von *PI*, daß die Funktionen f_j monoton seien. Dieser letzteren zufolge gilt nämlich offenbar $\Delta \sigma_j \Delta s_j < 0$ für jedes einzelne j, für welches $\Delta s_j \neq 0$, während 6') bloß verlangt, daß die Summe der n Zahlen $\Delta \sigma_j \Delta s_j$ negativ sei, falls mindestens eine der n Zahlen $\Delta s_j \neq 0$. Stellt man die beiden n-Tupel $(\Delta \sigma_1, \ldots, \Delta \sigma_n)$ und $(\Delta s_1, \ldots, \Delta s_n)$ als Vektoren des n-dimensionalen euklidischen Raumes dar, so besagt 6'), daß diese beiden Vektoren einen Winkel $> 90^0$ mit einander einschließen.

Die Eindeutigkeitsbehauptung des Theorems beruht auf folgendem Hilfssatz:

1. Ist $\ddot{u}_i, s_j, \rho_i, \sigma_j$ ein System von positiven Lösungen der obigen Gleichungen (II) und $\ddot{u}_i + \Delta \ddot{u}_i$, $s_j + \Delta s_j$, $\rho_i + \Delta \rho_i$, $\sigma_j + \Delta \sigma_j$ ein System von positiven Lösungen des Gleichungssystemes (II'), das aus (II) entsteht, wenn man r_i durch $r_i + \Delta r_i$ ($i=1, \ldots, m$)

[1]) Wie Herr **Menger** bemerkt hat, kann 5) dahin abgeschwächt werden, daß $\lim_{k=\infty} f_j(s_1^k, \ldots, s_n^k)$ größer ist als eine gewisse von den mn Koeffizienten a_{ij} abhängige und aus ihnen leicht berechenbare endliche Zahl.

und die $f_j(s_1, \ldots, s_n)$ durch irgend welche Funktionen $f_j^*(s_1, \ldots, s_n)$ ersetzt $(j=1, \ldots, n)$, so gilt:

$$\alpha)\ \sum_{i=1}^{m} \rho_i \Delta r_i = \sum_{j=1}^{n} \sigma_j \Delta s_j + \sum_{i=1}^{m} \rho_i \Delta \ddot{u}_i,$$

$$\beta)\ \sum_{i=1}^{m} (\rho_i + \Delta \rho_i) \Delta r_i = \sum_{j=1}^{n} (\sigma_j + \Delta \sigma_j) \Delta s_j + \sum_{i=1}^{m} (\rho_i + \Delta \rho_i) \Delta \ddot{u}_i,$$

$$\gamma)\ \sum_{i=1}^{m} \rho_i \Delta \ddot{u}_i \geqq 0 \geqq \sum_{i=1}^{m} (\rho_i + \Delta \rho_i) \Delta \ddot{u}_i.$$

Um α) und β) zu erhalten, betrachtet man die offenbar bestehenden Gleichungen

$$\Delta r_i = \sum_{j=1}^{n} a_{ij} \Delta s_j + \Delta \ddot{u}_i (i=1, \ldots, m),\ \sigma_j + \Delta \sigma_j = \sum_{i=1}^{m} a_{ij} (\rho_i + \Delta \rho_i) \quad (j=1, \ldots, n),$$

multipliziert die m ersten der Reihe nach mit $\rho_1, \ldots, \rho_m$ bzw. mit $\rho_1 + \Delta \rho_1, \ldots, \rho_m + \Delta \rho_m$ und addiert. Zum Beweise von γ) beachtet man, daß, wenn $\Delta \ddot{u}_i \geqq 0$, so auch $\rho_i \Delta \ddot{u}_i \geqq 0$, wenn dagegen $\Delta \ddot{u}_i < 0$, so $\ddot{u}_i > 0$ wegen $\ddot{u}_i + \Delta \ddot{u}_i \geqq 0$ und mithin $\rho_i = 0$ und $\rho_i \Delta \ddot{u}_i = 0$ gilt; und ferner, daß wenn $\Delta \ddot{u}_i > 0$, so auch $\ddot{u}_i + \Delta \ddot{u}_i > 0$ und mithin $\rho_i + \Delta \rho_i = 0$ gilt; wegen $\rho_i + \Delta \rho_i \geqq 0$ $(i=1, \ldots, m)$ ergibt sich dann die zweite Hälfte der Ungleichung γ). Aus **1** folgt nun: Sind s_j, $\ddot{u}_i$, ρ_i, σ_j und $s_j + \Delta s_j$, $\ddot{u}_i + \Delta \ddot{u}_i$, $\rho_i + \Delta \rho_i$, $\sigma_j + \Delta \sigma_j$ zwei Lösungssysteme von (II), so haben wir:

$$\sum_{j=1}^{n} \sigma_j \Delta s_j + \sum_{i=1}^{m} \rho_i \Delta \ddot{u}_i = 0,\ \sum_{j=1}^{n} (\sigma_j + \Delta \sigma_j) \Delta s_j + \sum_{i=1}^{m} (\rho_i + \Delta \rho_i) \Delta \ddot{u}_i = 0,$$ also nach

1 γ): $\sum_{j=1}^{n} \sigma_j \Delta s_j \leqq 0 \leqq \sum_{j=1}^{n} (\sigma_j + \Delta \sigma_j) \Delta s_j$. Wären nun die beiden Lösungen in s_j, $\ddot{u}_i$, σ_j nicht identisch und mithin $\Delta s_j \neq 0$ für mindestens ein j, so würden diese Ungleichungen der Bedingung 6) widersprechen, womit die Eindeutigkeitsbehauptung bewiesen ist. Dem Beweis der Existenzbehauptung schicken wir folgenden ganz wie in P I (S. 13) beweisbaren Satz voran:

2. Konvergieren die m-Tupel positiver Zahlen $\{r_1^{(k)}, \ldots, r_m^{(k)}\}$ $(k=1, 2, \ldots$, ad inf.) gegen das m-Tupel $\{r_1, \ldots, r_m\}$ und die Funktionenfolge $\{f_j^{(k)}(s_1, \ldots, s_n)\}$ gleichmäßig gegen die Funktion $f_j(s_1, \ldots, s_n)$ $(j=1, \ldots, n)$ und ist $\ddot{u}_i^{(k)}$, $s_j^{(k)}$, $\rho_i^{(k)}$, $\sigma_j^{(k)}$ ein System positiver Lösungen der Gleichungen (II^k), welche entstehen, wenn man $r_1, \ldots, r_m$ in (II) durch $r_1^{(k)}, \ldots, r_m^{(k)}$ und $f_j(s_1, \ldots, s_n)$ durch $f_j^{(k)}(s_1, \ldots, s_n)$ $(j=1, \ldots, n)$ ersetzt, dann sind die Zahlenfolgen $\{\ddot{u}_1^{(k)}\}, \ldots, \{\ddot{u}_m^{(k)}\}$, $\{s_1^{(k)}\}, \ldots, \{s_n^{(k)}\}$ $(k=1, 2, \ldots$, ad inf.) beschränkt, ebenso die Zahlenfolge $\{\rho_i^{(k)}\}$ $(k=1, 2, \ldots$ ad inf.) für jedes i, für welches $r_i \neq 0$. Konvergieren die Lösungssysteme $\ddot{u}_i^{(k)}$, $s_j^{(k)}$, $\rho_i^{(k)}$, $\sigma_j^{(k)}$ von (II^k) mit wachsendem k gegen ein $(2m+2n)$-Tupel $\ddot{u}_i$, s_j, ρ_i, σ_j, so ist das letztere ein Lösungssystem von (II).

1*

Die Existenz einer Lösung wird nun durch Induktion nach n bewiesen ähnlich wie beim Theorem in P I. Die Fälle $n=1$ beider Theoreme sind identisch. Um den Fall n auf Grund des Falles n-1 zu beweisen, definieren wir $\bar{\lambda}$ ganz wie in P I und für jedes $0 \leqq \lambda < \bar{\lambda}$ das Gleichungssystem (II^{λ}_{n-1}), das sich von (Sch^{λ}_{n-1}) nur dadurch unterscheidet, daß jetzt die n-1 letzten Gleichungen $\sigma_j = f_j(s_1, \ldots, s_{n-1}, \lambda)$ lauten. Wie in P I führen wir $\Pi(\lambda)$ ein. Aus 2 folgert man leicht, daß zu jedem $\lambda_0 < \bar{\lambda}$ eine endliche Zahl A existiert derart, daß die Zahlen der Menge $\Pi(\lambda)$ durchwegs $< A$ sind. Die Menge $\Pi(\lambda)$ ist also für jedes $\lambda < \bar{\lambda}$ beschränkt, wegen 2 abgeschlossen und ihre Konvexität beweist man so wie in der ersten Mitteilung. $\Pi(\lambda)$ ist also ein abgeschlossenes Intervall, dessen eventuell zusammenfallende Endpunkte $\underline{\pi}(\lambda)$ und $\bar{\pi}(\lambda)$ heißen mögen. Ist $\lim_{k=\infty} \lambda_k = \lambda$ und $0 \leqq \lambda < \bar{\lambda}$, dann ergibt sich mit Rücksicht auf 2 mühelos, der Satz (*): Die Menge $\operatorname{Limsup}_{i=\infty} \Pi(\lambda_i)$ ist Teil von $\Pi(\lambda)$. Wir definieren nun $\underline{\tau}(\lambda) = \underline{\pi}(\lambda) - f_n(s_1^{\lambda}, \ldots, s_{n-1}^{\lambda}, \lambda)$, wobei $s_1^{\lambda}, \ldots, s_{n-1}^{\lambda}$ die (auf Grund der induktiven Annahme existierende in $s_1, \ldots, s_{n-1}$ eindeutige) Lösung der Gleichungen (II^{λ}_{n-1}) sind, und $\bar{\tau}(\lambda)$ entsprechend. Wir nennen $T(\lambda)$ das abgeschlossene Intervall zwischen $\bar{\tau}(\lambda)$ und $\underline{\tau}(\lambda)$. Aus 2 folgt, daß $s_1^{\lambda}, \ldots, s_{n-1}^{\lambda}$ stetige Funktionen von λ sind. Also ist auch $f_n(s_1^{\lambda}, \ldots, s_{n-1}^{\lambda}, \lambda)$ eine stetige Funktion von λ. Aus dem obigen Satz (*) ergibt sich dann der in der Mitteilung P I mit (**) bezeichnete Satz.

Der Beweis der Existenz einer Lösung von (II) erfolgt nun wie im Falle 1 von P I (S. 16) gestützt auf Satz (**), wenn eine positive Zahl $\lambda_0 < \bar{\lambda}$ existiert, für welche $\bar{\pi}(\lambda_0) \geqq f_n(s_1^{\lambda_0}, \ldots, s_{n-1}^{\lambda_0}, \lambda_0)$ gilt, und so wie im Falle 2 von P I, wenn kein solches λ_0 existiert.

Wir fügen noch einige Bemerkungen über die ökonomische Bedeutung der Bedingung 6) hinzu. Sei W ein bestimmter Wirtschafter, der bei den Preisen $\sigma_1, \ldots, \sigma_n$ eine Nachfrage nach s_{W_1} Einheiten von $S_1, \ldots$ und s_{W_n} Einheiten von S_n hat. Die Zahl s_j der vom Gut S_j produzierten Einheiten ist die Summe der Zahlen s_{Wj} für alle Wirtschafter W. Man kann nun für jeden W eine, wie wir sehen werden, aus werttheoretischen Voraussetzungen beweisbare Annahme formulieren, welche zu 6) analog ist, nämlich

6_W). *Hat W bei den Preisen* $\sigma_1, \ldots, \sigma_m$ *die Nachfragen* $s_{W1}, \ldots, s_{Wn}$ *und bei den Preisen* $\sigma'_1, \ldots, \sigma'_n$ *die Nachfragen* $s_{W1} + \Delta s_{W1}, \ldots, s_{Wn} + \Delta s_{Wn}$, *wobei mindestens eine der Zahlen* $\Delta_{Wj} < 0$ *ist, und gilt* $\sum_{j=1}^{n} \sigma_j \Delta s_{Wj} \leqq 0$, *so gilt* $\sum_{j=1}^{n} \sigma'_j \Delta s_{Wj} < 0$.

Wir bemerken zunächst: Aus 6) folgt nicht, daß 6_W) für jeden Wirtschafter W gilt, wie auch umgekehrt aus der Annahme, daß 6_W) für jeden W gilt, nicht die Giltigkeit von 6) folgt. Aber wenn

6_W) für jeden W gilt und 6) nicht gilt, so müssen, wie in einer folgenden Mitteilung ausgeführt werden soll, besondere Beziehungen zwischen den Nachfragen der einzelnen Wirtschafter W bestehen, welche statistisch unwahrscheinlich sind, so daß *mit einer statistischen Wahrscheinlichkeit aus der Annahme, daß* 6_W) *für alle* W *gilt, die Giltigkeit von* 6) *folgt.* Die Annahme 6_W) ihrerseits aber folgt, wie wir nun zeigen wollen, aus dem werttheoretischen Satz:

α_W). *Der Grenznutzen des Gutes* S_j *für den Wirtschafter* W *hängt nur von der in seinem Besitz befindlichen Quantität des Gutes* S_j *ab und ist eine im schärferen Sinne monoton abnehmende Funktion* ($j=1, \ldots, n$)[1].

Aus α_W) folgt nämlich zunächst der Satz:

6^*_W). *Sind von den* n *Zahlen* $\Delta s_{W1}, \ldots, \Delta s_{Wn}$ *die* k *Zahlen* $\Delta s_{Wl_1}, \ldots, \Delta s_{Wl_k}$ ($k \geqq 1$) *negativ, die übrigen* $\Delta s_{Wr_1}, \ldots, \Delta s_{Wr_{n-k}} \geqq 0$, *so gilt* $\sigma'_{r_j} : \sigma'_{l_i} < \sigma_{r_j} : \sigma_{l_i}$ ($i=1, \ldots, k; j=1, \ldots, n-k$).

Bezeichnet nämlich $\varphi_{Wj}(s_{Wj})$ die Grenznutzenfunktion von W für das Gut S_j, so gilt bekanntlich

$$\sigma_1 : \varphi_{W1}(s_{W1}) = \sigma_2 : \varphi_{W2}(s_{W2}) = \ldots = \sigma_n : \varphi_{Wn}(s_{Wn}) \text{ und}$$

$$\sigma'_1 : \varphi_{W1}(s'_{W1}) = \sigma'_2 : \varphi_{W2}(s'_{W2}) = \ldots = \sigma'_n : \varphi_{Wn}(s'_{Wn}),$$

woraus wegen der durch α_W) gewährleisteten Monotonie der Funktionen φ_{Wj} offenbar 6^*_W) folgt.

Aus 6^*_W) folgt aber leicht die Bedingung 6_W), die man ja auch folgendermaßen formulieren kann: Ist $k \geqq 1$, so folgt aus

$$(\sigma_{r_1} \Delta s_{Wr_1} + \ldots + \sigma_{r_{n-k}} \Delta_{Wr_{n-k}}) : (\sigma_{l_1} |\Delta \sigma_{Wl_1}| + \ldots + \sigma_{l_k} |\Delta s_{Wl_k}|) \leqq 1$$

stets

$$(\sigma'_{r_1} \Delta s_{Wr_1} + \ldots + \sigma'_{r_{n-k}} \Delta s_{Wr_{n-k}}) : (\sigma'_{l_1} |\Delta s_{Wl_1}| + \ldots + \sigma'_{l_k} |\Delta s_{Wl_1}|) < 1.$$

Da aus 6^*_W) für jedes j, für welches $\Delta s_{Wr_j} \neq 0$ gilt,

$$\sigma_{r_j} \Delta s_{Wr} : (\sigma'_{l_1} |\Delta s_{Wl_1}| + \ldots + \sigma'_{l_k} |\Delta s_{Wl_k}|) < \sigma_{r_j} \Delta s_{Wr_j} : (\sigma_{l_1} |\Delta s_{Wl_1}| + \ldots$$
$$\ldots + \sigma_{l_k} |\Delta s_{Wl_k}|)$$

folgt, so ergibt sich unmittelbar unsere Behauptung.

Wir bemerken noch, daß zur Herleitung von 6_W) eine viel schwächere Annahme als α_W) genügen würde, nämlich daß für jedes j die Zahl $\frac{\partial \varphi_{Wj}}{\partial s_{Wi}}$ negativ und dem Betrage noch groß gegenüber den Zahlen $\frac{\partial \varphi_{Wj}}{\partial s_{Wk}}$ für $k \neq j$ ist, d. h., es ist nicht erforderlich anzunehmen,

[1]) Um Mißverständnissen vorzubeugen, sei ausdrücklich bemerkt, daß α_W) die in *PI* gemachte Annahme, daß die Nachfrage nach einem Gut nur vom Preis des betreffenden Gutes abhängt, keineswegs impliziert. Diese Implikation bestünde nur, wenn der Grenznutzen des Geldes konstant wäre.

daß der Grenznutzen des Gutes S_j für den Wirtschafter W von dessen Besitz an sonstigen Gütern unabhängig sei, sondern es genügt anzunehmen, *daß der Grenznutzen von S_j durch eine Änderung im Besitze an S_j viel mehr beeinflußt wird als durch eine Änderung im Besitze an sonstigen Gütern.*

Gödel (Diskussionsbemerkung): In Wirklichkeit hängt für jeden einzelnen Wirtschafter die Nachfrage auch von seinem Einkommen ab und dieses wieder vom Preis der Produktionsmittel. Man kann ein dementsprechendes Gleichungssystem formulieren und auf seine Lösbarkeit hin untersuchen.

Kurzer Beweis des Satzes von Menger, daß für ein pseudoeuklidisches $(n+3)$-Tupel $p_1, p_2, \ldots, p_{n+3}$ stets sgn $D(p_1, \ldots, p_{n+3}) = (-1)^n$ gilt. Von L. M. Blumenthal (National Research Fellow).

Es seien $p'_1, \ldots, p'_{n+2}, p'_{n+3}$ Punkte des R_n derart, daß $p'_{n+2}\, p'_{n+3} \neq p_{n+2}\, p_{n+3}$, hingegen $p'_i p'_j = p_i p_j$ für alle anderen Abstände gilt. Bezeichnen wir mit $D(x)$ die quadratische Funktion, welche entsteht, wenn wir in der Determinante $D(p_1, \ldots, p_{n+3})$ jeden der Terme $(p_{n+2} p_{n+3})^2$ und $(p_{n+3} p_{n+2})^2$ durch x ersetzen. Offenbar verschwindet $D(x)$ für $x = (p'_{n+2}\, p'_{n+3})^2$. Denn setzen wir diesen Wert in die Determinante an Stelle von $(p_{n+2} p_{n+3})^2$, so erhalten wir die Determinante der $n+3$ im R_n gelegenen Punkte $p'_1, \ldots, p_{n+2} p'_{n+3}$, also 0.

Da der R_{n-1} die Quasikongruenzordnung $n+1$ hat, muß die Determinante von mindestens einem unter den Punkten $p_1, \ldots, p_{n+3}$ enthaltenen $(n+1)$-Tupel $\neq 0$ sein. Ist etwa $D(p_1, \ldots, p_{n+1}) \neq 0$, so gilt nach einem Jacobischen Satz (vgl. Bôcher, Einführung in die höhere Algebra S. 36).

$$D(x) = \frac{D(p_1, \ldots, p_{n+1}, p_{n+2}) . D(p_1, \ldots, p_{n+1}, p_{n+3}) - [D(p_1, \ldots, p_{n+1}) . x + C]^2}{D(p_1, \ldots, p_{n+1})}$$

wo C von x unabhängig ist. Da aber $D(p_1, \ldots, p_{n+2}) = D(p'_1, \ldots, p'_{n+2}) = 0$, so ist also $D(x)$ ein vollständiges Quadrat und $(p'_{n+2} p'_{n+3})^2$ mithin eine *doppelte* Nullstelle. Setzen wir $(p_{n+2} p_{n+3})^2 \neq (p'_{n+2} p'_{n+3})^2$ in den obigen Ausdruck für $D(x)$ ein, so erhalten wir demnach sgn $D(p_1, \ldots, p_{n+3}) = -$ sgn $D(p_1, \ldots, p_{n+1})$; da $p_1, \ldots, p_{n+1}$ mit Punkten den R_n, aber nicht des R_{n+1} kongruent sind, gilt sgn $D(p_1, \ldots, p_{n+1}) = (-1)^{n+1}$, also sgn $D(p_1, \ldots, p_{n+3}) = (-1)^n$, was von Menger (Amer. Journ. *53*, S. 740) bloß mit Hilfe von Stetigkeitsüberlegungen bewiesen war.

81. Kolloquium (30. XI. 1934).

Die kontradiktorische Logik. Von T. Hirano.

Die Anfangsfiguren:

1) $X v X \rightarrow X$ 2) $X \rightarrow X v Y$ 3) $X v Y \rightarrow Y v X$

4) $(X \rightarrow Y) \rightarrow (Z v X \rightarrow Z v Y)$ 5) $(x) f(x) \rightarrow f(x)$

6) $f(y) \rightarrow (Ex) f(x)$

und Transformationsregeln:

α) die Einsetzungsregel,

β) das Schema für (x) und (Ex): aus einer ableitbaren Figur $A \rightarrow B(x)$ gewinnt man $A \rightarrow (x) B(x)$, aus $B(x) \rightarrow A$ gewinnt man $(Ex) B(X) \rightarrow A$.

γ) das Schlußschema

$$\frac{\begin{matrix} P \\ P \rightarrow Q \end{matrix}}{Q}$$

bestimmen ein System von Figuren.

Interpretiert man die Zeichen —, v, $\rightarrow$, (x), (Ex) wie gewöhnlich als bzw. nicht, oder, impliziert, alle, es gibt, so bedeuten die Figuren des Systems Tautologien. D. h. das System ist eine *tautologische Logik.*

Interpretiert man aber das Zeichen — wie oben, dagegen v, (x), (Ex) als bzw. und, es gibt, alle, und $P \rightarrow Q$ als: („Q impliziert P" ist falsch), so wird durch ganz analoge Überlegungen wie bei der tautologischen Logik gezeigt, daß alle Figuren dieses Systems Kontradiktionen bedeuten. Wir nennen das System in diesem Fall die *kontradiktorische Logik.*

Überlegen wir die Anwendung der kontradiktorischen Logik in unserer Sprache. Hat man einen wahren empirischen Satz P und in dem System der tautologischen Logik eine Tautologie $P \rightarrow Q$, so bekommt man bekanntlich durch das Anwendungsschema

$$\frac{\begin{matrix} P \\ P \rightarrow Q \end{matrix}}{Q}$$

einen neuen wahren Satz Q. Nehmen wir zur kontradiktorischen Logik als Anwendungsschema

$$\frac{\begin{matrix} P \\ Q \rightarrow P \end{matrix}}{Q},$$

so wird durch eine leichte Überlegung gezeigt, daß, wenn P ein empirisch richtiger Satz und $Q \rightarrow P$ ein Satz der kontradiktorischen Logik (d. h. also eine Kontradiktion) ist, dann auch Q ein richtiger Satz ist, wobei die in P, Q, $P \rightarrow Q$ vorkommenden logischen Zeichen im Sinne der oben an zweiter Stelle angegebenen Übersetzung zu interpretieren sind. Wenn man also aus einem wahren Satz P mittels der tautologischen Logik einen wahren Satz schließen kann, kann man dasselbe auch in der kontradiktorischen, d. h. die kontradiktorische Logik kann ganz dieselbe Rolle für unsere Sprache wie die tautologische spielen.

Remarks Concerning The Euclidean Four-Point Property.

By Leonard M. Blumenthal (National Research Fellow).

Consider a semi-metric space M and a real function $\varphi(x)$ defined for every value of $x = pq$, where p, q are points of M. A space M' is called the *metric transform of M by* $\varphi(x)$ provided (1) the points of M and M' are in one-to-one, reciprocal, correspondence, and (2) if p', q' are points of M' corresponding to p, q of M, respectively, then $p'q' = \varphi(pq)$. We write $M' = \varphi(M)$.

Lemma. Let $\sigma(x)$ denote a monotonic increasing concave[1] *function defined for all non-negative values of x and $=0$ for $x = 0$. If M is a metric space, its metric transform by the Function $\sigma(x)$ is a metric space.*

Proof. The metric transform M' of M is obviously a semi-metric space. Let, now, p_1', p_2', p_3' be three pairwise distinct points of M'. Then $p_i'p_j' = \sigma(p_ip_j)$, $(i, j = 1, 2, 3)$, where p_i are the points of M corresponding to p_i' $(i = 1, 2, 3)$, respectively. We may assume the labeling so that $p_1'p_2' \leqq p_2'p_3' \leqq p_1'p_3'$. From the definition of $\sigma(x)$, it follows readily that $\sigma(p_1p_2) + \sigma(p_2p_3) > \sigma(p_1p_2 + p_1p_3)$. Since M is a metric space and σ is monotonic increasing we have $\sigma(p_1p_2) + \sigma(p_2p_3) > \sigma(p_1p_3)$, which proves the lemma.

Remark 1. The space $M' = \sigma(M)$, where M is a metric space, is *not convex*; for, as shown in the Lemma, if p_1', p_2', p_3' are three pairwise distinct points of M', the sum of any two of the three distances they determine exceeds the third distance.

Remark 2. Let p_i, $(i = 1, 2, 3)$, be the pairwise distinct vertices of a triangle, degenerate or proper. Then for $0 \leqq \alpha < 1$, the points p_i', $(i = 1, 2, 3)$, such that $p_i'p_j' = +(p_ip_j)^\alpha$, $(i, j = 1, 2, 3)$, are the vertices of a *proper* triangle. This is evident since, for $\alpha = 0$, the points p_i' are equilateral, while for $0 < \alpha < 1$ it follows from the Lemma and Remark 1.

Remark 3. If in Remark 2 the exponent α is less than $1/2$, then each angle of the triangle with vertices p_i', $(i = 1, 2, 3)$, is acute. This is immediate upon application of the law of cosines and Remark 2.

A semi-matric space is said to have to have the *euclidean four-point property* provided that each quadruple of its points may be imbedded congruently in a euclidean space.

Theorem I. *The metric transform M' of any metric space M by the function $\sigma(x) = x^\alpha$, $0 \leqq \alpha \leqq 1/2$, has the euclidean four-point property.*

---

[1]) A function $\sigma(x)$ of a single variable x is called *concave* if for $x_1 < x < x_2$, the determinant

$$\begin{vmatrix} \sigma(x_1) & x_1 & 1 \\ \sigma(x) & x & 1 \\ \sigma(x_2) & x_2 & 1 \end{vmatrix}$$

is positive.

Proof. We shall first prove the theorem for the case $\alpha=1/2$. Since M' is a metric space (Lemma), it is sufficient[2]) to show that if p'_i, $(i=1, 2, 3, 4)$ are any four points of M', then

$$D(p'_1, \ldots, p'_4) = \begin{vmatrix} 0 & 1 \\ 1 & (p'_i p'_j)^2 \end{vmatrix}_{(i,\, j=1,\, 2,\, 3,\, 4)} = \begin{vmatrix} 0 & 1 \\ 1 & p_i p_j \end{vmatrix}_{(i,\, j=1,\, 2,\, 3,\, 4)}$$

is not negative. Now $D(\alpha) = \begin{vmatrix} 0 & 1 \\ 1 & (p_i p_j)^{2\alpha} \end{vmatrix}_{(i,\, j=1,\, 2,\, 3,\, 4)}$

is a continuous function of α which is positive for $\alpha=0$. To show that $D(1/2)$ is not negative, we suppose the contrary, and quickly deduce a contradiction. For since $D(\alpha)$ is continuous, positive for $\alpha=0$ and negative for $\alpha=1/2$, there exists a value α' such that $0<\alpha'<1/2$, $D(\alpha')=0$; that is, the quadruple of points p^*_i, $(i=1,2,3,4)$, such that $p^*_i p^*_j = +(p_i p_j)^{\alpha'}$, $(i, j=1, 2, 3, 4)$, is congruent with four points of a plane. But this is impossible, for by Remark 3 all the angles determined by the four points p^*_i, $(i=1, 2, 3, 4)$, are acute (since $0<\alpha'<1/2$), while at least one of the twelve angles determined by four points in a plane exceeds an acute angle. Hence $D(1/2)$ cannot be negative, and the theorem is proved for $\alpha=1/2$. Since, however, it has been shown impossible for the function $D(\alpha)$ to vanish for any value between $\alpha=0$ and $\alpha=1/2$, it follows that $D(\alpha)$ must be *positive* for $0\leqq\alpha<1/2$; that is, the elements p'_i, $(i=1, 2, 3, 4)$ with $p'_i p'_j=(p_i p_j)^\alpha$, $(i, j=1, 2, 3, 4)$, $0\leqq\alpha\leqq 1/2$, are congruent with four points of the euclidean space of three dimensions, R_3, which proves the theorem.

The following interesting determinant theorem is a direct consequence of the preceding theorem:

Theorem II. If the bordered, symmetric determinant $D=|r_{ij}|$, $r_{ij}=r_{ij}>0$, $(i\neq j)$. $r_{ii}=0$, $r_{0i}=1$, $(i\neq 0)$, $(i, j=0, 1, \ldots, 4)$, *of order* 5, *has each bordered fourth-order principal minor negative or zero, the for* $0\leqq k\leqq 1/2$, *the determinant* $D^{(k)}$, *each element of which is then positive k th root of the corresponding element of D, has each bordered fourth-order principal minor negative or zero, and the determinant* $D^{(k)}$ *itself is positive or zero.*

It is easily seen that $\alpha=1/2$ is the greatest exponent for which Theorem I is valid; for if the points p_i, $(i=1, 2, 3, 4)$, forming a pseudo-linear quadruple with $p_1p_2=p_2p_3=p_3p_4=p_1p_4=1$, $p_1p_3= =p_2p_4=2$, be metrically transformed by the function $\sigma(x)=x^{\frac{1}{2}(1+\varepsilon)}$, $0\leqq\varepsilon<1$, we find that $D(p'_1, \ldots, p'_4)=-32\cdot 2^{2\varepsilon}(2^\varepsilon-1)$, which vanishes for $\varepsilon=0$ and is negative for ε positive. Hence p'_i $(i=1,2,3,4)$, are congruent with four points of a plane for $\varepsilon=0$, while for ε positive the transformed set is not congruent with four points of any euclidean space.

---

[2]) Menger, Math. Annalen *100*, p. 136.

Finally, it may be remarked that there is reason to believe that if M is a *euclidean* space, the transform M' of M by the function $\sigma(x)=x^\alpha$, has the euclidean fourpoint property for $0\leqq\alpha\leqq1$.

82. Kolloquium (12. XII. 1934).

Sur une proposition de M. L. M. Blumenthal. Par A. Ville.

Soit E un espace métrique, où la distance de deux points p et q sera désigné par $\delta(p, q)$. Nous appellerons $E(x)$ l'espace semi-métrique formé des mêmes points que E, l'écart des points p et q étant pris égal au nombre $[\delta(p, q)]^x$, x étant un nombre réel positif ou nul. M. Blumenthal a démontré: Pour $0\leqq x\leqq 1/2$, tout quadruplet de points de $E(x)$ est congruent à un quadruplet euclidien. *Nous allons former un espace métrique E, formé de 5 points, tel que $E(1/2)$ ne soit pas congruent à un quintuplet euclidien.*

Considérons dans l'espace euclidien à 3 dimensions le quintruplet $Q(x)$ formé des 5 points $p_1(x), \ldots, p_5(x)$ variant continûment avec x, les distances $d_{ij}(x)$ des points $p_i(x)$ et $p_j(x)$ étant données par les relations:

$$d_{12}(x)=d_{13}(x)=d_{14}(x)=1,\quad d_{23}(x)=d_{34}(x)=d_{42}(x)=x,$$
$$d_{52}(x)=d_{53}(x)=d_{54}(x)=y,\quad d_{15}(x)=z,$$

où $0<x\leqq\sqrt{2}$, x, y, z étant liées par la relation:

$$\frac{1+y^2-z^2}{2y}=1-\frac{x^2}{2}, \text{ donc } y=\frac{2\sqrt{(3-x^2)(4-x^2)}-2(2-x^2)}{8-3x^2} \text{ pour } x>\frac{\sqrt{7}-1}{\sqrt{3}}$$

y et z sont deux fonctions continues de x entre $\frac{\sqrt{7}-1}{\sqrt{3}}$ et $\sqrt{2}$. Le quintuplet $Q(x)$ est formé de 10 triangles contenant en tout 30 angles qui sont fonctions continues de x. Pour $x=\sqrt{2}$ ceux des angles dont le cosinus est $1-x^2/2$, sont droits, les autres aigus Il existe donc une valeur $x_0<\sqrt{2}$ telle que tous les angles de $Q(x_0)$ sont aigus. Si y_0 et z_0 sont les valeurs correspondantes de y et z, elles satisfont, avec x_0, aux inégalités: $d^2_{ij}(x_0)+d^2_{jk}(x_0)-d^2_{ik}(x_0)>0$

Le quintuplet $Q(Z)$ des points $p'_1, \ldots, p'_5$, dont les distances sont données par $d'_{15}=Z$, et $d'_{ij}=d_{ij}(x_0)$ sauf pour $(i, j)=(1, 5)$, est congruent à un quintuplet euclidien si

$$D(Z)=\begin{vmatrix}0 & 1\\ 1 & (d_{ij})^2\end{vmatrix}<0,$$

et si ce déterminant est nul, à un quintuplet de l'espace à trois dimensions[1]). Soit $p^*_5(x_0)$ le symétrique de $p_5(x_0)$ par rapport au plan $[p_2(x_0), p_3(x_0), p_4(x_0)]$. En désignant par z^*_0 la distance de $p_1(x_0)$ et $p^*_5(x_0)$, on voit facilement que $D(z_0)=D(z^*_0)=0$. Puisque

[1]) Cf. Menger, Math. Annalen *100*, 136.

$D(Z)$ est un polynôme du 2e degré en Z^2, admettant deux racines, distinctes, et que $Q'(z_0)$ satisfait à $D(z_0)=0$, il existe une valeur z_1 telle que $Q'(z_1)$ soit métrique et que $D(z_1)>0$. Il suit que $Q'(z_1)$ n'est congruent à aucun quintuplet euclidien. L'espace E formé des points $a_1, \ldots, a_5$ avec les distances $(d'_{ij})^2$ entre a_i et a_j est métrique et $E(1/2)=Q'(z_1)$ ne peut pas être plongé dans un espace euclidien.

83. Kolloquium (18. I. 1935).

Algebra der Geometrie (Zur Axiomatik der projektiven Verknüpfungsbeziehungen). Von K. Menger.

Eine der abstrakten Algebra analoge und die Boolesche Algebra umfassende Algebra der projektiven Geometrie[1]) kann auf die Annahme gegründet werden, es liege eine Klasse von Dingen (zu deuten als lineare Gebilde eines projektiven Raumes) vor, für welche zwei Operationen + und . definiert sind, so daß je zwei Dingen A und B zwei Dinge $A+B$ und $A.B$ (zu deuten als Verbindungs- und Durchschnittsgebilde) zugeordnet sind gemäß den folgenden fünf Bedingungen: **I.** Jede der Operationen ist *assoziativ*. **II.** Jede der Operationen besitzt ein *indifferentes Element*, d. h. es existieren Elemente V und U, so daß $A+V=A=A.U$ für jedes A gilt. **III.** Die beiden Operationen sind durch folgendes *schwache Distributivgesetz* verknüpft:

$$A+(A+B).C=A+(A+C).B \text{ und } A.(A.B+C)=A.(A.C+B).$$

IV. Jede der Operationen ist *teilweise umkehrbar:* Wenn $A+B=B$, so existiert ein B', so daß $A+B'=B$ und $A.B'=V$. Wenn $A.B=B$, so existiert ein B'', so daß $A.B''=B$ und $A+B''=U$. **V.** *Jede Teilkette ist endlich*, d. h. aus $A_1 \supseteq A_2 \supseteq \ldots \supseteq A_n \supseteq \ldots$ folgt, daß fast alle A_n mit einander identisch sind. (Die Definition von $B \supseteq A$ siehe unten).

Aus **II** und **III** ergibt sich, daß jede der Operationen *kommutativ* ist. Aus **IV** folgt $A.V=V$ und $A+U=U$ für jedes A. Infolgedessen ergibt sich aus **III** die *Absorptionsformel* $A+A.B= =A=A.(A+B)$ und der *Absorptionssatz*, daß $A+B=B$ und $A.B=A$ äquivalent sind. Bestehen beide Beziehungen, so schreiben wir $A \subseteq B$ und nennen A *Teil* von B. Es ergibt sich weiter die Formel $A+A=A=A.A$ für jedes A. Wir nennen *Punkt* ein Ding P, für welches $P' \subseteq P$ nur für $P'=V$ und $P'=P$ gilt; wir nennen H eine *Hyperebene*, wenn $H \subseteq H'$ nur für $H'=U$ und $H'=H$ gilt. Aus den Axiomen **I** und **IV** ergibt sich dann folgender *Einschiebungssatz* von Bergmann: Ist P ein Punkt, so folgt für jedes A aus $A \subseteq B \subseteq A+P$ entweder $B=A$ oder $B=A+P$; und das duale

[1]) Vgl. Menger, Jahresber. DMV, *37*, 1928, S. 309, Bergmann, Monatshefte f. Math. u. Phys. *36*, 1929, S. 269, sowie die Noten von Bergmann, Menger, Schreiber und Wald in diesen Ergebnissen *1*, S. 28, *4*, S. 6, 14, 34, 38. Eine ausführliche Darstellung erscheint in den Annals of Mathematics 1936.

Gegenstück: Ist H eine Hyperebene, so folgt aus $A \supseteq B \supseteq A \,.\, H$ entweder $B = A$ oder $B = A \,.\, H$. Bezeichnet man k Punkte als *unabhängig*, wenn jeder mit der Summe der übrigen den Durchschnitt V hat, so ergibt sich aus dem Einschiebungssatz, wie in der Dissertation von Schreiber (Wien 1933) gezeigt wird, daß wenn k unabhängige Punkte und l unabhängige Punkte die gleiche Summe haben, stets $k = l$ gilt. Nennt man $\dim A$ die um 1 verminderte Zahl der unabhängigen Punkte, als deren Summe A darstellbar ist, so folgt aus den Axiomen **I** und **V**, daß $\dim A$ für jedes A endlich ist und der Funktionalgleichung $\dim A + \dim B = \dim (A+B) + \dim A \,.\, B$ genügt.

84. Kolloquium (6. II. 1935).

K. Popper: Über nachwirkungsfreie Folgen.

A. Wald: Über den Kollektivbegriff.

85. Kolloquium (20. II. 1935).

A. Wald: Allgemeines Existenztheorem für Kollektiva bezüglich beliebiger Auswahlvorschriften und Merkmalmengen. (Erscheint in Heft 8 dieser Ergebnisse. Ein Auszug erschien in den C. R. Paris 20. I. 1936.)

86. Kolloquium (1. III. 1935).

A. Wald: Ein Spielsystem für vorwirkende Angaben.

Axiomatik der affinen Verknüpfungsbeziehungen. Von F. Alt.

Wir zerlegen den Einschiebungssatz (vgl. 83. Kolloquium) in zwei Aussagen: *a)* Ist P ein Punkt, so folgt für jedes A aus $A \subseteq B \subseteq A + P \neq U$, daß $B = A$ oder $B = A + P$. *b)* Ist P ein Punkt, so folgt aus $A \subseteq B \subseteq A + P = U$, daß $B = A$ oder $B = U$. Bereits die Aussage *a)* genügt zur Definition einer Dimension und zum Beweis der Funktionalungleichung $\dim A + \dim B \geqq \dim (A+B) + \dim A \,.\, B$, falls $A + B \neq U$. Satz *b)* gestattet, diese Formel auch im Falle $A + B = U$ zu beweisen. Von der Aussage *a)* gilt in affinen Räumen auch das duale Gegenstück. Ist H eine Hyperebene, so folgt aus $A \supseteq B \supseteq A \,.\, H \neq V$ entweder $B = A$ oder $B = A \,.\, H$. Dies ermöglicht die Herleitung einer Theorie der linearen Abhängigkeit von Hyperebenen und die Einführung einer dualen Dimension, indem man als $\dim' A$ die um 1 verminderte Anzahl unabhängiger Hyperebenen bezeichnet, deren Durchschnitt A ist, wobei man beweisen kann: $\dim' A + \dim' B \geqq \dim' (A+B) + \dim' A \,.\, B$, falls $A \,.\, B \neq V$. Da nun für jedes A die beiden Dimensionen durch die Gleichung $\dim A + \dim' A = \dim U$ zusammenhängen, folgt aus den beiden erwähnten Ungleichungen die Funktionalgleichung $\dim A + \dim B = \dim (A+B) + \dim A \,.\, B$, falls $A \,.\, B \neq V$. Im projektiven Raum gilt auch das duale Gegenstück von *b)* und daher die Funktionalgleichung für die Dimension ausnahmslos. Im affinen Raum gilt statt dessen das Parallelenaxiom. Eine ausführliche Darstellung erscheint in der im 83. Kolloquium erwähnten Arbeit in den Annals of Math. 1936.

87. Kolloquium (13. III. 1935).

Über im kleinen zusammenhängende Kontinua. Von G. v. Alexits, Budapest (nach brieflicher Mitteilung).

Wir bezeichnen mit K^c die Menge aller irrationalen Punkte der Menge K, mit K^{cc} die Menge aller irrationalen Punkte der Menge K^c.

Satz 1. *Ist U offen im Kontinuum K und ist K in allen Punkten von U nicht zusammenhängend im kleinen, so ist K^{cc} dicht in U.* Ist V ein offener Teil von U so haben wir zu zeigen, daß V einen Punkt von K^{cc} enthält, oder, was dazu hinreichend ist, daß in V unabzählbar viele paarweise fremde Teilkontinua von K enthalten sind. Dazu betrachten wir die Komponenten von V. Jede ist in V abgeschlossen. Jede ist in V nirgends dicht, sonst enthielte sie einen relativ offenen Teil und daher gäbe es Punkte von V, in denen K zusammenhängend im kleinen wäre. Da V nicht von erster Kategorie ist, gibt es unabzählbar viele solcher Komponenten, die paarweise fremd sind. Jede enthält ein Kontinuum. Damit ist Satz 1 bewiesen.

Satz 2. *Ist K nur in den Punkten einer nirgendsdichten Menge N zusammenhängend im kleinen, so ist K^{cc} dicht in K.* Denn $K-\overline{N}$ ist eine in K dichte offene Menge, in deren Punkten K nicht zusammenhängend im kleinen ist, in der also K^{cc} nach Satz 1 dicht ist.

A. Ville: *Bemerkungen über vollständige Räume.*

88. Kolloquium (19. III. 1935).

A. Tarski: *Grundzüge des Systemenkalküls.* (Erscheint in Fundamenta Mathem. *25* und *26*). In der Diskussion wird die weitgehende Analogie des Systemenkalküls mit der *Algebra der Geometrie* (vgl. dieses Heft S. 11) festgestellt und auf Interpretationsmöglichkeiten für die Abschließungs- und Komplementbildungen des Systemenkalküls durch die mengentheoretische Geometrie der Ebene, insbesondere durch die Operation der Bildung der abgeschlossenen Hülle, hingewiesen.

89. Kolloquium (21. V. 1930).

Über die φ-Metrik und φ-Bogenlänge. Von K. Menger.

Sei R ein metrischer Raum, in dem p und q den Abstand $d(p, q)$ haben. Man kann den im allgemeinen bloß halbmetrischen (d. h. nicht notwendig der Dreiecksungleichung genügenden) Raum betrachten, in dem für eine gewisse monoton wachsende Funktion $\varphi(x)$ je zwei Punkte p und q von R den Abstand $\varphi(d(p, q))$ haben. Für die Grundlegung der Variationsrechnung ist speziell die Theorie der Bogenlänge in dieser φ-Metrik von einem gewissen Interesse. Ist R der euklidische Raum und $\varphi(x) = x^2$ so kommt man zu der von Denjoy eingeführten quadratischen Bogenlänge. Ist in einem euklidischen Raum C stetiges Bild der Strecke $a \leq t \leq b$, so nennen wir Teilpolygon P von C das geordnete Bild $\{p_1, p_2, \ldots, p_n\}$ einer geordneten endlichen Zahlenmenge $\{a=t_1<t_2<\ldots<t_n=b\}$. Wir nennen $\nu_c(P)$ die größte der Zahlen $t_{i+1}-t_i$, bezeichnen mit $l(P)$ die *φ-Länge* $\sum\varphi(d(p_i, p_{i+1}))$ von P und nennen eine Folge von Teilpolygonen $\{P_k\}$ von C ausgezeichnet, wenn $\lim\nu_c(P_k)=0$. Nun gibt es, wenn in der φ-Metrik die Dreiecksungleichung nicht gilt, im allgemeinen, z. B. schon für $\varphi(x)=x^2$, Streckenbilder C, welche verschiedene ausgezeichnete Folgen von Teilpolygonen enthalten, deren φ-Längen gegen verschiedene Grenzwerte konvergieren. Wir müssen daher *obere* und *untere φ-Länge* eines Streckenbildes C unterscheiden und man kann leicht in der euklidischen Ebene Streckenbilder C angeben, u. zw. Bogen, die in der Form $y=f(x)$ mit nirgends differenzierbarem $f(x)$ gegeben sind, für welche die untere φ-Länge $=0$, die obere >0 ist. Problem: *Gibt es Streckenbilder, für welche obere und untere quadratische Länge übereinstimmen und einen endlichen Wert >0 haben?* Es ist klar, daß derartige Untersuchungen eine Klassifikation der in der gewöhnlichen Metrik nicht rektifizierbaren Bogen ergeben würden.

Ein metrisches Quadrupel, das in genau zwei Kugeln einbettbar ist. Von J. Groiß.

S_k sei für $k<0$ die hyperbolische Ebene der Krümmung k, für $k=0$ die euklidische Ebene, für $k>0$ die Kugelfläche der Krümmung k, in der als Abstand zweier Punkte die Länge des kürzeren sie verbindenden Großkreisbogens gilt. Ist Q ein aus vier Punkten bestehender metrischer Raum, so bezeichnen wir mit $K(Q)$ die Menge aller Zahlen k derart, daß Q mit vier Punkten von S_k kongruent ist. Es ist klar, daß $K(Q)$ für ein lineares (d. h. mit vier Punkten der Geraden kongruentes) Quadrupel Q ein Intervall $0\leqq k\leqq k^*$ ist. Ferner weiß man[1]), daß $K(Q)$ leer für manche Q ist, und daß es auch nicht-lineare Q gibt, für welche $K(Q)$ aus mindestens zwei Zahlen besteht, von denen eine $=0$ ist.

Im folgenden gebe ich ein Quadrupel Q an, *für welches $K(Q)$ aus genau zwei (u. zw. positiven) Zahlen besteht:* Q sei der metrische Raum bestehend aus den Punkten p_1, p_2, p_3, p_4 mit den Abständen $p_1p_3=p_1p_4=p_2p_3=p_2p_4=\pi$, $p_1p_2=p_3p_4=3\pi/2$. Da $D(p_1,p_2,p_3,p_4)=-81\pi^6/16\neq 0$, wo D die Volumsdeterminante bezeichnet, so ist vor Q allem nicht in die euklidische Ebene S_0 einbettbar[2]). Notwendig und hinreichend für die Einbettbarkeit von Q in S_k mit $k>0$ ist[3]) das Verschwinden der Determinante $|\cos p_ip_j/r|=(1-\cos 3\pi/2r)^2.[(1+\cos 3\pi/2r)^2-4\cos^2\pi/r]$, wobei $r\geqq \operatorname{Max} p_ip_j/\pi=3/2$. Setzt man diesen Ausdruck $=0$, so liefern die zwei Gleichungen $1+\cos 3\pi/2r=\pm 2\cos\pi/r$ genau je eine endliche Nullstelle für $r\geqq 3/2$, nämlich $r=3$, bzw. eine zwischen 3/2 und 2. Ein mit Q kongruentes Quadrupel einer hyperbolischen Ebene existiert nicht. Denn wären p_1', p_2', p_3', p_4' mit Q kongruente Punkte in S_k mit $k<0$, so müßten

[1]) Vgl. Wald, diese Ergebnisse *6*, S. 31.
[2]) Vgl. Menger, diese Ergebnisse *1*, S. 20.
[3]) Vgl. Klanfer, diese Ergebnisse *4*, S. 43.

auf Grund der Kongruenz- und Anordnungssätze die Strecken p'_1, p'_2 und p'_3, p'_4 einander senkrecht in ihrem gemeinsamen Halbierungspunkt q' schneiden. (Dieser Schluß wäre für die Pseudosphäre von der Krümmung k nicht zulässig.) Also müßte in S_k ein bei q' rechtwinkeliges Dreieck q', p'_2, p'_3 existieren mit $q'p'_2 = q'p'_3 = 3\pi/4$ und $p'_2p'_3 = \pi$. Dies ist aber nach Formeln der hyperbolischen Trigonometrie für kein $k < 0$ möglich.

90. Kolloquium (5. VI. 1935).

F. Waismann: Bemerkungen von Freges und Russells Definition der Zahl; mit Diskussionsbemerkungen von Gödel, Menger, Tarski.

91. Kolloquium (12. VI. 1935).

Über die Beschränktheit der Ausdrucksmittel deduktiver Theorien. Von A. Lindenbaum und A. Tarski.

Die folgenden mathematischen Ergebnisse[1]) lassen sich verschiedenen Systemen der mathematischen Logik und verschiedenen diese Systeme der Logik „überbauenden" deduktiven Theorien anpassen. Um aber die Betrachtungen zu konkretisieren, nehmen wir als Logik ein beliebiges System an, welches

a) die folgendermaßen modifizierte und noch einer strengen Formalisierung zugeführte allgemeine Logik der *„Principia Mathematica"* enthält:

1⁰) es werde eine vereinfachte Typentheorie und das Extensionalitätsaxiom angenommen;

2⁰) man schalte alle definierten Operatoren, wie z. B. „$\hat{x}(\varphi x)$" oder „$(\imath x)(\varphi x)$" aus;

3⁰) als Sätze betrachte man nur diejenigen Satzfunktionen, die keine freien Variablen enthalten,

b) eventuell noch weitere logische Aussagen, wie z. B. Unendlichkeitsaxiom, Auswahlaxiom usw. enthält, dagegen mit keinen neuen undefinierten Konstanten operiert.

---

[1]) Über den Satz 9 hat Tarski auf dem II. Polnischen Philosophenkongreß in Warschau 1927 in einem Vortrag über die Begriffe der Widerspruchsfreiheit und der Vollständigkeit berichtet; auf Grund einer mündlichen Mitteilung erwähnt bei Fränkel: Einl. i. d. Mengenlehre, 3. Aufl., 1928, S. 352(3). Andere Ergebnisse der vorliegenden Mitteilung wurden auf den Sitzungen der Logischen Sektion der Warschauer Philosophischen Gesellschaft am 15. VI. 1932 (Lindenbaum: Über einige methodologische Fragen, die mit den Grundlagen der Geometrie zusammenhängen. I) und 1. II. 1933 (Lindenbaum und Tarski: Über einige metalogische Sätze) zum ersten Mal vorgetragen. — Zum selben Ideenkreis vgl.: Tarski u. Lindenbaum: Sur l'indépendance des notions primitives dans les systèmes mathématiques, Ann. de la Soc. Polon. Math. *5*, 1927, S. 111 bis 113; Tarski: Einige methodologische Untersuchungen über die Definierbarkeit der Begriffe, Erkenntnis *5*, 1935, S. 80—100 (etwas ausführlicher polnisch: Przeglad Filozoficzny *37*, 1934, S. 438—460).

Zu anderen im folgenden benutzten Begriffen und Symbolen siehe: Tarski, Erkenntnis, 5, 1935, S. 81—82. Insbesondere treten die (l. c. S. 90f. erklärten) symbolischen Ausdrücke der Form

$$R\frac{x', y', z' \ldots}{x'', y'', z'' \ldots}$$

auf, welche in inhaltlicher Deutung besagen, daß die Relation R die Klasse aller Individuen auf sich selbst eineindeutig abbildet, wobei die Individuen, Klassen, Relationen usw. $x', y', z', \ldots$ beziehungsweise auf $x'', y'', z'', \ldots$ abgebildet werden. Als allgemeines Schema einer Satzfunktion wird der symbolische Ausdruck „$\sigma(a, b, c\ldots; x, y, z, \ldots)$" gebraucht; hier sind „$a$", „$b$", „$c$" ... alle außerlogischen Konstanten und „x", „y", „z" ... alle freien Variablen der betrachteten Satzfunktion. Die Konjunktion aller Axiome eines deduktiven axiomatisierbaren Systems[2]), in dem als undefinierte außerlogische Konstanten „a", „b", „c" ... auftreten, wird dementsprechend in der schematischen Form „$\alpha(a, b, c, \ldots)$" dargestellt.

I.

Satz 1. Jeder Satz der Form

$$(x', x'', y', y'', z', z'' \ldots R) :. R\frac{x', y', z' \ldots}{x'', y'', z'' \ldots} . \supset : \sigma(x', y', z' \ldots) . \equiv$$
$$\equiv . \sigma(x'', y'', z'' \ldots)$$

ist in der Logik beweisbar.

Frei ausgesprochen, besagt Satz 1: Jede Beziehung zwischen den Gegenständen (Individuen, Klassen, Relationen usw.), die sich mit rein logischen Mitteln formulieren läßt, ist in Bezug auf jede eineindeutige Abbildung der „Welt" (d. h. der Klasse aller Individuen) auf sich selbst invariant und diese Invarianz ist logisch nachweisbar. — Der Satz ist sicher sehr plausibel und wurde schon in gewissen inhaltlichen Überlegungen als eine Prämisse verwendet, nichtsdestoweniger wurde er bisher nie streng bewiesen, ja nicht einmal präzis formuliert. — Der Beweis des Satzes 1 wird durch Rekursion in Bezug auf die Form der Satzfunktion „$\sigma(x', y', z' \ldots)$" geführt[3]).

Es ist zu beachten, daß das Symbol „σ", welches in den Schematen der Satzfunktionen auftritt, nicht als ein selbständiges Zeichen (etwa als ein Funktionszeichen) betrachtet werden darf. So sind z. B. im Gegensatz zu Satz 1 die Sätze von der Form

$$(x', x'', y', y'', z', z'' \ldots R, f) :. R\frac{x', y', z' \ldots}{x'', y'', z'' \ldots} . \supset : f(x', y', z' \ldots) . \equiv$$
$$\equiv . f(z'', y'', z'' \ldots)$$

---

[2]) Siehe Tarski, Fundamentale Begriffe der Methodologie der deduktiven Wissenschaften I, Monatsh. f. Math. u. Phys. 37, 1930, S. 369 u. 375.

[3]) Die genauen Beweise werden einer größeren Arbeit vorbehalten.

im allgemeinen in der Logik nicht beweisbar, sondern widerlegbar: beweisbar sind erst — auf Grund des Satzes 1 — die Sätze von der folgenden Form:

$$(x', x'', y', y'', z', z'' \ldots, R, f', f''): . R \frac{x', y', z' \ldots, f'}{x'', y'', z'' \ldots, f''} . \supset : f'(x', y', z' \ldots) . \equiv$$
$$\equiv . f''(x'', y'', z'' \ldots).$$

Wenden wir Satz 1 auf Satzfunktionen mit einer einzigen freien Variablen, und zwar einer Individuenvariablen an, so gewinnen wir als unmittelbares Korollar

Satz 2. Sind „x'" und „x''" Individuenvariablen, so ist jeder Satz der Form

$$(x', x''): \sigma(x') . \equiv . \sigma(x'')$$

in der Logik beweisbar.

Inhaltlich gedeutet: keine zwei Individuen lassen sich mit rein logischen Mitteln voneinander unterscheiden. — Durch eine leichte Umformung erhält man aus Satz 2

Satz 3. Ist „x" eine Individuenvariable, so ist jeder Satz der Form

$$(x) . \sigma(x) . \vee . (x) . \sim \sigma(x)$$

in der Logik beweisbar.

M. a. W. außer der Allklasse und der Nullklasse kann mit rein logischen Mitteln keine Klasse von Individuen definiert werden[4]). — Es sei bemerkt, daß folgende Verschärfung des Satzes 3 im allgemeinen falsch ist: *Ist „x" eine Individuenvariable, so ist für jede Satzfunktion „$\sigma(x)$" entweder der Satz „$(x) . \sigma(x)$" oder der Satz „$(x) . \sim \sigma(x)$" in der Logik beweisbar.* Diese Verschärfung gilt nämlich dann und nur dann, wenn das zugrundegelegte System der Logik vollständig ist.

Wenden wir nun Satz 1 auf Satzfunktionen mit zwei freien Individuenvariablen, so ergibt sich:

Satz 4. Sind „x" und „y" Individuenvariablen, so ist jeder Satz der Form

$$(x, y) . \sigma(x, y) . \vee . (x, y) . \sim \sigma(x, y) . \vee : (x, y) : \sigma(x, y) . \equiv . x = y : \vee :$$
$$(x, y) : \sigma(x, y) . \equiv . x \neq y$$

in der Logik beweisbar.

[4]) Die Worte „definieren", „definierbar" usw. werden in zwei verschiedenen Sinnen gebraucht: im ersten Sinne handelt es sich um eine formale Beziehung gewisser Ausdrücke zu anderen Ausdrücken einer Theorie (vgl. z. B. Tarski, Erkenntnis *5*); im zweiten Sinne — um eine semantische Beziehung zwischen den Gegenständen und Ausdrücken (vgl. Tarski, Fund. Math. *17*, 1931, S. 220). Hier meinen wir den zweiten Sinn.

Somit kann man mit rein logischen Mitteln nur vier zweigliedrige Relationen zwischen den Individuen definieren, und zwar die Allrelation, die leere Relation, die Identität und die Verschiedenheit (also die vier Moduli der **Schröder**schen Algebra der Relative).

Dieser Satz läßt sich auf mehrgliedrige Relationen ausdehnen: für jede natürliche Zahl n kann man mit rein logischen Mitteln nur eine angebbare endliche Anzahl n-gliedriger Relationen zwischen den Individuen definieren; dabei läßt sich jede dieser Relationen bloß mit Hilfe der Identität und der Begriffe aus dem Aussagenkalkül ausdrücken.

Es sei hier noch ein Spezialfall des Satzes 1 hervorgehoben, der sich auf Satzfunktionen mit einer freien Variable bezieht, wobei aber diese Variable eine Klasse von Individuen bezeichnet; „$\mathbf{Nc}(x)$“ soll hier die Kardinalzahl (im Sinne der „*Principia Mathematica*“) der Klasse x und „$\bar{x}$“ das Komplement von x bezeichnen.

Satz 5. Sind „x'“ und „x''“ zum Typus der Klassen von Individuen gehörende Variablen, so ist jeder Satz der Form

$$(x', x''):.\mathbf{Nc}(x')=\mathbf{Nc}(x'').\mathbf{Nc}(\bar{x}')=\mathbf{Nc}(\bar{x}''):\supset:\sigma(x').\equiv.\sigma(x'')$$

in der Logik beweisbar.

Also: zwei beliebige Klassen von Individuen, die gleichmächtig sind und deren Komplemente auch gleichmächtig sind, sind mit rein logischen Mitteln nicht unterscheidbar[5]).

Man pflegt zu sagen, daß unsere Logik eine Logik der Umfänge und nicht der Inhalte ist, da zwei Begriffe von verschiedenen Inhalten, aber gleichen Umfängen sich logisch nicht unterscheiden lassen. Im Lichte des Satzes 5 könnte man diese Behauptung verschärfen: unsere Logik ist nicht einmal Logik der Umfänge, sondern bloß Logik der Anzahlen, da zwei Begriffe schon dann in der Logik nicht unterscheidbar sind, wenn ihre Umfänge zwar verschieden sind, wenn jedoch die Anzahl der Elemente, die zu diesen Umfängen gehören, identisch ist und ebenso die Anzahl der Elemente, die zu den Umfängen der Komplementärbegriffe gehören.

Wir gehen zu den Anwendungen des Satzes 1 in **speziellen deduktiven Theorien** über, und zwar betrachten wir, um die Überlegungen zu konkretisieren, die **euklidische Geometrie**. Dieselbe läßt sich bekanntlich mit Hilfe eines einzigen Grundbegriffes aufbauen, z. B. mit Hilfe der viergliedrigen Relation der Kongruenz zweier Punktpaare. Wir wollen diese Relation mit „a“ bezeichnen; der Ausdruck „$a(x, y, z, t)$“ wird demnach gelesen: „*der Punkt x ist soweit entfernt vom Punkte y, wie der Punkt z vom Punkte t*“. Wir setzen noch voraus, daß das Axiomensystem der euklidischen Geometrie in einer gewissen Richtung verschärft ist, und zwar daß man in dieses System den Satz aufnimmt, daß jedes Individuum

---

[5]) Das ist die Entscheidung eines Problems, das von **Tarski** in Fund. Math. *16*, 1930, S. 182[(1)] gestellt wurde.

ein Punkt ist (also der Raum sich mit der Klasse aller Individuen deckt)[6]). Als einen Spezialfall des Satzes 1 gewinnen wir sofort:

Satz 6. Jeder Satz der Form

$$(x', x'', y', y'', z', z'' \ldots, R) : . R \frac{a, x', y', z' \ldots}{a, x'', y'', z'' \ldots} . \supset : \sigma(a; x', y', z' \ldots) . \equiv . \\ \sigma(a; x'', y'', z'' \ldots)$$

ist in der euklidischen Geometrie beweisbar.

Inhaltlich gedeutet: jede Beziehung zwischen den Gegenständen (Punkten, Punktmengen, Klassen um Punktmengen usw.), die sich mit den Mitteln der Logik und der Geometrie formulieren läßt, ist in Bezug auf jede eineindeutige Abbildung des Raumes auf sich selbst, bei der die Relation a in sich selbst übergeht, d. h. in Bezug auf jede ähnliche Abbildung invariant (nicht nur in Bezug auf jede isometrische Abbildung).

Wenn man die spezifischen Eigentümlichkeiten der Axiome der euklidischen Geometrie beachtet, so kann man leicht aus Satz 6 die Analoga der Sätze 2—4 ableiten. Also: mit den Mitteln der Logik und der euklidischen Geometrie kann man keine zwei Punkte von einander unterscheiden, kann man ferner keine konkreten Punktmengen außer dem ganzen Raum und der leeren Menge definieren, sowie keine zweigliedrige Relationen zwischen den Punkten außer der Allrelation, der leeren Relation, der Identität und der Verschiedenheit (insbesondere läßt sich somit in der Geometrie keine Richtung auf der Geraden auszeichnen). Der Unterschied zwischen der Logik und der Geometrie tritt aber schon bei der Betrachtung der dreigliedrigen Relationen zu Tage: auf Grund der euklidischen Geometrie kann man bekanntlich unendlich viele drei- (und mehr-) gliedrige Relationen definieren.

In zwei folgenden Korollaren des Satzes 6 treten die Formeln „$x' \sim x''$“ (*„die Punktmengen x' und x'' sind geometrisch ähnlich“*), „$x' \cong x''$“ (*„die Punktmengen x' und x'' sind kongruent“*) und „$S(x)$“ (*„x ist eine Strecke“*) auf, die sich bekanntlich mit Hilfe des Terminus „a“ definieren lassen.

Satz 7. Sind „x'“ und „x''“ zum Typus der Punktmengen gehörende Variablen, so ist jeder Satz der Form

$$(x', x'') : . x' \sim x'' . \supset : \sigma(a; x') . \equiv . \sigma(a; x'')$$

in der euklidischen Geometrie beweisbar.

Satz 8. Unter derselben Voraussetzung ist jeder Satz der Form

$$(x') : S(x') . \sigma(a; x') . \supset . (\exists x'') . S(x'') . \sigma(a; x'') . \sim (x' \cong x'')$$

in der euklidischen Geometrie beweisbar.

[6]) Vgl. die Schlußbemerkungen.

Also: keine zwei ähnlichen Punktmengen lassen sich mit den Mitteln der Logik und der euklidischen Geometrie voneinander unterscheiden; auf Grund der euklidischen Geometrie läßt sich keine absolute Einheit der Länge definieren.

Verschiedene derartige Ergebnisse kann man auch bei der Untersuchung anderer deduktiven Theorien, z. B. der projektiven Geometrie und der Topologie erreichen. Der Satz 6 gilt überhaupt für jede auf die Logik gestützte Theorie, weil man bei seinem Beweise von den spezifischen Eigenschaften der Axiome der Geometrie keinen Gebrauch macht. Auf die projektive Geometrie kann man nicht nur die Sätze 2—4, sondern auch das Analogon des Satzes 4 für dreigliedrige Relationen übertragen: in der projektiven Geometrie lassen sich nur endlich viele dreigliedrige und erst unendlich viele viergliedrige Relationen zwischen Punkten definieren (u. a. läßt sich in dieser Geometrie die Relation des Zwischenliegens nicht definieren). Auf Grund der allgemeinen Topologie kann man überhaupt nur diejenigen zwei- und mehrgliedrigen Punktrelationen definieren, zu deren Definition schon rein logische Mitteln ausreichen.

II.

Als eine andere Anwendung des Satzes 1 wollen wir hier zwei allgemeine mathematische Sätze angeben.

Ein Axiomensystem „$\alpha(a, b, c, \ldots)$" heißt bekanntlich kategorisch oder monomorph, wenn der Satz

$$(x', x'', y', y'', z', z'' \ldots) : \alpha(x', y', z' \ldots) \,.\, \alpha(x'', y'', z'' \ldots) \,.\supset .\, (\exists R) \,.\, R \frac{x', y', z' \ldots}{x'', y'', z'' \ldots}$$

logisch beweisbar ist. Das Axiomensystem wird nichtgabelbar genannt, wenn für jede Satzfunktion „$\sigma(x, y, z \ldots)$" die Alternative

$$(x, y, z \ldots) : \alpha(x, y, z \ldots) \,.\supset .\, \sigma(x, y, z \ldots) : \vee : (x, y, z \ldots) : \alpha(x, y, z \ldots) \,.\supset .\, \sim \sigma(x, y, z \ldots)$$

logisch beweisbar ist; wenn dagegen für jede Satzfunktion schon ein Glied dieser Alternative logisch beweisbar ist, so ist das System entscheidungsdefinit oder vollständig[7]). Durch Anwendung von Satz 1 beweist man nun

Satz 9. Jedes kategorische Axiomensystem ist nichtgabelbar.

Z. B. ist die axiomatisch aufgebaute Arithmetik natürlicher Zahlen kategorisch, also auch nichtgabelbar. Sie ist zugleich unvollständig, d. h. es gibt arithmetische Aussagen, die weder aus den Axio-

---

[7]) Zu diesen Begriffen vgl. Fränkel, op. cit., S. 347—354 (insbes. 353). — Zwischen dem Teil II unserer Mitteilung und der Arbeit von Carnap, Bericht üb. Unters. zur allgem. Axiomatik, Erkenntnis *1*, 1930, S. 303—307, bestehen gewisse Berührungspunkte, jedoch von mehr äußerlichem Charakter.

men der Arithmetik abgeleitet, noch widerlegt werden können[8]). Wegen der Nichtgabelbarkeit ist der Unabhängigkeitsbeweis für diese Sätze in keiner Richtung mittels einer Interpretation in der Logik durchführbar.

Satz 9 läßt sich unter einer zusätzlichen Voraussetzung umkehren. Wir definieren: ein Axiomensystem „$\varkappa(a, b, c, \ldots)$" heißt effektiv logisch interpretierbar, wenn es in der Logik eine solche Satzfunktion „$\tau(x, y, z \ldots)$" gibt, daß die drei Formeln

$$(x, y, z \ldots): \tau(x, y, z \ldots) . \supset . \varkappa(x, y, z \ldots),$$
$$(\exists x, y, z \ldots) . \tau(x, y, z \ldots),$$
$$(x', x'', y', y'', z', z'' \ldots): \tau(x', y', z' \ldots) . \tau(x'', y'', z'' \ldots) . \supset . x' = x'' . \\ . y' = y'' . z' = z'' \ldots$$

logisch beweisbar sind. Mit Rücksicht auf die in der Logik üblicherweise angenommenen Definitionsregeln kann man diese Formulierung erheblich vereinfachen: das Axiomensystem „$\varkappa(a, b, c \ldots)$" ist dann und nur dann effektiv logisch interpretierbar, wenn es (undefinierte oder definierte) logische Konstanten „a'", „b'", „c'" ... gibt, derart daß sich der Satz „$\varkappa(a', b', c' \ldots)$" (der durch die Einsetzung in das betrachtete Axiomensystem der logischen Konstanten an die Stelle von Grundzeichen gewonnen wird) in der Logik beweisen läßt.

Satz 10. Jedes nichtgabelbare effektiv logisch interpretierbare Axiomensystem ist kategorisch.

Es ist möglich, daß man eine effektive logische Interpretation für ein Axiomensystem erst an einem höheren Typus finden kann (so z. B. für das Axiomensystem der Arithmetik der reellen Zahlen): dann kann man auch — unter gewissen Bedingungen, die für gewöhnliche logische Systeme erfüllt sind — aus der Nichtgabelbarkeit auf die Kategorizität schließen. Es ist uns aber unbekannt, ob der Satz 9 ganz allgemein umkehrbar ist[9]).

In allen derartigen Betrachtungen ist es wichtig auf die Relativität der drei Begriffe: kategorisch, nichtgabelbar und vollständig — in Bezug auf das vorausgesetzte System der Logik zu achten. Ist z. B. die Logik vollständig, so haben die Begriffe: nichtgabelbar und vollständig für jedes Axiomensystem den gleichen Umfang, so daß jedes kategorische System nicht nur nichtgabelbar, sondern auch vollständig ist. — Die drei erörterten Begriffe wurden hier als *a*-Begriffe im Sinne Carnaps eingeführt; zieht man die entsprechenden *f*-Begriffe in Betracht[10]), so ergibt sich, daß sich die *f*-Vollständigkeit mit der *f*-Nichtgabelbarkeit ihrem Umfang nach deckt; das Analogon des Satzes 9 bleibt

[8]) Vgl. Gödel, Monatsh. f. Math. u. Phys. *38*, 1931, S. 174f. u. 187ff.

[9]) Weitere uns bekannte Sätze über den Zusammenhang der Begriffe der Kategorizität und der logischen Interpretation sollen in einer anderen Arbeit angegeben werden.

[10]) Über die *a*- und *f*-Begriffe s. Carnap, Logische Syntax der Sprache, 1934, S. 123ff.

gültig, und das Problem seiner Umkehrung ist auch hier noch nicht allgemein entschieden.

Zum Schluß wollen wir folgendes bemerken: Bei den Anwendungen des Satzes 1 auf konkrete deduktive Theorien und in der allgemeinen Mathematik muß man sich auf solche Axiomensysteme beschränken, aus denen folgt, daß es keine Individuen außerhalb des „Bereiches" der betrachteten Theorie gibt. Man kann sich von dieser Beschränkung befreien, und zwar in der Weise, daß man den Satz 1 entsprechend verallgemeinert und daß man nur diejenigen Satzfunktionen betrachtet, die — in einem gewissen, hier nicht näher zu präzisierendem Sinne — in Bezug auf die Grundzeichen der betrachteten Theorie „intern" sind. Auf diesem Wege kann man verschiedene neue Resultate erreichen; so kann man z. B. beweisen, daß jede Eigenschaft einer Punktmenge x, die sich mit den Mitteln der Logik und der Topologie ausdrücken läßt und dabei in Bezug auf x intern ist, zugleich gegenüber jeder homöomorpher Abbildung der Menge x auf eine andere Menge invariant bleibt[11]).

Ein Satz über quadratische Länge. Von A. Ville.

Jeder Bogen C (d. h. jedes topologische Streckenbild) *in der euklidischen Ebene, dessen zweidimensionales Lebesguesches Maß* 0 *ist, besitzt die untere quadratische Länge* 0, d. h. er enthält für jedes $\nu>0$ und jedes $\varepsilon>0$ ein Teilpolygon $P=\{p_1, p_2, \ldots, p_n\}$, für welches p_1 und p_2 mit den Endpunkten a und b von C identisch sind, dessen Kantenlängen $d(p_i, p_{i+1})\leqq\nu$ sind $(i=1, 2, \ldots, n-1)$ und dessen quadratische Länge $\sum d^2(p_i, p_{i+1})<\varepsilon$.

Vor allem gilt der einfache Hilfssatz: Liegen in der euklidischen Ebene die Punkte $p_1, p_2, \ldots p_6$ im Innern eines Kreises vom Radius r, so gilt für mindestens einen der 15 Abstände $d(p_i, p_j)\leqq r$. Denn ist 0 das Kreiszentrum, so gilt für mindestens einen der Winkel $\sphericalangle p_i 0 p_j \leqq \pi/3$. Sei etwa $d(0, p_i)\leqq d(0, p_j)$, so haben wir $d^2(p_i, p_j)\leqq d^2(0, p_i)+d^2(0, p_j)-d(0, p_i)d(0, p_j)\leqq d^2(0, p_i)\leqq r^2$. Sei nun $\nu>0$ irgend eine vorgelegte Zahl. Nach einem Satz von Menger (Math. Annalen, *103,* S. 468) enthält C ein Teilpolygon $P=\{a=p_1, p_2, \ldots p_n=b\}$, so daß $d(p_i\, p_{i+1})=\nu\ (i=1, 2, \ldots, n-2)$, $d(p_{n-1}, b)\leqq\nu$, $d(p_i, p_{i+1})>\nu$ für $|i-j|>1$. Bilden wir zu $p_1, p_2, \ldots, p_{n-1}$ den offenen Kreis vom Radius ν, so ist die Summe dieser Kreise in der Menge $U(C; \nu)$ aller Punkte der Ebene, die von C einen Abstand $<\nu$ haben, enthalten. Die Summe der Flächeninhalte dieser Kreise ist, da nach dem Hilfssatz kein Punkt der Ebene in mehr als 5 von diesen Kreisen liegt, kleiner als das 5fache des Lebesgueschen Maßes von $U(C; \nu)$. Also ist die quadratische Länge von P, die Zahl $\sum d^2(p_i, p_{i+1})$ kleiner als das $5/\pi$-fache des Maßes von $U(C; \nu)$. Da, wenn C das Lebesguesche Maß 0 hat, das Maß von $U(C; \nu)$ mit ν gegen 0 konvergiert, folgt daraus der eingangs behauptete Satz.

---

[11]) Vgl. Kuratowski, Topologie I, 1933, S. 13.

Der Satz kann leicht für Bogen in höherdimensionalen euklidischen Räumen verallgemeinert werden.

92 Kolloquium (19. VI. 1934).

Über die Länge von Beweisen. Von Kurt Gödel.

Unter der *Länge* eines Beweises in einem formalen System S möge die Anzahl der Formeln, aus denen er besteht, verstanden werden. Ist ferner jede natürliche Zahl in bestimmter Weise durch ein Symbol (Zahlzeichen) in S darstellt (z. B. durch ein Symbol der Form $1+1+\ldots+1$) so möge eine Funktion $\varphi(x)$ *berechenbar in* S heißen, falls es zu jedem Zahlzeichen m ein Zahlzeichen n gibt, so daß $\varphi(m)=n$ in S beweisbar ist. Insbesondere sind z. B. alle rekursiv definierten Funktionen schon in der klassischen Arithmetik (d. h. im System S_1 der unten definierten Folge) berechenbar.

Sei nun S_i das System der Logik i-ter Stufe, wobei die natürlichen Zahlen als Individuen betrachtet werden. D. h. genauer: S_i soll Variable und Quantifikatoren für nat. Zahlen, für Klassen nat. Zahlen, für Klassen von Klassen nat. Zahlen u. s. w. bis Klassen von i-tem Typus samt den zugehörigen logischen Axiomen enthalten, aber keine Variable von höherem Typus. Dann gibt es bekanntlich Sätze aus S_i, die in S_{i+1}, aber nicht in S_i beweisbar sind. Betrachten wir hingegen diejenigen Formeln f, die sowohl in S_i als in S_{i+1} beweisbar sind, so gilt folgendes: Zu jeder in S_i berechenbaren Funktion φ gibt es unendlich viele Formeln f von der Art, daß, wenn k die Länge eines kürzesten Beweises für f in S_i und l die Länge eines kürzesten Beweises für f in S_{i+1} ist, $k>\varphi(l)$. Setzt man z. B. $\varphi(n)=10^6 . n$, so folgt: Es gibt unendlich viele Formeln, deren kürzester Beweis in S_i mehr als 10^6 mal länger ist als ihr kürzester Beweis in S_{i+1}. *Der Übergang zur Logik der nächst höheren Stufe bewirkt also nicht bloß, daß gewisse früher unbeweisbare Sätze beweisbar werden, sondern auch daß unendlich viele der schon vorhandenen Beweise außerordentlich stark abgekürzt werden können.*

Die Formeln f, für welche die obige Ungleichung $k>\varphi(l)$ gilt, sind arithmetische Sätze von demselben Charakter wie die von mir konstruierten in S_i unentscheidbaren Sätze, d. h. sie können auf folgende Normalform gebracht werden:

$$(\mathfrak{P})[Q(x_1, x_2, \ldots, x_n)=0],$$

wobei $x_1, x_2, \ldots, x_n$ Variable für ganze Zahlen sind, Q ein bestimmtes Polynom mit ganzzahligen Koeffizienten in den n Variablen $x_1, x_2, \ldots, x_n$ ist und $\mathfrak{P}$ ein Präfix bezeichnet, d. h. eine bestimmte Folge von All- und Existenzzeichen für die Variabeln $x_1, x_2, \ldots x_n$. Ein solcher Satz spricht also eine Eigenschaft der Diophantischen Gleichung $Q=0$ aus. Z. B. besagt der Satz

$$(x_1)(x_2)(Ex_3)(Ex_4)\ [Q(x_1, x_2, x_3, x_4)=0],$$

daß die Diophantische Gleichung $Q(a, b, x, y)=0$ für alle Werte der Parameter a und b ganzzahlige Lösungen x, y besitzt. Enthält das Präfix die Allzeichen für $x_{i_1}, x_{i_2}, \ldots, x_{i_k}$ und die Existenzzeichen für $x_{j_1}, x_{j_2}, \ldots, x_{j_{n-k}}$, so hat man $x_{i_1}, \ldots, x_{i_k}$ als Parameter aufzufassen und die obige Formel besagt, daß für beliebige Werte der Parameter Lösungen $x_{j_1}, x_{j_2}, \ldots, x_{j_{n-k}}$ existieren, wobei der Wert von x_{j_r} nur abhängt von den Werten derjenigen Parameter, die x_{j_r} im Präfix vorangehen.

(Bemerkung während der Drucklegung):

Es läßt sich übrigens zeigen, das eine Funktion, die in einem der Systeme S_i oder auch in einem System transfiniter Stufe berechenbar ist, schon in S_1 berechenbar ist, so daß also der Begriff „berechenbar" in gewissem Sinn „absolut" ist, während fast alle sonst bekannten metamathematischen Begriffe (z. B. beweisbar, definierbar etc.) sehr wesentlich vom zu Grunde gelegten System abhängen.

93. Kolloquium (25. VI. 1935).

Begründung einer koordinatenlosen Differentialgeometrie der Flächen. Von A. Wald.

Im folgenden geben wir eine neue koordinatenlose Begründung der Flächentheorie im Sinne des Mengerschen Programms einer allgemeinen koordinatenlosen metrischen Geometrie. Menger hat in zahlreichen Abhandlungen und Vorträgen seit Jahren die Ansicht entwickelt, daß zur Untersuchung der lokalen Eigenschaften von Raumgebilden die Darstellung der Punkte und Gebilde durch Koordinaten bzw. differenzierbare Funktionen und Gleichungen ein Hilfsmittel bildet, das sich zwar als sehr fruchtbar erwiesen hat, aber weder den einzigen, noch den allgemeinsten, noch auch den dem geometrischen Kern des Problemes angemessensten Weg darstellt. Einer rein metrischen und koordinatenlosen Behandlung gelinge es *erstens,* die Theorie von vielen überflüssigen Voraussetzungen zu befreien, insbesondere von zahlreichen Differenzierbarkeitsannahmen, die vielfach lediglich zum Zwecke der Anwendbarkeit analytischer Methoden erforderlich sind, ohne mit dem geometrischen Kern des Problemes etwas zu tun zu haben, während in den Fällen, wo Differenzierbarkeitsannahmen erforderlich sind, durch die metrischen Methoden ihre geometrische Bedeutung aufgeklärt werde; *zweitens* gelinge ihr durch Schaffung neuer Methoden, für zahlreiche klassische Sätze einfache und anschauliche Beweise zu erbringen und darüber hinaus Probleme zu behandeln, die den klassischen Methoden prinzipiell unzugänglich sind.

Eine metrische Theorie der *Kurvenkrümmung* hat bekanntlich Menger selbst gegeben, indem er sich lediglich auf die Betrachtung von Punktetripeln und ihren Abstandstripeln stützt [1]). Im Zusammenhang damit hat Menger das Problem formuliert, die *Flächenkrümmung* durch bloße Betrachtung von Punktequadrupeln und

---

[1]) Mathem. Annalen *103.*

ihren Abstandssextupeln zu definieren und zu behandeln. Diese Aufgabe wird im folgenden vollständig gelöst [2]).

Wir betrachten *metrische Räume* im Sinne von Fréchet, d. h. Mengen von irgend welchen Elementen („Punkten"), in denen je zwei Punkten p und q ein Abstand $pq = qp \geqq 0$, zugeordnet ist, wobei $pq = 0$ dann und nur dann gilt, wenn $p = q$, und wobei die Dreiecksungleichung $pq + qr \geqq pr$ erfüllt ist. Jede Teilmenge eines metrischen Raumes ist ein metrischer Raum. Zwei metrische Räume heißen *kongruent*, wenn sie eineindeutig und abstandstreu aufeinander abbildbar sind. Ist R mit einer Teilmenge von R' kongruent, so nennen wir R in R' *einbettbar*. Mit Menger nennen wir die in die euklidische Gerade einbettbaren Mengen *linear*, die mit den Intervallen der Geraden kongruenten Mengen *Strecken*.

Von großer Bedeutung für diese Arbeit sind folgende Begriffsbildungen der Mengerschen Konvexitätstheorie [3]): Ein Punkt q liegt *zwischen* den Punkten p und r, wenn $p \neq q \neq r$ und $pq + qr = pr$. (Z. B. sind drei paarweise verschiedene Punkte dann und nur dann linear, wenn einer zwischen den beiden anderen liegt.) Ein metrischer Raum heißt *konvex*, wenn zu je zwei Punkten $p \neq r$ mindestens ein Punkt zwischen p und r existiert. Oft verwenden wir folgendes Theorem von Menger [4]): Ein konvexer vollständiger (z. B. kompakter) Raum enthält zu je zwei verschiedenen Punkten mindestens eine sie verbindende Strecke.

Ein besonders wichtiges Beispiel konvexer Räume ist für jede reelle Zahl k der S_k. So nennen wir für $k > 0$ die Oberfläche einer Kugel der Krümmung k, also vom Radius $1/\sqrt{k}$, in der je zwei Punkten als Abstand die Länge des kürzeren sie verbindenden Großkreisbogens zugeordnet wird, für $k = 0$ die euklidische Ebene, für $k < 0$ die hyperbolische Ebene der Krümmung k. Die linearen Teilmengen von S_k sind für $k \leq 0$ die Teilmengen der Geraden, für $k > 0$ die Teilmengen von halben Großkreisen.

Wir definieren nun: Der metrische Raum R hat im Punkt p die **Flächenkrümmung $k(p)$**, *wenn keine Umgebung von p linear ist und zu jedem $\varepsilon > 0$ ein $\delta > 0$ existiert derart, daß jedes Quadrupel von Punkten, die von p einen Abstand $< \delta$ haben, in einen S_k mit $|k - k(p)| < \varepsilon$ einbettbar ist.*

Nach einigen Hilfssätzen über den S_k im Kapitel I zeigen wir im Kapitel II, daß man, wenn ein kompakter konvexer Raum in einem Punkt p eine Flächenkrümmung besitzt, in einer Umgebung von p Polarkoordinaten einführen kann, durch welche die Umgebung auf eine Teilmenge der Ebene topologisch abgebildet wird. In Kapitel III zeigen wir, daß, wenn der Raum in jedem Punkt eine Flächenkrümmung besitzt, das Bogenelement eingeführt werden kann,

[2]) Hauptresultate finden sich in meiner Note C. R. Paris 18. XI. 1935. Der von mir im Heft 6 dieser Ergebnisse entwickelte Krümmungsbegriff wird im folgenden modifiziert, wodurch die Theorie beträchtlich vereinfacht wird.

[3]) Mathem. Annalen *100*, S. 76.

[4]) l. c., S. 89.

so daß der Raum als Gaußsche Fläche erwiesen ist. Die Gaußsche Krümmung derselben ist, wie wir sehen werden, in jedem Punkt mit der Flächenkrümmung des Raumes identisch. In Kapitel IV zeigen wir, daß umgekehrt eine Gaußsche Fläche in jedem Punkt eine Flächenkrümmung besitzt, die mit der Gaußschen Krümmung identisch ist. Diese Resultate lassen sich zusammengenommen aussprechen als folgendes

Fundamentaltheorem der Flächentheorie. *Damit ein metrischer Raum R eine Gaußsche Fläche sei, ist notwendig und hinreichend, daß R kompakt und im Mengerschen Sinne konvex sei und in jedem Punkt eine Flächenkrümmuug im oben definierten Sinn besitze. Dieselbe stimmt dann für jeden Punkt mit der Gaußschen Krümmung überein.*

Diesem Theorem zufolge dient also unsere Definition einerseits einer neuen einfachen und anschaulichen Einführung der Gaußschen Krümmung (nämlich einer metrischen, die nur auf die Betrachtung von Punktequadrupeln beruht), andererseits einer Grundlegung für den koordinatenlosen Aufbau der Differentialgeometrie der Flächen.

I. Hilfssätze über S_k.

Sind $p \neq q \neq r$ Punkte von S_k, so bezeichnen wir den kleineren der beiden Winkel, welche die Strecken qp und qr miteinander einschließen, mit $\sphericalangle pqr$. Es gilt dann also stets $0 \leqq \sphericalangle pqr \leqq \pi$. Sind p, q, r linear, so gilt $\sphericalangle pqr = 0$ oder $= \pi$. Umgekehrt sehen wir:

1. *Ist $\sphericalangle pqr = \pi$ für p, q, r von S_k, so liegt q zwischen p und r, es sei denn, daß $k > 0$ und $pq + qr + rp = u(k)$ gilt, wo $u(k)$ die Länge des Großkreises von S_k bezeichnet.*

2. *Sind p, a, b drei nicht lineare Punkte von S_k und ist c ein Punkt im Innern des Winkelraumes zwischen den Strecken pa und pb für den $cp = bp$, so ist $ac < ab$.*

3. *Ist a', b', c' ein nicht lineares Tripel in S_k, und a'', b'', c'' ein Tripel in $S_{k''}$, so daß $k' < k''$, $a'b' = a''b''$, $a'c' = a''c''$ und $\sphericalangle b'a'c' = \sphericalangle b''a''c''$ gilt, dann ist $b'c' > b''c''$.* Führen wir auf S_k geodätische Polarkoordinaten r, φ ein, wobei $r \geqq 0$ und, wenn $k > 0$, überdies $r < u(k)/2$ gilt, so hat das Bogenelement auf S_k bekanntlich die Gestalt $ds^2 = dr^2 + G_k^2(r, \varphi)\, d\varphi^2$, wo $G_k(r, \varphi) = G_k(r) = \sin\sqrt{k}\, r/\sqrt{k}$. Die Ableitung von $G_k(r)$ nach k ist negativ. Ist $\sphericalangle b'a'c' = \pi$, so ist nach **1:** $a'b' + b'c' + c'a' = u(k) > u(k'') = a''b'' + b''c'' + c''a''$, also $b'c' > b''c''$. Zum Beweis von **3** genügt es daher offenbar zu zeigen, daß zu jedem nichtlinearen Tripel a, b, c von S_k, wobei $\sphericalangle bac \neq \pi$ ist, ein Zahl $\varepsilon > 0$ existiert mit folgender Eigenschaft: Ist $k - \varepsilon < k^* < k + \varepsilon$, und a^*, b^*, c^* ein Tripel von S_{k^*}, so daß $a^*b^* = ab$, $a^*c^* = ac$, $\sphericalangle b^*a^*c^* = \sphericalangle bac$, dann gilt $bc \gtreqless b^*c^*$, je nachdem $k \lesseqgtr k^*$. Um dies einzusehen, führen wir in S_k und S_{k^*} Polarkoordinaten mit den Zentren a bzw. a^* ein, so daß b und c dieselben Koordinaten in S_k haben, wie b^* und c^* in S_{k^*}. Da a, b, c nicht linear sind, so sind keine zwei dieser drei Punkte diametral gegenüberliegend. Ist

$r=r(\varphi)$ $(\varphi_1 \leqq \varphi \leqq \varphi_2)$ die Gleichung des geodätischen Bogens B zwischen b und c in S_k, so existiert daher, wenn ε hinlänglich klein ist, in S_{k^*} der b^* und c^* verbindende Bogen mit der Gleichung $r=r(\varphi)$ und hat, da G_k monoton abnimmt, eine größere bzw. kleinere Länge als B, je nachdem $k>k^*$ oder $<k^*$. Daraus folgt unmittelbar die Behauptung.

4. *Sind die nichtlinearen Tripel a', b', c' von $S_{k'}$, und a'', b'', c'' von $S_{k''}$ kongruent und ist $k'<k''$, so ist jeder der drei Winkel des ersten Tripels kleiner als der entsprechende Winkel des zweiten Tripels.* Kein Winkel des Tripels ist $=\pi$, da sonst **1** zufolge, $k'>0$ und $a'b'+b'c'+c'a'=u(k')$ gelten würde und kein zu a', b', c' kongruentes Tripel in $S_{k''}$ existieren könnte. Wir wählen nun einen Punkt d'' in $S_{k''}$ derart, daß $b''d''=b''c''$ und $\sphericalangle a''b''d''=\sphericalangle a'b'c'$. Nach **3** ist dann $a'c'>a''d''$. Da $a''c''=a'c'$, also $a''c''>a''d''$, so folgt aus **2**, daß $\sphericalangle a''b''c''>\sphericalangle a''b''d''=\sphericalangle a'b'c'$.

5. *Sei $Q'=\{a', b', c', d'\}$ ein echtes*[5]) *nichtlineares Quadrupel auf S_k, für welches $a'b'+b'c'=a'c'$ gilt, und $Q''=\{a'', b'', c'', d''\}$ ein Quadrupel $S_{k''}$, wobei $k'<k''$ gilt und a'', b'', c'' mit a', b', c' kongruent sind. Dann behaupten wir:*

α_1) *Aus $a'd'=a''d''$ und $b'd'=b''d''$ folgt $c'd'>c''d''$.*

α_2) *Aus $c'd'=c''d''$ und $b'd'=b''d''$ folgt $a'd'>a''d''$.*

β) *Aus $a'd'=a''d''$ und $c'd'=c''d''$ folgt $b'd'<b''d''$.*

α_1) Ist a', b', d' nicht linear, dann ist $\sphericalangle a'b'd'<\sphericalangle a''b''d''$ nach **4**. Für die Nebenwinkel gilt also $\sphericalangle c'b'd'>\sphericalangle c''b''d''$, woraus nach **3** und **2** sich $c'd'>c''d''$ ergibt. Ist zweitens a', b', d' linear, so muß, da Q' nicht linear ist, a' zwischen b' und d' liegen. Dann sieht man, daß $\sphericalangle c'a'd'=\sphericalangle b'a'd'=\pi$, also gilt nach **1**:

$$(+) \qquad k'>0 \text{ und } a'c'+c'd'+d'a'=u(k').$$

Da andererseits $a'c'+c''d''+d'a'=a''c''+c''d''+d''a''\leqq u(k'')<u(k')$ gilt, so folgt wiederum $c'd'>c''d''$, wie α_1) behauptet. Analog beweist man α_2). Um β) zu beweisen, bemerken wir zunächst, daß weder a', b', d' noch a', c', d' linear ist. Denn wäre a', b', d' linear, so müßte a' zwischen b' und d' liegen und (+) gelten, was wegen $a'c'+c'd'+d'a'=a''c''+c''d''+d''a''\leqq u(k'')<u(k')$ unmöglich ist. Wäre a', c', d' linear, so müßte, da Q' nicht linear ist, d' zwischen a' und c' liegen. Es wäre $k'>0$ und a' und c' wären diametral gegenüberliegend in $S_{k'}$, was wegen $a'c'=a''c''$ und $k'<k''$ unmöglich ist. Ist nun erstens a'', b'', d'' linear, so muß, da das mit a', c', d' kongruente Tripel a'', c'', d'' nicht linear ist, a'' zwischen b'' und d'' liegen, d. h. $b''a''+a''d''=b''d''$ gelten. Da a', b', d' nicht linear ist, also $b'a'+a'd'>b'd'$ gilt, ist $b'd'<b''d''$. Ist zweitens a'', b'', d'' nicht linear, so sei d^* ein Punkt von $S_{k'}$, so daß $a'd^*=a''d''$ und $b'd^*=b''d''$. Wegen des bereits bewiesenen Satzes **5**α) ist dann $c'd^*>c''d''=c'd'$. Wegen **2** ist daher $\sphericalangle b'a'd^*$

---

[5]) Ein n-Tripel von Punkten heiße **echt**, falls je zwei seiner Punkte einen Abstand $\neq 0$ haben.

$> \sphericalangle b' a' d'$ und mithin $b' d' < b' d^* = b'' d''$, womit **5** in allen Stücken bewiesen ist. Es folgt als Korollar:

6. *Ein echtes nichtlineares Quadrupel, das ein lineares Tripel enthält, ist für höchstens eine Zahl k in S_k einbettbar.*

Gilt für die Punkte a, b, c, d von S_k, daß $\sphericalangle bac + \sphericalangle bad + \sphericalangle cad = 2\pi$, so wollen wir sagen, a liege auf S_k *zwischen* b, c, d. Wir beweisen dann:

7. *Es sei $Q = \{a, b, c, d\}$ ein echtes Quadrupel des S_k. Liegt einer der vier Punkte zwischen den drei anderen, so ist Q für kein $k'' > k$ in $S_{k''}$ einbettbar. Liegt keiner der vier Punkte zwischen den drei anderen, so ist Q für kein $k' < k$ in $S_{k'}$ einbettbar*[6]. Wäre nämlich im ersten Fall a'', b'', c'', d'' ein zu Q kongruentes Quadrupel in $S_{k''}$ mit $k'' > k$, dann enthielte Q nach **6** kein lineares Tripel und es müßte wegen 4 offenbar $\sphericalangle b'' a'' c'' + \sphericalangle b'' a'' d'' + \sphericalangle c'' a'' d'' > 2\pi$ gelten, was unmöglich ist. Wäre im zweiten Fall a', b', c', d' ein zu Q kongruentes Quadrupel in $S_{k'}$ mit $k' < k$, so kann Q nach **6** kein lineares Tripel enthalten; es gäbe ferner, da keiner der vier Punkte von Q zwischen den drei anderen liegt, zwei gegenüberliegende Strecken, etwa ab und cd, die sich in einem inneren Punkt e schneiden würden. Es sei e' ein Punkt von $S_{k'}$ zwischen a' und b' so, daß $a' e' = ae$. Wegen **5** β) wäre dann $c' e' < ce$ und $d' e' < de$, also $c' e' + e' d' < ce + ed = cd$. Da $c' d' \leqq c' e' + e' d'$ ist, so gilt erst recht $c' d' < cd$ im Widerspruch zur vorausgesetzten Kongruenz. Die damit bewiesene Behauptung **7** enthält offenbar als Korollar:

8. *Ein echtes nichtlineares metrisches Quadrupel Q ist für höchstens zwei verschiedene Zahlen $k' < k''$ sowohl in $S_{k'}$ als auch in $S_{k''}$ einbettbar. Gibt es zwei solche Zahlen, so enthält jedes zu Q kongruente Quadrupel von $S_{k''}$ einen Punkt zwischen den drei übrigen, dagegen enthält kein zu Q kongruentes Quadrupel in $S_{k'}$ einen zwischen den drei übrigen Punkten liegenden Punkt.*

II. Über metrische Räume, welche in einem Punkt eine Flächenkrümmung besitzen.

Wir ziehen vor allem einige unmittelbare Folgerungen aus den bewiesenen Hilfssätzen auf Grund des bekannten Theorems von Menger[7]), daß ein mehr als vier Punkte enthaltender metrischer Raum, von dem je drei Punkte linear sind, linear ist. Als *Umgebung* eines Punktes p eines metrischen Raumes werden wir im folgenden stets die Menge aller Punkte des Raumes bezeichnen, die für irgend eine Zahl $\rho > 0$ von p einen Abstand $< \rho$ haben; genauer nennen wir diese Menge die *Umgebung von p vom Radius ρ*.

Ein konvexer metrischer Raum R besitzt in einem Punkt p höchstens eine Flächenkrümmung. Sei U irgend eine Umgebung von p. Ist U linear, dann hat R in p keine Flächenkrümmung. Ist U nicht linear, so existieren dem erwähnten Mengerschen Theorem

[6]) Die Beweise der Sätze 7 und 8 hat mir J. Groiß geliefert.

[7]) Mathem. Annalen *100*, S. 128.

zufolge drei nicht lineare Punkte q, r, s von U. Wegen der Konvexität von R existiert ein Punkt t zwischen r und s. Dann sind q, r, s, t ein nicht lineares, aber ein lineares Tripel enthaltendes Quadrupel. Die Behauptung folgt dann aus **6** [8]).

Wir nennen die Umgebung U von p vom Radius ρ *regulär*, wenn jedes Quadrupel von U einbettbar ist in S_k mit $|k| < \pi^2/16\rho^2$. Die untere Schranke K aller Zahlen k, für welche jedes Quadrupel von U in $S_{k'}$ mit $|k'| \leqq k$ einbettbar ist, nennen wir die *Krümmungsschranke* von U. Aus der Definition der Flächenkrümmung folgt unmittelbar: *Wenn R in p eine Flächenkrümmung besitzt, so ist jede Umgebung von p mit hinlänglich kleinem Radius regulär.* Wir zeigen weiter: In einem kompakten konvexen Raum hat jede reguläre Umgebung U eines Punktes p, in dem R eine Flächenkrümmung besitzt, folgende Eigenschaften:

1. *Jedes echte Quadrupel Q von U, das zwei lineare Tripel enthält, ist linear.* Q ist mit einem Quadrupel Q' von S_k mit $k < \pi^2/16\rho^2$ kongruent, das zwei lineare Tripel enthält. Ist $k \leqq 0$, so liegt Q' auf einer Geraden. Ist $k > 0$, so liegt Q' auf einem halben Großkreis von S_k, da je zwei Punkte von Q' als Bilder von zwei Punkten der Umgebung U vom Radius ρ einen Abstand $< 2\rho < \pi/2\sqrt{k}$ haben. Jedenfalls ist also Q' linear.

2. *Es existieren zwei Punkte a, b von U, so daß p, a, b nicht linear sind.* Anderenfalls wäre wegen Eigenschaft 1. jedes Tripel von U linear. Da U wegen der Konvexität von R mehr als vier Punkte enthält, wäre dann nach dem eingangs erwähnten Theorem von Menger U linear, was unmöglich ist, da R in p eine Flächenkrümmung besitzt.

3. *U ist konvex.* Seien $a \neq b$ zwei Punkte von U. Sind p, a, b linear, so liegt jeder Punkt zwischen a und b in U. Es seien p, a, b nicht linear; wäre x ein Punkt zwischen a und b, der nicht in U liegt, für den also $px \geqq \rho$, so existierten wegen der Konvexität von R zwei Punkte $y_1 \neq y_2$ zwischen a und b derart, daß $pa < py_1 = py_2 < \rho$. Das Quadrupel $Q = \{p, a, y_1, y_2\}$ wäre mit einem Quadrupel $Q' = \{p', a', y_1', y_2'\}$ von S_k mit $|k| < \pi^2/16\rho^2$ kongruent. Da a, y_1, y_2 linear sind und a nicht zwischen y_1 und y_2 liegt, so ist dies für $k \leqq 0$ wegen $pa < py_1$ unmöglich; für $k > 0$ wäre dies nur möglich, wenn das gleichschenkelige Dreieck p', y_1', y_2' eine Höhe $\geqq \pi/2\sqrt{k}$ hätte, was aber nicht der Fall ist, da diese Höhe $< 2\rho$ ist. Aus den Eigenschaften 1. und 3. folgt nach bekannten Sätzen von Menger [9]) als *Korrolar:*

Je zwei Punkte von U sind durch genau eine Strecke (d. h. mit einem Intervall der Geraden kongruente Menge) *verbunden und dieselbe verläuft in U.*

[8]) Erwähnen wir noch, daß analog aus 8 folgt, *daß jeder beliebige metrische Raum in keinem seiner Punkte mehr als zwei Flächenkrümmungen besitzen kann.*

[9]) Mathem. Annalen *100*, S. 103—113.

Einführung des Winkelbegriffes. R habe im Punkt p die Flächenkrümmung $k(p)$ und es seien a und b zwei Punkte $\neq p$ in einer regulären Umgebung U von p. Bezeichnen wir dann für jede Zahl λ, die kleiner als pa und pb ist, mit $a(\lambda)$ bzw. $b(\lambda)$ den zwischen p und a bzw. b im Abstand λ von p liegenden Punkt und nennen wir $\delta(\lambda)$ den Abstand von $a(\lambda)$ und $b(\lambda)$, so behaupten wir: *Es existiert lim* $\delta(\lambda)/\lambda$ *für* $\lambda \to 0$.

Vor allem sehen wir, daß, wenn p, a, b linear sind, $\delta(\lambda) = 2\lambda$ bzw. $= 0$ für jedes λ gilt, je nachdem ob p zwischen a und b liegt oder nicht, daß dann also der fragliche Limes existiert und $= 2$ bzw. $= 0$ ist. Wir nehmen nun an, p, a, b seien nicht linear. K sei die Krümmungsschranke von U. Wir wählen zwei Zahlen $k' < -K$ und $k'' > K$ und irgend ein λ_0, das kleiner als pa und pb ist. Wir bilden zu p, $a(\lambda_0)$, $b(\lambda_0)$ zwei kongruente Tripel p', $a'(\lambda_0)$, $b'(\lambda_0)$ und p'', $a''(\lambda_0)$, $b''(\lambda_0)$ auf $S_{k'}$ bzw. auf $S_{k''}$. Wir bezeichnen sodann für jedes $\lambda \leqq \lambda_0$ mit $a'(\lambda)$ bzw. $b'(\lambda)$ den zwischen p' und $a'(\lambda_0)$ bzw. $b'(\lambda_0)$ im Abstand λ von p' liegenden Punkt von $S_{k'}$, nennen $\delta'(\lambda)$ den Abstand der Punkte $a'(\lambda)$ und $b'(\lambda)$ und definieren analog $a''(\lambda)$, $b''(\lambda)$, $\delta''(\lambda)$. Wir zeigen nun, daß $\delta'(\lambda) < \delta(\lambda)$ für jedes $\lambda < \lambda_0$. Da das Quadrupel p, $a(\lambda_0)$, $b(\lambda_0)$, $a(\lambda)$ von U in eine S_{k_1} mit $k_1 > k'$ einbettbar ist, während p', $a'(\lambda_0)$, $b'(\lambda_0)$, $a'(\lambda)$ in $S_{k'}$ liegen, so folgt aus **5** β, daß $b'(\lambda_0)a'(\lambda) < b(\lambda_0)a(\lambda)$ gilt. Das Quadrupel p, $b(\lambda_0)$, $a(\lambda)$, $b(\lambda)$ von U ist in eine S_{k_2} mit $k_2 > k'$ einbettbar. Es sei p^*, $b^*(\lambda_0)$, $a^*(\lambda)$ ein zu p, $b(\lambda_0)$, $a(\lambda)$ kongruentes Tripel $S_{k'}$ und $b^*(\lambda_0)$ der zwischen p und $b^*(\lambda_0)$ im Abstand λ von p^* liegende Punkt. Dann folgt aus **5** β, daß $a^*(\lambda)b^*(\lambda) < a(\lambda)b(\lambda) = \delta(\lambda)$. Wegen $b'(\lambda_0)a'(\lambda) < b(\lambda_0)a(\lambda) = b^*(\lambda_0)a^*(\lambda)$ folgt aus **2**, daß $\sphericalangle b^*(\lambda_0)p^*a^*(\lambda) > \sphericalangle b'(\lambda_0)p'a'(\lambda)$, mithin $a^*(\lambda)b^*(\lambda) > a'(\lambda)b'(\lambda) = \delta'(\lambda)$ und demnach $\delta'(\lambda) < \delta(\lambda)$ gilt. Analog beweist man $\delta''(\lambda) > \delta(\lambda)$ und hat also $\delta'(\lambda) < \delta(\lambda) < \delta''(\lambda)$ für jedes $\lambda < \lambda_0$. Nach bekannten Formeln der sphärischen Trigonometrie gilt:

$$\frac{\sin\sqrt{k'}\,\delta'(\lambda)/2}{\sin\sqrt{k'}\,\lambda} = \frac{\sin\sqrt{k'}\,\delta'(\lambda_0)/2}{\sin\sqrt{k'}\,\lambda_0} = L'(\lambda_0)$$ für jedes $\lambda \leqq \lambda_0$, also $\lim \delta'(\lambda)/\lambda = 2L'(\lambda_0)$ für $\lambda \to 0$. Analog definiert man $L''(\lambda_0)$ und hat $\lim \delta''(\lambda)/\lambda = 2L''(\lambda_0)$. Zwischen den Zahlen $2L'(\lambda_0)$ und $2L''(\lambda_0)$, die zwischen 0 und 2 liegen, müssen alle Häufungswerte von $\delta(\lambda)/\lambda$ liegen. Da, wie man leicht sieht $\lim (L'(\lambda_0) - L''(\lambda_0)) = 0$ für $\lambda_0 \to 0$, so muß $\delta(\lambda)/\lambda$ konvergieren, w. z. b. w.

Wir definieren nun:

$$\sphericalangle apb = 2 \text{ Arc sin } \frac{1}{2} \lim_{\lambda=0} \frac{\delta(\lambda)}{\lambda},$$ wo Arc sin den zwischen 0 und $\pi/2$ gelegenen Wert von arc sin bedeutet, dann geht aus unseren Überlegungen hervor:

9. *Für je zwei Punkte* $a \neq p \neq b$ *von* U *ist* $0 \leqq \sphericalangle apb \leqq \pi$, *wobei eines der Gleichheitszeichen dann und nur dann gilt, wenn* p, a, b *linear sind. Ist* $pa = pb$, *so gilt* $\sphericalangle apb = 0$ *dann und nur dann, wenn* $a = b$.

10. *Es seien $a \neq p \neq b$ Punkte der regulären Umgebung U von p mit der Krümmungsschranke K und a', p', b' bzw. a'', p'', b'' Tripel in $S_{k'}$ bzw. $S_{k''}$, wobei $k' < -K$, $k'' > K$, so gilt, wenn die Tripel mit p, a, b kongruent sind, $\sphericalangle a'p'b' \leqq \sphericalangle apb \leqq \sphericalangle a''p''b''$; wenn dagegen $pa = p'a' = p''a''$, $pb = p'b' = p''b''$, $\sphericalangle apb = \sphericalangle a'p'b' = \sphericalangle a''p''b''$, so gilt $a'b' \geqq ab \geqq a''b''$.*

Einführung von Polarkoordinaten. Ist U eine reguläre Umgebung von p, so können wir wegen Eigenschaft 2. der Regularität zwei Punkte p_1 und p_2 wählen, so daß p, p_1, p_2, nicht linear sind. Wir ordnen nun jedem Punkt $s \neq p$ von U die zwei Zahlen zu: $r(s) = ps$ und $\varphi(s) = \sphericalangle spp_1$ bzw. $2\pi - \sphericalangle spp_1$ je nachdem $|\sphericalangle p_1pp_2 - \sphericalangle spp_1| = \sphericalangle spp_2$ oder $\neq \sphericalangle spp_2$ ist und beweisen, daß U folgende wichtige Eigenschaft hat:

4. *Bildet man den Punkt p der regulären Umgebung U auf das Zentrum p' eines Polarkoordinatensystems der euklidischen Ebene ab und jeden Punkt $s \neq p$ auf den Punkt s' der Ebene mit den Polarkoordinaten $r(s)$, $\varphi(s)$, so wird hiedurch U topologisch abgebildet auf eine Teilmenge der Ebene. Sind a' und b' zwei Punkte dieser ebenen Menge U', so daß p', a', b' nicht linear sind, so existiert ein $\lambda > 0$ derart, daß jeder Punkt im Winkelraum zwischen den Strecken $p'a'$ und $p'b$, der von p' einen Abstand $> \lambda$ hat, in U' liegt.*

Wir beginnen mit einer Hilfsüberlegung: $p_1, \ldots p_n$ seien n Punkte $\neq p$ von U. Für jedes λ, das nicht größer als die kleinste der Zahlen $pp_1, \ldots, pp_n$ ist, bezeichnen wir mit $p_i(\lambda)$ den zwischen p und p_i im Abstand λ von p liegenden Punkt $(i = 1, 2, \ldots, n)$ und setzen $p_0(\lambda) = p$ für jedes λ. Dann ist für jedes λ das Quadrupel $p_{i_1}(\lambda), \ldots, p_{i_4}(\lambda)$ mit einem Quadrupel Q^* von S_k kongruent, wo k von λ und i_1, i_2, i_3, i_4 abhängt und mit $\lambda \to 0$ gegen $k(p)$ konvergiert. Da Q^* von S_k mit $\lambda \to 0$ unbegrenzt klein wird, so nähern sich die Abstandsverhältnisse der Punkte von Q^* unbegrenzt den Abstandsverhältnissen eines Quadrupels der euklidischen Ebene. Dasselbe gilt für vier Punkte, deren Abstände gleich dem λ-ten Teil der Abstände der entsprechenden Punkte von Q^* sind. Setzen wir $d_{ij} = \lim p_i(\lambda)\, p_j(\lambda)/\lambda$ $(i - 0, 1, \ldots, n)$, so existiert also für je vier Zahlen $i_1, \ldots, i_4$ ein Quadrupel $p''_{i_1}, \ldots, p''_{i_4}$ der euklidischen Ebene, in welchem erstens p''_i und p''_j den Abstand d_{ij} haben und zweitens $\sphericalangle p''_i p'' p''_j = \sphericalangle p_i\, p\, p_j$ gilt. Aus der ersten Bemerkung folgt auf Grund Mengerscher Sätze über die Ebene[10]), daß $n+1$ Punkte $p''_0, p''_1, \ldots, p''_n$ der Ebene existieren, von denen p''_i und p''_j den Abstand d_{ij} haben $(i, j = 0, 1, \ldots, n)$ aus der zweiten, daß $\sphericalangle p''_i p''_0 p''_j = \sphericalangle p_i p_0 p_j$ $(i, j = 1, \ldots, n)$ gilt. Wenden wir dies auf die Punkte p_1 und p_2 von U an, welche zur Einführung der Polarkoordinaten in U verwendet wurden, und auf irgend zwei von p verschiedene Punkte p_3 und p_4 von U, wobei wir in p''_0 das Zentrum eines ebenen Koordinatensystems, durch

---

[10]) Mathem. Annalen *100*, S. 128.

p_1'' dessen Halbstrahl $\varphi = 0$ legen und das wir so orientieren, daß $\varphi(p_2'') < \pi$, so folgt aus der zweiten Bemerkung: Es gilt $\varphi(p_i'') = \varphi(p_i)$ $(i = 1, 2, 3, 4)$ und $\sphericalangle p_3'' p_0'' p_4'' = \sphericalangle p_3 p p_4$. Hieraus ergibt sich für die Abbildung von U auf die ebene Menge U', wenn man in der Ebene das eben betrachtete Koordinatensystem wählt, der *Hilfssatz:*

$$\varphi(p_i') = \varphi(p_i'') \ (i = 1, 2, 3, 4) \text{ und } \sphericalangle p_3' p_0' p_4' = \sphericalangle p_3'' p_0'' p_4'' = \sphericalangle p_3 p_0 p_4.$$

Hieraus folgt: Die Abbildung von U auf U' ist *eineindeutig*. Denn sind s und t zwei verschiedene Punkte von U, so ist entweder $ps \neq pt$, also $r(s') \neq r(t')$ oder es ist $ps = pt \neq 0$, also nach **9**, da $s \neq t$ gilt, $\sphericalangle spt > 0$, woraus nach dem Hilfssatz $\varphi(s') \neq \varphi(t')$ folgt. Jedenfalls also folgt $s' \neq t'$ aus $s \neq t$. Ferner ist die Abbildung von U auf U' *stetig*. Dies ist evident im Punkte p. Ist $\{q_i\}$ eine Folge von Punkten von U mit $\lim q_i = q \neq p$, so gilt wegen der Stetigkeit der Metrik $\lim r(q_i) = r(q)$. Wir zeigen ferner: $\lim \varphi(q_i) = \varphi(q)$, abgesehen davon, daß $\varphi(q) = 0$ mit $\lim \varphi(q_i) = 2\pi$ verträglich ist. Setzen wir $\Delta\varphi_i = \sphericalangle q_i p q$, so genügt es nach dem Hilfssatz, zu zeigen: $\lim \Delta\varphi_i = 0$. Sei s ein fester Punkt zwischen p und q, und s_i der Punkt zwischen p und q_i, für den $ps_i = ps$, dann gilt wegen der Eindeutigkeit der Strecken zwischen den Punkten von U offenbar $\lim s_i = s$. Betten wir r, s_i, s in $S_{k'}$ bzw. $S_{k''}$ ein, wo $k' < K$ und $k'' > K$, und bezeichnen wir die Winkel der Bildtripel, deren Scheitel der Bildpunkt von p ist, mit $\Delta\varphi_i'$ bzw. $\Delta\varphi_i''$ so gilt nach **10** $\Delta\varphi_i' \leqq \Delta\varphi_i \leqq \Delta\varphi_i''$. Aus $\lim s_i = s$, folgt $\lim \Delta\varphi_i' = \lim \Delta\varphi_i'' = 0$, also $\lim \Delta\varphi_i = 0$. Da R kompakt ist, ist die Abbildung von U auf U' also *topologisch*.

Um noch die behauptete Eigenschaft von U' zu erweisen, bezeichnen wir, wenn p', a', b' in U' nicht kollinear sind, mit a und b die Urbildpunkte in U von a' und b'. Es ist $0 \neq \sphericalangle a' p' b' \neq \pi$, also ist $0 \neq \sphericalangle apb \neq \pi$ und p liegt somit nicht zwischen a und b. Die Strecke S zwischen a und b hat somit von p einen Abstand $\lambda > 0$. Das Bild S' von S in U' hat den Abstand λ von p. Jede Strecke zwischen p' und einem Punkt von S' liegt in U'. Zum Beweis der behaupteten Eigenschaft von U' haben wir also nur zu zeigen, daß S' im Winkelraum zwischen den Strecken $p'a'$ und $p'b'$ verläuft. Mit diesen beiden Strecken hat S', da je zwei Punkte von U durch genau eine Strecke verbunden sind, nur a' bzw. b' gemein. Läge S' nicht im Winkelraum zwischen den Strecken $p'a'$ und $p'b'$, so läge S' also ganz im äußeren Winkelraum. Dann enthielte S' aber einen Punkt c' (mit einem Urbild c in U), so daß p' zwischen a' und c' läge. Dann läge aber p zwischen a und c, was, da a, c, b linear und a, p, b nicht linear sind, der Eigenschaft 1 der Regularität von U widerspräche.

Wir bezeichnen mit Menger[11]) einen Punkt q des Raumes R als *Schalenpunkt* bezüglich p, wenn kein Punkt r von R existiert,

[11]) Mathem. Annalen *100*, S. 102.

so daß q zwischen p und r liegt, und zeigen, daß die reguläre Umgebung U folgende Eigenschaft besitzt:

5) *Sind a und b Punkte von U, von denen mindestens einer (etwa b) kein Schalenpunkt bezüglich p ist, so ist kein Punkt der Strecke zwischen a und b Schalenpunkt bezüglich p.* Ist p, a, b linear, so ist 5) evident. Sind p, a, b nicht linear, so existiert ein Punkt b^* von U, so daß b zwischen p und b^* liegt. Ist c irgend ein Punkt zwischen a und b, so existiert auf der Strecke zwischen a und b^* genau ein Punkt c^*, so daß $\sphericalangle c^*pa = \sphericalangle cpa$ gilt. Wegen Eigenschaft 1) der regulären Umgebung U ist $c^* \neq c$. Läge c^* zwischen p und c, so müßten wegen Eigenschaften 3) und 4) von U die Strecken b^*c^* und bc sich schneiden, also hätten die Strecken ab und ab^* einen Punkt $\neq a$ gemein, was wegen Eigenschaft 1) von U unmöglich ist. Also liegt c zwischen p und c^* und ist daher kein Schalenpunkt w. z. b. w.

Daß jede Umgebung eines Punktes p, deren Radius hinlänglich klein ist, die Eigenschaften 1)—5) besitzt, hat sich aus der bloßen Annahme ergeben, daß R in p eine Flächenkrümmung besitzt. Herr Menger hat bemerkt, daß sich alle Schlüsse sogar schon aus folgender schwächeren Annahme ziehen lassen; es existiert eine Zahl $k(p)$ derart, daß jedes Quadrupel von U, *das ein lineares Tripel enthält,* in S_k einbettbar ist, wo $|k - k(p)|$ beliebig klein ist, wenn das Quadrupel hinreichend nahe an p liegt. (Ein solches Quadrupel ist nach **6** (S. 28), wenn es nicht linear ist, in höchstens ein S_k einbettbar.) Die Frage von Herrn Menger, ob auch die Sätze des folgenden Kapitels giltig bleiben, wenn man in jedem Punkt die Existenz einer Flächenkrümmung in diesem schwächeren Sinne annimmt, bleibt unentschieden.

III. Einführung des Bogenelementes in Räumen, die in jedem Punkt eine Flächenkrümmung besitzen.

Wir bezeichnen als *Gaußsche Fläche* einen kompakten konvexen metrischen Raum, zu dem eine Zahl $\rho_0 > 0$ existiert derart, daß jede Umgebung jedes Punktes, deren Radius $< \rho_0$ ist, die Eigenschaften 1)—5) besitzt, und in dem eine stetige Funktion $\varkappa(p)$, genannt die *Gaußsche Krümmung*, definiert ist, derart, daß, wenn U eine Umgebung eines Punktes mit einem Radius $< \rho_0$ ist und die stetig differenzierbaren Funktionen $r(t)$, $\varphi(t)$ $(t_1 \leq t \leq t_2)$ eine Kurve in U darstellen, die Länge dieser Kurve gegeben ist durch

$$\int_{t_1}^{t_2} \sqrt{dr^2 + G^2(r, \varphi)\, d^2\varphi},$$

wo $G(r, \varphi)$ die den Bedingungen $G(0, \varphi) = 0$ und $\partial G(0, \varphi)/\partial r = 1$ genügende Lösung der Differentialgleichung $\frac{\partial^2 G}{\partial r^2}(r, \varphi) + \varkappa(r, \varphi)\, G(r, \varphi) = 0$ ist. In diesem Kapitel beweisen wir das Theorem: *Jeder kompakte konvexe, metrische Raum, der in jedem Punkt eine Flächenkrümmung*

3

besitzt, ist eine Gaußsche Fläche, für welche in jedem Punkt die Gaußsche Krümmung mit der Flächenkrümmung identisch ist.

Wir setzen also voraus, R sei ein kompakter konvexer Raum, der in jedem Punkt p eine Flächenkrümmung $k(p)$ besitzt und beweisen vor allem[12]):

$k(p)$ ist eine stetige Funktion von p. Sei $\varepsilon > 0$ gegeben. Es existiert ein $\delta > 0$ derart, daß jedes nicht lineare Quadrupel, das ein lineares Tripel enthält und von p einen Abstand $< \delta$ hat, in S_k mit $|k - k(p)| < \varepsilon/2$ einbettbar ist. Ist q ein Punkt mit $qp < \delta/2$, so existiert wegen der Eigenschaften 1), 2), 3) der regulären Umgebungen in einem Abstand $< \delta/2$ von q nicht lineares Quadrupel, das ein lineares Tripel enthält und in S_k mit $|k - k(q)| < \varepsilon/2$ einbettbar ist. Mit Rücksicht auf **6** folgt aus $qp < \delta/2$, daß $|k(p) - k(q)| < \varepsilon$. — Wegen der Kompaktheit von R, existiert also eine endliche obere Schranke K der Zahlen $|k(p)|$ für alle Punkte p von R.

11. *Ist in einer regulären Umgebung U von p eine Folge $\{a_i, b_i\}$ von Punktpaaren gegeben, $a_i \neq b_i$, $pa_i = pb^i = r_i \neq 0$ für jedes i und setzen wir $\psi_i = \sphericalangle a_i p b_i$, so hat $a_i b_i / r_i \psi_i$ eine endliche obere und eine positive untere Schranke.* Bilden wir nämlich zwei Tripel p', a_i' b_i' und p'', a_i'', b_i'' auf S_{-K} bzw. S_K derart, daß $p'a_i' = p'b_i' = p''a_i'' = p''b_i'' = r_i$ und $\sphericalangle a_i' p' b_i' = \sphericalangle a_i'' p'' b_i'' = \psi_i$, so gilt nach **10**, daß $a_i' b_i' \geqq a_i b_i \geqq a_i'' b_i''$. Da sowohl $a_i' b_i' / r_i \psi_i$ als auch $a_i'' b_i'' / r_i \psi_i$ eine endliche obere und eine positive untere Schranke besitzt, folgt **11.**

Wir werden ferner die zwei folgenden Bemerkungen über S_k verwenden, wobei wir den Flächeninhalt des Dreiecks a, b, c in S_k mit $F(a, b, c)$ bezeichnen.

12. *Liegt a_i, b_i, c_i in S_{k_i} $(i = 1, 2, \ldots$ ad inf.), und konvergiert der Durchmesser von a_i, b_i, c_i gegen 0, so ist, wenn $|k_i|$ beschränkt ist, auch $\dfrac{F(a_i, b_i, c_i)}{a_i b_i \cdot b_i c_i}$ beschränkt.*

12'. *Sind a_i, b_i, c_i in S_{k_i} und a_i', b_i', c_i' in $S_{k_i'}$ für jedes natürliche i kongruent, konvergiert der Durchmesser von a_i, b_i, c_i gegen 0, ist $|k_i|$ beschränkt und $\lim |k_i - k_i'| = 0$, dann ist $\lim F(a_i, b_i, c_i) / F(a_i', b_i', c_i') = 1$.*

Ist nun U eine reguläre Umgebung von p, so nennen wir *Polartrapez* ein nicht lineares echtes geordnetes Quadrupel $T = \{a, b, c, d\}$ derart, daß 1) a zwischen p und d liegt und b zwischen p und c liegt und 2) $ap = bp$, $cp = dp$ gilt. Wir bezeichnen diese Zahlen stets mit r bzw. $r + \rho$. Offenbar ist $r > 0$ und wegen 1) gilt $ad = bc = \rho > 0$. Wir nennen $\sphericalangle apb$ den *Polwinkel* von T und bezeichnen ihn stets mit ψ. Als Quadrupel einer regulären Umgebung kann T für mindestens ein k in S_k, nach Hilfssatz **8** für höchstens

---

[12]) Der Beweis ist analog einem Beweis von F. Alt dafür, daß die Mengersche Kurvenkrümmung eine stetige Funktion ist.

zwei k' und k'' in $S_{k'}$ und $S_{k''}$ eingebettet werden. Wir bezeichnen als die T zugeordnete Krümmung k diejenige[13]), von den Zahlen k' und k'', die an $k(a)$ näher liegt, wo $k(a)$ die Flächenkrümmung von R im Punkt a von T bezeichnet. Als *Bildquadrupel* von T werden wir stets ein mit T konkruentes Quadrupel $T' = \{a', b', c', d'\}$ in diesem S_k bezeichnen. Jedes Bildquadrupel eines Polartrapezes ist echt, nicht linear und $a'd' = b'c'$. Die Winkel

$$\alpha = \sphericalangle b'a'd' - 90^0,\ \beta = \sphericalangle a'b'c' - 90^0,\ \gamma = 90^0 - \sphericalangle b'c'd',\ \delta = 90^0 - \sphericalangle c'd'a'$$

nennen wir die vier *T zugeordneten Winkel.* Von besonderer Wichtigkeit wird der Winkel $\eta = |\alpha - \beta|$ sein. Man beachte, daß α, β, γ, δ, η in S_k erklärt sind, dagegen ψ in R. Die Abstände ab und und cd werden wir stets mit σ bzw. τ bezeichnen.

13. *Ist T_i $(i = 1, 2, \ldots, ad\ inf.)$ eine Folge von Polartrapezen, für welche $\lim \rho_i = \lim r_i = 0$ so gilt*

1) $$\lim \frac{\xi_i - \psi_i/2}{\psi_i} = 0 \ \textit{für}\ \xi_i = \alpha_i,\ \beta_i,\ \gamma_i,\ \delta_i.$$

2) $$\eta_i = \lambda_i\,(r_i + \rho_i)^2 \psi_i,\ \textit{wo}\ \lim \lambda_i = 0.$$

Wegen **10** gibt es für jedes i offenbar eine Zahl k_i^*, so daß $\lim k_i^* = k(p)$ und $\sphericalangle a_i^* p^* b_i^* = \psi_i$ gilt, wobei a_i^*, b_i^*, p^* ein zu a_i, b_i, p kongruentes Tripel in $S_{k_i^*}$ bedeutet. Es sei

a_i, b_i, d_i, p kongruent mit a_i^1, b_i^1, d_i^1, p^1 in S_k mit $k = k_i^1$;

$a_i, b_i, c_i\, p$ kongruent mit a_i^2, b_i^2, c_i^2, p^2 in S_k mit $k = k_i^2$.

Nach trigonometrischen Formeln haben wir, wenn wir $\sphericalangle a_i^l p_i^l b_i^l = \psi_i^l\ (l = 1, 2)$ setzen und mit a_i', b_i', c_i', d_i' das Bildquadrupel T_i' von T_i bezeichnen:

$$2\sphericalangle b_i^1 a_i^1 d_i^1 = \pi + \psi_i^1 - k_i^1 F(p^1, a_i^1, b_i^1),$$

$$2\sphericalangle a_i^2 b_i^2 c_i^2 = \pi + \psi_i^2 - k_i^2 F(p^2, a_i^2, b_i^2),$$

$$|\sphericalangle b_i' a_i' d_i' - \sphericalangle b_i^1 a_i^1 d_i^1| \leq |k_i F(a_i' b_i' d_i') - k_i^1 F(a_i^1 b_i^1 d_i^1)|,$$

$$|\sphericalangle a_i' b_i' c_i' - \sphericalangle a_i^2 b_i^2 c_i^2| \leq |k_i F(a_i', b_i', c_i') - k_i^2 F(a_i^2, b_i^2, c_i^2)|,$$

$$|\psi_i - \psi_i^l| \leq |k_i^* F(p^*, a_i^*, b_i^*) - k_i^l F(p^l, a_i^l, b_i^l)|;\ (l = 1, 2).$$

Aus **11** folgt, daß $a_i b_i / r_i \psi_i$ beschränkt ist. Wegen **12** sind die Ausdrücke $F(p^l, a_i^l, b_i^l)/r_i^2 \psi_i$, $F(a_i', b_i', d_i')/r_i \rho_i \psi_i$ und $F(a_i', b_i', c_i')$, $r_i \rho_i \psi_i$ beschränkt. Unter Berücksichtigung von **12'** folgen nun, da $\lim k_i = \lim k_i^1 = \lim k_i^2 = \lim k_i^* = k(p)$ gilt, die Behauptungen 2) und 1) für $\xi_i = \alpha_i$, β_i. Für $\xi_i = \gamma_i$, δ_i beweist man 1 analog.

[13]) Bzw. irgend eine der Zahlen k', k'', falls $k(a)$ ihr arithmetisches Mittel ist.

3*

Wir wollen eine reguläre Umgebung U von p *trapezregulär* nennen, wenn für jedes Polartrapez T in U der Betrag jedes der vier zugeordneten Winkel, α, β, γ, δ zwischen dem $^3/_8$fachen und dem $^5/_8$fachen des Polwinkels ψ liegt. Aus Behauptung 1) von **13** ergibt sich dann unmittelbar: *Jede reguläre Umgebung von p mit hinlänglich kleinem Radius ist trapezregulär.*

14. *Ist U eine trapezreguläre Umgebung von p und T_i $(i=1, 2, \ldots$, ad inf.$)$ eine nicht gegen die Begrenzung von U sich häufende Folge von Polartrapezen für welche 1) $\rho_i < r_i$, 2) ρ_i/ψ_i eine positive untere und eine endliche obere Schranke hat, 3) $\lim \rho_i = \lim \psi_i = 0$ gilt, dann ist $\lim \eta_i/r_i^2\psi_i = 0$.*

Es genügt **14** für den Fall zu beweisen, daß die T_i gegen einen Punkt a von U konvergieren. Ist $a=p$, so folgt **14** aus Behauptung 2) von **13**. Sei also $a \neq p$. Für jedes natürliche i betrachten wir auf der Strecke zwischen p und dem Punkt a_i von T_i endlich viele Punkte $a_{i1}, \ldots, a_{in(i)-1}, a_{in(i)} = a_i$, so daß $pa_{i1} \leqq \rho_i$ und $a_{ij}a_{ij+1} = \rho_i$ $(j=1, \ldots, n(i)-1)$. Wir setzen $r_{ij} = pa_{ij}$, nennen b_{ij} den Punkt zwischen p und b_i für den $pb_{ij} = pa_{ij}$, setzen $c_{ij} = b_{ij+1}$, $d_{ij} = a_{ij+1}$ $(j=1, \ldots, n(i)-1)$, $c_{in(i)} = c_i$, $d_{in(i)} = d$ und nennen T_{ij} das Polartrapez $\{a_{ij}, b_{ij}, c_{ij}, d_{ij}\}$ $(j=1, \ldots, n(i))$. Es sei nun $j(1), j(2), \ldots, j(i), \ldots$ irgend eine Folge natürlicher Zahlen $1 \leqq j(i) \leqq n(i)$. Wir betrachten die Folge $T_{ij(i)}$, $(i=1, 2, \ldots$, ad inf.$)$. Das Bildquadrupel $\mathrm{T}'_{ij(i)} = \{a'_{ij(i)}, \ldots\}$ liegt für $k = k_{ij(i)}$ in S_k. Dann gilt

1') $a'_{ij(i)} d'_{ij(i)} = b'_{ij(i)} c'_{ij(i)} = \rho_i$ für jedes natürliche i,

2') $|\xi' - \psi_i/2| \leqq \psi_i/8$, wenn ξ' irgend ein $T_{ij(i)}$ zugeordneter Winkel ist, d. h. für die Winkel $\sphericalangle b'_{ij(i)}, a'_{ij(i)} d'_{ij(i)} - 90^0$ usw. von S_k, da $T_{ij(i)}$ in einer trapezregulären Umgebung von p liegt.

Nun existiert für jedes i ein $\lambda'_i > 0$, so daß

$$(+) \quad \lim \lambda'_i = 0, \quad \eta_{ij} \leqq \lambda'_i r_{ij}^2 \psi_i \quad \text{für } 2 \leqq j \leqq 1/\sqrt{\rho_i} \text{ und jedes } i.$$

Denn für $2 \leqq j(i) \leqq 1/\sqrt{\rho_i}$ erfüllt die Folge $T_{ij(i)}$ offenbar die Voraussetzungen von **13**, so daß Behauptung 2) von **13** gilt. Wir behaupten weiter, daß für jedes i ein $\mu_i > 0$ und ein $\lambda''_i > 0$ existiert, so daß

$$(++) \quad \lim \mu_i = \lim \lambda''_i = 0, \quad \eta_{ij+1} - \eta_{ij} \leqq (\mu_i \eta_{ij} + \lambda''_i r_{ij} \psi_i)\rho_i \quad \text{für } j \geqq 1/\sqrt{\rho_{ij}}$$

und jedes i. Um dieses zu beweisen, setzen wir, wenn $j(i)$ eine Zahlenfolge mit $j(i) \geqq 1/\sqrt{\rho_i}$ ist, $a_{ij(i)} b_{ij(i)} = \sigma_{ij(i)}$, $c_{ij(i)} d_{ij(i)} = \tau_{ij(i)}$ und haben dann mit Rücksicht auf **11**:

3') $\sigma_{ij(i)}/\psi_i$ und $\tau_{ij(i)}/\psi_i$ sind beschränkt; $\tau_{ij(i)}/\sigma_{ij(i)}$ hat eine positive untere und eine endliche obere Schranke und $\lim \rho_i \psi_i/\sigma_{ij(i)} = 0$.

Nach einem elementaren Satz über Quadrupelfolgen von S_k, (dessen Beweis sich S. 40 findet) ergibt sich aus 1'), 2'), 3') und 2) für jedes i die Existenz eines $\varkappa_i$, so daß

$\lim \varkappa_i = 0$; $|\delta_{ij} - \gamma_{ij}| - \eta_{ij} \leqq \varkappa_i \eta_{ij} \rho_i$ für $j \geqq 1/\sqrt{\rho_i}$ und jedes i.

Hieraus folgt (+ +) offenbar mühelos, sobald man weiß, daß

$$(*) \qquad |\delta_{ij} - \alpha_{ij+1}| + |\gamma_{ij} - \beta_{ij+1}| < \nu_{ij} r_{ij} \psi_i \rho_i,$$

wo ν_{ij} mit wachsenden i gleichmäßig gegen 0 konvergiert. Um den ersten Summanden von (*) abzuschätzen, betrachten wir das (ein lineares Tripel enthaltende Quadrupel c_{ij}, a_{ij}, a_{ij+1}, a_{ij+2}, das nach **6** für genau ein (von i und j abhängiges) k^* mit einem Quadrupel c^*, a_0^*, a_1^*, a_2^* von S_{k^*} kongruent ist. Wegen der Linearität von a_0^*, a_1^*, a_2^* gilt, wenn wir $\delta^* = 90^0 - \sphericalangle c^* a_1^* a_0^*$, $\alpha^* = \sphericalangle c^* a_1^*, a_2^* - 90^0$ setzen, $a^* = \delta^*$ und

$$|\alpha^* - \alpha_{ij+1}| \leqq |k_{ij+1} F(a'_{ij+1}, b'_{ij+1}, d'_{ij+1}) - k^* F(a_1^*, a_2^*, c^*)|$$

$$|\delta^* - \delta_{ij}| \leqq |k_{ij} F(a'_{ij}, c'_{ij}, d'_{ij}) - k^* F(a_0^*, a_1^*, c^*)|.$$

Da nach **11** $a_{ij+1} c_{ij} / r_{ij} \psi_i$ beschränkt ist, so folgt aus **12**, daß $F(a'_{ij}, d'_{ij}, c'_{ij})/r_{ij}\psi_i\rho_i$ und $F(a'_{ij+'}, b'_{ij+'}, d'_{ij+'})/r_{ij}\psi_i\rho_i$ beschränkt sind. Da $k_{ij} - k^*$ und $k_{ij+1} - k^*$ mit wachsendem i gleichmäßig in j gegen 0 konvergieren, folgen wegen **12'** Abschätzungen für $|\alpha^* - - \alpha_{ij+1}|$ und $|\delta^* - \delta_{ij}|$, welche wegen $\alpha^* = \delta^*$ die gewünschte Abschätzung des ersten Summanden von (*) ergeben. Die des zweiten ergibt sich analog, womit (+ +) bewiesen ist.

Wir nennen nun λ_i die größere der Zahlen λ'_i und λ''_i und zeigen zunächst: Ist $y_i(r)$ die der Bedingung $y_i(0) = 0$ genügende Lösung der Differentialgleichung $y'_i(r) = \mu_i y_i(r) + 2\lambda_i \psi_i r$ $(i = 1, 2, \ldots,$ ad inf.), so gilt $\eta_{ij} \leqq y_i(r_{ij})$ für jedes i und j. Da die der Bedingung $z_i(0) = 0$ genügende Lösung der Differentialgleichung $z'_i(r) = 2\lambda_i \psi_i r$ eine Minorante von $y_i(r)$ ist und $z_i(r) = \lambda_i \psi_i r^2 \geqq \eta_{ij}$ für $2 \leqq j \leqq 1/\sqrt{\rho_i}$, so gilt erst recht $y_i(r_{ij}) \geqq \eta_{ij}$. Für $j > 1/\sqrt{\rho_i}$ sieht man nach dem Mittelwertsatz, daß $y_i(r_{ij+1}) - y_i(r_{ij}) = y'(r^*_{ij})\rho_i$, wo $r_{ij} \leqq r^*_{ij} \leqq r_{ij+1}$. Aus der Differentialgleichung folgt unmittelbar, daß $y_i(r)$ und $y'_i(r)$ monoton wachsende Funktionen sind, also ist $y'_i(r^*_{ij})\rho_i \geqq y'_i(r_{ij})\rho_i = (\mu_i y_i(r_{ij}) + 2\lambda_i \psi_i r_{ij})\rho_i \geqq \eta_{ij+1} - \eta_{ij}$, woraus durch Induktion nach j die Behauptung $y_i(r_{ij}) \geqq \eta_{ij}$ für alle i und j folgt.

Um **14** zu beweisen, genügt es offenbar zu zeigen, daß $y_i(r)/r^2\psi_i$ mit wachsendem i gleichmäßig in r gegen 0 konvergiert. Es gilt:

$$y_i(r) = e^{\mu_i r} \int_0^r 2\lambda_i r \psi_i e^{-\mu_i r}\, dr < y_i^*(r) = e^{\mu_i r} \int_0^r 2\lambda_i r \psi_i\, dr = \lambda_i r^2 e^{\mu_i r} \psi_i.$$

Da $\lim \mu_i = \lim \lambda_i = 0$, so ist **14** bewiesen.

Ist U eine reguläre, also mit Polarkoordinaten r, φ beschreibbare Umgebung von p, so bezeichnen wir den Abstand der Punkte $a = (r, \varphi)$ und $b = (r, \varphi + \Delta\varphi)$ mit $\sigma(r, \varphi, \Delta\varphi)$ und setzen:

$$\sigma(r+\Delta r, \varphi, \Delta\varphi)-\sigma(r, \varphi, \Delta\varphi)=\Delta\sigma(r, \varphi, \Delta\varphi, \Delta r),$$
$$\Delta\sigma(r+\Delta r, \varphi, \Delta\varphi, \Delta r)-\Delta\sigma(r, \varphi, \Delta\varphi, \Delta r)=\Delta^2\sigma(r, \varphi, \Delta\varphi, \Delta r).$$

15. *Sei U eine trapezreguläre Umgebung von p; die Punktefolge $a_i=\{r_i, \varphi_i\}$ $(i=1, 2, \ldots,$ ad inf.) konvergiere gegen einen Punkt $a=(r, \varphi)\neq p$. Es seien $\Delta r_i>0$ und $\Delta\varphi_i$ zwei Zahlenfolgen, so daß $\lim \Delta r_i=0$ und wenn wir $b_i=(r_i, \varphi_i+\Delta\varphi_i)$ und $\psi_i=\sphericalangle a_i p b_i$ setzen, folgende Bedingungen erfüllt sind: $0\neq\psi_i\neq\pi$, $\lim \psi_i=0$ und $\Delta r_i/\psi_i$ hat eine positive untere und eine endliche obere Schranke. Dann gilt:*

$$\lim \frac{\Delta^2\sigma(r_i, \varphi_i, \Delta\varphi_i, \Delta r_i)}{\sigma(r_i, \varphi_i, \Delta\varphi_i)\,\Delta r_i^2}=-k(r, \varphi)$$

wobei wir $k(r, \varphi)=k(a)$ schreiben.

Wir setzen $c_i=b_i^*=(r_i+\Delta r_i, \varphi_i+\Delta\varphi_i)$, $d_i=a_i^*=(r_i+\Delta r_i, \varphi_i)$, $c_i^*=(r_i+2\Delta r_i, \varphi_i+\Delta\varphi_i)$, $d_i^*=(r_i+2\Delta r_i, \varphi_i)$ und $T_i=\{a_i, b_i, c_i, d_i\}$, $T_i^*=\{a_i^*, \ldots\}$. Die Bildquadrupel $T_i'=\{a_i', \ldots\}$ von T_i genügen, wie man aus **13** und **14** entnimmt, folgenden Bedingungen:

1) $a_i'd_i'=b_i'c_i'=\Delta r_i$ für jedes i

2) $|\xi'-\psi_i/2|\leqq\varphi_i/8$, wenn ξ' ein T_i ungeordneter Winkel ist.

3) σ_i/ψ_i und τ_i/ψ_i haben eine positive untere und eine endliche obere Schranke, da $a\neq p$.

4) $\lim \eta_i/\psi_i=0$ nach **14**.

5) $\lim k_i=k(a)$.

Aus einem elementaren Satz über Quadrupelfolgen von S_k, dessen Beweis sich S. 41 findet, folgt aus 1)—5):

6) $$\lim \frac{\gamma_i-\beta_i}{\sigma(r_i, \varphi_i, \Delta\varphi_i)\,.\,\Delta r_i}=\lim \frac{\delta_i-\alpha_i}{\sigma(r_i, \varphi_i, \Delta\varphi_i)\,\Delta r_i}=-\frac{1}{2}k(a);$$

und

$$\Delta\sigma(r_i, \varphi_i, \Delta\varphi_i, \Delta r_i)=\tau_i-\sigma_i=(\beta_i+\delta_i)\Delta r_i+\varepsilon_i\sigma(r_i, \varphi_i, \Delta\varphi_i)\Delta r_i^2,$$
wo $\lim \varepsilon_i=0$.

Ebenso sieht man, da die Bildquadrupel von T_i^* analoge Voraussetzungen 1*)—5*) erfüllen, daß

$$\Delta\sigma(r_i+\Delta r_i, \varphi_i, \Delta\varphi_i, \Delta r_i)=\tau_i^*-\sigma_i^*=(\beta_i^*+\delta_i^*)\Delta r_i+\varepsilon_i^*\sigma(r_i+\Delta r_i, \varphi_i, \Delta\varphi_i)\Delta r_i^2,$$
wo $\lim \varepsilon_i^*=0$.

Da $\alpha_i, \beta_i, \gamma_i, \delta_i$ gegen 0 konvergieren und $\Delta r_i/\sigma_i$ eine positive untere und eine endliche obere Schranke hat, so ist $\lim \tau_i/\sigma_i=1$.

Durch Subtraktion der letzten Formeln ergibt sich dann:

$$\Delta^2\sigma(r_i, \varphi_i, \Delta\varphi_i, \Delta r_i)=[(\beta_i^*-\beta_i)+(\delta_i^*-\delta_i)]\Delta r_i+\varepsilon_i'\sigma(r_i, \varphi_i, \Delta\varphi_i)\Delta r_i^2;$$
$\lim \varepsilon_i'=0$.

Daher gilt:

$$\lim \frac{\Delta^2\sigma(r_i, \varphi_i, \Delta\varphi_i, \Delta r_i)}{\sigma(r_i, \varphi_i, \Delta\varphi_i)\, \Delta r_i^2} = \lim \frac{(\beta_i^* - \beta_i) + (\delta_i^* - \delta_i)}{\sigma(r_i, \varphi_i, \Delta\varphi_i)\, \Delta r_i}.$$

Wegen 6) und 6*) folgt die Behauptung **15** sobald wir wissen, daß

$$\lim\, [(\alpha_i^* - \delta_i)/\sigma(r_i, \varphi_i, \Delta\varphi_i)\Delta r_i + (\gamma_i - \beta_i^*)/\sigma(r_i, \varphi_i, \Delta\varphi_i)\Delta r_i] = 0,$$

was man genau so, wie Formel (*) von S. 37 beweist,

Wir setzen nun: $\sigma(r, \varphi, \Delta\varphi)/\psi = G(r, \varphi, \Delta\varphi)$,

$G(r+\Delta r, \varphi, \Delta\varphi) - G(r, \varphi, \Delta\varphi) = \Delta G(r, \varphi, \Delta\varphi, \Delta r)$,

$\Delta G(r+\Delta r, \varphi, \Delta\varphi, \Delta r) - \Delta G(r, \varphi, \Delta\varphi, \Delta r) = \Delta^2 G(r, \varphi, \Delta\varphi, \Delta r)$.

Dann lautet die Behauptung von **15**

$$\lim \frac{\Delta^2 G(r_i, \varphi_i, \Delta\varphi_i, \Delta r_i)}{G(r_i, \varphi_i, \Delta\varphi_i)\, \Delta r_i^2} = -k(r, \varphi),$$

wobei $G(0, \varphi, \Delta\varphi) = 0$ und für $r \to 0$ gilt: $\partial G(0, \varphi, \Delta\varphi)/\partial r = \lim G(r, \varphi, \Delta\varphi)/r = \lim \sigma(r, \varphi, \Delta\varphi)/r\psi = 2/\psi \,.\, \sin \psi/2$. Für $\psi \to 0$ konvergiert dies gegen 1. Aus diesen Tatsachen folgt leicht:

16. *Ist $a_i = (r_i, \varphi_i)$ $(i = 1, 2, \ldots$ ad inf.) eine gegen den Punkt $a_0 = (r_0, \varphi_0)$ konvergente Folge von Punkten der trapezregulären Umgebung U von p und ist $\Delta\varphi_i$ eine gegen 0, 2π oder -2π konvergente Folge von Zahlen $\neq 0$, $\neq 2\pi$, $\neq -2\pi$, so gilt lim $G(r_i, \varphi_i, \Delta\varphi_i) = G(r_0, \varphi_0)$, wo $G(r, \varphi)$ die den Bedingungen $G(0, \varphi) = 0$, $\partial G(0, \varphi)/\partial r = 1$ genügende Lösung der Differentialgleichung*

$$\frac{\partial^2 G}{\partial r^2}(r, \varphi) + k(r, \varphi)\, G(r, \varphi) = 0 \ \textit{ist.}$$

Daß R eine Gaußsche Fläche ist, für welche $\varkappa(p) = k(p)$ gilt, ist also offenbar erwiesen, sobald noch gezeigt ist:

17. *Sind $a_i = (r_i, \varphi_i)$ und $c_i = (r_i + \Delta r_i, \varphi_i + \Delta\varphi_i)$ zwei gegen einen Punkt $a_0 = (r_0, \varphi_0)$ konvergente Punktefolgen einer hinlänglich kleinen Umgebung U, wo $r_i > 0$, $\Delta r_i \geqq 0$, $\Delta r_i^2 + \Delta\varphi_i^2 \neq 0$, lim $\Delta\varphi_i = 0$, dann gilt lim $a_i c_i / \sqrt{\Delta r_i^2 + G^2(r_i, \varphi_i)\Delta\varphi_i^2} = 1$.*

Zum Beweise wählen wir U so klein, daß U regulär ist und, wenn K eine Majorante aller Zahlen $k(p)$ auf R ist, einen Radius $< \pi/\sqrt{|K|}$ besitzt, da dann aus Sturmschen Sätzen folgt, daß für $r \neq 0$ stets $G(r, \varphi) > 0$ gilt. Falls $\Delta\varphi_i = 0$, so ist **17** evident. Sei also $\Delta\varphi_i \neq 0$. Dann gilt zunächst lim $G(r_i, \varphi_i, \Delta\varphi_i)/G(r_i, \varphi_i) = 1$; denn ist $a_0 \neq p$, so folgt dies wegen der gleichmäßigen Stetigkeit von $G(r, \varphi)$ daraus, daß lim $G(r_i, \varphi_i, \Delta\varphi_i) = G(r_0, \varphi_0) \neq 0$; wenn $a_0 = p$ daraus, daß lim $G(r_i, \varphi_i)/r_i = 1$ und wegen **10** lim $G(r_i, \varphi_i, \Delta\varphi_i)/r_i = 1$ gilt. Auf Grund der damit bewiesenen Beziehung genügte also für den Beweis von **17** zu zeigen, daß

$\lim a_i c_i / \sqrt{\Delta r_i^2 + G^2(r_i, \varphi_i, \Delta\varphi_i)\Delta\varphi_i^2} = 1$ ist. Für $\Delta r_i = 0$ ist dies evident. Wenn $\Delta r_i \neq 0$, so betrachten wir die Punkte $b_i = (r_i, \varphi_i + \Delta\varphi_i)$ und $d_i = (r_i + \Delta r_i, \varphi_i)$ und die Polartrapeze $T_i = \{a_i, b_i, c_i, d_i\}$. Da ihre zugeordneten Winkel gegen 0 konvergieren, also $\lim \sphericalangle a'_i, b'_i c'_i = 90^0$, und $a_i c_i / \sqrt{\Delta r^2 + G(r_i, \varphi_i, \Delta\varphi_i)\, \Delta\varphi_i^2} = a'_i c'_i / \sqrt{(a'_i, b'_i)^2 + (b'_i c'_i)^2}$ ist, so ergibt sich **17** nach dem Pythagoräischen Satz.

Wir beweisen nun als Anhang zu diesem Kapitel zwei in ihm verwendete elementare Hilfssätze über Quadrupelfolgen.

Hilfssatz 1. Voraussetzungen. Für jedes natürliche i seien gegeben:
erstens eine Zahl k_i, *so daß* $|k_i|$ *beschränkt ist;*
zweitens in S_{k_i} *ein echtes Quadrupel* $\{a_i, b_i, c_i, d_i\}$, *für welche* $a_i d_i = b_i c_i = \rho_i$ und für welches wir setzen: $a_i b_i = \sigma_i$; $c_i d_i = \tau_i$; $\sphericalangle a_i b_i c_i = 90^0 + \alpha_i$; $\sphericalangle a_i b_i c_i = 90^0 + \beta_i$; $\sphericalangle b_i c_i d_i = 90^0 - \gamma_i$; $\sphericalangle c_i d_i a_i = 90^0 - \delta_i$; $|\alpha_i - \beta_i| = \eta_i$;
drittens ein $\psi_i > 0$, *so daß* $\lim \psi_i = 0$ *und folgendes gilt:*

1) ρ_i / ψ_i *hat eine positive untere und eine endliche obere Schranke,*

2) $|\xi_i - \psi_i / 2| \leqq \psi_i / 8$ *für* $\xi_i = \alpha_i, \beta_i, \gamma_i, \delta_i$,

3) σ_i / ψ_i *und* τ_i / ψ_i *sind beschränkt;* τ_i / σ_i *hat eine positive untere und eine endliche obere Schranke.*

4) $\lim \rho_i \psi_i / \sigma_i = 0$.

Behauptung: I) $|\delta_i - \gamma_i| - \eta_i \leqq \varepsilon_i \eta_i \rho_i$, *wo* $\lim \varepsilon_i = 0$.

II) Wenn σ_i / ψ_i *eine positive untere Schranke hat, so gilt außerdem* $\gamma_i = \beta_i - \lambda_i k_i \sigma_i \rho_i / 2 + \mu_i \eta_i \rho_i$ *und* $\delta_i = \alpha_i - \lambda'_i k_i \sigma_i \rho_i / 2 + \mu'_i \eta_i \rho_i$, *wo* $\lim \lambda_i = \lim \lambda'_i = 1$ *und* μ_i, μ'_i *beschränkt sind.*

Sei zum Beweis etwa $\alpha_i \geqq \beta_i$. Aus 1), 2), 3) folgt leicht, daß d_i und c_i für fast alle i auf derselben Seite von $a_i b_i$ liegen. Sei d'_i ein Punkt von S_{k_i}, der mit d_i auf derselben Seite von $a_i b_i$ liegt und so, daß $a_i d'_i = \rho_i$ und $\sphericalangle b_i a_i d'_i = 90^0 + \beta_i$ gilt. Dann haben wir $\sphericalangle d_i a_i d'_i = \eta_i$, also

5) $\sphericalangle b_i c_i d'_i = \sphericalangle a_i d'_i c_i = 90^0 - \beta_i - k_i F(a_i, b_i, c_i, d'_i) / 2$, wo $F(a_i, b_i, c_i, d'_i)$ den Flächeninhalt von (a_i, b_i, c_i, d'_i) bezeichnet, und

6) $\sphericalangle a_i d'_i d_i = 90^0 - \eta_i / 2 + k_i F(a_i, d_i, d'_i) = 90^0 - \nu_i \eta_i / 2$, wo $\lim \nu_i = 1$. Wegen $\lim F(a_i, b_i, c_i, d'_i) / \psi_i = 0$ und $|\beta_i - \psi_i / 2| \leqq \psi_i / 8$ folgt aus 5) und 6)

7) $\sphericalangle a_i d'_i c_i + \sphericalangle a_i d'_i d_i < 180^0$ für fast alle i, also

8) $\sphericalangle a_i d'_i c_i + \sphericalangle a_i d'_i d_i = \sphericalangle c_i d'_i d_i$.

Aus 7) folgt, daß d'_i nicht im Innern des Vierecks (a_i, b_i, c_i, d_i) liegt, so daß

9) $\sphericalangle b_i c_i d_i = \sphericalangle b_i c_i d'_i - \sphericalangle d'_i c_i d_i$.

Anwendung des Sinussatzes auf das Dreieck (d_i, c_i, d'_i) ergibt $\sin \sphericalangle d'_i c_i d_i = \sin \sphericalangle c_i d'_i d_i \sin (d_i d'_i \sqrt{k_i}) / \sin (c_i d_i \sqrt{k_i})$. Hieraus folgt wegen 2) und wegen der Beschränktheit von $(180^0 - \sphericalangle c_i d'_i d_i)'/ \psi_i > 0$

10) $\sphericalangle d'_i c_i d_i = \zeta'_i \psi_i \eta_i \rho_i / \tau_i = \zeta_i \psi_i \eta_i \rho_i / \sigma_i$, wo $\zeta_i \geqq 0$ und beschränkt ist.

Aus 5), 9) und 10) ergibt sich $90^0 - \gamma_i = \sphericalangle b_i c_i d_i = 90^0 - \beta_i + k_i F(a_i, b_i, c_i, d'_i)/2 - \zeta_i \eta_i \rho_i \psi_i / \sigma_i$, also

11) $\gamma_i = \beta_i - k_i F(a_i, b_i, c_i, d'_i)/2 + \zeta_i \psi_i \eta_i \rho_i / \sigma_i$.

Hieraus folgt wegen $\alpha_i + \beta_i - \gamma_i - \delta_i = k_i F(a_i, b_i, c_i, d_i)$

12) $\delta_i = \alpha_i + k_i F(a_i, b_i, c_i, d'_i)/2 - \zeta_i \psi_i \eta_i \rho_i / \sigma_i - k_i F(a_i, b_i, c_i, d_i)$.

Subtrahiert man 11) und 12), so ergibt sich

13) $\delta_i - \gamma_i = \eta_i + k_i [F(a_i, b_i, c_i, d'_i) - F(a_i, b_i, c_i, d_i)] - 2\zeta_i \psi_i \eta_i \rho_i / \sigma_i$.

Nun ist offenbar $|F(a_i, b_i, c_i, d'_i) - F(a_i, b_i, c_i, d_i)| \leqq F(d_i, c_i, d'_i) + F(d_i, a_i, d'_i)$ und $F(d_i, a_i, d'_i)/\eta_i \rho_i$ konvergiert gegen 0. Aus 10) folgt, daß auch $\lim F(d_i, c_i, d'_i)/\eta_i \rho_i = 0$. Aus 13) ergibt sich also

$$\delta_i - \gamma_i = \eta_i + \varepsilon'_i \eta_i \rho_i - 2\zeta_i \psi_i \eta_i \rho_i / \sigma_i, \text{ wo } \lim \varepsilon'_i = 0.$$

Nach 4) folgt hieraus

$$|\delta_i - \gamma_i| - \eta_i = \varepsilon_i \eta_i \rho_i - 2\zeta_i \psi_i \eta_i \rho_i / \sigma_i, \text{ wo } \lim \varepsilon_i = 0$$

und daraus wegen $\zeta_i \geqq 0$ die Behauptung I).

Hat σ_i / ψ_i eine positive untere Schranke, so gilt offenbar $\lim \tau_i / \sigma_i = 1$ und $\lim F(a_i, b_i, c_i, d'_i)/\sigma_i \rho_i = 1$. Aus 11) und 12) ergibt sich unmittelbar die Behauptung II).

Hilfssatz 2. Außer den Voraussetzungen von Hilfssatz 1 gelte 1′) $\lim k_i = k$; 2′) $\lim \eta_i / \psi_i = 0$; 3′) σ_i / ψ_i hat eine positive untere Schranke. Dann behaupten wir:

III) $\lim (\gamma_i - \beta_i)/\sigma_i \rho_i = \lim (\delta_i - \alpha_i)/\sigma_i \rho_i = -k/2$,

IV) $\tau_i - \sigma_i = \rho_i(\beta_i + \delta_i) + \varepsilon_i \sigma_i \rho_i^2$, wo $\lim \varepsilon_i = 0$.

Wegen 2′) und 3′) gilt $\lim \eta_i / \sigma_i = 0$. Hieraus und aus 1′) ergibt sich III) unmittelbar auf Grund der Behauptung II) von Hilfssatz 1. Nach dem Kosinussatz erhalten wir, wenn wir $k_i = 1/r_i^2$ setzen:

$$\cos a_i c_i / r_i = \cos \tau_i / r_i \cos \rho_i / r_i + \cos (90^0 - \delta_i) \sin \tau_i / r_i \sin \rho_i / r_i,$$
$$\cos a_i c_i / r_i = \cos \sigma_i / r_i \cos \rho_i / r_i + \cos (90^0 + \beta_i) \sin \sigma_i / r_i \sin \rho_i / r_i.$$

Hieraus ergibt sich

$$\cos \tau_i / r_i - \cos \sigma_i / r_i = -tg\, \rho_i / r_i [\sin \beta_i \sin \sigma_i / r_i + \sin \delta_i \sin \tau_i / r_i]$$

und daraus wegen $\lim \tau_i / \sigma_i = 1$

$$(\tau_i^2 - \sigma_i^2) k_i (1 + \mu_i \sigma_i^2)/2 = tg\, \rho_i / r_i [\sin \beta_i \sin \sigma_i / r_i + \sin \delta_i \sin \tau_i / r_i],$$

wobei μ_i beschränkt ist, also

$$(\tau_i^2 - \sigma_i^2)/2 = \rho_i(\beta_i \sigma_i + \delta_i \tau_i) + \nu_i \sigma_i^2 \rho_i^3, \text{ wo } \nu_i \text{ beschränkt ist.}$$

Hieraus ergibt sich

$$\tau_i = \rho_i \delta_i \pm 1/2 \,.\, \sqrt{4\rho_i^2 \delta_i + 4(\sigma_i^2 + 2\rho_i \beta_i \sigma_i + 2\nu_i \sigma_i^2 \rho_i^3)}.$$

Da $\tau_i > 0$ gilt das Zeichen $+$ und man erhält daher leicht:

$$\tau_i = \rho_i \delta_i + \sigma_i [1 + \rho_i \beta_i / \sigma_i + \rho_i^2 \delta_i^2 / 2\sigma_i^2 - \rho_i^3 \beta_i^2 / 2\sigma_i^2] + \nu_i' \sigma_i \rho_i^2,$$

wo $\lim \nu_i' = 0$, also

$$(+) \qquad \tau_i - \sigma_i = \rho_i (\delta_i + \beta_i) + \rho_i^2 (\delta_i^2 - \beta_i^2) / 2\sigma_i + \nu_i' \sigma_i \rho_i^2.$$

Aus der Behauptung III) folgt $\lim (\delta_i - \alpha_i) / \sigma_i = 0$.

Wegen 2') und 3') gilt $\lim (\alpha_i - \beta_i) / \psi_i = \lim (\alpha_i - \beta_i) / \sigma_i = 0$, also wegen $(+)$ die Behauptung IV).

IV. Die Existenz der Flächenkrümmung auf Gaußschen Flächen.

Sei F eine Gaußsche Fläche im Sinne der Definition von S. 33. Wir beweisen in diesem Kapitel: *In jedem Punkt p von F existiert eine Flächenkrümmung $k(p)$ und dieselbe ist in jedem Punkt mit der Gaußschen Krümmung $\varkappa(p)$ identisch.*

Sei U eine Umgebung von p in F, deren Radius ρ kleiner ist, als die auf S. 33 erwähnte Zahl ρ_0. Wegen Eigenschaft 4) von U können wir Polarkoordinaten mit dem Zentrum p einführen. U' sei die ebene Bildmenge von U. Sie ist Teil des Kreisinnern C mit dem Zentrum p' und dem Radius ρ. Wird jedem Punkt q' von U', wenn q das Urbild von q' bezeichnet, die Zahl $\varkappa(q)$ zugeordnet, so liegt eine gleichmäßig stetige Funktion in U' vor, welche wir zu einer gleichmäßig stetigen Funktion $\varkappa^*(q)$ auf ganz C erweitern können. Sei $G^*(r, \varphi)$ die den Bedingungen $G^*(0, \varphi) = 0$ und $\partial G^*(0, \varphi) / \partial r = 1$ genügende Lösung der Differentialgleichung $\partial^2 G^*(r, \varphi) / \partial r^2 + \varkappa^*(r, \varphi) G^*(r, \varphi) = 0$. Wir bezeichnen mit U^* den metrischen Raum bestehend aus den Punkten von C, wobei je zwei Punkten q_1^* und q_2^* von C als Abstand die untere Schranke aller Zahlen $\int \sqrt{dr^2 + G^{*2}(r, \varphi) d\varphi^2}$, gebildet für alle q_1^* und q_2^* verbindenden stetig differenzierbaren in C verlaufenden Bogen: $r = r(t)$, $\varphi = \varphi(t)$, zugeordnet wird. Jeder Punkt von U ist Punkt von U^*. Wir zeigen nun

18. *Es existiert ein $\rho^* > 0$ derart, daß der Abstand je zweier Punkte q_1 und q_2 von U, die von p einen Abstand $< \rho^*$ haben, identisch ist mit dem Abstand, den diese Punkte aufgefaßt als Punkte von U^* haben, und diese Zahl ρ^* kann gleichmäßig für alle Umgebungen mit Radien $< \rho_0$ in F gewählt werden.* Vor allem ist klar, daß jeder in U liegende Bogen die gleiche Länge hat, ob wir ihn als Bogen von U oder von U^* auffassen. Nach klassischen Sätzen existiert eine Zahl $\rho^* > 0$ (u. zw. gleichmäßig für die verschiedenen Umgebungen), so daß in der Menge $U^*(\rho^*)$ der Punkte von U^*, die von p einen Abstand $< \rho^*$ haben, zu je zwei Punkten a^* und b^* genau ein sie innerhalb $U^*(\rho^*)$ verbindender Bogen B^* existiert, dessen Länge ein relatives Minimum ist, d. h. nicht unterschritten wird von den Längen aller hinreichend benachbarten, a^* und b^* verbindenden Bogen. Die Länge dieses Bogens ist zugleich ein absolutes Minimum. Zum Beweise der Behauptung **18** genügt es, dieselbe für Punktepaare

q_1, q_2 zu beweisen, die nicht zur Schale von U bezüglich p gehören. Nach Eigenschaft 5) von U enthält die q_1 und q_2 verbindende Strecke S in U keine Schalenpunkte. S aufgefaßt als Teilbogen S^* von U^* liefert offenbar ein relatives Minimum, welches zugleich ein absolutes Minimum ist. Da S^* und S dieselbe Länge haben, so ist unsere Behauptung bewiesen.

Wir bezeichnen als *ausgezeichnet* eine Umgebung von U von p in F vom Radius ρ, wenn $\rho<\rho_0$, $\rho<\rho^*$ und $\varkappa(q)<\pi^2/16\rho^2$ für jeden Punkt q von U. Jede ausgezeichnete Umgebung U hat offenbar folgende Eigenschaften:

a) Ist k'' die obere Schranke von $\varkappa(q)$ in U und k eine beliebige Zahl $\leqq k''$, so ist jede Kugelumgebung in S_k, deren Radius $\leqq$ als der Radius von U, konvex.

b) Ist $\{a, b, c\}$ ein beliebiges Punktetripel in U, k eine beliebige Zahl $\leqq$ als die obere Schranke von $\varkappa(q)$ in U, dann gibt es in S_k ein zu $\{a, b, c\}$ kongruentes Tripel.

In Analogie zu den Hilfssätzen **2**, **3**, **4** über S_k gelten ferner folgende Sätze:

19. *Sind p, a, b drei nicht lineare Punkte einer ausgezeichneten Umgebung von p und ist c ein Punkt, im Innern des Winkelraumes zwischen den Strecken pa und pb, für den $cp=bp$, so ist $ac<ab$.* Es existiert ein Punkt x zwischen a und b, der wegen der Konvexität von U in U liegt, so daß p, c, x linear sind. Ist $x=c$, so ist $ac=ax<ab$. Liegt c zwischen p und x, dann sind, da p, a, b, x, sowie p, a, c, x nicht linear sind, wegen Eigenschaft 1) von U die Tripel a, c, x und p, b, x nicht linear. Also gilt $ax+xc>ac$, $xb+bp>xp=xc+cp=xc+bp$ mithin $xb>xc$ und daher $ab=ax+$ $+xb>ax+xc>ac$. Analog behandelt man den Fall, daß x zwischen p und c liegt.

Bezeichnet man mit k' die untere und mit k'' die obere Schranke der Zahlen $\varkappa(q)$ für alle Punkte q der ausgezeichneten Umgebung U von p, so gilt, wenn man in U Polarkoordinaten einführt nach bekannten Sätzen von Sturm

$$G_{k'}(r, \varphi)\geqq G(r, \varphi)\geqq G_{k''}(r, \varphi),$$

wo $G_k(r, \varphi)$ und $G(r, \varphi)$ die Funktionen von S. 26 und S. 33 sind.

20. *Ist die Umgebung $U(p; \rho)$ von p mit dem Radius ρ ausgezeichnet, k' die untere, k'' die obere Schranke der Gaußschen Krümmungen in $U(p; \rho)$, ferner $\{a, b, c\}$ ein nicht lineares Tripel in $U(p; \rho/4)$, $\{a', b', c'\}$ bzw. $\{a'', b'', c''\}$ ein Tripel in $S_{k'}$ bzw. $S_{k''}$, so daß $ab=a'b'=a''b''$, $ac=a'c'=a''c''$ und $\sphericalangle bac=\sphericalangle b'a'c'=$ $\sphericalangle b''a''c''$ gilt, dann ist $b'c'\geqq bc\geqq b''c''$.* Aus der Dreiecksungleichung folgt unmittelbar, daß $U(a; \rho/2)$ Teil von $U(p; \rho)$ ist. Also ist k' nicht größer als die untere und k'' nicht kleiner als die obere Schranke der Zahlen $\varkappa(q)$ für alle Punkte q von $U(a; \rho/2)$. Da $U(p; \rho)$ ausgezeichnet ist und mithin $k''<\pi^2/16\rho^2$ gilt, ist daher auch $U(a; \rho/2)$ ausgezeichnet. Wir betrachten eine Erweiterung

$U^*(a; \rho/2)$ von $U(a; \rho/2)$, wie sie auf S. 42 betrachtet wurde, wobei für die Erweiterung $\varkappa^*(q)$ von $\varkappa(q)$ die Ungleichung $k' \leqq \varkappa^*(q) \leqq k''$ für alle Punkte q^* von $U^*(a; \rho/2)$ gelte. Es ist daher $G_{k'} \geqq G^* \geqq G_{k''}$ für jeden Punkt von $U^*(a; \rho/2)$. Wir führen in $S_{k'}$ Polarkoordinaten ein mit dem Zentrum a' und so daß b' und c' dieselben Koordinaten haben, wie b, c in $U(a; \rho/2)$. Die Gleichung der Strecke $b'c'$ sei $r = f(\varphi)\,(\varphi_1 \leqq \varphi \leqq \varphi_2)$. Da $U(a; \rho/2)$ ausgezeichnet ist, so ist wegen Eigenschaft *a)* die Umgebung $U(a'; \rho/2)$ in $S_{k'}$ konvex; mithin liegt der Bogen $r = f(\varphi)$ in $U(a'; \rho/2)$. Daraus folgt, daß ein Bogen B^* mit der Gleichung $r = f(\varphi)\,(\varphi_1 \leqq \varphi \leqq \varphi_2)$ in $U^*(a; \rho/2)$ existiert. Wegen $G^* \leqq G_{k'}$ hat B^* eine Länge $l(B^*) \leqq b'c'$. Wegen $bc = b^*c^* \leqq l(B^*)$ gilt $bc \leqq b'c'$. Analog zeigt man, daß $bc \geqq b''c''$ ist.

Aus **19** und **20** folgert man mühelos

21. *Ist $U(p; \rho)$ eine ausgezeichnete Umgebung, k' die untere, k'' die obere Schranke der Gaußschen Krümmungen in $U(p; \rho)$, ferner $\{a, b, c\}$ ein nicht lineares Tripel in $U(p; \rho/4)$, $\{a', b', c'\}$ bzw. $\{a'', b'', c''\}$ ein zu $\{a, b, c\}$ kongruentes Tripel in $S_{k'}$ bzw. $S_{k''}$, so sind die Winkel in $\{a, b, c\}$ $\geqq$ bzw. $\leqq$ als die entsprechenden Winkel in $\{a', b', c'\}$ bzw. $\{a'', b'', c''\}$.* Wir beweisen nun:

22. *Ist $U(p; \rho)$ eine ausgezeichnete Umgebung, k' die untere, k'' die obere Schranke der Gaußschen Krümmungen in $U(p; \rho)$, ferner $Q = \{a_1, a_2, a_3, a_4\}$ ein beliebiges Punktquadrupel in $U(p; \rho/4)$, dann existiert eine Zahl k, so daß $k' \leqq k \leqq k''$ und Q in S_k einbettbar ist.* Wir unterscheiden zum Beweis fünf Fälle:

I) Q ist *nicht echt;* II) Q ist *linear.* In diesen Fällen folgt **22** unmittelbar aus Eigenschaft *b)* von $U(p; \rho)$.

III) *Q ist ein echtes nicht lineares Quadrupel, das ein lineares Tripel enthält.* Es liege etwa a_2 zwischen a_1 und a_3. Es sei $Q' = \{a'_1, \ldots\}$ ein Quadrupel in $S_{k'}$, $Q'' = \{a''_1, \ldots\}$ ein Quadrupel in $S_{k''}$, so daß $a_i a_j = a'_i a'_j = a''_i a''_j$ für alle Indexpaare (i, j) außer $(3, 4)$ gilt. Die Existenz von Q' und Q'' folgt unmittelbar aus Eigenschaft *b).* Wegen der Kongruenz der Tripel: $\{a_1, a_2, a_4\}$, $\{a'_1 a'_2 a'_4\}$, $\{a''_1, a''_2, a''_4\}$ gilt nach **21** $\sphericalangle a'_1 a'_2 a'_4 \leqq \sphericalangle a_1 a_2 a_4 \leqq \sphericalangle a''_1 a''_2 a''_4$. Für die Nebenwinkel gilt dann: $\sphericalangle a'_3 a'_2 a'_4 \geqq \sphericalangle a_3 a_2 a_4 \geqq \sphericalangle a''_3 a''_2 a''_4$. Aus **21** und **2** folgert man dann mühelos $a'_3 a'_4 \geqq a_3 a_4 \geqq a''_3 a''_4$ und daraus die Behauptung.

IV) *Q enthält kein lineares Tripel und es gibt einen Punkt, etwa a_2, so daß $\sphericalangle a_1 a_2 a_4 + \sphericalangle a_1 a_2 a_3 + \sphericalangle a_3 a_2 a_4 = 2\pi$ gilt,* wofür wir auch kurz sagen werden, daß *a_2 zwischen den Punkten a_1, a_3, a_4 liegt.* Es sei $Q'' = \{a''_1, \ldots\}$ ein Quadrupel in $S_{k''}$, so daß $a''_i a''_j = a_i a_j$ für alle Indexpaare außer $(3, 4)$ gilt und a''_3, a''_4 auf verschiedenen Seiten der Geraden $a''_1 a''_2$ liegen. Die Existenz von Q'' folgt unmittelbar aus Eigenschaft *b).* Nach **21** gilt dann $\sphericalangle a''_1 a''_2 a''_3 \geqq$

$\geqq \sphericalangle a_1 a_2 a_3$ und $\sphericalangle a''_1 a''_2 a''_4 \geqq \sphericalangle a_1 a_2 a_4$. Da a_2 zwischen a_1, a_3, a_4 liegt, so gilt $\sphericalangle a_1 a_2 a_3 + \sphericalangle a_1 a_2 a_4 > \pi$, also erst recht $\sphericalangle a''_1 a''_2 a''_3 + + \sphericalangle a''_1 a''_2 a''_4 > \pi$. Daraus folgt: $\sphericalangle a''_1 a''_2 a''_3 + \sphericalangle a''_1 a''_2 a''_4 + \sphericalangle a''_3 a''_2 a''_4 = = 2\pi$ und mithin $\sphericalangle a''_3 a''_2 a''_4 < \sphericalangle a_3 a_2 a_4$. Aus **21** und **2** folgt dann:

$$(+) \qquad a''_3 a''_4 \leqq a_3 a_4.$$

Wegen Eigenschaft *b)* gibt es ein Quadrupel $Q' = (a'_1, \ldots)$ in $S_{k'}$, so daß $a_i a_j = a'_i a'_j$ für alle Indexpaare außer (3, 4) gilt und a'_3, a'_4 auf verschiedenen Seiten der Geraden $a'_1 a'_2$ liegen. Ist nun *erstens* $\sphericalangle a'_1 a'_2 a'_3 + \sphericalangle a'_1 a'_2 a'_4 > \pi$, dann ist $\sphericalangle a'_1 a'_2 a'_3 + \sphericalangle a'_1 a'_2 a'_4 + \sphericalangle a'_3 a'_2 a'_4 = = 2\pi$; nach **21** haben wir $\sphericalangle a'_1 a'_2 a'_3 \leqq \sphericalangle a_1 a_2 a_3$ und $\sphericalangle a'_1 a'_2 a'_4 \leqq \leqq \sphericalangle a_1 a_2 a_4$, also $\sphericalangle a'_3 a'_2 a'_4 \geqq \sphericalangle a_3 a_2 a_4$. Aus **21** und **2** folgt dann $a'_3 a'_4 \geqq a_3 a_4$, woraus sich wegen (+) die Behauptung ergibt. Sei *zweitens* $\sphericalangle a'_1 a'_2 a'_3 + \sphericalangle a'_1 a'_2 a'_4 \leqq \pi$. Wegen $\sphericalangle a''_1 a''_2 a''_3 + \sphericalangle a''_1 a''_2 a''_4 > \pi$ gibt es eine Zahl k^* ($k' \leqq k^* < k''$), so daß S_{k^*} ein Quadrupel $Q^* = \{a^*_1, \ldots\}$ enthält, wobei $a^*_i a^*_j = a_i a_j$ für alle Indexpaare außer (3, 4) gilt und a^*_2 zwischen a^*_3 und a^*_4 liegt. Daraus folgt $a^*_3 a^*_4 > > a_3 a_4$, woraus sich wegen (+) und wegen $k' \leqq k^*$ die Behauptung ergibt.

V) *Keiner der Fälle* I)—IV) *liegt vor.* Man kann die Punkte von Q so numerieren, daß die Strecken $a_1 a_2$ und $a_3 a_4$ sich in einem inneren Punkte x schneiden. Da a_1 und a_2 in $U(p; \rho/4)$ liegen und $U(p; \rho/4)$ konvex ist, liegt x in $U(p; \rho/4)$. Wegen Eigenschaft *b)* existiert ein Quadrupel $Q' = \{a'_1, \ldots\}$ in $S_{k'}$, so daß $a'_i a'_j = a_i a_j$ für alle Indexpaare außer (3, 4) gilt und a'_3, a'_4 auf verschiedenen Seiten von der Geraden $a'_1 a'_2$ liegen. Es sei x' ein Punkt der Strecke $a'_1 a'_2$, so daß $a'_1 x' = a_1 x$ gilt. Wir beweisen: $a'_3 x' \leqq a_3 x$. Zu diesem Zweck sei a^*_3 ein Punkt in $S_{k'}$, so daß $a^*_3 a'_1 = a_3 a_1$ und $a^*_3 x' = a_3 x$ gilt. Nach **21** ist $\sphericalangle a'_1 x' a^*_3 \leqq \sphericalangle a_1 x a_3$ und mithin $\sphericalangle a'_2 x' a^*_3 \geqq \geqq \sphericalangle a_2 x a_3$. Daraus folgt wegen **21** und **2**, daß $a'_2 a^*_3 \geqq a_2 a_3 = a'_2 a'_3$ ist. Wegen **2** ist dann $\sphericalangle a'_2 a'_1 a^*_3 \geqq \sphericalangle a'_2 a'_1 a'_3$, woraus ebenfalls wegen **2** folgt: $a'_3 x' \leqq a^*_3 x' = a_3 x$. Ebenso zeigt man: $a'_4 x' \leqq a_4 x$. Also gilt

$$(++) \qquad a_3 a_4 = a_3 x + x a_4 \geqq a'_3 x' + x' a'_4 \geqq a'_3 a'_4.$$

Wegen Eigenschaft *b)* existiert ein Quadrupel $Q'' = \{a''_1, \ldots\}$ in $S_{k''}$, so daß $a''_i a''_j = a_i a_j$ für alle Indexpaare außer (3, 4) gilt und a''_3, a''_4 auf verschiedenen Seiten von der Geraden $a''_1 a''_2$ liegen. Es sind zwei Fälle möglich:

1. *Die Strecken* $a''_1 a''_2$ *und* $a''_3 a''_4$ *schneiden sich in einem inneren Punkte* x''. Es sei $\bar{x}$ ein Punkt der Strecke $a_1 a_2$, so daß $a_1 \bar{x} = = a''_1 x''$ gilt. Um zu zeigen, daß $a''_3 x'' \geqq a_3 \bar{x}$ ist, sei a^*_3 ein Punkt

in $S_{k''}$, so daß $a''_1 a^*_3 = a_1 a_3$ und $x'' a^*_3 = \bar{x} a_3$ gilt. Nach **21** ist dann $\sphericalangle a''_1 x'' a^*_3 \geqq \sphericalangle a_1 x a_3$, also $\sphericalangle a''_2 x'' a^*_3 \leqq \sphericalangle a_2 \bar{x} a_3$, woraus wegen **21** und **2** folgt, daß $a''_2 a^*_3 \leqq a_2 a_3 = a''_2 a''_3$. Daraus ergibt sich ebenfalls wegen **2**, daß $\sphericalangle x'' a''_1 a^*_3 \leqq \sphericalangle x'' a''_1 a^*_3$ und mithin $x'' a''_3 \geqq x'' a^*_3 = \bar{x} a_3$. Analog zeigt man: $x'' a''_4 \geqq \bar{x} a_4$. Also gilt $a''_3 a''_4 = a''_3 x'' + x'' a''_4 \geqq$ $\geqq a_3 \bar{x} + \bar{x} a_4 \geqq a_3 a_4$. woraus wegen $(++)$ die Behauptung folgt.

2. *Die Strecken* $a''_1 a''_2$ *und* $a''_3 a''_4$ *schneiden sich nicht im Inneren.* Dann muß entweder $\sphericalangle a''_2 a''_1 a''_3 + \sphericalangle a''_2 a''_1 a''_4 > \pi$ oder $\sphericalangle a''_1 a''_2 a''_3 +$ $+ \sphericalangle a''_1 a''_2 a''_4 > \pi$ gelten. Es gelte etwa die letztere Ungleichung. Da nach Voraussetzung die Strecken $a_1 a_2$ und $a_3 a_4$ sich im Inneren schneiden, so ist $\sphericalangle a_1 a_2 a_3 + \sphericalangle a_1 a_2 a_4 < \pi$, also wegen **21** erst recht $\sphericalangle a'_1 a'_2 a'_3 + \sphericalangle a'_1 a'_2 a'_4 < \pi$. Es gibt also eine Zahl k^* $(k' \leqq k^* \leqq k'')$, so daß S_{k^*} ein Quadrupel $Q^* = \{a^*_1, \ldots\}$ enthält, wobei $a^*_i a^*_j = a_i a_j$ für alle Indexpaare außer (3, 4) gilt, und a^*_2 zwischen a^*_3 und a^*_4 liegt. Es gilt daher $a^*_3 a^*_4 = a^*_3 a^*_2 + a^*_2 a^*_4 = a_3 a_2 + a_2 a_4 > a_3 a_4$, woraus wegen $(++)$ und wegen $k' \leqq k^* \leqq k''$ die Behauptung des Satzes **22** folgt, der damit in allen Fällen bewiesen ist.

Aus **22** folgert man mühelos die zu Beginn dieses Kapitels formulierte Behauptung.

94. Kolloquium (26. VI. 1935).

Über mehrfache Erreichbarkeit eines unzerlegbaren Kontinuums. Von J. Novák, Brno.

Satz. *Es sei F ein unzerlegbares Kontinuum in der Ebene, $G_1, G_2, \ldots, G_n$ $(n \geqq 2)$ seien paarweise verschiedene Komplementärgebiete von F. Sind dann r paarweise fremde Teilkontinua $C_1, \ldots, C_r$ von F gegeben, von denen jedes aus mehreren von den Gebieten G_i erreichbar ist, darunter r_λ Kontinua, deren jedes aus λ von den Gebieten G_i erreichbar ist $(2 \leqq \lambda \leqq n)$, so daß $r = \sum_{\lambda=2}^{n} r_\lambda$, dann gilt*

$$\sum_{\lambda=2}^{n} (\lambda - 1) r_\lambda \leqq n - 1.$$

Korollar. *Ein unzerlegbares Kontinuum F, das in der Ebene n Gebiete bestimmt, besitzt höchstens $n-1$ mehrfach erreichbare Punkte. Das Kontinuum B_n von* K n a s t e r (Fund. Math., 7, S. 274) *zeigt, daß die Grenze $n-1$ scharf ist.*

Zum Beweis wählen wir in jedem Gebiete G_i einen Punkt p_i $(i = 1, \ldots, n)$. Jedes der r_λ Kontinua C_j, das aus λ von den Gebieten G_i erreichbar ist, verbinden wir mit den in diesen Gebieten G_i gewählten Punkten p_i durch je einen einfachen Bogen, der mit Ausnahme eines Endpunktes ganz in G_i verläuft. Im ganzen liegen also $\sum_{\lambda=2}^{n} \lambda r_\lambda$ Bogen vor, von denen wir voraussetzen dürfen, daß sie paarweise höchstens Endpunkte gemein haben. Wir bezeichnen nun mit Ω einen Graphen, dessen Strecken die eben konstruierten $\sum \lambda r_\lambda$ einfachen Bogen und dessen „Endpunkte" die n Punkte $p_1, \ldots, p_n$, sowie die r Kontinua $C_1, \ldots, C_r$ sind. Es genügt zu beweisen, daß jede Komponente dieses Graphen Ω ein Baum ist. Denn dann ist ja bekanntlich die Anzahl $\sum_{\lambda=2}^{n} \lambda r_\lambda$ der Strecken von Ω kleiner als die Anzahl $n + \sum_{\lambda=2}^{n} r_\lambda$ der Endpunkte von Ω und das besagt gerade die zu beweisende Ungleichung. Machen wir also die Annahme, daß nicht jede Komponente von Ω ein Baum sei! Dann gibt es in Ω einen einfachen geschlossenen „Streckenzug", der offenbar mindestens zwei der Punkte $p_1, \ldots, p_n$ enthalten muß. Unter diesen Punkten existieren also zwei — wir bezeichnen sie mit p und p' — so daß es zwei Teilkontinua Ω_1 und Ω_2 von Ω gibt, deren Durchschnitt aus den beiden Punkten p und p' besteht. Bekanntlich gibt es zwischen p und p' irreduzible Kontinua $K_1 \subset \Omega_1$ und $K_2 \subset \Omega_2$. Nach Rosenthal (Sitzb. Münch. Akad. 1919, S. 91) ist das Kontinuum $K_1 + K_2$ gemeinsame Begrenzung von zwei Gebieten H und H'. Da $F.(K_1 + K_2) \subset C_1 + \ldots + C_r$ gilt und die C_j echte Teilkontinua des unzerlegbaren Kontinuums F sind, so ist die Menge $F - F.(K_1 + K_2)$ bekanntlich zusammenhängend, so daß entweder H oder H' zu F punktfremd ist. Dann ist aber $H + p + p'$ oder $H' + p + p'$ eine zusammenhängende Teilmenge des Komplementes von F und das ist ein Widerspruch, da ja p und p' in verschiedenen Komplementärgebieten von F liegen.

Über die Bettischen Gruppen kompakter Räume[1]. Von E. Čech (Brno).

R sei ein kompakter Raum, $\mathfrak{G}$ eine Abelsche Gruppe. Ein *$(n, \mathfrak{G})$-Grenzzyklus* in R ist eine Folge $\{\gamma_i\}_1^\infty$ von algebraischen n-Zykeln in R, deren Koeffizienten $\mathfrak{G}$ entnommen sind, und für die es zu jedem $\varepsilon > 0$ ein m gibt, so daß: (1) für $i > m$ γ_i ein (n, ε)-Zyklus[2]) ist; (2) für $i > m$ und $j > m$ γ_i und γ_j ε-homolog[3]) in R sind. Zwei $(n, \mathfrak{G})$-Grenzzykeln $\{\gamma_i\}$ und $\{\gamma'_i\}$ sind äquivalent, wenn es zu jedem $\varepsilon > 0$ ein m gibt, so daß für $i > m$ γ_i und γ'_i ε-homolog in R sind. Äquivalente $(n, \mathfrak{G})$-Grenzzykeln werden immer identifiziert. Die $(n, \mathfrak{G})$-Grenzzykeln in R bilden eine (additive) Abelsche Gruppe $B_n(R, \mathfrak{G})$, die *volle n^{te} Bettische Gruppe* des Raumes R bei Koeffizientenbereich $\mathfrak{G}$.

---

[1]) Ein sehr wesentlicher Teil der hier mitgeteilten Sätze wurde etwa gleichzeitig und völlig unabhängig auch von N. E. Steenrod gefunden; s. seine demnächst in den Proc. Nat. Acad. erscheinende Note „On universal homology groups", die ich in Manuskriptform zu lesen Gelegenheit hatte.

[2]) D. h. die Durchmesser der Simplexe sind sämtlich $< \varepsilon$.

[3]) D. h. es gibt in R eine algebraische $(n+1, 2)$-Kette (vgl. Anm.[2])) mit Koeffizienten aus $\mathfrak{G}$, deren Rand gleich $\gamma_i - \gamma_j$ ist.

Γ sei die additive Gruppe der ganzen Zahlen, $\mathfrak{P}$ sei die additive Gruppe der modulo Eins reduzierten reellen Zahlen. In $B_n(R, \mathfrak{P})$ kann man nach Pontrjagin[4]) einen Stetigkeitsbegriff einführen, wodurch aus $B_n(R, \mathfrak{P})$ eine hier mit $P_n(R)$ bezeichnete *kompakte* Gruppe entsteht.

Ein algebraischer n-Zyklus γ mit Koeffizienten aus $\mathfrak{G}$ heiße ein *Torsionszyklus*, wenn $\gamma = g_i \tau_i$[5]), wo $g_i \varepsilon \mathfrak{G}$, und es eine positive Zahl $m \varepsilon \Gamma$ gibt, so daß $m\tau_i$ gleich dem Rande einer $(n+1)$-Kette mit Koeffizienten aus Γ ist. Ein $(n, \mathfrak{G})$-Grenzzyklus $\{\gamma_i\}$ heiße ein $(n, \mathfrak{G})$-*Torsionsgrenzzyklus*, wenn γ_i für fast alle i ein Torsionszyklus ist. Die $(n, \mathfrak{G})$-Torsionsgrenzzykeln bilden eine Untergruppe $T_n(R, \mathfrak{G})$ von $B_n(R, \mathfrak{G})$, die n^{te} *Torsionsgruppe* des Raumes R bei Koeffizientenbereich $\mathfrak{G}$. In $T_n(R, \Gamma)$ kann man wieder einen Stetigkeitsbegriff einführen, wodurch aus $T_n(R, \Gamma)$ eine hier mit $Q_n(R)$ bezeichnete *kompakte* Gruppe entsteht.

Die Faktorgruppe in

$$B_n(R, \mathfrak{G}) / T_n(R, \mathfrak{G}) = B'_n(R, \mathfrak{G})$$

heiße die *reduzierte n^{te} Bettische Gruppe des Raumes R* bei Koeffizientenbereich $\mathfrak{G}$. Es gelten die Sätze.

I. Die volle n^{te} Bettische Gruppe $B_n(R, \mathfrak{G})$ ist direkte Summe der n^{ten} Torsionsgruppe $T_n(R, \mathfrak{G})$ und einer anderen — natürlich mit der reduzierten n^{ten} Bettischen Gruppe $B'_n(R, \mathfrak{G})$ isomorphen — Untergruppe.

II. Die n^{te} Torsionsgruppe $T_n(R, \mathfrak{G})$ ist durch $\mathfrak{G}$ und $Q_n(R)$ eindeutig bestimmt wie folgt. Nach Pontrjagin[6]) kann man die kompakte Gruppe $Q_n(R)$ festlegen durch eine Folge $\{Q_n^{(i)}\}_{i=1}^{\infty}$ von Abelschen Gruppen mit je endlich vielen Erzeugenden (gegebenenfalls sogar endlichen Gruppen), wobei für jedes i eine homomorphe Abbildung π_i von $Q_n^{(i+1)}$ auf $Q_n^{(i)}$ vorliegt. Für jedes i sei die Gruppe $Q_n^{(i)}$ durch Erzeugende α_{ih} und definierende Relationen $a_{ihk}\alpha_{ih} = 0$ $(\alpha_{ihk} \varepsilon \Gamma)$ festgelegt; sei $\pi_i(\alpha_{i+1,\lambda}) = c_{\lambda h}^{(i)} \alpha_{ih}$[7]) $(c_{\lambda h}^{(i)} \varepsilon \Gamma)$. Man bilde die (additive) Gruppe $\mathfrak{Q}_n^{(i)}(\mathfrak{G})$ der Vektoren $g_h X_{ih} (g_h \varepsilon \mathfrak{G})$ mit definierenden Relationen $a_{ihk} g X_{ih} = 0$ $(g \varepsilon \mathfrak{Q})$. Durch $\overset{*}{\pi}_i(g_\lambda X_{i+1,\lambda}) = c_{\lambda h}^{(i)} X_{ih}$[7]) wird $\mathfrak{Q}_n^{(i+1)}(\mathfrak{G})$ auf eine Untergruppe von $\mathfrak{Q}_n^{(i)}(\mathfrak{G})$ homomorph bezogen. Die Folgen $\{v_i\}_1^{\infty}$ mit $v_i \varepsilon \mathfrak{Q}_n^{(i)}(\mathfrak{G})$, $\overset{*}{\pi}_i v_{i+1} = v_i$ bilden eine mit $T_n(R, \mathfrak{G})$ isomorphe Gruppe.

III. Die reduzierte n^{te} Bettische Gruppe $B'_n(R, \mathfrak{G})$ ist durch $\mathfrak{G}$ und $P_n(R)$ eindeutig bestimmt wie folgt. Sei $M_n(R)$ die nach Pontrjagin[6]) definierte abzählbare Gruppe der Charaktere der kompakten Gruppe $P_n(R)$. Die Gruppe $M_n(R)$ sei gegeben durch

---

[4]) Annals of Math., *35* (1934), S. 909.
[5]) Über doppelt vorkommende Indizes wird immer summiert.
[6]) Annals of Math., *35*, 361—388 (1934).
[7]) Über i wird hier nicht summiert.

eine aufsteigende Folge $\{M_n^{(i)}\}_{i=1}^{\infty}$ von Untergruppen mit je endlich vielen Erzeugenden. Für jedes i sei die Gruppe $M_n^{(i)}$ durch Erzeugende β_{ih} und definierende Relationen $b_{ihk}\beta_{ih}=0$ $(b_{ihk}\varepsilon\Gamma)$ festgelegt. Da $M_n^{(i)}$ Untergruppe von $M_n^{(i+1)}$ ist, hat man noch Relationen der Form $\beta_{ih}=e_{h\lambda}^{(i)}\beta_{i+1,\lambda}$ $(e_{h\lambda}^{(i)}\varepsilon\Gamma)$. Man bilde die (additive) Gruppe $\mathfrak{M}_n^{(i)}(\mathfrak{G})$ der Vektoren $g_h Y_{ih}$, wobei g_h den Relationen $b_{ihk}g_h= =0$ unterworfene Elemente der Gruppe $\mathfrak{G}$ sind. Durch $\rho_i(g_\lambda Y_{i+1,\lambda})= =e_{h\lambda}^{(i)}g_\lambda Y_{ih}$[7]) wird $\mathfrak{M}_n^{(i+1)}(\mathfrak{G})$ auf eine Untergruppe von $\mathfrak{M}_n^{(i)}(\mathfrak{G})$ homomorph bezogen.

Die Folgen $\{w_i\}_1^{\infty}$ mit $w_i\varepsilon\mathfrak{M}_n^{(i)}(\mathfrak{G})$, $\rho_i(w_{i+1})=w_i$ bilden eine mit $B'_n(R, \mathfrak{G})$ isomorphe Gruppe.

Ein algebraischer n-Zyklus γ mit Koeffizienten aus $\mathfrak{G}$ heiße *rein*, wenn $\gamma=g_i\vartheta_i$, wo $g_i\varepsilon\mathfrak{G}$ und die ϑ_i algebraische n-Zykeln mit Koeffizienten aus Γ sind. Ein $(n, \mathfrak{G})$-Grenzzyklus $\{\gamma_i\}$ heiße *rein*, wenn die algebraischen n-Zykeln γ_i für fast alle i rein sind. Die reinen $(n, \mathfrak{G})$-Grenzzykeln bilden eine Untergruppe $C_n(R, \mathfrak{G})$ von $B_n(R, \mathfrak{G})$ die *reine* n^{te} *Bettische Gruppe* von R bei Koeffizientenbereich $\mathfrak{G}$. Die Torsionsgruppe $T_n(R, \mathfrak{G})$ ist eine Untergruppe von $C_n(R, \mathfrak{G})$. Die Faktorgruppe

$$C_n(R, \mathfrak{G})\,/\,T_n(R, \mathfrak{G})=C'_n(R, \mathfrak{G})$$

heiße die *reduzierte reine* n^{te} *Bettische Gruppe* von R bei Koeffizientenbereich $\mathfrak{G}$.

Als Untergruppe der *kompakten* Gruppe $B_n(R, \mathfrak{P})=P_n(R)$ ist $C_n(R, \mathfrak{P})$ in $P_n(R)$ abgeschlossen und folglich kompakt. In diesem Sinne (als kompakte Gruppe) wollen wir $C_n(R, \mathfrak{P})$ mit $Z_n(R)$ bezeichnen. Auch die Faktorgruppe $P_n(R)\,/\,Z_n(R)$ ist kompakt.

IV. Die Gruppe $Z_n(R)$ besteht aus denjenigen Elementen x der Gruppe $P_n(R)$, für die $\chi(x)=0$ bei jedem Charakter *endlicher Ordnung* der kompakten Gruppe $P_n(R)$.

V. Die Gruppe $Q_n(R)$ ist mit der Faktorgruppe $P_{n+1}(R)/ /Z_{n+1}(R)$ stetig isomorph.

VI. Die reine n^{te} Bettische Gruppe $C_n(R, \mathfrak{G})$ ist direkte Summe der n^{ten} Torsionsgruppe $T_n(R, \mathfrak{G})$ und einer anderen — natürlich mit der reduzierten reinen n^{ten} Bettischen Gruppe $C'_n(R, \mathfrak{G})$ isomorphen — Untergruppe.

VII. Die reduzierte reine n^{te} Bettische Gruppe $C'_n(R, \mathfrak{G})$ ist durch $\mathfrak{G}$ und $Z_n(R)$ eindeutig bestimmt. Man braucht nur in der im Satz III beschriebenen Konstruktion die Gruppe $P_n(R)$ durch $Z_n(R)$ zu ersetzen.

VIII. Die Faktorgruppe

$$B_n(R, \mathfrak{G})\,/\,C_n(R, \mathfrak{G})=B'_n(R, \mathfrak{G})\,/\,C'_n(R, \mathfrak{G})$$

ist durch $\mathfrak{G}$ und $Q_{n-1}(R)$ eindeutig bestimmt. Man braucht wieder nur in der im Satz III beschriebenen Konstruktion die Gruppe $P_n(R)$ durch $Q_{n-1}(R)$ zu ersetzen.

Der kompakte Raum R heiße *n-normal,* wenn es zu jedem $\eta>0$ ein $\varepsilon>0$ gibt, so daß, wenn γ ein algebraischer (n, ε)-Zyklus[2]) in R mit beliebigem Koeffizientenbereiche $\mathfrak{G}$ ist, ein $(n, \mathfrak{G})$-Grenzzyklus $\{\gamma_i\}$ in R existiert, so daß γ_i η-homolog γ für fast alle i ist. Ähnlich werden *n-seminormale* und *n-quasinormale* kompakte Räume eingeführt; man erhält die Definition, indem man γ und $\{\gamma_i\}$ auf *reine* Zykeln bzw. auf Torsionszykeln einschränkt. Ist der Koeffizientenbereich $\mathfrak{G}$ vorgegeben, so spreche man von $(n, \mathfrak{G})$-*Normalität* usw.

IX. Jeder kompakte Raum ist n-quasinormal.

X. Ein (n, Γ)-seminormaler kompakter Raum ist n-seminormal.

XI. Damit der kompakte Raum R n-seminormal sei, ist notwendig und hinreichend, daß die Charaktergruppe der kompakten Gruppe $Z_n(R)$ direkte Summe von höchstens abzählbar vielen zyklischen Gruppen sei.

Ist p eine Primzahl, so sei Δ_p die direkte Summe von abzählbar vielen zyklischen Gruppen $\varphi_i (i=1, 2, 3, \ldots)$, wobei φ_i die endliche zyklische Gruppe der Ordnung p^i ist. Δ_p ist also eine abzählbare Abelsche Gruppe.

XII. Wenn ein kompakter R gleichzeitig (n, Γ)-seminormal[8]) und (n, Δ_p)-normal für jede Primzahl p ist, so ist R n-normal.

XIII. Damit der kompakte Raum R n-normal sei, ist notwendig und hinreichend, daß die Charaktergruppe der kompakten Gruppe $P_n(R)$ direkte Summe von höchstens abzählbar vielen zyklischen Gruppen sei.

Wir nennen eine Untergruppe $\mathfrak{H}$ einer Abelschen Gruppe *speziell*, wenn: (1) $\tau(\mathfrak{H})\subset\mathfrak{H}$ bei jeder homomorphen Abbildung τ der Gruppe $\mathfrak{G}$ auf eine Untergruppe von $\mathfrak{G}$, (2) $\mathfrak{H}$ immer abgeschlossen in $\mathfrak{G}$ ist, wenn man irgend einen Stetigkeitsbegriff in $\mathfrak{G}$ einführt, der $\mathfrak{G}$ zu einer kompakten Gruppe macht.

XIV. Jeder kompakte Raum R ist $(n, \mathfrak{G})$-normal, wenn die Abelsche Gruppe $\mathfrak{G}$ dem folgenden Teilerkettensatz genügt: jede absteigende Folge von speziellen Untergruppen bricht im Endlichen ab. Dem Teilerkettensatz genügen z. B. alle endlichen Abelschen Gruppen, dann die additive Gruppe der rationalen Zahlen und die Gruppe $\mathfrak{P}$, dagegen nicht z. B. die Gruppe Γ und die Gruppen Δ_p.

G. v. Alexits (Budapest): Metrische Behandlung der Torsion von Raumkurven.

95. Kolloquium (27. VI. 1935).

Programmatisches zur Anwendung der metrischen Geometrie auf die Variationsrechnung. Von Karl Menger.

Die Theorie der geodätischen Linien in metrischen Räumen (Mathem. Annalen *103*) legt es nahe, die eindimensionalen Probleme der Variationsrechnung in eine allgemeine metrische Theorie der

---

[8]) Oder (n, Γ)-normal; die beiden Begriffe sind ja äquivalent.

geodätischen Linien einzuordnen. Man kann von einem metrischen Raum R im Sinne von Fréchet ausgehen und annehmen, daß in ihm eine Funktion definiert ist, welche jedem Punkt p und jedem Paar verschiedener Punkte $q \neq r$ eine Zahl $\varphi(p; q, r)$ zuordnet. Das Kurvenintegral ist dann die Bogenlänge im Sinne der durch diese Funktion φ verzerrten Metrik von R, in welcher man je zwei Punkten q und r als Abstand den mit $\varphi(q; q, r)$ multiplizierten Abstand, den sie in R haben, zuordnet. In einem Raum R, in welchem Richtungen definiert sind, wie z. B. in einem Vektorraum, kann die Funktion von Punkt und Punktepaar durch eine Funktion von Punkt und Richtung ersetzt werden. Es sind in dieser Theorie Eigenschaften der Funktion φ zu formulieren, welche die Existenz geodätischer Linien in der durch φ verzerrten Metrik von R gewährleisten. Insbesondere kann man Eigenschaften, welche dieser verzerrten Metrik zukommen und schwächer sind als die Eigenschaften eines metrischen Raumes, aber doch zum Beweis der Existenz geodätischer Linien hinreichen, z. B. eine abgeschwächte Symmetriebedingung und eine abgeschwächte Dreiecksungleichung, zur Definition *„fast metrischer Räume"* verwenden. Eine ausführliche Darstellung der Theorie der Bogenlänge in solchen fast metrischen Räumen erscheint in Band *25* der Fundamenta Mathematicae.

96. Kolloquium (3. VII. 1935).

Über die Erweiterungen der unvollständigen Systeme des Aussagenkalküls. Von A. Tarski.

Nach einem bekannten Satz von A. Lindenbaum[1]) läßt sich jede widerspruchsfreie Aussagenmenge einer beliebigen deduktiven Theorie zu einem widerspruchsfreien und vollständigen System erweitern. Es entsteht nun die Frage, wie viele derartige Erweiterungen es gibt, m. a. W. wie groß die Anzahl aller widerspruchsfreien und vollständigen Systeme ist, die eine gegebene Aussagenmenge enthalten. Ich will hier einen allgemeinen Satz aus dem Gebiet des „Metaaussagenkalküls" beweisen, demgemäß im Bereich des Aussagenkalküls unter gewissen Voraussetzungen nur eine einzige Erweiterung existiert; weiterhin will ich zeigen, daß die meisten bis jetzt behandelten unvollständigen Systeme des Aussagenkalküls diesen Voraussetzungen genügen, so daß das übliche (zweiwertige) System des Aussagenkalküls für alle diese Systeme die einzige widerspruchsfreie und vollständige Erweiterung darstellt.

In den Ausdrücken des Aussagenkalküls treten bekanntlich spezielle Variablen, sog. *Aussagevariablen p, q, r,...* auf, überdies das *Implikationszeichen C* und eventuell noch andere Konstante, z. B. die *Zeichen der Negation N, der Alternative A, der Konjunktion*

[1]) Vgl. A. Tarski, *Fundamentale Begriffe der Methodologie der deduktiven Wissenschaften. I.*, Mh. f. Math. u. Phys. *37*, 1930 (unten als *T* zitiert), S. 394 f., Satz I. 56.

4*

K usw.[2]) Jeder Konstanten wird eine sog. *Grundoperation an Ausdrücken* zugeordnet, durch die aus einfacheren Ausdrücken zusammengesetztere gebildet werden können; so entspricht z. B. dem Implikationszeichen die Operation der *Implikationsbildung,* mit deren Hilfe aus zwei gegebenen Ausdrücken a und b die Implikation Cab gebildet wird. Der Durchschnitt aller Mengen von Ausdrücken, die alle Aussagevariablen enthalten und in bezug auf jede Grundoperation an Ausdrücken abgeschlossen sind, wird *Menge der (sinnvollen) Aussagen* genannt und mit s bezeichnet.

Ferner werden noch zwei Operationen ausgezeichnet, die gewissen Aussagen andere Aussagen zuordnen, und zwar die Operation der *Einsetzung* und die der *Abtrennung.* Es wäre etwas umständlich und überflüssig, den anschaulich evidenten Sinn des Begriffs der Einsetzung hier näher zu präzisieren. Die Menge aller Aussagen, die durch Einsetzung aus den Aussagen einer gegebenen Menge x gewonnen werden können, wird mit „$sb(x)$" bezeichnet; allgemeiner bezeichnen wir mit „$sb_y(x)$" die Menge aller Aussagen, die man dadurch erhält, daß man in den Aussagen der Menge x alle Variablen durch Aussagen der Menge y ersetzt (wobei gleichgestaltete Variable, die in derselben Aussage vorkommen, durch gleichgestaltete Aussagen ersetzt werden). — Die Operation der Abtrennung ordnet zwei beliebigen Aussagen von der Form a und Cab die Aussage b zu. Ist x eine beliebige Aussagenmenge, so heißt der Durchschnitt aller Aussagenmengen, die die Menge $sb(x)$ umfassen und in bezug auf die Operation der Abtrennung abgeschlossen sind, die *Folgerungsmenge von x,* symbolisch $fl(x)$.

Mit Hilfe der Begriffe der Aussage und der Folgerungsmenge lassen sich bekanntlich andere metalogische Begriffe definieren. So wird z. B. eine Aussagenmenge x *deduktives System* oder einfach *System* genannt, wenn $fl(x) = x$. Die Aussagenmenge x heißt *widerspruchsvoll* oder *widerspruchsfrei,* je nachdem die Formel $fl(x) = s$ gilt oder nicht. Die Aussagenmenge x heißt *vollständig,* wenn jede beliebige Aussage a entweder eine Folgerung der Menge x ist oder, zu x hinzugefügt, eine widerspruchsvolle Menge $x + \{a\}$ ergibt; allgemeiner nennen wir die Aussagenmenge x *vollständig in bezug auf die Aussagenmenge y,* wenn für jede Aussage b der Menge y entweder $b \varepsilon fl(x)$ gilt oder die Menge $x + \{b\}$ widerspruchsvoll ist.

Verschiedene elementare Eigenschaften obiger Begriffe werden hier als bekannt vorausgesetzt[3]). Neben den oben angeführten Sym-

---

[2]) Wegen der Symbolik vgl. J. Łukasiewicz und A. Tarski, *Untersuchungen über den Aussagenkalkül,* C. R. des séances de la Soc. d. Sc. et d. L. de Varsovie, *23*, 1930, Classe III, S. 30—50 (unten als *LT.* zitiert); was die Terminologie und die Begriffsbildung betrifft, vgl. *T.* und *LT.* — Es sei bemerkt, daß die Zeichen „*p*", „*q*",. . ., „*C*", „*N*" . . . sowie die aus diesen Zeichen zusammengesetzten Ausdrücke hier nicht als Ausdrücke des Aussagenkalküls, sondern als Bezeichnungen entsprechender Ausdrücke verwendet werden.

[3]) Mehrere solche Eigenschaften wurden in *T.* festgestellt.

bolen werde ich die übliche Symbolik aus dem Gebiet der Mengenlehre verwenden. Der anfangs erwähnte allgemeine Satz lautet:

Satz 1. *Es sei x eine widerspruchsfreie Aussagenmenge, die folgender Bedingung genügt: es gibt eine Aussagenmenge z, derart, daß x in bezug auf die Menge $sb_z(s)$ vollständig ist und daß dabei die Menge $x+sb_z(s)$ widerspruchsvoll ist. Dann existiert genau éin widerspruchsfreies und vollständiges deduktives System y, das die Menge x umfaßt.*

Dem Beweise dieses Satzes schicken wir einige Hilfssätze voran.

Hilfssatz 1. *Für jede Aussagenmenge x ist $sb_x(s)$ die kleinste aller Aussagenmengen, die x umfassen und in bezug auf jede Grundoperation an Ausdrücken abgeschlossen sind.* Der Beweis — auf Grund einer genauen Definition des Einsetzungsbegriffs — bietet keine Schwierigkeiten. Als Korollar gewinnen wir leicht

Hilfssatz 2. $x \subset sb_x(s) = sb_{sb_x(s)}(s)$ *für beliebige Aussagenmenge x.*

Hilfssatz 3. *Für beliebige Aussagenmengen x und y gilt* $sb_y\big(fl(x)\big) \subset fl\Big(sb_y\big(sb(x)\big)\Big)$. Beweis: Es sei z die Menge aller Aussagen a, die die Formel $sb_y(\{a\}) \subset fl\Big(sb_y\big(sb(x)\big)\Big)$ erfüllen. Man zeigt leicht, daß $sb(x) \subset z$ und daß z in bezug auf die Operation der Abtrennung abgeschlossen ist (d. h. wenn $a \varepsilon z$ und $Cab \varepsilon z$, so auch $b \varepsilon z$). Daraus ergibt sich sofort, daß $fl(x) \subset z$, m. a. W. daß $sb_y(\{a\}) \subset fl\Big(sb_y\big(sb(x)\big)\Big)$ für jede Aussage $a \varepsilon fl(x)$; schließlich also $sb_y\big(f(x)\big) \subset fl\Big(sb_y\big(sb(x)\big)\Big)$, w. z. b. w.

Setzt man im Hilfssatz 3: $y = s$, so erhält man nach einfachen Umformungen die Formel: $sb\big(fl(x)\big) \subset fl(x)$ für jede Aussagenmenge x; dieser Formel zufolge läßt sich die oben angegebene Definition der Folgerungsmenge durch folgende äquivalente Definition ersetzen: $fl(x)$ ist der Durchschnitt aller Aussagenmengen, die die Menge x umfassen und in bezug auf die Operationen der Einsetzung und der Abtrennung abgeschlossen sind.

Hilfssatz 4. *Sind die Aussagenmengen $x+y$ und $x+z$ widerspruchsvoll, so ist auch die Menge $x + sb_z\big(sb(y)\big)$ widerspruchsvoll.* Beweis: Wenn wir in Hilfssatz 3 „x" durch „$x+y$" und „y" durch „z" ersetzen, so erhalten wir: $sb_z\big(fl(x+y)\big) \subset fl\Big(sb_z\big(sb(x+y)\big)\Big)$; da nach der Voraussetzung $fl(x+y) = s$ ist, so folgt hieraus, daß $sb_z(s) \subset fl\Big(sb_z\big(sb(x+y)\big)\Big)$, wonach, mit Rücksicht auf Hilfssatz 2, $z \subset fl\Big(sb_z\big(sb(x+y)\big)\Big)$ und ferner (1) $x + z \subset x + fl\Big(sb_z\big(sb(x+y)\big)\Big)$. Wir haben ersichtlich: $sb_z(sb(x+y)) = sb_z(sb(x)) + sb_z(sb(y)) \subset sb(x) + sb_z(sb(y)) \subset fl(x) + sb_z(sb(y))$, woraus $fl\Big(sb_z\big(sb(x+y)\big)\Big) \subset fl\Big(fl(x) + sb_z\big(sb(y)\big)\Big) = fl\Big(x + sb_z\big(sb(y)\big)\Big)$ [nach der bekannten Formel: $fl(fl(x)+y) = fl(x+y)$[4]]; da hiebei $x \subset x + sb_z(sb(y)) \subset fl\Big(x + sb_z\big(sb(y)\big)\Big)$, so gilt: (2) $x + fl\Big(sb_z\big(sb(x+y)\big)\Big) \subset fl\Big(x + sb_z\big(sb(y)\big)\Big)$. Die Inklusionen (1) und (2) ergeben sofort: $x + z \subset fl\Big(x + sb_z\big(sb(y)\big)\Big)$; da die Menge $x+z$ widerspruchsvoll ist, so ist auch $fl\Big(x + sb_z\big(sb(y)\big)\Big)$ widerspruchsvoll [5]. Daraus folgt aber unmittelbar, daß auch $x + sb_z(sb(y))$ widerspruchsvoll ist w. z. b. w.

Hilfssatz 5: *Sind x und y widerspruchsfreie deduktive Systeme und ist $x.y$ in bezug auf die Aussagenmenge z vollständig, so gilt:* $x.y = y.x$. Beweis: Es sei a eine beliebige Aussage der Menge $x.z$. Es gilt: $x.y + \{a\} \subset x$, die Menge $x.y + \{a\}$ ist also widerspruchsfrei; da $x.y$ in Bezug auf z vollständig ist und $a \varepsilon z$, so folgt hieraus, daß $a \varepsilon fl(x.y)$. Die Menge $x.y$ ist als Durchschnitt zweier Systeme selbst ein System: $fl(x.y) = x.y$[6]; folglich: $a \varepsilon x.y$. Die Formel $a \varepsilon x.z$ impliziert also stets $a \varepsilon x.z$; demnach $x.z \subset x.y$, $x.z = x.y.z$. Ganz analog beweist man $y.z = x.y.z$, woraus schließlich $x.z = y.z$, w. z. b. w.

Wir wenden uns nun dem Beweis von Satz 1 zu. Nach dem schon zitierten Satz von Lindenbaum läßt sich die widerspruchsfreie Aussagenmenge x mindestens zu einem widerspruchsfreien und vollständigen System y erweitern. Nehmen wir an, es gäbe noch ein anderes derartiges System y_1, des von y verschieden wäre. Da y und y_1 vollständig sind, so muß die Aussagenmenge $y + y_1$ widerspruchsvoll sein[7]; da andrerseits $x + sb_z(s) \subset y + sb_z(s)$ und $x + sb_z(s)$ widerspruchsvoll ist, so muß auch $y + sb_z(s)$ widerspruchsvoll sein. Auf Grund von Hilfssatz 4 folgt hieraus, daß auch (1) die Menge $y + sb_{sb_z(s)}(y_1)$ widerspruchsvoll ist. Es gilt ferner: $sb_{sb_z(s)}(y_1) \subset sb(y_1) \subset fl(y_1) = y_1$ [y_1 ist ja ein deduktives System] und $sb_{sb_z(s)}(y_1) \subset sb_{sb_z(s)}(s) = sb_z(s)$ [Hilfssatz 2], wonach $sb_{sb_z(s)}(y_1) \subset y_1 . sb_z(s)$ und $y + sb_{sb_z(s)}(y_1) \subset y + y_1 . sb_z(s)$; diese Inklusion zusammen mit (1) ergibt, daß (2) die Menge $y + y_1 . sb_z(s)$ widerspruchsvoll ist. Wir haben noch; $x \subset y$ und $x \subset y_1$, folglich $x \subset y . y_1$; die Menge $y . y_1$ ist demnach — ebenso wie Menge x — vollständig in Bezug auf $sb_z(s)$.

---

[4] Vgl. *T.*, S. 365 f., Satz I. 2.
[5] Vgl. *T.*, S. 388, Satz I. 46.
[6] Vgl. *T.*, S. 370, Satz I. 11.
[7] Vgl. *T.*, S. 393, Satz I. 53.

Mit Hilfe von Hilfssatz 5 schließt man daraus, daß $y \,.\, sb_z(s) = y_1 \,.\, sb_z(s)$, daß also (3) $y = y + y_1 \,.\, sb_z(s)$.

Aus (2) und (3) ergibt sich sofort, daß y widerspruchsvoll ist, was unserer Annahme über y widerspricht. Das System y_1 kann also nicht von y verschieden sein. Damit ist Satz 1 bewiesen.

Um ihn auf konkrete Systeme anzuwenden, betrachten wir zuerst gewisse Systeme des sog. beschränkten Aussagenkalküls, d. h. der Theorie, in der das Implikationszeichen als einzige Konstante auftritt[8]).

Satz 2. *Wenn x eine widerspruchsfreie Aussagenmenge ist, die die drei Sätze: $CpCqp$, $CpCCpqq$ und $CCqrCCpqCpr$ als Elemente enthält, so ist das übliche (zweiwertige) System des beschränkten Aussagenkalküls das einzige widerspruchsfreie und vollständige deduktive System, das die Menge x umfaßt.*

Hilfssatz 6. *Wenn $x_0 = \{CpCqp, CpCCpqq, CCqrCCpqCpr\}$ und $z = \{p\}$, so gilt für jede Aussage $a \varepsilon sb_z(s)$ entweder $a \varepsilon fl(x_0)$ oder zugleich $Cap \varepsilon fl(x_0)$ und $Cpa \varepsilon fl(x_0)$.*

Beweis: Es sei y die Menge aller Aussagen, die die Behauptung des Hilfssatzes erfüllen. Durch die Anwendung der üblichen Schlußweisen des Aussagenkalküls zeigt man mühelos, daß die Aussage Cpp eine Folgerung der Menge x_0 ist. Daraus ergibt sich unmittelbar, daß für $a \varepsilon z$, also für $a = p$, zugleich $Cap \varepsilon fl(x_0)$ und $Cpa \varepsilon fl(x_0)$ gilt, wonach $a \varepsilon y$; folglich ist $z \subset y$. — Betrachten wir ferner zwei beliebige Aussagen a und b der Menge y; wir wollen beweisen, daß auch $Cab \varepsilon y$. Hier sind drei Fälle zu unterscheiden: (1) $b \varepsilon fl(x_0)$; (2) $Cap \varepsilon fl(x_0)$, $Cpa \varepsilon fl(x_0)$, $Cbp \varepsilon fl(x_0)$ und $Cpb \varepsilon fl(x_0)$; (3) $a \varepsilon fl(x_0)$, $Cbp \varepsilon fl(x_0)$ und $Cpb \varepsilon fl(x_0)$.

Der Kürze halber soll hier nur der erste Fall näher untersucht werden. Da $CpCqp \varepsilon x_0$, so $CbCab \varepsilon sb(x_0)$ und demnach $CbCab \varepsilon fl(x_0)$; da hiebei $b \varepsilon fl(x_0)$ und die Menge $fl(x_0)$ in Bezug auf die Operation der Abtrennung abgeschlossen ist, so $Cab \varepsilon fl(x_0)$. Analog zeigt man, daß auch im Falle (2) $Cab \varepsilon fl(x_0)$, dagegen im Falle (3) gilt zugleich $CCabp \varepsilon fl(x_0)$ und $CpCab \varepsilon fl(x_0)$; jedenfalls also $Cab \varepsilon y$. — Damit haben wir festgestellt, daß y die Menge z umfaßt und in Bezug auf die Operation der Implikationsbildung, d. i. auf die einzige Grundoperation des beschränkten Aussagenkalküls, abgeschlossen ist. Nach Hilfssatz 1 folgt hieraus, daß $sb_z(s) \subset y$; für jede Aussage $a \varepsilon sb_z(s)$ gilt also entweder $a \varepsilon fl(x_0)$ oder $Cap \varepsilon fl(x_0)$ und $Cpa \varepsilon fl(x_0)$, w. z. b. w.

Hilfssatz 7. *Wenn $x_0 = \{CpCqp, CpCCpqq, CCqrCCpqCpr\}$ und $z = \{p\}$, so gilt: a) die Menge x_0 ist in Bezug auf die Menge $sb_z(s)$ vollständig, b) die Menge $x_0 + sb_z(s)$ ist widerspruchsvoll.* Beweis: *a)* ergibt sich leicht aus dem vorigen Hilfssatz: für jede Aussage $a \varepsilon sb_z(s)$ gilt nämlich entweder $a \varepsilon fl(x_0)$ oder $Cap \varepsilon fl(x_0)$; wenn also a zu der Menge $fl(x)$ nicht gehört, so ist die Menge $x_0 + \{a\}$ widerspruchsvoll, da ihre Folgerungsmenge zugleich die

[8]) Vgl. *LT.*, S. 42 ff.

Aussagen a und Cap, demnach auch die Aussagenvariable p und folglich alle möglichen Aussagen als Elemente enthält (jede Aussage läßt sich ja durch Einsetzung aus der Variable p gewinnen). Aus analogem Grunde gilt auch *b)*.

Beweis des Satzes 2: Es ist eine wohl bekannte Tatsache, daß das übliche System L des Aussagenkalküls (das z. B. auf Grund der üblichen zweiwertigen Matrix definiert werden kann) widerspruchsfrei und vollständig ist[9]) und die Aussagen $CpCqp$, $CpCCpqq$ und $CCqrCCpqCpr$ als Elemente enthält. Daraus folgt, daß die Menge x_0, die aus diesen drei Aussagen besteht, widerspruchsfrei ist; dieser Umstand zusammen mit dem Hilfssatz 7 gestattet uns, den Satz 1 anzuwenden und auf diesem Wege zu schließen, daß L das einzige widerspruchsfreie und vollständige System ist, das X_0 umfaßt. Ist nun x eine beliebige widerspruchsfreie Aussagenmenge derart, daß $x_0 \subset x$, so läßt x sich sicher nach dem Satz von Lindenbaum zu einem widerspruchsfreien und vollständigen System y erweitern; da $x_0 \subset y$, so kann y nicht von L verschieden sein. Demnach ist L das einzige widerspruchsfreie und vollständige System, das x als Teilmenge enthält w. z. b. w.

Aus Satz 2 ergibt sich unmittelbar das folgende

Korollar. *Das übliche System des beschränkten Aussagenkalküls ist das einzige widerspruchsfreie und vollständige System, das die Aussagen $CpCqp$, $CpCCpqq$ und $CCqrCCpqCpr$ als Elemente enthält.*

Analoge Korollare lassen sich auch aus den unten angeführten Sätzen 3 und 4 ableiten.

Im Zusammenhang mit dem Satz 2 sei bemerkt, daß die Klasse aller Systeme des Aussagenkalküls, die die Aussagen $CpCqp$, $CpCCpqq$ und $CCqrCCpqCpr$ enthalten, ziemlich umfassend ist; zu dieser Klasse gehören z. B. alle *mehrwertigen Systeme* des Aussagenkalküls, die von J. Łukasiewicz stammen[10]), sowie das sog. *positive System* des Aussagenkalküls, das von D. Hilbert stammt[11]). Jedes dieser Systeme läßt sich also nur auf eine einzige Weise zu einem widerspruchsfreien und vollständigen Systeme erweitern.

Wenden wir uns nun zu solchen Systemen des Aussagenkalküls, in denen zwei Konstanten: das Implikations- und das Negationszeichen auftreten. Hier gilt der

Satz 3. *Wenn x eine widerspruchsfreie Aussagenmenge ist, die die vier Aussagen: $CpNNp$, $CqCpq$, $CNpCpq$ und $CpCNqNCpq$ als Elemente enthält, so ist das übliche (zweiwertige System des Aussagenkalküls das einzige widerspruchsfreie und vollständige System, das die Menge x umfaßt.*

---

[9]) Vgl. *LT.*, S. 42 f.

[10]) Vgl. *LT.*, S. 38 ff. und S. 43 f.

[11]) Als positives System des Aussagenkalküls bezeichnet man die Menge aller Folgerungen derjenigen Hilbertschen Axiome für den Aussagenkalkül (Math. Ann., *88*, 1923, S. 153), die kein Negationszeichen enthalten.

Der Beweis ist dem des Satzes 2 vollkommen analog und stützt sich auf den folgenden Hilfssatz, der den Hilfssätzen 6 und 7 genau entspricht:

Hilfssatz 8. *Wenn* $x_0 = \{CpNNp, CpCqp, CNpCpq, CpCNqNCpq\}$ *und* $z = \{CpNCNq\}$, *so gilt: a) für jede Aussage* $a \varepsilon sb_z(s)$ *entweder* $a \varepsilon fl(x_0)$ *oder* $Na \varepsilon fl(x_0)$, *b)* x_0 *ist in Bezug auf die Menge* $sb_z(s)$ *vollständig, c) die Menge* $x_0 + sb_z(s)$ *ist widerspruchsvoll.*

Nehmen wir schließlich eine Theorie des Aussagenkalküls in Betracht, in der neben dem Implikations- und dem Negationszeichen noch zwei Konstante, Zeichen der Alternative und der Konjunktion, auftreten. Satz 3 läßt sich dann folgendermaßen ergänzen:

Satz 4. *Wenn* x *eine widerspruchsfreie Aussagenmenge ist, die folgende zehn Aussagen:* $CpNNp$, $CqCpq$, $CNpCpq$, $CpCNqNCpq$, $CpApq$, $CqApq$, $CNpCNqNApq$, $CpCqKpq$, $CNpNKpq$ *und* $CNqNKpq$ *als Elemente enthält, so ist das übliche (zweiwertige) System des Aussagenkalküls das einzige widerspruchsfreie und vollständige System, das die Menge* x *umfaßt.*

Zu den Systemen x, die der Voraussetzung des Satzes 4 (bzw. 3) genügen, gehören wiederum alle mehrwertigen Systeme von Łukasiewicz und daneben u. a. das *intuitionistische System* des Aussagenkalküls von A. Heyting[12]). Für jedes dieser Systeme stellt also das übliche zweiwertige System die einzige widerspruchsfreie und vollständige Erweiterung dar[13]).

---

[12]) Heyting, S.-B. Preuß. Akad. d. Wiss., Phys.-math. Kl., 1930, II, S. 42—56.

[13]) Für die mehrwertigen Systeme des Aussagenkalküls wurde dieses Ergebnis schon früher auf einem anderen Wege gewonnen; vgl. LT., S. 40, Satz 21 und Anmerkung.

Mitteilungen des Jahres 1934/35.

Concerning rationality bases for curves. By G. T. Whyburn (University, Virginia, U. S. A.).

In a recent paper[1]) Knaster shows that in any stably regular curve[2]), i. e., a regular curve which remains regular upon the addition to it of an arbitrary regular curve, every rationality basis[3]) is also a regularity basis[3]); and he raises the question as to whether this property characterizes the stably regular curves among the rational curves. This question may be answered in the negative since it is possible to show (see Corollary 3 below) that *every dendrite* has this property and it is known[4]) that not all dendrites are stably regular. However, in this paper we shall prove a theorem characterizing the rational curves which do have this property; and from our theorem it follows (see Corollary 1 below) that *among the rational curves having no cut point, Knasters property does characterize the stably regular curves.*

Theorem. *In order that a rational curve H should have the property that every rationality basis in H is also a regularity basis in H it is necessary and sufficient that H be a regular curve and that every true cyclic element of H be stably regular.*

Proof. The conditions are necessary. For since H is rational, it therefore has a rationality basis and thus also a regularity basis so that H is a regular curve. Now let C be any true cyclic element of H. If we suppose, contrary to what we wish to show, that C is not stably regular, then[5]) C contains an arc t such that the points of order >2 in C are dense on t. Since C is cyclicly connected it follows that[6]) if N denotes the set of all points of t which are not local separating points of C, then N is dense on t. Let B denote the set of points $H-t+N$. We shall show that B is a rationality basis in H.

---

[1]) Monatsh. f. Math. u. Phys. *42* (1935), pp. 37—44.

[2]) For this and other notions related to the theory of curves the reader is referred to *K.* Menger, *Kurventheorie*, B. G. Teubner 1932.

[3]) If H is a continuum, a subset B of H is said to be a rationality basis (resp. a regularity basis) for H provided every point of L is contained in arbitrarily small neighborhoods whose boundaries intersect H in a countable (resp. finite) subset of B. See Knaster, loc. cit., in this connection see also G. E. Schweigert, abstract in Bulletin of the Amer. Math. Soc. *40* (1934), p. 535.

[4]) See my paper in Monatsh. f. Math. u. Phys. *38* (1931), pp. 1—4.

[5]) See my paper, loc. cit., or *Kurventheorie*, p. 361.

[6]) See my paper in Math. Ann., *102* (1929), p. 320.

To this end let p be any point of H. If $p \subset H-t$, obviously p is contained in arbitrarily small neighborhoods whose boundaries intersect H in a set which is countable and contained in B. Thus we may suppose $p \subset t$. Since N is dense in t, it follows that for any given $\varepsilon > 0$ there exists a neighborhood U of p of diameter $< \varepsilon$ whose boundary A intersects t in just two points a and b, where $a+b \subset N$. For each $x \varepsilon [A.H-(a+b)]$, let U_x be a neighborhood of x of diameter $< 1/2 \rho(x, t)$ such that if A_x denotes the intersection of the boundary of U_x with H then A_x is countable and of course $A_x \subset H-t \subset B$. Now, using the Borel Theorem, we can cover the set of all such points x of $A.H$ with $\rho(x, t) \geq 1$ by a finite number of these neighborhoods $[U_x]$, say with $U_1, U_2, U_3, \ldots, U_{k_1}$. Similarily we can cover the set of points x of $A.H$ with $1 \geq \rho(x, t) \geq \geq 1/2$ by a finite number of these neighborhoods, say U_{k_1+1}, $U_{k_1+2}, \ldots, U_{k_1+k_2}$, and so on. Continuing this process indefinitely we obtain a sequence $U_1, U_2, U_3, \ldots$ of the neighborhoods $[U_x]$ covering $A.H-(a+b)$ in such a way that if Q denotes the set $a+b+\Sigma A_i$, where A_i is the intersection of the boundary of U_i with H, then clearly Q is closed, countable and contained in N. Let R denote the component of $H-Q$ containing p. Then the boundary (rel. H) of R is a countable subset of Q and therefore of B. Also $R \subset U$. For if not, R would contain an arc pq where $q \subset A.H-$ $-(a+b)$; and this would give $q \subset U_i$ for some i, which is impossible since pq does not intersect A_i. Thus R is an ε-neighborhood of p rel. H whose boundary is countable and is contained in B; and accordingly B is a rationality basis for H.

But now by hypothesis B must also be a regularity basis for H. Hence if x and y are two points of t, there exists a neighborhood V of x such that $\bar{V}$ does not contain y and whose boundary D intersects H in just a finite number of points. But clearly D intersects t; and if q denotes the first point of D on the arc yx of t in the order y, x, then clearly x is a local separating point of C (since $H.D$ is finite); and this is impossible since $q \subset B.t = N$ and no point of N is a local separating point of C. Thus C is stably regular and the necessity of the condition is proven.

To establish the sufficiency of the conditions, we show first that if the conditions are satisfied in H and if B is any rationality basis in H, then any two points a and b in H are separated in H by a finite number of points of B. To this end let a and b be any two such points and consider the cyclic chain[7]) $C(a, b)$ in H. Take an arc ab in H. Then if $C(a, b) = ab$, any single point of $ab-$ $-(a+b)$ will separate[7]) a and b in H and clearly B must contain points of $ab-(a+b)$. If $C(a, b) \neq ab$, then there exists[7]) a true cyclic element C_1 of H which belongs to $C(a, b)$ and which intersects ab in a subarc $a_1 b_1$ where we have the order a, a_1, b_1, b on ab (a may coincide with a_1 and b_1 may coincide with b). Now

---

[7]) See Kuratowski et Whyburn, Fund. Math. *16* (1930), pp. 305—331.

since C_1 is stably regular, it follows by known results[8]) that there exist in C_1 a finite number of open „free" arc segments $\alpha_1, \alpha_2, \ldots, \alpha_k$ whose sum separates a_1 and b_1 in C_1. Now since clearly $B.C_1$ must be dense in C_1, for each i, we can find a point x_i on α_i which belongs to B; and clearly $x_1+x_2+\ldots+x_k$ also separates a_1 and b_1 in C_1 an thus likewise[9]) in H. Since we have the order a, a_1, b_1, b on ab it follows that this finite subset $x_1+x_2+\ldots+x_k$ of B also separates a and b in H and our statement is proven.

The rest of the proof now follows by a simple application of the Borel Theorem. For let $p \subset H$ and let $\varepsilon > 0$ be given. Let U be an ε-neighborhood of p. By what we have just shown it follows that for each $x \varepsilon H - U.H$ there exists a finite subset F_x of B and open subsets P_x and Q_x of H such that $P_x \supset p$, $Q_x \supset x$ and $P_x + Q_x + F_x = H$. Using the Borel Theorem we can find a finite number of the sets $[Q_x]$, say $Q_{x_1}, Q_{x_2}, \ldots, Q_{x_k}$ which cover $H - U.H$. Then it is clear that the product R of $P_{x_1}, P_{x_2}, \ldots, P_{x_k}$ is an open subset of H containing p, $R \subset U$ and the boundary (rel. H) of R is a subset of $F_{x_1}+F_{x_2}+\ldots+F_{x_k}$, and this is a finite subset of B. Accordingly B is a regularity basis in H and our theorem is proven.

Corollaries. (1) *If C is a rational curve and has no cut point, then in order that C be a stably regular curve it is necessary and sufficient that every rationality basis in H be a regularity basis in H.*

(2) *The property of a continuum H that every rationality basis in H be a regularity basis in H is cyclicly extensible.*

(3) *In any dendrite every rationality basis is also a regularity basis.*

Zur topologischen Algebra. Von O. Taussky.

Während sich die Nullteiler in topologischen Ringen[1]) nicht durch topologische Eigenschaften charakterisieren lassen, ist dies für die annullierenden Elemente in lokal kompakten, separabeln zusammenhängenden Ringen möglich. Dabei wird das Element a annullierend genannt, wenn $ax = 0$ für alle Elemente x des Ringes.

Ein lokal kompakter separabler zusammenhängender Ring ist — als abelsche Gruppe hinsichtlich der Addition betrachtet — direkte Summe einer endlichdimensionalen Vektorgruppe und einer kompakten Gruppe, welche jede kompakte Untergruppe der Gesamtgruppe enthält[2]). Diese kompakte Gruppe besteht nur aus annullierenden Elementen[3]). Umgekehrt gilt: Läßt sich eine lokal kompakte, separable zusammenhängende Abelsche Gruppe G als additive Gruppe

[8]) See my paper in Math. Ann., loc. cit., p. 322.

[9]) See Kuratowski et Whyburn loc. cit.

[1]) Die Definitionen für die Begriffe der topologischen Algebra stammen von D. von Dantzig, Math. Annalen *107*.

[2]) Siehe L. Pontrjagin, Ann. of Math. *33*, 163—174.

[3]) Siehe N. Jacobson and O. Taussky, Proc. Nat. Acad. *21*, 106—108.

eines topologischen Ringes interpretieren, so erfüllen die dabei notwendigerweise auftretenden annullierenden Elemente eine kompakte Untergruppe. Denn G ist direkte Summe einer Vektorgruppe und einer kompakten Gruppe C, wobei die Vektorgruppe immer als additive Gruppe eines Ringes ohne annullierende Elemente aufgefaßt werden kann, z. B. eines hyperkomplexen Systems, dessen Basiselemente e_i die Relationen $e_i^2 = e_i$, $e_i e_k = 0$, $(i \neq k)$, erfüllen, während die kompakte Gruppe C aus lauter annullierenden Elementen besteht.

Insbesondere entspricht der Tatsache, daß jeder Ring mindestens ein annullierendes Element, nämlich die 0, enthält, die Tatsache, daß die additive Gruppe jedes topologischen Ringes wenigstens eine kompakte Untergruppe, nämlich $\{0\}$ besitzt.

Angelegenheiten des Kolloquiums.

Im Studienjahre 1934/35 arbeiteten mit ausländischen Stipendien im Kolloquium A. Ville (Paris), im Wintersemester Dr. L. M. Blumenthal (National Research Fellow), A. Flores (Madrid), Prof. T. Hirano (Tokio), ab Jänner 1935 Prof. A. Tarski (Warschau), überdies während kürzerer Perioden Dr. L. Geymonat (Torino) und Frau Dr. Lutmann (Lwów).

Druck von Gottlieb Gistel & Cie., Wien, III., Münzgasse 6.

Die metrische Methode in der Variationsrechnung[1]).

Von Karl Menger.

Einleitung.

Die Probleme der Variationsrechnung gehen davon aus, daß drei Daten vorliegen: 1) ein *Raum* R oder ein Teil eines Raumes; 2) eine *Funktion* φ, die in R definiert ist und aus der durch einen Integrationsprozeß ein Funktional hergeleitet wird, das jedem Gebilde C einer gewissen Klasse $\mathfrak{C}(\varphi)$ von Gebilden des Raumes R eine Zahl zuordnet, die wir $\lambda_\varphi(C)$ nennen wollen; 3) die *Klasse der Vergleichsgebilde,* eine Teilklasse $\mathfrak{C}$ von $\mathfrak{C}(\varphi)$. Eines der Hauptprobleme ist: *Existiert in* $\mathfrak{C}$ *ein Gebilde* C_0, *welches das Funktional* λ_φ *auf* $\mathfrak{C}$ *zu einem Extremum macht, z. B. eine Minimante* C_0 *für* λ_φ *auf* $\mathfrak{C}$, *so daß* $\lambda_\varphi(C_0) \leqq \lambda_\varphi(C)$ *für alle Elemente* C *von* $\mathfrak{C}$ *gilt?* Dabei ist bisweilen wünschenswert, daß C_0 einer vorweg gegebenen engeren Teilklasse $\mathfrak{C}_0$ von $\mathfrak{C}$ angehört.

Es ist klar, daß diese Aufgabe umso allgemeiner gelöst ist, je allgemeiner R, φ und $\mathfrak{C}$ gewählt werden können und je spezieller die Teilklasse $\mathfrak{C}_0$ von $\mathfrak{C}$. Tatsächlich war es auch die Tendenz der Variationsrechnung, die Existenz von Minimanten vor allem für immer allgemeinere Integranden φ und für immer allgemeinere Klassen $\mathfrak{C}$ von Vergleichsgebilden zu beweisen. Was bisher systematisch erreicht worden ist, ist für eindimensionale Probleme (bei denen die betrachteten Raumgebilde *Kurven* sind) und Integranden, die nicht höhere als erste Ableitungen enthalten, der Beweis der Existenz von Minimanten unter folgenden Voraussetzungen:

1) Der Raum R ist eine *Teilmenge eines euklidischen Raumes,* in dem jeder Punkt als ein geordnetes n-Tupel reeller Zahlen $\xi_1, \ldots, \xi_n$, und der Abstand je zweier Punkte durch die Pythagoräische Abstandformel gegeben ist.

2) Der Integrand φ, der von Punkt und Richtung abhängt, also als eine Funktion von $2n$ Variablen $\xi_1, \ldots, \xi_n$ und $\xi_1', \ldots \xi_n'$,

---

1) Der Inhalt der vorliegenden Abhandlung wurde im Mathematischen Kolloquium vorgetragen am 3. X. 1935 und in vier Sitzungen im Sommersemester 1936, in denen auch die anschließenden Ergebnisse von Alt, Mayer, Novák und Wald mitgeteilt wurden. Den Herren Alt und Wald bin ich für zahlreiche Bemerkungen zu dieser Arbeit und für viele selbstlose Mitarbeit zu aufrichtigem Dank verpflichtet, Herrn Wald insbesondere für eine Abkürzung meiner ursprünglichen Bedingung V*, durch welche der Beweis des zweiten Fundamentallemmas in § 30 so einfach wurde, und für Ratschläge zu Kapitel VI.

gegeben wurde, die in $\xi_1', \ldots, \xi_n'$ positiv homogen vom ersten Grade ist, mußte besitzen:

a) Stetigkeits- und Differenzierbarkeitseigenschaften; das mindeste, was diesbezüglich bisher in der systematischen Theorie jedenfalls verlangt wurde, war *Stetigkeit von* φ *im ganzen Definitionsbereich.*

b) Regularitätseigenschaften; im Falle der Existenz zweiter partieller Ableitungen von φ nach den ξ_i' wurde die *Legendresche Bedingung* vorausgesetzt, mindestens aber die *Konvexität der Indikatrix* von φ in jedem Punkt $\xi_1, \ldots, \xi_n$.

Als Funktional $\lambda_\varphi(C)$ wird für die Kurve $C = \{\xi_1(\gamma), \ldots, \xi_n(\gamma)\}$ $(\alpha \leqq \gamma \leqq \beta)$ das zwischen α und β erstreckte

$$\int \varphi(\xi_1(\gamma), \ldots, \xi_n(\gamma), \xi_1'(\gamma), \ldots, \xi_n'(\gamma))\, d\gamma$$

betrachtet. Damit die Integration durchführbar sei, muß natürlich der Integrand (zumindest fast überall) einen Sinn haben, d. h. die $\xi_i(\gamma)$ müssen fast überall Ableitungen $\xi_i'(\gamma)$ besitzen. Tonelli führte für $\int$ das Lebesguesche Integral ein, brauchte also nur die Existenz von Ableitungen $\xi_i'(\gamma)$ abgesehen von einer Nullmenge, was auf Grund eines bekannten Satzes von Lebesgue durch die von Tonelli auch sonst benutzte Voraussetzung der Rektifizierbarkeit von C gewährleistet wird. Also:

3). Die Klasse $\mathfrak{C}$ der Vergleichskurven wurde *als Klasse von rektifizierbaren Kurven* angenommen.

Das Resultat der vorliegenden Arbeit ist ein Beweis der Existenz von Minimanten und des Osgoodschen Satzes unter Voraussetzungen, die erheblich allgemeiner sind als alle bisherigen Lösungen, u. zw. sowohl was den Raum R, also auch was den Integranden φ, als auch was die Klasse der Vergleichskurven $\mathfrak{C}$ betrifft. Wir werden nämlich zulassen:

1) daß *R ein beliebiger kompakter metrischer Raum* im Sinne von Fréchet sei, ja sogar das, was wir einen *Längenraum* nennen.

2) Die Funktien φ lassen wir, da Richtungen in einem allgemeinen metrischen Raum nicht erklärt sein müssen, von den Punkten und den Paaren verschiedener Punkte des Raumes abhängen. Sind im Raume Richtungen definiert und besitzt φ gewisse Eigenschaften, so erhalten wir den klassischen Fall einer Punkt-Richtung-Funktion als Spezialfall (vgl. § 44). Die Punkt-Punktepaar-Funktion φ brauchen wir weder als überall stetig noch als überall regulär vorauszusetzen, sondern lassen, wenn φ *beschränkt* ist, im wesentlichen (über die genauen Voraussetzungen vgl. § 21) zu:

a) Jede rektifizierbare Kurve C enthält eine Nullmenge von Punkten, in denen φ *hinsichtlich der Punkt-Veränderlichen unstetig ist, wenn nur in der Nachbarschaft dieser Unstetigkeitspunkte das Integral entlang geschlossenen Kurven nicht negativ ist.*

b) Jede rektifizierbare Kurve enthält eine Nullmenge von Punkten, in denen φ *nicht quasiregulär ist,* wobei Quasiregularität in einem Punkt rein metrisch als eine Eigenschaft von φ in der Punktepaar-Veränderlichen definiert werden kann (vgl. § 12).

Unter diesen Voraussetzungen zeigen wir (§§ 26, 27), daß das Riemannsche Integral von φ entlang C für jede rektifizierbare Kurve C existiert und für jedes λ in der Klasse der Kurven mit Längen $\leqq \lambda$ unterhalb stetig ist. Bezeichnen wir dieses Integral mit $\lambda_\varphi(C)$, so beweisen wir (§ 40) für dieses Funktional:

Wenn für jedes λ die Längen der Kurven C, für welche $\lambda_\varphi(C) \leqq \lambda$ gilt, beschränkt sind, *so enthält jede abgeschlossene Klasse* $\mathfrak{C}$ *von rektifizierbaren Kurven eine Minimante.*

Dabei ist zu beachten: Auch wenn man sich auf Probleme in euklidischen Räumen, z. B. in der Ebene beschränkt, so ist durch die erwähnten Ergebnisse die Klasse der Integranden, für welche Minimanten existieren, gegenüber dem bisher Bekannten erheblich vergrößert, wobei unstetige Integranden bekanntlich auch in der Physik eine Rolle spielen.

3) Machen wir über φ noch eine Zusatzvoraussetzung, welche im wesentlichen darauf hinausläuft, *daß für jeden Punkt* p *eine Zahl* $\omega(p) > 0$ *existiert, so daß für jede geschlossene Kurve* C, *die hinlänglich nahe an* p *liegt,* $\lambda_\varphi(C) \geqq \omega(p)\,\lambda(C)$ *gilt, wo* $\lambda(C)$ *die Länge von* C *ist,* so existiert $\lambda_\varphi(C)$ für jede stetige Kurve C, ob C rektifizierbar ist oder nicht (§ 31). Über $\mathfrak{C}$ brauchen wir dann bloß vorauszusetzen, daß es eine *abgeschlossene Klasse von stetigen Kurven* ist, und können die Existenz einer Minimante C_0 für λ_φ auf $\mathfrak{C}$ beweisen, die selbst *rektifizierbar* ist, so daß $\mathfrak{C}_0$ als die Klasse der rektifizierbaren Kurven von $\mathfrak{C}$ vorgegeben werden kann (§ 40).

3′) Machen wir über φ bloß die schwächere Zusatzvoraussetzung, daß für jeden Punkt p *abgesehen höchstens von den Punkten einer abzählbaren abgeschlossenen Menge* ein $\omega(p)$ der erwähnten Art existiert, so gilt, wenn $\varphi \geqq 0$ ist, immer noch: *Jede abgeschlossene Klasse* $\mathfrak{C}$ *von stetigen Kurven enthält eine Minimante* C_0 in dem Sinn, daß zumindest ein uneigentliches Riemannsches Integral von φ entlang C_0 existiert und der unteren Schranke der Zahlen $\lambda_\varphi(C)$ für alle C von $\mathfrak{C}$ gleich ist. Aber nun ist *die Minimante* C_0 *nicht notwendig rektifizierbar.*

Daß es Variationsprobleme mit nichtrektifizierbaren Minimanten gibt, hat Hahn entdeckt (Wien. Ber. *134*, S. 446). Daß schon für sehr einfache Integranden in der Ebene, z. B. wenn φ Quadratwurzel eines Polynoms 8-ten Grades in $\xi_1, \xi_2, \xi_1', \xi_2'$ ist) nicht rektifizierbare Extremanten auftreten können, zeigte Carathéodory. (Variationsrechnung, 1935 S. 310). Unser erwähnter Satz gibt nun eine umfassende Klasse von Problemen an, für welche die Lösbarkeit, d. h. die Existenz von Minimanten, nur wenn man nichtrektifizierbare Vergleichskurven zuläßt, behauptet werden kann, dann aber auch in der Tat gewährleistet ist.

Wieder ist zu beachten: Auch wenn man sich auf überall stetige und quasireguläre Integranden in einem euklidischen Raum beschränkt, so gestatten die erwähnten Theoreme bemerkenswerte, bisher unbekannte Folgerungen: Bilden wir nämlich für jeden Punkt $\xi_1, \ldots, \xi_n$ das Minimum $\chi(\xi_1, \ldots, \xi_n)$ der Zahlen $\varphi(\xi_1, \ldots, \xi_n, \xi_1', \ldots, \xi_n') + \varphi(\xi_1, \ldots, \xi_n, -\xi_1', \ldots, -\xi_n')$ für alle $\xi_1', \ldots, \xi_n'$ mit der Quadratsumme 1, so sind die Zusatzvoraussetzungen 3. bzw. 3'., damit gleichbedeutend (vgl. § 56), daß $\chi(\xi_1, \ldots, \xi_n) > 0$ für alle Punkte $\xi_1, \ldots, \xi_n$, bzw. für alle Punkte abgesehen höchstens von denen einer abzählbaren abgeschlossenen Menge gilt. 3. umfaßt insbesondere alle Integranden, die Tonelli *seminormal* nennt.

Selbstverständlich können alle Ergebnisse dieser Arbeit mutatis mutandis für Maximanten ausgesprochen werden.

Was die Methoden betrifft, mit denen wir die erwähnten Ergebnisse (und noch viel allgemeinere) im folgenden gewinnen, so wird ihr Wesen vielleicht am klarsten durch einen Vergleich mit den Methoden Tonellis. Der außerordentliche Fortschritt, den die Variationsrechnung diesem Gelehrten hinsichtlich der uns hier beschäftigenden Fragen verdankt, beruht auf zwei methodisch ganz verschiedenen Leistungen: *Erstens* hat Tonelli durch Analyse der Arbeiten von Hilbert und Lebesgue entdeckt, daß es die Unterhalbstetigkeit des Funktionals sei, was die Anwendung der direkten Methoden ermögliche, ein bei aller Einfachheit sehr fruchtbarer Gedanke, der methodisch der *allgemeinen Analysis* angehört. *Zweitens* ist es Tonelli durch sehr tiefliegende Untersuchungen, die methodisch der *Theorie der reellen Funktionen* verwandt sind, gelungen, in euklidischen Räumen für eine sehr große Klasse von Integranden und für Klassen von rektifizierbaren Kurven zu beweisen, daß das Funktional tatsächlich unterhalbstetig ist.

Was den ersten Gedanken betrifft, so übernehmen wir ihn im wesentlichen unverändert von Tonelli, indem wir die Schlüsse Tonellis explizite als einen Satz der allgemeinen Analysis unter dem Namen „*Tonellisches Prinzip*" formulieren; die einzige vielleicht erwähnenswerte Verallgemeinerung, die wir in dieser Hinsicht vornehmen, besteht darin, daß wir einen Satz über zwei Funktionen einer Limesklasse aussprechen, was sich in der Schlußbemerkung von § 40 als fruchtbar erweist, während Tonelli als die eine Funktion in der Limesklasse der Kurven stets speziell die Kurvenlänge betrachtet hat.

Was den zweiten Teil, den Nachweis der Unterhalbstetigkeit, betrifft, so gehen wir ganz andere Wege als Tonelli, wie ja schon daraus hervorgeht, daß wir unsere Untersuchungen in äußerst allgemeinen Räumen durchführen, in denen nicht einmal Richtungen erklärt sein müssen, daß wir die Kurven nicht durch Funktionen definieren müssen und, wenn sie so definiert sind, uns nie auf die Existenz von Ableitungen dieser Funktionen berufen, daß wir unsere Sätze ohne Einführung des Lebesgueschen Integrales gewinnen, daß Existenz und Unterhalbstetigkeit des Funktionals als unmittelbare Folgerungen eines und desselben Satzes (des ersten Fundamentallemmas und des Satzes von § 25) gewonnen werden usw. Auch dürfen wir vielleicht die besondere Kürze der Beweise unserer allgemeinen Sätze betonen. All dies hängt damit zusammen, daß wir die Methoden unserer *Geometrie der Abstände* oder *metrischen Geometrie* verwenden in Weiterführung unserer früheren Untersuchungen über geodätische Bogen in allgemeinen metrischen Räumen (vgl. insbesondere Mathem. Annalen *104*). Das Funktional $\lambda_\varphi(C)$, das Riemannsche Integral von φ entlang C, ist für uns nichts anderes als die Länge von C in einem anderen Raum mit einer durch φ verzerrten Metrik. In jedem Punkte des Raumes läßt sich eine „Tangential"-Metrik definieren, in welcher die Quasiregularität sich als eine Dreiecksungleichung ausdrückt. Die Riemannschen Summen, welche dieses Integral approximieren, sind für uns die Längen der C eingeschriebenen Polygone, wie wir sie für $\varphi = 1$ in der Theorie der Länge von Kurven in metrischen Räumen (Math. Ann. *104*) herangezogen haben. Für den Fall, daß R Teil eines euklidischen Raumes ist, hat inzwischen Bouligand in einer (mir erst nach Abfassung meiner Arbeit in den Fundam. Mathem. *25* bekannt gewordenen) Abhandlung (Mém. Soc. Sc. Liége, 3. ser. t. *19*) den Gedanken entwickelt, das Funktional, das Riemannsche Integral von φ entlang C, direkt als Limes seiner Riemannschen Summen aufzufassen, und wenn er sich auch bloß auf die Durchführung im Falle positiv semidefiniter Integranden beschränkt, so geht aus seiner Arbeit doch hervor, daß er sich einer großen Tragweite dieses Gedankens bewußt war.

Allgemeine metrische Räume werden auch in den tiefliegenden neuesten Untersuchungen von Morse betrachtet, aber in ganz anderem Sinn als in der vorliegenden Arbeit. Der metrische Raum, in dem Morse Funktionen untersucht, entspricht dem, was wir mit $\mathfrak{C}$ (φ) bezeichnen, d. h. der Menge der *Kurven* eines anderen Raumes, für welche ein Funktional λ_φ definiert ist, wobei für je zwei Kurven ein passender Abstand definiert wird. Morse studiert unterhalb stetige Funktionen in seinem Raum, d. h. in unserer Ausdrucksweise: unterhalb stetige Funktionale in der (geeignet metrisierten) Menge der Kurven eines Raumes. Während das Tonellische Prinzip aus der Unterhalbstetigkeit eines Funktionals lediglich die Existenz von *Minimanten* folgert, zieht Morse durch Anwendung topologischer Methoden aus der Unterhalbstetigkeit viel weitergehende Schlüsse hinsichtlich der viel tieferen Probleme der Existenz und Anzahl von Elementen, für welche das Funktional in gewissem Sinne *stationär* ist. Insoferne Morse seine Theorie bloß auf die Unterhalbstetigkeit des Funktionals gründet, kann sie auch an die allgemeinen Voraussetzungen der vorliegenden Arbeit angeschlossen werden, d. h. für die allgemeinen von uns betrachteten Punkträume und die allgemeinen von uns zugelassenen Integranden kann durch Anwendung der Morseschen Theorie (statt des Tonellischen Prinzips) auf die

Existenz und Anzahl von stationären Kurven bestimmter Art geschlossen werden. Für manche seiner Überlegungen benötigt Morse übrigens bloß, daß das Funktional auf der Menge der Kurven beschränkter Länge unterhalb stetig sei — eine Eigenschaft, welche im folgenden unter ganz besonders allgemeinen Voraussetzungen über den Integranden als erfüllt nachgewiesen wird. Die Grundgedanken und Hauptergebnisse der vorliegenden Arbeit wurden in drei Noten in C. R. Paris, t. *201*, p. 705, t. *202* p. 1009 und p. 1648 veröffentlicht (21. X. 1935, 23. III. und 18. V. 1936).

I. Die metrischen Grundlagen.

Allgemeine Abstandsräume und ihre Polygone.

§ 1. Als *Raum mit reellen Abständen* bezeichnen wir eine Menge R von irgendwelchen Elementen, die *Punkte* genannt werden, wenn jedem geordneten Paar p, q von Elementen eine reelle Zahl $\delta(p, q)$ zugeordnet ist gemäß der einzigen Bedingung: $\delta(p, p)=0$ für jeden Punkt p. Eine geordnete endliche Menge $P=\{p_1, \ldots, p_n\}$ von Punkten von R nennen wir ein *Polygon*. Ist $p_1=p_n$, so heißt P *geschlossen*. Jedes Polygon $\{p_h, p_{h+1}, \ldots, p_i\}$ heißt ein *Abschnitt* von P, für $h=1$ ein *Anfangs*-, für $i=n$ ein *End*abschnitt. Wir bilden zu P die $n-1$ Abstände $\delta(p_i, p_{i+1})$ und führen die folgenden Zahlen ein:

$\lambda(P)$, die *Länge* von P, als die Summe der $n-1$ Zahlen $\delta(p_i, p_{i+1})$;

$\mu(P)$, die *Spanne* von P, als die Zahl $\delta(p_1, p_n)$; ein geschlossenes Polygon hat also stets die Spanne 0;

$\xi(P)$, den *Defekt* von P, als den Quotienten $\frac{\mu(P)-\lambda(P)}{|\mu(P)|}$, falls $\mu(P)\neq 0$, während wir, wenn $\mu(P)=0$ ist, $\xi(P)=\infty, -\infty, 0$ setzen je nachdem $\lambda(P)<0, >0, =0$ ist. Für jedes aus zwei Punkten bestehende Polygon $P=\{p, q\}$ gilt $\mu(P)=\lambda(P)=\delta(p, q)$, also $\xi(P)=0$.

§ 2. Gilt $\delta(p, q)=\delta(q, p)$, bzw. $\delta(p, q)>0$ für je zwei verschiedene Punkte p und q, so nennen wir R einen *Raum mit symmetrischen*, bzw. *mit positiven Abständen*. Die Räume mit symmetrischen *und* positiven Abständen sind die von Fréchet eingeführten *halbmetrischen* Räume. In einem halbmetrischen Raum bezeichnen wir in üblicher Weise für jede Teilmenge M und jede Zahl $\rho>0$ mit $U(M, \rho)$ die Menge aller Punkte q, die von mindestens einem Punkt von M einen Abstand $<\rho$ haben; insbesondere mit $U(p, \rho)$ die *Kugel* mit dem Zentrum p und dem Radius ρ, d. h. die Menge aller Punkte q, für welche $\delta(p, q)<\rho$ gilt. Wir nennen *Durchmesser* einer Menge M die obere Schranke der Zahlen $\delta(p, q)$ für alle Punkte p, q von M. Als *Norm* $\nu(P)$ des Polygons $P=\{p_1, \ldots, p_n\}$ bezeichnen wir die größte der $n-1$ Zahlen $\delta(p_i, p_{i+1})$ $(i=1, \ldots, n)$.

§ 3. Wir sagen, ein halbmetrischer Raum habe *gleichmäßig stetige Abstände*, wenn für jedes $\varepsilon > 0$ ein $\delta > 0$ existiert, so daß aus $\delta(p, p') + \delta(q, q') < \delta$ stets $|\delta(p, q) - \delta(p', q')| < \varepsilon$ folgt. Man überzeugt sich leicht davon, daß in diesen Räumen ein Großteil der für metrische Räume bewiesenen Sätze der Theorie der reellen Funktionen und der Topologie giltig bleiben.

§ 4. Wir nennen einen Raum mit reellen Abständen (die nicht notwendig positiv oder symmetrisch sein müssen) einen *Dreiecksraum im Punkte* p, wenn $\delta(p, q) + \delta(q, r) \geqq \delta(p, r)$ für je zwei Punkte q und r des Raumes gilt. Ist R ein Dreiecksraum im Punkt p, so gilt, wie wir durch Betrachtung der Punkte q und $r = p$ sehen, stets $\delta(p, q) + \delta(q, p) \geqq 0$; ferner, wie sich durch vollständige Induktion ergibt, für ein in p beginnendes Polygon $P = \{p_1 = p, p_2, \ldots, p_n\}$ stets $\lambda(P) \geqq \mu(P)$, also $\xi(P) \leqq 0$. Ein Raum mit reellen Abständen, der in jedem Punkt Dreiecksraum ist, heißt ein *Dreiecksraum* schlechthin. In einem Dreiecksraum hat jedes Polygon eine die Länge nicht übertreffende Spanne und daher einen nicht positiven Defekt. Die Dreiecksräume mit symmetrischen und positiven Abständen sind die von Fréchet eingeführten *metrischen Räume*. Da in einem metrischen Raum offenbar stets $|\delta(p, q) - \delta(p', q')| \leqq \delta(p, p') + \delta(q, q')$ gilt, hat er gleichmäßig stetige Abstände.

Über die Länge von Streckenbildern in Räumen mit reellen Abständen.

§ 5. Es sei R ein Raum mit reellen Abständen. Ist jeder Zahl γ des abgeschlossenen Intervalles $[\alpha, \beta]$ genau ein Punkt $p(\gamma)$ von R zugeordnet, so sagen wir, es liege ein Streckenbild vor und bezeichnen dasselbe mit $p[\alpha, \beta]$. Ist $\alpha \leqq \alpha_1 < \beta_1 \leqq \beta$, so bezeichnen wir die Zuordnung des Punktes $p(\gamma)$ zu jeder Zahl γ des Intervalles $[\alpha_1, \beta_1]$ als den *Abschnitt* $p[\alpha_1, \beta_1]$ von $p[\alpha, \beta]$. Ist $\alpha < \alpha_1 < \beta_1 < \beta$, so heißt $p[\alpha_1, \beta_1]$ ein *innerer* Abschnitt von $p[\alpha, \beta]$.

Als *Teilpolygon* von $C = p[\alpha, \beta]$ bezeichnen wir jedes Polygon $P = \{p(\gamma_1), \ldots, p(\gamma_n)\}$, für welches $\alpha = \gamma_1 < \gamma_2 < \ldots < \gamma_{n-1} < \gamma_n = \beta$ gilt. Wir nennen *C-Norm* von P und bezeichnen mit $\nu_C(P)$ die größte der Zahlen $\gamma_{i+1} - \gamma_i$. Ein und dasselbe Polygon P kann Teilpolygon verschiedener Streckenbilder C und C' sein, wobei $\nu_C(P)$ und $\nu_{C'}(P)$ natürlich keineswegs gleich sein müssen. Eine Folge $P_1, P_2, \ldots$ von Teilpolygonen des Streckenbildes C heißt *ausgezeichnet*, wenn $\lim \nu_C(P_n) = 0$ gilt.

§ 6. Wir bezeichnen als *obere* bzw. *untere Länge* von C die (eventuell unendliche) obere bzw. untere Schranke der Zahlen λ, für welche eine ausgezeichnete Folge $P_1, P_2, \ldots$ von Teilpolygonen von C mit $\lim \lambda(P_n) = \lambda$ existiert. Stimmen obere und untere Länge von C überein, so nennen wir ihren gemeinsamen Wert die *Länge* von C, bezeichnen ihn mit $\lambda(C)$ und nennen C *im weiteren Sinne rektifizierbar*. Kurven mit endlicher Länge heißen *rektifizierbar*. Legen wir zwei ausgezeichnete Folgen von Teilpolygonen von C

zusammen, so ist die entstehende Folge ausgezeichnet. Daraus ergibt sich in üblicher Weise: *Damit C rektifizierbar, bzw. im weiteren Sinn rektifizierbar sei, ist notwendig und hinreichend, daß für jede ausgezeichnete Folge* $P_1, P_2, \ldots$ *von Teilpolygonen von C die Zahlenfolge* $\lambda(P_1), \lambda(P_2), \ldots$ *gegen eine endliche Zahl, bzw. gegen eine endliche oder unendliche Zahl konvergiere.* In einem metrischen Raum ist bekanntlich jedes Streckenbild im weiteren Sinn rektifizierbar.

§ 7. Aus der Definition von $\lambda(C)$ folgert man mühelos: 1) Ist das Streckenbild $C = p[\alpha, \beta]$ rektifizierbar, so auch jeder Abschnitt $p[\alpha_1, \beta_1]$ von C. 2) Ist $C = p[\alpha, \beta]$ sowie alle Abschnitte von C rektifizierbar im weiteren Sinn und ist γ eine Zahl aus $[\alpha, \beta]$, so gilt, wenn wir $C' = p[\alpha, \gamma]$ und $C'' = p[\gamma, \beta]$ setzen, $\lambda(C) = \lambda(C') + \lambda(C'')$. 3) Ist $C = p[\alpha, \beta]$ sowie alle Abschnitte von C im weiteren Sinn rektifizierbar und konvergieren die Zahlen α_n und β_n aus $[\alpha, \beta]$ gegen α bzw. β, so gilt, wenn wir $C_n = p[\alpha_n, \beta_n]$ setzen, $\lim \lambda(C_n) = \lambda(C)$.

Die Verzerrung einer Metrik durch eine Funktion und die Tangentialräume.

§ 8. Es sei in einem halbmetrischen Raum mit gleichmäßig stetigen Abständen jedem Punkt p und jedem Paar von untereinander verschiedenen Punkten q, r eine Zahl $\varphi(p; q, r)$ zugeordnet. Wir bezeichnen mit R_φ die Menge der Punkte von R, wenn je zwei Punkten q, r als Abstand nicht wie in R die Zahl $\delta(q, r)$, sondern die Zahl

$$\delta_\varphi(q, r) = \varphi(q; q, r)\, \delta(q, r) \text{ für } q \neq r \text{ und } \delta_\varphi(q, q) = 0$$

zugeordnet wird. Wir nennen den Raum R_φ mit reellen (nicht notwendig symmetrischen oder positiven) Abständen *den durch* φ *verzerrten Raum R.* Da die Punkte von R_φ mit denen von R übereinstimmen, so können wir zu jedem Polygon $P = \{p_1, \ldots, p_n\}$ in R das Polygon $\{p_1, \ldots, p_n\}$ in R_φ bilden, das wir gleichfalls mit P bezeichnen wollen. Länge, Spanne, Defekt und Norm von P sind aber in R und in R_φ im allgemeinen verschieden. Wir bezeichnen Länge und Spanne von P in R_φ mit $\lambda_\varphi(P)$, bzw. $\mu_\varphi(P)$, haben also

$$\lambda_\varphi(P) = \sum \delta_\varphi(p_i, p_{i+1}) = \sum \varphi(p_i; p_i, p_{i+1})\, \delta(p_i, p_{i+1}),$$
$$\mu_\varphi(P) = \delta_\varphi(p_1, p_n) = \varphi(p_1; p_1, p_n)\, \delta(p_1, p_n),$$

falls wir alle Terme $=0$ setzen, die einen Faktor $\delta(q, q)$ enthalten.

Es ist für das Folgende nicht notwendig, anzunehmen, daß $\varphi(p; q, r)$ für *alle* Paare verschiedener Punkte q, r definiert sei. Es würde genügen, daß eine Zahl $\rho > 0$ existiert derart, daß für jeden Punkt p die Funktion φ für alle Paare verschiedener Punkte q, r aus $U(p, \rho)$ definiert ist. Nur der Einfachheit halber machen wir die erwähnte weitergehende Annahme, die übrigens unter gewissen Voraussetzungen (vgl. § 44) gar nicht weitergehend ist.

§ 9. Wir bezeichnen für jeden Punkt p von R mit $R_{\varphi p}$ die Menge aller Punkte von R_φ (oder, was dasselbe ist, von R), wenn je zwei Punkten q, r als Abstand die Zahl

$$\delta_{\varphi p}(q, r) = \varphi(p; q, r)\, \delta(q, r) \text{ für } q \neq r \text{ und } \delta_{\varphi p}(q, q) = 0$$

zugeordnet ist. Wir nennen $R_{\varphi p}$ den *Tangentialraum* an R_φ im Punkte p, welcher auch der *Berührungspunkt* von $R_{\varphi p}$ mit R_φ heißt. Aus den Definitionen von $\delta_\varphi(q, r)$ und $\delta_{\varphi p}(q, r)$ erkennt man, daß dieselben im Falle $q=p$ übereinstimmen, d. h. *vom Berührungspunkt p hat im Tangentialraum $R_{\varphi p}$ jeder Punkt r den gleichen Abstand wie im Raume R_φ.* Jedes Polygon $P=\{p_1, \ldots, p_n\}$ in R oder R_φ kann auch als Polygon in $R_{\varphi p}$ aufgefaßt werden. Länge, Spanne und Defekt von P in $R_{\varphi p}$ bezeichnen wir mit $\lambda_{\varphi p}(P)$, $\mu_{\varphi p}(P)$, $\xi_{\varphi p}(P)$, haben also:

$$\lambda_{\varphi p}(P)=\sum \delta_{\varphi p}(p_i, p_{i+1})=\sum \varphi(p; p_i, p_{i+1})\,\delta(p_i, p_{i+1})$$

$$\mu_{\varphi p}(P)=\delta_{\varphi p}(p_1, p_n)=\varphi(p; p_1, p_n)\,\delta(p_1, p_n)$$

$$\xi_{\varphi p}(P)=[\mu_{\varphi p}(P)-\lambda_{\varphi p}(P)] : |\mu_{\varphi p}(P)|, \text{ bzw. } \infty, -\infty, 0,$$
$$\text{wenn } \mu_{\varphi p}(P)\neq 0, \text{ bzw. } =0 \text{ und } \lambda_{\varphi p}(P)<0, >0, =0.$$

Für jedes im Berührungspunkt p beginnende Polygon $P=\{p_1=p, p_2, \ldots, p_n\}$ gilt dem eben Bemerkten zufolge $\mu_{\varphi p}(P)=\mu_\varphi(P)$, d. h. es hat die gleiche Spanne in $R_{\varphi p}$ wie in R_φ, während seine Längen in $R_{\varphi p}$ und R_φ verschieden sein können.

Schwankung und Tangentialkontraktion der Funktion φ.

§ 10. Für jeden Punkt p und jede Zahl $\rho>0$ bezeichnen wir mit $\sigma(p, \rho)$ die (endliche oder unendliche) obere Schranke der Zahlen $|\varphi(p'; q, r)-\varphi(p''; q, r)|$ für alle Punkte p', p'', q, r von $U(p, \rho)$. Wir nennen $\sigma(p, \rho)$ die *ρ-Schwankung von φ in p.* Für festes p ist $\sigma(p, \rho)$ offenbar eine monoton wachsende Funktion von ρ. Falls $\varphi(p; q, r)$ in einer Umgebung von p beschränkt ist, so besitzt also $\sigma(p, \rho)$ für $\rho\to 0$ einen Limes, den wir mit $\sigma(p)$ bezeichnen und die *Schwankung von φ in p* nennen wollen. $\sigma(p)=0$ ist offenbar damit gleichbedeutend, daß die Punkt-Punktepaar-Funktion φ in p hinsichtlich ihrer Punktvariablen stetig ist. Aus der Definition von $\sigma(p)$ erhalten wir unmittelbar, bzw. durch Anwendung des Borelschen Theorems:

Bemerkung 1. Ist $\sigma(p)<\sigma$, so existiert ein $\rho(p)>0$, so daß $\sigma(p, \rho(p))<\sigma$.

Bemerkung 2. Für jedes $\sigma>0$ ist die Menge aller Punkte p mit $\sigma(p)\geqq\sigma$ abgeschlossen.

Bemerkung 3. Gilt $\sigma(p)<\sigma$ für jeden Punkt einer kompakten Menge A, so existiert ein $\rho>0$ derart, daß $\sigma(p, \rho)<\sigma$ für alle Punkte p von A.

§ 11. Für jeden Punkt p und jede Zahl $\rho>0$ bezeichnen wir mit $\tau(p, \rho)$ die (endliche oder unendliche) obere Schranke der Zahlen $\xi_{\varphi p}(P)$ für alle in p beginnenden Polygone $P=\{p_1=p, p_2, \ldots, p_n\}$, für die $\delta(p, p_i)<\rho$, welche also in R innerhalb $U(p, \rho)$ liegen. Wir nennen $\tau(p, \rho)$ die *ρ-Kontraktion* der Funktion φ in p. Da jedes aus zwei Punkten bestehende Polygon $P=\{p, q\}$ den Defekt $\xi_{\varphi p}(P)=0$ hat (vgl. § 1), so ist $\tau(p, \rho)\geqq 0$ für jedes p und ρ. Für festes p ist $\tau(p, \rho)$ offenbar eine monoton wachsende Funktion von ρ. Wir bezeichnen mit $\tau_0(p)$, bzw. $\tau(p)$ den (eventuell unendlichen) nicht negativen Limes von $\tau(p, \rho)$ für $\rho\to 0$, bzw. $\rho\to\infty$. Natürlich gilt für $\tau_0(p)$ ein Analogon der Bemerkung 1 über $\sigma(p)$. Hingegen ist das Analogon der Bemerkungen 2 und 3 weder für $\tau_0(p)$ noch für $\tau(p)$ allgemein gültig.

§ 12. Ist $\tau(p, \rho)=0$, so hat insbesondere jedes Polygon $\{p, q, r\}$ in $U(p, \rho)$ in $R_{\varphi p}$ einen Defekt von $\leqq 0$, d. h. $U(p, \rho)$ ist in $R_{\varphi p}$ ein Dreiecksraum in p. Ist umgekehrt $R_{\varphi p}$ ein Dreiecksraum in p, so folgt, daß jedes Polygon $\{p, q, r\}$ in $U(p, \rho)$ einen Defekt $\leqq 0$ hat, und durch Induktion, daß jedes in p beginnende Polygon in $U(p, \rho)$ einen Defekt $\leqq 0$ hat, mithin $\tau(p, \rho)=0$ gilt. Also: *Damit* $\tau(p, \rho)=0$, *bzw.* $\tau(p)=0$ *gelte, ist notwendig und hinreichend, daß* $U(p, \rho)$ *in* $R_{\varphi p}$ *bzw.* $R_{\varphi p}$ *selbst ein Dreiecksraum in* p *sei.* Gilt $\tau(p)=0$, so sagen wir, φ sei in p *quasiregulär.*

Über Streckenbilder in halbmetrischen Räumen mit gleichmäßig stetigen Abständen.

§ 13. Es sei R ein halbmetrischer Raum mit gleichmäßig stetigen Abständen. Als *Streckenbild* in R bezeichnen wir von nun an eine Zuordnung eines Punktes $p(\gamma)$ zu jeder Zahl γ eines Intervalles $[\alpha, \beta]$ nur dann, wenn sie *stetig* ist, d. h. wenn aus $\lim \gamma_n=\gamma$ stets $\lim \delta(p(\gamma_n), p(\gamma))=0$ folgt. Man beweist mühelos die Bemerkung: *Ist* C *ein Streckenbild, so existiert zu jedem* $\nu>0$ *ein* $\nu_C>0$ *derart, daß für jedes Teilpolygon* P *von* C, *für welches* $\nu_C(P)<\nu_C$ *gilt, die Norm* (vgl. § 2) *der Ungleichung* $\nu(P)<\nu$ *genügt.*

§ 14. Sind $C=p[\alpha, \beta]$ und $C'=p'[\alpha', \beta']$ zwei Streckenbilder, so bezeichnen wir als ihren *Abstand* $\delta(C, C')$ die untere Schranke aller Zahlen δ, für welche eine eineindeutige und ordnungstreue (also beiderseits stetige) Abbildung zwischen den Intervallen $[\alpha, \beta]$ und $[\alpha', \beta']$ existiert derart, daß für zwei einander vermöge dieser Abbildung entsprechende Zahlen γ aus $[\alpha, \beta]$ und γ' aus $[\alpha', \beta']$ der Punkt $p(\gamma)$ von C und der Punkt $p'(\gamma')$ von C' einen Abstand $<\delta$ von einander haben. Offenbar gilt $\delta(C, C')=\delta(C', C)$; es gilt $\delta(C, C')=0$ dann und nur dann, wenn $[\alpha, \beta]$ und $[\alpha', \beta']$ ordnungstreu derart auf einander abbildbar sind, daß $p(\gamma)=p'(\gamma')$ für je zwei einander vermöge dieser Abbildung entsprechende Zahlen γ aus $[\alpha, \beta]$ und γ' aus $[\alpha', \beta']$ gilt.

Ist $\delta(C. C')=0$, so entspricht also jedem Punkt $p(\gamma)$ von $C=p[\alpha, \beta]$ ein Punkt $p'(\gamma')$ von $C'=p'[\alpha', \beta']$, der mit $p(\gamma)$ identisch ist, mithin jedem Teilpolygon P von C ein Teilpolygon P' von C', das aus den nämlichen Punkten in der nämlichen Anordnung besteht wie P. Während P und P' in Länge, Spanne, Defekt und Norm offenbar übereinstimmen, können $\nu_C(P)$ und $\nu_{C'}(P')$ natürlich verschieden sein. Doch beweist man mühelos die Bemerkung: *Ist*

$\delta(C, C') = 0$ *und ist* $P_1, P_2, \ldots$ *eine ausgezeichnete Folge von Teilpolygonen von* C, *so sind die entsprechenden (mit* $P_1, P_2, \ldots$ *identischen) Polygone* $P'_1, P'_2, \ldots$ *eine ausgezeichnete Folge von Teilpolygonen von* C', d. h. aus $\lim \nu_C(P_n) = 0$ folgt stets $\lim \nu_{C'}(P'_n) = 0$.

§ 15. Hieraus ergibt sich nach § 6 unmittelbar: Ist $\delta(C, C') = 0$, so stimmen die oberen (die unteren) Längen von C und C' überein. Ist C rektifizierbar (im weiteren Sinne rektifizierbar), so auch C' und es gilt $\lambda(C) = \lambda(C')$.

§ 16. Von Wichtigkeit für das Weitere ist folgender Hilfsbegriff: Ist $P = \{p_1, \ldots, p_n\} = \{p(\gamma_1), \ldots, p(\gamma_n)\}$ ein Teilpolygon des Streckenbildes $C = p[\alpha, \beta]$, das durch P in $n-1$ Abschnitte $C_i = = p[\gamma_i, \gamma_{i+1}]$ $(i = 1, \ldots, n-1)$ zerlegt wird, und ist $P' = \{p'_1, \ldots, p'_{n'}\}$ irgend ein Polygon des Raumes, so sagen wir, P' *habe von* P *mit Bezug auf* C *einen Abstand* $< \delta$ und schreiben $\delta(P', P_C) < \delta$, wenn P' sich in $n-1$ aneinander schließende Abschnitte $P'_i = = \{p'_{h_i}, p'_{h_i+1}, \ldots, p'_{h_{i+1}}\}$ $(i = 1, \ldots, n-1)$ zerlegen läßt, so daß

1) $h_1 = 1$ und $h_n = n'$, d. h. der erste Abschnitt P'_1 ein Anfangs-, der letzte P'_{n-1} ein Endabschnitt von P' ist;

2) $\delta(p_i, p'_{h_i}) < \delta$ $(i = 1, \ldots, n)$, d. h. der Anfangs- und der Schlußpunkt des Abschnittes P'_i liegt in einem Abstand $< \delta$ vom Anfangs- bzw. Schlußpunkt des Abschnittes C_i;

3) P'_i ist in $U(C_i, \delta)$ enthalten.

Die untere Schranke aller Zahlen δ, für welche $\delta(P', P_C) < \delta$ gilt, können wir mit $\delta(P', P_C)$ bezeichnen und den Abstand des Polygones P' von P mit bezug auf C nennen.

Es ist klar und wird ja auch durch die Bezeichnung zum Ausdruck gebracht, daß $\delta(P', P_C) < \delta$ keine symmetrische Beziehung zwischen P und P' ist, denn P muß dabei Teilpolygon von C sein, P' nicht notwendig. Aber selbst wenn P' ebenfalls Teilpolygon von C ist, folgt aus $\delta(P', P_C) < \delta$, wie man leicht einsieht, keineswegs notwendig $\delta(P, P'_C) < \delta$. Zum Beispiel muß ja P' mindestens so viele Punkte enthalten wie P.

§ 17. Hingegen zeigen wir: *Es seien* $C = p[\alpha, \beta]$ *und* $C' = p'[\alpha', \beta']$ *zwei Streckenbilder,* δ *eine Zahl* $> \delta(C, C')$. *Dann existiert zu jedem Teilpolygon* $P = \{p(\gamma_1), \ldots, p(\gamma_n)\}$ *von* C *ein* $\nu' > 0$ *derart, daß für jedes Teilpolygon* P' *von* C' *mit* $\nu_{C'}(P') < \nu'$ *die Ungleichung gilt:* $\delta(P', P_C) < \delta$.

Wir setzen $\delta(C, C') = \delta_1$. Dann existiert zu jedem Punkt p_i von P ein Punkt $p^*_i = p'(\gamma^*_i)$ von C' $(i = 1, \ldots, n)$, so daß $\delta(p_i, p^*_i) < \delta$ und daß, wenn wir $C^*_i = p'[\gamma^*_i, \gamma^*_{i+1}]$ setzen, C^*_i Teilmenge von $U(C_i, \delta)$ ist, wobei wir $\alpha' = \gamma^*_1 < \gamma^*_2 < \ldots < \gamma^*_{n-1} < \gamma^*_n = \beta'$ annehmen können Ist nun $\nu' > 0$ hinreichend klein und bezeichnen wir den Teil des Intervalles $[\gamma^*_i - \nu', \gamma^*_i + \nu']$, der in $[\alpha', \beta']$ liegt, mit $J^*_i = [\gamma'_i, \gamma''_i]$, so sind die Intervalle $J^*_1, J^*_2, \ldots, J^*_n$ paarweise fremd und es gilt:

Wie immer wir die Zahl γ_i^0 in J_i^* wählen, stets hat $p'(\gamma_i^0)$ von p_i einen Abstand $<\delta$, da $\delta>\delta_1$, und stets ist $p'[\gamma_i^0, \gamma_{i+1}^0]$ Teilmenge von $U(C_i, \delta)$. Ist dann $P'=\{p_1', \ldots, p_{n'}'\}$ irgend ein Teilpolygon von C' mit $\nu_{C'}(P')<\nu'$, so können wir offenbar in jedem der Intervalle J_i^* genau ein γ_{h_i}' wählen, insbesondere γ_1' in J_1^* und γ_n' in J_n^*. Setzen wir dann $P_i'=\{p'(\gamma_{h_i}'), \ldots, p'(\gamma_{h_{i+1}}')\}$, so ist klar, daß $\delta(P', P_C)<\delta$.

§ 18. Wir führen ohne seinen einfachen Beweis folgenden Satz an: *Ist C ein Streckenbild mit endlicher oberer Länge, so existiert zu jedem $\varepsilon>0$ ein $\beta>0$ derart, daß für jedes System von endlich vielen Kugeln mit einer Durchmessersumme $<\beta$ ein $\nu>0$ existiert mit folgender Eigenschaft: Für jedes Teilpolygon $P=\{p_1, \ldots, p_n\}$ von C mit $\nu_C(P)<\nu$ ist die Summe aller Zahlen $\delta(p_i, p_{i+1})$, für welche p_i in einer der Kugeln liegt, $<\varepsilon$*[2]).

II. Die Grundlagen aus der allgemeinen Analysis.

§ 19. Es sei L eine Limesklasse im Sinne von Fréchet. Ist jedem Element e der Teilmenge M von L ein $\alpha(e)$ zugeordnet, wo $\alpha(e)$ entweder eine endliche reelle Zahl oder ∞ oder $-\infty$ ist, so sagen wir: Auf M ist die *Funktion* $\alpha(e)$ definiert; dieselbe heißt *endlichwertig*, wenn keine unendlichen Funktionswerte auftreten, das heißt $\alpha(e)$ für jedes e von M eine endliche Zahl ist. Die Funktion $\alpha(e)$ heißt *nach oben beschränkt* auf der Teilmenge M_1 von M, wenn eine reelle Zahl α^* existiert, so daß $\alpha(e)\leqq\alpha^*$ für jedes Element e von M_1. Sie heißt *unterhalb stetig auf* M_1, wenn für jedes Element e von M_1 und für jede Elementefolge $e_1, e_2, \ldots$ aus M_1 mit $\lim e_n=e$ die Ungleichung $\liminf \alpha(e_n)\geqq\alpha(e)$ gilt.

Eine Folge $e_1, e_2, \ldots$ von Elementen von M, für welche die Zahlen $\alpha(e_1), \alpha(e_2), \ldots$ gegen die untere Schranke der Zahlen $\alpha(e)$ für alle Elemente e von M konvergieren, heißt eine *Minimalfolge für* $\alpha(e)$ *auf* M. Ein Element e_M, für welches $\alpha(e_M)\leqq\alpha(e)$ für alle Elemente e von M gilt, nennen wir *minimant für* $\alpha(e)$ *auf* M.

§ 20. Es ist zweckmäßig, für jede Teilmenge M von L und jede reelle Zahl ρ mit $M[\alpha(e)\leqq\rho]$ die Menge aller Elemente e von M zu bezeichnen, für welche $\alpha(e)\leqq\rho$ gilt, mit $M[\alpha(e)<\infty]$ die Menge aller e von M, für welche $\alpha(e)$ endlich ist. In dieser Ausdrucksweise gilt folgender Satz, der von Tonelli für die Limesklasse der rektifizierbaren Streckenbilder euklidischer Räume und für den Fall, daß α die Länge ist, verwendet wurde, und den wir bezeichnen als

[2]) Anmerkung bei der Korrektur: Die Behauptung von § 18 gilt ohne weitere Einschränkung nur für rektifizierbare Streckenbilder, sodaß auch die Theoreme des Abschnitts III zunächst nur für rektifizierbare Streckenbilder ausgesprochen werden können. Vom Abschnitt IV angefangen bleiben alle Resultate hiedurch unberührt.

Tonellisches Prinzip. *Es sei auf der Limesklasse L eine (nicht notwendig endlichwertige) Funktion $\alpha(e)$ definiert. Auf $L[\alpha(e) < \infty]$ sei ferner eine endlichwertige Funktion $\beta(e)$ definiert, so daß für jede endliche Zahl ρ folgende Bedingungen erfüllt sind:*

1) *Die Menge $L[\alpha(e) \leqq \rho]$ ist kompakt.*

2) *$\beta(e)$ ist auf $L[\alpha(e) \leqq \rho]$ unterhalb stetig.*

3) *$\alpha(e)$ ist auf $L[\beta(e) \leqq \rho]$ nach oben beschränkt.*

Dann enthält jede abgeschlossene Teilmenge A von $L[\alpha(e) < \infty]$ mindestens ein Element e_A, welches Limes einer Minimalfolge für $\beta(e)$ auf A ist. Jedes derartige Element ist minimant für $\beta(e)$ auf A.

Zusatz 1. *Gilt außer* 1), 2), 3) *noch*

4) *$\beta(e) = \infty$ für jedes e, für welches $\alpha(e) = \infty$,*

dann enthält jede abgeschlossene Teilmenge A von L mindestens ein Element e_A, das Limes einer Minimalfolge für $\beta(e)$ auf A ist. Dabei ist $\alpha(e_A)$ endlich, es sei denn, daß A Teilmenge von $L[\alpha(e) = \infty] = L[\beta(e) = \infty]$ ist.

Zusatz 2. *Gilt außer* 1), 2) *die Voraussetzung*

5) *Aus $\lim e_n = e$ und $\lim \alpha(e_n) = \infty$ folgt stets $\lim \beta(e_n) = \infty$,*

die zusammen mit 2) *für die Unterhalbstetigkeit von $\beta(e)$ hinreichend ist, so folgt, daß* 3) *äquivalent ist mit*

3') *Jede Folge $e_1, e_2, \ldots,$ für welche $\beta(e_n) \leqq \beta^* < \infty$ für $n = 1, 2, \ldots$ gilt, enthält eine konvergente Teilfolge.*

Zum Beweis des Tonellischen Prinzipes setzen wir

6) β = untere Schranke der $\beta(e)$ für alle e von A.

Wenn nicht der im Zusatz 1 ausgeschlossene Fall vorliegt, daß $A \subseteq L[\beta(e) = \infty]$, so haben wir: 6') $\beta < \infty$. Sei $e_1, e_2, \ldots$ eine Minimalfolge für $\beta(e)$ auf A. Setzen wir $\beta' = \beta + 1$, wenn $\beta > -\infty$, und $\beta' = 1$, wenn $\beta = -\infty$, so ist β' wegen 6') endlich und

7) $\beta(e_n) \leqq \beta'$ für fast alle n, etwa für $n = n_0, n_0 + 1, \ldots$

Da β' endlich ist, ist $\alpha(e)$ auf $A[\beta(e) \leqq \beta']$ wegen 3) nach oben beschränkt, d. h. es existiert eine Zahl $\alpha^* < \infty$, so daß $\alpha(e) \leqq \alpha^*$ für alle e von $A[\beta(e) \leqq \beta']$, insbesondere

8) $\alpha(e_n) \leqq \alpha^*$ für $n = n_0, n_0 + 1, \ldots$

Nun ist nach 1) die Menge $L[\alpha(e) \leqq \alpha^*]$ kompakt. Die Folge $e_{n_0}, e_{n_0+1}, \ldots$ enthält also eine Teilfolge $e_1', e_2', \ldots,$ die offenbar gleichfalls eine Minimalfolge für $\beta(e)$ auf A ist, und die gegen ein Element von $L[\alpha(e) \leqq \alpha^*]$, das e_A heißen möge, konvergiert. Da $e_1', e_2', \ldots$ Elemente der abgeschlossenen Menge A sind, liegt e_A in A. Damit ist der erste Teil der Behauptung bewiesen. Sei zweitens e_A Limes irgend einer Minimalfolge für $\beta(e)$ auf A, etwa von $e_1', e_2', \ldots$ mit $\lim \beta(e_n') = \beta$. Da e_A und fast alle e_n' für ein $\alpha^* < \infty$ zu $L[\alpha(e) \leqq \alpha^*]$ gehören, so folgt aus 2), daß $\beta(e_A) \leqq \beta$, also wegen

6), daß e_A minimant für $\beta(e)$ auf A ist. Damit ist das Tonellische Prinzip bewiesen. Zusatz 1 folgt aus ihm unmittelbar.

Zum Beweis von Zusatz 2 setzen wir 1), 2), 5) voraus und zeigen: *Erstens* folgt 3′) aus 3). Denn liegt eine Folge $e_1, e_2, \ldots$ mit $\beta(e_n) \leqq \beta^* < \infty$ für alle n vor, so existiert nach 3) ein $\alpha^* < \infty$, so daß $\alpha(e_n) \leqq \alpha^*$ für alle n, so daß also die Folge nach 1) eine konvergente Teilfolge enthält. *Zweitens* folgt 3) aus 3′). Denn existierte eine Folge $e_1, e_2, \ldots$ und eine Zahl $\beta^* < \infty$, so daß $\lim \alpha(e_n) = \infty$ und $\beta(e_n) \leqq \beta^*$ für alle n, so enthielte sie wegen 3′) eine konvergente Teilfolge, etwa $e_{i1}, e_{i2}, \ldots$ mit $\lim e_{i_n} = e$. Das widerspricht 5).

Zusatz 3. *Ist auf L eine (nicht notwendig endlichwertige) Funktion $\alpha(e)$ definiert und unterhalb stetig und ist für jede endliche Zahl ρ die Menge $L[\alpha(e) \leqq \rho]$ kompakt, dann enthält jede abgeschlossene Teilmenge A von L mindestens ein Element e_A, welches $\alpha(e)$ auf A minimisiert.* Dies ergibt sich aus dem Tonellischen Prinzip, wenn wir $\alpha(e) = \beta(e)$ setzen, da dann die Bedingungen 3), 4) sicher erfüllt sind.

III. Existenz und Unterhalbstetigkeit der Länge in $\mathbf{R}_\varphi$.

§ 21. Wir machen nun über die Funktion $\varphi(p; q, r)$ von § 8 folgende Voraussetzungen:

I. *Zu jedem Streckenbild C existiert eine Umgebung U, so daß die obere Schranke der Zahlen $|\varphi(p; q, r)|$ für alle Punkte p, q, r von U endlich ist.* Wir bezeichnen diese obere Schranke mit φ_U.

II. *Für jedes Streckenbild C mit endlicher oberer Länge und jedes $\sigma > 0$ ist die Menge $C[\sigma(p) \geqq \sigma]$ aller Punkte p von C, für welche $\sigma(p) \geqq \sigma$ gilt, eine Nullmenge im engeren Sinn.*

III. *Für jedes Streckenbild C mit endlicher oberer Länge und jedes $\tau > 0$ existiert eine Zahl $\rho_\tau > 0$ derart, daß die Menge $C[\tau(p, \rho_\tau) \geqq \tau]$ aller Punkte von C, für welche $\tau(p, \rho_\tau) \geqq \tau$ gilt, eine Nullmenge im engeren Sinn ist.*

IV. *Für jedes Streckenbild C mit endlicher oberer Länge existiert ein $\rho_C > 0$, so daß die obere Schranke der Zahlen $\tau(p, \rho_C)$ endlich ist.* Wir bezeichnen diese obere Schranke mit τ_C.

V. *Zu jedem Punkt p, für den $\sigma(p) > 0$ gilt, existiert eine Umgebung U und eine Zahl $\nu > 0$, so daß $\lambda_\varphi(P) \geqq 0$ gilt für jedes geschlossene Polygon P, das in U enthalten ist und eine Norm $\nu(P)$ besitzt, die $< \nu$ ist.*

Dabei sagen wir im Sinne von Carathéodorys Theorie des linearen Maßes, die Teilmenge M eines halbmetrischen Raumes mit gleichmäßig stetigen Abständen sei eine *Nullmenge im engeren* bzw. *im weiteren Sinne,* wenn für jede beschränkte Teilmenge M' von M und für jedes $\beta > 0$ endlich bzw. unendlich viele Kugeln mit einer Durchmessersumme $< \beta$ existieren, welche M' überdecken. Jede kompakte Nullmenge im weiteren Sinne ist nach dem Borelschen Überdeckungssatz eine Nullmenge im engeren Sinne.

§ 22. Nach Bemerkungen 1, 2, 3 von § 10 ist **II** äquivalent mit einer **III** analogen Voraussetzung: Für jedes C und $\sigma > 0$ existiert ein $\rho_\sigma > 0$ derart, daß $C[\sigma(p, \rho_\sigma) \geqq \sigma]$ eine Nullmenge im engeren Sinn ist. Nach der Schlußbemerkung von § 21 ist **II** äquivalent mit der Voraussetzung: *Für jedes rektifizierbare Streckenbild C ist die Menge $C[\sigma(p) > 0]$ aller Unstetigkeitspunkte von φ eine Nullmenge im weiteren Sinne.* Hingegen ist **III**, da die Analoga der Bemerkungen 2 und 3 von § 10 für $\tau(p)$ nicht gelten, im allgemeinen weder äquivalent mit einer **II** analogen Voraussetzung, noch damit, daß die Menge $C[\tau(p) > 0]$ eine Nullmenge im weiteren Sinn sei. — Aus einem Ergebnis von Novák in diesem Heft geht hervor, daß schon in der Ebene die Voraussetzung „*Jedes rektifizierbare Streckenbild* einer offenen Menge enthält eine Nullmenge von Punkten bestimmter Art" schwächer ist als die Voraussetzung „*Die offene Menge* enthält eine Nullmenge von Punkten der betreffenden Art".

§ 23. Es sei in einem halbmetrischen Raum mit gleichmäßig stetigen Abständen $C = p[\alpha, \beta]$ ein Streckenbild mit endlicher oberer Länge. U sei eine nach **I** existierende offene Menge $\supseteq C$, so daß gilt

1) Die obere Schranke φ_U der Zahlen $|\varphi(p;\, q, r)|$ für alle Punkte p, q, r von U ist endlich.

Es sei σ irgend eine Zahl > 0. Wir setzen $C_\sigma = C[\sigma(p) \geqq \sigma]$. Da C_σ abgeschlossen in C, also kompakt ist, existieren wegen **V** nach dem Borelschen Theorem endlichviele C_σ überdeckende Umgebungen $U_1, \ldots, U_k$ und Zahlen $\nu_1, \ldots, \nu_k$, so daß $\lambda_\varphi(P) \geqq 0$ gilt für jedes geschlossene Polygon P, das für eine der Zahlen $j = 1, \ldots, k$ den Bedingungen $P \subseteq U_j$ und $\nu(P) \leqq \nu_j$ oder gar $\nu(P) \leqq \nu$ genügt, wo ν die kleinste der Zahlen $\nu_1, \ldots \nu_k$ ist. Für jedes $\beta > 0$ existieren nach **II** endlich viele C_σ überdeckende Kugeln mit einer Durchmessersumme $< \beta$; daraus folgt:

2) Für jedes $\beta > 0$ existieren endlichviele offene Kugeln $K_1, \ldots, K_l$ mit einer Durchmessersumme $< \beta$, so daß *a)* $C_\sigma \subseteq K_1 + \ldots + K_l$ und daß *b)* ein $\rho_1 > 0$ existiert mit folgender Eigenschaft: Für jede der Zahlen $j = 1, \ldots, l$ ist $U(K_j, \rho_1)$ Teil einer Umgebung $U_1, \ldots, U_k$, so daß also nach dem Obigen: *c)* $\lambda_\varphi(P) \geqq 0$ für jedes geschlossene Polygon P, das Teil einer der Umgebungen $U(K_j, \rho_1)$ $(j = 1, \ldots, l)$ ist und für das $\nu(P) < \nu$ gilt.

Für jeden Punkt p der Menge $C'_\sigma = C - C.(K_1 + \ldots + K_l)$ haben wir wegen 2 *a)* und der Definition von C_σ offenbar $\sigma(p) < \sigma$. Da C'_σ kompakt ist, folgt aus Bemerkung 3 von § 10:

3) Es existiert ein $\rho_2 > 0$, so daß $\sigma(p, \rho_2) < \sigma$ für jeden Punkt p von C'_σ.

Es sei nun τ irgend eine Zahl >0. Nach **III** existiert ein $\rho_\tau>0$, so daß für jedes $\beta>0$, wenn wir $C_\tau=C[\tau(p, \rho_\tau)\geqq\tau]$ setzen, gilt:

4) Es existieren endlichviele Kugeln $K_1', \ldots, K_m'$ mit einer Durchmessersumme $<\beta$, so daß $C_\tau \subseteq K_1'+\ldots+K_m'$. Hieraus folgt auf Grund der Definition von C_τ:

5) Es gilt $\tau(p, \rho_\tau)<\tau$ für jeden Punkt p der Menge $C_\tau'= =C-C.(K_1'+\ldots+K_m')$.

6) Wir nennen ρ die kleinste der Zahlen $\rho_1, \rho_2, \rho_\tau, \nu$ und ρ_C, wo ρ_C die nach **IV** existierende Zahl ist.

7) Es existiert nach § 13 ein $\nu_C>0$, so daß, wenn $\alpha\leqq\gamma\leqq\gamma'\leqq\beta$ und $\gamma'-\gamma\leqq\nu_C$ gilt, der Abschnitt $p[\gamma, \gamma']$ einen Durchmesser $<\rho$ hat, also $\subseteq U(p(\gamma), \rho)$ ist.

Es sei nun $P=\{p_1, \ldots, p_n\}=\{p(\gamma_1), \ldots, p(\gamma_n)\}$ irgend ein Teilpolygon von C mit $\nu_C(P)<\nu_C$. Aus 7) folgt $p[\gamma_i, \gamma_{i+1}]\subseteq U(p_i, \rho)$ und weiter:

8) Es existiert ein $\delta>0$, so daß *a)* $\delta<\nu$ und *b)* $U(C_i, \delta)\subseteq \subseteq U(p_i, \rho)$, wo $C_i=p[\gamma_i, \gamma_{i+1}]$ $(i=1, \ldots, n-1)$.

Es sei weiter $P'=\{p_1', \ldots, p_{n'}'\}$ irgend ein Polygon, für das $\nu(P')<\nu$ und $\delta(P', P_C)<\delta$ gilt. P' zerfällt also in $n-1$ Abschnitte $P_i'=\{p_{h_i}', \ldots, p_{h_{i+1}}'\}$ $(i=1, \ldots, n-1)$ mit $h_1=1$ und $h_n=n'$ und es gilt

9*a*) $\delta(p_{h_i}', p_i)<\delta$ $(i=1, \ldots, n)$ und $P_i'\subseteq U(C_i, \delta)$; wegen 8) ist also $P_i'\subseteq U(p_i, \rho)$. Da P_i' und wegen 8*b*) auch p_{i+1}, in $U(p_i, \rho)$ liegen, so haben wir:

9*b*) Die Punkte $p_i, p_{h_i}', \ldots, p_{h_{i+1}}', p_{i+1}$ liegen in $U(p_i, \rho)$ $(i=1, \ldots, n-1)$. Aus 9*a*) und 1) folgt:

10) $\delta_\varphi(p_i, p_{hi}')+\delta_\varphi(p_{h_{i+1}}', p_{i+1})<2\varphi_U\delta$.

Wir suchen nun zu den zwei betrachteten Polygonen P und P' für den Ausdruck

$$\lambda_\varphi(P)-\lambda_\varphi(P')=\sum_{i=1}^{n-1}[\delta_\varphi(p_i, p_{i+1})-\lambda_\varphi(P_i')]$$

eine Abschätzung nach oben. Wir betrachten zu diesem Zweck die einzelnen Summanden und unterscheiden zwei Fälle:

Erster Fall: p_i liegt in $K_1+\ldots+K_l$. Dann setzen wir $P_i''=\{p_i, p_{h_i}', \ldots, p_{h_{i+1}}', p_{i+1}, p_i\}$ und haben

11) $\delta_\varphi(p_i, p_{i+1})-\lambda_\varphi(P_i')=\delta_\varphi(p_i, p_{i+1})-\lambda_\varphi(P_i'')+\delta_\varphi(p_i, p_{h_i}')+ +\delta_\varphi(p_{h_{i+1}}', p_{i+1})+\delta_\varphi(p_{i+1}, p_i)$.

Wegen 9*b*) und 6) gilt: $P_i''\subseteq U(p_i, \rho_1)$, also, da p_i in einer der Kugeln K_j liegt, $P_i''\subseteq U(K_j, \rho_1)$. Nun ist P_i'' ein geschlossenes

Polygon und es gilt $\nu(P''_i) < \nu$; denn wegen $8a$) und $9a$) ist $\delta(p_i, p'_{h_i}) < \nu$ und $\delta(p'_{h_{i+1}}, p_{i+1}) < \nu$, wegen $\nu(P') < \nu$ ist $\delta(p'_j, p'_{j+1}) < \nu$ für $j = h_i, \ldots, h_{i+1} - 1$ und wegen $\nu_C(P) < \nu_C$ im Verein mit 7) und 6) gilt $\delta(p_{i+1}, p_i) < \nu$. Aus $2c$) ergibt sich also $\lambda_\varphi(P''_i) \geqq 0$, so daß wir wegen 1) und 10) aus 11) erhalten:

11') $\delta_\varphi(p_i, p_{i+1}) - \lambda_\varphi(P'_i) < 2\varphi_U \delta(p_i, p_{i+1}) + 2\varphi_U \delta$, falls p_i in $K_1 + \ldots + K_l$.

Zweiter Fall: p_i liegt nicht in $K_1 + \ldots + K_l$. Dann liegt p_i in C'_σ. Setzen wir $P''_i = \{p_i, p'_{h_i}, \ldots, p'_{h_{i+1}}, p_{i+1}\}$, so haben wir mit Rücksicht auf $9a$):

$$12a)\quad \lambda(P''_i) = \lambda(P'_i) + \delta(p_i, p'_{h_i}) + \delta(p'_{h_{i+1}}, p_{i+1}) \leqq \lambda(P'_i) + 2\delta.$$

$$12b)\quad \delta_\varphi(p_i, p_{i+1}) - \lambda_\varphi(P'_i) = [\delta_\varphi(p_i, p_{i+1}) - \lambda_{\varphi p_i}(P''_i)] + $$
$$+ [\lambda_{\varphi p_i}(P''_i) - \lambda_\varphi(P''_i)] + \delta_\varphi(p_i, p'_{h_i}) + \delta_\varphi(p'_{h_{i+1}}, p_{i+1}).$$

Wegen $9b$) und 6) gilt $P''_i \subseteq U(p_i, \rho_2)$. Daraus folgt, da p_i in C'_σ liegt, auf Grund der Definition von $\sigma(p, \rho)$ und wegen 3): $|\varphi(p_i; p'_j, p'_{j+1}) - \varphi(p'_j; p'_j, p'_{j+1})| \leqq \sigma(p_i, \rho_2) < \sigma$, also wegen $12a$) $\lambda_{\varphi p_i}(P''_i) - \lambda_\varphi(P''_i) < \sigma\lambda(P''_i) < \sigma\lambda(P'_i) + 2\sigma\delta$[3]. Wegen $\delta_{\varphi p_i}(p_i, p_{i+1}) = \delta_\varphi(p_i, p_{i+1})$ folgt aus der Definition von $\tau(p, \rho)$, daß $\delta_\varphi(p_i, p_{i+1}) - \lambda_{\varphi p_i}(P''_i) \leqq \tau(p_i, \rho) |\delta_\varphi(p_i, p_{i+1})| \leqq \tau(p_i, \rho)\varphi_U \delta(p_i, p_{i+1})$, so daß wir aus $12b$) erhalten:

$$12')\quad \delta_\varphi(p_i, p_{i+1}) - \lambda_\varphi(P'_i) < \tau(p_i, \rho)\varphi_U \delta(p_i, p_{i+1}) + \sigma\lambda(P'_i) + 2(\sigma + \varphi_U)\delta,$$
wenn p_i nicht in $K_1 + \ldots + K_l$.

Addieren wir die Ungleichungen 11') und 12'), so ergibt sich, wenn λ_σ die Summe der Zahlen $\delta(p_i, p_{i+1})$ bezeichnet, für welche p_i in $K_1 + \ldots + K_l$ liegt:

---

[3]) *Zusatz während der Drucklegung:* Herr N. Aronszajn (Paris) hat mich darauf aufmerksam gemacht, daß die obige Schlußfolgerung schon dann gezogen werden kann, wenn man weiß, daß $\varphi(p_i; p'_j, p'_{j+1}) - \varphi(p'_j; p'_j, p'_{j+1}) < \sigma$ gilt. Um dessen sicher zu sein, genügt es, mit $K_1 + \ldots + K_l$ einen Teil von $C[\sigma(p) \geqq \sigma]$ zu überdecken, nämlich die Menge $C[\sigma^+(p) \geqq \sigma]$ bestehend aus den Punkten von C, für welche die Zahl $\sigma^+(p)$, die folgendermaßen definiert ist, $\geqq \sigma$ ist: $\sigma^+(p)$ ist Limes für $\rho \to 0$ von $\sigma^+(p, \rho)$, wobei $\sigma^+(p, \rho)$ die obere Schranke der Zahlen $\varphi(p; q, r) - \varphi(p'; q, r)$ für alle Punkte p', q, r von $U(p, \rho)$ ist. $\sigma^+(p) = 0$ ist gleichbedeutend mit der *Unterhalbstetigkeit* von φ im Punkt p, so wie $\sigma(p) = 0$ mit der Stetigkeit gleichbedeutend ist. Daraus ergibt sich die bemerkenswerte Konsequenz, daß die Voraussetzungen **II** und **V** durch schwächere Voraussetzungen **II**$^+$ und **V**$^+$ ersetzt werden können, nämlich *daß für jedes Streckenbild C mit endlicher oberer Länge und jedes $\sigma > 0$ die Menge $C[\sigma^+(p) \geqq \sigma]$ eine Nullmenge im engeren Sinn ist* und daß die Voraussetzung **V** erfüllt ist *für jeden Punkt p, für den $\sigma^+(p) > 0$ gilt oder für den in jeder Umgebung Punkte q mit $\sigma^+(q) > 0$ existieren.*

2

$$13)\ \lambda_\varphi(P)-\lambda_\varphi(P')<2\,\varphi_U\lambda_\sigma+\sum\tau(p_i,\rho)\varphi_U\delta(p_i,p_{i+1})+\sigma\lambda(P')+ +2(n-1)(\sigma+\varphi_U)\delta.$$

Um den zweiten Summanden rechts, in dem über alle i zu summieren ist, abzuschätzen, berücksichtigen wir: Wegen 6) ist $\rho<\rho_\tau$, also gilt wegen 5), wenn p_i nicht in $K'_1+\ldots+K'_m$ liegt, $\tau(p_i,\rho)<\tau$. Wegen 6) ist ferner $\rho<\rho_C$, also $\tau(p_i,\rho)\leqq\tau_C$, wenn p_i in $K'_1+\ldots+K'_m$ liegt. Nennen wir also λ_τ die Summe der Zahlen $\delta(p_i,p_{i+1})$, für die p_i in $K'_1+\ldots+K'_m$ liegt, so ist der zweite Summand in 13) offenbar $\leqq\tau_C\varphi_U\lambda_\tau+\tau\varphi_U\lambda(P)$. Wir erhalten so aus 13) die **Fundamentale Ungleichung:**

$$14)\ \lambda_\varphi(P)-\lambda_\varphi(P')< <2\,\varphi_U\lambda_\sigma+\tau_C\varphi_U\lambda_\tau+\tau\,\varphi_U\lambda(P)+\sigma\lambda(P')+2(n-1)(\sigma+\varphi_U)\delta.$$

§ 24. **Erstes Fundamentallemma.** *Sei C ein Streckenbild, dessen obere Länge $\lambda^*(C)$ endlich ist, und seien $\varepsilon>0$ und $\lambda^*>\lambda^*(C)$ gegebene Zahlen. Dann existiert ein $\nu>0$ und ein $\nu_C>0$ derart, daß zu jedem Teilpolygon P von C mit $\nu_C(P)<\nu_C$ ein $\delta>0$ existiert mit folgender Eigenschaft: Ist P' irgend ein Polygon mit $\nu(P')<\nu$, $\lambda(P')\leqq\lambda^*$ und $\delta(P',P_C)<\delta$, dann gilt $\lambda_\varphi(P')>\lambda_\varphi(P)-\varepsilon$.*

Wir setzen $\sigma=\varepsilon/5\lambda^*$ und $\tau=\varepsilon/5\,\varphi_U\lambda^*$. Dann ist auf der rechten Seite von 14) der vierte Summand $<\varepsilon/5$ und, wenn ν_C hinlänglich klein ist, für alle Teilpolygone P von C mit $\nu_C(P)<\nu_C$ wegen $\lambda^*(C)<\lambda^*$ auch der dritte Summand $<\varepsilon/5$. Nach § 18 können wir β in 2) so klein wählen, daß für jedes System von endlich vielen Kugeln mit einer Durchmessersumme $<\beta$ ein $\nu(\beta)>0$ existiert, so daß für jedes Teilpolygon $P^*=\{p^*_1,\ldots,p^*_h\}$ von C mit $\nu_C(P^*)<$ $<\nu(\beta)$ gilt: die Summe aller Zahlen $\delta(p^*_i,p^*_{i+1})$, für welche p^*_i in den endlichvielen Kugeln liegt, ist beliebig klein, z. B. kleiner als $\varepsilon/10\,\varphi_U$ und als $\varepsilon/5\,\tau_C\varphi_U$. Ist $\nu_C(P)<\nu_C(\beta)$, so ist also wegen 2) und 4) und der Definition von λ_σ und λ_τ auch der erste und zweite Summand in 14) je $<\varepsilon/5$. Wird schließlich für das betrachtete Polygon $P=\{p_1,\ldots,p_n\}$ das δ, das in 8) auftritt, $=\varepsilon/10\,n(\sigma+\varphi_U)$, so ist in 14) auch der fünfte Summand $<\varepsilon/5$, also $\lambda_\varphi(P)-\lambda_\varphi(P')<\varepsilon$, wie das Fundamentallemma behauptet.

§ 25. Sind C, $\varepsilon>0$ und $\lambda^*>\lambda^*(C)$ gegeben und P ein Teilpolygon von C mit $\nu_C(P)<\nu_C$, so existiert nach der Bemerkung von § 17 ein $\nu'>0$ mit folgender Eigenschaft: Ist C' ein Streckenbild mit $\delta(C,C')<\delta$, so gilt $\delta(P',P_C)<\delta$ für jedes Teilpolygon P' von C' mit $\nu_{C'}(P')<\nu'$. Ist $\lambda^*(C')<\lambda^*$ und ν' hinlänglich klein, so ist $\lambda(P')<\lambda^*$ für jedes Teilpolygon P' von C' mit $\nu_{C'}(P')<\nu'$. Wir erhalten also aus dem Fundamentallemma den

Satz. *Ist C ein Streckenbild mit endlicher oberer Länge $\lambda^*(C)$ und sind $\varepsilon>0$ und $\lambda^*>\lambda^*(C)$ vorgelegt, so existiert ein $\nu_C>0$ derart, daß zu jedem Teilpolygon P von C, für welches $\nu_C(P)<\nu_C$ gilt, ein $\delta>0$ existiert mit folgender Eigenschaft: Zu jedem Streckenbild*

C', für welches $\lambda^*(C') < \lambda^*$ und $\delta(C, C') < \delta$ gilt, existiert ein $\nu' > 0$ derart, daß für jedes Teilpolygon P' von C' mit $\nu_{C'}(P') < \nu'$ die Ungleichung gilt: $\lambda_\varphi(P') > \lambda_\varphi(P) - \varepsilon$.

§ 26. Es sei C ein Streckenbild mit endlicher oberer Länge und $\lambda^* > \lambda^*(C)$. Es sei $P_1, P_2, \ldots$ eine ausgezeichnete Folge von Teilpolygonen von C. Wir geben ein beliebig kleines $\varepsilon > 0$ vor und wenden den Satz von § 25 auf $C, \varepsilon, \lambda^*$ und $C' = C$ an. Es ergibt sich die Existenz eines $\nu_C > 0$ und eines $\nu' > 0$, so daß für jedes Teilpolygon P von C mit $\nu_C(P) < \nu_C$, also für fast jedes P_n der ausgezeichneten Folge, und für jedes Teilpolygon P' von C' mit $\nu_C(P') < \nu'$, also für fast alle P_m, die Ungleichung gilt: $\lambda_\varphi(P_m) > > \lambda_\varphi(P_n) - \varepsilon$. Da überdies $|\lambda_\varphi(P_i)| \leqq \varphi_U \cdot \lambda(P_i) < \varphi_U \lambda^*$ für fast alle i gilt, d. h. die Zahlenfolge $\lambda_\varphi(P_1), \lambda_\varphi(P_2), \ldots$ beschränkt ist, so konvergiert sie gegen eine endliche Zahl und nach § 6 haben wir folgendes

Erstes Theorem über die Länge in R_φ. *Ist R ein halbmetrischer Raum mit gleichmäßig stetigen Abständen, φ eine Funktion, die den Voraussetzungen* **I**—**V** *genügt, C ein Streckenbild mit endlicher oberer Länge, so hat C im Raum R_φ eine endliche Länge $\lambda_\varphi(C)$. Ist $\delta(C, C') = 0$, so gilt $\lambda_\varphi(C) = \lambda_\varphi(C')$.* Aus Beispielen von Alt und Wald in diesem Heft geht hervor, daß die Voraussetzung der Endlichkeit der oberen Länge von C für die Existenz von $\lambda_\varphi(C)$ nicht entbehrlich ist.

§ 27. Wählen wir im Streckenbild C mit endlicher oberer Länge, wenn $\varepsilon > 0$ und $\lambda^* > \lambda^*(C)$ gegeben sind, ein Teilpolygon P, für das $\lambda_\varphi(P) > \lambda_\varphi(C) - \varepsilon/3$ gilt und $\nu_C(P)$ hinreichend klein ist, so folgt nach dem Satz vom § 25 für jedes Streckenbild C', für welches $\lambda^*(C') < \lambda^*$ und $\delta(C, C')$ hinreichend klein ist, für jedes Teilpolygon P', für welches $\nu_{C'}(P')$ hinreichend klein ist, $\lambda_\varphi(P') > > \lambda_\varphi(P) - \varepsilon/3$. Dabei können wir P' so wählen, daß $\lambda_\varphi(P') < < \lambda_\varphi(C') + \varepsilon/3$ gilt, woraus hervorgeht, daß $\lambda_\varphi(C') > \lambda_\varphi(C) - \varepsilon$ ist. Wir haben also folgendes

Erstes Unterhalbstetigkeitstheorem. *Unter den Voraussetzungen des ersten Theorems über die Länge existiert zu jedem $\lambda^* > \lambda^*(C)$ und jedem $\varepsilon > 0$ ein $\delta > 0$, so daß für jedes Streckenbild C', für welches $\delta(C, C') < \delta$ und $\lambda^*(C') < \lambda^*$ gilt, die Ungleichung besteht $\lambda_\varphi(C') > \lambda_\varphi(C) - \varepsilon$.*

§ 28. Wir ersetzen nun **V** durch die stärkere Voraussetzung **V***. *Zu jedem Punkt p existiert eine Umgebung $V(p)$, und zwei Zahlen $\nu_p > 0$ und $\omega(p) > 0$, so daß für jedes geschlossene Polygon P, für das $P \subseteq V(p)$ und $\nu(P) < \nu_p$ gilt, die Ungleichung $\lambda_\varphi(P) \geqq \omega(p)\lambda(P)$ besteht.*

§ 29. Nach dem Borelschen Überdeckungstheorem ergibt sich hieraus unmittelbar: Zu jeder kompakten Menge C existieren ein $\rho > 0$, ein $\nu > 0$ und ein $\omega > 0$, so daß für jedes geschlossene Polygon P, das für irgend einen Punkt p von C den Bedingungen $P \subseteq U(p, \rho)$ und $\nu(P) < \nu$ genügt, $\lambda_\varphi(P) \geqq \omega \lambda(P)$ gilt.

2*

§ 30. **Zweites Fundamentallemma.** *Ist C ein Streckenbild, so existiert ein $\nu_C > 0$ derart, daß zu jedem Teilpolygon P von C mit $\nu_C(P) < \nu_C$ und zu jedem $\lambda > 0$ ein $\lambda^* > 0$ und ein $\delta > 0$ existieren mit folgender Eigenschaft: Ist P' irgend ein Polygon mit $\delta(P', P_C) < \delta$ und $\lambda(P') > \lambda^*$, so ist $\lambda_\varphi(P') > \lambda$.*

Wir bestimmen nach der Bemerkung von § 29 zur kompakten Menge C die Zahlen ρ, ν und ω. Ist $\nu_C > 0$ hinlänglich klein, so hat für jedes Teilpolygon $P = \{p_1, \ldots, p_n\} = \{p(\gamma_1), \ldots, p(\gamma_n)\}$ von C der Abschnitt $C_i = p[\gamma_i, \gamma_{i+1}]$ $(i = 1, \ldots, n-1)$ einen Durchmesser $< \rho$ und $< \nu$ und es existiert ein positives $\delta < \nu$, so daß gilt:

1) $U(C_i, \delta)$ hat einen Durchmesser $< \rho$.

Es sei nun $P' = \{p'_1, \ldots, p'_n\}$ irgend ein Polygon mit $\nu(P') < \nu$ und $\delta(P', P_C) < \delta$, so daß P' in die Abschnitte $P'_i = \{p'_{h_i}, \ldots, p'_{h_{i+1}}\}$ $(i = 1, \ldots, n-1)$ zerfällt, wobei

2) $\delta(p'_{h_i}, p_i) < \delta$ $(i = 1, \ldots, n)$ und 2') $P'_i \subseteq U(C_i, \delta)$ $(i = 1, \ldots, n-1)$.

Wir setzen $P''_i = \{p_i, p'_{h_i}, \ldots, p'_{h_{i+1}}, p_{i+1}, p_i\}$, haben $\nu(P''_i) < \nu$ und $\lambda_\varphi(P''_i) \leqq \lambda_\varphi(P'_i) + 2\varphi_U\delta + \varphi_U\delta(p_i, p_{i+1})$, also

3) $\lambda_\varphi(P') = \sum \lambda_\varphi(P'_i) \geqq \sum \lambda_\varphi(P''_i) - 2(n-1)\varphi_U\delta - \varphi_U\lambda(P)$.

Wegen 1), 2') und der Bestimmung von ρ und ω gilt $\lambda_\varphi(P''_i) \geqq \geqq \omega\lambda(P''_i)$, also

4) $\sum \lambda_\varphi(P''_i) \geqq \omega \sum \lambda(P''_i) \geqq \omega(\lambda(P') - 2(n-1)\delta - \lambda(P))$.

Kombiniert man 3) und 4), so wird klar, daß für das feste P, wenn $\lambda(P')$ hinreichend groß ist, $\lambda_\varphi(P') > \lambda$ für jedes gegebene λ gilt, wie das Lemma behauptet. Aus ihm ergibt sich unter Berücksichtigung der Ungleichung aus § 23 folgendes:

§ 31. **Zweites Theorem über die Länge in R_φ.** *Ist R ein halbmetrischer Raum mit gleichmäßig stetigen Abständen, φ eine Funktion welche die Voraussetzungen* **I—IV, V*** *erfüllt, und C ein Streckenbild mit unendlicher oberer Länge, so hat C in R_φ die Länge $\lambda_\varphi(C) = \infty$. Aus $\delta(C, C') = 0$ folgt $\lambda_\varphi(C') = \infty$.*

§ 32. Sei nun C ein Streckenbild mit endlicher oberer Länge. 1) Zu jedem λ, z. B. zu $\lambda = \lambda_\varphi(C)$, existiert ein λ^*, so daß für jedes Streckenbild C', für welches $\delta(C, C')$ hinreichend klein und die obere Länge $> \lambda^*$ ist, $\lambda_\varphi(C') > \lambda_\varphi(C)$ gilt. 2) Im System der Streckenbilder einer oberen Länge $\leqq \lambda^*$ gilt aber nach § 27 für jedes $\varepsilon > 0$, wenn $\delta(C, C')$ hinreichend klein ist, $\lambda_\varphi(C') > \lambda_\varphi(C) - \varepsilon$. Nimmt man 1) und 2) zusammen und beweist man, was keine Schwierigkeit bereitet, das Analogon von 1) für Streckenbilder mit unendlicher oberer Länge, so erhält man folgendes

Zweites Unterhalbstetigkeitstheorem. *Unter den Voraussetzungen des zweiten Theorems über die Länge existiert zu jedem Streckenbild C von endlicher oberer Länge und zu jedem $\varepsilon > 0$ ein $\delta > 0$ derart, daß für jedes Streckenbild C' mit $\delta(C, C') < \delta$ stets $\lambda_\varphi(C') > \lambda_\varphi(C) - \varepsilon$ gilt. Zu jedem Streckenbild C mit unendlicher* *oberer Länge, und jedem* λ *existiert ein* $\delta > 0$, *so daß für alle* C', *für die* $\delta(C, C') < \delta$ *gilt und die obere Länge hinreichend groß ist, die Ungleichung* $\lambda_\varphi(C') > \lambda$ *besteht.*

Ferner ergibt sich aus dem zweiten Fundamentallemma unmittelbar: *Ist* C *ein Streckenbild,* $C_1, C_2 \ldots$ *eine gegen* C *konvergente Folge von Streckenbildern mit* $\lim \lambda(C_n) = +\infty$, *so gilt* $\lim \lambda_\varphi(C_n) = +\infty$.

IV. Minimanten in Längenräumen.

§ 33. Sei R ein halbmetrischer Raum mit gleichmäßig stetigen Abständen. Wir betrachten die konstante Funktion $\varphi(p; q, r) = 1$. Sie ist beschränkt und hat in jedem Punkt p für jedes $\rho > 0$ die ρ-Schwankung $\sigma(p, \rho) = 0$, erfüllt also die Voraussetzungen **I** und **II**, ferner **V***, wo wir $V(p) = R$ und $\omega(p) = 1$ für jeden Punkt p setzen können. Damit auch **III** und **IV** erfüllt seien, ist offenbar hinreichend die Voraussetzung, daß zu jedem $\tau > 0$ ein $\rho_\tau > 0$ existiert, so daß $\tau(p, \rho_\tau) < \tau$ für alle Punkte p gilt. Diese Voraussetzung für $\varphi = 1$ besagt, daß für alle hinreichend kleinen Polygone P der Defekt $\xi(P)$ beliebig klein ist, d. h. *daß zu jedem* $\xi > 0$ *ein* $\delta > 0$ *existiert, so daß* $\lambda(P) > (1 - \xi)\mu(P)$ *für alle* P *mit einem Durchmesser* $< \delta$, wo $\lambda(P)$ die Länge, $\mu(P)$ die Spanne von P ist. Einen halbmetrischen Raum mit gleichmäßig stetigen Abständen, für welchen diese Bedingung erfüllt ist, nennen wir einen *Längenraum.*

§ 34. Aus den beiden Theoremen über die Länge in R_φ ergibt sich in einem Längenraum für $\varphi = 1$ offenbar, daß für jedes Streckenbild die Länge mit der oberen Länge übereinstimmt, also der (sehr leicht auch direkt beweisbare) Satz: *Jedes Streckenbild eines Längenraumes ist im weiteren Sinne rektifizierbar.*

§ 35. Aus der Definition des Längenraumes folgt unmittelbar der Satz: *Zu jedem* $\xi > 0$ *existiert ein* $\delta > 0$, *so daß* $\lambda(C) > (1 - \xi)\mu(C)$ *für jedes Streckenbild* C *mit einem Durchmesser* $< \delta$ *gilt, wo* $\mu(C)$ *der Abstand der Endpunkte von* C *ist.*

§ 36. Aus den beiden Unterhalbstetigkeitstheoremen ergibt sich ferner unmittelbar: In einem Längenraum existiert zu jedem Streckenbild, das rektifizierbar ist, d. h. für welches $\lambda(C) < \infty$ gilt, und zu jedem $\varepsilon > 0$ ein $\delta > 0$, so daß für alle Streckenbilder C' mit $\delta(C, C') < \delta$ die Ungleichung $\lambda(C') > \lambda(C) - \varepsilon$ besteht. Zu jedem nicht rektifizierbaren Streckenbild C, d. h. wenn $\lambda(C) = \infty$, existiert, wie man überdies unschwer beweist, zu jeder noch so großen Zahl λ ein $\delta > 0$, so daß für jedes Streckenbild C' mit $\delta(C, C') < \delta$ stets $\lambda(C') > \lambda$ gilt. Zusammenfassend hat man den Satz: *Die Länge ist in Längenräumen ein unterhalbstetiges Funktional der Streckenbilder.*

§ 37. Es sei nun R ein kompakter Längenraum, $C_1, C_2, \ldots$ eine Folge von Streckenbildern mit Längen $\leqq \lambda$. Wir bezeichnen für jeden Punkt $p_n = p_n(\gamma)$ der Kurve $C_n = p_n[\alpha_n, \beta_n]$ den Quotienten der Länge des Abschnittes $p_n[\alpha_n, \gamma]$ durch die Länge des Streckenbildes $p_n[\alpha_n, \beta_n]$ mit $\zeta_n(p_n)$. Dies tun wir für jedes natürliche n. Die Punkte

C_n werden dadurch eineindeutig und umkehrbar stetig auf das Intervall [0, 1] abgebildet. Dies beruht darauf, daß die Länge ein für eine gewisse Menge von Streckenbildern definiertes Funktional ist, das für die Anfangsabschnitte jedes rektifizierbaren Streckenbildes stetig und in schärferem Sinne monoton wächst. Dies wiederum wird durch folgende Eigenschaften des Funktionals gewährleistet:

1) Für je zwei verschiedene Punkte p und q ist die untere Schranke der Funktionalwerte aller p und q verbindenden Streckenbilder > 0.

2) Ist das Funktional für ein Streckenbild definiert, so auch für alle seine Abschnitte.

3) Entlang jedem Streckenbild $C = p[\alpha, \beta]$ ist das Funktional additiv, d. h. wenn $\alpha \leqq \gamma \leqq \beta$, so ist der Wert des Funktionals für $p[\alpha, \beta]$ gleich der Summe seiner Werte für $p[\alpha, \gamma]$ und $p[\gamma, \beta]$.

4) Ist $p[\alpha, \beta]$ ein Streckenbild mit endlichem Funktionalwert und ist $\gamma_1, \gamma_2, \ldots$ eine monoton wachsend gegen γ konvergente Folge von Zahlen, so konvergieren die Funktionalwerte von $p[\alpha, \gamma_n]$ gegen den Funktionalwert von $p[\alpha, \gamma]$.

Aus 3) und 4) folgt, daß auch das Analogon von 4) für monoton abnehmende Folgen $\gamma_1, \gamma_2, \ldots$, also für alle gegen γ konvergente Folgen gilt.

§ 38. Man greift nun auf Grund der Kompaktheit des Raumes nach dem bekannten Hilbertschen Verfahren aus $C_1, C_2, \ldots$ zunächst eine Teilfolge $C_{11}, C_{12}, \ldots$ heraus, für welche die Endpunkte, d. h. die Punkte p mit $\zeta_{1n}(p) = 0$ und $= 1$ gegen zwei Punkte $p(0)$ und $p(1)$ des Raumes konvergieren. Aus dieser Teilfolge greift man sodann eine Teilfolge $C_{21}, C_{22}, \ldots$ heraus, für welche auch die Mittelpunkte, d. h. die Punkte mit $\zeta_{2n}(p) = 1/2$ gegen einen Punkt $p(1/2)$ des Raumes konvergieren; aus ihr eine Teilfolge $C_{31}, C_{32}, \ldots$, für welche auch die Punkte mit $\zeta_{3n}(p) = 1/4$ und $= 3/4$ gegen Punkte $p(1/4)$ und $p(3/4)$ konvergieren usw. Man ordnet so schließlich jeder dyadisch rationalen Zahl ζ aus [0, 1] einen Punkt $p(\zeta)$ des Raumes zu, gegen den die Punkte der Diagonalfolge $C_{11}, C_{22}, \ldots$ mit $\zeta_{nn}(p) = \zeta$ konvergieren. Es handelt sich nun nur darum zu zeigen, daß diese Zuordnung *gleichmäßig stetig* ist. Denn, wenn sie dies ist, so kann man sie zu einer stetigen Abbildung des Intervalles [0, 1], d. h. zu einem Streckenbild $C = p[0, 1]$ erweitern, von dem man leicht zeigt, daß $\lim \delta(C_{nn}, C) = 0$. Wegen der Unterhalbstetigkeit des Funktionals der Länge, gilt $\lambda(C) \leqq \lambda$. Es ist dann also bewiesen: *In einem kompakten Längenraum ist für jedes λ die Limesklasse aller Streckenbilder mit Längen $\leqq \lambda$ kompakt.*

Wäre nun die Abbildung der dyadisch rationalen Zahlen ζ aus [0, 1] auf die Punkte $p(\zeta)$ nicht gleichmäßig stetig, so gäbe es eine konvergente Folge von Paaren dyadisch rationaler Zahlen ζ_n, ζ'_n mit $\lim (\zeta_n - \zeta'_n) = 0$, für welche die Punkte $p(\zeta_n)$ und $p(\zeta'_n)$ gegen zwei verschiedene Punkte p und p' konvergieren würden.

Man sieht sofort, daß dann p und p', obwohl ihr Abstand >0 wäre, durch Polygone beliebig kleiner Länge verbindbar wären, im Widerspruch zu § 33.

§ 38*a*. Ganz analog ergibt sich: *Ist in einem kompakten Raum R irgend ein unterhalb stetiges Funktional $\lambda(C)$ für alle Streckenbilder definiert, das die Eigenschaften 1), 2), 3), 4) von § 37 besitzt, und hat der Raum R die Eigenschaft*

5) *Jeder Punkt p ist für jedes $\varepsilon>0$ mit jedem Punkt einer hinlänglich kleinen Umgebung durch ein Streckenbild C verbunden, für welches $\lambda(C)<\varepsilon$ gilt, — dann ist für jedes $\lambda<\infty$ die Menge aller Streckenbilder C mit $\lambda(C)\leqq\lambda$ kompakt.* Dabei sind nach § 7 die Eigenschaften 2), 3), 4) für $\lambda(C)$ sicher erfüllt, wenn $\lambda(C)$ das Riemannsche Integral einer Funktion entlang C ist. Daß wir im Beispiel der Länge kein Analogon der Voraussetzung 5) über den Raum benötigten, kommt daher, daß wir in § 36 mit einer Funktion der Punktepaare operierten und die Grenzpunkte p und p' durch ein *Polygon* beliebig kleiner Länge verbinden können, auch wenn p und p' mit den gegen sie konvergierenden Punkten $p(\zeta_n)$ und $p(\zeta_n')$ nicht durch Streckenbilder beliebig kleiner Länge verbunden sind.

§ 39. Es liege nun ein kompakter Längenraum vor und eine Funktion $\varphi(p;\, q, r)$. Wir bezeichnen mit L die Limesklasse der Streckenbilder des Raumes, indem wir $\lim C_n = C$ gleichbedeutend mit $\lim \delta(C_n, C)=0$ erklären, und wenden die Ergebnisse von § 20 auf $\alpha(C)=\lambda(C)$ und $\beta(C)=\lambda_\varphi(C)$ an. Voraussetzung 2) von § 20 ist auf Grund des Theorems von § 27 durch die Voraussetzungen **I—V** über φ gewährleistet; die Voraussetzungen 4) und 5) von § 20 sind auf Grund der Theoreme von §§ 31, 32 durch die Voraussetzung **V*** über φ gewährleistet. Voraussetzung 1) ist nach dem Satz von § 38 erfüllt. Formulieren wir die Voraussetzungen 3) und 3') von § 20 als Voraussetzungen über φ, so lauten sie beziehungsweise:

VI. *Für jede Zahl λ ist $\lambda(C)$ für alle rektifizierbaren Streckenbilder C mit $\lambda_\varphi(C)\leqq\lambda$ beschränkt.*

VI'. *Jede Folge $C_1, C_2, \ldots$ von Streckenbildern, für welche die Zahlen $\lambda_\varphi(C_n)$ unter einer endlichen Schranke liegen, enthält eine konvergente Teilfolge.*

§ 40. Dann folgt aus den Sätzen von § 20:

Erstes Existenztheorem für Minimanten. *Liegt in einem kompakten Längenraum eine Funktion φ vor, die den Voraussetzungen* **I—VI** *genügt, so enthält jede abgeschlossene Menge $\mathfrak{C}$ von rektifizierbaren Streckenbildern mindestens ein (rektifizierbares) Streckenbild, das Limes einer Minimalfolge von $\lambda_\varphi(C)$ auf $\mathfrak{C}$ ist. Jedes Streckenbild, das Limes einer solchen Minimalfolge ist, ist eine Minimante für $\lambda_\varphi(C)$ auf $\mathfrak{C}$.*

Zweites Existenztheorem für Minimanten. *Genügt in einem kompakten Längenraum die Funktion* φ *den Voraussetzungen* **I—IV, V*** *und einer der beiden (dann unter einander aquivalenten) Voraussetzungen* **VI** *oder* **VI'**, *so enthält jede abgeschlossene Menge* $\mathfrak{C}$ *von Streckenbildern mindestens ein Streckenbild, das Limes einer Minimalfolge für* $\lambda_\varphi(C)$ *auf* $\mathfrak{C}$ *ist. Dasselbe ist rektifizierbar, es sei denn, daß* $\mathfrak{C}$ *ausschließlich nicht rektifizierbare Streckenbilder enthält. Jedes Streckenbild, das Limes einer solchen Minimalfolge ist, ist eine Minimante für* $\lambda_\varphi(C)$ *auf* $\mathfrak{C}$.

Nach § 38*a* spielt in diesen Theoremen die Länge keine ausgezeichnete Rolle und kann durch jedes unterhalbstetige Funktional $\lambda(C)$ ersetzt werden, das für jedes Streckenbild C durch Integration einer Funktion entlang C definiert ist, wofern 1) für je zwei verschiedene Punkte p und q die untere Schranke von $\lambda(C)$ für alle p und q verbindenden Streckenbilder >0 ist und 2) der Raum R die in § 38*a* erwähnte Eigenschaft 5) hat.

§ 41. Ist $C=p[\alpha, \beta]$ ein Streckenbild, so bezeichnen wir als *Teilstreckenbild* von C jedes Streckenbild $p'[\alpha', \beta']$ derart, daß jedem γ' aus $[\alpha', \beta']$ eindeutig (aber nicht notwendig eineindeutig) ein $\gamma(\gamma')$ aus $[\alpha, \beta]$ zugeordnet ist, für welches $p'(\gamma')=p(\gamma(\gamma'))$ gilt. Einer bekannten Konstruktion von Kaluszay zufolge (vgl. Kerékjártó, Vorl. üb. Topologie 1923, p. 103) enthält jedes (stetige) Streckenbild einen *konfinalen Bogen* (d. h. ein eineindeutiges stetiges Streckenbild mit den gleichen Endpunkten) als Teil.

§ 42. **Theorem über minimante Bogen.** *Genügt die Klasse* $\mathfrak{C}$ *der Vergleichskurven der Voraussetzung*

VIII. *Ist C ein Streckenbild der Klasse* $\mathfrak{C}$, *so auch jedes konfinale Teilstreckenbild von C, die z. B. impliziert wird durch*

VIII*. $\mathfrak{C}$ *ist die Klasse aller Streckenbilder, deren Anfangspunkt und Endpunkt auf zwei gegebenen Mengen A und B liegen,*

dann existiert unter den Minimanten der Theoreme von § 40 mindestens ein Bogen.

Denn ist C eine Minimante, so gilt für jedes geschlossene Teilstreckenbild C' von C offenbar $\lambda_\varphi(C')\geqq 0$. Also gilt für jeden konfinalen Teilbogen B von C, wie er nach § 41 existiert, $\lambda_\varphi(B)\leqq \leqq\lambda_\varphi(C)$. Da B nach **VIII** zu $\mathfrak{C}$ gehört und C Minimante ist, ist auch B eine Minimante.

V. Die Variationsrechnung in euklidischen Räumen.

§ 43. Wir nennen *Richtungsraum* einen halbmetrischen Raum R, wenn jedem geordneten Paar p, q von verschiedenen Punkten ein Element $d(p, q)$ einer Menge D zugeordnet ist und aus $d(p, q)= =d(p' q')$ stets $d(q, p)=d(q', p')$ folgt. Wir schreiben dann $d(p, q)= —d(q, p)$. Wir nennen jedes Paar p, d, wo p Punkt von R und d Element von D ist, ein *Linienelement* und bezeichnen die Menge aller Linienelemente mit R, D.

§ 44. Eine Punkt-Punktepaar-Funktion $\varphi(p; q, r)$ eines Richtungsraumes soll eine *Punkt-Richtungs-Funktion* heißen, wenn 1) aus $d(q, r) = d(q', r')$ für jeden Punkt p folgt, $\varphi(p; q, r) = \varphi(p; q', r')$, so daß wir φ als eine Funktion $\varphi(p, d)$ der Linienelemente symbolisieren können, deren Definitionsbereich eine Teilmenge von R, D ist, und wenn 2) zu je zwei Elementen (p, d) und (p', d') des Definitionsbereiches von φ auch (p, d') und (p', d) zum Definitionsbereich von φ gehören.

§ 45. Beispiele von Richtungsräumen sind die *Vektorräume*, wenn man jedem geordneten Paar verschiedener Punkte p, q die Menge $d(p, q)$ aller positiven Multipla des Vektors $p \rightarrow q$ zuordnet. Der n-dimensionale Cartesische Raum ist ein Richtungsraum, in dem wir für D auch ein Büschel orientierter Geraden oder die Oberfläche einer Kugel wählen können.

§ 46. Ist R Teilmenge eines Vektorraumes, in dem jedem Vektor v eine Länge $|v|$ zugeordnet ist, und ist eine Punkt-Richtungs-Funktion $\varphi(p, d)$ in R definiert, so wird für jeden Punkt p im Tangentialraum $R_{\varphi p}$ jedem Vektor v, wenn $d(v)$ seine Richtung bezeichnet, die Länge $\varphi(p, d(v))|v|$ zugeordnet. Ist z. B. in einer Teilmenge R eines euklidischen Raumes eine Punkt-Richtungs-Funktion definiert, so ist für jeden Punkt p von R der Tangentialraum $R_{\varphi p}$ ein *Raum mit verallgemeinerter Strahlenmetrik* im Sinne von Minkowski (verallgemeinert: da auch negative und verschwindende Abstände auftreten können). $\tau(p) = 0$ ist (vgl. § 12) damit gleichbedeutend, daß diese Abstände der Dreiecksungleichung genügen, also *daß der Tangentialraum ein Dreiecksraum sei.*

§ 47. Sind die Abstände eines Raumes mit Strahlenmetrik positiv, so weiß man seit Minkowski, daß die Giltigkeit der Dreiecksungleichung für die Abstände gleichbedeutend mit der *Konvexität der Eichfläche* (der Menge aller Punkte im Abstand 1 vom Ursprung) ist. $R_{\varphi p}$ ist ein Raum mit positiven Abständen dann und nur dann, wenn $\varphi(p, d)$ stets > 0 ist. In einem Raum mit verallgemeinerter Strahlenmetrik ist, wie Alt in diesem Heft zeigt, die Giltigkeit der Dreiecksungleichung im wesentlichen mit *projektiver Konvexität der Eichfläche* äquivalent. Dies liefert uns ein Mittel, im Falle eines Cartesischen Raumes die Punkte p mit $\tau(p) = 0$ oder, wie wir sie nennen wollen, die Punkte, in denen φ *quasiregulär* ist, *durch projektive Konvexität der Indikatrix von φ zu definieren.* Zugleich zeigt Alt, daß für jedes $\tau > 0$ die Punkte p mit $\tau(p) \leqq \tau$ durch *nicht allzustarke Konkavität der Indikatrix* gekennzeichnet sind.

§ 48. Zwei Polygone $P = \{p_1, \ldots, p_n\}$ und $P' = \{p'_1, \ldots, p'_n\}$ sollen *kovektoriell* heißen, wenn die $n-1$ Vektoren $p_1 \rightarrow p_2, p_2 \rightarrow \rightarrow p_3, \ldots, p_{n-1} \rightarrow p_n$ eine Permutation der Vektoren $p'_1 \rightarrow p'_2, p'_2 \rightarrow \rightarrow p'_3, \ldots, p'_{n-1} \rightarrow p'_n$ sind. Kovektorielle Polygone haben offenbar gleiche Länge und, wenn sie gleiche Anfangspunkte haben, so auch gleiche Endpunkte.

§ 49. Wir sagen, der Vektorraum R besitze eine polygonale Streckbarkeit $\geqq \varkappa$, wenn zu jedem geschlossenen Polygon P von R ein kovektorielles Polygon P' existiert, dessen Durchmesser $\geqq \varkappa \lambda(P) = \varkappa \lambda(P')$ ist. Die untere Schranke aller Zahlen $\varkappa$, für die R eine polygonale Streckbarkeit $\geqq \varkappa$ besitzt, nennen wir die *polygonale Streckbarkeit* von R.

§ 50. Man sieht leicht, daß der n-dimensionale euklidische Raum R_n eine Streckbarkeit $\geqq 1/2n$ besitzt. Denn sind $v_i = \{\xi_{i1}, \ldots, \xi_{in}\}$ $(i = 1, \ldots, k)$ die k Vektoren eines geschlossenen Polygons P in R_n, so gilt für dessen Länge offenbar $\lambda(P) \leqq \sum_{i=1}^{k} [|\xi_{i1}| + \ldots + |\xi_{in}|]$. Für mindestens eine der Zahlen $j = 1, \ldots, n$ gilt also $\sum_{i=1}^{k} |\xi_{ij}| \geqq \lambda(P)/n$. Da die betreffenden Zahlen $\xi_{1j}, \ldots, \xi_{kj}$ wegen der Geschlossenheit von P die Summe 0 haben, so gilt, wenn $\xi_{i_1 j}, \ldots, \xi_{i_g j}$ die nicht negativen unter ihnen sind und $\xi_{h_1 j}, \ldots, \xi_{h_{k-g} j}$ die übrigen negativen sind: $\xi_{i_1 j} + \ldots + \xi_{i_g j} \geqq \lambda(P)/2n$. Wir bilden dann das zu P kovektorielle Polygon P', dessen Vektoren folgende Permutation der Vektoren von P sind: $v_{i_1}, \ldots, v_{i_g}, v_{h_1}, \ldots, v_{h_{k-g}}$. Vergleichen wir Anfangs- und Endpunkt des Vektors $v_{i_1} + \ldots + v_{i_g}$, so sehen wir, daß ihre j-te Koordinaten sich um mindestens $\lambda(P)/2n$ unterscheiden. Also ist die Länge eines Vektors und daher der Durchmesser von P' mindestens $\lambda(P)/2n$, womit die Behauptung bewiesen ist.

§ 51. Ist $\varphi(p, d)$ eine Punkt-Richtungs-Funktion in einem Richtungsraum, so nennen wir die Funktion $\varphi(p, d) + \varphi(p, -d)$ die *Symmetrisierung von* φ. Die untere Schranke der Werte, welche die symmetrisierte Funktion im Punkte p für alle Richtungen d annimmt, werden wir *Symmetrisierungsminimum* von φ in p nennen und mit $\chi(p)$ bezeichnen.

§ 52. Ist $\chi(p) < 0$, so ist $\tau(p) = \infty$. Denn sei q ein Punkt, so daß für die Richtung $d = d(p, q)$ die Zahl $\varphi(p, d) + \varphi(p, -d)$ negativ ist. Für das Polygon $P = \{p, q, p\}$ ist dann $\mu_{\varphi p}(P) = 0$, $\lambda_{\varphi p}(P) < 0$, also $\xi_{\varphi p}(P) = \infty$ und daher $\tau(p) = \infty$.

§ 53. In einem Vektorraum existiert zu jedem Polygon P und zu jedem Punkt p ein in p beginnendes, mit P kovektorielles Polygon; ferner zu jedem Polygon ein beliebig kleines ähnliches Polygon. Daraus folgt: Ist φ eine Punkt-Richtungs-Funktion in einer Teilmenge eines Vektorraumes und p ein Punkt mit $\tau(p, \rho) \leqq \tau$, so gilt $\xi(P) \leqq \tau$ für *jedes* Polygon P von $R_{\varphi p}$, nicht nur für die in p beginnenden Polygone $\subseteq U(p, \rho)$.

§ 54. Lemma. *Ist* φ *eine Punkt-Richtungs-Funktion in einer Teilmenge eines Vektorraumes mit einer polygonalen Streckbarkeit* $\geqq \varkappa$, *so ist hinreichend für die Giltigkeit von* **V*** *(§ 28) die Voraussetzung*

$V^*_\varkappa$. *Zu jeder kompakten Menge C existiert ein $\omega_0 > 0$ und ein $\rho > 0$ derart, daß für jeden Punkt p von C*

$$\chi(p)/2 \geqq \varphi_U \tau(p) + \sigma(p, \rho)/\varkappa + \omega_0.$$

Sei nämlich P ein geschlossenes Polygon, das für einen Punkt p von C und für die in $V^*_\varkappa$ auftretende Zahl ρ in $U(p, \rho)$ liegt. Dann gilt:

$$1)\ \lambda_\varphi(P) \geqq \lambda_{\varphi p}(P) - \sigma(p, \rho)\lambda(P).$$

Da der Raum eine Streckbarkeit $\geqq \varkappa$ besitzt, existiert ein mit P kovektorielles (und daher geschlossenes, mit P gleichlanges) Polygon $P' = \{p'_1, \ldots, p'_n\}$ mit $p'_1 = p'_n = p$, dessen Durchmesser $\geqq \varkappa\lambda(P)$ ist. P' enthält mindestens einen Punkt $p'_j = q$, dessen Abstand von p mindestens dem halben Durchmesser von P' gleich ist, so daß also

$$2)\ \delta(p, q) \geqq \varkappa\lambda(P)/2.$$

Setzen wir $P'_1 = \{p'_1, \ldots, p'_j\}$, $P'_2 = \{p'_j, \ldots, p'_n\}$, so haben wir:

$$3)\ \lambda_{\varphi p}(P') = \lambda_{\varphi p}(P'_1) + \lambda_{\varphi p}(P'_2) \geqq \delta_{\varphi p}(p, q) + \delta_{\varphi p}(q, p) - \\ - \tau(p)[\,|\delta_{\varphi p}(p, q)| + |\delta_{\varphi p}(q, p)|\,],$$

$$4)\ \delta_{\varphi p}(p, q) + \delta_{\varphi p}(q, p) = [\varphi(p; p, q) + \varphi(p; q, p)]\,\delta(p, q) \geqq \\ \geqq \chi(p)\delta(p, q).$$

Ferner gilt $|\delta_{\varphi p}(p, q)| + |\delta_{\varphi p}(q, p)| \leqq 2\varphi_U \delta(p, q)$, so daß aus 3) und 4) wegen 2) folgt

$$\lambda_{\varphi p}(P') \geqq \chi(p)\delta(p, q) - 2\tau(p)\varphi_U\delta(p, q) \geqq (\chi(p) - 2\tau(p)\varphi_U)\varkappa\lambda(P)/2.$$

Wegen 1) folgt daher aus $V^*_\varkappa$, daß $\lambda_\varphi(P) > \omega_0\varkappa\lambda(P)$, also der Satz von § 29 gilt.

§ 55. *Verallgemeinerter Satz von Hahn. Existiert in einem kompakten Raum eine Folge rektifizierbarer Streckenbilder $C_1, C_2, \ldots$ mit $\lim \lambda(C_n) = \infty$ und $\lambda_\varphi(C_n) \leqq \lambda < \infty$, so existiert auch entweder eine Folge rektifizierbarer Streckenbilder $D_1, D_2, \ldots$ mit $\lim \lambda(D_n) = \infty$ und $\lambda_\varphi(D_n) \leqq 0$ oder ein Punkt p, von dem jede Umgebung ein geschlossenes Polygon P enthält, für das $\lambda(P)$ beliebig groß und $\nu(P)$ und $\lambda_\varphi(P)$ beliebig klein sind.*

Die von Tonelli verallgemeinerte Konstruktion Hahns zum Beweise dieses Satzes kann folgendermaßen angeordnet werden: Zunächst kann für jedes $\lambda_1 > 0$ auf Grund der Folge $C_1, C_2, \ldots$ leicht eine Folge $C'_1, C'_2, \ldots$ hergeleitet werden, so daß $\lambda(C'_n) \geqq n\lambda_1$ und $\lambda_\varphi(C'_n) < \lambda/n$ gilt. Die Folge $C'_1, C'_2, \ldots$ enthält eine Teilfolge, deren Anfangsabschnitte der Länge λ_1 gegen ein Streckenbild B_1 konvergieren. Diese Teilfolge enthält eine Teilfolge, deren zweite Abschnitte der Länge λ_1 gegen ein an B_1 anschließendes Streckenbild B_2 konvergieren usw. So erhalten wir eine Folge $B_1, B_2, \ldots$ von Streckenbildern, von denen der Endpunkt eines jeden mit dem Anfangspunkt des nächsten übereinstimmt, von denen jedes eine Länge $\leqq \lambda_1$ besitzt und so daß für jedes k eine Teilfolge $C_1^k, C_2^k, \ldots$ von $C_1', C_2', \ldots$ existiert, deren Anfangsabschnitt $A_1^k, A_2^k, \ldots$ der Länge $k\lambda_1$ gegen $A_k = B_1 + \ldots + B_k$ konvergieren. Existiert nun *erstens* ein k, so daß $\lambda_\varphi(A_k) > 0$, so bilden die Endabschnitte $D_n = C_n^k - A_n^k$, wie man aus der Unterhalbstetigkeit von λ_φ leicht folgert, eine Folge von Streckenbildern mit $\lim \lambda(D_n) = \infty$ und $\lambda_\varphi(D_n) < 0$ für fast alle n. Gilt *zweitens* $\lambda_\varphi(A_k) \leqq 0$ für alle k und $\lim \lambda(A_k) = \infty$, so sind wir wiederum am Ziel. Gilt *drittens* $\lambda_\varphi(A_k) \leqq 0$, während $\lambda(A_k)$ beschränkt ist, dann ist $A = \Sigma A_k$ nach Hinzufügung eines Schlußpunktes p ein Streckenbild. Jede Umgebung dieses Punktes p enthält dann, wie man leicht einsieht, beliebig lange Polygone P, die man zu geschlossenen ergänzen kann, für welche $\nu(P)$ und $\lambda_\varphi(P)$ beliebig klein sind.

§ 56. Sprechen wir die Resultate der §§ 40, 41 für den euklidischen Raum aus, so können wir unter Berücksichtigung der Ergebnisse dieses Kapitels sagen:

Erstes Existenztheorem für Minimanten. *Es sei in einem Teil eines euklidischen Raumes eine Punkt-Richtung-Funktion $\varphi(x, x')$ definiert, welche folgenden Voraussetzungen genügt:*

I. *φ sei in jedem beschränkten Teil des Raumes beschränkt.*

II. Auf jedem rektifizierbaren Streckenbild sei für jedes $\sigma > 0$ die Menge aller Punkte, in denen φ eine Schwankung $\geqq \sigma$ besitzt, eine Nullmenge im engeren Sinn oder, was gleichbedeutend ist, *auf jedem rektifizierbaren Streckenbild sei die Menge aller Unstetigkeitspunkte eine Nullmenge im weiteren Sinn.*

III. Auf jedem rektifizierbaren Streckenbild sei für jedes $\tau > 0$ die Menge aller Punkte, in denen die Indikatrix eine Abweichung $\geqq \tau$ von der projektiven Konvexität hat, eine Nullmenge im engeren Sinn, oder, was hiefür hinreichend ist, *in jedem rektifizierbaren Streckenbild sei die Menge der Punkte, in denen φ nicht quasiregulär ist (d. h. nicht eine projektiv konvexe Indikatrix besitzt) eine Nullmenge im engeren Sinn.*

IV. *Auf jedem rektifizierbaren Streckenbild sei die Abweichung der Indikatrizes von der projektiven Konvexität beschränkt.*

V. *Zu jedem Unstetigkeitspunkt von φ existiert eine Umgebung U, so daß $\lambda_\varphi(P) \geqq 0$ für jedes hinreichend feine geschlossene Polygon $P \subseteq U$ gilt.*

VI. *Für jede Zahl λ ist die Länge aller rektifizierbaren Streckenbilder C mit $\lambda_\varphi(C) \leqq \lambda$ beschränkt.*

Dann enthält jede abgeschlossene Klasse $\mathfrak{C}$ von rektifizierbaren Streckenbildern mindestens ein (rektifizierbares) Streckenbild, das Limes einer Minimalfolge von $\lambda_\varphi(C)$ auf $\mathfrak{C}$ ist. Jedes Streckenbild, *das Limes einer solchen Minimalfolge ist, ist eine Minimante für* $\lambda_\varphi(C)$ *auf* $\mathfrak{C}$. Dabei ist $\lambda_\varphi(C)$ das Riemannsche Integral von φ entlang C.

Zweites Existenztheorem für Minimanten. φ *genüge außer* **I—IV** *den beiden Voraussetzungen*

$V_\varkappa^*$. *Zu jeder kompakten Menge* C *existiert ein* $\omega_0 > 0$ *und ein* $\rho > 0$ *derart, daß für jeden Punkt* p *von* C

$$\chi(p)/2 \geqq \varphi_U \tau(p) + \sigma(p, \rho)/\varkappa + \omega_0,$$

wo $\chi(p)$ *die untere Schranke der Zahlen* $\varphi(p, d) + \varphi(p, -d)$ *für alle Richtungen* d *ist und* $\varkappa = 1/2n$ *gesetzt werden kann,* eine Bedingung, die sich im Falle, daß φ überall stetig und quasiregulär ist, so daß $\sigma(p) = \tau(p) = 0$ für jeden Punkt p gilt, auf folgendes reduziert: *Zu jeder kompakten Menge* C *existiert ein* $\omega_0' > 0$, *so daß* $\chi(p) \geqq \omega_0'$ *für alle Punkte* p *von* C.

VI_0. *Die Länge aller rektifizierbaren Streckenbilder* C *mit* $\lambda_\varphi(C) \leqq 0$ *ist beschränkt.*

Dann enthält jede abgeschlossene Klasse $\mathfrak{C}$ *von Streckenbildern mindestens ein Streckenbild, das Limes einer Minimalfolge für* $\lambda_\varphi(C)$ *auf* $\mathfrak{C}$ *ist. Dasselbe ist rektifizierbar, es sei denn, daß* $\mathfrak{C}$ *ausschließlich nicht rektifizierbare Streckenbilder enthält. Jedes solche Streckenbild ist Minimante für* $\lambda_\varphi(C)$ *auf* $\mathfrak{C}$.

Entsprechend ist das Theorem von § 42 anwendbar.

Im Hilbertschen Raum ist, wie Wald in diesem Heft zeigt, $\chi(p) \geqq \varepsilon > 0$ für alle Punkte p einer Kurve C, nicht immer hinreichend für die Existenz von $\lambda_\varphi(C)$.

VI. Nicht rektifizierbare Minimanten.

§ 57. Wir beschränken uns nun auf Funktionen φ, für die gilt:

VII. φ *ist positiv semidefinit,* d. h. $\geqq 0$ für alle Argumente.

Ist $C = p[\alpha, \beta]$ ein Streckenbild, für das jeder innere Abschnitt $C' = p[\alpha', \beta']$ $(\alpha < \alpha' < \beta' < \beta)$ eine Länge $\lambda_\varphi(C')$ in R_φ hat, so bezeichnen wir mit $\lambda_\varphi^{\sim}(C)$ und nennen *die uneigentliche Länge* von C in R_φ die Zahl $\lim \lambda_\varphi(C_n)$ für eine Folge $C_n = p[\alpha_n, \beta_n]$ von inneren Abschnitten mit $\lim \alpha_n = \alpha$ und $\lim \beta_n = \beta$. Diese Definition ist möglich, da, wie man leicht sieht, $\lim \lambda_\varphi(C_n)$ von der Wahl der Folge $C_1, C_2, \ldots$ unabhängig ist. Aus Bemerkung 3) von § 7 folgt: Wenn $\lambda_\varphi(C)$ existiert, so gilt $\lambda_\varphi(C) = \lambda_\varphi^{\sim}(C)$. Doch ist, wie Wald in diesem Heft zeigt, möglich, daß $\lambda_\varphi(C)$ nicht existiert, obwohl $\lambda_\varphi(C')$ für jeden inneren Abschnitt C' von C existiert.

§ 58. Sind alle inneren Abschnitte $p[\alpha', \beta']$ eines Streckenbildes $C = p[\alpha, \beta]$ rektifizierbar, so nennen wir C *grenzrektifizierbar.* Das Bild $p(\alpha, \beta)$ des offenen Intervalles (α, β) nennen wir dann ein grenzrektifizierbares offenes Streckenbild.

§ 59. Ein Bogen $B=p[\alpha, \beta]$ heiße *reduzibel-rektifizierbar*, wenn eine abzählbare abgeschlossene Teilmenge A von B existiert, so daß $C-A$ Summe von endlich oder abzählbar vielen grenzrektifizierbaren offenen Bogen $B_1, B_2, \ldots$, ist. Wir definieren als *uneigentliche Länge* von B in R_φ die Zahl $\lambda_\varphi^{\sim}(B)=\Sigma\lambda_\varphi^{\sim}(B_i)$, falls diese Summe absolut konvergiert. Hinreichend dafür ist offenbar **VII**.

§ 60. Wir nehmen nun zu **V** folgende Voraussetzung, die minder scharf als **V*** ist, hinzu:

$\mathbf{V}^*_A$. *Ist C ein Bogen, so existieren zu jedem Punkt p von C, abgesehen höchstens von den Punkten einer abzählbaren abgeschlossenen Menge A, eine Umgebung $U(p)$ und zwei Zahlen $\nu(p)$ und $\omega(p)$ derart, daß für jedes geschlossene Polygon P, für das $P \subseteq U(p)$ und $\nu(P)<\nu(p)$ gilt, die Ungleichung besteht: $\lambda_\varphi(P) \geqq \omega(p)\lambda(P)$.*

§ 61. Wir machen ferner über die abgeschlossene Klasse $\mathfrak{C}$ von Streckenbildern die Voraussetzung **VIII***, daß sie aus den zwei Mengen verbindenden Streckenbildern bestehe, und über den Raum R die Voraussetzung (vgl. § 38*a*):

IX. *Zu jedem Punkt p von R, zu jeder Umgebung U von p und zu jedem $\varepsilon>0$ existiert eine Umgebung U', so daß p mit jedem Punkt von U' durch ein Streckenbild C verbunden ist, für das $C \subset U$ und $\lambda_\varphi(C)<\varepsilon$ gilt.*

§ 62. Sei $\mathfrak{C}$ eine abgeschlossene Klasse von Streckenbildern, μ die untere Schranke der Zahlen $\lambda_\varphi(C)$ für alle Streckenbilder C von $\mathfrak{C}$. Wir bilden eine Minimalfolge $C_1, C_2, \ldots$ für λ_φ auf $\mathfrak{C}$. Sie enthält, wenn wir die Voraussetzung **VI′** machen, eine Teilfolge, die gegen ein Streckenbild konvergiert, das wegen der Abgeschlossenheit von $\mathfrak{C}$ zu $\mathfrak{C}$ gehört. Wegen der Voraussetzung **VIII*** können wir annehmen, daß das Grenzstreckenbild ein Bogen $B=p[\alpha, \beta]$ ist. Wir können jedes der Streckenbilder C_n, indem wir es nötigenfalls durch ein anderes im Abstand 0 ersetzen, wodurch $\lambda(C_n)$ und $\lambda_\varphi(C_n)$ sich nicht ändern, in der Form $p_n[\alpha, \beta]$ und so annehmen, daß $p(\gamma)=\lim p_n(\gamma)$ für jedes γ aus $[\alpha, \beta]$ gilt. Es sei A die auf B nach $\mathbf{V}^*_A$ auftretende abzählbare abgeschlossene Menge. $B-A$ kann also als Summe von endlich oder abzählbar vielen offenen Bogen angenommen werden. Wir bezeichnen dieselben in irgend einer Reihenfolge mit $B^1, B^2, \ldots$

Sei D^i irgend ein abgeschlossener Abschnitt von B^i. Wir behaupten: D^i ist *rektifizierbar*. Sei nämlich $\mathfrak{D}$ die abgeschlossene Klasse aller mit D^i konfinalen Streckenbilder. Ist $D^i_1, D^i_2, \ldots$ die gegen D^i konvergente Folge von Abschnitten von $C_1, C_2, \ldots$ und verbinden wir für jedes n die Endpunkte von D^i_n mit den Endpunkten von D^i durch kleine Streckenbilder, so können wir dies wegen **IX** so einrichten, daß für die so entstehenden mit D^i konfinalen Streckenbilder $E^i_1, E^i_2, \ldots$ der lim inf der Zahlen $\lambda_\varphi(E^i_1), \lambda_\varphi(E^i_2), \ldots$ gleich $\lim\inf \lambda_\varphi(D^i_n)$ ist. Die Folge $E^i_1, E^i_2, \ldots$ ist eine Minimalfolge für λ_φ in $\mathfrak{D}$. Denn andernfalls könnte man offenbar die Streckenbilder $C_1, C_2, \ldots$ so modifizieren (indem man ihre Abschnitte $D_1^i, D_2^i, \ldots$ durch konfinale Abschnitte ersetzte), daß lim inf $\lambda_\varphi(C_n)$ verkleinert würde, was unmöglich ist, da die modifizierten Streckenbilder wegen **VIII*** zu $\mathfrak{C}$ gehören und $C_1, C_2, \ldots$ eine Minimalfolge für λ_φ auf $\mathfrak{C}$ ist. D^i ist mithin Grenzstreckenbild einer Minimalfolge für λ_φ auf $\mathfrak{D}$ und, da auf D^i die Voraussetzung **V*** gilt, ist D^i nach dem zweiten Theorem von § 40 rektifizierbar. Daraus folgt, daß B^i *grenz*rektifizierbar und B *reduzibel rektifizierbar* ist.

Für jeden abgeschlossenen Abschnitt D^i von B^i folgt nach dem Theorem von § 32 aus **V***

1) $\lambda_\varphi(D^i) \leqq \liminf \lambda_\varphi(D_n^i)$.

Es bezeichne C_n^i den offenen Abschnitt von C_n, der dem offenen Bogen B^i entspricht. Da $\lambda_\varphi(C_n)$ existiert, so existiert nach § 7 auch $\lambda_\varphi(C_n^i)$ und aus 1) folgt

2) $\tilde{\lambda}_\varphi(B^i) \leqq \liminf \lambda_\varphi(C_n^i)$.

Da $C_n^1, C_n^2, \ldots$ für jedes n paarweise fremde offene Abschnitte von C_n sind, so folgt aus 2)

3) $\Sigma \tilde{\lambda}_\varphi(B^i) \leqq \liminf \lambda_\varphi(C_n) = \mu$, also $\tilde{\lambda}_\varphi(B) \leqq \mu$.

Um zu zeigen, daß $\tilde{\lambda}_\varphi(B) = \mu$ gilt, müssen wir nur noch ausschließen, daß $\tilde{\lambda}_\varphi(B) < \mu$. Existiert $\lambda_\varphi(B)$, so gilt $\tilde{\lambda}_\varphi(B) = \lambda_\varphi(B)$ und es kann wegen der Definition von μ nicht $\lambda_\varphi(B) < \mu$ gelten. Existiert $\lambda_\varphi(B)$ nicht, so würde, wenn $\tilde{\lambda}_\varphi(B) < \mu$ wäre, für irgend einen abgeschlossenen Abschnitt D^i eines Bogens B^i die Ungleichung $\lambda_\varphi(D^i) < \liminf \lambda_\varphi(D_n^i)$ gelten, wo D_n^i die gegen D^i konvergenten Abschnitte der C_n sind. Dann könnte man wieder wegen **IX** die C_n so modifizieren, daß $\lambda_\varphi(C_n)$ vermindert wird und doch die modifizierten Kurven zu $\mathfrak{C}$ gehören, was der Voraussetzung, daß $C_1, C_2, \ldots$ eine Minimalfolge für λ_φ auf $\mathfrak{C}$ ist, widerspräche. Man brauchte nämlich nur die Abschnitte D_n^i durch konfinale Streckenbilder zu ersetzen bestehend aus D_i und kleinen Streckenbildern, welche die Endpunkte von D^i mit den Endpunkten von D_n^i verbinden. Damit haben wir folgendes

Drittes Existenztheorem für Minimanten. *Liegt in einem kompakten Längenraum eine Funktion φ vor, die den Voraussetzungen* **I—V, V$_A^*$, VI′, VII** *genügt, und genügt der Raum der Voraussetzung* **IX**, *so enthält jede abgeschlossene Klasse $\mathfrak{C}$ aller zwei Mengen verbindenden Streckenbilder einen nicht notwendig rektifizierbaren Bogen, der eine Minimante für $\lambda_\varphi(C)$ auf $\mathfrak{C}$ ist in dem Sinne, daß $\tilde{\lambda}_\varphi(C)$ gleich der unteren Schranke der Zahlen $\lambda_\varphi(C)$ für alle C von $\mathfrak{C}$ ist.*

Wir überlassen es dem Leser, die Anwendung auf den euklidischen Raum zu ziehen. Dieselbe umfaßt insbesondere die Beispiele von Hahn und Carathéodory, in welchen die Menge A der „punti eccessionali“ im Sinne von Tonelli aus einem einzigen Punkt besteht und rektifizierbare Minimanten nicht existieren. Unser Theorem lehrt z. B., daß immer, wenn die Menge A endlich oder abzählbar und abgeschlossen ist, sicherlich *zumindest nicht rektifizierbare Minimanten existieren.*

§ 63. Ein interessanter Spezialfall des Theorems liegt vor, wenn über φ folgende Voraussetzung gemacht wird:

VII*. *Für je zwei verschiedene Punkte p und q ist die untere Schranke der Zahlen $\lambda_\varphi(C)$ für alle p und q verbindenden Streckenbilder* > 0.

Da dann 1) von § 37 erfüllt ist, ergibt sich aus § 38*a*, daß die Voraussetzung **VI′** von selbst erfüllt ist. Unter gewissen Stetigkeitsvoraussetzungen ist dann auch **VII** erfüllt. Ganz insbesondere können wir sagen, daß für überall stetiges und quasireguläres φ, wenn $\mathbf{V}_A^*$ und **VII*** erfüllt sind, die Aussage des dritten Existenztheoremes für Minimanten gilt.

Dreiecksungleichung und Eichkörper in verallgemeinerten Minkowskischen Räumen. Von F. Alt.

Sei R ein n-dimensionaler affiner Raum, der durch Hinzufügung einer unendlich fernen Hyperebene zu einem projektiven Raum erweitert ist. Wir definieren in R eine verallgemeinerte Minkowskische Metrik, vermöge welcher jedem Vektor $\mathfrak{p}$ eine Länge $|\mathfrak{p}|$ zugeordnet wird, indem wir jeder durch einen festen Punkt o hindurchgehenden orientierten Geraden H einen ihrer von o verschiedenen Punkte e_H zuordnen, und für jeden Vektor $\mathfrak{p}$ der Richtung H festsetzen $|\mathfrak{p}| = \mathfrak{p} : \mathfrak{e}_H$, wo $\mathfrak{e}_H$ der Vektor $o e_H$ ist. (Dieser Abstand ist im Allgemeinen weder positiv noch symmetrisch.) Dadurch ist die Metrik festgelegt.

$E(H)$ sei die Menge aller Punkte von H, die im Sinne der Orientierung von H zwischen o und e_H liegen, diese Punkte mit eingeschlossen. Als *Eichkörper* E der Metrik bezeichnen wir die Summe der Mengen $E(H)$ für alle H durch o. Man beweist mühelos

1. Sind $\mathfrak{p}$, $\mathfrak{q}$, $\mathfrak{r}$ drei Vektoren, so daß $\mathfrak{p}+\mathfrak{q}=\mathfrak{r}$ gilt, und liegen auf ihren orientierten Trägergeraden P, Q, R die Punkte e_P, e_Q, e_R auf einer Geraden, so gilt $|\mathfrak{p}|+|\mathfrak{q}|=|\mathfrak{r}|$.

2. Ersetzt man e_H durch einen anderen Einheitspunkt e_H', so wird die Länge aller Vektoren in der Richtung von H dann und nur dann vergrößert, wenn e_H' in $E(H)$ liegt. — Aus **1** und **2** folgt unmittelbar

3. Ist $\mathfrak{p}+\mathfrak{q}=\mathfrak{r}$, wo $\mathfrak{p}$ und $\mathfrak{q}$ unabhängig sind, so gilt $|\mathfrak{p}|+|\mathfrak{q}|\geqq|\mathfrak{r}|$ dann und nur dann, wenn der Schnittpunkt der Trägergeraden R von $\mathfrak{r}$ mit der Verbindungsgeraden von e_P und e_Q in $E(R)$ liegt.

Sind $\mathfrak{p}$ und $\mathfrak{q}$ zwei Vektoren (nicht Nullvektoren) mit den orientierten Trägergeraden P und Q, so bezeichnen wir mit $H(P, Q)$

die Menge aller orientierten Geraden H, die einen Vektor $\lambda \mathfrak{p} + \mu \mathfrak{q}$ mit $\lambda > o$ und $\mu > o$ tragen, mit $E(P, Q)$ die Summe aller $E(H)$ für die Geraden H aus $H(P, Q)$. Aus **3.** folgt leicht:

4. In einer verallgemeinerten Minkowskischen Metrik gilt die Dreiecksgleichung dann und nur dann, wenn für je zwei orientierte Gerade P, Q der Durchschnitt von $H(P, Q)$ mit der Geraden durch e_P und e_Q in $E(P, Q)$ liegt. — Da alle Begriffe in dieser Aussage projektiv invariant sind, gilt:

5. Die Eigenschaft einer verallgemeinerten Minkowskischen Metrik, der Dreiecksungleichung zu genügen, ist invariant gegenüber projektiven Transformationen von E.

Nach Minkowski gilt die Dreiecksungleichung stets und im Falle durchwegs positiver Abstände nur dann, wenn der Eichkörper E konvex ist. Wir haben also

6. *Ist der Eichkörper E projektives Bild einer konvexen Menge, so gilt die Dreiecksungleichung.* Wir beweisen umgekehrt:

7. *Gilt in einer verallgemeinerten Minkowskischen Metrik die Dreiecksungleichung, so gibt es eine projektive Transformation des Raumes, die den Eichkörper in eine konvexe Menge überführt.*

Aus der Dreiecksungleichung folgt die Stetigkeit der Abstände. Wir führen eine Hilfsmetrik ein, indem wir je zwei Punkten p und q ihren euklidischen Abstand zuordnen. Ist $\mathfrak{i}_H$ der euklidische Einheitsvektor in der Richtung H, so ist $|\mathfrak{i}_H|$ eine stetige Funktion von H. Es sei H_0 eine Gerade, für welche $|\mathfrak{i}_H|$ sein Minimum annimmt. J_0 sei die auf H_0 senkrechte Hyperebene durch e_{H_0}. Es sei T eine projektive Transformation, welche o und alle durch o hindurchgehenden orientierten Geraden festläßt und J_0 in die unendlich ferne Hyperebene überführt. T erzeugt auf jeder Geraden durch o eine Projektivität, welche, wie man leicht nachrechnet, etwa durch die Gleichung $1/x' = 1/x - |\mathfrak{i}_{H_0}| \cos(H, H_0)$ dargestellt wird. Daraus ergibt sich leicht, daß $T(e_H)$ stets auf der positiven Hälfte von $T(H)$ liegt. Wird E der Transformation T unterworfen, so geht also die Metrik in eine solche mit durchwegs nicht negativen Abständen über, für die nach **5.** die Dreiecksungleichung gilt. Also ist nach Minkowski $T(E)$ konvex, womit **7.** bewiesen ist.

Die Giltigkeit der Dreiecksungleichung für eine verallgemeinerte Minkowskische Metrik ist nach Menger gleichbedeutend mit dem Verschwinden der von ihm eingeführten nichtnegativen Zahl $\tau(p)$ für einen und daher für jeden Punkt p des Raumes. $\tau(p) \leqq \tau$ ist in einem Vektorraum, wie man leicht einsieht, damit gleichbedeutend, daß der Eichkörper nicht allzu konkav sei, d. h. daß, wenn E^* die konvexe Hülle des Eichkörpers ist, die Schnittpunkte der Halbstrahlen durch o mit der Begrenzung von E^* nicht allzu große Minkowskische Abstände von o haben.

3

Ein Streckenbild, für welches λ_φ nicht existiert, obwohl φ beschränkt, stetig und quasiregulär ist. Von F. Alt.

Der zugrundeliegende Raum sei die Gerade. Jedem Punkt ξ sei für jede der beiden Richtungen, die wir durch + und — bezeichnen, die Zahl $\varphi(\xi, +) = -\xi^2 \cos 2\pi/\xi^3 = -\varphi(\xi, -)$ zugeordnet, wobei noch $\varphi(0, +) = \varphi(0, -) = 0$ sei. Wir betrachten das stetige Streckenbild C, das durch $\xi(\gamma) = \gamma^2 \cos 2\pi/\gamma^6$ für $-\sqrt[3]{2} \leqq \gamma \leqq 0$ gegeben ist. C enthält sowohl beliebig feine Teilpolygone, für welche $\lambda_\varphi(P)$ beliebig klein ist, als auch beliebig feine Teilpolygone, für welche $\lambda_\varphi(P)$ beliebig groß ist. $\lambda_\varphi(C)$ existiert daher nicht.

Ein Streckenbild, für welches λ_φ nicht existiert, obwohl λ_φ für jeden Anfangsabschnitt existiert. Von A. Wald.

Es sei B irgend ein Bogen eines euklidischen Raumes mit den Endpunkten p und q, der nicht rektifizierbar ist, während jeder in p beginnende echte Abschnitt rektifizierbar ist. (Beispiel: $\xi \sin 1/\xi$). Es gibt sicherlich in B eine Punktfolge $\{p_i\}$ ($i = 1, 2, \ldots$, ad inf.), so daß $p_1 = p$, daß p_{i+1} im Inneren des Teilbogens $B(p_i, q)$ mit den Endpunkten p_i und q liegt, daß $\lim p_i = q$ und daß $\sum_{i=1}^{\infty} p_i p_{i+1} = \infty$ gilt. Wir bezeichnen den Teilbogen $B(p_i, p_{i+1})$ mit B_i und wählen im Inneren von B_i zwei Punkte q_i und q_i' so, daß q_i' zwischen q_i und p_{i+1} liegt und die Länge der Teilbogen $p_i\, q_i$ und $q_i'\, p_{i+1}$ kleiner als $1/i^2$ sind. Den Teilbogen $B(q_i, q_i')$ bezeichnen wir mit B_i'. Es seien $\{\alpha_i\}$ und $\{\beta_i\}$ ($i = 1, 2, \ldots$, ad inf.) zwei gegen 0 konvergente Folgen positiver Zahlen, so daß $\sum \alpha_i \cdot p_i\, p_{i+1} = \infty$ und $\beta_i < 1/i^2 \lambda(B_i)$ für jedes i gilt, wobei $\lambda(B_i)$ die Bogenlänge von B_i bezeichnet. Wegen $\sum p_i\, p_{i+1} = \infty$ existiert offenbar eine solche Folge $\{\alpha_i\}$.

Wir definieren nun die nur vom Punkt abhängige Funktion φ folgendermaßen: Im Punkte p_i habe φ den Wert α_i. Auf dem Teilbogen $B(p_i, q_i)$ sei φ eine lineare Funktion der Bogenlänge, derart daß $\varphi(q_i) = \beta_i$. Auf dem Teilbogen B_i' sei φ konstant gleich β_i. Auf dem Teilbogen $B(q_i', p_{i+1})$ sei φ wiederum eine lineare Funktion der Bogenlänge, derart daß $\varphi(p_{i+1}) = \alpha_{i+1}$ ist. Ferner gelte $\varphi(q) = 0$. Die Funktion φ ist offenbar stetig. Man bestätigt mühelos, daß $\int \varphi\, ds$ entlang des Teilbogens $B(p, p_i)$ existiert und mit wachsendem i gegen eine endliche Zahl λ konvergiert. Dagegen existiert dieses Integral nicht entlang ganz B. Wegen $\varphi(p_i) = \alpha_i$ und $\sum \alpha_i \cdot p_i\, p_{i+1} = \infty$ besitzt nämlich B einerseits offenbar ausgezeichnete Folgen $\{P_i\}$ ($i = 1, 2, \ldots$, ad inf.) von Teilpolygonen, für welche $\lambda_\varphi(P_i)$ mit wachsendem i gegen ∞ konvergiert (z. B. Folgen, für welche mit wachsendem i immer längere Abschnitte der Folge $p_1, p_2, \ldots$ als konsekutive Punkte von P_i auftreten), andererseits auch ausgezeichnete Folgen $\{Q_i\}$ von Teilpolygonen, für welche $\lambda_\varphi(Q_i)$ mit wachsendem i gegen λ konvergiert (z. B. Folgen, für welche Q_i ein

Teilpolygon einer Norm $< 1/i$ von $B(p, p_i) = C_i$ ist, für welches $|\lambda_\varphi(Q_i) - \lambda_\varphi(C_i)| < 1/i$ gilt).

Eine Bemerkung über Linearmaß und Rektifizierbarkeit. Von J. Novák.

In Beantwortung einer Frage von Menger zeigen wir: *Es existiert in der Ebene eine Menge M, die Teilmenge eines Streckenbildes C ist und ein äußeres lineares Maß > 0 besitzt, aber mit jedem rektifizierbaren Streckenbild C' einen Durchschnitt vom linearen Maß 0 hat.*

Sei D ein gleichseitiges Dreieck der Seitenlänge 1. Wir bilden in jeder der drei Ecken ein zu D seitenparalleles Dreieck der Seitenlänge 1/3 und iterieren dieses Verfahren ad inf., so daß beim k-ten Schritt 3^k gleichseitige Dreiecke der Seitenlänge $1/3^k$ entstehen. Wir bezeichnen mit M die abgeschlossene Hülle der Menge der Eckpunkte aller dieser Dreiecke, mit C die abgeschlossene Hülle der Summe der Kanten aller dieser Dreiecke. C ist offenbar ein im kleinen zusammenhängendes Kontinuum, also ein stetiges Streckenbild. Die Menge M hat nicht das lineare Maß 0. Denn wenn $U_1, \ldots, U_n$ irgend ein System von endlich vielen M überdeckenden konvexen offenen Mengen ist, so existiert ein k, so daß die 3^k Dreiecke des k-ten Schrittes von $U_1, \ldots, U_n$ überdeckt werden. Die orthogonale Projektion dieser Dreiecke auf eine Seite von D hat die Länge 1. Also haben die Durchmesser von $U_1, \ldots, U_n$ eine Summe $\geqq 1$. Man sieht übrigens leicht, daß M das äußere lineare Maß 1 hat.

Wir behaupten weiter: Ist C' irgend ein Streckenbild, dessen Durchschnitt mit M nicht das lineare Maß 0 hat, dann ist C' nicht rektifizierbar. Für jedes k seien nämlich $D_1^k, \ldots, D_{n_k}^k$ die zu $C' \cdot M$ gehörigen Dreiecke des k-ten Schrittes, d. h. diejenigen, welche mit $C' \cdot M$ Punkte gemein haben und $C' \cdot M$ überdecken. Die Summe ihrer Durchmesser ist offenbar $n_k/3^k$. Da $C' \cdot M$ nicht das lineare Maß 0 hat, können die Zahlen $n_k/3^k$ mit wachsendem k nicht gegen 0 konvergieren. Da die Zahlenfolge $n_k/3^k$ offenbar monoton abnimmt, so gilt also $\lim n_{k+1}/n_k = 3$ und $\lim n_{k+3}/n_k = 27$. Es sei C_k' der Durchschnitt von C' mit dem Innern der zu $C' \cdot M$ gehörigen Dreiecke $D_1^k, \ldots, D_{n_k}^k$ und $E_k = C_k' - C_{k+3}'$. Offenbar sind die Mengen

$$(*) \qquad E_1, E_4, E_7, E_{10}, \ldots$$

paarweise fremd. Wegen $\lim n_{k+3}/n_k = 27$ existiert eine natürliche Zahl k_0, so daß für jedes $k \geqq k_0$ mindestens $n_k/2$ Dreiecke des k-ten Schrittes je 27 zu $C' \cdot M$ gehörige Dreiecke des $(k+3)$-ten Schrittes enthalten. C' hat mit jedem zu $C' \cdot M$ gehörigen Dreieck mindestens einen Punkt gemein. Betrachten wir nun ein Dreieck D_i^k, das 27 zu $C' \cdot M$ gehörige Dreiecke des $(k+3)$-ten Schrittes enthält. Unter ihnen gibt es auch solche, die zum Rand von D_i^k fremd sind. Der Abschnitt von C', der im Innern von D_i^k, aber außerhalb

3*

der Dreiecke des $(k+3)$-ten Schrittes verläuft, hat also eine Länge $\geqq 1/3^{k+3}$. Also hat E_k eine Gesamtlänge $\geqq n_k/2 \cdot 3^{k+3} = n_k/3^k \cdot 54$. Da die Mengen (*) paarweise fremd sind, hat also C' eine Länge $\geqq \sum n_{1+3k}/3^{1+3k} \cdot 54$. Da $n_k/3^k$ nicht gegen 0 konvergiert, ist diese Summe $= \infty$, also ist C' nicht rektifizierbar.

Die φ-Länge im Hilbertschen Raum. Von A. Wald.

In §§ 51—56 seiner Abhandlung in diesem Heft zeigt Menger, daß in einem euklidischen Raum für eine stetige und quasireguläre Funktion φ von Punkt und Richtung, für welche in jedem Punkt das Symmetrisierungsminimum $\varphi(p, d) + \varphi(p, -d)$ größer als 0 ist, jede (auch nicht rektifizierbare) stetige Kurve eine φ-Länge besitzt. Wir zeigen an einem Beispiel, daß im Hilbertschen Raum die entsprechende Behauptung nicht zutrifft.

Wir bezeichnen im Hilbertschen Raum für $n = 1, 2, \ldots$ mit p_n den Punkt mit den Koordinaten $(1, 1/2, 1/3, \ldots, 1/n, 0, 0, \ldots)$ und mit p_0 den Punkt $(0, 0, \ldots)$. Es sei C_{n+1} die Strecke zwischen p_n und p_{n+1} $(n = 0, 1, \ldots)$ und p der Punkt $(1, 1/2, 1/3, \ldots, 1/n, 1/n+1, \ldots)$. Es sei r_n der Mittelpunkt von C_n, ferner q_n ein Punkt zwischen p_{n-1} und r_n, und s_n ein Punkt zwischen r_n und p_n, so daß $q_n r_n = r_n s_n = 1/n^2$ gilt. Schließlich sei d_n die Richtung der positiven n-ten Koordinate und $\varepsilon_1, \varepsilon_2, \ldots$ eine gegen 0 konvergente Folge von Zahlen > 0, so daß $\varepsilon_n > 1/n$ und $\sum \varepsilon_n/n = \infty$ gilt.

Es sei C die um den Schlußpunkt p vermehrte Summe $C_1 + C_2 + \ldots$ Wir definieren $\varphi(q, d)$ für alle Punkte q von C und alle Richtungen d. Wir setzen $\varphi(q, -d_n) = \varepsilon > 0$ für jedes n. Liegt q in C_n, so setzen wir $\varphi(q, d_m) = 1/m$ für $m \neq n$, ferner $\varphi(q, d_n) = 1/n$, falls q nicht zwischen q_n und s_n liegt. Es sei $\varphi(r_n, d_n) = \varepsilon_n$. Von q_n bis r_n nehme φ linear zu, von r_n bis s_n linear ab. Setzen wir noch $\varphi(p, d_n) = 1/n$, so ist für jeden Punkt q von C für alle Koordinatenrichtungen $\varphi(q, d)$ definiert. Wir tragen auf jedem durch q gehenden zu einer Koordinatenachse parallelem Halbstrahl den entsprechenden Wert von $1/\varphi$ auf und bilden den kleinsten durch diese Punkte bestimmten konvexen Körper K_q. Für jede Richtung d bilden wir den in dieser Richtung von q ausgehenden Halbstrahl und setzen $\varphi(q, d)$ gleich dem reziproken Abstand zwischen q und dem Schnittpunkt dieses Halbstrahles mit der Begrenzung von K_q.

Die so für alle Punkte q von C und alle Richtungen d definierte Funktion ist offenbar beschränkt, stetig und quasiregulär. Wir behaupten: Für jeden Punkt q von C und für jede Richtung d gilt $\varphi(q, d) + (q, -d) \geqq \varepsilon$. Vor allem gilt dies, wenn d in dem von den n ersten Koordinaten aufgespannten linearen Raum R_n liegt. Für $n = 1$ ist dies trivial. Nehmen wir an, es gelte für n und d sei eine Richtung von R_{n+1}. Ist d senkrecht zu R_n, so gilt $d = d_{n+1}$ und die Behauptung ist offenbar erfüllt. Ist d nicht senkrecht auf R_n, so sei d' die Projektion von d auf R_n. In der von d' und d_{n+1} bestimmten Ebene erkennt man leicht: Aus $\varphi(q, -d_{n+1}) = \varepsilon$ und der nach induktiver Annahme geltenden Ungleichung $\varphi(q, d') +$

$+\varphi(q, -d') \geqq \varepsilon$ folgt $\varphi(q, d) + \varphi(q, -d) \geqq \varepsilon$. Wegen der Stetigkeit von φ folgt die Giltigkeit dieser Ungleichung für jede Richtung des Raumes.

Dennoch hat C keine φ-Länge. Zunächst bestätigt man mühelos, daß jeder echte Anfangsabschnitt von C eine φ-Länge hat und daß diese gegen eine endliche Zahl λ konvergiert, wenn der Endpunkt des Abschnittes gegen den Schlußpunkt p von C konvergiert. Daraus folgt die Existenz einer ausgezeichneten Folge von Teilpolygonen von C, deren φ-Längen gegen λ konvergieren. Anderseits existiert auch eine ausgezeichnete Folge $P_1, P_2, \ldots$ von Teilpolygonen, deren λ-Längen gegen ∞ konvergieren. Wir haben zu diesem Zweck bloß zwei monoton wachsende Folgen $m_1, m_2, \ldots$ und $n_1, n_2, \ldots$ von natürlichen Zahlen geeignet zu wählen und P_i so zu bestimmen, daß der m_i-te Eckpunkt von P_i p_{m_i} ist und daß die n_i darauffolgenden Eckpunkte von P_i der Reihe nach $r_{m_i}, p_{m_i+1}, r_{m_i+1}, p_{m_i+2}, \ldots$ sind.

Über den größten Durchmesser kovektorieller Polygone im R_k. Von Anton E. Mayer.

Im Anschluß an §§ 48—50 seiner Abhandlung hat Menger folgende Frage aufgeworfen: Wie groß ist die polygonale Streckbarkeit des R_k? M. a. W., wenn wir für jedes Polygon P des R_k mit $\Delta(P)$ die obere Schranke der Durchmesser aller mit P kovektoriellen Polygone und mit $\lambda(P)$ deren gemeinsame Länge bezeichnen, — wie groß ist die untere Schranke $\rho(k)$ der Zahlen $\Delta(P)/\lambda(P)$ für alle Polygone des R_k? In § 50 l. c. wird bewiesen, daß $\rho(k) \geqq 1/2k$ gilt. Diese Abschätzung ist für $k=1$ offenbar scharf. Für beliebiges k gilt, wie wir anderen Ortes ausführen werden:

$$\rho(k) = \frac{1}{2\sqrt{\pi}} \frac{\Gamma(k/2)}{\Gamma((k+1)/2)}.$$

Zum Beweise ziehen wir den Begriff der Längenmenge heran (vgl. diese Ergebnisse, Heft 5, S. 1 ff.).

Die Widerspruchsfreiheit des Kollektivbegriffes der Wahrscheinlichkeitsrechnung.

Von A. Wald.

R. v. Mises[1]) führt zur Grundlegung der Wahrscheinlichkeitsrechnung den Begriff des Kollektivs ein. Gegen seine Formulierung wurden jedoch verschiedene Einwände erhoben, welche hauptsächlich gegen das Axiom der Regellosigkeit sich richteten, das in der Tat nicht genügend präzis formuliert wurde. In dieser Note werden wir den Kollektivbegriff genau präzisieren und dessen Widerspruchsfreiheit beweisen.

Die Arbeit besteht aus drei Teilen. In dem ersten Teil geben wir die Definition des Kollektivs (§ 1), behandeln ihre Beziehungen zum Misesschen Kollektivbegriff (§ 2), ferner die Einwände gegen die Misessche Definition (§ 3) und geben schließlich eine Übersicht über die erhaltenen Ergebnisse, die einen vollkommen einwandfreien und auch der Misesschen Auffassung gerecht werdenden Aufbau der Wahrscheinlichkeitsrechnung ermöglichen. Im zweiten Teil geben wir die Beweise der Theoreme an. Der dritte Teil behandelt die Frage, ob der im ersten Teil definierte Kollektivbegriff nicht in gewissem Sinne modifiziert werden könnte. Die Ergebnisse fallen jedoch negativ aus.

Einen Teil der in dieser Arbeit publizierten und bewiesenen Ergebnisse, habe ich bereits in den Comptes Rendus Acad. sciences, Paris, t. 202, 1936 p. 180—183, ohne Beweise veröffentlicht.

Erster Teil. Der Kollektivbegriff.

§ 1. Definition des Kollektivs.

Es sei M eine Menge von Elementen, die wir auch *Merkmale* oder *Merkmalpunkte* nennen werden. Die Menge M selbst wird *Merkmalraum* genannt. Ist n eine natürliche Zahl (bzw. 0) und f_n eine Funktion, die jedem geordneten n-Tupel von Merkmalen (bzw. der leeren Menge) entweder 0 oder 1 zuordnet, so nennen wir f_n eine *n-stellige Auswahlfunktion*. Besteht z. B. der Merkmalraum M bloß aus den zwei Elementen 0 und 1, so ist eine 2-stellige

[1]) R. v. Mises: Wahrscheinlichkeitsrechnung und ihre Anwendungen in der Statistik und theor. Physik. Deuticke, Leipzig Wien 1931. Die wesentlichen Grundzüge seiner Theorie hat Mises bereits 1919 in der Mathematischen Zeitschrift Bd. 5 publiziert.

Auswahlfunktion eine Funktion, die jedem geordneten Merkmalpaar, also jedem der folgenden vier Merkmalpaare: (1,1); (1,0), (0,1); (0,0) entweder den Wert 0 oder 1 zuordnet. Es gibt also in diesem Falle insgesamt $2^4 = 16$ verschiedene 2-stellige Auswahlfunktionen. Besteht der Merkmalraum aus k Elementen, wobei k eine natürliche Zahl ist, dann gibt es für jede natürliche Zahl n offenbar 2^{k^n} verschiedene n-stellige Auswahlfunktionen.

Es sei $\{f_n\}$ ($n = 0, 1, 2, \ldots$, ad inf.) eine Folge von Auswahlfunktionen, wobei f_n für jedes n eine n-stellige Auswahlfunktion bedeutet. *Wir können dann jeder Folge* $\{m_i\}$ ($i = 1, 2, \ldots$, *ad inf.*) *von Merkmalen eine Teilfolge* $\{m_{i_j}\}$ ($j = 1, 2, \ldots$) *nach folgender Vorschrift zuordnen: wir bilden die Folge der Funktionswerte*

$$f_0, f_1(m_1), f_2(m_1, m_2), \ldots, f_n(m_1, \ldots, m_n), \ldots, \text{ad inf.},$$

nennen i_1 *die kleinste natürliche Zahl, für welche* $f_{i_1-1}(m_1, \ldots, m_{i_1-1}) = 1$ *ist, und allgemein* i_h ($h = 2, 3, \ldots$, *ad inf.*) *die kleinste natürliche Zahl* $> i_{h-1}$, *für welche* $f_{i_h-1}(m_1, \ldots, m_{i_h-1}) = 1$ *gilt. Eine solche Vorschrift nennen wir „Auswahlvorschrift“.* Eine Auswahlvorschrift ist also durch eine Folge $\{f_n\}$ ($n = 1, 2, \ldots$, ad inf.) von Auswahlfunktionen definiert und ordnet jeder Folge $\{m_i\}$ ($i = 1, 2, \ldots$, ad inf.) von Merkmalen eine Teilfolge $\{m_{i_j}\}$ ($j = 1, 2 \ldots$) zu. Die zugeordnete Teilfolge nennen wir *„Auswahlfolge“*. Sie kann unendlich, endlich oder auch leer sein. Im ersten Falle sagen wir, daß die Auswahlfolge *eigentlich* ist. Die Auswahlvorschrift, welche jeder Folge von Merkmalen die Folge selbst zuordnet, nennen wir die Identität.

Ist $\mathfrak{S}$ ein die Identität enthaltendes System von Auswahlvorschriften (jede Auswahlvorschrift ist durch eine Folge $f_1, f_2, \ldots$, ad inf. von Auswahlfunktionen definiert), $\mathfrak{M}$ ein System von Teilmengen von M, ferner $K = \{m_i\}$ ($i = 1, 2, \ldots$, ad inf.) eine Folge von Merkmalen, so sagen wir, daß K *ein Kollektiv bezüglich* $\mathfrak{S}$ *und* $\mathfrak{M}$ *ist*, im Zeichen $K(\mathfrak{S}, \mathfrak{M})$, falls folgende Bedingung erfüllt ist: *Ist* L *ein Element von* $\mathfrak{M}$, *so existiert eine Zahl* W_L, *genannt die Wahrscheinlichkeit von* L *im Kollektiv* $K(\mathfrak{S}, \mathfrak{M})$, *derart, daß für jede nach einer Vorschrift des Systems* $\mathfrak{S}$ *gebildete unendliche Auswahlfolge* $A = \{m_{i_j}\}$ ($j = 1, 2, \ldots$, *ad inf.*) *gilt:* $\lim_{n=\infty} H_n(L, A) = W_L$, *wobei* $H_n(L, A)$ *die relative Häufigkeit von* L *im Anfangsabschnitt* $(m_{i_1}, m_{i_2}, \ldots, m_{i_n})$ *ist, d. h. die durch* n *dividierte Anzahl der zu* L *gehörigen Elemente dieses Anfangsabschnittes.* Die Wahrscheinlichkeit ist also eine nicht negative, additive Mengenfunktion, definiert für alle Elemente aus $\mathfrak{M}$. Diese Mengenfunktion nennt man auch *„Verteilungsfunktion“* des Kollektivs $K(\mathfrak{S}, \mathfrak{M})$.

§ 2. Die Beziehungen zum Misesschen Kollektivbegriff.

Es sei zunächst ein zwischen unserer und der üblichen Formulierung des Kollektivbegriffes bestehender Unterschied erwähnt, der rein darstellerischer Natur ist. In der üblichen Darstellung ist das Kollektiv eine Folge $\{a_i\}$ ($i=1, 2, \ldots$, ad inf.) von Elementen (genannt Beobachtungen), wobei jedem Element a_i ein Merkmal m_i aus dem Merkmalraum M zugeordnet ist. Da für die Theorie die Elemente a_i gar keine Bedeutung haben und nur die zugeordneten Merkmale m_i eine Rolle spielen, so habe ich nach einem Vorschlag von K. Menger und A. Tarski die Merkmalfolge $\{m_i\}$ selbst als Kollektiv betrachtet. Besteht der Merkmalraum bloß aus den zwei Elementen 0 und 1, so ist in unserer Darstellung ein Kollektiv nichts anderes als eine Merkmalfolge $\{m_i\}$ ($i=1, 2, \ldots$, ad inf.), wobei m_i für jedes i entweder gleich 1 oder gleich 0 ist.

Unser Begriff der „Auswahlvorschrift" ist offenbar nichts anderes als eine präzise Formulierung des Misesschen Begriffs der „Stellenauswahl". So bezeichnet nämlich Mises eine Auswahl, bei der über die Zugehörigkeit, bzw. Nichtzugehörigkeit eines Elements zur Auswahlfolge höchstens mit Benützung der Merkmalwerte der vorangegangenen Elemente entschieden wird, dagegen der Merkmalwert des betreffenden Elementes dabei nicht verwendet wird. Jede „Stellenauswahl" ist eine „Auswahlvorschrift" in unserem Sinne, da sie durch eine Folge $\{f_n\}$ ($n=1, 2, \ldots$, ad inf.) von Auswahlfunktionen dargestellt werden kann und umgekehrt definiert jede Folge $\{f_n\}$ ($n=1, 2, \ldots$, ad inf.) von Auswahlfunktionen eine „Stellenauswahl". Wir wollen dies an einigen Beispielen veranschaulichen: Die Stellenauswahl, „jedes gradzahlige Element wird ausgewählt", läßt sich durch die Folge $\{f_n\}$ ($n=0, 1, 2, \ldots$) von Auswahlfunktionen darstellen, wobei $f_0=0$ und f_n eine n-stellige Auswahlfunktion ist, die identisch (d. h. für jedes n-Tupel von Merkmalen) gleich 0 oder gleich 1 ist, je nachdem ob n gerade oder ungerade ist. Ein zweites Beispiel: Der Merkmalraum bestehe bloß aus den zwei Merkmalen 0 und 1 und wir betrachten die Stellenauswahl: „Ein Element m_i ($=0$ oder 1) wird aus der Merkmalfolge $\{m_n\}$ ($n=1, 2, \ldots$, ad inf.) ausgewählt, falls n eine Primzahl und $m_{n-2}=m_{n-1}=1$ ist". Diese Stellenauswahl ist nichts anderes als die Auswahlvorschrift definiert durch die Folge $\{f_n\}$ von Auswahlfunktionen, in welcher f_n für jede natürliche Zahl n eine n-stellige Auswahlfunktion ist, die, falls $n+1$ keine Primzahl ist, identisch gleich 0 ist und falls $n+1$ eine Primzahl ist, für jedes geordnete n-Tupel von Merkmalen, in welchem die zwei letzten Elemente gleich 1 sind, den Wert 1, für alle übrigen n-Tupel den Wert 0 hat. In diesem Beispiel hängt also die Auswahl eines Elementes m_i aus der Merkmalfolge $\{m_n\}$ auch von den zwei unmittelbar vorangehenden Merkmalwerten m_{i-1} und m_{i-2} ab. Wir wollen noch ein Beispiel betrachten, wo die Auswahl eines Elementes von sämtlichen vorangehenden Merkmalen abhängt. Der Merkmalraum M bestehe wiederum aus den zwei Elementen 0 und 1 und die Stellenauswahl sei folgendermaßen definiert: ein Element m_i aus der Merkmalfolge $\{m_n\}$ wird ausgewählt, falls i eine Quadratzahl und die Anzahl der Einsen in dem Anfangsabschnitt $m_1, \ldots, m_{i-1}$, eine gerade Zahl ist. Diese Stellenauswahl ist identisch mit der Auswahlvorschrift, gegeben durch die Folge $\{f_n\}$ von Auswahlfunktionen, wobei die n-stellige Auswahlfunktion f_n folgendermaßen definiert wird: ist $n+1$ keine Quadratzahl, so ist f_n identisch gleich 0; ist $n+1$ eine Quadratzahl, so ist f_n für jedes geordnete n-Tupel von Zeichen 0 und 1 gleich 1 oder 0, je nachdem, ob die Anzahl der Einsen in dem betrachteten n-Tupel gerade oder ungerade ist.

In unserer Ausdrucksweise lautet dann die Misessche Definition der Kollektivs folgendermaßen: Ist M der Merkmalraum und $\{m_i\}$ $(i=1, \ldots,$ ad inf.) eine Folge von Merkmalen, so ist $K=\{m_i\}$ ein Kollektiv, falls folgende zwei Axiome erfüllt sind:

I. Grenzwertaxiom. Ist L eine Teilmenge von M, so konvergieren mit wachsenden n die relativen Häufigkeiten H_n (L, K) *gegen einen Grenzwert* W_L.

II. Regellosigkeitsaxiom. Es seien A und B zwei zueinander fremde Teilmengen von M und bezeichne W_A *bzw.* W_B *den Grenzwert der relativen Häufigkeit von A bzw. B in der Merkmalfolge* $\{m_i\}$, *wobei* $W_B \neq 0$ *sei: Es sei* m_i' $(i=1, 2, \ldots$ *ad inf.) die Folge derjenigen Elemente der Folge* $\{m_i\}$, *welche in* $A+B$ *enthalten sind. Wählen wir nach irgendeiner „Auswahlvorschrift" aus* $\{m_i'\}$ *eine unendliche Auswahlfolge* $\{m_i''\}$ $(i=1, 2, \ldots,$ *ad inf.) aus, dann existiert der Grenzwert* W_A'' *bzw.* W_B'' *der relativen Häufigkeit von A bzw. B in* $\{m_i''\}$ *und es gilt* $W_A'' : W_B'' = W_A : W_B$.

Mises behauptet[2]) versehentlich, daß das Axiom II *nicht* äquivalent sei mit dem folgenden einfachen Axiom:

II'. Es sei A eine Teilmenge von M und es bezeichne W_A *den Grenzwert der relativen Häufigkeit von A in der Folge* $\{m_i\}$. *Ist* $\{m_i'\}$ $(i=1, 2, \ldots,$ *ad inf.) eine nach irgendeiner Auswahlvorschrift gebildete unendliche Auswahlfolge aus* $\{m_i\}$, *dann existiert der Grenzwert der relativen Häufigkeit von A in* $\{m_i'\}$ *und es gilt* $W_A' = W_A$.

Wir werden beweisen, daß II und II' äquivalent sind.

a) Aus II folgt II'. Es sei A eine beliebige Teilmenge von M und $B=M-A$. Es gilt dann offenbar $W_A + W_B = 1$, also sind W_A und W_B nicht beide $=0$. Es sei etwa $W_B \neq 0$. Wegen II existiert dann offenbar der Grenzwert W_A' bzw. W_B' der relativen Häufigkeit von A bzw. B in jeder nach einer Auswahlvorschrift gebildeten unendlichen Auswahlfolge $\{m_i'\}$ aus $\{m_i\}$ und es gilt $W_A' : W_B' = W_A : W_B$. Da wegen $A+B=M$ offenbar auch die Beziehung $W_A' + W_B' = 1$ gilt, so muß $W_A' = W_A$ und $W_B' = W_B$ gelten, womit II' bewiesen ist.

b) Aus II' folgt II. Es seien A und B zwei zueinander fremde Teilmengen von M, sodaß $W_B \neq 0$ ist. Wir heben aus der Folge $\{m_i\}$ $(i=1, 2, \ldots,$ ad inf.) diejenigen Elemente heraus, die in $A+B$ enthalten sind. Es sei die so herausgehobene Teilfolge $\{m_i'\}$ $(i=1, 2, \ldots,$ ad inf.). Es sei $\{m_i''\}$ $(i=1, 2, \ldots,$ ad inf.) eine unendliche Auswahlfolge, ausgewählt aus $\{m_i'\}$ nach irgend einer

---

[2]) l. c. Seite 18.

Auswahlvorschrift V. Die Auswahlvorschrift V sei gegeben durch die Folge $\{f_n\}$ ($n=0, 1, \ldots,$ ad inf.) von Auswahlfunktionen, wobei f_n eine n-stellige Auswahlfunktion bedeutet. Wir definieren nun eine Auswahlvorschrift V^* durch die Folge $\{g_n\}$ von Auswahlfunktionen, wobei g_n für jedes n eine n-stellige Auswahlfunktion bedeutet, die folgendermaßen gegeben ist: Ist $(p_1, \ldots, p_n)$ ein geordnetes n-Tupel von Merkmalen und bezeichnet $(p_{i_1}, \ldots, p_{i_k})$ das geordnete k-Tupel, welches aus $(p_1, \ldots, p_n)$ durch Streichung der in $C=M-(A+B)$ enthaltenen Merkmale entsteht, so sei $g_n(p_1, \ldots, p_n)=f_k(p_{i_1}, \ldots, p_{i_k})$. Wir wählen nach Vorschrift V^* aus $\{m_i\}$ eine Teilfolge $\{m_i^*\}$ aus. Es folgt unmittelbar aus der Definition der Funktion g_n, daß, wenn man in der Folge $\{m_i^*\}$ diejenigen Elemente streicht, die in C enthalten sind, die Folge $\{m_i^*\}$ in die Folge $\{m_i''\}$ übergeht. Nach II' existiert in der Folge $\{m_i^*\}$ für eine beliebige Teilmenge N von M der Grenzwert W_N^* der relativen Häufigkeit von N und ist $W_N^*=W_N$. Insbesondere gilt daher

$$W_A^*=W_A;\quad W_B^*=W_B;\quad W_C^*=W_C.$$

Da die Folge $\{m_i^*\}$ nach Streichung der in C enthaltenen Elemente in $\{m_i''\}$ übergeht, so existieren offenbar die Grenzwerte W_A'' und W_B'' der relativen Häufigkeiten von A und B in der Teilfolge $\{m_i''\}$ und es gelten folgende Beziehungen:

$$W_A''=W_A^*:(W_A^*+W_B^*)=W_A:(W_A+W_B);$$
$$W_B''=W_B^*:(W_A^*+W_B^*)=W_B:(W_A+W_B).$$

Daraus folgt unmittelbar $W_A'':W_B''=W_A:W_B$, womit II bewiesen ist.

Wir können also die Misesschen Axiome I und II durch die Axiome I und II' ersetzen. Der Unterschied zwischen dem Misesschen und dem in § 1 definierten Kollektivbegriff besteht daher lediglich darin, *daß wir den Kollektivbegriff sowohl in bezug auf ein System $\mathfrak{S}$ von Auswahlvorschriften, als auch in bezug auf ein System $\mathfrak{M}$ von Teilmengen des Merkmalraumes M relativieren.* Das Misessche Kollektiv ist ein Kollektiv K ($\mathfrak{S}$, $\mathfrak{M}$), wobei $\mathfrak{S}$ das System aller Auswahlvorschriften und $\mathfrak{M}$ das System aller Teilmengen von M bedeutet.

§ 3. Einwände gegen den Misesschen Kollektivbegriff und verschiedene Versuche zur Einschränkung der Regellosigkeitsforderung.

Die Einwände richten sich hauptsächlich gegen die Missessche Formulierung des Regellosigkeitsaxioms, u. zw. gegen die Forderung, daß $\mathfrak{S}$ das System *aller* Auswahlvorschriften sei. In der Tat, wenn man das Wort *alle* gar keiner Einschränkung unterwirft, führt dies bekanntlich zu einen Widerspruch, was man wie folgt einsieht: eine Folge $\{i_j\}$ ($j=1, 2, \ldots,$ ad inf.) von natürlichen Zahlen definiert offenbar eine Auswahlvorschrift, indem man festsetzt: das

n-te Element der Merkmalfolge $\{m_i\}$ ($i=1, 2, \ldots$, ad inf.) soll dann und nur dann ausgewählt werden, falls n ein Element der Folge $\{i_j\}$ ($j=1, 2, \ldots$, ad inf.) ist. Da $\mathfrak{S}$ nach Voraussetzung das System aller Auswahlvorschriften bedeutet, so enthält $\mathfrak{S}$ das System $\mathfrak{S}'$ von Auswahlvorschriften, das dem System aller natürlichen Zahlenfolgen entspricht. Ist nun $K=\{m_i\}$ ($i=1, 2, \ldots$, ad inf.) ein Kollektiv bezüglich $\mathfrak{S}$ und irgend eines Systems $\mathfrak{M}$ von Teilmengen von M, so betrachten wir ein Element A aus $\mathfrak{M}$, für welches $W_A \neq 0$ und $\neq 1$ gilt. (Den trivialen Fall, daß W_A entweder $=0$ oder $=1$ für jedes Element A von $\mathfrak{M}$ ist, wollen wir ausschließen.) Aus $W_A \neq 0$ folgt, daß unendlich viele Elemente aus $\{m_i\}$ in A enthalten sind. Daraus ergibt sich aber unmittelbar, daß es eine Auswahlvorschrift in $\mathfrak{S}'$ und erst recht in $\mathfrak{S}$ gibt, die aus $\{m_i\}$ eine unendliche Auswahlfolge liefert, deren sämtliche Elemente in A enthalten sind. Wir sind also zu einem Widerspruch gelangt und haben gezeigt, daß kein Kollektiv bezüglich $\mathfrak{S}'$ und erst recht bezüglich $\mathfrak{S}$ existiert.

Offenbar meinte jedoch Mises, daß nicht alle beliebigen Auswahlvorschriften zugelassen seien, sondern nur solche, die durch mathematische Gesetze definierbar sind. Dann müßte allerdings der Begriff des mathematischen Gesetzes präzisiert werden. Dies tut aber Mises nicht und solange dies nicht geschieht, kann man die Frage der Widerspruchsfreiheit des Misesschen Kollektivbegriffes gar nicht sinnvoll stellen. Es sei hier noch bemerkt, daß Mises die mit dem Wort „alle" verbundenen Schwierigkeiten auch dadurch zu beseitigen versucht, daß er seiner Regellosigkeitsforderung wörtlich folgende Deutung gibt[3]): „Die Festsetzung, daß in einem Kollektiv jede Stellenauswahl (in unserer Ausdrucksweise, Auswahl nach einer „Auswahlvorschrift") die Grenzhäufigkeit unverändert läßt, besagt nichts anderes als dies: Wir verabreden, daß, wenn in einer konkreten Aufgabe ein Kollektiv einer bestimmten Stellenauswahl unterworfen wird, wir annehmen wollen, diese Stellenauswahl ändere nichts an den Grenzwerten der relativen Häufigkeiten". Gibt man der Regellosigkeitsforderung diese Deutung, so wird der Kollektivbegriff sozusagen bezüglich der vorliegenden Aufgabe relativiert, indem $\mathfrak{S}$ als das System der in der Aufgabe tatsächlich vorkommenden Auswahlvorschriften angenommen wird. Liegt dann eine konkrete Aufgabe vor, so kann man zwar die Frage, ob Kollektivs dieser Art existieren, sinnvoll stellen, sie muß jedoch noch gelöst werden, d. h. es muß für jede konkret vorliegende Aufgabe gezeigt werden, daß entsprechende Kollektivs existieren[4]).

---

[3]) Mises: „Wahrscheinlichkeit, Statistik und Wahrheit" vierte Auflage, Wien, Julius Springer 1936, S. 119.

[4]) Es sei hier vorwegnehmend bemerkt (ausführlicher sprechen wir darüber in § 4), daß in dieser Note ein Widerspruchsfreiheitbeweis für den Misesschen Kollektivbegriff gegeben wird, sowohl wenn man den Begriff des mathematischen Gesetzes im Sinne der modernen Logik präzisiert, als auch wenn man der Regellosigkeitsforderung die obige Deutung gibt. Im letzteren Falle muß man noch eine Voraussetzung machen, die jedoch in jeder konkreten Aufgabe der Wahrscheinlichkeitsrechnung zweifellos erfüllt ist.

Die erwähnten Schwierigkeiten haben einige Mathematiker veranlaßt, eine weitgehend eingeschränkte Regellosigkeitsforderung der Theorie zugrundezulegen. Merkmalfolgen, die einer stark eingeschränkten Regellosigkeitsforderung genügen, sind die von Reichenbach [5]) und neuerdings von K. Popper [6]) untersuchten sogenannten „normalen" bzw. „nachwirkungsfreien" Folgen. Ist M endlich, so ist eine nachwirkungsfreie Folge $\{m_i\}$ in unserer Ausdrucksweise ein Kollektiv K ($\mathfrak{S}$, $\mathfrak{M}$), wobei $\mathfrak{M}$ das System aller Teilmengen von M bedeutet und $\mathfrak{S}$ außer der Identität die Auswahlvorschriften von folgender und nur von folgender Art enthält: ein Element m_n aus $\{m_i\}$ wird dann und nur dann ausgewählt, falls das diesem Element unmittelbar vorangehende k-Tupel (m_{n-k}, $m_{n-k+1}, \ldots, m_{n-1}$) von Merkmalen einem festgegebenen k-Tupel von Merkmalen ($a_1, \ldots, a_k$) gleich ist. Besteht der Merkmalraum M bloß aus den zwei Elementen 0 und 1, so sind Auswahlvorschriften solcher Art z. B. die folgenden: ein Element wird dann und nur dann ausgewählt, falls ihm das Element 0 unmittelbar vorangeht; oder: ein Element wird dann und nur dann ausgewählt, falls ihm die Kombination 0 0 unmittelbar vorangeht, usw. Reichenbach formuliert l. c. als ungelöstes Problem die Frage, ob nachwirkungsfreie Folgen durch eine Rechenvorschrift konstruktiv definiert werden können. Diese Frage hat K. Popper l. c. für den Fall, daß der Merkmalraum M bloß aus zwei Elementen besteht, positiv entschieden, indem er ein Konstruktionsverfahren für nachwirkungsfreie Folgen angibt. Noch weiter gehende Untersuchungen über die Existenz von Kollektivs bei verschiedenen Einschränkungen der Regellosigkeitsforderung finden sich bei A. Copeland [7]). Er betrachtet sogenannte „admissible numbers", die auf das hinauslaufen, was Reichenbach eine normale Folge nennt. Auf die von Copeland erreichten Ergebnisse werden wir im § 4 noch zurückkommen.

Die Bemühungen, die mit der Regellosigkeitsforderung verbundenen Schwierigkeiten zu überwinden, laufen also auf eine entsprechende Einschränkung der Regellosigkeitsforderung hinaus, die zumindest so weitgehend sein muß, daß die Existenz von Kollektivs beweisbar sei. Manche stellen sogar die Forderung auf, daß die Regellosigkeitsforderung so weitgehend eingeschränkt sein muß, daß Kollektivs durch Rechenvorschriften konstruktiv definiert werden können. Dies ist von Wichtigkeit, da für viele Anwendungen der Wahrscheinlichkeitsrechnung auch eine weitgehend eingeschränkte Regellosigkeitsforderung ausreicht, anderseits die konstruktive Definierbarkeit des Kollektivs sicherlich einen Vorteil bedeutet. Jedoch eignet sich eine solche Theorie, wie von Mises betont wird [8]), nicht zur Beschreibung der Glücksspiele und ähnlicher Erscheinungs-

---

[5]) Reichenbach, Mathem. Zeitschrift, Bd. *34*.

[6]) K. Popper, Logik der Forschung, Wien 1935.

[7]) A. H. Copeland: Point set theory applied to the random selection of the digits of an admissible number. Amer. Journ. of math. *58*, 1936.

[8]) Vgl. etwa Mises: Wahrscheinlichkeit, Statistik und Wahrheit, S. 117.

reihen. Ist nämlich das System $\mathfrak{S}$ der zulässigen Auswahlvorschriften so weitgehend eingeschränkt, daß es möglich ist, Kollektivs durch Rechenvorschriften konstruktiv zu definieren, dann wäre es möglich, auch eine durch eine Rechenvorschrift konstruktiv definierte Auswahlvorschrift anzugeben, so daß in der Auswahlfolge der Grenzwert der relativen Häufigkeit eines Merkmals von dem Grenzwert in der ursprünglichen Reihe verschieden wäre, d. h. es gäbe dann ein Spielsystem, welches die Gewinnmöglichkeiten wesentlich beeinflussen würde.

In der vorliegenden Arbeit wird daher ganz allgemein untersucht, *erstens,* wie weitgehend die Regellosigkeitsforderung gefaßt werden kann, damit zumindest die Existenz von Kollektivs beweisbar sei, insbesondere, ob der Misessche Kollektivbegriff, falls man den Begriff des mathematischen Gesetzes im Sinne der modernen Logik präzisiert, widerspruchsfrei ist; *zweitens*, welche weiteren Einschränkungen man der Regellosigkeitsforderung auferlegen muß, damit es möglich sei, Kollektivs durch Rechenvorschriften konstruktiv zu definieren.

Weiter behandeln wir die bisher nicht genügend untersuchte Frage, wie das System $\mathfrak{M}$ von Teilmengen von M, die eine Wahrscheinlichkeit besitzen sollen, gewählt werden darf, damit der Kollektivbegriff widerspruchsfrei sei. Wir werden zeigen, daß, falls M aus unendlich vielen Elementen besteht, auch über $\mathfrak{M}$ gewisse Annahmen gemacht werden müssen. Wir geben in dem nächsten Abschnitt eine genaue Formulierung der Probleme und eine kurze Übersicht über die erzielten wichtigsten Resultate.

§ 4. Formulierung der Hauptergebnisse.

Wir knüpfen an unsere im § 1 gegebene Definition des Kollektivbegriffs an.

Problem 1. *Sei M ein gegebener Merkmalraum. Welchen Bedingungen muß das System $\mathfrak{S}$ von Auswahlvorschriften, das System $\mathfrak{M}$ von Teilmengen von M, und die für alle Elemente von $\mathfrak{M}$ erklärte Mengenfunktion μ genügen, damit ein Kollektiv $K(\mathfrak{S}, \mathfrak{M})$ existiert, dessen Verteilungsfunktion mit μ identisch ist?*

Die Lösung dieses Problems wird durch die folgenden vier Theoreme gegeben:

Theorem I. *Ist der Merkmalraum M eine endliche Menge (von einer Mächtigkeit > 1), $\mathfrak{S}$ ein System von abzählbar vielen Auswahlvorschriften, $\mathfrak{M}$ das System aller Teilmengen von M und μ eine nichtnegative additive Mengenfunktion definiert für alle Elemente von $\mathfrak{M}$, wobei $\mu(M) = 1$ gilt, so gibt es kontinuierlich viele Kollektivs $K(\mathfrak{S}, \mathfrak{M})$, deren Verteilungsfunktion gleich μ ist.*

Theorem II. *Ist der Merkmalraum M eine unendliche (abzählbare oder unabzählbare) Menge, $\mathfrak{S}$ ein System von abzählbar vielen Auswahlvorschriften, $\mathfrak{M}$ das System aller Teilmengen von M,* *und μ eine nichtnegative additive Mengenfunktion definiert für alle Elemente von $\mathfrak{M}$, wobei $\mu(M)=1$ gilt, dann ist die notwendige und hinreichende Bedingung dafür, daß ein Kollektiv K ($\mathfrak{S}$, $\mathfrak{M}$) mit der Verteilungsfunktion μ existiere, daß es eine Folge $\{m_i\}$ ($i=1, 2, \ldots$, ad inf.) von paarweise verschiedenen Merkmalen gibt, für welche $\sum_{i=1}^{\infty}\mu(m_i)=1$ ist.*

In der klassischen Theorie wird im Falle der sogen. geometrischen Wahrscheinlichkeiten (wo also der Merkmalraum aus kontinuierlich vielen Punkten besteht) stets die Annahme gemacht, daß die einzelnen Merkmalpunkte die Wahrscheinlichkeit 0 haben. Aus Theorem II ergibt sich daher, daß in allen diesen Fällen *keine* Kollektivs bezüglich aller Teilmengen von M existieren. Wir können also höchstens von Kollektivs bezüglich gewisser Teilmengen von M sprechen. Das folgende Theorem zeigt, daß es jedoch möglich ist, sehr umfassende für die Anwendungen völlig ausreichende Systeme von Teilmengen von M anzugeben, bezüglich welcher Kollektivs existieren.

Theorem III. *Voraussetzungen:*

1. M ist ein unendlicher Merkmalraum von beliebiger Mächtigkeit.

2. $\mathfrak{K}$ ist ein abzählbarer Mengenkörper, dessen Elemente Teilmengen von M sind und der M selbst als Element enthält.

3. μ ist eine nichtnegative additive Mengenfunktion, erklärt für alle Elemente aus $\mathfrak{K}$, wobei $\mu(M)=1$ gilt.

4. $\mathfrak{M}$ sei das System aller in bezug auf $\mathfrak{K}$ μ-meßbaren Teilmengen von M. Dabei heißt eine Teilmenge A von M in bezug auf $\mathfrak{K}$ μ-meßbar, falls das äußere μ-Maß von A, d. i. die untere Schranke der Werte von μ für alle Elemente aus $\mathfrak{K}$, die A enthalten, und das innere μ-Maß von A, d. i. die obere Schranke der Werte von μ für alle Elemente aus $\mathfrak{K}$, die in A enthalten sind, einander gleich sind; dieser gemeinsame Wert heißt $\mu(A)$, das μ-Maß von A.

5. $\mathfrak{S}$ sei ein System von abzählbar vielen Auswahlvorschriften.

Behauptung: Es gibt kontinuierlich viele Kollektivs K ($\mathfrak{S}$, $\mathfrak{M}$), deren Verteilungsfunktion gleich μ ist.

In den Anwendungen werden nur Merkmalmengen M betrachtet, die entweder endlich, oder im Peanoschen Sinne meßbare Teilmengen des n-dimensionalen euklidischen Raumes R_n sind. Im letzteren Fall sind nur die Wahrscheinlichkeiten der im Peanoschen Sinne meßbaren Teilmengen von M für die Anwendungen von Bedeutung. Ferner handelt es sich hier ausschließlich um Verteilungsfunktionen μ, für welche folgendes gilt: ist $\{M_i\}$ ($i=1, 2, \ldots$, ad inf.) eine Folge von Teilmengen von M, wobei das Peanosche Maß $\mu_P(M_i)$ von M_i mit wachsendem i gegen 0 konvergiert, so gilt auch lim $\mu(M_i)=0$. Für die Zwecke der Anwendungen reicht daher vollkommen aus das

Theorem IV. *Ist der Merkmalraum M eine beschränkte im Peanoschen Sinne meßbare Teilmenge des n-dimensionalen euklidischen Raumes R_n, ist ferner μ eine nichtnegative additive Mengenfunktion, erklärt für alle im Peanoschen Sinne meßbaren Teilmengen von M, wobei $\mu(M)=1$ gilt und lim $\mu(M_i)=0$ ist, falls $\{M_i\}$ eine Folge von Peanosch-meßbaren Teilmengen bedeutet, für welche das Peanosche Maß $\mu_P(M_i)$ mit wachsendem i gegen 0 konvergiert, und ist schließlich $\mathfrak{S}$ ein System von abzählbar vielen Auswahlvorschriften, so gibt es kontinuierlich viele Kollektivs bezüglich $\mathfrak{S}$ und aller im Peanoschen Sinne meßbaren Teilmengen von M, deren Verteilungsfunktion identisch gleich μ ist.*

Theorem IV ist als Spezialfall in Theorem III enthalten, denn wählt man den abzählbaren Mengenkörper $\mathfrak{K}$ so, daß er jede Teilmenge von M, die Durchschnitt von M mit einem rationalen Intervall des R_n ist, enthält, so ist jede im Peanoschen Sinne meßbare Teilmenge von M zugleich μ-meßbar in bezug auf den Körper $\mathfrak{K}$.

Die Theoreme I—IV geben Bedingungen über $\mathfrak{S}$, $\mathfrak{M}$ und μ an, die hinreichend, aber nicht notwendig sind, damit Kollektivs $K(\mathfrak{S}, \mathfrak{M})$ mit der Verteilungsfunktion μ existieren. Doch sind diese Bedingungen schon so wenig einschränkend, daß eine weitere Abschwächung derselben kaum von Interesse wäre. Hinsichtlich der Bedingungen für $\mathfrak{M}$ und μ haben wir dies bereits gesehen. Ist nämlich M endlich, so haben wir über $\mathfrak{M}$ und μ gar keine einschränkende Annahmen gemacht (Theorem I). Ist M unendlich, so haben wir bloß die Bedingung gestellt, daß die Elemente von $\mathfrak{M}$ in gewissem Sinne meßbare Teilmengen von M seien (Theorem III und IV). Nun wollen wir klar machen, daß auch die an $\mathfrak{S}$ gestellte Bedingung, nämlich daß $\mathfrak{S}$ nur abzählbar viele Auswahlvorschriften enthält, so schwach ist, daß eine weitere Abschwächung für die Anwendungen nicht von Interesse wäre. Sogar der Misessche Kollektivbegriff genügt nämlich, wenn man den Begriff des mathematischen Gesetzes im Sinne der modernen Logik präzisiert, dieser Bedingung. Der Begriff des mathematischen Gesetzes kann ja nur innerhalb einer formalisierten Logik sinnvoll definiert werden und mithin gibt es offenbar nur abzählbar viele mathematische Gesetze. Man kann z. B. die von Whitehead und Russell in den Principia Mathematica entwickelte formalisierte Logik (innerhalb deren das gesamte Gebiet der Mathematik dargestellt werden kann) zugrunde legen, und $\mathfrak{S}$ als das System aller Auswahlvorschriften annehmen, die durch mathematische Gesetze dieser formalisierten Logik definiert werden können. Man sieht also, daß die Bedingung der Abzählbarkeit von $\mathfrak{S}$ so wenig einschränkend ist, daß eine weitere Abschwächung für die Anwendungen kaum von Bedeutung wäre. Besonders klar wird dies, wenn man der Regellosigkeitsforderung die erwähnte Misessche

Deutung gibt, welche die Regellosigkeitsforderung bezüglich der vorliegenden Aufgabe relativiert, indem bloß gefordert wird, daß die Auswahlen, die nach in der Aufgabe tatsächlich vorkommenden Auswahlvorschriften gebildet werden, an den Grenzwerten der relativen Häufigkeiten nichts ändern. Denn es ist klar, daß in jeder konkreten Aufgabe der Wahrscheinlichkeitsrechnung höchstens abzählbar viele Auswahlvorschriften tatsächlich verwendet werden. Die Theoreme I—IV liefern also auch dann einen Widerspruchsfreiheitsbeweis für den Misesschen Kollektivbegriff, wenn man der Regellosigkeitsforderung die obige Deutung gibt. Bezeichnet man ein Kollektiv K ($\mathfrak{S}$, $\mathfrak{M}$) als Misessches Kollektiv bezüglich einer formalisierten Logik L, falls $\mathfrak{S}$ das System aller Auswahlvorschriften bedeutet, die durch mathematische Gesetze der formalisierten Logik L definiert sind, und als Misessches Kollektiv bezüglich einer Aufgabe A, falls $\mathfrak{S}$ das System der in der Aufgabe A tatsächlich vorkommenden Auswahlvorschriften bedeutet, so gilt folgendes

Korollar: *Ist M ein Merkmalraum, μ eine nichtnegative additive Mengenfunktion definiert in einem abzählbaren Körper $\mathfrak{K}$ von Teilmengen von M (falls M endlich ist, so sei μ für alle Teilmengen von M definiert), wobei $\mathfrak{K}$ die Menge M als Element enthält und $\mu(M)=1$ gilt, so gibt es Misessche Kollektivs sowohl bezüglich jeder formalisierten Logik L, als auch bezüglich jeder Aufgabe A, die nur abzählbar viele Auswahlvorschriften verwendet (und nur solche kommen in den Anwendungen vor), derart, daß jede in bezug auf den Körper $\mathfrak{K}$ μ-meßbare Teilmenge A (falls M endlich ist, so jede Teilmenge A) eine Wahrscheinlichkeit W_A besitzt, die gleich $\mu(A)$ ist.*

Die Theoreme I—IV stellen also eine für die Anwendungen völlig ausreichende Lösung des Problems 1 dar. Es sei hier noch bemerkt, daß, falls die Wahrscheinlichkeitsrechnung als ein Teil der Mathematik in einer formalisierten Logik L entwickelt und dargestellt wird, es am zweckmäßigsten ist, das Misessche Kollektiv bezüglich der formalisierten Logik L an die Spitze zu stellen. Das Misessche Kollektiv bezüglich der formalisierten Logik L wird in dieser Theorie gleichzeitig auch ein Misessches Kollektiv bezüglich jeder Aufgabe A sein, da in jeder Aufgabe nur Auswahlvorschriften auftreten, die durch mathematische Gesetze der formalisierten Logik L definiert sind.

Ein weiteres Problem knüpft an folgende Definition: Die Merkmalfolge $\{m_i\}$ ($i=1, 2, \ldots$, ad inf.) heiße *konstruktiv definiert, falls ein Verfahren vorliegt, das für jede natürliche Zahl i den Merkmalwert m_i in endlich vielen Schriften tatsächlich zu berechnen gestattet.* Z. B. ist die Merkmalfolge $\{m_i\}$ ($i=1, 2, \ldots$, ad inf.), wo $m_n=1$ oder 0 ist, je nachdem ob n eine Primzahl ist oder nicht, konstruktiv definiert, da man für jede natürliche Zahl n in endlich vielen Schritten berechnen kann, ob n eine Primzahl ist oder nicht. Dagegen ist die Folge $\{m_i\}$ ($i=1, 2, \ldots$, ad inf.), wo m_n gleich 1 oder 0 ist, je nachdem der große Fermatsche Satz für n gilt oder nicht, nicht konstruktiv definiert, da wir nicht für jedes n in endlich vielen Schritten feststellen können, ob der große Fermatsche Satz gilt oder nicht.

Problem 2. *Welchen Bedingungen muß das abzählbare System $\mathfrak{S}$ von Auswahlvorschriften, das System $\mathfrak{M}$ von Teilmengen des Merkmalraumes M, und die für die Elemente von $\mathfrak{M}$ erklärte Mengenfunktion μ genügen, damit Kollektivs $K(\mathfrak{S}, \mathfrak{M})$ mit der Verteilungsfunktion μ konstruktiv definiert werden können? Ein entsprechendes Konstruktionsverfahren für solche Kollektivs $K(\mathfrak{S}, \mathfrak{M})$ soll effektiv angegeben werden.*

Wir beginnen mit Hilfsbegriffen. Eine Auswahlvorschrift V ist durch eine Folge $\{f_n\}$ von Auswahlfunktionen gegeben. Da $\mathfrak{S}$ abzählbar ist, so kann man die Auswahlvorschriften von $\mathfrak{S}$ in eine Folge $\{V_i\}$ $(i=1, 2, \ldots)$ anordnen. Für jede natürliche Zahl i sei die Auswahlvorschrift V_i durch die Folge $\{f_n^i\}$ $(n=0, 1, 2, \ldots, \text{ad inf.})$ von Auswahlfunktionen gegeben, wobei f_n^i eine n-stellige Auswahlfunktion bedeutet. Wir sagen, daß das System $\mathfrak{S}=\{V_i\}$ *konstruktiv definiert* ist, *falls für jedes natürliche Zahlenpaar i und n der Funktionswert von f_n^i für jedes geordnete n-Tupel von Merkmalen in endlich vielen Schritten berechnet werden kann.*

Ist $\mathfrak{A}=\{A_i\}$ $(i=1, 2, \ldots)$ eine endliche oder unendliche Folge von Teilmengen von M, so sagen wir, daß die Mengenfunktion μ für die Elemente von $\mathfrak{A}$ *konstruktiv definiert ist, falls für jede natürliche Zahl i der Funktionswert $\mu(A_i)$ in endlich vielen Schritten berechnet werden kann.* Ist der Merkmalraum endlich, so ist μ stets konstruktiv definiert, da man für jeden Merkmalpunkt von M den Wert von μ explizit angibt. — Die Lösung des Problems 2 wird dann durch die folgenden zwei Theoreme gegeben:

Theorem V. *Ist M ein endlicher Merkmalraum, ist μ eine nichtnegative additive Mengenfunktion, definiert für alle Teilmengen von M, wobei $\mu(M)=1$ ist, und ist $\mathfrak{S}$ ein konstruktiv definiertes abzählbares System von Auswahlvorschriften, so kann man Kollektivs bezüglich $\mathfrak{S}$ und aller Teilmengen von M mit der Verteilungsfunktion μ konstruktiv definieren.*

Theorem VI. *Ist M ein unendlicher Merkmalraum, ist μ eine nichtnegative additive Mengenfunktion, die konstruktiv definiert ist für die Elemente eines abzählbaren Körpers $\mathfrak{K}$ von Teilmengen von M, wobei M ein Element von $\mathfrak{K}$ ist und $\mu(M)=1$ gilt, und ist $\mathfrak{S}$ ein konstruktiv definiertes abzählbares System von Auswahlvorschriften, so kann man Kollektivs $K(\mathfrak{S}, \mathfrak{M})$ mit der Verteilungsfunktion μ konstruktiv definieren, wobei $\mathfrak{M}$ das System aller in bezug auf $\mathfrak{K}$ μ-meßbaren Teilmengen von M bedeutet.*

Für Kollektivs, die den Voraussetzungen des Theorem V bzw. VI genügen, geben wir im zweiten Teil dieser Untersuchung ein Konstruktionsverfahren effektiv an. Damit ist das Problem 2 vollständig gelöst.

In den konkreten Aufgaben werden wohl nur Kollektivs betrachtet, bei denen die Verteilungsfunktion μ und das System der Auswahlvorschriften, die in der gegebenen Aufgabe zur Bildung von Auswahlfolgen tatsächlich verwendet werden, konstruktiv definiert sind. Unsere Theorie liefert also ein Konstruktionsverfahren für Misessche Kollektivs in bezug auf jede praktisch vorkommende Aufgabe. Damit fallen auch alle Einwände der Intuitionisten weg, da nach dem intuitionistischen Standpunkt erst recht nur Aufgaben gestellt werden, bei denen die obgenannten Bedingungen erfüllt sind.

Alle bis jetzt betrachteten und näher untersuchten Kollektivs mit eingeschränkter Regellosigkeitsforderung genügen den Bedingungen der Theoreme V und VI; mithin liefert unsere Theorie ein Konstruktionsverfahren für solche Kollektivs. Wir wollen dies etwa für die nachwirkungsfreien Folgen klar machen. Wir betrachten zunächst den Fall, daß der Merkmalraum M bloß aus den zwei Elementen 0 und 1 besteht. Wir ordnen die Menge aller endlichen Zeichenkombinationen von 0 und 1 in die Folge

(†) 0, 1, 00, 01, 10, 11, 000, 001, 010, 011, 100, 101, 110, 111, 0000, ...

Wenn man diese Zeichenkombinationen als dyadische Zahlen auffaßt, so kommen zuerst die einstelligen, dann die zweistelligen, dann die dreistelligen Zahlen usw., wobei die 2^n verschiedenen n-stelligen Zahlen monoton wachsend angeordnet werden. Wir bezeichnen das k-te Element der Folge (†) mit λ_k ($k=1, 2, \ldots$, ad inf.). Z. B. ist λ_3 die Kombination 00, λ_8 die Kombination 001. Wir definieren nun eine Folge $\mathfrak{S}_0 = \{V_i\}$ ($i=1, 2, \ldots$, ad inf.) von Auswahlvorschriften:

$$V_1 = f_0^1, f_1^1, f_2^1, \ldots, f_n^1, \ldots, \text{ ad inf.},$$
$$V_2 = f_0^2, f_1^2, f_2^2, \ldots, f_n^2, \ldots, \text{ ad inf.}$$
$$\ldots\ldots\ldots\ldots\ldots\ldots\ldots\ldots\ldots\ldots$$
$$V_i = f_0^i, f_1^i, f_2^i, \ldots, f_n^i, \ldots, \text{ ad inf.}$$

wobei f_n^i für eine n-stellige Zeichenkombination den Wert 1 oder 0 hat, je nachdem ob diese n-stellige Zeichenkombination mit λ_i endet oder nicht. Z. B. ist f_5^3 für die 5-stellige Kombination 01100 gleich 1, da die Kombination mit $\lambda_3 = 00$ endet, dagegen ist $f_5^3 = 0$ für die Kombination 00101, da diese nicht mit λ_3 endet. Die Auswahlfunktion f_3^2 ist z. B. für die Kombinationen 001, 011, 101, 111 gleich 1 und für die Kombinationen 000, 010, 100, 110 gleich 0. Offenbar bedeutet V_i für jede natürliche Zahl i die Auswahlvorschrift: ein Element m_k der aus 0 und 1 gebildeten Zeichenfolge $\{m_n\}$ ($n=1, 2, \ldots$, ad inf.) wird dann und nur dann ausgewählt, falls in der Folge $\{m_n\}$ dem Element m_k die Zeichenkombination λ_i unmittelbar vorangeht. Z. B. bedeutet V_4 die Auswahlvorschrift: jedes

Element wird ausgewählt, dem die Zeichenkombination 01 unmittelbar vorangeht. Dieses System $\mathfrak{S}_0$ von Auswahlvorschriften ist konstruktiv definiert, da für jedes natürliche Zahlenpaar (i, n) der Wert von f_n^i für jede n-stellige Zeichenkombination in endlich vielen Schritten berechnet werden kann.

Die Kollektivs bezüglich $\mathfrak{S}_0$ sind offenbar mit den nachwirkungsfreien Folgen identisch. Wir haben also im Falle, daß der Merkmalraum M bloß aus 2 Elementen besteht, gezeigt, daß die nachwirkungsfreien Folgen Kollektivs sind, die den Bedingungen des Theorem V genügen. Man kann analog zeigen, daß auch, wenn M mehr als zwei aber endlich viele Elemente enthält, die nachwirkungsfreien Folgen mit einer beliebigen Verteilungsfunktion μ Kollektivs sind, die den Bedingungen des Theorem V genügen. *Unsere Theorie liefert daher, falls M endlich ist, ein Konstruktionsverfahren für nachwirkungsfreie Folgen mit beliebig vorgegebener Verteilungsfunktion μ.*

Ist der Merkmalraum M unendlich, so muß zunächst präzisiert werden, was man unter einer nachwirkungsfreien Merkmalfolge versteht, denn im allgemeinen kann nicht für jede Teilmenge von M der Grenzwert der relativen Häufigkeit existieren. Ist $\mathfrak{M}$ ein System von Teilmengen von M, μ eine Mengenfunktion erklärt für die Elemente von $\mathfrak{M}$, so sagen wir, daß eine Merkmalfolge $\{m_n\}$ $(n = 1, 2, \ldots, \text{ad inf.})$ nachwirkungsfrei in bezug auf $\mathfrak{M}$ sei und die Verteilungsfunktion μ habe, falls der Grenzwert der relativen Häufigkeit für jedes Element A aus $\mathfrak{M}$ gleich $\mu(A)$ ist und dieser Grenzwert durch Auswahlen folgender Art nicht geändert wird: ein Element m_i wird aus der Folge $\{m_n\}$ dann und nur dann ausgewählt, falls m_{i-k} in A_1, m_{i-k+1} in $A_2, \ldots$ und schließlich m_{i-1} in A_k enthalten ist, wobei $(A_1, \ldots, A_k)$ ein beliebiges aber fest gegebenes k-Tupel von Elementen aus $\mathfrak{M}$ bedeutet.

Es sei $\mathfrak{K}$ ein abzählbarer Mengenkörper, dessen Elemente Teilmengen von M sind und in welchem M als Element enthalten ist. Ferner sei μ eine nichtnegative additive Mengenfunktion definiert für alle Elemente von $\mathfrak{K}$, wobei $\mu(M) = 1$ gilt. Ist μ konstruktiv definiert für die Elemente aus $\mathfrak{K}$, so kann man leicht (analog zum Falle endlicher Merkmalräume) zeigen, daß die in bezug auf den Körper $\mathfrak{K}$ nachwirkungsfreien Folgen mit der Verteilungsfunktion μ Kollektivs sind, die den Bedingungen des Theorems VI genügen. Hat die in bezug auf den Körper $\mathfrak{K}$ nachwirkungsfreie Folge $\{m_n\}$ die Eigenschaft, daß für jedes Element A des Körpers $\mathfrak{K}$, für welches der Grenzwert der relativen Häufigkeit 0 ist, auch die relative Häufigkeit in jedem Anfangsabschnitt von $\{m_n\}$ gleich 0 ist, so ist die Folge $\{m_n\}$ — wie man leicht zeigen kann — auch in bezug auf das System $\mathfrak{M}$ aller (in bezug auf $\mathfrak{K}$) μ-meßbaren Teilmengen von M nachwirkungsfrei. Da unser Konstruktionsverfahren für Kollektivs, die den Bedingungen des Theorems III bzw. VI genügen, nur Merkmalfolgen liefert, welche die genannte Eigenschaft haben, so gilt: *Ist M ein unendlicher Merkmalraum, μ*

4*

eine nichtnegative additive Mengenfunktion, die konstruktiv definiert ist für die Elemente eines abzählbaren Körpers $\mathfrak{K}$ von Teilmengen von M, wobei M ein Element von $\mathfrak{K}$ ist und $\mu(M)=1$ gilt, dann kann man Merkmalfolgen konstruktiv definieren, die in bezug auf das System $\mathfrak{M}$ aller (in bezug auf $\mathfrak{K}$) μ-meßbaren Teilmengen von M nachwirkungsfrei sind und die vorgegebene Verteilungsfunktion μ haben. Dieses Resultat geht weit über das hinaus, was bis jetzt erzielt worden war, da bisher ein Konstruktionsverfahren für nachwirkungsfreie Folgen nur für den Fall gegeben wurde, daß der Merkmalraum bloß aus zwei Elementen besteht.

Es sei noch bemerkt, daß die erwähnte Arbeit von A. H. Copeland gewisse Berührungspunkte mit unserem Theorem I aufweist, in welchem insbesondere Copelands Theorem I (l. c. S. 185) als Spezialfall enthalten ist. Copeland betrachtet speziell einen Merkmalraum, der bloß aus zwei Merkmalen besteht, und Auswahlvorschriften von folgender speziellen Art: Es wird das i_k-te Element ($k=1, 2, \ldots$ ad inf.) ausgewählt, wobei $\{i_k\}$ eine monoton wachsende Folge von natürlichen Zahlen ist. (Insbesondere hängt also in diesem Falle die Auswahl des Merkmales m_i aus einer Merkmalfolge bloß von i ab, nicht von den m_i vorangehenden Merkmalwerten.) Copelands Theorem I besagt dann, ohne übrigens Konstruktionsvorschriften zu enthalten: Ist D ein abzählbares System von Auswahlvorschriften obiger Art, so existieren kontinuierlich viele „admissible numbers", von denen jede die Eigenschaft hat, daß jede aus ihr nach einer Vorschrift des Systems D gebildete Auswahlfolge ebenfalls ein „admissible number" ist. Nun ist ein „admissible number" ein Kollektiv bezüglich eines abzählbaren (und sogar konstruktiv definierten) Systems $\mathfrak{S}_0$ von Auswahlvorschriften. Um zu zeigen, daß Copelands Theorem I in unserem Theorem I enthalten ist, definieren wir zunächst als *Produkt* von endlich vielen Auswahlvorschriften $V_1, \ldots, V_k$ folgende Vorschrift V: Es wird aus der Merkmalfolge $\{m_i\}$ ($i=1, 2, \ldots$ ad inf.) zuerst eine Auswahlfolge $\{m_i^{(1)}\}$ nach Vorschrift V_k gebildet, dann aus der Folge $\{m_i^{(1)}\}$ nach Vorschrift V_2 eine Auswahlfolge $\{m_i^{(2)}\}$ gebildet, ... und schließlich wird aus $\{m_i^{(k-1)}\}$ ($i=1, 2, \ldots$ ad inf.) eine Auswahlfolge $\{m_i^{(k)}\}$ ($i=1, 2, \ldots$ ad inf.) nach Vorschrift V_k gebildet. Die Vorschrift V ordnet also einer Merkmalfolge $\{m_i\}$ ($i=1, 2, \ldots$ ad inf.) die Auswahlfolge $\{m_i^{(k)}\}$ ($i=1, 2, \ldots$ ad inf.) zu [9]. Wir nennen ein

[9]) Sind $V=\{f_n\}$ und $V^*=\{f_n^*\}$, ($n=0, 1, \ldots$ ad inf.) zwei Auswahlvorschriften, wo f_n und f_n^* zwei n-stellige Auswahlfunktionen sind, so ist das Produkt VV^* selbst eine Auswahlvorschrift, da es offenbar mit der Auswahlvorschrift $\{g_n\}$ identisch ist, wo g_n für jedes n folgende n-stellige Auswahlfunktion bezeichnet: Ist $m_1, \ldots, m_n$ ein geordnetes n-Tupel von Merkmalen, so sei $m_{i_1}, \ldots, m_{i_j}$ das geordnete j-Tupel von Merfmalen, das aus $(m_1, \ldots m_n)$ durch die Vorschrift V ausgewählt wird ($j \geqq 0$); es sei nun $g_n(m_1, \ldots, m_n)=1$ bzw. $=0$, je nachdem ob $f_n(m_1, \ldots, m_n)=f_j^*(m_{i_1}, \ldots, m_{i_j})=1$ gilt oder nicht.

System von Auswahlvorschriften, das zu je zwei Auswahlvorschriften auch ihr Produkt enthält, eine *Gruppe*. Es sei nun $\mathfrak{S}_1$ die kleinste Gruppe, die die Auswahlvorschriften von $\mathfrak{S}_0$ und D enthält. Offenbar ist $\mathfrak{S}_1$ abzählbar und jedes Kollektiv bezüglich $\mathfrak{S}_1$ genügt der Behauptung des Copelandschen Theorem I, das also in unserem Theorem I als Spezialfall enthalten ist.

Unsere bisherigen Ausführungen bezogen sich auf ein einziges Kollektiv. In den Aufgaben der Wahrscheinlichkeitsrechnung geht man öfters von mehreren Grundkollektivs K_1, K_2, ... aus. Aus diesen Grundkollektivs werden Auswahlfolgen nicht nur durch Auswahlvorschriften, sondern auch durch Auswürfelung mittels eines anderen Kollektivs gebildet; dabei sagen wir, daß aus einem Kollektiv $K=\{m_i\}$ die Auswahlfolge durch Auswürfelung mittels des Kollektivs $K'=\{m_i'\}$ gebildet ist, falls ein Element m_i aus $\{m_i\}$ dann und nur dann ausgewählt wird, wenn m_i' in einer vorgegebenen Teilmenge N' des Merkmalraumes M' von K' enthalten ist. Aus den so gewonnenen Auswahlfolgen werden wiederum Auswahlfolgen entweder durch Auswahlvorschriften oder durch Auswürfelung mittels eines Grundkollektivs bzw. einer bereits gebildeten Auswahlfolge gebildet, usw. Die Forderungen, denen die so gewonnenen Auswahlfolgen genügen müssen, hat K. Dörge formuliert[10]). Er geht von einem System von Grundkollektivs K_1, K_2, ... und von einem Anfangssystem $\mathfrak{S}$ von Auswahlvorschriften aus. Dabei nimmt Dörge an, daß $\mathfrak{S}$ nur Auswahlvorschriften enthält, bei denen die Auswahl eines Elementes m_i aus der Merkmalfolge $\{m_n\}$ ($n=1, 2, \ldots$ ad inf.) nur von dem Index i, aber nicht von den dem Element m_i vorangehenden Merkmalwerten abhängt. Er betrachtet das System $\mathfrak{A}$ aller Auswahlfolgen, die aus den Grungkollektivs durch endlich oftmalige Anwendung von Auswahloperationen gebildet werden können, wobei jede Auswahloperation entweder eine Auswahl durch Auswürfelung, oder eine Auswahl nach einer Vorschrift des Ausgangssystems $\mathfrak{S}$ bedeutet. Das System $\mathfrak{A}$ ist in bezug auf die genannten Auswahloperationen abgeschlossen, d. h. wird aus einem Element des System $\mathfrak{A}$ eine Teilfolge nach einer Auswahlvorschrift des Systems $\mathfrak{S}$ oder durch Auswürfelung mittels eines Elements von $\mathfrak{A}$ gebildet, so ist diese Teilfolge ebenfalls ein Element von $\mathfrak{A}$. Dörge unterscheidet l. c. reguläre, singuläre und irreguläre Auswahlfolgen und formuliert drei Axiome: 1. daß jedes Grundkollektiv K_ν ($\nu=1, 2, \ldots$) eine Verteilung hat, d. h. daß für jede Teilmenge N_ν des Merkmalraumes M_ν von K_ν der Grenzwert $W(N_\nu)$ der relativen Häufigkeit in K_ν existiert. 2. daß jede reguläre Auswahlfolge, gebildet aus einem Grundkollektiv K_ν, dieselbe Verteilung hat, wie K_ν. 3. ein Axiom über singuläre Auswahlfolgen. Über die irregulären Auswahlfolgen wird nichts postuliert. Dörge zeigt l. c., daß im Falle, daß das Anfangssystem $\mathfrak{S}$ bloß die Identität enthält, seine Axiome trivialerweise erfüllt sind. Für den Fall, daß $\mathfrak{S}$ außer der Identität

[10]) Mathem. Zeitschrift *32* (1930).

noch weitere Auswahlvorschriften enthält, gibt Dörge keinen Widerspruchsfreiheitsbeweis an. Auf Grund der Theoreme I—VI kann auch dieses Problem gelöst werden. Man kann nämlich unter der Voraussetzung, daß aus einer Teilfolge eines Grundkollektivs K_ν keine Auswahlfolge durch Auswürfelung mittels einer Teilfolge desselben Kollektivs K_ν gebildet wir, allgemein beweisen:

VII. *Ist* M_ν ($\nu = 1, 2, \ldots$, *ad inf.*) *ein Merkmalraum,* $\mathfrak{K}_\nu$ ($\nu = 1, 2, \ldots$, *ad inf.*) *ein abzählbarer Mengenkörper von Teilmengen von* M_ν[11]*, wobei* M_ν *ein Element von* $\mathfrak{K}_\nu$ *ist, ist ferner* μ_ν ($\nu = 1, 2, \ldots$, *ad inf.*) *eine nichtnegative additive Mengenfunktion definiert für die Elemente von* $\mathfrak{K}_\nu$*, wobei* $\mu_\nu(M_\nu) = 1$ *gilt, und ist schließlich* $\mathfrak{S}$ *ein System von abzählbarvielen Auswahlvorschriften, dann gibt es für jede natürliche Zahl* ν *ein Kollektiv* K_ν *bezüglich* $\mathfrak{S}$ *und des Systems* $\mathfrak{M}_\nu$ *aller* μ_ν*-meßbaren Teilmengen von* M_ν *mit der vorgeschriebenen Verteilungsfunktion* μ_ν*, so daß in bezug auf die Grundkollektivs* $\{K_\nu\}$ ($\nu = 1, 2, \ldots$, *ad inf.*) *und das Anfangssystem* $\mathfrak{S}$ *von Auswahlvorschriften die Dörgeschen Axiome erfüllt sind*[12]. *Sind* $\{\mu_\nu\}$ ($\nu = 1, 2, \ldots$, *ad inf.*) *und* $\mathfrak{S}$ *konstruktiv definiert, so kann man auch die Folge* $\{K_\nu\}$ ($\nu = 1, 2, \ldots$, *ad inf.*) *von Grundkollektivs konstruktiv definieren.*

Damit ist auch die Frage der Widerspruchsfreiheit der auf den Kollektivbegriff aufgebauten Wahrscheinlichkeitsrechnung in befriedigender Weise erledigt. Von den Intuitionisten dürften kaum Einwände dagegen erhoben werden, da in jeder von ihnen zugelassenen konkreten Aufgabe der Wahrscheinlichkeitsrechnung nur abzählbar viele Grundkollektivs betrachtet werden, deren Verteilungsfunktionen konstruktiv definiert sind, und die in der Aufgabe vorkommenden Auswahlvorschriften ebenfalls ein abzählbares konstruktiv definiertes System bilden. Man kann daher nach Satz VII die Grundkollektivs mit den vorgeschriebenen Verteilungen konstruktiv definieren, so daß die Dörgeschen Axiome erfüllt sind, falls als Ausgangssystem $\mathfrak{S}$ das System der in der Aufgabe tatsächlich vorkommenden Auswahlvorschriften gewählt wird.

Eine eingehende Untersuchung der Dörgeschen Axiomatik und den Beweis des obigen Satzes VII werden wir wegen Raummangel in einer folgenden Mitteilung durchführen.

Zweiter Teil. Beweis der Theoreme I—VI.

§ 5. Zwei Definitionen.

Definition 1. Ist M ein Merkmalraum, $\{m_i\}$ ($i = 1, 2, \ldots$, ad inf.) eine Folge von Merkmalen, $\{V_i\}$ ($i = 1, 2, \ldots$, ad inf.) eine Folge

[11]) Ist M_ν endlich, so sei $\mathfrak{K}_\nu$ das System aller Teilmengen von M_ν.

[12]) Dörge betrachtet nur den Fall, daß die Merkmalräume M_1, $M_2, \ldots$ endlich sind, und fordert die Gültigkeit seiner Axiome für beliebige Teilmengen der Merkmalräume. Im Falle unendlicher Merkmalräume, muß man sich auf die μ-meßbaren (bzw. bei gewissen Operationen auf abzählbare Systeme von) Teilmengen derselben beschränken.

von Auswahlvorschriften und $(r_1, r_2, \ldots, r_n)$ ein geordnetes n-Tupel von Zeichen 0 und 1 (r_i ist $=0$ oder $=1$ für $i=1, \ldots, n$), so bezeichne $A_{r_1 r_2 \ldots r_n}$ die Teilfolge von $\{m_i\}$ bestehend aus den und nur den Elementen m_j, welche für $k=1, 2, \ldots, n$ nach der Auswahlvorschrift V_k ausgewählt werden oder nicht, je nachdem $r_k=1$ oder $=0$. (Eine solche Teilfolge $A_{r_1 r_2 \ldots r_n}$ kann endlich oder sogar leer sein.) Die 2^n Folgen $A_{r_1 r_2 \ldots r_n}$, die wir erhalten, wenn $(r_1, r_2, \ldots, r_n)$ sämtliche n-Tupel von Zeichen 0 und 1 durchläuft, sind paarweise fremd und haben zur Summe die ganze Folge $\{m_i\}$. Jedes Element m_r der Folge $\{m_i\}$ bestimmt daher eindeutig eine Folge von Teilfolgen

(*) $$A_{r_1}, A_{r_1 r_2}, \ldots, A_{r_1 r_2 \ldots r_n}, \ldots$$

wo $A_{r_1 r_2 \ldots r_n}$ für jedes n diejenige der 2^n Folgen $A_{x_1 x_2 \ldots x_n}$ bezeichnet, die m_r enthält. Ist $n > m$ und $r_1, r_2, \ldots, r_m, r_{m+1}, \ldots, r_n$ ein n-Tupel von Zeichen 0 und 1, so ist $A_{r_1 r_2 \ldots r_n}$ eine Teilfolge der Folge $A_{r_1 r_2 \ldots r_m}$.

Definition 2. Wir zerlegen nun die Folge $\{m_i\}$ in paarweise fremde endliche Mengen $B_{r_1 \ldots r_n}$, die für jedes natürliche n und jedes geordnete n-Tupel $(r_1, \ldots, r_n)$ von Zeichen 0 und 1 definiert werden. Wir wählen eine im schärferen Sinn monoton wachsende Folge $\{n_i\}$ von natürlichen Zahlen. m_1 gehöre zu B_1 oder B_0, je nachdem ob m_1 zu A_1 oder zu A_0 gehört. Ist bereits für jedes der Elemente $m_1, \ldots, m_{r-1}$ bestimmt, zu welcher B-Menge es gehört, so bilden wir die m_r zugeordnete Folge.

(*) $$A_{r_1}, A_{r_1 r_2}, \ldots, A_{r_1 r_2 \ldots r_n}, \ldots$$

und die entsprechende Folge der Zeichenkombinationen von 0 und 1

(**) $$r_1, r_1 r_2, r_1 r_2 r_3, \ldots, r_1 r_2 \ldots r_n, \ldots$$

Ist k die kleinste Zahl, für welche der Menge $B_{r_1 r_2 \ldots r_k}$ weniger als n_k von den Elementen $m_1, m_2, \ldots, m_{r-1}$ zugeteilt worden sind, dann ordnen wir m_r der Menge $B_{r_1 r_2 \ldots r_k}$ zu.

Offenbar enthält eine B-Menge mit k Indizes höchstens n_k Elemente. Ist $B_{r_1 r_2 \ldots r_k}$ nicht leer, so enthält $B_{r_1 r_2 \ldots r_{k-1}}$ genau n_{k-1} Elemente.

Bemerkung. *Für jedes natürliche r sind die Indizes der das Element m_r enthaltenden B-Menge und die vor m_r stehenden Elemente dieser B-Menge schon durch $m_1, m_2, \ldots, m_{r-1}$ eindeutig bestimmt.* In der Tat genügt zur Feststellung der fraglichen Indizes und Elemente die Kenntnis der Folge (**) und der Indizes der B-Mengen, in denen m_j für $j < r$ enthalten ist. Für $r=1$ ist die Bemerkung evident. Ist sie für jede natürliche Zahl $\leq r-1$ richtig, so folgt ihre Gültigkeit für r daraus, daß (**) durch die Merkmalwerte $m_1, m_2, \ldots, m_{r-1}$ bestimmt ist.

§ 6. Ein Hilfssatz.

Voraussetzungen: $M=\{p_j\}$ ($j=1, 2, \ldots$, evt. ad inf.) *sei ein mehr als ein Element enthaltender abzählbarer (evt. endlicher) Merkmalraum. Jedem* p_j *sei eine Zahl* $\alpha_j \geqq 0$ *zugeordnet, so daß* $\Sigma \alpha_j = 1$ *und* $\alpha_{j+1} \leqq \alpha_j$ *gilt.* $\{m_i\}$ *sei eine folgendermaßen definierte Merkmalfolge:* m_1 *sei* p_1 *oder* p_2*; sind* $m_1, \ldots, m_{k-1}$ *bereits definiert, so sei* $m_k = p_j$*, wo* j *die kleinste natürliche Zahl ist, für die* p_j *in* $m_1, \ldots, m_{k-1}$ *eine relative Häufigkeit* $H_{k-1}(p_j) \leqq \alpha_j$ *besitzt.* Ein solches j existiert, denn wäre $H_{k-1}(p_j) > \alpha_j$ für jedes j, so wäre die über j erstreckte Summe $\Sigma H_{k-1}(p_j) > \Sigma \alpha_j = 1$, was unmöglich ist.

Behauptung: *Für jedes* n *und* j *gilt:*

$$|H_n(p_j) - \alpha_j| \leqq 1 / \sqrt{n}.$$

Wir beweisen zunächst

$$(*) \qquad H_n(p_j) - \alpha_j \leqq 1 / n,$$

indem wir annehmen, daß diese für $n=1$ triviale Ungleichung für $n-1$ gelte. Sei erstens $m_n \neq p_j$ und s die Anzahl derjenigen von den Elementen $m_1, \ldots, m_{n-1}$, die $= p_j$ sind. Dann ist

$$H_n(p_j) - H_{n-1}(p_j) = s / n - s / (n-1) = -s / n(n-1).$$

Da nach Induktionsvoraussetzung $H_{n-1}(p_j) - \alpha_j \leqq 1 / (n-1)$, so gilt $H_n(p_j) - \alpha_j \leqq (n-s) / n(n-1)$, also $\leqq 1/n$ für $s \geqq 1$, während für $s=0$ gilt: $H_n(p_j) - \alpha_j = -\alpha_j < 1/n$. Sei zweitens $m_n = p_j$. Zufolge der Konstruktionsvorschrift der Folge $\{m_i\}$ kann dies nur eintreten, wenn $H_{n-1}(p_j) \leqq \alpha_j$. Nun ist $H_n(p_j) = (s+1)/n$, also $H_n(p_j) - H_{n-1}(p_j) = (n-s-1) / n(n-1)$, also gilt (*) auch in diesem Fall.

Sei nun A die Teilmenge von M, welche alle Punkte von M außer $p_1, p_2, \ldots, p_j$ enthält. Wir behaupten

$$(**) \qquad H_n(A) = n_A / n \leqq \Sigma \alpha_\nu + 1 / n,$$

wobei diese und die folgenden Summationen über alle Indizes $\nu > j$ zu erstrecken ist und n_A die Anzahl der in A enthaltenen Elemente der Reihe $m_1, m_2, \ldots, m_n$ bezeichnet. Wir nehmen an, daß die für $n=1$ triviale Behauptung (**) für $n-1$ gelte. Ist erstens m_n nicht in A enthalten, so gilt $(n-1)_A = n_A$, also $H_n(A) - H_{n-1}(A) = -n_A / n(n-1)$; folglich $H_n(A) \leqq \Sigma \alpha_\nu + 1 / (n-1) - n_A / n(n-1) = \Sigma \alpha_\nu + (n - n_A) / n(n-1)$. Ist $n_A \geqq 1$, so folgt also (**). Ist $n_A = 0$, so gilt $H_n(A) = 0$ und ebenfalls (**). Ist zweitens m_n in A enthalten, so muß wegen der Konstruktionsvorschrift der Folge $\{m_i\}$ für jedes $s \leqq j$ die Ungleichung $H_{n-1}(p_s) > \alpha_s$ gelten, also wegen

$$H_{n-1}(p_1) + \ldots + H_{n-1}(p_j) + H_{n-1}(A) = 1$$

die Ungleichung $H_{n-1}(A) < \Sigma \alpha_\nu$. Wegen $(n-1)_A = n_A - 1$ und

$$H_n(A)-H_{n-1}(A)=n_A/n-(n_A-1)/(n-1)\leqq 1/n$$

folgt hieraus (**). Aus (*) und (**) folgt mühelos

$$H_n(p_j)=1-H_n(p_1)-\ldots-H_n(p_{j-1})-H_n(A)\geqq 1-\alpha_1-\alpha_2-\ldots$$
$$-\alpha_{j-1}-\Sigma\alpha_\nu-j/n,$$

also $H_n(p_j)\geqq\alpha_j-j/n$, also wegen (*) $\alpha_j-j/n\leqq H_n(p_j)\leqq\alpha_j+j/n$. Für $j\leqq\sqrt{n}$ folgt daraus die Behauptung des Hilfssatzes. Sei $j>\sqrt{n}$. Aus $\alpha_{\nu+1}\leqq\alpha_\nu$ und $\Sigma\alpha_\nu=1$ folgt $\alpha_j\leqq 1/\sqrt{n}$. Ist $H\ (p_j)\geqq\alpha_j$, so folgt aus (*), daß $|H_n(p_j)-\alpha_j|\leqq 1/n$, also $\leqq 1/\sqrt{n}$. Ist $H_n(p_j)<\alpha_j$, so ist $|H_n(p_j)-\alpha_j|\leqq\alpha_j\leqq 1/\sqrt{n}$. Damit ist der Hilfssatz in allen Stücken bewiesen.

§ 7. Beweis der Theoreme I, II, V.

Satz 1. *Ist M ein abzählbarer (endlicher oder unendlicher) Merkmalraum einer Mächtigkeit $\geqq 2$, $\mathfrak{M}$ die Menge aller Teilmengen von M, $\mathfrak{S}$ ein die Identität enthaltendes System von abzählbar vielen Auswahlvorschriften, μ eine σ-additive nicht negative Mengenfunktion erklärt für alle Teilmengen von M, wobei $\mu(M)=1$ ist, so gibt es kontinuierlich viele Kollektivs bezüglich $\mathfrak{S}$ und $\mathfrak{M}$, deren Verteilungsfunktion mit μ identisch ist.*

Da M nach Voraussetzung abzählbar ist, so kann man die Elemente von M in eine Folge anordnen. Es sei $M=\{p_i\}$ und $\mu(p_i)=\alpha_i\geqq 0\ (i=1, 2, \ldots$ evt. ad inf.). Wegen der σ-Additivität von μ und $\mu(M)=1$ gilt offenbar $\sum\limits_i\alpha_i=1$, also, wenn $M=\{p_i\}$ eine unendliche Folge ist, $\lim\alpha_i=0$. Man kann daher die Zahlen $\{\alpha_i\}\ (i=1, 2, \ldots)$ monoton abnehmend anordnen. Wir nehmen an, daß bereits in der gegebenen Anordnung $\alpha_{i+1}\leqq\alpha_i$ gilt. Wir ordnen die abzählbar vielen Auswahlvorschriften des Systems $\mathfrak{S}$ in eine Folge $\{V_i\}\ (i=1, 2, \ldots$, ad inf.) an, und setzen die monoton wachsende Zahlenfolge $\{n_k\}\ (k=1, 2, \ldots$, ad inf.), die in § 5 bei der Definition der B-Mengen verwendet wird, gleich $\{2^{4k}\}\ (k=1, 2, \ldots$, ad inf.), also $n_k=2^{4k}$.

Wir ordnen nun jeder dyadischen Zahl $\lambda\ (0\leqq\lambda\leqq 1)$ eine Folge $K=\{m_i\}$ von Merkmalen nach folgender rekursiven Vorschrift zu:

a_λ. *Es sei $m_1=p_1$ oder $=p_2$ je nachdem λ als erste Ziffer* 0 *oder* 1 *besitzt. Sind $m_1, \ldots, m_{n-1}$ bereits bestimmt, so sind nach der Schlußbemerkung von § 5 die Indizes der das Element m_n enthaltenden B-Menge und die vor m_n stehenden Elemente dieser B-Menge eindeutig bestimmt. Ist m_n nicht das erste Element der betreffenden B-Menge, so sei $m_n=p_j$, wo j die kleinste natürliche Zahl ist mit der Eigenschaft, daß die relative Häufigkeit von p_j in dem vor m_n stehenden Abschnitt der betreffenden B-Menge $\leqq\alpha_j$ ist.* (Ein solches j existiert offenbar, sonst wäre $\Sigma H(p_i)>\Sigma\alpha_i=1$, was unmöglich ist.) *Ist m_n das erste Element der betreffenden B-Menge, so sei $m_n=p_1$ oder p_2 je nachdem, ob λ als k-te Ziffer* 0 *oder* 1 *besitzt, wo k die Anzahl der B-Mengen ist, die mit $m_1, \ldots, m_n$ einen nichtleeren Durchschnitt haben.* Aus der Schlußbemerkung von § 5 folgt, daß auch k durch $m_1, \ldots, m_{n-1}$ eindeutig bestimmt ist.

Wir zeigen nun: *Die durch Vorschrift a_λ definierte Folge $K=\{m_i\}$ ist ein Kollektiv $K(\mathfrak{S}, \mathfrak{M})$ mit der Verteilungsfunktion μ.* Zu diesem Zweck genügt es zu zeigen, daß für jede Teilmenge M' von M der Grenzwert $H(M')$ der relativen Häufigkeit von M' in jeder unendlichen A-Folge gleich $\mu(M')$ ist. Man kann nämlich die ursprüngliche Folge $K=\{M_i\}$ $(i=1, 2, \ldots, \text{ad inf.})$ und jede Auswahlfolge, die nach einer Vorschrift des Systems $\mathfrak{S}$ gebildet wird, als Summe von endlich vielen paarweise fremden A-Folgen darstellen. Die Folge $\{m_i\}$ ist z. B. gleich $\Sigma A_{x_1 \ldots x_n}$, wobei $x_1 x_2 \ldots x_n$ die 2^n verschiedenen n-Tupel von 0 und 1 durchläuft, und ist $\{m_i'\}$ $(i=1, 2, \ldots)$ eine Teilfolge von $\{m_i\}$ gebildet nach der Auswahlvorschrift V_i, so ist $\{m_i'\}=\Sigma A_{x_1 x_2 \ldots x_{i-1} 1}$, wobei $x_1 x_2 \ldots x_{i-1}$ alle $i-1$-Tupel der Zeichen 0 und 1 durchläuft. Es genügt sogar zu zeigen, daß für jedes i der Grenzwert $H(p_i)$ von p_i in jeder unendlichen A-Folge gleich $\mu(p_i)=\alpha_i$ ist. Ist M' eine endliche Teilmenge von M, so folgt aus $H(p_i)=\mu(p_i)$ unmittelbar, daß $H(M')= =\mu(M')$ ist. Ist $M'=\{p_{i_j}\}$ $(j=1, 2, \ldots, \text{ad. inf.})$ eine unendliche Teilmenge von M, so ist zunächst klar, daß die Folge der relativen Häufigkeiten von M' nur Häufungswerte $\geqq \alpha_{i_1}+\alpha_{i_2}+\alpha_{i_3}+\ldots=\mu(M')$ haben kann. Nun kann es aber keinen Häufungswert $>\mu(M')$ geben, denn wäre ν ein solcher, so könnte man wegen $\Sigma\alpha_i=1$ in $M-M'$ endlich viele Punkte $p_{k_1}, \ldots, p_{k_r}$ derart angeben, daß

$$\alpha_{k_1}+\alpha_{k_2}+\ldots+\alpha_{k_r}+\nu>1$$

ist. Daraus würde aber wegen $H(p_i)=\alpha_i$ folgen, daß in einem entsprechend großen Anfangsabschnitt die relative Häufigkeit von $M''= =M'+\{p_{k_1}, \ldots, p_{k_r}\}$ größer als 1 wäre, was unmöglich ist.

Wir haben also nur zu zeigen, daß in jeder unendlichen A-Folge $H(p_i)=\alpha_i$ ist. Es sei $A_{r_1 r_2 \ldots r_l}$ eine unendliche A-Folge, ferner seien k und s zwei natürliche Zahlen, so daß $n_{l+k} \leqq s \leqq \leqq n_{l+k+1}$ gilt. Wir betrachten die Folge $A^*_{r_1 \ldots r_l}$, die aus $A_{r_1 \ldots r_l}$ entsteht, wenn man jene Elemente, die in $(B_{r_1}+B_{r_1 r_2}+\ldots+B_{r_1 r_2 \ldots r_l})$ enthalten sind, streicht. Es genügt offenbar zu zeigen, daß $H(p_i)= =\alpha_i$ in $A^*_{r_1 \ldots r_l}$ gilt. Wir betrachten den aus den ersten s Elementen bestehenden Anfangsabschnitt T der Folge $A^*_{r_1 \ldots r_l}$ und behaupten zunächst: Es gibt höchstens 2^{k+2} verschiedene B-Mengen, die mit T einen nicht leeren Durchschnitt haben. Jede B-Menge, die mit $A^*_{r_1 \ldots r_l}$ einen nicht leeren Durchschnitt hat, muß offenbar mehr als l Indizes haben und die ersten l Indizes müssen der Reihe nach mit $r_1, \ldots, r_l$ übereinstimmen. Daraus folgt, daß die Anzahl der B-Mengen mit höchstens $l+k+1$ Indizes, die mit $A^*_{r_1 \ldots r_l}$ einen nicht leeren Durchschnitt haben, nicht größer als $2+2^2+\ldots+2^{k+1}=2^{k+2}-2$ sein kann. Würden daher mehr als 2^{k+2} B-Mengen existieren, die mit T einen nicht leeren Durchschnitt haben, so müßte es unter diesen B-Mengen auch eine mit mindestens $l+k+2$ Indizes geben. Sei etwa $B_{r_1 \ldots r_l x_{l+1} \ldots x_{l+j}} (j \geqq k+2)$ eine solche. Da $B_{r_1 \ldots r_l x_{l+1} \ldots x_{l+j}}$ nicht leer ist, so muß $B_{r_1 \ldots r_l x_{l+1} \ldots x_{l+j-1}}$ genau n_{l+j-1} Elemente haben. Da $B_{r_1 \ldots r_l x_{l+1} \ldots x_{l+j-1}}$ in $A^*_{r_1 \ldots r_l}$ enthalten ist, ferner sämtliche Elemente von $B_{r_1 \ldots r_l x_{l+1} \ldots x_{l+j-1}}$ einen kleineren Index haben, als ein beliebiges Element von $B_{r_1 \ldots r_l x_{l+1} \ldots x_{l+j}}$, so folgt daraus, daß $B_{r_1 \ldots r_l x_{l+1} \ldots x_{l+j-1}}$ schon in T enthalten ist und sogar eine echte Teilmenge von T sein muß, da auch $B_{r_1 \ldots r_l x_{l+1} \ldots x_{l+j}}$ mit T einen nicht leeren Durchschnitt hat. Also gilt $n_{l+j-1} < s$. Wegen $j \geqq k+2$ und der Monotonie der Zahlenfolge $\{n_i\}$ ist aber $n_{l+j-1} \geqq \geqq n_{l+k+1} \geqq s$, womit ein Widerspruch hergeleitet und unsere Behauptung bewiesen ist.

Da jede B-Menge, die nicht fremd zu $A^*_{r_1 \ldots r_l}$ ist, in $A^*_{r_1 \ldots r_l}$ enthalten ist, so ist der Durchschnitt einer B-Menge mit dem Anfangsabschnitt T ein Anfangsabschnitt der betreffenden B-Menge.

Es seien diese Durchschnitte der Reihe nach

$$D_1, D_2, \ldots, D_j,$$

wobei $j \leqq 2^{k+2}$ ist, da, wie bereits gezeigt wurde, höchstens 2^{k+2} B-Mengen existieren, die mit T einen nicht leeren Durchschnitt haben. Die Anzahl der in $D_\nu (\nu = 1, \ldots, j)$ enthaltenen Elemente sei d_ν. Es gilt offenbar $d_1 + d_2 + \ldots + d_j = s$. Da jede B-Menge offenbar den Voraussetzungen des Hilfssatzes von § 6 genügt, so ist die relative Häufigkeit von p_i in $D_\nu (\nu = 1, \ldots, j)$ gleich $\alpha_i + \varepsilon_\nu$, wobei $|\varepsilon_\nu| \leqq 1/\sqrt{d_\nu}$. Also kommt p_i in D_ν $\alpha_i d_\nu + \varepsilon_\nu d_\nu = \alpha_i d_\nu + \delta_\nu$ mal vor, wo $|\delta_\nu| \leqq \sqrt{d_\nu}$. Ist also $H_s(p_i)$ die relative Häufigkeit von p_i im betrachteten Anfangsabschnitt T von $A^*_{r_1 \ldots r_l}$, so ist

$$H_s(p_i) = (\alpha_i d_1 + \alpha_i d_2 + \ldots + \alpha_i d_j + \delta_1 + \delta_2 + \ldots + \delta_j) / s \, (j \leqq 2^{k+2}),$$

also

$$(\dagger) \qquad H_s(p_i) = \alpha_i + (\delta_1 + \delta_2 + \ldots + \delta_j) / s.$$

Es ist, wobei alle Summen von 1 bis j zu erstrecken sind,

$$|\Sigma \delta_\nu / s| \leqq \Sigma |\delta_\nu| / s \leqq \Sigma \sqrt{d_\nu} / s \leqq 2^{k+2} \sqrt{s} / s = 2^{k+2} / \sqrt{s} \leqq 2^{k+2} / \sqrt{n_{l+k}}.$$

Wegen $n_{l+k} = 2^{4(l+k)}$ geht der letzte Ausdruck mit wachsendem k gegen 0. Aus (†) folgt also $H(p_i) = \alpha_i$. Es ist also $\{m_i\}$ ein $K(\mathfrak{S}, \mathfrak{M})$ mit der Verteilungsfunktion μ.

Zum Beweis von Satz 1 haben wir nur noch zu zeigen: Sind λ und λ' zwei verschiedene dyadische Zahlen, so ist die nach Vorschrift a_λ definierte Merkmalfolge $\{m_i\}$ von der nach Vorschrift $a_{\lambda'}$

definierten Merkmalfolge $\{m'_i\}$ verschieden. Es sei n die kleinste natürliche Zahl, so daß die n-te Ziffer von λ mit der n-ten Ziffer von λ' nicht übereinstimmt. Es ist klar, daß $m_i = m'_i$ für jedes i gilt, für welches die Anzahl der B-Mengen, die mit $(m_1, m_2, \ldots, m_i)$ einen nicht leeren Durchschnitt haben, $\leqq n-1$ ist. Es sei s die kleinste natürliche Zahl derart, daß mit $(m_1, \ldots, m_s)$ genau n verschiedene B-Mengen einen nicht leeren Durchschnitt haben. Dann ist m_s erstes Element einer B-Menge. Da $m'_1 = m_1, \ldots, m'_{s-1} = m_{s-1}$ gilt, so gibt es ebenfalls genau n verschiedene B-Mengen, die mit $(m'_1, \ldots, m'_s)$ einen nicht leeren Durchschnitt haben, und m'_s ist erstes Element einer B-Menge. Da die n-te Ziffern von λ und λ' verschieden sind, so ist $m'_s \neq m_s$. Damit ist Satz 1 in allen Stücken bewiesen.

Theorem I von § 4 ist offenbar als Spezialfall in Satz 1 enthalten.

Zum Beweis von Theorem V von § 4 genügt es zu zeigen, daß es, falls der Merkmalraum endlich ist und das System $\mathfrak{S} = \{V_i\}$ $(i = 1, \ldots,$ ad inf.) von Auswahlvorschriften konstruktiv definiert ist, dyadische Zahlen λ gibt, für welche die rekursive Vorschrift a_λ das Kollektiv $K = \{m_i\}$ konstruktiv definiert, d. h. so daß für jede natürliche Zahl n der Merkmalwert m_n auf Grund der Vorschrift a_λ in endlich vielen Schritten berechnet werden kann. Ist λ eine konstruktiv definierte dyadische Zahl, so daß für jedes natürliche n die n-te Ziffer von λ in endlich vielen Schritten berechnet werden kann, so zeigen wir durch vollständige Induktion, daß für jede natürliche Zahl n der Merkmalwert m_n und die Indizes der B-Menge, in welcher m_n enthalten ist, in endlich vielen Schritten berechnet werden kann. Für $n = 1$ ist dies evident. Sind für $i = 1, \ldots, r-1$ der Merkmalwert m_i und die Indizes der B-Menge, in welcher m_i enthalten ist, bereits berechnet, so kann man offenbar m_r in endlichen vielen Schritten berechnen, falls die Indizes der B-Menge, welche m_r enthält, bekannt sind. Wir haben also nur zu zeigen, daß diese Indizes in endlich vielen Schritten berechnet werden können, falls für $i = 1, \ldots, r-1$ sowohl der Merkmalwert m_i, als auch die Indizes der B-Menge, in welcher m_i enthalten ist, bekannt sind. Da die Folge $\{V_i\}$ $(i = 1, 2, \ldots,$ ad inf.) konstruktiv definiert ist, so kann man offenbar für jedes natürliche i in endlich vielen Schritten berechnen, ob m_r nach V_i ausgewählt wird oder nicht. Wir können also einen beliebig langen Anfangsabschnitt der Folge $\{r_i\}$ $(i = 1, 2, \ldots,$ ad inf.) in endlich vielen Schritten berechnen, wobei $r_i = 1$ oder $= 0$ ist, je nachdem ob m_r nach V_i ausgewählt wird oder nicht. Da nach Voraussetzung für $j < r$ die Indizes der B-Menge, welche m_j enthält, bekannt sind, so kann man für jedes natürliche i die Anzahl der Elemente aus dem Anfangsabschnitt $(m_1, \ldots, m_{r-1})$, die in $B_{r_1 r_2 \ldots r_i}$ enthalten sind, in endlich vielen Schritten berechnen. Man kann daher die kleinste Zahl j mit der Eigenschaft, daß $B_{r_1 \ldots r_j}$ weniger als n_j Elemente aus $(m_1, \ldots, m_{r-1})$ enthält, ebenfalls in endlich vielen Schritten berechnen. m_r ist nach Definition in $B_{r_1 r_2 \ldots r_j}$ enthalten. Damit ist gezeigt, daß die Indizes der m_r enthaltenden B-Menge und mithin auch der Merkmalwert von m_r in endlich vielen Schritten berechnet werden können, wie Theorem V behauptet. *Die Vorschrift* a_λ *stellt ein allgemeines Konstruktionsverfahren für Kollektivs dar.*

Satz 2. *Ist* M *ein (abzählbar oder unabzählbar) unendlicher Merkmalraum und* $\mu(M)$ *eine für alle Teilmengen von* M *erklärte nichtnegative additive Mengenfunktion, wobei* $\mu(M)=1$ *ist und für keine Folge* $\{p_i\}$ $(i=1, 2, \ldots$, ad inf.) *von paarweise verschiedenen Merkmalpunkten* $\Sigma\,\mu(p_i)=1$ *gilt, dann gibt es keine Merkmalfolge* $\{m_i\}$ $(i=1, 2, \ldots$, ad inf.), *so daß für jede Teilmenge* M' *von* M *der Grenzwert der relativen Häufigkeit von* M' *gleich* $\mu(M')$ *ist.*

Wir machen die Annahme, es sei $\{m_i\}$ $(i=1, 2, \ldots$, ad inf.) eine Merkmalfolge mit der obigen Eigenschaft, und führen diese Annahme ad absurdum. Wir betrachten die abzählbare Teilmenge M^* von M, die alle in der Folge $\{m_i\}$ auftretenden Merkmalpunkte (und nur diese) enthält. Da nach Annahme für jede Teilmenge M' von M der Grenzwert der relativen Häufigkeit von M' gleich $\mu(M')$ ist, so muß $\mu(M-M^*)=0$ sein, da die relative Häufigkeit von $M-M^*$ in jedem Abschnitt der Folge $\{m_i\}$ gleich 0 ist. Wegen $\mu(M)=1$ und der Additivität der Mengenfunktion μ folgt dann, daß $\mu(M^*)=1$ ist. Wir behaupten zunächst, daß M^* unendlich ist. Denn wäre M^* etwa die endliche Menge $(p_1, \ldots, p_n)$ von paarweise verschiedenen Merkmalpunkten, so gäbe es — da M unendlich ist — in $M-M^*$ eine unendliche Folge $\{p_i\}$ $(i=n+1, n+2, \ldots$, ad inf.) von paarweise verschiedenen Merkmalpunkten und es würde offenbar $\Sigma\,\mu(p_i)=1$ gelten entgegen unserer Voraussetzung. M^* muß also abzählbar unendlich sein. Wir ordnen die Elemente von M^* in eine Folge an, also $M^*=\{p_i\}$ $(i=1, 2, \ldots$, ad inf.). Wir bezeichnen $\mu(p_i)$ mit α_i. Nach Voraussetzung gilt $\Sigma\,\alpha_i \neq 1$. Wegen $\mu(M^*)=1$ und der Additivität von μ gilt dann offenbar $\Sigma\,\alpha_i=\alpha<1$.

Wir betrachten eine Folge $\{A_i\}$ $(i=1, 2, \ldots$, ad inf.) von aneinander anschließenden Abschnitten von $\{m_i\}$ mit den Längen $\{n_i\}$ $(i=1, 2, \ldots$, ad inf.), d. h. es ist

$$A_1=(m_1, \ldots, m_{n_1}),$$

$$A_2=(m_{n_1+1}, m_{n_1+2}, \ldots, m_{n_1+n_2}),$$

usw. Wir bezeichnen mit M_i die Menge der Merkmalpunkte, die in A_i, aber nicht in A_j $(j=1, \ldots, i-1)$ auftreten. Die Mengen M_i $(i=1, 2, \ldots$, ad inf.) sind offenbar endlich, paarweise fremd und es gilt $\Sigma\,M_i=M^*$, und $\Sigma\,\mu(M_i)=\Sigma\,\alpha_i=\alpha<1$.

Es seien ε und δ zwei positive Zahlen, so daß

1) $$\alpha+\delta<1-\delta-\varepsilon.$$

Wir wählen n_1 so groß, daß $\mu(M_1) \geqq \alpha - \varepsilon$. Dann gilt

2) $$\mu(M_2) + \mu(M_3) + \ldots \leqq \varepsilon.$$

Wir wählen n_2 so groß, daß $|H_{n_1+n_2}(M_1) - \mu(M_1)| \leqq \delta$ sei. Es ist offenbar möglich, eine solche Zahl n_2 zu bestimmen, da nach Voraussetzung für jede Teilmenge von M, also auch für M_1, der Grenzwert der relativen Häufigkeit in $\{m_i\}$ existiert und gleich $\mu(M_1)$ ist. Wir wählen ferner n_3 so groß, daß $|H_{n_1+n_2+n_3}(M_2) - \mu(M_2)| \leqq \delta$ sei, und allgemein n_{2k} und n_{2k+1} so groß, daß

3) $$|H_{n_1+n_2}+\ldots+n_{2k}(M_1 + M_3 + \ldots + M_{2k-1}) - \\ - \mu(M_1 + M_3 + \ldots + M_{2k-1})| \leqq \delta,$$

4) $$|H_{n_1+n_2}+\ldots+n_{2k+1}(M_2 + M_4 + \ldots + M_{2k}) - \\ - \mu(M_2 + M_4 + \ldots + M_{2k})| \leqq \delta.$$

Es sei nun $M' = M_1 + M_3 + \ldots + M_{2k-1} + \ldots$ Dann gilt offenbar:

5) $$H_{n_1+n_2}+\ldots+n_{2k}(M') = H_{n_1+n_2}+\ldots+n_{2k}(M_1 + M_3 + \ldots + M_{2k-1})$$

und

6) $$H_{n_1+n_2}+\ldots+n_{2k+1}(M') = \\ = 1 - H_{n_1+n_2}+\ldots+n_{2k+1}(M_2 + M_4 \ldots + M_{2k}).$$

Wegen 2) gilt $\mu(M_2 + M_4 + \ldots + M_{2k}) \leqq \varepsilon$, also ist mit Rücksicht auf 4)

$$H_{n_1+n_2}+\ldots+n_{2k+1}(M_2 + M_4 + \ldots + M_{2k}) \leqq \delta + \varepsilon$$

und daher wegen 6)

7) $$H_{n_1+n_2}+\ldots+n_{2k+1}(M') \geqq 1 - \delta - \varepsilon.$$

Dagegen ist mit Rücksicht auf 5) und 3)

8) $$H_{n_1+n_2}+\ldots+n_{2k}(M') \leqq \alpha + \delta.$$

Aus 1), 7) und 8) ergibt sich, daß die relative Häufigkeit von M' in $\{m_i\}$ keinen Grenzwert hat. Wir sind also zu einem Widerspruch gelangt und Satz 2 ist bewiesen.

Aus den Sätzen 1 und 2 folgt nun mühelos das Theorem II. Es sei nämlich M ein Merkmalraum, μ eine Mengenfunktion, $\mathfrak{S}$ ein System von Auswahlvorschriften, die den Voraussetzungen des Theorem II genügen. Gibt es in M eine Folge $N = \{p_i\}$ $(i = 1, 2, \ldots,$ ad inf.) von paarweise verschiedenen Merkmalpunkten, so daß $\Sigma \mu(p_i) = 1$, so gilt offenbar $\mu(M - N) = 0$ und μ ist eine σ-additive Mengenfunktion. Betrachtet man statt M die Teilmenge N als Merkmalraum, so sind offenbar alle Voraussetzungen des Satzes 1 erfüllt. Es gibt daher Kollektivs bezüglich $\mathfrak{S}$ und alle Teilmengen von N, deren Verteilungsfunktion μ ist. Diese Kollektivs sind aber wegen $\mu(M - N) = 0$ gleichzeitig auch Kollektivs bezüglich alle Teilmengen von M. Damit ist die eine Hälfte des Theorems II bewiesen. Gibt es in M keine Folge $\{p_i\}$ $(i=1, 2, \ldots,$ ad inf.) von paarweise verschiedenen Merkmalpunkten, für welche $\Sigma \mu(p_i)=1$ gilt, so folgt aus Satz 2, daß kein Kollektiv bezüglich aller Teilmengen von M existiert. Damit ist Theorem II bewiesen.

§ 8. Beweis der Theoreme III, IV und VI.

Es sei M ein Merkmalraum, $\mathfrak{K}$ ein abzählbarer Mengenkörper von Teilmengen von M, μ eine Mengenfunktion, $\mathfrak{M}$ ein System von Teilmengen von M und $\mathfrak{S}$ ein System von Auswahlvorschriften, für welche die Bedingungen 1)—5) des Theorems III erfüllt sind.

Man sieht zunächst, daß ein Kollektiv $K(\mathfrak{S}, \mathfrak{K})$ mit der Verteilungsfunktion μ zugleich ein Kollektiv $K(\mathfrak{S}, \mathfrak{M})$ ist. Es sei nämlich L ein beliebiges Element von $\mathfrak{M}$, d. h. eine μ-meßbare Menge. Es ist klar, daß die Häufungswerte der relativen Häufigkeiten von L in der Merkmalfolge $K(\mathfrak{S}, \mathfrak{K})$ und in einer Auswahlfolge aus $K(\mathfrak{S}, \mathfrak{K})$, die nach einer Vorschrift des Systems $\mathfrak{S}$ ausgewählt wurde, nicht größer als das äußere μ-Maß und nicht kleiner als das innere μ-Maß von L sein können. Da L μ-meßbar ist, so muß die Folge der relativen Häufigkeiten von L gegen das μ-Maß von L konvergieren, d. h. das Kollektiv $K(\mathfrak{S}, \mathfrak{K})$ ist zugleich ein Kollektiv $K(\mathfrak{S}, \mathfrak{M})$. Um das Theorem III zu beweisen, genügt es also zu zeigen, daß kontinuierlich viele Kollektivs $K(\mathfrak{S}, \mathfrak{K})$ existieren, deren Verteilungsfunktion gleich μ ist. Dabei können wir wegen Theorem I annehmen, daß der Körper $\mathfrak{K}$ unendlich viele Elemente enthält.

Wir ordnen die Elemente von $\mathfrak{K}$ in eine Folge $\{A_i\}$ $(i=1, 2, \ldots,$ ad inf.). Wir bezeichnen das Element A_1 mit M_0 und $M-M_0$ mit M_1. Den Durchschnitt von A_2 und M_0 bezeichnen wir mit M_{00} und M_0-M_{00} mit M_{01}, den Durchschnitt von A_2 und M_1 mit M_{10} und M_1-M_{10} mit M_{11}. Im allgemeinen bezeichnen wir für jede natürliche Zahl n den Durchschnitt der Mengen A_{n+1} und $M_{x_1 x_2 \ldots x_n}$, wobei jeder der Indizes $x_1, x_2, \ldots, x_n$ gleich 0 oder 1 ist, mit $M_{x_1 x_2 \ldots x_n 0}$ und $M_{x_1 \ldots x_n} - M_{x_1 \ldots x_n 0}$ mit $M_{x_1 \ldots x_n 1}$. Es bestehen dann offenbar die folgenden vier Eigenschaften:

1) $M_0+M_1=M$.

2) $M_{x_1 \ldots x_n}$ ist zu $M_{y_1 \ldots y_n}$ fremd, falls es ein $i \leqq n$ gibt, für welches $y_i \neq x_i$ ist.

3) Jedes Element A_n von $\mathfrak{K}$ ist als Summe von endlich vielen M-Mengen mit n Indizes darstellbar.

4) $M_{x_1 \ldots x_n 0}+M_{x_1 \ldots x_n 1}=M_{x_1 \ldots x_n}$.

Manche Mengen $M_{x_1 \ldots x_n}$ können auch leer sein, doch existiert für jedes natürliche n unter den 2^n M-Mengen mit n Indizes mindestens eine, die nicht leer ist, da die Summe dieser 2^n Mengen offenbar gleich M ist.

Für die folgenden Untersuchungen erscheint es zweckmäßig, auch der leeren Menge ein μ-Maß zuordnen, und zwar das μ-Maß 0.

Hilfssatz: Voraussetzungen: Es sei $\{m_i\}(i=1, 2, \ldots)$ *eine Folge von Merkmalen, die folgendermaßen gegeben ist: das Element* m_1 *sei ein Punkt der Menge* $M_{x_1 \ldots x_n}$, *wobei* n *eine gegebene natürliche Zahl ist,* $x_1=1$ *oder* 0 *ist, je nachdem ob* $\mu(M_1) \neq 0$ *oder* $=0$ *ist, ferner* $x_i (i=2, \ldots, n)=1$ *oder* 0 *ist, je nachdem ob* $\mu(M_{x_1 \ldots x_{i-1} 1}) \neq 0$ *oder* $=0$ *ist*[13]. *Sind bereits die ersten* $j-1$ *Elemente* $m_1, \ldots, m_{j-1}$ *gegeben, so sei* m_j *ein Punkt der Menge* $M_{x_1 \ldots x_n}$, *wobei die Indizes* $x_1, \ldots, x_n$ *folgendermaßen bestimmt werden: der Index* x_1 *ist* $=1$ *falls* $\mu(M_1) \neq 0$ *und die relative Häufigkeit* $H_{j-1}(M_1)$ *von* M_1 *in dem Anfangsabschnitt* $m_1, \ldots, m_{j-1}$ *nicht größer als* $\mu(M_1)$ *ist, andernfalls sei* $x_1=0$. *Sind die ersten* $i (i<n)$ *Indizes bereits bestimmt, so sei* $x_{i+1}=1$ *falls* $\mu(M_{x_1 \ldots x_i 1}) \neq 0$ *und*

$$H_{j-1}(M_{x_1 \ldots x_i 1}) / H_{j-1}(M_{x_1 \ldots x_i}) \leqq \mu(M_{x_1 \ldots x_i 1}) / \mu(M_{x_1 \ldots x_i}),$$

andernfalls[14] $x_{i+1}=0$. *Ist* $H_{j-1}(M_{x_1 \ldots x_i})=0$ *und mithin auch* $H_{j-1}(M_{x_1 \ldots x_i 1})=0$, *so setzen wir* $H_{j-1}(M_{x_1 \ldots x_i 1}) / H_{j-1}(M_{x_1 \ldots x_i})=0$.

Behauptung: Für jedes natürliche $k \leqq n$, *jedes natürliche* s *und jedes* k*-Tupel* $(y_1, \ldots, y_k)$ *von Zeichen* 0 *und* 1 *gilt:*

$$1) \qquad |H_s(M_{y_1 \ldots y_k}) - \mu(M_{y_1 \ldots y_k})| \leqq k/s.$$

---

[13]) Die so bestimmte Menge $M_{x_1 \ldots x_n}$ ist sicherlich nicht leer, vielmehr gilt $\mu(M_{x_1 \ldots x_n})>0$. Es ist zunächst klar, daß $\mu(M_{x_1})>0$ ist, denn x_1 ist dann und nur dann gleich 1, wenn $\mu(M_1) \neq 0$ und mithin ist im Falle, daß $x_1=0$ ist, $\mu(M_1)=0$ und daher $\mu(M_0)=\mu(M)=1$. Wir machen nun die Voraussetzung, daß $\mu(M_{x_1 \ldots x_i})>0$ (für $i<n$) und zeigen, daß auch $\mu(M_{x_1 \ldots x_i x_{i+1}})>0$ ist. Es gilt nämlich $x_{i+1}=1$ dann und nur dann, wenn $\mu(M_{x_1 \ldots x_i 1}) \neq 0$ ist. Ist daher $x_{i+1}=0$, so ist $\mu(M_{x_1 \ldots x_i 1})=0$ und mithin $\mu(M_{x_1 \ldots x_i 0})=\mu(M_{x_1 \ldots x_i}) \neq 0$. Es gilt daher für jeden Fall $\mu(M_{x_1 \ldots x_i x_{i+1}}) \neq 0$ und mithin ist $M_{x_1 \ldots x_n}$ nicht leer.

[14]) Sicher ist $\mu(M_{x_1 \ldots x_i}) \neq 0$. Dies trifft für $i=1$ zu, denn im Falle $x_1=1$ ist $\mu(M_1) \neq 0$ nach Voraussetzung und im Falle $x_1=0$ ist $\mu(M_0) \neq 0$, da sonst $\mu(M_1)=1$, also $H_{j-1}(M_1) \leqq \mu(M_1)$ wäre, was nach unserer Vorschrift nicht möglich ist. Setzen wir $\mu(M_{x_1 \ldots x_{i-1}}) \neq 0$ voraus, so ist im Falle $x_i=1$ nach Voraussetzung $\mu(M_{x_1 \ldots x_{i-1} 1}) \neq 0$. Aber auch für $x_i=0$ ist $\mu M_{x_1 \ldots x_{i-1} 0}) \neq 0$, denn wäre $\mu(M_{x_1 \ldots x_{i-1} 0})=0$, so wäre $\mu(M_{x_1 \ldots x_{i-1} 1})=\mu(M_{x_1 \ldots x_{i-1}}) \neq 0$ und

$$H_{j-1}(M_{x_1 \ldots x_{i-1} 1}) / H_{j-1}(M_{x_1 \ldots x_{i-1}}) \leqq \mu(M_{x_1 \ldots x_{i-1} 1}) / \mu(M_{x_1 \ldots x_{i-1}})=1.$$

Dies ist aber ein Widerspruch, da in diesem Fall nach unserer Vorschrift $x_i=1$ gesetzt werden muß.

Für $k=1$ beweisen wir den für $s=1$ trivialen Hilfssatz für $s+1$ unter der Annahme seiner Gültigkeit für s. Sei *erstens* m_{s+1} nicht in M_{y_1} enthalten. Ist i die Anzahl der in M_{y_1} enthaltenen Elemente der Reihe $m_1, \ldots, m_s$, so ist $H_s(M_{y_1})=i/s$, $H_{s+1}(M_{y_1})= =i/(s+1)$, also

$$2) \qquad H_{s+1}(M_{y_1})-H_s(M_{y_1})=-i/s(s+1).$$

Auf Grund der Induktionsvoraussetzung folgt hieraus

$$H_{s+1}(M_{y_1})-\mu(M_{y_1})\leqq(s+1-i)/s(s+1).$$

Falls $i\geqq 1$, so ergibt sich hieraus

$$3) \qquad H_{s+1}(M_{y_1})-\mu(M_{y_1})\leqq 1/(s+1).$$

Wir beweisen weiter

$$4) \qquad H_s(M_{y_1})\geqq\mu(M_{y_1}).$$

Ist $y_1=1$, so wäre das Gegenteil von 4) nach der Bildungsvorschrift für die Folge $\{m_j\}$ nur damit verträglich, daß m_{s+1} in M_{y_1} enthalten ist, was nicht der Fall ist. Ist $y_1=0$, so ist m_{s+1} nicht in M_0, also in M_1 enthalten und nach der Bildungsvorschrift für $\{m_j\}$ muß $H_s(M_1)\leqq\mu(M_1)$ gelten, woraus 4) wegen

$$H_s(M_0)+H_s(M_1)=\mu(M_0)+\mu(M_1)=1$$

folgt. Aus 2), 3), 4) folgt 1) für $k=1$. Falls $i=0$, so behaupten wir: $\mu(M_{y_1})=0$. Ist $y_1=1$, so muß $\mu(M_1)=0$ gelten, sonst wäre m_{s+1} nach unserer Vorschrift in M_1 enthalten, da $H_s(M_1)=0$, also $<\mu(M_1)$. Ist $y_1=0$, so ist m_{s+1} nicht in M_0, also in M_1 enthalten. Aus der Bildungsvorschrift für $\{m_j\}$ folgt dann $H_s(M_1)\leqq\mu(M_1)$. Wegen $H_s(M_{y_1})=H_s(M_0)=0$ ist $H_s(M_1)=1$, also auch $\mu(M_1)=1$ und daher $\mu(M_0)=\mu(M_{y_1})=0$. Aus $H_{s+1}(M_{y_1})=\mu(M_{y_1})=0$ folgt 1) für $k=1$.

Sei *zweitens* m_{s+1} in M_{y_1} enthalten. Dann gilt

$$5) \qquad H_s(M_{y_1})\leqq\mu(M_{y_1}).$$

Für $y_1=1$ folgt 5) aus der Bildungsvorschrift für $\{m_j\}$. Für $y_1=0$ muß dieser Vorschrift zufolge $\mu(M_1)=0$ oder $H_s(M_1)>\mu(M_1)$ gelten. Im ersten Falle folgt 5) aus $\mu(M_0)=1$, im zweiten daraus, daß $H_s(M_0)+H_s(M_1)=\mu(M_0)+\mu(M_1)=1$.

Aus 5) folgt $H_{s+1}(M_{y_1})-\mu(M_{y_1})\leqq(s-i)/s(s+1)\leqq 1/(s+1)$. Aus der Induktionsvoraussetzung folgt

$$H_{s+1}(M_{y_1})-\mu(M_{y_1})\geqq(-i-1)/s(s+1),$$

also $\geqq -1/s+1$, falls $i<s$; für $i=s$ haben wir $H_s(M_{y_1})=H_{s+1}(M_{y_1})=1$ und wegen $H_s(M_{y_1})-\mu(M_{y_1})\leqq 0$ auch $\mu(M_{y_1})=1$, also $|H_{s+1}(M_{y_1})-\mu(M_{y_1})|=0$. Jedenfalls gilt also 1) für $k=1$.

5

Wir beweisen nun den Hilfssatz unter der Voraussetzung, daß er für $k-1$ gilt, d. h. daß

6) $$H_s(M_{y_1\ldots y_{k-1}})=\mu(M_{y_1\ldots y_{k-1}})+\varepsilon_s, \quad \text{wo } |\varepsilon_s|\leqq(k-1)/s.$$

Ist *erstens* $H_s(M_{y_1\ldots y_{k-1}})=0$, dann ist auch $H_s(M_{y_1\ldots y_k})=0$ und $|\mu(M_{y_1\ldots y_{k-1}})|\leqq(k-1)/s$ wegen 6). Also gilt erst recht $|\mu(M_{y_1\ldots y_k})|=|H_s(M_{y_1\ldots y_k})-\mu(M_{y_1\ldots y_k})|\leqq(k-1)/s<k/s$, wie 1) behauptet. Sei *zweitens* $H_s(M_{y_1\ldots y_{k-1}})\neq 0$. Es seien $m'_1, m'_2, \ldots, m'_j$ die in $M'=M_{y_1\ldots y_{k-1}}$ enthaltenen Elemente der Reihe $m_1, \ldots, m_s$, wobei $j=s\,H_s(M_{y_1\ldots y_{k-1}})$. Wir setzen $M'_{z_k}=M_{y_1\ldots y_{k-1}z_k}$ und

$$\mu'(M'_{z_k})=\mu(M'_{z_k})/\mu(M').$$

Aus der Bildungsvorschrift für $\{m_j\}$ folgt, daß m'_r für $r\leqq j$ in M'_1 enthalten ist, falls $\mu(M'_1)\neq 0$ und $H'_{r-1}(M'_1)\leqq\mu'(M'_1)$; andernfalls ist m'_r in M'_0 enthalten. Die Folge $\{m'_1, \ldots, m'_j\}$ erfüllt offenbar in bezug auf die M'-Mengen mit einem Index und die Mengenfunktion μ' die Voraussetzungen des für $k=1$ bewiesenen Hilfssatzes. Derselbe ergibt $|H'_j(M'_{y_k})-\mu'(M'_{y_k})|\leqq 1/j$, woraus wegen

$$j=s\,H_s(M_{y_1\ldots y_{k-1}})$$

folgt, daß

$$H_s(M_{y_1\ldots y_k})/H_s(M_{y_1\ldots y_{k-1}})=\mu(M_{y_1\ldots y_k})/\mu(M_{y_1\ldots y_{k-1}})+\varepsilon^*,$$

wo $|\varepsilon^*|\leqq 1/s\,H_s(M_{y_1\ldots y_{k-1}})$. Hieraus im Verein mit 6) folgt

$$H_s(M_{y_1\ldots y_k})=\mu(M_{y_1\ldots y_k})+\varepsilon_s\,\mu(M_{y_1\ldots y_k})/\mu(M_{y_1\ldots y_{k-1}})+$$
$$+H_s(M_{y_1\ldots y_{k-1}})\,\varepsilon^*, \text{ also}$$

$$|H_s(M_{y_1\ldots y_k})-\mu(M_{y_1\ldots y_k})|\leqq$$
$$\leqq|\varepsilon_s|+|H_s(M_{y_1\ldots y_{k-1}})\,\varepsilon^*|\leqq(k-1)/s+1/s=k/s,$$

wie 1) behauptet, womit der Hilfssatz bewiesen ist.

Wir definieren nun eine Merkmalfolge $K=\{m_i\}\,(i=1, 2, \ldots,$ ad inf.) nach folgender rekursiver Vorschrift:

a) Das Element m_1 sei ein Punkt von M_1 oder M_0, je nachdem ob $\mu(M_1)\neq$ oder $=0$ ist. Sind bereits die ersten $j-1$ Elemente bestimmt, so bestimmen wir m_j nach folgender Vorschrift: Zunächst werden auf Grund der vorangehenden Merkmalwerte $m_1, \ldots, m_{j-1}$ die Indizes der B-Menge festgesetzt, in der m_j enthalten ist, so wie der m_j vorangehende Anfangsabschnitt T dieser B-Menge. (Ist m_j erstes Element der B-Menge, so ist T die leere Menge). *m_j sei ein Punkt der Menge $M_{x_1\ldots x_n}$, wobei n die Anzahl der Indizes der betrachteten B-Menge ist und $x_1, \ldots, x_n$, so wie im Hilfssatz, folgendermaßen bestimmt werden: Ist m_j erstes Element, so sei $x_1=1$*

oder $=0$, *je nachdem* $\mu(M_1) \neq 0$ *oder* $=0$; *für* $i=2, \ldots, n$ *sei* $x_i=1$ *oder* $=0$, *je nachdem* $\mu(M_{x_1 \ldots x_{i-1} 1}) \neq 0$ *oder* $=0$. *Ist* m_j *nicht erstes Element, so sei* $x_1=1$, *falls* $\mu(M_1) \neq 0$ *und* M_1 *in* T *eine relative Häufigkeit* $H(M_1) \leqq \mu(M_1)$ *besitzt; andernfalls sei* $x_1=0$. *Für* $i=1, \ldots, n-1$ *sei* $x_{i+1}=1$, *falls* $\mu(M_{x_1 \ldots x_i 1}) \neq 0$ *und*

$$H(M_{x_1 \ldots x_i 1}) / H(M_{x_1 \ldots x_i}) \leqq \mu(M_{x_1 \ldots x_i 1}) / \mu(M_{x_1 \ldots x_i});$$

andernfalls sei $x_{i+1}=0$. *Dabei bedeutet* $H(M_{x_1 \ldots x_i 1})$ *bzw.* $H(M_{x_1 \ldots x_i})$ *die relative Häufigkeit von* $M_{x_1 \ldots x_i 1}$ *bzw.* $M_{x_1 \ldots x_i}$ *in* T. *Ist* $H(M_{x_1 \ldots x_i})$ $=0$, *so wird* $H(M_{x_1 \ldots x_i 1})/H(M_{x_1 \ldots x_i})=0$ *gesetzt.* (Daß $\mu(M_{x_1 \ldots x_i}) \neq 0$ für $i=1, \ldots, n$ und mithin $M_{x_1 \ldots x_n}$ nicht leer ist, folgt aus der Fußnote zum Hilfssatz).

Wir beweisen, daß die so definierte Merkmalfolge $K=\{m_i\}$ ein Kollektiv $K(\mathfrak{S}, \mathfrak{K})$ mit der Verteilungsfunktion μ ist und daher nach der Bemerkung von S. 63 auch ein Kollektiv $K(\mathfrak{S}, \mathfrak{M})$ ist. Zu diesem Zweck genügt es zu zeigen, daß für ein beliebiges ν-Tupel $y_1, \ldots, y_\nu$ der Zeichen 0 und 1 der Grenzwert $H(M_{y_1 \ldots y_\nu})$ der relativen Häufigkeit von $M_{y_1 \ldots y_\nu}$ in jeder unendlichen A-Folge gleich $\mu(M_{y_1 \ldots y_\nu})$ ist, da die Folge $K=\{m_i\}$ und jede nach einer Vorschrift des Systems $\mathfrak{S}$ gebildete Auswahlfolge Summe von endlich vielen paarweise fremden A-Folgen und jedes Element des Körpers $\mathfrak{K}$ Summe von endlich vielen paarweise fremden M-Mengen ist. Es sei $A_{i_1 \ldots i_l}$ eine unendliche A-Folge.

Für $\nu \leqq l$ sei $A^* = A_{i_1 \ldots i_l} - (B_{i_1} + B_{i_1 i_2} + \ldots + B_{i_1 \ldots i_l})$.

Für $\nu > l$ sei $A^* = A_{i_1 \ldots i_l} - (B_{i_1} + B_{i_1 i_2} + \ldots + B_{i_1 \ldots i_l}) - \Sigma B_{i_1 \ldots i_l y_{l+1} \ldots y_\nu}$

wobei die Summe über sämtliche $(\nu - l)$-Tupel $y_{l+1}, \ldots, y_\nu$ von 0 und 1 erstreckt wird.

Es sei $\{n_i\}$ die im schärferen Sinn monoton wachsende Folge von natürlichen Zahlen, welche zur Definition der B-Mengen in § 5 verwendet wurden. Es seien s und k zwei natürliche Zahlen, so daß $n_{l+k} \leqq s \leqq n_{l+k+1}$. Sei T der aus den s ersten Elementen bestehende Abschnitt von A^*. Da jede B-Menge, die mit A^* einen nichtleeren Durchschnitt hat, mehr als l Indizes haben muß, von denen die l ersten der Reihe nach $i_1, \ldots, i_l$ sein müssen, ist die Anzahl der zu A^* nicht fremden B-Mengen mit höchstens $l+k+\nu+1$ Indizes, $\leqq 2+2^2+\ldots+2^{k+\nu+1}=2^{k+\nu+2}-2$. Existierten mehr als $2^{k+\nu+2}$ zu T nicht fremde B-Mengen, so hätte mindestens eine von ihnen mindestens $l+k+\nu+2$ Indizes, etwa $B_{i_1 \ldots i_l x_{l+1} \ldots x_{l+j}}$ $(j \geqq k+\nu+2)$. Da sie nicht leer wäre, müßte $B_{i_1 \ldots i_l x_{l-1} \ldots x_{l+j-1}}$ genau n_{l+j-1} Elemente enthalten. Die letztere Menge wäre in A^* enthalten und da ihre sämtlichen Elemente einen kleineren Index haben, als die Elemente von $B_{i_1 \ldots i_l x_{l+1} \ldots x_{l+j}}$, so wäre sie schon in T enthalten, und zwar

5*

als echte Teilmenge, da auch $B_{i_1 \ldots i_l x_{l+1} \ldots x_{l+j}}$ mit T einen nicht leeren Durchschnitt hat. Es wäre also $n_{l+j-1} < s$, was wegen $j \geqq k+\nu+2$ und $n_{l+j-1} \geqq n_{l+k+\nu+1} \geqq s$ unmöglich ist. Also existieren höchstens $2^{k+\nu+2}$ zu T nicht fremde B-Mengen.

Ihre Durchschnitte mit T seien der Reihe nach $N_1, N_2, \ldots, N_j$ $(j \leqq 2^{k+\nu+2})$. Die Mächtigkeit von N_i sei $a_i (i=1, \ldots, j)$, so daß $a_1 + \ldots + a_j = s$. Da $N_1, \ldots, N_j$ offenbar Anfangsabschnitte von B-Mengen sind, so ergibt der Hilfssatz: Die relative Häufigkeit von $M_{y_1 \ldots y_\nu}$ in der Menge $N_\lambda (\lambda = 1, \ldots, j)$ ist $\mu(M_{y_1 \ldots y_\nu}) + \delta_\lambda / a_\lambda$, wobei $|\delta_\lambda| \leqq \nu$. Es gilt daher $H_s(M_{y_1 \ldots y_\nu}) = \mu(M_{y_1 \ldots y_\nu}) + (\delta_1 + \ldots + \delta_j)/s$, wobei $H_s(M_{y_1 \ldots y_\nu})$ die relative Häufigkeit von $M_{y_1 \ldots y_\nu}$ in T bedeutet. Es ist $\Sigma \delta_i / s \leqq \nu\, 2^{k+\nu+2}/s \leqq 2^\nu \nu \,.\, (2^{k+2}/n_{l+k})$.

Wählt man $n_i = 3^i (i=1, 2, \ldots, \text{ad inf.})$, so konvergiert der letzte Ausdruck mit wachendem k gegen 0. Damit ist die Behauptung, daß $K = \{m_i\}$ ein Kollektiv $K(\mathfrak{S}, \mathfrak{K})$ mit der Verteilungsfunktion μ ist, bewiesen.

Um Theorem III zu beweisen, haben wir noch zu zeigen, daß kontinuierlich viele Kollektivs $K(\mathfrak{S}, \mathfrak{M})$ mit der Verteilungsfunktion μ existieren. Zu diesem Zweck werden wir die obige Vorschrift *a)* für die Bildung eines Kollektivs $K(\mathfrak{S}, \mathfrak{M})$ etwas modifizieren. Es seien p_1 und p_2 zwei verschiedene Merkmalpunkte und λ eine gegebene dyadische Zahl zwischen 0 und 1, also eine unendliche Zeichenfolge von 0 und 1. Die modifizierte Bildungsvorschrift für ein Kollektiv $K = \{m_i\}$ lautet dann:

a_λ. *Das erste Element m_1 sei gleich p_1 oder p_2, je nachdem ob die erste Ziffer von λ gleich 0 oder 1 ist. Sind die ersten $j-1$ Elemente bereits bestimmt, so bestimmen wir m_j folgendermaßen: zunächst werden auf Grund von $m_1, \ldots, m_{j-1}$ die Indizes der B-Menge, in der m_j enthalten ist, und die vor m_j stehenden Elemente dieser B-Menge bestimmt. Ist m_j erstes Element der B-Menge, so sei $m_j = p_1$ oder p_2, je nachdem ob die k-te Ziffer von λ gleich 0 oder 1 ist, wobei k die Anzahl der B-Mengen bedeutet, die mit dem Anfangsabschnitt $(m_1, \ldots, m_j)$ einen nicht leeren Durchschnitt haben. (Diese Anzahl k kann offenbar bereits auf Grund von $m_1, \ldots, m_{j-1}$, festgestellt werden). Ist m_j nicht erstes Element der B-Menge, so bestimmen wir m_j nach Vorschrift a).*

Man kann nun leicht zeigen, daß auch die nach $a_\lambda)$ definierte Merkmalfolge $\{m_i\} (i=1, 2, \ldots, \text{ad inf.})$ ein Kollektiv $K(\mathfrak{S}, \mathfrak{M})$ mit der Verteilungsfunktion μ ist und daß die nach a_λ definierte Merkmalfolge von der nach $a_{\lambda'}$ definierten Merkmalfolge verschieden ist, falls $\lambda \neq \lambda'$ ist. Damit ist das Theorem III in allen Stücken bewiesen.

Das Theorem IV folgt unmittelbar aus Theorem III, falls folgender Satz gilt:

Ist M eine beschränkte im Peanoschen Sinne meßbare Teilmenge des n-dimensionalen euklidischen Raumes R_n, ferner μ eine nicht-

negative additive Mengenfunktion erklärt für alle im Peanoschen Sinne meßbaren Teilmengen von M, wobei $\mu(M)=1$ *und* $\lim \mu(M_i)=0$ *ist für jede Folge* $\{M_i\}(i=1, 2, \ldots$, ad inf.) *von Peanosch-meßbaren Teilmengen von M, für welche das Peanosche Maß* $\mu_P(M_i)$ *mit wachendem i gegen 0 konvergiert, dann gibt es einen abzählbaren Mengenkörper $\mathfrak{K}$, dessen Elemente im Peanoschen Sinne meßbare Teilmengen von M sind und wobei für jede im Peanoschen Sinne meßbare Teilmenge A von M das äußere und das innere μ-Maß in bezug auf den Körper $\mathfrak{K}$ einander gleich sind, und zwar gleich* $\mu(A)$.

Um die Richtigkeit des obigen Satzes einzusehen, sei $\mathfrak{K}$ der kleinste Körper über dem System der Teilmengen von M, die Durchschnitte eines rationalen Intervalls mit M sind. Der Körper $\mathfrak{K}$ ist offenbar abzählbar und alle Elemente von $\mathfrak{K}$ sind im Peanoschen Sinne meßbar. Daß das innere und das äußere μ-Maß in bezug auf den Körper $\mathfrak{K}$ für jede im Peanoschen Sinne meßbare Teilmenge A von M gleich $\mu(A)$ ist, folgt unmittelbar aus der Annahme, daß $\lim \mu(M_i)=0$ für jede Folge $\{M_i\}(i=1, 2, \ldots$, ad inf.) von Peanosch-meßbaren Teilmengen gilt, für welche $\lim \mu_P(M_i)=0$ ist. Damit ist der obige Satz und mithin auch das Theorem IV bewiesen.

Um das Theorem VI zu beweisen, haben wir bloß zu zeigen, daß, falls μ für die Elemente des Körpers $\mathfrak{K}$ konstruktiv definiert ist und $\mathfrak{S}$ ein konstruktiv definiertes abzählbares System von Auswahlvorschriften ist, durch die rekursive Vorschrift *a)* das Kollektiv $K=\{m_i\}$ konstruktiv definiert wird, d. h. daß dann auf Grund der Vorschrift *a)* für jede natürliche Zahl n der Merkmalwert m_n in endlich vielen Schritten tatsächlich berechnet werden kann. Wir beweisen dies durch vollständige Induktion und dazu noch, daß auch die Indizes der B-Menge, in der m_n enthalten ist, in endlich vielen Schritten berechnet werden können. Zur Berechnung von m_n genügt es, die Indizes der M-Menge zu berechnen, in der m_n enthalten ist, denn nach Vorschrift *a)* kann dann m_n einem beliebigen Element dieser M-Menge gleich gesetzt werden. Für $n=1$ ist die Behauptung trivial. Wir beweisen sie für $n+1$ unter der Annahme ihrer Gültigkeit für alle natürliche Zahlen $\leq n$. Wir zeigen zunächst, daß die Indizes der B-Menge, welche m_{n+1} enthält, in endlich vielen Schritten berechnet werden können. Da nach Voraussetzung $\mathfrak{S}$ ein konstruktiv definiertes abzählbares System ist, so kann man die Auswahlvorschriften von $\mathfrak{S}$ in eine Folge $\{V_i\}(i=1, 2, \ldots$, ad inf.) anordnen und für jede natürliche Zahl i in endlich vielen Schritten berechnen, ob m_{n+1} nach V_i ausgewählt wird oder nicht, da $m_1, \ldots, m_n$ nach Induktionsvoraussetzung berechnet werden können. Wir können also einen beliebig langen Anfangsabschnitt der Folge $\{r_i\}(i=1, 2, \ldots$, ad inf.) in endlich vielen Schritten berechnen, wobei $r_i=1$ oder 0 ist, je nachdem ob m_{n+1} nach V_i ausgewählt wird, oder nicht. Da nach Induktionsvoraussetzung für $j<n+1$ die Indizes der B-Menge, die m_j enthält, in endlich vielen Schritten berechnet werden können, so kann man für jede natürliche Zahl i die Anzahl der Elemente

aus dem Anfangsabschnitt $(m_1, \ldots, m_n)$, die in $B_{r_1 \ldots r_i}$ enthalten sind, in endlich vielen Schritten berechnen. Man kann daher die kleinste Zahl j mit der Eigenschaft, daß $B_{r_1 \ldots r_j}$ weniger als n_j Elemente aus dem Abschnitt $(m_1, \ldots, m_n)$ enthält, ebenfalls in endlich vielen Schritten berechnen. m_{n+1} ist nach Definition (S. 55) in $B_{r_1 \ldots r_j}$ enthalten und somit ist gezeigt, daß die Indizes der B-Menge, die m_{n+1} enthält, in endlich vielen Schritten berechnet werden können. Nach Induktionsvoraussetzung können die vor m_{n+1} stehenden Elemente von $B_{r_1 \ldots r_j}$ in endlich vielen Schritten berechnet werden. Um m_{n+1} nach Vorschrift *a)* berechnen zu können, genügt daher offenbar, wenn man für jede vorgegebene Menge $M_{y_1 \ldots y_k}$ den Wert $\mu(M_{y_1 \ldots y_k})$ berechnen kann. Nun ist aber $M_{y_1 \ldots y_k}$ offenbar ein Element des Körpers $\mathfrak{K}$ und da μ nach Voraussetzung in $\mathfrak{K}$ konstruktiv definiert ist, so kann $\mu(M_{y_1 \ldots y_k})$ in endlich vielen Schritten berechnet werden. Damit ist das Theorem VI bewiesen *und die rekursive Vorschrift a) stellt ein allgemeines Konstruktionsverfahren für Kollektivs dar.*

Dritter Teil. Untersuchungen über Auswahlverfahren, wo die Auswahl eines Elementes auch von nachfolgenden Elementen abhängt.

Nach unserer in § 1 gegebenen Definition des Kollektivs, wird die Invarianz der Grenzhäufigkeiten nur gegenüber Auswahlen gefordert, bei denen die Auswahl eines Elementes aus einer Merkmalfolge höchstens von den diesem Element vorangehenden Elementen abhängt. Es ist jedoch von Interesse zu untersuchen, ob es Merkmalfolgen gibt, in denen die Grenzhäufigkeiten auch gegenüber Auswahlen invariant bleiben, bei denen die Auswahl eines Elementes auch von den diesem Element nachfolgenden Merkmalwerten abhängig gemacht wird. Wir werden eine Vorschrift zur Bildung von Auswahlfolgen als *„Auswahlverfahren"* bezeichnen, falls die Auswahl eines Elementes nicht von dem Merkmalwert des betreffenden Elementes abhängt. Insbesondere sind die in § 1 definierten Auswahlvorschriften zugleich auch Auswahlverfahren, doch ist nicht jedes Auswahlverfahren eine Auswahlvorschrift. Ein Auswahlverfahren ist z. B. die Vorschrift: ein Element m_i wird aus der Merkmalfolge $\{m_n\}$ $(n = 1, 2, \ldots,$ ad inf.) ausgewählt, falls i eine Primzahl und $m_{i-1} = m_{i+1} = 1$ ist, oder die Vorschrift: ein Element m_i wird ausgewählt, falls in dem unmittelbar nachfolgenden i-Tupel $(m_{i+1}, m_{i+2}, \ldots, m_{2i})$ die Anzahl der Einsen gerade ist. Man könnte nun die Definition des Kollektivs $K(\mathfrak{S}, \mathfrak{M})$ in § 1 etwas modifizieren, indem man zuläßt, daß $\mathfrak{S}$ auch Auswahlverfahren enthalte, die keine Auswahlvorschriften sind. Wir werden aber sehen, daß dann unsere Theoreme nicht mehr gelten. Wir werden in § 9 sogar zeigen, daß es ein Auswahlverfahren V gibt, so daß kein Kollektiv bezüglich V existiert.

§ 9. Ein Auswahlverfahren, in bezug auf welches kein Kollektiv existiert.

Der Merkmalraum M bestehe aus den Elementen 0 und 1. Ist f_n für jede natürliche Zahl n eine Funktion, die jeder unendlichen Zeichenfolge von 0 und 1 entweder die Zahl 1 oder die Zahl 0 als Funktionswert zuordnet, so definiert die Folge $\{f_n\}(n=1, 2, \ldots,$ ad inf.) ein Auswahlverfahren, indem man festsetzt: das i-te Element m_i der Merkmalfolge $\{m_j\}(j=1, 2, \ldots,$ ad inf.) wird dann und nur dann ausgewählt, falls die Funktion f_i für die unendliche Zeichenfolge $\{m_{i+k}\}(k=1, 2, \ldots,$ ad inf.) den Wert 1 hat. Um ein Auswahlverfahren $V=\{f_n\}(n=1, 2, \ldots,$ ad inf.) mit der behaupteten Eigenschaft zu definieren, denken wir die Menge Z aller unendlichen Zeichenfolgen von 0 und 1 in Klassen eingeteilt, wobei die folgenden vier Bedingungen erfüllt sind:

1) Jede Klasse enthält mindestens ein Element von Z.

2) Ist die Zeichenfolge z ein Element einer Klasse A, so ist auch jede Zeichenfolge z', die nur an endlich vielen Stellen von z verschieden ist, ein Element von A.

3) Sind z und z' zwei Zeichenfolgen, die an unendlich vielen Stellen voneinander verschieden sind, so sind sie nicht in einer Klasse enthalten.

4) Jedes Element von Z ist in mindestens einer Klasse enthalten.

Es ist klar, daß eine solche Klasseneinteilung existiert, ferner sieht man unmittelbar, daß eine Zeichenfolge nicht in zwei verschiedenen Klassen enthalten sein kann, d. h. die Klassen sind paarweise fremd. Jede Klasse enthält nur abzählbar viele Elemente. Es gibt daher kontinuierlich viele Klassen und man kann die Menge aller Klassen auf die Menge aller reellen Zahlen eineindeutig abbilden. Wir bezeichnen die der reellen Zahl α zugeordnete Klasse mit A_α. Aus jeder Klasse A_α wählen wir ein bestimmtes Element z_α aus. Es sei nun z eine beliebige Zeichenfolge von 0 und 1; wir bilden die Zeichenfolge z', die aus z entsteht, wenn man das Zeichen 0 n-mal vor z anschreibt. Ist z. B. $n=4$ und $z=101010\ldots$, so ist $z'=0000101010\ldots$ Die Zeichenfolge z' bestimmt eindeutig die reelle Zahl α, für welche z' in A_α enthalten ist. Wir setzen nun $f_n(z)$ gleich 1 oder 0, je nachdem ob die n-te Stelle in z_α 1 oder 0 ist. Ist nun $\{m_i\}(i=1, 2, \ldots,$ ad inf.) eine beliebige Merkmalfolge und bildet man aus $\{m_i\}$ eine Auswahlfolge $\{m'_i\}(i=1, 2, \ldots)$ nach dem Auswahlverfahren $V=\{f_n\}(n=1, 2, \ldots,$ ad inf.), so bestätigt man mühelos, daß fast alle Elemente der Auswahlfolge $\{m'_i\}$ gleich 1 sind. Es gibt also kein Kollektiv in bezug auf V, abgesehen von dem trivialen Fall, daß fast alle Elemente des Kollektivs gleich 1 (bzw. gleich 0) sind.

Das obige Auswahlverfahren V ist nicht konstruktiv definiert. Es wurde bloß seine Existenz bewiesen. In den Beweis geht sogar das

Auswahlaxiom der Mengenlehre ein, indem wir aus jeder Klasse A_α ein bestimmtes Element z_α auswählen. In § 10 werden wir ein sehr einfaches und konstruktiv definiertes Auswahlverfahren angeben, welches die merkwürdige Eigenschaft hat, daß, wenn man aus einer beliebigen Bernoullischen Folge nach diesem Auswahlverfahren eine Auswahlfolge bildet, diese Auswahlfolge keine Bernoullische Folge ist.

§ 10. Ein Auswahlverfahren, das aus jeder Bernoullischen Folge eine nicht-Bernoullische Folge auswählt.

Es bestehe der Merkmalraum wieder bloß aus den zwei Elementen 0 und 1. Eine Merkmalfolge $\{m_i\}(i=1, 2, \text{ad inf.})$ heißt Bernoullisch, falls für jede natürliche Zahl n folgendes gilt: Man denke die Merkmalfolge in aufeinanderfolgende Abschnitte der Länge n geteilt. Jeder solche Abschnitt ist ein geordnetes n-Tupel von 0 und 1. Für ein beliebiges n-Tupel $(a_1, \ldots, a_n)$ von 0 und 1, hat die relative Häufigkeit der Abschnitte, die mit dem n-Tupel $(a_1, \ldots, a_n)$ identisch sind, einen Grenzwert, der gleich $p^k(1-p)^{n-k}$ ist, wobei p eine feste Zahl >0 und <1, und k die Anzahl der Nullen in dem n-Tupel $(a_1, \ldots, a_n)$ bedeutet.

Wir definieren ein Auswahlverfahren V mit der behaupteten Eigenschaft folgendermaßen: Es sei i eine fest gegebene ungerade Zahl >3; wir denken die Merkmalfolge $\{m_n\}(n=1, 2, \ldots, \text{ad inf.})$ in aufeinander folgende Abschnitte der Länge i geteilt und bezeichnen dieselben der Reihe nach mit $A_1, A_2, \ldots$, ad inf. Ein Element m_j eines Abschnittes A_r wird dann und nur dann ausgewählt, falls die Anzahl der Einsen, die in $A'_r = A_r - m_j$ enthalten sind, gerade ist, dabei wird die Zahl 0 als gerade betrachtet.

Es ist klar, daß vermöge dieses Auswahlverfahreus aus jedem Abschnitt A_ν $(\nu=1, 2, \ldots, \text{ad inf.})$ entweder alle Einsen, oder alle Nullen ausgewählt werden. Daraus folgt mühelos, daß dieses Auswahlverfahren aus jeder Bernoullischen Folge eine nicht-Bernoullische Folge auswählt.

Über ein ökonomisches Gleichungssystem und eine Verallgemeinerung des Brouwerschen Fixpunktsatzes.

Von J. v. Neumann (Princeton, N. J.).

Der Gegenstand dieser Note ist die Auflösung eines typischen ökonomischen Gleichungssystems. Dasselbe hat die folgenden Eigenschaften:

(1) Die Güter werden nicht bloß aus den „natürlichen Produktionsfaktoren", sondern in erster Linie *auseinander* hergestellt, u. zw. können die Produktionsprozesse *zirkelhaft* sein, d. h. Gut G_1 wird mit Hilfe von Gut G_2 erzeugt, und G_2 mit Hilfe von G_1.

(2) Es sind unter Umständen mehr technisch mögliche Produktionsprozesse als Güter da. Die landläufige Methode des „Gleichungen-Abzählens" versagt also. Entscheidend ist vielmehr, herauszufinden, welche Prozesse wirklich Verwendung finden werden, und welche (als „unrentabel") nicht.

Um (1), (2) ganz rein diskutieren zu können, werden wir andere Elemente der Situation weitgehend idealisieren. (Vgl. §§ **1** bis **2**) Von diesen Idealisationen sind die meisten unwesentlich, aber wir wollen darauf hier nicht eingehen.

Unsere Fragestellung führt zwingend zu einem System von Ungleichheiten (3)—(8') in § **3**, dessen Lösbarkeit gar nicht evident ist, d. h. durch *keinerlei qualitatives Argument bewiesen werden kann.* Der mathematische Beweis gelingt vielmehr erst mit Hilfe einer Verallgemeinerung des Brouwerschen Fixpunktssatzes, d. h. durch Verwendung recht tief liegender *topologischer* Tatsachen. Dieser verallgemeinerte Fixpunktssatz (der „Satz" von § **7**) ist auch an sich von Interesse.

Der Zusammenhang mit der Topologie mag zunächst recht überraschend sein, aber der Verf. glaubt, daß er in Problemen dieser Art naturgemäß ist. Das Auftreten eines gewissen „Minimax" Problems, wie es aus der Variationsrechnung wohlbekannt ist, ist der direkte Anlaß dazu. In unserem Problem ist das „Minimax" Problem in § **5** formuliert, es ist eng verwandt mit einem anderem, das in der Theorie der Gesellschaftsspiele auftritt. [Vgl. [2]) in § **6.**]

Eine direkte Interpretation der hierbei auftretenden Funktion $\Phi(X, Y)$ wäre sehr erwünscht. Ihre Rolle scheint jener der thermodynamischen Potentiale in der phänomenologischen Thermodynamik ähnlich zu sein und es ist zu vermuten, daß ihr auch in voller phänomenologischer Allgemeinheit (unabhängig von unseren unnatürlich einschränkenden Idealisationen) eine ähnliche Rolle zukommen wird.

Ein anderer vorläufig uninterpretierter Zug in unserer Theorie ist die bemerkenswerte Dualität (Symmetrie) zwischen den monetären Variablen (Preise y_j, Zinsfaktor β) und den technischen (Produktionsintensitäten x_i, Expansionskoeffizient der Wirtschaft α). Diese ist sowohl in § **3** (3)—(8'), als auch in § **4** (7*)—(8*), sowie in der „Minimax"-Formulierung von § **5** (7**)—(8**) ungemein auffallend.

Zum Schluß sei noch auf die Resultate von § **11** hingewiesen, aus denen u. a. hervorgeht, daß (falls unsere Annahmen gelten) der normale Preis-Mechanismus die rein technisch zweckmäßigste Verteilung der Produktionsintensitäten herbeiführt. Da wir alle monetären Komplikationen ausgeschaltet haben, ist dies nicht unvernünftig.

Die vorliegenden Untersuchungen wurden vom Verfasser zum erstenmal im Winter 1932 im mathematischen Kolloquium der Universität Princeton vorgetragen. Der Anlaß zur gegenwärtigen Veröffentlichung ergab sich aus einer Aufforderung von Herrn K. Menger, dem der Verfasser auch an dieser Stelle seinen Dank aussprechen möchte.

1. Wir betrachten das folgende Problem: Es existieren n Güter $G_1, \ldots, G_n$, die durch m Prozesse $P_1, \ldots, P_m$ hergestellt werden können. Welche Prozesse werden (als „rentabel") Verwendung finden, und welche Preise der Güter werden Gültigkeit haben? Das Problem ist offenbar nicht-trivial, da eine jede seiner beiden Hälften bloß dann beantwortet werden kann, wenn es die andere Hälfte schon ist — d. h. es ist implizit. Wir bemerken insbesondere:

(a) Da $m > n$ sein kann, ist es gewiß nicht durch die sonst übliche Methode des „Gleichungs-Abzählens" lösbar.

Um andersgeartete Komplikationen auszuschließen, nehmen wir an:

(b) Daß der Ertrag der Produktion konstant ist. — Ferner:

(c) Daß die natürlichen Produktionsfaktoren, einschließlich der Arbeit, unbeschränkt zur Verfügung stehen.

Das wesentliche Phänomen, das wir zu erfassen wünschen, ist dieses: Die Güter werden mittels der Produktionsprozesse auseinander hergestellt (vgl. w. u. Gleichung (7)), und wir wollen ermitteln, *(i)* welche Prozesse verwendet werden und mit welchen Intensitäten, *(ii)* mit welcher relativ-Geschwindigkeit die Gesamt-Gütermenge anwächst, *(iii)* welche Preise bestehen werden, *(iv)* welcher Zinsfuß gültig ist. Um dieses Phänomen ganz zu isolieren, nehmen wir weiter an:

(d) Der einzige stattfindende Konsum ist der Güterverbrauch der Produktionsprozesse, in den auch der lebensnotwendige Güter-

verbrauch der Arbeiter und Angestellten eingeschlossen sein muß; d. h. wir nehmen an, daß jedes über das Lebensminimum hinausgehende Einkommen vollkommen reinvestiert wird.

Es ist wohl klar, welchen modellmäßigen Vorstellungen über die Wirtschaft diese Annahmen entsprechen.

2. Jeder Prozeß P_i, $i=1, \ldots, m$, hat die folgende Natur: Er verbraucht die (in irgendwelchen Einheiten ausgedrückten) Quantitäten a_{ij} der bzw. Güter $G_j (j=1, \ldots, n)$ und er erzeugt die Quantitäten b_{ij} derselben. Er kann also symbolisch so dargestellt werden:

$$(1) \qquad P_i\colon \sum_{j=1}^{n} a_{ij}\, G_j \rightarrow \sum_{j=1}^{n} b_{ij}\, G_j.$$

Hierbei ist zu bemerken:

(e) Kapitalgüter sind einfach auf beiden Seiten von (1) anzusetzen. Abnutzung eines Kapitalgutes ist dadurch zu beschreiben, daß man seinen verschiedenen Abnutzungs-Stadien als verschiedene Güter einführt, und für jedes derselben P_i besonders ansetzt.

(f) Jeder Prozeß P_i habe die Zeiteinheit als Zeitdauer. Längere Prozesse zerlege man in Einzelprozesse dieser Länge, nötigenfalls die Zwischenprodukte als besondere Güter einführend.

(g) (1) kann insbesondere auch den Fall beschreiben, daß ein Gut G_j nur zusammen mit gewissen anderen, seinen permanenten Nebenprodukten, gewonnen werden kann.

Im tatsächlichen Prozeß der Gesamt-Wirtschaft werden diese Prozesse P_i, $i=1, \ldots, m$, mit gewissen *Intensitäten* x_i, $i=1, \ldots, m$, verwendet werden. D. h.: im Gesamt-Prozeß sind die Quantitäts-Angaben der Gleichungen (1) mit x_i zu multiplizieren. Wir schreiben symbolisch:

$$(2) \qquad W = \sum_{i=1}^{m} x_i P_i.$$

$x_i=0$ bedeutet, daß der Prozeß P_i unbenutzt bleibt.

Wir interessieren uns für jene Zustände der Gesamt-Wirtschaft, in denen sie sich bei unveränderter Struktur ausdehnt, d. h. wo die Verhältnisse der Intensitäten, $x_1 : \ldots : x_m$, unverändert bleiben, aber die $x_1, \ldots, x_m$ selbst sich ändern dürfen. Dann werden die $x_1, \ldots, x_m$ mit einem gemeinsamen Faktor α pro Zeiteinheit multipliziert. Dieser Faktor α ist der *Expansionskoeffizient der Gesamt-Wirtschaft.*

3. Die numerischen Unbekannten unseres Problems sind: *(i)*. Die *Intensitäten* $x_1, \ldots, x_m$ der Prozesse $P_1, \ldots, P_m$, *(ii)* der *Expansionskoeffizient der Gesamt-Wirtschaft* α, *(iii)* die *Preise* $y_1, \ldots, y_n$ der Güter $G_1, \ldots, G_n$, *(iv)* der *Zinsfaktor* β. ($\beta = 1 + \frac{z}{100}$, z ist der Zinsfuß in $^0/_0$ pro Zeiteinheit).

Offenbar ist stets

(3) $x_i \geqq 0$, (4) $y_j \geqq 0$,

und da eine Lösung mit $x_1 = \ldots = x_m = 0$ oder $y_1 = \ldots y_n = 0$ sinnlos wäre

(5) $\sum_{i=1}^{m} x_i > 0$, (6) $\sum_{j=1}^{n} y_j > 0$.

Die ökonomischen Gleichungen lauten nunmehr:

(7) $$\alpha \sum_{i=1}^{m} a_{ij} x_i \leqq \sum_{j=1}^{m} b_{ij} x_j,$$

(7′) und falls in (7) $<$ gilt, so ist $y_j = 0$.

(8) $$\beta \sum_{j=1}^{n} a_{ij} y_j \geqq \sum_{i=1}^{n} b_{ij} y_j,$$

(8′) und falls in (8) $>$ gilt, so ist $x_i = 0$.

(7), (7′) besagen: Von einem Gute G_j kann im Gesamt-Prozesse (2) unmöglich mehr verbraucht werden, als erzeugt wird. Wird aber weniger verbraucht, d. h. wird G_j überproduziert, so wird G_j ein freies Gut und sein Preis y_j wird 0. Und (8), (8′) besagen: Im Gleichgewichtszustande kann bei keinem Prozeß P_i ein Profit erzielt werden (da sonst die Preise oder der Zinsfuß steigen würden — es ist klar wie diese Idealisation zu verstehen ist). Ist aber ein Verlust da, d. h. ist P_i unrentabel, so wird P_i unbenützt bleiben, seine Intensität x_i wird 0.

Die Koeffizienten a_{ij}, b_{ij} sind als gegebene Größen anzusehen, während die x_i, α, y_j, β die Unbekannten sind. Dies sind $m+n+2$ Unbekannte, da aber an den x_i, y_j offenbar nur die Verhältnisse $x_1 : \ldots : x_m$, $y_1 : \ldots : y_n$ wesentlich sind, sind es eigentlich bloß $m+n$. Dem stehen $m+n$ Bedingungen (7)+(7′) und (8)+(8′) gegenüber. Da aber dieselben keine Gleichungen, sondern ziemlich komplizierte Ungleichheiten sind, garantiert die Gleichheit dieser Anzahlen die Lösbarkeit des Gleichungssystems keineswegs.

Die duale Symmetrie der Gleichungen (3), (5), (7), (7′), der Variablen x_i, α und des Begriffes „unbenützter Prozeß" einerseits und der Gleichungen (4), (6), (8), (8′), der Variablen y_j, β und des Begriffes „freies Gut" andererseits, scheint bemerkenswert.

4. Unsere Aufgabe ist, (3)—(8′) zu lösen. Wir werden zeigen: *Lösungen von* (3)—(8′) *existieren immer.* Dagegen können mehrere Lösungen mit verschiedenen $x_1 : \ldots : x_m$ oder $y_1 : \ldots : y_n$ existieren. Ersteres ist möglich, da wir nicht einmal den Fall ausgeschlossen haben, daß mehrere P_i denselben Prozeß beschreiben, oder daß ein P_i durch Kombinieren anderer entsteht. Letzteres ist möglich, da möglich ist, daß gewisse Güter G_j in jedem Prozeß P_i nur in einem festen Verhältnis mit gewissen anderen eingehen.

Aber auch wenn diese trivialen Umstände nicht eintreten, können, aus weniger direkten Gründen, mehrere Lösungen $x_1 : \ldots : x_m$, $y_1 : \ldots : y_n$ existieren. — Demgegenüber ist es von Wichtigkeit, daß α, β in allen Lösungen dieselben Werte haben. D. h.

α, β *sind eindeutig bestimmt.*

Wir werden sogar sehen, daß α und β in einfacher Weise direkt charakterisiert werden können. (Vgl. §§ **10** u. **11**.)

Um unsere Betrachtungen zu vereinfachen, nehmen wir an, daß stets

(9) $$a_{ij} + b_{ij} > 0$$

ist (natürlich ist immer $a_{ij}, b_{ij} \geqq 0$). Da die a_{ij}, b_{ij} beliebig klein sein dürfen, ist diese Beschränkung nicht sehr weitgehend. Sie ist jedenfalls notwendig, um die Eindeutigkeit von α, β zu sichern, da sonst W in unzusammenhängende Teile zerfallen könnte. Wir wollen aber diese Fragen bei einer anderen Gelegenheit untersuchen.

Betrachten wir nun eine (hypothetische) Lösung x_i, α, y_j, β von (3)—(8′).

Hätten wir in (7) stets $<$, so würde wegen (7′) stets $y_j = 0$ gelten, entgegen (6). Hätten wir in (8) stets $>$, so würde wegen (8′) stets $x_i = 0$ gelten, entgegen (5). Also: In (7) gilt stets $\leqq$, aber mindesten einmal $=$; in (8) gilt stets $\geqq$, aber mindestens einmal $=$. Infolgedessen gilt[1]):

(10) $$\alpha = \operatorname*{Min}_{j=1,\ldots,n} \left[\sum_{i=1}^{m} b_{ij}\, x_i \Big/ \sum_{i=1}^{m} a_{ij}\, x_i \right],$$

(11) $$\beta = \operatorname*{Max}_{i=1,\ldots,m} \left[\sum_{j=1}^{n} b_{ij}\, y_j \Big/ \sum_{j=1}^{n} a_{ij}\, y_j \right].$$

Somit bestimmen x_i, y_j die α, β eindeutig. (Die rechten Seiten von (10), (11) können wegen (3)—(6) und (9) niemals die sinnlose Form $\frac{0}{0}$ annehmen.) Wir können daher (7)+(7′) und (8)+(8′) als Bedingungen für x_i, y_j allein aussprechen:

(7*) Für jedes $j = 1, \ldots, n$, für welches

$$\sum_{i=1}^{m} b_{ij} x_i \Big/ \sum_{i=1}^{m} a_{ij} x_i$$

nicht seinen Minimalwert (aller $j = 1, \ldots, n$) annimmt, ist $y_j = 0$.

8*) Für jedes $i = 1, \ldots, m$ für welches

$$\sum_{j=1}^{n} b_{ij} y_j \Big/ \sum_{j=1}^{n} a_{ij} y_j$$

nicht seinen Maximalwert (aller $i = 1, \ldots, m$) annimmt, ist $x_i = 0$.

---

[1]) Statt der Bruchstriche wird aus drucktechnischen Gründen stets / gesetzt.

(In (7*) sind die $x_1, \ldots, x_m$ als fest gegeben anzusehen, in (8*) sind es die $y_1, \ldots, y_n$.) Wir müssen somit (3)—(6), (7*), (8*) in Bezug auf x_i, y_j lösen.

5. Wir bezeichnen eine Variablenreihe $(x'_1, \ldots, x'_m)$, die die Analoga von (3), (5)

$$(3') \qquad x'_i \geqq 0, \qquad (5') \qquad \sum_{i=1}^{m} x'_i > 0$$

erfüllt, mit X'; und eine Variablenreihe $(y'_1, \ldots, y'_n)$, die die Analoga von (4), (6)

$$(4') \qquad y'_j \geqq 0, \qquad (6') \qquad \sum_{j=1}^{n} y_j > 0$$

erfüllt, mit Y'. Ferner setzen wir

$$(12) \qquad \Phi(X', Y') = \sum_{i=1}^{m} \sum_{j=1}^{n} b_{ij} x'_i y'_j \Big/ \sum_{i=1}^{m} \sum_{j=1}^{n} a_{ij} x'_i y'_j.$$

Seien $X = (x_1, \ldots, x_m)$, $Y = (y_1, \ldots, y_n)$ die (hypothetische) Lösung, $X' = (x', \ldots, x'_m)$, $Y' = (y'_1, \ldots, y'_n)$ frei variabel, aber so, daß (3)—(6), bzw. (3′)—(6′) erfüllt sind, dann können, wie man leicht verifiziert, (7*), (8*) folgendermaßen formuliert werden:

(7**) $\Phi(X, Y')$ nimmt bei $Y' = Y$ seinen Y'-Minimalwert an.

(8**) $\Phi(X', Y)$ nimmt bei $X' = X$ seinen X'-Maximalwert an.

Die Frage nach der Lösbarkeit von (3)—(8′) geht infolgedessen in die Frage nach der Lösbarkeit von (7**), (8**) über und diese kann so formuliert werden:

(*) *Betrachten wir $\Phi(X', Y')$ in dem durch (3′)—(6′) abgegrenzten Bereiche. Gesucht wird ein Sattelpunkt $X' = X$, $Y' = Y$, d. h. einer, indem $\Phi(X, Y')$ ein Y'-Minimum hat, und gleichzeitig $\Phi(X', Y)$ ein X'-Maximum.*

(7), (7*), (10) bzw. (8), (8*), (11) ergeben:

$$\alpha = \sum_{j=1}^{n} \left[\sum_{i=1}^{m} b_{ij} x_i \right] y_j \Big/ \sum_{j=1}^{n} \left[\sum_{i=1}^{m} a_{ij} x_i \right] y_j = \Phi(X, Y), \text{ bzw.}$$

$$\beta = \sum_{i=1}^{m} \left[\sum_{j=1}^{n} b_{ij} y_j \right] x_i \Big/ \sum_{i=1}^{m} \left[\sum_{j=1}^{n} a_{ij} y_j \right] x_i = \Phi(X, Y).$$

D. h.:

(**) *Falls unser Problem lösbar ist, d. h. falls $\Phi(X', Y')$ einen Sattelpunkt $X' = X$, $Y' = Y$ besitzt (vgl. oben), so ist*

$$(13) \qquad \alpha = \beta = \Phi(X\,Y) = \textit{der Wert im Sattelpunkte.}$$

6. Wegen der Homogenität von $\Phi(X', Y')$ (in X', Y', d. h. in $x'_1, \ldots, x'_m$ und $y'_1, \ldots, y'_n$) bleibt unser Problem unbeeinflußt, wenn wir (5′), (6′) (und entsprechend (5), (6)) durch die Normierungen

$$(5^*) \qquad \sum_{i=1}^{m} x'_i = 1, \qquad (6^*) \qquad \sum_{j=1}^{n} y'_j = 1$$

ersetzen. Dies wollen wir tun und nennen die durch

$$(3') \quad x'_i \geqq 0, \qquad (5^*) \quad \sum_{i=1}^{m} x'_i = 1$$

beschriebene X'-Menge S, die durch

$$(4') \quad y'_i \geqq 0, \qquad (6^*) \quad \sum_{j=1}^{n} y'_j = 1$$

beschriebene Y'-Menge T. (S, T sind $m-1$- bzw. $n-1$-dimensionale Simplices).

Um (*) zu lösen[2]), kehren wir zur direkteren Formulierung (7*), (8*) zurück, die wir mit

$$(3) \quad x_i \geqq 0, \qquad (5^*) \quad \sum_{i=1}^{m} x_i = 1,$$

$$(4) \quad y_j \geqq 0, \qquad (6^*) \quad \sum_{j=1}^{n} y_j = 1,$$

kombinieren — d. h. damit, daß $X=(x_1, \ldots, x_m)$ in S liegt und $Y=(y_1, \ldots, y_n)$ in T liegt.

7. Wir werden einen etwas allgemeineren Satz beweisen:

Sei R_m der m-dimensionale Raum aller Punkte $X=(x_1, \ldots, x_m)$, R_n der n-dimensionale Raum aller Punkte $Y=(y_1, \ldots, y_n)$, R_{m+n} der $m+n$-dimensionale Raum aller Punkte $(X, Y)=(x_1, \ldots, x_m, y_1, \ldots, y_n)$.

Eine Menge (in R_m oder R_n oder R_{m+n}), die *nicht leer, konvex abgeschlossen* und *beschränkt* ist, nennen wir eine Menge C.

Seien S^0, T^0 Mengen C in R_m bzw. R_n. Sei $S^0 \times T^0$ die Menge aller (X, Y) (in R_{m+n}), wo X ganz S^0 durchläuft und Y ganz T^0. Seien V, W zwei abgeschlossene Teilmengen von $S^0 \times T^0$. Für jedes X in S^0 sei die Menge $Q(X)$ aller Y mit (X, Y) in V eine Menge C, für jedes Y in T^0 sei die Menge $P(Y)$ aller X mit (X, Y) in W eine Menge C. Dann gilt der

---

[1]) Die Lösbarkeit unseres Problems hängt sonderbarerweise mit jener eines in der Theorie der Gesellschaftsspiele auftretenden Problems zusammen, das der Verf. anderwärtig behandelt hat. (Math. Annalen, *100*, 1928, S. 295—320; insbesondere S. 305 und 307—311.) Jenes Problem ist ein Spezialfall von (*) und wird durch unsere Lösung von (*) (vgl. w. u.) auf eine neue Weise miterledigt. In der Tat: Für $a_{ij}=1$ wird wegen (5*), (6*) $\sum_{i=1}^{m} \sum_{j=1}^{n} a_{ij} x'_i y'_j = 1$, und daher $\Phi(X', Y') = \sum_{i=1}^{m} \sum_{j=1}^{n} b_{ij} x'_i y'_j$, und daher fällt unser (*) mit loc. cit., S. 307, zusammen. (Unsere $\Phi(X', Y')$, b_{ij}, x'_i, y'_j, m, n entsprechen den dortigen $h(\xi, \eta)$, α_{pq}, ξ_p, η_q, $M+1$, $N+1$.)

Es ist übrigens bemerkenswert, daß wir durch (*) nicht, wie es sonst meistens der Fall ist, auf ein einfaches Maximum- oder Minimum-Problem geführt worden sind, dessen Lösbarkeit evident wäre, sondern auf ein Problem vom Sattelpunkt- oder Minimax-Typus, bei dem die Lösbarkeitsfrage viel tiefer liegt.

Satz. *Unter den obigen Annahmen besitzen V, W (mindestens) einen gemeinsamen Punkt.*

Unser Problem entsteht hieraus, indem man $S^0 = S$, $T^0 = T$ setzt, und V = Menge aller $(X, Y) = (x_1, \ldots, x_m, y_1, \ldots, y_n)$ die (7*) erfüllen, W = Menge aller $(X, Y) = (x_1, \ldots, x_m, y_1, \ldots, y_n)$ die (8*) erfüllen. Man erkennt mühelos, daß V, W abgeschlossen sind, und die Mengen $S^0 = S$, $T^0 = T$, $Q(X)$, $P(Y)$ sind alle Simplices, also Mengen C. Der gemeinsame Punkt dieser V, W ist natürlich unsere gesuchte Lösung $(X, Y) = (x_1, \ldots, x_m, y_1, \ldots, y_n)$.

8. Zum Beweis des obigen Satzes seien S^0, T^0, V, W wie vor dem Satz beschrieben.

Betrachten wir zunächst V. Für jedes X von S^0 wählen wir einen Punkt $Y^0(X)$ aus $Q(X)$ aus (z. B. den Schwerpunkt dieser Menge). Es wird im allgemeinen nicht möglich sein, $Y^0(X)$ als stetige Funktion von X zu wählen. — Sei $\varepsilon > 0$; wir definieren

$$(14) \qquad w^\varepsilon(X, X') = \mathit{Max}\left(0, 1 - \frac{1}{\varepsilon}\, \mathit{Distanz}\,(X, X')\right).$$

Nun sei $Y^\varepsilon(X)$ der Schwerpunkt der $Y^0(X')$ mit der (relativen) Gewichtsfunktion $w^\varepsilon(X, X')$, wo X' ganz S^0 durchläuft. D. h.: Wenn

$$Y^0(X) = \left(y_1^0(X), \ldots, y_n^0(X)\right), \quad Y^\varepsilon(X) = \left(y_1^\varepsilon(X), \ldots, y_n^\varepsilon(X)\right),$$

dann

$$(15) \qquad y_j^\varepsilon(X) = \int_{S^0} w^\varepsilon(X, X')\, y_j^0(X')\, d\,X' \Big/ \int_{S^0} w^\varepsilon(X, X')\, d\,X'.$$

Wir leiten nun eine Reihe von Eigenschaften von $Y^\varepsilon(X)$ (gültig für alle $\varepsilon > 0$) her:

(*i*) $Y^\varepsilon(X)$ liegt in T^0. Beweis: $Y^0(X')$ liegt in $Q(X')$, also in T^0, und da $Y^\varepsilon(X)$ ein Schwerpunkt von Punkten $Y^0(X')$ ist und T^0 konvex ist, so liegt auch $Y^\varepsilon(X)$ in T^0.

(*ii*) $Y^\varepsilon(X)$ ist eine (in ganz S^0) stetige Funktion von X. Beweis: Es genügt dies für jedes $y_j^\varepsilon(X)$ zu beweisen. Nun ist $w^\varepsilon(X, X')$ eine überall stetige Funktion von X, X'; $\int_{S^0} w^\varepsilon(X, X')\, d\,X'$ ist stets > 0; und alle $y_j^0(X)$ sind beschränkt (es sind Koordinaten von Punkten der beschränkten Menge S^0). Somit ergibt (15) die Stetigkeit der $y_j^\varepsilon(X)$.

(*iii*) Zu jedem $\delta > 0$ gibt es ein $\varepsilon_0 = \varepsilon_0(\delta) > 0$, so daß für $0 < \varepsilon < \varepsilon_0$ jeder Punkt $(X, Y^\varepsilon(X))$ eine Distanz $< \delta$ von V hat. Beweis: Nehmen wir das Gegenteil an. Dann gäbe es ein $\delta > 0$ und eine Folge von $\varepsilon_\nu > 0$ mit $\lim_{\nu \to \infty} \varepsilon_\nu = 0$, so daß für jedes $\nu = 1, 2, \ldots$ ein X_ν in S^0 existiert, für welches $(X_\nu, Y^{\varepsilon_\nu}(X_\nu))$ eine Distanz $\geqq \delta$ von V hat A fortiori hat dann $Y^{\varepsilon_\nu}(X_\nu)$ eine Distanz $\geqq \delta/2$ von jedem $Q(X')$ mit Distanz $(X_\nu, X') \leqq \delta/2$.

Alle X_ν, $\nu=1, 2, \ldots$, liegen in S^0, daher besitzen sie einen Häufungspunkt X^* in S^0. Somit gibt es eine Teilfolge von $X_\nu, \nu=1, 2, \ldots$, welche gegen X^* konvergiert, und in der stets Distanz $(X_\nu, X^*) \leq \delta/2$ gilt. Ersetzen wir die ε_ν, X_ν durch diese Teilfolge, so sehen wir, daß wir annehmen dürfen: $\lim X_\nu = X^*$, Distanz $(X_\nu, X^*) \leq \delta/2$. Also können wir für jedes $\nu=1, 2, \ldots$ $X'=X^*$ setzen, und somit haben wir stets: $Y^{\varepsilon_\nu}(X_\nu)$ hat eine Distanz $\geq \delta/2$ von $Q(X^*)$.

$Q(X^*)$ ist konvex, und die Menge aller Punkte mit einer Distanz $< \delta/2$ von $Q(X^*)$ ist es daher auch. Da $Y^{\varepsilon_\nu}(X_\nu)$ nicht zu dieser Menge gehört, und da es ein Schwerpunkt von Punkten $Y^0(X')$ mit Distanz $(X_\nu, X') \leq \varepsilon_\nu$ ist (denn für Distanz $(X_\nu, X') > \varepsilon_\nu$ ist nach (14) $w^{\varepsilon_\nu}(X_\nu X')=0$), so gehören auch diese Punkte nicht alle zur genannten Menge. Also existiert ein $X'=X_\nu'$, für welches Distanz $(X_\nu, X_\nu') \leq \varepsilon_\nu$ ist, und $Y^0(X_\nu')$ eine Distanz $\geq \delta/2$ von $Q(X^*)$ hat.

Da $\lim X_\nu = X^*$, $\lim$ Distanz $(X_\nu, X_\nu')=0$ ist, ist $\lim X_\nu'=X^*$. Alle $Y^0(Y_\nu')$ gehören zu T^0, also besitzen sie einen Häufungspunkt Y^*. Somit ist (X^*, Y^*) ein Häufungspunkt der $(X_\nu', Y^0(X_\nu'))$, und da diese alle zu V gehören, gehört es zu V. Also ist Y^* in $Q(X^*)$. Nun hat jedes $Y^0(Y_\nu')$ eine Distanz $\geq \delta/2$ von $Q(X^*)$, also auch der Häufungspunkt Y^*. Dies ist ein Widerspruch, und der Beweis ist damit durchgeführt.

(*i*)—(*iii*) besagen zusammen: Für jedes $\delta > 0$ existiert eine stetige Abbildung $Y_\delta(X)$ von S^0 auf eine Teilmenge von T^0, wobei jeder Punkt $(X, Y_\delta(X))$ eine Distanz $< \delta$ von V hat. (Man setze $Y_\delta(X) = Y^\varepsilon(X)$ mit $\varepsilon = \varepsilon_0 = \varepsilon_0(\delta)$.)

9. Vertauschung von S^0 und T^0 sowie von V und W ergibt nun: Für jedes $\delta > 0$ existiert eine stetige Abbildung $X_\delta(Y)$ von T^0 auf eine Teilmenge von S^0, wobei jeder Punkt $(X_\delta(Y), Y)$ eine Distanz $< \delta$ von W hat.

Setzen wir $f_\delta(X) = X_\delta(Y_\delta(X))$, so ist $f_\delta(X)$ eine stetige Abbildung von S^0 auf eine Teilmenge von S^0. Da S_0 eine Menge C, also topologisch ein Simplex ist[3]), können wir nun den Fixpunktssatz von L. E. J. Brouwer anwenden[4]); $f_\delta(X)$ besitzt einen Fixpunkt. D. h. ein X^δ in S^0 existiert, für welches $X^\delta = f_\delta(X^\delta) = X_\delta(Y_\delta(X^\delta))$. Sei $Y^\delta = Y_\delta(X^\delta)$, dann haben wir $X^\delta = X_\delta(Y^\delta)$. Somit hat der Punkt (X^δ, Y^δ) in R_{m+n} Distanzen $< \delta$ sowohl von V als auch von W. Also haben V und W eine Distanz $< 2\delta$.

Da dies für jedes $\delta > 0$ gilt, haben V, W die Distanz 0. Da V, W beschränkt und abgeschlossen sind, müssen sie somit einen gemeinsamen Punkt besitzen. Damit ist unser Satz restlos bewiesen.

---

[3]) Bezügl. dieser sowie der anderen hier benützten Eigenschaften der konvexen Mengen vgl. z. B. P. Alexandroff und H. Hopf, „Topologie", Bd. *1*, J. Springer, Berlin 1935, S. 598—609.

[4]) Vgl. z. B. l. c. [1]) S. 480.

10. Wir haben somit (7*), (8*) aus § **4** gelöst, also auch das äquivalente Problem (*) aus § **5**, sowie die ursprüngliche Aufgabe aus § **3**: Die Auflösung von (3)—(8'). Falls die x_i, y_j (die wir in §§ **7—9** X, Y nannten) bestimmt sind, ergeben sich α, β aus (13) in (**) in § **5**. Insbesondere ist $\alpha = \beta$.

Wie wir bereits in § **4** hervorhoben, kann es durchaus mehrere Lösungen x_i, y_j (d. h. X, Y) geben, wir wollen aber nun zeigen, daß es bloß einen Wert von α (d. i. von β) gibt. Seien in der Tat X_1, Y_1, α_1, β_1 und X_2, Y_2, α_2, β_2 zwei Lösungen. Dann ergeben (7**), (8**) und (13):

$$\alpha_1 = \beta_1 = \Phi(X_1, Y_1) \leqq \Phi(X_1, Y_2),$$
$$\alpha_2 = \beta_2 = \Phi(X_2, Y_2) \geqq \Phi(X_1, Y_2),$$

also $\alpha_1 = \beta_1 \leqq \alpha_2 = \beta_2$. Aus Symmetriegründen ist ebenso $\alpha_2 = \beta_2 \leqq \alpha_1 = \beta_1$, also ist $\alpha_1 = \beta_1 = \alpha_2 = \beta_2$.

Wir sehen also:

Es existiert mindestens eine Lösung X, Y, α, β. Für alle Lösungen gilt

$$\alpha = \beta = \Phi(X, Y), \tag{13}$$

und hat für alle Lösungen denselben numerischen Wert, m. a. W.: *Der Zinsfaktor und der Expansionskoeffizient der Wirtschaft sind einander gleich und durch die technisch möglichen Prozesse $P_1, \ldots, P_m$ eindeutig bestimmt.*

Wegen (13) ist $\alpha > 0$. Es kann aber $\alpha \gtreqless 1$ sein. Man sollte $\alpha > 1$ erwarten, aber $\alpha \leqq 1$ ist offenbar bei unserer allgemeinen Fragestellung nicht auszuschließen: Die Prozesse $P_1, \ldots, P_m$ können wirklich *unproduktiv* sein.

11. Wir wollen α noch auf zwei independente Weisen charakterisieren.

Betrachten wir zunächst einen Zustand der Wirtschaft, der rein technisch möglich ist und mit dem Faktor α' per Zeiteinheit expandiert. D. h. für die Intensitäten $x_1', \ldots, x_m'$ gilt

$$(3')\quad x_i' \geqq 0, \qquad\qquad (5')\quad \sum_{i=1}^{m} x_i' > 0$$

und

$$(7'')\quad \alpha' \sum_{i=1}^{m} a_{ij}\, x_i' \leqq \sum_{i=1}^{m} b_{ij}\, x_i'.$$

Auf Preise nehmen wir gar keine Rücksicht. Seien x_i, y_j, $\alpha = \beta$ eine Lösung unseres ursprünglichen Problems (3)—(8') in § **3** Multiplikation von (7'') mit y_j und Addition $\sum_{j=1}^{n}$ ergibt:

$$\alpha' \sum_{i=1}^{m} \sum_{j=1}^{n} a_{ij}\, x_i'\, y_j \leqq \sum_{i=1}^{m} \sum_{j=1}^{n} b_{ij}\, x_i'\, y_j,$$

also $\alpha' \leqq \Phi(X', Y)$. Wegen (8**) und (13) in § 5 ist also:

(15) $$\alpha' \leqq \Phi(X', Y) \leqq \Phi(X, Y) = \alpha = \beta.$$

Betrachten wir zweitens ein Preissystem, bei dem der Zinsfaktor β' keinen Profit mehr zuläßt. D. h. für die Preise $y_1', \ldots, y_n'$ gilt

(4') $y_j' \geqq 0,$ (6') $\sum_{j=1}^{n} y_j' > 0$

und

(8'') $$\beta' \sum_{j=1}^{n} a_{ij} y_j' \geqq \sum_{j=1}^{n} b_{ij} y_j'.$$

Auf die Produktions-Intensitäten nehmen wir gar keine Rücksicht. Seien x_i, y_j, $\alpha = \beta$ wie oben. Multiplikation von (8'') mit x_i und Addition $\sum_{i=1}^{m}$ ergibt:

$$\beta' \sum_{i=1}^{m} \sum_{j=1}^{n} a_{ij} x_i y_j' \geqq \sum_{i=1}^{m} \sum_{j=1}^{n} b_{ij} x_i y_j',$$

also $\beta' \geqq \Phi(X, Y')$. Wegen (7**) und (13) in § 5 ist also:

(16) $$\beta' \geqq \Phi(X, Y') \geqq \Phi(X, Y) = \alpha = \beta.$$

Diese zwei Resultate können wir auch so formulieren:

Der größte Expansionsfaktor α' der Gesamtwirtschaft, der rein technisch möglich ist, ist $\alpha' = \alpha = \beta$. Hierbei wird auf Preise keine Rücksicht genommen.

Der niedrigste Zinsfaktor β' bei dem ein profitloses Preissystem möglich ist, ist $\beta' = \alpha = \beta$. Hierbei wird auf die Produktionsintensitäten gar keine Rücksicht genommen.

Man beachte, daß diese Charakterisierungen erst auf Grund unseres Wissens von der Existenz von Lösungen des ursprünglichen Problems möglich werden — obwohl sie selber nicht direkt auf unser Problem Bezug nehmen. Ferner ist die Gleichheit des Maximums in der ersten und des Minimums in der zweiten Form bloß auf Grund dieser Existenz beweisbar.

Angelegenheiten des Kolloquiums:

Im Studienjahr 1935/36 nahmen von ausländischen Gästen am Kolloquium teil: Dr. J. Novák (Brno), Herr und Frau Chr. Pauc (Paris) und im Sommersemester Frau Dr. J. v. Schwarz (München).

Außer den in diesem Hefte enthaltenen Mitteilungen wurden unter anderem folgende Vorträge gehalten:

Chr. Pauc: *Structure d'un continu plan au voisinage d'un point où il admet une courbure de Alt ou de Menger. Dérivée seconde généralisée liée à la courbure de Alt.* (Erschienen in C. R. Paris, t. 203, 1936.)

Chr. Pauc: *Structure d'un continu métrique au voisinage d'un point où il admet une courbure de Alt ou de Menger.* (Erschienen in Rendiconti della R. Accademia nazionale dei Lincei, vol. XXIV, serie 6^a, 2^0 sem., fasc. 5–6, 1936.)

Chr. Pauc: *Introduction de directions dans un espace distancié. Analyse du contingent et du paratingent de Bouligand du point de vue topologique.* (Erschienen in C. R. Paris, t. 203, 1936; Bull. de la classe des Sc. de l'Ac. r. de Belgique, XXII, 1936.)

A. Wald: *Beweis für die Lösbarkeit der Tauschgleichungen der Ökonomie.*

Die Veröffentlichung dieser Arbeit, die wegen Raummangels nicht in diesem Heft erfolgen konnte, wird demnächst vorgenommen werden.

Druck von Gottlieb Gistel & Cie., Wien, III., Münzgasse 6.

Biographical notes on the editors of the *Ergebnisse*

Alt, Franz, born 1910 in Vienna, studied mathematics and received his Ph.D. with Menger. Alt had to emigrate in 1938. He contributed to the development of the computer, working at the Computing Laboratory in Aberdeen and the National Bureau of Standards. In 1967 he became departmental leader at the American Institute of Physics in New York. He was a founding member and a president of the Association for Computing Machinery and the first editor of Advances in Computers.

Gödel, Kurt (1906–1978), born in Brünn (today Brno in Czechia), studied in Vienna, where he achieved early fame with his incompleteness theorems which revolutionized mathematical logic. He received his Ph.D in 1930 and his habilitation in 1932. After two extended stays in the US, he emigrated in 1940 to the Institute for Advanced Study in Princeton, where he became professor in 1953 in recognition of his towering work in set theory and the foundations of mathematics.

Menger, Karl (1902-1985), born in Vienna as son of the noted economist Carl Menger, studied mathematics in Vienna, where he graduated in 1924. After two years with L.E.J. Brouwer in Amsterdam, Menger became extraordinary professor for geometry in Vienna in 1929. Visiting professor 1930/31 in Harvard, 1937/38 in Notre Dame. After the annexation of Austria, Menger remained at Notre Dame until 1946 and then moved to the Illinois Institute of Technology in Chicago. Many visiting positions. Menger was one of the pioneers of curve and dimension theory. He also contributed to geometry, algebra, probability theory, mathematical economics and didactics.

Nöbeling, Georg, born 1907 in Lüdenscheid (Westphalia), studied in Göttingen and Vienna. In 1933 he lost his appointment as assistant in Vienna and had to return to Germany. In 1935, he received his habilitation, in 1940 he became professor in Erlangen. Nöbeling held many distinguished positions (dean and rector in Erlangen, president of the German Mathematicians' Association). He greatly contributed to topology and geometry.

Wald, Abraham (1902-1950), born in Klausenburg (today Cluj in Romania), studied with Menger from 1927 on, became tutor of the banker Karl Schlesinger and consultant at Oskar Morgenstern's Institut für Konjunkturforschung. Forced to emigrate in 1938, he became professor at Columbia University and worked with the Statistical Research group. Fundamental contributions to general equilibrium theory, mathematical statistics and sequential analysis. Wald died in a plane crash in India.

Editorial notices

A report on the year 1929/30 which is similar, but not identical to the content of Volume I of the Ergebnisse has appeared in "Bericht über ein Mathematisches Kolloquium 1929/30" by Karl Menger, *Monatshefte für Mathematik und Physik* **38**, 1931, 17–38.

English translations of all papers by Kurt Gödel can be found in Volume I of his *Collected Works* (Eds. S. Feferman et al.), Oxford University Press, 1986.

The paper by Abraham Wald "Über Produktionsgleichungen der ökonomischen Wertlehre (II)" has been translated by Otto Eckstein and appeared under the title "On some systems of equations of mathematical economics", *Econometrica* **19** (1951), p. 368–403.

A translation of A. Wald's paper "Uber die eindeutige positive Lösbarkeit der neuen Produktionsgleichungen" has appeared in *Precursors in mathematical economics* (W.J. Baumol and S.M. Goldfeld, LSE Series Reprints of Scarce Works on Political Economy) **19**, London 1968.

The paper by John von Neuman "Über ein ökonomisches Gleichungssystem und eine Verallgemeinerung des Brouwerschen Fixpunktsatzes" has been translated by G. Morton and appeared under the title "A model of general economic equilibrium" in the *Review of Economic Studies* **13**, p. 1–9.

A revised and translated version of the paper by A. Tarski and A. Lindenbaum "Über die Beschränktheit der Ausdrucksmittel deduktiver Theorien" **7**, 1936, p. 15–21, has appeared under the title "On the Limitations of the Means of Expression of Deductive Theories" in *Logic, Semantics, Metamathematics. Papers from 1923–1938* by A. Tarski, Clarendon Press, Oxford 1956, p.334–392. The same volume contains on pp. 393–400 a revised and translated version of "Über die Erweiterungen der unvollständigen Systeme des Aussagenkalküls", **7** (1936), p. 51–57 with the title "On extensions of incomplete systems of the sentential calculus".

Afterword

Franz Alt

When I, one of the last survivors of the group of participants in Menger's Colloquium, heard about the plan to republish the *Ergebnisse,* my first reaction was "Why? Who would want to read that jumble of disconnected unfocussed trifles?"

But in glancing through the eight issues, the last of which I myself have helped see through to publication, I found a different picture. While some of the articles do indeed look out of context, isolated soundings that found no resonance, many more group themselves into a few clear areas of interest so that, out of many snippets of questions, conjectures, partial answers, there emerges a rounded body of knowledge. Differential geometry in spaces without coordinates, for instance; or axiomatic definitions of geometric incidence relations. Karl Menger, after all, occupied a chair intended for geometry.

And then mixed in with these, there are some shining gems, self-contained articles which have proved seminal through the twentieth century. What appears to have been Kurt Gödel's first oral presentation of a proof of his incompleteness theorem, was given at the 24th Colloquium in less than an hour, reported in *Ergebnisse* so concisely that it takes up little more than a page. From a group of articles by Karl Schlesinger, Abraham Wald and John von Neumann in issues 6, 7 and 8, has sprung a leading theory in mathematical economics.

Yet the *Ergebnisse* are only a pale reflection of what it meant to be present at the Colloquium meetings, to experience the give and take, the absorbing interest, the earnest or sometimes humorous exchange of words. There was the unforgettable quiet after Gödel's presentation, ended by what must be the understatement of the century: "That is very interesting. You should publish that." Then a question: "You use Peano's system of axioms. Will it work for other systems?" Gödel, after a few seconds of thought: “Yes, any system broad enough to define the field of the integers.” Olga Taussky (half-smiling): “The integers do not constitute a field!” Gödel, who knew this as well as anyone, and had only spoken carelessly: “Well, the ... the ... the domain of integrity of the integers.” And final relaxing laughter.

The reprinting of the *Ergebnisse* may not wholly recreate these experiences, but may give the reader an inkling of what Karl Menger accomplished by bringing this group of people together. How much more might have been accomplished if the group had not been so brutally broken up?

PS: Excerpts from two letters (originally in German) between Karl Menger, then at Notre Dame University, and Franz Alt:

December 31, 1937
My dear Franz

...

Since we expected the addition to our family for December, it was clear that I would not be able to return to Vienna in February. I have applied for an extension of my leave of absence.

...

I am deeply saddened by the fact that I can do so little for the beautiful Viennese circle of mathematicians which is so dear to me. I believe you should get together from time to time and especially see to it that Gödel takes part in the colloquium. It would not only be of greatest benefit for all the other participants, but also for himself, though he may not realize it. Heaven knows into what he could entangle himself [„in was er sich einspinnen könnte“] if he does not talk to you and the other friends in Vienna from time to time. If necessary, be pushy, on my say-so.

Vienna, March 1, 1938
Dear Karl:

...

A few days ago Wald told me of his imminent departure for America. You can imagine that I envy him. The Colloquium will hardly exist much longer – since my last letter there has not been a meeting. But that is not the most serious concern we have here. Until now I always closed my letters expressing my hope to see you again soon in Vienna. At present I have to hope to find some way to get together with you over there.

SPRINGER NATURE

GPSR Compliance

The European Union's (EU) General Product Safety Regulation (GPSR) is a set of rules that requires consumer products to be safe and our obligations to ensure this.

If you have any concerns about our products, you can contact us on ProductSafety@springernature.com

In case Publisher is established outside the EU, the EU authorized representative is:

Springer Nature Customer Service Center GmbH
Europaplatz 3
69115 Heidelberg, Germany

FSC
www.fsc.org
MIX
Papier | Fördert gute Waldnutzung
FSC® C083411

Kolibri 360 GmbH
Ferdinand-Jühlke-Straße 7
99095 Erfurt, Deutschland
produktsicherheit@kolibri360.de